Nanosafety

Ernesto Alfaro-Moreno · Fiona Murphy
Editors

Nanosafety

A Comprehensive Approach
to Assess Nanomaterial Exposure
on the Environment and Health

# Springer

Editors
Ernesto Alfaro-Moreno
Nanosafety Group
International Iberian Nanotechnology
Laboratory (INL)
Braga, Portugal

Fiona Murphy
Strathclyde Institute of Pharmacy
and Biomedical Sciences
University of Strathclyde
Glasgow, UK

ISBN 978-3-031-93870-2   ISBN 978-3-031-93871-9   (eBook)
https://doi.org/10.1007/978-3-031-93871-9

© The Editor(s) (if applicable) and The Author(s), under exclusive license to Springer Nature Switzerland AG 2025. This book is an open access publication.

Open Access This book is licensed under the terms of the Creative Commons Attribution-NonCommercial-NoDerivatives 4.0 International License (http://creativecommons.org/licenses/by-nc-nd/4.0/), which permits any noncommercial use, sharing, distribution and reproduction in any medium or format, as long as you give appropriate credit to the original author(s) and the source, provide a link to the Creative Commons license and indicate if you modified the licensed material. You do not have permission under this license to share adapted material derived from this book or parts of it.

The images or other third party material in this book are included in the book's Creative Commons license, unless indicated otherwise in a credit line to the material. If material is not included in the book's Creative Commons license and your intended use is not permitted by statutory regulation or exceeds the permitted use, you will need to obtain permission directly from the copyright holder.

This work is subject to copyright. All commercial rights are reserved by the author(s), whether the whole or part of the material is concerned, specifically the rights of translation, reprinting, reuse of illustrations, recitation, broadcasting, reproduction on microfilms or in any other physical way, and transmission or information storage and retrieval, electronic adaptation, computer software, or by similar or dissimilar methodology now known or hereafter developed. Regarding these commercial rights a non-exclusive license has been granted to the publisher.

The use of general descriptive names, registered names, trademarks, service marks, etc. in this publication does not imply, even in the absence of a specific statement, that such names are exempt from the relevant protective laws and regulations and therefore free for general use.

The publisher, the authors and the editors are safe to assume that the advice and information in this book are believed to be true and accurate at the date of publication. Neither the publisher nor the authors or the editors give a warranty, expressed or implied, with respect to the material contained herein or for any errors or omissions that may have been made. The publisher remains neutral with regard to jurisdictional claims in published maps and institutional affiliations.

This Springer imprint is published by the registered company Springer Nature Switzerland AG
The registered company address is: Gewerbestrasse 11, 6330 Cham, Switzerland

If disposing of this product, please recycle the paper.

Foreword

It's easy to write a foreword to a good book, so I have a simple task ahead of me. The book you are about to read is not the first nor the last to address Nanosafety, but one worth having. Edited by Ernesto Alfaro-Moreno (INL) and Fiona Murphy (Strathclyde), both established scholars in the field, it comes out at a time when our understanding of Nanosafety—the science of identifying and countering hazards linked uniquely to the novel properties of the nanoscale—has made great advances. We now have a good grasp of nanomaterial toxicity thanks to developing robust methodologies to study the illusive nanoscale, whilst models of nanomaterial mode of action have become more reliable. We have assessed toxicity on many environmental model species and human in vitro models and have begun tackling new pollutants such as nanoplastics. Our in silico capability has also advanced. These are just some examples of recent progress that the book addresses well.

Starting from the basics, the field of nanotechnology represents one of the most exciting frontiers in modern science, promising advancements in everything from medicine to environmental sustainability, agriculture to computing, climate mitigation to disease eradication. Yet, with these advances comes a profound responsibility: as we integrate nanomaterials into countless applications, it is important to rigorously investigate their effects on health and the environment. This critical balance between innovation and precaution is what nanosafety is all about. And the challenge is that the science of nanosafety must keep up with the advances of nanotechnology.

Nanosafety's role is twofold. On one hand, it allows us to harness the positive potential of nanomaterials—enabling groundbreaking progress in drug delivery, sustainable energy solutions, and environmental remediation. On the other hand, it seeks to prevent and manage potential risks associated with nanomaterials' unique properties, which can affect human and ecological systems in unforeseen ways. Nanoparticles, for instance, can more readily penetrate biological membranes due to their small size, crossing barriers within the body or dispersing into ecosystems, reaching parts of the biological and environmental machinery where they can cause harm. Such impacts have been shown to occur but remain hard to predict.

This volume, *Nanosafety – A Comprehensive Approach to Assess Nanomaterial Exposure on the Environment and Health*, provides a timely and thorough exploration of the multifaceted nature of nanosafety research. By dividing its focus into four essential domains—Toxicology, Advanced Models, Bridging Models, and In

Silico Approaches—it underscores that a comprehensive approach is necessary to fully understand the risks and safety mechanisms associated with nanomaterials. Each section is a reminder that no single methodology or model is sufficient on its own. Instead, a layered approach that combines experimental, computational, and environmental assessments is needed to capture the complexity of nanosafety.

As we look to the future, it is clear that nanosafety will play a pivotal role in defining how nanotechnology integrates with society. The knowledge, methods, and insights shared in this book contribute to a legacy of responsible development, ensuring that we prioritize both safety and sustainability as we continue to explore the extraordinary possibilities of the nanoscale.

I invite readers to immerse themselves in this volume and to be guided by the contributions to the ongoing dialogue on nanosafety. The book brings a fresh, thoughtful, and timely perspective of the science of nanosafety recognizing the complexity and providing informed guidance. To quote Sir William Bragg: **"The important thing in science is not so much to obtain new facts as to discover new ways of thinking about them."** And it is this approach to rethinking nanosafety by using innovative methods and models to address the complexities of nanomaterials that the book has done so well.

University of Birmingham Éva Valsami-Jones
Birmingham, UK

Preface

Nanotechnology is rapidly emerging as one of the most dynamic and transformative industries of our time, influencing a diverse range of sectors from energy to environmental remediation, healthcare to cosmetics. As we harness the power of nanomaterials, it is crucial to consider their potential impact on both human health and the environment. The field of Nanosafety has been dedicated to evaluating the release of nanomaterials into the environment and the subsequent exposure risks, striving to identify the safest materials while minimizing potential hazards.

History has taught us that groundbreaking advancements often come with unforeseen consequences. Innovations that promise to enhance our quality of life can also lead to significant challenges if not carefully managed. The dual-faced nature of technological advances—where a beneficial purpose can concomitantly result in unintended detrimental consequences—underscores the necessity for thorough investigation and regulation in the realm of nanotechnology.

In this book, we aim to compile and discuss essential tasks that must be addressed to ensure prompt and reliable evaluations of nanomaterial safety. A primary focus is related to the toxicity of these materials, a field now known as nanotoxicology. The unique physical and chemical properties exhibited at the nanoscale require specialized approaches that extend beyond traditional toxicity assessment methods. As the number and diversity of available nanomaterials continues to grow at an unprecedented pace, it is imperative that we develop faster, more efficient testing methods. This includes the use of animal-free yet biologically relevant models, particularly highlighted in our discussions on organ-on-a-chip technologies and advanced in vitro models.

Additionally, some contributions explore the potential utility of simple animal models such as planarias, *C. elegans*, fruit fly, and zebra fish embryos, and how the findings from these systems can be translated into toxicologically relevant effects on humans. This critical examination is presented in the section dedicated to pioneering the use of bridging models in nanotoxicology studies.

The environmental impact of nanomaterials is another significant theme of this book. A couple of chapters provide a comprehensive overview of how these novel nanoscale materials, alongside micro- and nanoplastics, affect our ecosystems. The intricate relationships between nanomaterials and environmental health are complex, and understanding these dynamics is essential for developing sustainable practices in nanotechnology.

Finally, various in silico approaches are presented, serving as crucial preliminary actions for rapid screening for potential adverse effects of nanomaterials. By leveraging computational models, we can identify risks based on characteristics shared with well-characterized materials, enabling more informed decision-making in the early stages of material development.

We hope that this book serves as a valuable resource for readers, drawing the attention of decision-makers to advance the field of nanosafety. Our goal is to inspire curiosity among those who are interested in these topics but may not have found a pathway to engage with the fascinating and rapidly evolving field of nanosafety. By fostering a deeper understanding of the risks and benefits associated with nanomaterials, we can work together to ensure that the promise of nanotechnology is realized in a safe and responsible manner.

As we stand at the intersection of innovation and caution, it is our collective responsibility to navigate the complexities of nanotechnology. Through informed discussions, rigorous research, and collaborative efforts, we can unlock the full potential of nanomaterials while safeguarding our health and the environment for future generations. This book is a step toward achieving that balance, and we invite you to join us on this journey.

Braga, Portugal	Ernesto Alfaro-Moreno
Glasgow, UK	Fiona Murphy

Acknowledgments

We would like to extend our heartfelt gratitude to all the authors for their invaluable contributions to this book. A special thanks goes to the members of INL's Nanosafety group—Ana Ribeiro, Nivedita Chatterjee, Vânia Vilas-Boas, Flipa Lebre, and Michael González-Durruthy—whose proactive efforts in inviting experts from various branches of nanosafety were instrumental in shaping this book.

This endeavor would not have been possible without the invaluable support of the European Commission through the ERA-Chair program. The funding provided for the Sinfonia Project (Horizon 2020, Grant Agreement No. 857253) has significantly enhanced our nanosafety capabilities at INL, enabling us to tackle the challenges that lie ahead. We deeply appreciate the collaborative spirit and dedication of all those involved in this project—Marina Brito, Marina Dias, and Paula Galvao—which has laid the groundwork for future advancements in the field. Thank you for your commitment to advancing nanosafety research and promoting a safer future.

Contents

1 **Nanosafety: Why Do We Need It?** 1
Ernesto Alfaro-Moreno and Fiona Murphy

Part I Nanotoxicology

2 **Methodological Considerations for Setting Up
Human-Relevant In Vitro Nanotoxicology
Experiments—A Practical Guide** 27
Vânia Vilas-Boas, Emma Arnesdotter, Félix Carvalho,
and Ernesto Alfaro-Moreno

3 **Exploring the Interplay Between Particles
and the Immune System**.. 55
Thais S. M. Lima and Filipa Lebre

4 **Nanoparticle Mediated Epigenetic and Immunological Modulation**.. 87
Ayse Basak Engin and Atilla Engin

5 **Engineered Nanomaterials in Rodent Models
of Allergic Lung Disease**....................................... 115
Logan J. Tisch, Ryan D. Bartone, and James C. Bonner

**Part II Advanced In Vitro Models and Organ-on-a-Chip
Used in Nanotoxicology**

6 **Advanced *In Vitro* Airway Models for NM Hazard Assessment**...... 141
Katie McAllister, Ernesto Alfaro-Moreno, and Fiona Murphy

7 **Advanced Skin Models for Nanomaterials Safety Assessment**........ 161
A. R. Ribeiro, S. Costa, S. Nogueira, M. González-Durruthy,
H. Colley, N. Oliva, R. De Vecchi, and E. Alfaro-Moreno

8 **Advanced Gut-on-a-Chip *In Vitro* Models
for ENMs Safety Assessment**................................... 193
M. D. Neto, L. Pastrana, and C. Gonçalves

9 BBB-on-a-Chip: Microfluidic Tools as an Alternative to *In Vivo* Experiments for Nanosafety Studies 217
M. C. Lefevre, M. C. Ceccarelli, M. Bernardeschi, and G. Ciofani

Part III In Vitro-In Vivo Bridging Models

10 Planarians as an Alternative Standard Invertebrate Model to Assess the Eco-safety of Nanoparticles/Nanomaterials 251
M. Bernardeschi, M. C. Lefevre, M. C. Ceccarelli, A. Salvetti, and G. Ciofani

11 *Caenorhabditis elegans*: A Bridging Model to Assess the Safety of Nanomaterials 275
Nivedita Chatterjee

12 Galleria Mellonella as a Potential Bridging Model for Nanotoxicology 313
Aaron Curtis, Kevin Kavanagh, and Fiona Murphy

13 *Drosophila* Model Unveils Nanoparticle Interactions: Implications for Safety 337
Ghada Tagorti and Bülent Kaya

14 Zebrafish as a Model to Investigate Nanoparticles 365
Lina Lundin and Steffen H. Keiter

Part IV Nanosafety Related to Nanomaterials in the Environment

15 Environmental Nanosafety..................................... 403
Begoña Espiña and Laura Rodriguez-Lorenzo

16 The Environmental Impacts of Nanoplastics in Marine Ecosystems... 439
Giulia Galani, Aline Nunes, Isadora Piccinin, Lucas Fazardo, Suelen Goulart, Thais Alberti, Vânia Vilas-Boas, and Marcelo Maraschin

Part V In Silico Approaches to Assess Nanosafety

17 Molecular Docking in Nanotoxicology 481
Michael González-Durruthy, Ana S. Moura, and M. Natália D. S. Cordeiro

18 Carbon and GaN Nanomaterials for Environmental Contaminant Removal Through Ab Initio Simulations 511
Laura F. O. Vendrame, Mariana Z. Tonel, Paulo B. O. Lira Junior, Silvete Guerini, Mirkos O. Martins, Ivana Zanella, and Solange B. Fagan

19 **Quasi-SMILES as a Tool for Simulation of Endpoints Related to Nanomaterials**.. 531
Andrey A. Toropov, Alla P. Toropova, Alessandra Roncaglioni, and Emilio Benfenati

20 **Life Cycle Assessment Towards Safe and Sustainable Biorefinery Systems of Marine Biomass—Focus on Engineered Nanoparticles**... 559
Carla Lopes, Véronique Adam, Luis Mauricio Ortiz-Galvez, Beatrice Salieri, Blanca Suarez Merino, Cyrille Durand, and Luis Taboada Antelo

Index.. 595

Nanosafety: Why Do We Need It?

Ernesto Alfaro-Moreno and Fiona Murphy

Abstract

Nanosafety has become a pivotal area of study amidst the rapid advancements in nanotechnology. Nanomaterials (NMs), defined by their nanoscale dimensions, exhibit unique properties that enable diverse applications across industries, including electronics, medicine, and consumer goods. However, these properties also raise significant health and environmental concerns. Nanosafety encompasses the investigation and management of risks linked to the lifecycle of NMs—from production to disposal. As NMs interact with biological systems, they can present altered toxicological profiles, prompting concerns over their cytotoxicity, immunotoxicity, and genotoxicity. Additionally, understanding the environmental behavior of NMs, such as their transport and potential for bioaccumulation, is essential for assessing their ecological impact. A multidisciplinary approach, integrating toxicology, materials science, and computational modeling, is crucial for developing robust risk assessment frameworks.

Keywords

Nanosafety · Nanotechnology · Nanomaterials · Nanoparticles

E. Alfaro-Moreno (✉)
Nanosafety Group, International Iberian Nanotechnology Laboratory, Braga, Portugal
e-mail: ernesto.alfaro@inl.int

F. Murphy
Strathclyde Institute of Pharmacy and Biomedical Sciences, University of Strathclyde, Glasgow, UK

© The Author(s) 2025
E. Alfaro-Moreno, F. Murphy (eds.), *Nanosafety*,
https://doi.org/10.1007/978-3-031-93871-9_1

1 Introduction: Definition and Importance of Nanosafety

Nanosafety has emerged as a critical area of research and concern in the rapidly evolving field of nanotechnology. Nanomaterials (NMs), defined as materials with at least one dimension in the size range of 1–100 nanometers (nm), possess unique physical, chemical, and biological properties that can be utilized for a wide range of applications, from electronics and energy to medicine and consumer products. However, these same unique properties that make NMs appealing also may raise potential health and environmental concerns (Lebre et al. 2022). Nanosafety covers the study and management of the potential risks associated with the production, use, and disposal of NMs. As NMs interact with biological systems, they may exhibit altered toxicological profiles compared to their bulk counterparts, leading to concerns about their cytotoxicity, immunotoxicity, and genotoxicity (Donaldson et al. 2004; Nel et al. 2006). Furthermore, the environmental fate and behavior of NMs, such as their transport, transformation, and bioaccumulation, are crucial considerations for assessing their potential impact on ecosystems and human health (Maynard et al. 2006; Wiesner et al. 2006). The relevance of nanosafety research lies in its ability to ensure the responsible development and application of nanotechnology. By understanding the fundamental mechanisms of nanomaterial-biological system and nanomaterial-environment interactions, researchers can develop robust frameworks for risk assessment, design safer NMs, and implement effective control measures to mitigate potential hazards (Krug and Wick 2011; Oberdörster et al. 2005). This knowledge is essential for regulatory agencies, policymakers, and industry stakeholders to make informed decisions and maintain public trust in the safe use of NMs (Gottschalk and Nowack 2011). Advancing the field of nanosafety requires a multidisciplinary approach, integrating expertise from toxicology, materials science, environmental health, computational modeling, and other relevant disciplines. The combination of experimental investigations, advanced in vitro models, and predictive in silico tools can provide a comprehensive understanding of the complex and dynamic nature of nanomaterial interactions, ultimately paving the way for the responsible and sustainable development of nanotechnology. In the present chapter, we aim to provide a general overview of the different topics that are addressed in this book and can be used as an introductory document to grasp the basics of much more complex information provided in the other chapters.

2 Toxicology of Nanomaterials

In the following section, we will explore three critical aspects of nanosafety in relation to toxicity: cytotoxicity, immunotoxicity, and genotoxicity. These aspects are fundamental for understanding the potential health risks and safety considerations. Firstly, we will examine cytotoxicity, which refers to the ability of NMs to cause damage to cells, potentially leading to cell death or dysfunction. Next, we will discuss the basics of immunotoxicity. The immune system's primary role is to protect the body from harmful entities, and any disruption caused by NMs can lead to a

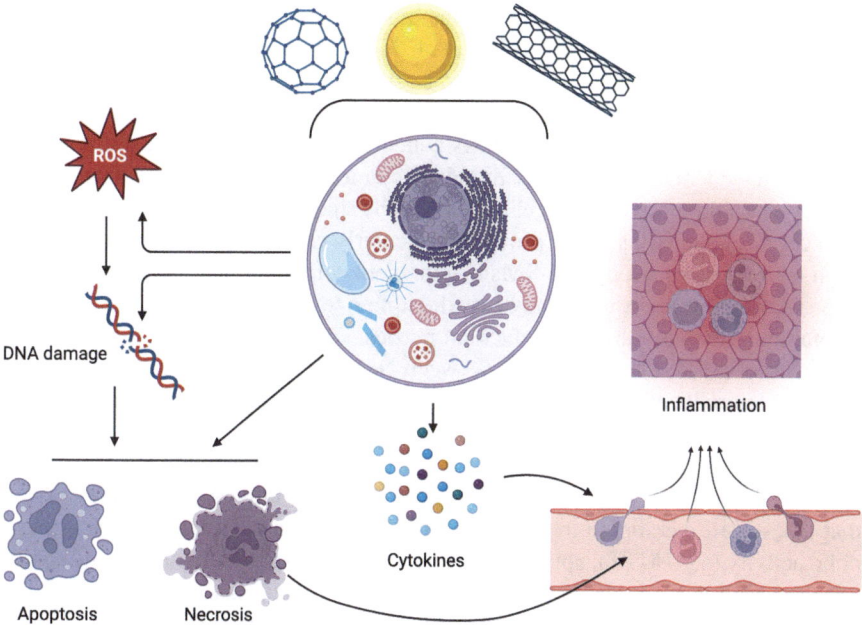

Fig. 1.1 General overview of potential toxic effects triggered by nanoparticles. After exposure, nanoparticles can trigger directly or indirectly, a number of effects, from oxidative stress increasing the concentration of reactive oxygen species (ROS), and the secretion of cytokines. As a consequence, we may induce cellular alterations including DNA damage, cellular death (i.e. apoptosis, necrosis) or induce inflammation. Created with Biorender

wide variety of effects. Finally, we will address genotoxicity, which concerns the potential of NMs to induce genetic damage. Such damage can result in mutations, cancer, and other type of alterations, making it a critical area of study for long-term health impacts. A summaraized view of these aspects is presented in Fig. 1.1.

2.1 Cytotoxicity

The unique physicochemical properties of NMs, such as their small size, high surface-to-volume ratio, and tailored surface chemistry, have the potential to elicit distinct biological responses compared to their bulk counterparts. One of the primary concerns in nanosafety research is the cytotoxicity of NMs, which refers to their ability to induce adverse effects on cellular function and viability.

The mechanisms by which NMs can exert cytotoxic effects are complex and can involve various pathways. At the cellular level, NMs may interfere with normal physiological processes, such as membrane integrity, organelle function, and signaling cascades. For example, certain NMs can generate reactive oxygen species (ROS), which can cause oxidative stress and damage to cellular macromolecules, leading to apoptosis or necrosis (Vilas-Boas et al. 2024). NMs may also disrupt mitochondrial function, impair cellular respiration, and trigger inflammatory responses (Donaldson et al. 2004; Nel et al. 2006; Vilas-Boas et al. 2024).

The cytotoxicity of NMs is influenced by a range of physicochemical properties, including size, shape, surface chemistry, and aggregation state (Albanese et al. 2012; Gatoo et al. 2014). Smaller NMs generally exhibit increased cellular uptake and may have a higher propensity to induce cytotoxic effects due to their increased surface area and reactivity (Nel et al. 2009). The shape of NMs can also play a role, with certain geometries, NMs proving particularly challenging for cellular uptake. High-aspect ratio of elongated NMs may prevent complete encapsulation into lysosomes during uptake by cells and cause disruption to lysosomal and cell membranes (Nagai and Toyokuni 2012). Surface properties, such as charge and functionalization, can modulate the interactions between NMs and cellular components, affecting their internalization, intracellular trafficking, and ultimately, their cytotoxicity (Saptarshi et al. 2013; Zhao et al. 2011).

To evaluate the cytotoxicity of NMs, there are a range of in vitro assays that measure different aspects of cellular function and viability. These include assays that assess cell membrane integrity (lactate dehydrogenase [LDH] release), metabolic activity (e.g., MTT), apoptosis (Annexin V/PI, caspase activation), and oxidative stress (ROS generation, glutathione depletion) (Guadagnini et al. 2015). The choice of cell lines (related to different species and different organs), exposure conditions (submerged, air-liquid interface), and endpoint measurements can significantly influence the outcomes of these assays and the interpretation of cytotoxicity data (Hirsch et al. 2011; Kroll et al. 2012). It is important to note that in vitro cytotoxicity studies provide valuable insights into the potential hazards of NMs, but their extrapolation to in vivo scenarios and human health risk assessment needs a thouroughful evaluation of the limitations and complexities involved. In vivo studies, along with the integration of computational modeling and advanced analytical techniques, remain necessary to develop a comprehensive understanding of nanomaterial toxicology and inform the safe design and application of NMs (Fadeel et al. 2018; Kagan et al. 2005).

2.2 Immunotoxicity

The immune system play a central role in the homeostasis of complex organisms. The original function of the immune system is to protect organisms from exogenous and non-self objects/organisms. When the immune system comes in contact with a potential pathogen a series of processes are triggered in the body to neutralise the exogenous agent. Overactivation, suppression or interference with immune processes are all forms of immunotoxicity that may lead to adverse outcomes in a population.

The unique physicochemical properties of NMs have raised concerns about their potential interactions with the human immune system and the potential immunotoxicity risks (Hofer et al. 2022). Immune system interactions with NMs have been a subject of extensive research in recent years. Studies have demonstrated that NMs can interact with various immune cells, such as macrophages, neutrophils, dendritic cells, lymphocytes, and natural killer cells, leading to a complex array of immunological responses (Cronin et al. 2020). The nature and extent of these interactions are heavily influenced by factors like nanoparticle size, surface properties, and material composition (Ray et al. 2021). Smaller nanoparticles, have been found to be more readily internalized by immune cells, potentially leading to altered cell function and signaling pathways (Dobrovolskaia and McNeil 2007). Surface charge and hydrophobicity of NMs can also play a significant role in their interactions with the immune system, as these properties can affect protein adsorption, cell recognition, and therefore cellular responses (Ray et al. 2021).

It has been described that the chemical composition of NMs, including the presence of certain functional groups or the release of ionic species, may trigger specific immune responses, ranging from inflammation to immunosuppression (Alsaleh and Brown 2018). Understanding these complex interactions is crucial for assessing the potential immunotoxicity of NMs and developing safe and effective NMs for biomedical applications. Alongside the exploration of immune system interactions, research efforts have also focused on the dual-edged nature of NMs' effects on the immune system, namely immunostimulation and immunosuppression (Aljabali et al. 2023). Certain NMs have been found to possess the ability to stimulate the immune system, which can be exploited for the development of effective vaccine adjuvants and immunotherapies (Ray et al. 2021). In contrast, the immunosuppressive potential of NMs may be desirable in specific applications, such as reducing inflammatory responses or modulating autoimmune disorders (Dobrovolskaia and McNeil 2007). The underlying mechanisms behind these immunomodulatory effects involve complex signaling pathways and interactions with various immune cells (Pondman et al. 2023). For instance, NMs can trigger the activation of pattern recognition receptors, leading to the release of proinflammatory cytokines and the recruitment of immune cells (Aljabali et al. 2023). Conversely, NMs may also induce the production of anti-inflammatory mediators or interfere with the activation and proliferation of immune cells, resulting in immunosuppression (Ray et al. 2021). Exposure to NMs have also been shown to affect immune cell functions such as phagocytosis and bacterial killing when challenged by secondary pathogens (DeLoid et al. 2016).

NMs have been shown to activate the complement system through the classical, alternative and lectin pathways with the prime consequence being surface opsonization, which subsequently leads to particle clearance. It has also been suggested that NMs could interfere with adaptive immune responses through physical interactions with immunoglobulins, T-cells or antigen presenting cells and as a result initiate or exacerbate hypersensitivity reactions (Alsaleh and Brown 2018).

Comprehensive in vitro and in vivo testing protocols have been established to assess the potential immunotoxicity of NMs (Boraschi et al. 2021). In vitro assays, such as cell-based assays and cytokine/chemokine secretion profiles, provide valuable insights into the direct effects of NMs on immune cells. These assays allow for the evaluation of parameters like cell viability, proliferation, activation, and cytokine production, which can serve as early indicators of potential immunotoxicity.

Significant progress in understanding the complex and diverse interactions between NMs and the immune system on a cellular level has been achieved however modulation of the immune system is a complex and dynamic process dependent on intercellular interactions and temporal responses and therefore incredibly difficult to model with simple in vitro systems. In vivo studies using animal models, on the other hand, allow for the evaluation of the systemic immune responses and potential adverse effects of NMs on the entire immune system (Boraschi et al. 2021). These studies can assess the impact of NMs on various immune functions, such as antibody production, T-cell responses, and the function of innate immune cells, providing a more comprehensive understanding of the immunotoxicity profile. Sumarasing, the study of nanomaterial-immune system interactions and the associated immunotoxicity risks is a critical area of focus in the field of nanosafety. However the molecular mechanisms of NM-induced immunological responses are not yet fully understood. Novel approach methodologies such as multicell in vitro tissue models and in vitro-in vivo bridging models may serve as suitable low-cost and medium/high through put models to shed further light on the immunomodulatory effects of NMs and underlying mechanisms of action.

2.3 Genotoxicity

The widespread and rapidly expanding use of engineered NMs in various applications has raised significant concerns about their potential to induce genotoxicity, which refers to the ability of a substance to cause damage to genetic material (DNA) (Chatterjee and Alfaro-Moreno 2023). Understanding the mechanisms underlying nanomaterial-induced genotoxicity and the development of reliable assessment methods are crucial for ensuring the safe and responsible use of these emerging technologies.

At the mechanistic level, NMs can interact with genetic material through multiple pathways, leading to various types of DNA damage (Shukla et al. 2021). One of the primary mechanisms is the direct interaction of NMs with DNA, where they can intercalate or bind to the DNA, causing structural alterations and disrupting normal DNA functions (Thongkumkoon et al. 2014). This can result in the formation of DNA adducts, single-strand breaks, and double-strand breaks, which can ultimately lead to genetic instability and potentially mutagenic effects (Ishino et al. 2015). Additionally, NMs can induce indirect genotoxicity through the generation of reactive oxygen species (ROS) and the subsequent induction of oxidative stress (Vilas-Boas et al. 2024). The high surface area and reactive nature of certain NMs can lead to the increased production of ROS, which can damage DNA through the formation of oxidized DNA bases, DNA strand breaks, and chromosomal abnormalities (Chatterjee and Alfaro-Moreno 2023). Another mechanism of genotoxicity induced by particles that has been described involves the disruption of cellular processes, such as mitosis and cell division (Moreno et al. 1997). Similar action has been described for NMs that can interfere with the normal functioning of the mitotic spindle, leading to alterations including chromosomal aberrations, aneuploidy, and other forms of genomic instability (Xiao et al. 2023).

To assess the potential genotoxicity of NMs, a range of in vitro and in vivo testing approaches have been developed and refined (Verdon et al. 2022). These testing methods aim to evaluate various endpoints of genetic damage, including gene mutations, chromosomal aberrations, and DNA strand breaks. In vitro testing methods, such as the micronucleus assay, and the comet assay, are among the more commonly employed to screen for nanoparticles genotoxic potential (Landsiedel et al. 2022). The Ames assay is a classic and robust method for identifying and evaluating chemical mutagens. However it is not generally recommended for NM genotoxicity studies as NMs may be unable to penetrate the rigid outer double-membrane of gram negative bacteria, therefore mutagenic events observed may not be nano-object-specific (Elespuru et al. 2018). These cell-based assays provide a rapid and cost-effective means of evaluating potential genotoxicity, as they allow for the direct exposure of cells to NMs and the subsequent analysis of specific genetic endpoints. In vivo testing approaches, on the other hand, involve the use of animal models to assess the genotoxic effects of NMs in a more comprehensive and physiologically relevant manner. These studies can evaluate the systemic distribution and biodistribution of NMs, as well as their impacts on various organs and tissues, including the genetic material. Assays such as the comet assay are readily employed on cells and tissue isolated after in vivo exposure (Di Ianni et al. 2022).

3   Advanced In Vitro Models for Nanosafety Assessment

The forthcoming section discuss the utilization of advanced in vitro models, specifically focusing on organ-on-a-chip systems, organoids, and 3D cellular models. These cutting-edge technologies represent a significant leap forward in our ability to simulate human physiology and evaluate the safety of NMs with unprecedented precision. Firstly, we will explore organ-on-a-chip devices, which integrate microfluidic technology with living cells to mimic the microarchitecture and functions of human organs. Then, we will cover organoids, which could be understood as miniature, simplified versions of organs grown in vitro. Additionally, we will examine 3D cellular models, which provide a more realistic representation of the cellular environment compared to conventional monolayer cultures. Figure 1.2 presents a condensed image of these approaches.

3.1   Organ-on-a-Chip Models

Traditional in vitro cell-based assays have been widely used for initial nanosafety screening, but these simplistic models often fail to accurately predict the complex biological responses observed in vivo. To address this challenge, sophisticated organ-on-a-chip systems, that better recapitulate the physiological and functional characteristics of human tissues and organs, have been developed.

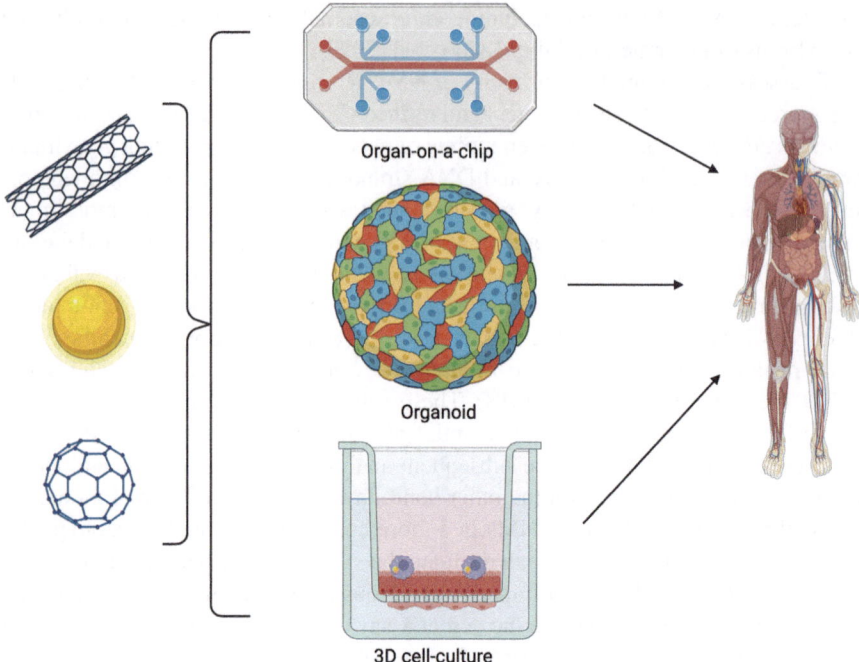

Fig. 1.2 Condensed overview of potential different advanced in vitro models. Among the many strategies to evaluate the potential effects of NMs, organ-on-a-chip, organoids and 3D cell cultures are the most promising and commonly used. Created with Biorender

Organ-on-a-chip models are devices that integrate living human cells, often in a three-dimensional (3D) configuration, and may include microfluidic channels that mimic the dynamic flow and mechanical aspects found in the human body (Brassard and Lutolf 2019). These advanced in vitro platforms have several advantages over traditional cell culture models, including the ability to recreate organ-specific physiological functions, maintain long-term cell viability and differentiation, and facilitate the study of cell-cell and cell-matrix interactions (Rossi et al. 2018).

Organ-on-a-chip systems can be used to evaluate the transport, fate, and biological impact of NMs, providing a more relevant and predictive assessment of nanosafety compared to conventional 2D cell culture models (Costa et al. 2023). One of the most well-studied organ-on-a-chip models for nanosafety assessment is the lung-on-a-chip, which recapitulates the alveolar-capillary interface of the human lung (Sengupta et al. 2023). Lung-on-a-chip models have been used to investigate the translocation (Braakhuis et al. 2015) and toxicity of various engineered NMs. Similarly, gut-on-a-chip and liver-on-a-chip models have been employed to study the absorption, metabolism, and cytotoxicity of NMs (Ashammakhi et al. 2020; Messelmani et al. 2022). More recently, researchers have developed multi-organ-on-a-chip systems that can simulate the interactions between different tissues, providing a more comprehensive platform for evaluating the systemic effects of NMs

(Sung 2021). Despite the significant advances in organ-on-a-chip technology, several challenges remain in the application of these models for nanosafety assessment. One key challenge is the need for standardized protocols and validation procedures to ensure the reproducibility and reliability of the experimental data. Additionally, the complexity of these systems can make it difficult to isolate the specific mechanisms underlying the observed biological responses to NMs.

Further research is needed to better understand the relationship between the physicochemical properties of NMs and their interactions with the various cell types and tissue structures present in organ-on-a-chip models (Chen et al. 2021). Looking to the future, the integration of organ-on-a-chip technology with other advanced in vitro techniques, such as high-throughput screening, computational modeling, and multi-omics analysis, holds great promise for improving the efficiency and accuracy of nanosafety assessment (Kumar et al. 2024). By combining these complementary approaches, a more comprehensive understanding of the biological effects of NMs could be achieved, which will be crucial for the safe and responsible development of NMs for various applications.

3.2 Organoid Models

Organoid models are three-dimensional (3D) multicellular structures derived from stem cells or organ-specific progenitor cells that self-organize and recapitulate the key structural and functional characteristics of their corresponding native tissues (Jackson and Lu 2016). These self-organizing, self-renewing, and highly differentiated in vitro models have several advantages over traditional two-dimensional (2D) cell cultures, including the ability to mimic the complex cellular interactions, extracellular matrix, and physiological microenvironment found in vivo (Rossi et al. 2018).

One of the key characteristics of organoid models that makes them particularly well-suited for nanosafety assessment is their ability to accurately represent the target organ's cellular composition and architecture. For instance, liver organoids can contain hepatocytes, Kupffer cells, endothelial cells, and stellate cells, allowing for the evaluation of nanomaterial interactions with the various cell types involved in hepatic function and toxicity (Takebe et al. 2013). Similarly, intestinal organoids can recapitulate the polarized epithelial layer, goblet cells, and Paneth cells, providing a physiologically relevant model for studying the absorption, metabolism, and toxicity of NMs in the gastrointestinal tract (Sato et al. 2009). Beyond organ-specific organoid models, researchers have also developed multi-organoid systems that integrate multiple organ-derived organoids, such as liver, intestine, and kidney, to simulate the complex interorgan interactions that occur in the human body (Shariati et al. 2021). These multi-organoid platforms can be used to investigate the systemic effects of NMs, including their biodistribution, metabolism, and potential for causing multi-organ toxicity.

While organoid models offer significant advantages over traditional in vitro approaches, they also face several limitations and considerations that must be addressed for their effective use in nanosafety assessment. One key challenge is the inherent variability in organoid formation and maturation, which can lead to inconsistencies in experimental results (Zhao et al. 2022). To overcome this, the development of standardized protocols and quality control measures to ensure the reproducibility and reliability of organoid-based studies will be critical. Another important consideration is the complexity of organoid models, which can make it challenging to isolate the specific mechanisms underlying the observed biological responses to NMs. To address this, the integration of advanced analytical techniques, such as single-cell omics and high-content imaging, can provide valuable insights into the cell-level interactions and signaling pathways involved in the nanomaterial-induced effects (Qin and Tape 2021).

3.3 3D Cellular Models

Both organ-on-a-chip and stem-cell derived organoid systems are resource and time-intensive and may require a significant level of technical expertise and specialised equipment e.g., microfluidic system, which present a barrier to the widespread adoption and use for screening high numbers of NMs. As the field of nanotechnology continues to expand, the need for reliable and physiologically relevant in vitro models for the rapid assessment of hazard potential has become increasingly pressing. Traditional two-dimensional (2D) cell culture systems, while valuable for initial screening, often fail to capture the complex cellular interactions and microenvironmental cues present in the human body. To address this limitation, researchers have turned to the development of three-dimensional (3D) cellular models, which offer a more realistic platform for evaluating the potential toxicity and biological effects of NMs.

The importance of 3D culture in nanosafety assessment lies in its ability to better mimic the in vivo cellular organization, extracellular matrix (ECM) composition, and signaling gradients that influence how cells respond to NMs (Alfaro-Moreno et al. 2008; Ravi et al. 2017). In 2D cultures, cells are typically grown as a monolayer, resulting in a flattened morphology and altered gene expression profiles compared to their in vivo counterparts. In contrast, 3D models, can be designed to combine relevant cells (primary cells or cell-lines) in a physiologically relevant manner, better recapitulating the complex tissue architecture and cellular microenvironment (Duval et al. 2017).

Several techniques have been employed to generate 3D cellular models for nanosafety assessment, each with its own advantages and limitations. Spheroid formation, for example, can be achieved through the use of hanging drop cultures, non-adherent surfaces, or microfluidic devices, resulting in compact, self-organizing aggregates of cells (Ryu et al. 2019). Hydrogel-based cultures, which incorporate cells within a 3D extracellular matrix, provide a more biomimetic environment for studying the interactions between NMs and the cellular microenvironment (Giobbe et al. 2019).

The validation and correlation of 3D cellular models with in vivo data is crucial for their effective use in nanosafety assessment. A variety of techniques to evaluate the predictive power of 3D models, including comparative analyses of gene expression profiles, cytotoxicity responses, and physiological functions are currently used (Ravi et al. 2017). For instance, studies have shown that 3D liver models better recapitulate the metabolic capabilities and sensitivity to hepatotoxic compounds compared to 2D cultures, highlighting their potential for improved prediction of induced liver toxicity (Serras et al. 2021). Furthermore, the integration of advanced analytical techniques, such as high-content imaging, single-cell omics, and computational modeling, has enabled a more comprehensive understanding of the cellular and molecular mechanisms underlying the interactions between NMs and 3D cellular models (Diamante et al. 2024). These approaches can provide valuable insights into the specific pathways and biological responses that are altered in response to nanomaterial exposure, informing the development of safer and more effective NMs. Despite the promising progress in the field, the widespread adoption of 3D cellular models for nanosafety assessment faces several challenges. One key issue is the inherent variability in 3D model generation and maturation, which can lead to inconsistencies in experimental results. To address this, researchers are working on developing standardized protocols and quality control measures to ensure the reproducibility and reliability of 3D model-based studies.

Another challenge is the scalability and throughput of 3D cellular models, which can be limited compared to traditional 2D cultures. To overcome this, development of automated, high-throughput platforms for the generation, maintenance, and analysis of 3D cellular models can be used, enabling their more efficient integration into the nanosafety assessment pipeline (Ranga et al. 2014).

As the field of nanotechnology continues to evolve, the integration of advanced 3D cellular models with other in vitro and in silico approaches will be crucial for addressing the complex safety and toxicity concerns associated with engineered NMs. By providing a more physiologically relevant platform for studying the interactions between NMs and human tissues, 3D cellular models can contribute to a better understanding of the potential health and environmental risks, ultimately supporting the safe and responsible development of nanotechnology.

4   In Vivo-In Vitro Bridging Models

Traditional toxicity testing methods relying solely on in vitro cell culture models or in vivo animal studies each have their own limitations, leading scientists to explore alternative "bridging" approaches that can help extrapolate findings between the two. The use of organisms that allow a quick evaluation of toxicity within the dynamic environment of a whole organism with reduced ethical concerns related to the use of animals gaining momentum in the field of nanosafety. Of special interest are invertebrate models such as the nematode Caenorhabditis elegans, the planarian Schmidtea mediterranea, the fruit fly Drosophila melanogaster, Galleria mellonella

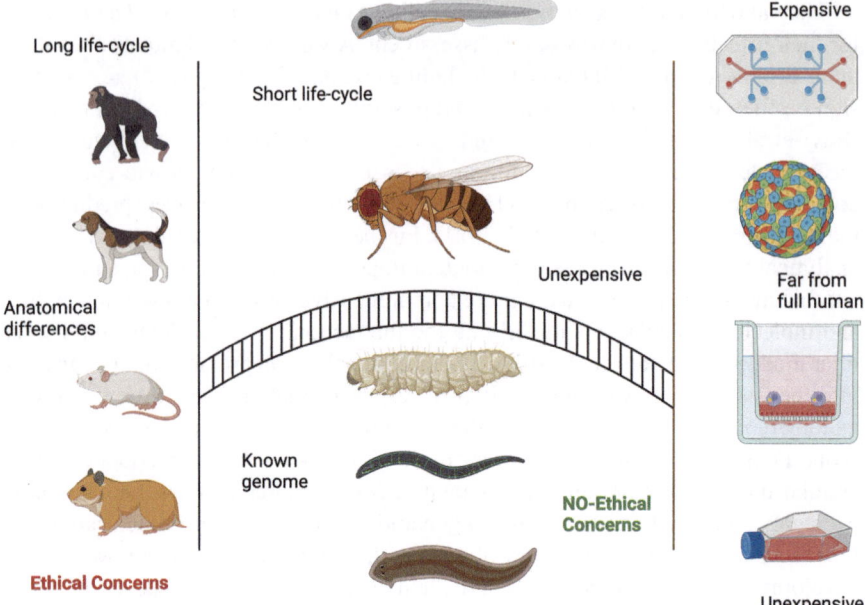

Fig. 1.3 Zebra fish embryos, Drosophila melanogaster, Galleria mellonella, Caenobarditis elegans and Planaria are among the most popular models that can be considered as a bridge between in vivo and in vitro approaches. On one side, the in vivo models are commonly criticized due to costs, anatomical and physiological differences between species but of most importance, ethical concerns. On the other side, the in vitro models have lower physiological relevance but avoid the ethical concerns and high costs associated with animal models.. The bridging models have the positive aspects of having a short life, relatively unexpensive, the genome of the animals is well known, and there are no ethical concerns. Created with Biorender

larvae (Leynen et al. 2024), and embryo fish models. In Fig. 1.3 we exemplify some of the bridging models that are more extensively explored in other chapters.

4.1 Bridging the Gap Between In Vivo and In Vitro

These small, genetically tractable model organisms offer several advantages for nanotoxicology testing, including rapid life cycles, transparent bodies for real-time imaging, and a high degree of evolutionary conservation with humans (Wu et al. 2019).

For example, the nematode C. elegans has been extensively used to study the toxicity and biodistribution of a wide range of engineered NMs, including metal oxides, quantum dots, and carbon-based NMs (Roh et al. 2009). It has been described that C. elegans exhibits dose-dependent responses to different types of particulate matter, including altered locomotive behavior, reproductive defects, and oxidative stress—findings that often align with those seen in mammalian models

(Chatterjee et al. 2024; Kong et al. 2017; Pluskota et al. 2009). Similarly, planarians have been employed to evaluate the toxicity and regenerative effects of NMs, taking advantage of their remarkable ability to regenerate entire body parts (Leynen et al. 2019).

4.2 Limitations and Future Developments

One of the primary limitations of these models is the inherent physiological and anatomical differences between the organisms and humans. For example, C. elegans, a nematode, has a relatively simple nervous system compared to the complexity of the human brain (Markaki and Tavernarakis 2020). Similarly, Planaria, possesses a nervous system that may not accurately reflect the integrated neural networks found in vertebrates (Agata et al. 1998). These fundamental differences can introduce challenges in directly translating findings from these models to human biology and disease, therefore highlighting the need for careful interpretation of results and the integration of multiple model systems to improve the predictive power of these bridging models (Wu et al. 2019).

These bridging models not only provide valuable insights into the mechanisms of nanotoxicity but also serve as cost-effective, high-throughput screening platforms to prioritize candidate NMs for further in-depth study. By combining the strengths of both in vitro and in vivo systems, these bridging models hold promise to accelerate the safe development and responsible implementation of nanotechnology in the years to come.

5 Environmental Health Aspects of Nanomaterials

The interaction of NMs with the environment is very complex, and here we will introduce some of the key factors including fate, bioacummulation, trophic transfer, and ecotoxicology which influence the risk assessment related to these materials. The environmental fate of NMs involves tracking their journey from release into the environment to their ultimate desposition and/or degradation. This includes examining NM persistence, transformation, and mobility in different environmental compartments such as air, water, and soil. Studies have shown that the environmental fate of NMs is governed by a complex interplay of factors, including the specific properties of the NM, the characteristics of the receiving environment, and the presence of other chemical and biological components. For instance, the size, shape, and surface chemistry of NMs can affect their interactions with natural organic matter, leading to the formation of heteroaggregates that may alter their mobility and bioavailability (Lowry et al. 2012). Similarly, the pH, ionic strength, and presence of other pollutants in the environment can influence the transformation and persistence of NMs, with processes such as dissolution, aggregation, and sedimentation playing crucial roles (Metreveli et al. 2016).

Bioaccumulation and trophic transfer are critical processes to consider, as NMs can accumulate in organisms and transfer through food webs, potentially leading to adverse effects on wildlife and human health. NMs can be taken up by organisms at the base of the food chain, such as algae and bacteria, and subsequently transferred to higher-level consumers through the process of trophic transfer (Maharramov et al. 2019). This can result in the bioaccumulation of NMs in the tissues of higher-level organisms, potentially leading to adverse effects. Metal and metal oxide nanoparticles have been reported to be transported through aquatic and terrestrial food webs (Deng et al. 2017). The extent of bioaccumulation and trophic transfer is influenced by factors such as the physico-chemical properties of the NMs, the feeding behavior and metabolic processes of the organisms, and the complexity of the food web (Croteau et al. 2014; Holbrook et al. 2008).

Ecotoxicology studies the effects of NMs on living organisms within ecosystems. The ecotoxicological effects of NMs on various aquatic and terrestrial organisms have been extensively studied, revealing a range of potential impacts on survival, growth, reproduction, and other key physiological processes (Gambardella and Pinsino 2022). In aquatic ecosystems, studies have reported adverse effects of ENMs on the growth, behaviour, and survival of organisms such as algae, crustaceans, fish, and amphibians (do Amaral et al. 2019; Prajitha et al. 2019). For example, silver nanoparticles have been shown to impair the development of zebrafish (Xin et al. 2015), while carbon nanotubes can cause respiratory distress and mortality in aquatic species (Sohn et al. 2015). In terrestrial environments, the ecotoxicological impacts of NMs have been observed in organisms ranging from soil microbes, invertebrates and plants (Judy and Bertsch 2014). Nanoparticles can alter soil microbial community structure and function, affect the growth and reproduction of earthworms and other soil organisms, and interfere with the uptake and translocation of nutrients in plants (Tripathi et al. 2017).

These effects can vary widely depending on the characteristics of the materials, and therefore ecotoxicological testing methods have incorporating advanced techniques to evaluate their toxicity accurately. Standardized protocols and guidelines are essential to ensure consistent and reliable results across different studies. A comprehensive approach integrating data from environmental fate, bioaccumulation, and ecotoxicology studies will be required to evaluate the environmental risks posed by NMs and mitigate potential adverse impacts. As nanotechnology continues to advance, ongoing research and updated regulatory frameworks are crucial to safeguard environmental health while harnessing the benefits of NMs (Fig. 1.4).

5.1 Standardized Testing Methods

To ensure the reliable and reproducible assessment of the (eco)toxicological impacts of NMs, the development and validation of standardized testing methods is crucial. Several international organizations, such as the Organization for Economic Co-operation and Development (OECD) and the International Organization for

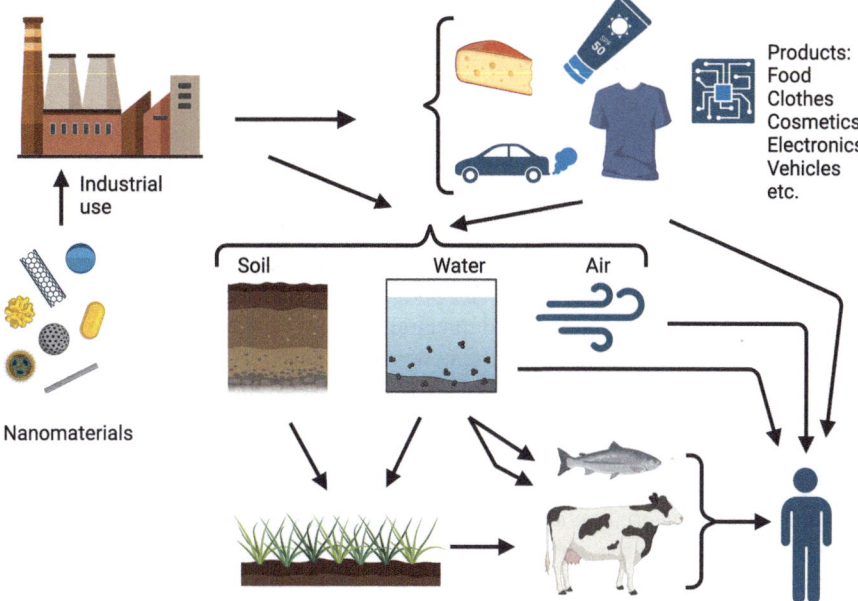

Fig. 1.4 The use of NMs in industry, to produce a wide variety of products, may have different life cycles and along this cycle, may end in different compaitments, including the trophic chain. The effects on the environment and related organisms, including humans, is the center of the evaluation and riskassessment of NMs impact in the environment

Standardization (ISO), have been actively working on the development of guidelines and protocols for nanomaterial toxicity testing (OECD 2021; ISO 2018).

These standardized methods address key challenges in nanomaterial toxicology, such as the characterization of NMs, the preparation of test media, the selection of appropriate test organisms and models, and the evaluation of endpoints relevant to human and environmental risk assessment. By establishing these frameworks, researchers can generate high-quality data that can be compared across studies and used to support regulatory decision-making.

6 Cheminformatics and In Silico Models for Nanosafety

6.1 Cheminformatics Approaches

As the field of nanotechnology continues to rapidly advance, there is a growing need to develop reliable methods for predicting the physicochemical properties and potential toxicity of novel NMs. Cheminformatics techniques have emerged as a powerful tool in this effort, allowing researchers to leverage large datasets and computational modeling to screen and prioritize NMs for further testing (Fourches et al. 2010). By applying machine learning algorithms to nanomaterial characterization data, it is possible to build predictive models that can estimate key properties such as size, shape, surface chemistry, and reactivity (Tao et al. 2021). These models can then be used to guide the design of safer NMs and accelerate the risk assessment process.

A key focus of cheminformatics research in nanosafety has been the development of quantitative structure-activity relationship (QSAR) models that can relate the physicochemical features of NMs to their biological effects (Singh et al. 2024). These models leverage statistical techniques to identify the structural descriptors that best explain the observed toxicity data, enabling the prediction of toxicity for new NMs. QSAR modeling has been applied to a wide range of NMs, including metal oxides, carbon nanotubes, and quantum dots, with promising results (Kumar and Kumar 2021; Meneses et al. 2023; Wyrzykowska et al. 2022). However, the complex and often unique properties of NMs present challenges in developing robust and generalizable QSAR models, requiring continued research in this area.

In addition to QSAR modeling, cheminformatics researchers have explored the use of other data mining and machine learning techniques to enhance our understanding of nanomaterial toxicity. These approaches leverage large datasets of nanomaterial characterization and biological response data to identify patterns and relationships that may not be easily discernible through traditional statistical analysis (Gajewicz et al. 2012; Lebre et al. 2022). For example, clustering algorithms can be used to group NMs with similar properties, while feature selection methods can help identify the most important predictors of toxicity (Pathakoti et al. 2014). Machine learning algorithms, such as artificial neural networks and support vector machines, have also been applied to develop predictive models for nanomaterial hazards (Zhou et al. 2023). As the availability of high-quality nanomaterial data continues to improve, these computational techniques will become increasingly valuable in advancing our understanding of nanosafety.

6.2 In Silico Models for Nanosafety

In parallel with the development of cheminformatics approaches, researchers have also made significant progress in the field of computational toxicology, leveraging in silico models to predict the potential hazards of NMs. These models, which are often based on physiologically-based pharmacokinetic (PBPK) or adverse outcome pathway (AOP) frameworks, can simulate the absorption, distribution, metabolism, and excretion (ADME) of NMs, as well as the biological responses that may lead to adverse effects (Chen et al. 2022; Halappanavar et al. 2020; Utembe et al. 2020). By incorporating data on nanomaterial properties, exposure scenarios, and biological mechanisms, these computational models can provide valuable insights into the potential toxicity of NMs, helping to prioritize materials for further testing and inform the design of safer nanoproducts.

In addition to toxicity prediction, in silico models have also been developed to simulate the environmental fate and transport of NMs, as well as human and ecological exposures (Lebre et al. 2022). These exposure and fate models can integrate information on nanomaterial physicochemical properties, environmental transformations, and transport processes to estimate concentrations in various environmental compartments and exposure pathways (Gottschalk et al. 2013; Ortiz-Galvez et al. 2024). By coupling exposure models with toxicity data, it is possible to perform risk assessments and develop safe exposure limits for NMs (Wigger et al. 2020). Furthermore, these models can be used to inform the design of NMs with improved environmental compatibility and reduced ecological impacts (Corsi et al. 2023).

To fully explore the potential of computational approaches in nanosafety, researchers have begun developing integrative in silico frameworks that combine cheminformatics, computational toxicology, and exposure modeling capabilities. These holistic platforms can integrate data from various sources, including experimental studies, literature, and high-throughput screening, to provide a comprehensive assessment of nanomaterial risks. By seamlessly linking nanomaterial properties, biological responses, and exposure scenarios, these frameworks can support decision-making throughout the product development lifecycle, from early-stage material design to post-market surveillance. As the field of computational nanotoxicology continues to evolve, these integrative in silico tools will become increasingly valuable in the safe and sustainable development of NMs.

6.3 Challenges and Future Perspectives

One of the key challenges in the development of cheminformatics and in silico models for nanosafety is the availability and quality of experimental data. NMs can exhibit a wide range of physicochemical properties and biological effects, and the experimental characterization of these materials can be time-consuming and resource-intensive (Krug 2014). Furthermore, the lack of standardized testing protocols and reporting guidelines has resulted in a heterogeneous and often incomplete dataset, which can limit the ability to develop robust and generalizable predictive models (Oomen et al. 2014). Addressing these data gaps will require concerted efforts to establish comprehensive nanomaterial databases, harmonize experimental methods, and promote data sharing among the research community.

Another key challenge in the field of computational nanotoxicology is the validation and regulatory acceptance of in silico models. Regulatory agencies have traditionally relied on experimental toxicology data to support risk assessment and decision-making, and there is a need to demonstrate the reliability and predictive accuracy of computational models before they can be widely adopted (Burden et al. 2017). This will require the development of robust validation protocols, the use of high-quality experimental data for model training and testing, and the establishment of clear guidelines for the appropriate use and interpretation of in silico tools (Toropov et al. 2012). As these validation and acceptance processes are established, computational approaches will become increasingly integrated into the risk assessment and regulation of NMs.

Looking to the future, the continued advancement of computational nanotoxicology will depend on the integration of multiple disciplines, including materials science, biology, and computer science. Researchers will need to work collaboratively to develop novel experimental techniques for comprehensive nanomaterial characterization, as well as innovative computational algorithms and modeling frameworks that can accurately capture the complex interactions between NMs and biological systems (Valsami-Jones and Lynch 2015). Additionally, the integration of in silico models with high-throughput screening and other advanced experimental methods will be crucial in generating the large, high-quality datasets needed to support the development of predictive models (Verma et al. 2023). As these multidisciplinary efforts progress, computational approaches will play an increasingly central role in the safe and sustainable development of NMs, enabling researchers and regulators to make more informed decisions and accelerate the translation of NMs into real-world applications.

7 Final Remarks and Conclusions

Summarizing, Nanosafety is essential as we navigate the complexities introduced by nanomaterials across various industries. The critical aspects of nanosafety encompass its definition and significance in ensuring public health and environmental protection. A thorough examination of the toxicological profiles of nanomaterials reveals important concerns regarding cytotoxicity, immunotoxicity, and genotoxicity. Advanced in vitro models, such as organ-on-a-chip and 3D cellular systems, play a vital role in enhancing safety assessments. The integration of in vivo-in vitro bridging models represents a significant advancement in understanding nanomaterial behavior and potential risks, addressing the limitations of traditional approaches. Additionally, the environmental health implications of nanomaterials, including their fate, behavior, and ecotoxicological effects, highlight the importance of standardized testing methods for effective risk assessment. Finally, chemoinformatics and in silico models offer promising avenues for nanosafety research. A collaborative and multidisciplinary approach is crucial for developing safe nanotechnology and fostering public trust as the field progresses.

Acknowledgements This work was partially funded by the European Union's H2020 project Sinfonia (N.857253) and Horizon Europe Projects LEARN (N.101057510), iCare (N.101092971).

References

Agata K, Soejima Y, Kato K et al (1998) Structure of the planarian central nervous system (CNS) revealed by neuronal cell markers. Zool Sci 15(3):433–440. https://doi.org/10.2108/zsj.15.433

Albanese A, Tang PS, Chan WCW (2012) The effect of nanoparticle size, shape, and surface chemistry on biological systems. Annu Rev Biomed Eng 14(1):1–16. https://doi.org/10.1146/annurev-bioeng-071811-150124

Alfaro-Moreno E, Nawrot TS, Vanaudenaerde BM et al (2008) Co-cultures of multiple cell types mimic pulmonary cell communication in response to urban PM10. Eur Respir J 32(5). https://doi.org/10.1183/09031936.00044008

Aljabali AA, Obeid MA, Bashatwah RM et al (2023) Nanomaterials and their impact on the immune system. Int J Mol Sci 24(3):2008. https://doi.org/10.3390/ijms24032008

Alsaleh NB, Brown JM (2018) Immune responses to engineered nanomaterials: current understanding and challenges. Curr Opin Toxicol 10:8–14. https://doi.org/10.1016/j.cotox.2017.11.011

Ashammakhi N, Nasiri R, de Barros NR et al (2020) Gut-on-a-Chip: current progress and future opportunities. Biomaterials 255:120196. https://doi.org/10.1016/j.biomaterials.2020.120196

Boraschi D, Li D, Li Y et al (2021) In vitro and in vivo models to assess the immune-related effects of nanomaterials. Int J Environ Res Public Health 18(22):11769. https://doi.org/10.3390/ijerph182211769

Braakhuis HM, Kloet SK, Kezic S et al (2015) Progress and future of in vitro models to study translocation of nanoparticles. Arch Toxicol 89(9):1469–1495. https://doi.org/10.1007/s00204-015-1518-5

Brassard JA, Lutolf MP (2019) Engineering stem cell self-organization to build better organoids. Cell Stem Cell 24(6):860–876. https://doi.org/10.1016/j.stem.2019.05.005

Burden N, Aschberger K, Chaudhry Q et al (2017) The 3Rs as a framework to support a 21st century approach for nanosafety assessment. Nano Today 12:10–13. https://doi.org/10.1016/j.nantod.2016.06.007

Chatterjee N, Alfaro-Moreno E (2023) In vitro cell transformation assays: a valuable approach for carcinogenic potentiality assessment of nanomaterials. Int J Mol Sci 24(9). https://doi.org/10.3390/ijms24098219

Chatterjee N, González-Durruthy M, Costa MD et al (2024) Differential impact of diesel exhaust particles on glutamatergic and dopaminergic neurons in caenorhabditis elegans: a neurodegenerative perspective. Environ Int 186:108597. https://doi.org/10.1016/j.envint.2024.108597

Chen X, Zhang YS, Zhang X et al (2021) Organ-on-a-Chip platforms for accelerating the evaluation of nanomedicine. Bioact Mater 6(4):1012–1027. https://doi.org/10.1016/j.bioactmat.2020.09.022

Chen Q, Riviere JE, Lin Z (2022) Toxicokinetics, dose–response, and risk assessment of nanomaterials: methodology, challenges, and future perspectives. WIREs Nanomed Nanobiotechnol 14(6). https://doi.org/10.1002/wnan.1808

Corsi I, Venditti I, Trotta F et al (2023) Environmental safety of nanotechnologies: the eco-design of manufactured nanomaterials for environmental remediation. Sci Total Environ 864:161181. https://doi.org/10.1016/j.scitotenv.2022.161181

Costa S, Vilas-Boas V, Lebre F et al (2023) Microfluidic-based skin-on-chip systems for safety assessment of nanomaterials. Trends Biotechnol. https://doi.org/10.1016/j.tibtech.2023.05.009

Cronin JG, Jones N, Thornton CA et al (2020) Nanomaterials and innate immunity: a perspective of the current status in nanosafety. Chem Res Toxicol 33(5):1061–1073. https://doi.org/10.1021/acs.chemrestox.0c00051

Croteau M-N, Misra SK, Luoma SN et al (2014) Bioaccumulation and toxicity of CuO nanoparticles by a freshwater invertebrate after waterborne and dietborne exposures. Environ Sci Technol 48(18):10929–10937. https://doi.org/10.1021/es5018703

DeLoid G, Casella B, Pirela S et al (2016) Effects of engineered nanomaterial exposure on macrophage innate immune function. NanoImpact 2:70–81. https://doi.org/10.1016/j.impact.2016.07.001

Deng R, Lin D, Zhu L et al (2017) Nanoparticle interactions with co-existing contaminants: joint toxicity, bioaccumulation and risk. Nanotoxicology 11(5):591–612. https://doi.org/10.1080/17435390.2017.1343404

Di Ianni E, Møller P, Cholakova T et al (2022) Assessment of primary and inflammation-driven genotoxicity of carbon black nanoparticles *in vitro* and *in vivo*. Nanotoxicology 16(4):526–546. https://doi.org/10.1080/17435390.2022.2106906

Diamante G, Ha SM, Wijaya D et al (2024) Single cell multiomics systems biology for molecular toxicity. Curr Opin Toxicol 39:100477. https://doi.org/10.1016/j.cotox.2024.100477

do Amaral DF, Guerra V, Motta AGC et al (2019) Ecotoxicity of nanomaterials in amphibians: a critical review. Sci Total Environ 686:332–344. https://doi.org/10.1016/j.scitotenv.2019.05.487

Dobrovolskaia MA, McNeil SE (2007) Immunological properties of engineered nanomaterials. Nat Nanotechnol 2(8):469–478. https://doi.org/10.1038/nnano.2007.223

Donaldson K, Stone V, Tran CL et al (2004) Nanotoxicology. Occup Environ Med 61(9):727–728. https://doi.org/10.1136/oem.2004.013243

Duval K, Grover H, Han L-H et al (2017) Modeling physiological events in 2D vs. 3D cell culture. Physiology 32(4):266–277. https://doi.org/10.1152/physiol.00036.2016

Elespuru R, Pfuhler S, Aardema MJ et al (2018) Genotoxicity assessment of nanomaterials: recommendations on best practices, assays, and methods. Toxicol Sci 164(2):391–416. https://doi.org/10.1093/toxsci/kfy100

Fadeel B, Bussy C, Merino S et al (2018) Safety assessment of graphene-based materials: focus on human health and the environment. ACS Nano 12(11):10582–10620. https://doi.org/10.1021/acsnano.8b04758

Fourches D, Pu D, Tassa C et al (2010) Quantitative nanostructure–activity relationship modeling. ACS Nano 4(10):5703–5712. https://doi.org/10.1021/nn1013484

Gajewicz A, Rasulev B, Dinadayalane TC et al (2012) Advancing risk assessment of engineered nanomaterials: application of computational approaches. Adv Drug Deliv Rev 64(15):1663–1693. https://doi.org/10.1016/j.addr.2012.05.014

Gambardella C, Pinsino A (2022) Nanomaterial ecotoxicology in the terrestrial and aquatic environment: a systematic review. Toxics 10(7):393. https://doi.org/10.3390/toxics10070393

Gatoo MA, Naseem S, Arfat MY et al (2014) Physicochemical properties of nanomaterials: implication in associated toxic manifestations. Biomed Res Int 2014:1–8. https://doi.org/10.1155/2014/498420

Giobbe GG, Crowley C, Luni C et al (2019) Extracellular matrix hydrogel derived from decellularized tissues enables endodermal organoid culture. Nat Commun 10(1):5658. https://doi.org/10.1038/s41467-019-13605-4

Gottschalk F, Nowack B (2011) The release of engineered nanomaterials to the environment. J Environ Monit 13(5):1145. https://doi.org/10.1039/c0em00547a

Gottschalk F, Kost E, Nowack B (2013) Engineered nanomaterials in water and soils: a risk quantification based on probabilistic exposure and effect modeling. Environ Toxicol Chem 32(6):1278–1287. https://doi.org/10.1002/etc.2177

Guadagnini R, Halamoda Kenzaoui B, Walker L et al (2015) Toxicity screenings of nanomaterials: challenges due to interference with assay processes and components of classic *in vitro* tests. Nanotoxicology 9(sup1):13–24. https://doi.org/10.3109/17435390.2013.829590

Halappanavar S, van den Brule S, Nymark P et al (2020) Adverse outcome pathways as a tool for the design of testing strategies to support the safety assessment of emerging advanced materials at the nanoscale. Part Fibre Toxicol 17(1):16. https://doi.org/10.1186/s12989-020-00344-4

Hirsch C, Roesslein M, Krug HF et al (2011) Nanomaterial cell interactions: are current *in vitro* tests reliable? Nanomedicine 6(5):837–847. https://doi.org/10.2217/nnm.11.88

Hofer S, Hofstätter N, Punz B et al (2022) Immunotoxicity of nanomaterials in health and disease: current challenges and emerging approaches for identifying immune modifiers in susceptible populations. WIREs Nanomed Nanobiotechnol 14(6). https://doi.org/10.1002/wnan.1804

Holbrook RD, Murphy KE, Morrow JB et al (2008) Trophic transfer of nanoparticles in a simplified invertebrate food web. Nat Nanotechnol 3(6):352–355. https://doi.org/10.1038/nnano.2008.110

International Organization for Standardization (2018) ISO 19007:2018 – Nanotechnologies — In vitro MTS assay for measuring the cytotoxic effect of nanoparticles. ISO. https://www.iso.org/standard/63698.html

Ishino K, Kato T, Kato M et al (2015) Comprehensive DNA adduct analysis reveals pulmonary inflammatory response contributes to genotoxic action of magnetite nanoparticles. Int J Mol Sci 16(2):3474–3492. https://doi.org/10.3390/ijms16023474

Jackson EL, Lu H (2016) Three-dimensional models for studying development and disease: moving on from organisms to organs-on-a-Chip and Organoids. Integr Biol 8(6):672–683. https://doi.org/10.1039/C6IB00039H

Judy JD, Bertsch PM (2014) Bioavailability, toxicity, and fate of manufactured nanomaterials in terrestrial ecosystems. Adv Agron 123:1–64. https://doi.org/10.1016/B978-0-12-420225-2.00001-7

Kagan VE, Bayir H, Shvedova AA (2005) Nanomedicine and nanotoxicology: two sides of the same coin. Nanomedicine 1(4):313–316. https://doi.org/10.1016/j.nano.2005.10.003

Kong L, Gao X, Zhu J et al (2017) Reproductive toxicity induced by nickel nanoparticles in *Caenorhabditis Elegans*. Environ Toxicol 32(5):1530–1538. https://doi.org/10.1002/tox.22373

Kroll A, Pillukat MH, Hahn D et al (2012) Interference of engineered nanoparticles with in vitro toxicity assays. Arch Toxicol 86(7):1123–1136. https://doi.org/10.1007/s00204-012-0837-z

Krug HF (2014) Nanosafety research—Are we on the right track? Angew Chem Int Ed 53(46):12304–12319. https://doi.org/10.1002/anie.201403367

Krug HF, Wick P (2011) Nanotoxicology: an interdisciplinary challenge. Angew Chem Int Ed 50(6):1260–1278. https://doi.org/10.1002/anie.201001037

Kumar A, Kumar P (2021) Cytotoxicity of quantum dots: use of QuasiSMILES in development of reliable models with index of ideality of correlation and the consensus modelling. J Hazard Mater 402:123777. https://doi.org/10.1016/j.jhazmat.2020.123777

Kumar D, Nadda R, Repaka R (2024) Advances and challenges in organ-on-Chip Technology: toward mimicking human physiology and disease in vitro. Med Biol Eng Comput 62(7):1925–1957. https://doi.org/10.1007/s11517-024-03062-7

Landsiedel R, Honarvar N, Seiffert SB et al (2022) Genotoxicity testing of nanomaterials. WIREs Nanomed Nanobiotechnol 14(6). https://doi.org/10.1002/wnan.1833

Lebre F, Chatterjee N, Costa S et al (2022) Nanosafety: an evolving concept to bring the safest possible nanomaterials to society and environment. Nano 12(11). https://doi.org/10.3390/nano12111810

Leynen N, Van Belleghem FGAJ, Wouters A et al (2019) *In vivo* toxicity assessment of silver nanoparticles in homeostatic versus regenerating planarians. Nanotoxicology 13(4):476–491. https://doi.org/10.1080/17435390.2018.1553252

Leynen N, Tytgat JS, Bijnens K et al (2024) Assessing the in vivo toxicity of titanium dioxide nanoparticles in Schmidtea Mediterranea: uptake pathways and (neuro)developmental outcomes. Aquat Toxicol 270:106895. https://doi.org/10.1016/j.aquatox.2024.106895

Lowry GV, Espinasse BP, Badireddy AR et al (2012) Long-term transformation and fate of manufactured Ag nanoparticles in a simulated large scale freshwater emergent wetland. Environ Sci Technol 46(13):7027–7036. https://doi.org/10.1021/es204608d

Maharramov AM, Hasanova UA, Suleymanova IA et al (2019) The engineered nanoparticles in food chain: potential toxicity and effects. SN Appl Sci 1(11):1362. https://doi.org/10.1007/s42452-019-1412-5

Markaki M, Tavernarakis N (2020) Caenorhabditis Elegans as a model system for human diseases. Curr Opin Biotechnol 63:118–125. https://doi.org/10.1016/j.copbio.2019.12.011

Maynard AD, Aitken RJ, Butz T et al (2006) Safe handling of nanotechnology. Nature 444(7117):267–269. https://doi.org/10.1038/444267a

Meneses J, González-Durruthy M, Fernandez-de-Gortari E et al (2023) A Nano-QSTR model to predict Nano-cytotoxicity: an approach using human lung cells data. Part Fibre Toxicol 20(1). https://doi.org/10.1186/s12989-023-00530-0

Messelmani T, Morisseau L, Sakai Y et al (2022) Liver organ-on-chip models for toxicity studies and risk assessment. Lab Chip 22(13):2423–2450. https://doi.org/10.1039/D2LC00307D

Metreveli G, Frombold B, Seitz F et al (2016) Impact of chemical composition of Ecotoxicological test media on the stability and aggregation status of silver nanoparticles. Environ Sci Nano 3(2):418–433. https://doi.org/10.1039/C5EN00152H

Moreno EA, Rojas GF, Frenk FH et al (1997) In vitro induction of abnormal anaphases by contaminating atmospheric dust from the City of Mexicali, Baja California, Mexico. Arch Med Res:28(4)

Nagai H, Toyokuni S (2012) Differences and similarities between carbon nanotubes and asbestos fibers during mesothelial carcinogenesis: shedding light on fiber entry mechanism. Cancer Sci 103(8):1378–1390. https://doi.org/10.1111/j.1349-7006.2012.02326.x

Nel A, Xia T, Mädler L et al (2006) Toxic potential of materials at the nanolevel. Science (1979) 311(5761):622–627. https://doi.org/10.1126/science.1114397

Nel AE, Mädler L, Velegol D et al (2009) Understanding biophysicochemical interactions at the nano–bio interface. Nat Mater 8(7):543–557. https://doi.org/10.1038/nmat2442

Oberdörster G, Oberdörster E, Oberdörster J (2005) Nanotoxicology: an emerging discipline evolving from studies of ultrafine particles. Environ Health Perspect 113(7):823–839. https://doi.org/10.1289/ehp.7339

OECD (2021) Interlaboratory comparison testing for the guidance document to support implementation of Test Guideline No. 312 for nanomaterial safety (Series on Testing and Assessment No. 338; ENV/CBC/MONO(2021)16). Organisation for Economic Co-operation and Development. https://one.oecd.org/document/ENV/CBC/MONO(2021)16/en/pdf

Oomen AG, Bos PMJ, Fernandes TF et al (2014) Concern-driven integrated approaches to nanomaterial testing and assessment—report of the nanosafety cluster working group 10. Nanotoxicology 8(3):334–348. https://doi.org/10.3109/17435390.2013.802387

Ortiz-Galvez LM, Caballero-Guzman A, Lopes C et al (2024) Probabilistic material flow analysis of released nano titanium dioxide in Mexico. NanoImpact 35:100516. https://doi.org/10.1016/j.impact.2024.100516

Pathakoti K, Huang M-J, Watts JD et al (2014) Using experimental data of Escherichia Coli to develop a QSAR model for predicting the photo-induced cytotoxicity of metal oxide nanoparticles. J Photochem Photobiol B 130:234–240. https://doi.org/10.1016/j.jphotobiol.2013.11.023

Pluskota A, Horzowski E, Bossinger O et al (2009) In Caenorhabditis Elegans nanoparticle-biointeractions become transparent: silica-nanoparticles induce reproductive senescence. PLoS One 4(8):e6622. https://doi.org/10.1371/journal.pone.0006622

Pondman K, Le Gac S, Kishore U (2023) Nanoparticle-induced immune response: health risk versus treatment opportunity? Immunobiology 228(2):152317. https://doi.org/10.1016/j.imbio.2022.152317

Prajitha N, Athira SS, Mohanan PV (2019) Bio-interactions and risks of engineered nanoparticles. Environ Res 172:98–108. https://doi.org/10.1016/j.envres.2019.02.003

Qin X, Tape CJ (2021) Deciphering organoids: high-dimensional analysis of biomimetic cultures. Trends Biotechnol 39(8):774–787. https://doi.org/10.1016/j.tibtech.2020.10.013

Ranga A, Gjorevski N, Lutolf MP (2014) Drug discovery through stem cell-based organoid models. Adv Drug Deliv Rev 69–70:19–28. https://doi.org/10.1016/j.addr.2014.02.006

Ravi M, Ramesh A, Pattabhi A (2017) Contributions of 3D cell cultures for cancer research. J Cell Physiol 232(10):2679–2697. https://doi.org/10.1002/jcp.25664

Ray P, Haideri N, Haque I et al (2021) The impact of nanoparticles on the immune system: a gray zone of nanomedicine. J Immunol Sci 5(1):19–33

Roh J, Sim SJ, Yi J et al (2009) Ecotoxicity of silver nanoparticles on the soil nematode *Caenorhabditis Elegans* using functional ecotoxicogenomics. Environ Sci Technol 43(10):3933–3940. https://doi.org/10.1021/es803477u

Rossi G, Manfrin A, Lutolf MP (2018) Progress and potential in organoid research. Nat Rev Genet 19(11):671–687. https://doi.org/10.1038/s41576-018-0051-9

Ryu N-E, Lee S-H, Park H (2019) Spheroid culture system methods and applications for mesenchymal stem cells. Cells 8(12):1620. https://doi.org/10.3390/cells8121620

Saptarshi SR, Duschl A, Lopata AL (2013) Interaction of nanoparticles with proteins: relation to bio-reactivity of the nanoparticle. J Nanobiotechnol 11(1):26. https://doi.org/10.1186/1477-3155-11-26

Sato T, Vries RG, Snippert HJ et al (2009) Single Lgr5 stem cells build crypt-villus structures in vitro without a mesenchymal niche. Nature 459(7244):262–265. https://doi.org/10.1038/nature07935

Sengupta A, Dorn A, Jamshidi M et al (2023) A multiplex inhalation platform to model in situ like aerosol delivery in a breathing lung-on-chip. Front Pharmacol:14. https://doi.org/10.3389/fphar.2023.1114739

Serras AS, Rodrigues JS, Cipriano M et al (2021) A critical perspective on 3D liver models for drug metabolism and toxicology studies. Front Cell Dev Biol:9. https://doi.org/10.3389/fcell.2021.626805

Shariati L, Esmaeili Y, Haghjooy Javanmard S et al (2021) Organoid technology: current standing and future perspectives. Stem Cells 39(12):1625–1649. https://doi.org/10.1002/stem.3379

Shukla RK, Badiye A, Vajpayee K et al (2021) Genotoxic potential of nanoparticles: structural and functional modifications in DNA. Front Genet 12. https://doi.org/10.3389/fgene.2021.728250

Singh AV, Varma M, Rai M et al (2024) Advancing predictive risk assessment of chemicals via integrating machine learning, computational modeling, and chemical/Nano-quantitative structure-activity relationship approaches. Adv Intell Syst 6(4). https://doi.org/10.1002/aisy.202300366

Sohn EK, Chung YS, Johari SA et al (2015) Acute toxicity comparison of single-walled carbon nanotubes in various freshwater organisms. Biomed Res Int 2015:1–7. https://doi.org/10.1155/2015/323090

Sung JH (2021) Multi-organ-on-a-chip for pharmacokinetics and toxicokinetic study of drugs. Expert Opin Drug Metab Toxicol 17(8):969–986. https://doi.org/10.1080/17425255.2021.1908996

Takebe T, Sekine K, Enomura M et al (2013) Vascularized and functional human liver from an IPSC-derived organ bud transplant. Nature 499(7459):481–484. https://doi.org/10.1038/nature12271

Tao H, Wu T, Aldeghi M et al (2021) Nanoparticle synthesis assisted by machine learning. Nat Rev Mater 6(8):701–716. https://doi.org/10.1038/s41578-021-00337-5

Thongkumkoon P, Sangwijit K, Chaiwong C et al (2014) Direct nanomaterial-DNA contact effects on DNA and mutation induction. Toxicol Lett 226(1):90–97. https://doi.org/10.1016/j.toxlet.2014.01.036

Toropov AA, Toropova AP, Benfenati E et al (2012) Novel application of the CORAL software to model cytotoxicity of metal oxide nanoparticles to bacteria Escherichia Coli. Chemosphere 89(9):1098–1102. https://doi.org/10.1016/j.chemosphere.2012.05.077

Tripathi DK, Shweta SS et al (2017) An overview on manufactured nanoparticles in plants: uptake, translocation, accumulation and phytotoxicity. Plant Physiol Biochem 110:2–12. https://doi.org/10.1016/j.plaphy.2016.07.030

Utembe W, Clewell H, Sanabria N et al (2020) Current approaches and techniques in physiologically based pharmacokinetic (PBPK) modelling of nanomaterials. Nanomaterials 10(7):1267. https://doi.org/10.3390/nano10071267

Valsami-Jones E, Lynch I (2015) How safe are nanomaterials? Science (1979) 350(6259):388–389. https://doi.org/10.1126/science.aad0768

Verdon R, Stone V, Murphy F et al (2022) The application of existing genotoxicity methodologies for grouping of nanomaterials: towards an integrated approach to testing and assessment. Part Fibre Toxicol 19(1):32. https://doi.org/10.1186/s12989-022-00476-9

Verma SK, Nandi A, Simnani FZ et al (2023) In silico nanotoxicology: the computational biology state of art for nanomaterial safety assessments. Mater Des 235:112452. https://doi.org/10.1016/j.matdes.2023.112452

Vilas-Boas V, Chatterjee N, Carvalho A et al (2024) Particulate matter-induced oxidative stress—mechanistic insights and antioxidant approaches reported in in vitro studies. Environ Toxicol Pharmacol 110:104529. https://doi.org/10.1016/J.ETAP.2024.104529

Wiesner MR, Lowry GV, Alvarez P et al (2006) Assessing the risks of manufactured nanomaterials. Environ Sci Technol 40(14):4336–4345. https://doi.org/10.1021/es062726m

Wigger H, Kägi R, Wiesner M et al (2020) Exposure and possible risks of engineered nanomaterials in the environment—current knowledge and directions for the future. Rev Geophys 58(4). https://doi.org/10.1029/2020RG000710

Wu T, Xu H, Liang X et al (2019) Caenorhabditis Elegans as a complete model organism for biosafety assessments of nanoparticles. Chemosphere 221:708–726. https://doi.org/10.1016/j.chemosphere.2019.01.021

Wyrzykowska E, Mikolajczyk A, Lynch I et al (2022) Representing and describing nanomaterials in predictive nanoinformatics. Nat Nanotechnol 17(9):924–932. https://doi.org/10.1038/s41565-022-01173-6

Xiao L, Pang J, Qin H et al (2023) Amorphous silica nanoparticles cause abnormal cytokinesis and multinucleation through dysfunction of the centralspindlin complex and microfilaments. Part Fibre Toxicol 20(1):34. https://doi.org/10.1186/s12989-023-00544-8

Xin Q, Rotchell JM, Cheng J et al (2015) Silver nanoparticles affect the neural development of zebrafish embryos. J Appl Toxicol 35(12):1481–1492. https://doi.org/10.1002/jat.3164

Zhao F, Zhao Y, Liu Y et al (2011) Cellular uptake, intracellular trafficking, and cytotoxicity of nanomaterials. Small 7(10):1322–1337. https://doi.org/10.1002/smll.201100001

Zhao Z, Chen X, Dowbaj AM et al (2022) Organoids. Nat Rev Methods Primers 2(1):94. https://doi.org/10.1038/s43586-022-00174-y

Zhou Y, Wang Y, Peijnenburg W et al (2023) Using machine learning to predict adverse effects of metallic nanomaterials to various aquatic organisms. Environ Sci Technol 57(46):17786–17795. https://doi.org/10.1021/acs.est.2c07039

Open Access This chapter is licensed under the terms of the Creative Commons Attribution-NonCommercial-NoDerivatives 4.0 International License (http://creativecommons.org/licenses/by-nc-nd/4.0/), which permits any noncommercial use, sharing, distribution and reproduction in any medium or format, as long as you give appropriate credit to the original author(s) and the source, provide a link to the Creative Commons license and indicate if you modified the licensed material. You do not have permission under this license to share adapted material derived from this chapter or parts of it.

The images or other third party material in this chapter are included in the chapter's Creative Commons license, unless indicated otherwise in a credit line to the material. If material is not included in the chapter's Creative Commons license and your intended use is not permitted by statutory regulation or exceeds the permitted use, you will need to obtain permission directly from the copyright holder.

Part I

Nanotoxicology

Methodological Considerations for Setting Up Human-Relevant In Vitro Nanotoxicology Experiments—A Practical Guide

Vânia Vilas-Boas, Emma Arnesdotter, Félix Carvalho, and Ernesto Alfaro-Moreno

Abstract

Nanotoxicology is a rapidly evolving field dedicated to assessing the safety and potential hazards of nanomaterials on human health. This practical guide outlines essential methodological considerations for designing human-relevant in vitro nanotoxicology experiments. A primary focus is placed on the comprehensive characterization of the nanomaterial in question, as properties such as size, shape, surface charge, and solubility significantly influence biological activity. The guide discusses the selection of appropriate in vitro models, including various cell sources, to ensure relevance to human exposure scenarios.

It is crucial to exercise caution when choosing test methods to account for potential nanoparticle interference with the selected assays; however, the use of suitable controls can help mitigate the impact of these interactions. The guide also emphasizes accurate practices for nanomaterial sample preparation and the importance of dosimetry, facilitating the translation of in vitro findings to realistic human exposure conditions. Guidance on exposure concentrations is provided to ensure that testing remains biologically and environmentally relevant. Furthermore, the guide includes reflections and perspectives on addressing

V. Vilas-Boas (✉) · E. Alfaro-Moreno
International Iberian Nanotechnology Laboratory, Nanosafety Research Group, Braga, Portugal
e-mail: vania.vilasboas@inl.int

E. Arnesdotter
Environmental Sustainability Assessment and Circularity (SUSTAIN) Unit, Luxembourg Institute of Science and Technology, Esch-sur-Alzette, Luxembourg

F. Carvalho
UCIBIO—Applied Molecular Biosciences Unit, Laboratory of Toxicology, Department of Biological Sciences, Faculty of Pharmacy, University of Porto, Porto, Portugal

Institute for Health and Bioeconomy (i4HB), Laboratory of Toxicology, Faculty of Pharmacy, University of Porto, Porto, Portugal

© The Author(s) 2025
E. Alfaro-Moreno, F. Murphy (eds.), *Nanosafety*,
https://doi.org/10.1007/978-3-031-93871-9_2

common challenges and enhancing reproducibility in nanotoxicology studies. By adhering to these guidelines, researchers can generate more reliable and human-relevant in vitro nanotoxicology data, thereby supporting the risk assessment of nanomaterials.

Graphical Abstract

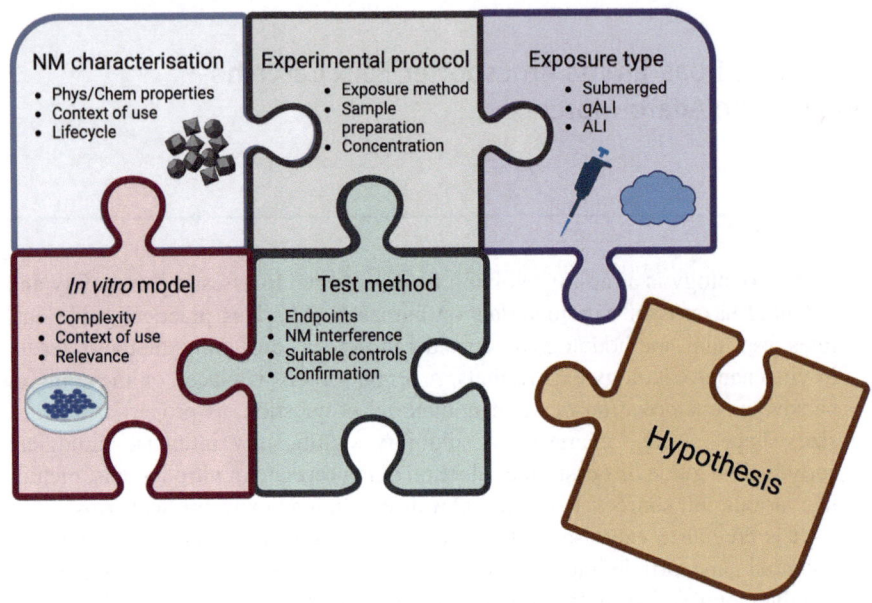

Keywords

Nanotoxicology · Advanced in vitro models · Nanomaterials · Test method · Experimental protocol

Abbreviations

ADME	Absorption, distribution, metabolism, excretion
ALI	Air-liquid interface
AO	Adverse outcome
AOP	Adverse outcome pathway
ELISA	Enzyme-linked immunosorbent assay
FBS	Fetal bovine serum
FDA	Food and Drug Administration
GIVIMP	Good in vitro method practices

IATA	Integrated approaches to testing and assessment
iPSC	Induced pluripotent stem cells
KE	Key event
LAL	Limulus amebocyte lysate
LBP	Lipopolysaccharide-binding protein
LDH	Lactate dehydrogenase
LPS	Lipopolysaccharide
MIE	Molecular initiating event
MTT	3-(4,5-dimethylthiazol-2-yl)-2,5-diphenyltetrazolium bromide
NM	Nanomaterial
NP	Nanoparticle
OECD	Organization for Economic Co-operation and Development
OoC	Organ-on-chip
qALI	*Quasi* air-liquid interface
ROS	Reactive oxygen species
SOP	Standard operating procedures
TNFα	Tumor necrosis factor alpha
TRL	Toll-like receptor

1 Introduction

Since the emergence of nanotechnology in the 1980's, a growing number of nanomaterials (NMs) has been discovered and developed for use in different sectors, including healthcare, agriculture, environmental, and energy applications (Xuan et al. 2023). An increasing number of consumer products, such as electronic devices, cosmetics, textiles, paints, and even food items, now incorporate NMs. This widespread usage raises the rate of human exposure, both directly and indirectly, through environmental accumulation. As exposure to potentially hazardous NMs rises, so does the associated risk. Therefore, it is crucial to evaluate the safety of NMs before their incorporation into consumer products or significant environmental accumulation, to prevent unwarranted public concern. While we possess some understanding of the general adverse effects associated with various types of NMs, our grasp of the underlying molecular mechanisms remains limited (Malakar et al. 2021). In recent years, adverse outcome pathways (AOPs)—schematic representations of the sequence of key events (KEs) from a molecular initiating event (MIE) to a measurable adverse outcome (AO)—have emerged as valuable tools for designing toxicity testing strategies in (nano)toxicology. These frameworks support the development of new testing methods and guidelines, as well as integrated approaches to testing and assessment (IATAs) (Gerloff et al. 2017). However, it is important to recognize that NM interactions with cells (or cellular targets, such as membrane receptors) may involve specific mechanisms, such as mechanical damage, which cannot be categorized as purely "molecular" interactions. This complexity may complicate the identification of a "molecular" initiating event, making it more appropriate to designate an initial KE that describes non-specific interactions. Nonetheless, the progress made in this area demonstrates the feasibility of using AOPs to articulate the mechanistic knowledge of the toxicological pathways induced by poorly soluble NMs (Gerloff et al. 2017). Consequently, further research should focus on advancing nano-related AOPs and their application to predict the AO of new materials.

There has been a paradigm shift in toxicology in recent years, driven by increasing ethical constraints and acknowledged species differences between laboratory animals and humans. Studies indicate that data derived from rodents, the most commonly used animal models in preclinical toxicology, can predict or replicate less than 50% of human clinical toxicological outcomes for 150 compounds (Olson et al. 2000). While conventional in vitro models (2D monolayers) have significantly contributed to our understanding of nanotoxicology, they are not suited to address all research questions due to their simplistic and inaccurate representation of human physiology.

In recent years, remarkable advancements have been made in the development of models that more closely replicate the in vivo human situation when compared to traditional 2D monolayer cultures. These advanced models mitigate some limitations of 2D monolayers by employing strategies such as co-culturing different cell types on transwell inserts with permeable membranes to allow aerosol exposure (Klein et al. 2013; Ramos-Godínez et al. 2013) or assembling cells into spherical 3D structures known as spheroids (Bell et al. 2016; Vilas-Boas et al. 2021). Others have designed chip-like structures, termed organs-on-chip (OoCs), where cells are seeded and the culture medium is perfused using microfluidics (Huh et al. 2007), or mechanical cues are induced to simulate physiological movements (Stucki et al. 2018). OoCs mimicking the lung, liver, or heart have already been successfully applied in nanotoxicology (Lu and Radisic 2021). Additionally, organoids—3D heterogenous structures that grow from pluripotent stem cells, recapitulating real organs by recreating various cell types within an organ system—show significant potential to replace animal studies in the future (Clevers 2016). To leverage the strengths of both OoCs and organoids, researchers have recently combined these technologies to create 3D tissue replacements that self-organize from stem cells in a dynamically controlled environment, continuously monitored by integrated sensors (Zhao et al. 2024). The different complexities of these models allow researchers to address varied research questions, each with its own advantages and limitations, and not all complex models will be suitable for every study.

Nanotoxicology primarily applies principles of chemical toxicology to NMs. However, due to their small size, NMs possess increased surface area and reactivity, which not only contribute to NMs-related toxicity but also interfere with many cytotoxicity assays used for conventional chemicals. This interference presents a significant challenge in nanotoxicology, emphasizing the need for researchers to consider this aspect when designing experiments. Methodological choices should rely on preliminary tests to exclude NM interference with reagents and readouts (Karlsson et al. 2015).

To date, numerous studies have demonstrated that NMs can induce oxidative stress, leading to inflammation and cell death (Vilas-Boas and Vinken 2021). The emergence of new methods, such as diverse omics techniques, has facilitated the identification of novel AOs triggered by NMs, as well as the underlying mechanisms involved.

Once NMs, in vitro models, and methodological approaches are selected, it is crucial to meticulously design the experimental procedure to address the research question and objectives while considering the challenges, possibilities, and limitations of the chosen methods. Careful design is essential, as inadequate experimental planning may lead to erroneous conclusions. Considerations must include the type of exposure, NM concentrations, the selection of appropriate positive and negative controls, and the necessity of confirmation assays.

In this chapter, we will discuss the major NM- and in vitro model-related aspects and key elements to consider when setting up an in vitro nanotoxicology study. Additionally, we will present the main methodological challenges associated with designing a robust and reliable nanotoxicological assessment.

2 The NM

2.1 Detailed Physical-Chemical Characterization of the NM

Factors such as size, surface area, charge, shape, and composition are crucial in determining the behavior and toxicity of nanomaterials (NMs). These characteristics often correlate with the cytotoxicity observed for specific NMs (Awashra and Młynarz 2023; Gerloff et al. 2017). For instance, positively charged NMs (i.e., those with a positive zeta potential) are generally more likely to interact with negatively charged cell membranes compared to negatively charged or neutral NMs. Additionally, different adverse outcomes or severities may be expected from NMs with average primary particle sizes of 5 nm versus 100 nm (Gerloff et al. 2017). Smaller NMs are typically more readily taken up by cells, which may result in more pronounced toxicity. Likewise, if a NM is partially soluble, toxicity related to released ions should also be considered (Drasler et al. 2017). A recent review by Ruijter et al. (2023) summarizes how NM characteristics influence their toxicity in in vitro studies.

While there is now a greater awareness of the importance of understanding the characteristics and behavior of NMs, the lack of such knowledge has historically been identified as a significant drawback. Even today, most available physicochemical data pertain to pristine NMs and do not account for changes that cell culture media (or other exposure vehicles) may induce in the inherent properties of the NMs. Moreover, the expected biotransformation of NMs in their environment before reaching their target—such as the transformations occurring in the gastrointestinal tract upon ingestion—will likely impact their kinetics in the body, including absorption, distribution, metabolism, and excretion (ADME) (Gerloff et al. 2017). Therefore, these factors must be considered when designing experiments intended to reflect real-life scenarios.

While data on particle composition, size, and shape can be obtained from pristine materials, it is essential to characterize the NM when dispersed in the exposure medium, as this is how it will encounter cells or tissues (Drasler et al. 2017). In this context, information on hydrodynamic size, solubility, aggregation, and the presence of contaminants, such as endotoxins, becomes equally essential (Swartzwelter et al. 2021). It is also vital to note the distinction between aggregates—particles comprising strongly bonded or fused entities—and agglomerates, which are collections of weakly bound particles, aggregates, or mixtures of the two (OECD 2012). For detailed information on the main techniques used for NM characterization, refer to Drasler et al. (2017).

2.2 Endotoxin: A Biological Component that Must Be Kept in Mind

Endotoxin is a molecule found on the surface of many materials; it is a heat-stable, pyrogenic complex component of the membrane of Gram-negative bacteria (Raetz and Whitfield 2002). Specifically, endotoxin is a lipopolysaccharide (LPS) heteropolymer composed of three elements: lipid A, core oligosaccharide, and O-specific polysaccharide, also known as antigen-O (Erridge et al. 2002). While LPS is a component of endotoxin, it is common for these terms to be used interchangeably, even though they are not synonymous. The heat-stable nature of endotoxin makes it difficult to eliminate from a wide range of materials (e.g., autoclaving does not effectively remove endotoxin). Thus, it is essential to ascertain the presence or absence of endotoxin when evaluating the toxicity of NMs.

Endotoxin binds to various cellular receptors, including toll-like receptor 4 (TLR4), LPS-binding protein (LBP), and CD14 (Ulevitch and Tobias 1995). The interaction of these receptors with endotoxin triggers a cascade of cellular events that leads to the release of proinflammatory mediators such as TNFα, IL-1β, IL-6, prostaglandins, and leukotrienes (Beutler and Rietschel 2003). For decades, evaluating endotoxin presence has been a standard practice when assessing respirable particles (Griwatz and Seemayer 1995), and its role in complex mixtures has been clarified using specific inhibitors such as LBP (Bonner et al. 1998) or Polymyxin B (Alfaro-Moreno et al. 2007). In the context of assessing NM toxicity, several studies have underscored the importance of evaluating endotoxin presence. For example, Dobrovolskaia et al. (2010) demonstrated that the presence of endotoxin in carbon nanotubes could lead to an overestimation of the NMs' cytotoxicity (Dobrovolskaia et al. 2010). Similarly, a review by Deng et al. (2009) emphasized the necessity for standardized protocols and guidelines to assess endotoxin contamination in NMs, as inconsistent or inadequate evaluations can yield unreliable toxicity data (Deng et al. 2009).

To address these concerns, various methods have been developed for detecting and quantifying endotoxin in NMs. Historically, the Limulus Amebocyte Lysate (LAL) assay, which utilizes the clotting cascade of the horseshoe crab Limulus polyphemus, has been the most commonly used method for endotoxin detection and quantification (Colas et al. 2014). However, a comparable method produced without the need for animal-derived raw material is now available, called Recombinant

Factor C, which is based on the same reaction principle as the LAL assay (Bolden and Smith 2017). Other techniques, such as mass spectrometry and nuclear magnetic resonance spectroscopy, have also been explored for identifying and characterizing endotoxin in NMs (Bergstrand et al. 2006; Gorbet and Sefton 2005).

In addition to analytical methods, it is crucial to develop strategies for removing or minimizing endotoxin contamination in NMs. These strategies may include thermal treatment, chemical modifications, and affinity-based purification techniques (Bacher et al. 2001; Sharma et al. 2014); however, no standardized methods for endotoxin removal from NMs currently exist (Hannon and Prina-Mello 2021). Consequently, evaluating the presence of endotoxin in NMs is a critical aspect of toxicity assessment, as this immunogenic molecule can significantly influence the interpretation of experimental results. By implementing robust protocols for endotoxin detection and employing strategies to prevent contamination, researchers and regulatory agencies can enhance the reliability and accuracy of NM toxicity data. This approach ultimately contributes to the safe development and application of these emerging technologies. More details on the relevance and impact of endotoxin on nanosafety assessments can be found in Chap. 3.

2.3 The Context of Use and Life Cycle

Knowing the intended use of NMs helps anticipate possible exposure scenarios and routes, as well as the main target organs or tissues. The primary routes of exposure to NMs are ingestion, dermal contact, inhalation, and occasionally injection (Fig. 2.1). Different exposure routes impact various organs in distinct ways. For example, when inhaled, NMs enter the nasal cavity, travel through the upper airways, and eventually reach the lungs. The size of the inhaled particles determines their final destination, while their surface charge and functionalization may also affect absorption, emphasizing the relevance of detailed physicochemical characterization of the NM (Savage et al. 2019).

Some organs or tissues, such as blood and liver, are generally at risk of exposure due to their interaction with the exposure route. The liver often accumulates and transforms NMs (Li et al. 2022) making it a crucial to study cytotoxic effects in liver models. The brain, fetus, reproductive system, and endocrine system can also be exposed once NMs enter the bloodstream, increasing the importance of assessing the potential effects on these organs. There has been growing interest in studying the endocrine disruption ability of NMs, revealing that ubiquitous materials, such as diesel exhaust NPs, may disrupt hormonal regulation in both men and women, although the effects might stem from the chemical mixture they carry (Iavicoli et al. 2013).

If the NM is intended for health applications (e.g., as a drug carrier), it is advisable to consult the literature and market for qualified models that have undergone regulatory acceptance for toxicological assessment. This approach will increase confidence in the results and expedite the review process by regulatory agencies.

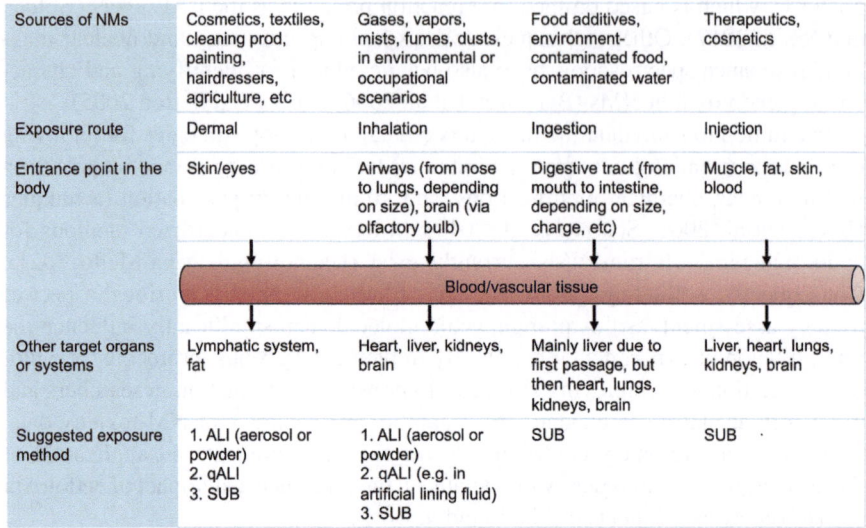

Sources of NMs	Cosmetics, textiles, cleaning prod, painting, hairdressers, agriculture, etc	Gases, vapors, mists, fumes, dusts, in environmental or occupational scenarios	Food additives, environment-contaminated food, contaminated water	Therapeutics, cosmetics
Exposure route	Dermal	Inhalation	Ingestion	Injection
Entrance point in the body	Skin/eyes	Airways (from nose to lungs, depending on size), brain (via olfactory bulb)	Digestive tract (from mouth to intestine, depending on size, charge, etc)	Muscle, fat, skin, blood
		Blood/vascular tissue		
Other target organs or systems	Lymphatic system, fat	Heart, liver, kidneys, brain	Mainly liver due to first passage, but then heart, lungs, kidneys, brain	Liver, heart, lungs, kidneys, brain
Suggested exposure method	1. ALI (aerosol or powder) 2. qALI 3. SUB	1. ALI (aerosol or powder) 2. qALI (e.g. in artificial lining fluid) 3. SUB	SUB	SUB

Fig. 2.1 Examples of NMs sources, their exposure routes, and main target organs implicated in the exposure. Based on this information, an in vitro exposure scenario is suggested

3 The In Vitro Model

The choice of the in vitro model used for experiments is a critical component of the setup. A careless selection may result in significant project costs without justifiable outcomes. For a thorough toxicity assessment, the selected in vitro model must express the cellular targets involved in the suspected pathways or mechanisms of toxicity, regardless of the model's complexity. In this subsection, we will outline the key considerations for selecting an appropriate in vitro model for a predefined study.

3.1 The Context of Use

The context of use refers to the applicability domain and specific purpose of an in vitro model or method, as outlined by the FDA (FDA 2017). Understanding this context, along with a thorough knowledge of the limitations, sensitivity, reproducibility, and relevance of the models and assays, is essential for obtaining high-quality results, especially in a regulatory framework. These parameters should be clearly defined for both the in vitro model and the selected test method to build confidence in the results.

Exploring the wide range of available complex in vitro models may help identify one that meets your specific needs. Collaborating with model developers can enhance focus and optimize collective efforts. If you choose to develop a new model, consider conducting experiments not only centered on your primary research question but also specifically aimed at validating or qualifying the new model. This approach will ensure its acceptance and adoption by end-users, such as pharmaceutical companies (FDA 2017).

To facilitate this process, it is crucial to adhere to the official guidelines set forth by the OECD for the validation of methods and models (OECD 2005). Note that the 2005 document is currently under revision and will be updated soon.

3.2 Cell Source

Various cell types can be employed to establish in vitro models, ranging from cell lines to primary cells and stem cells. The choice of cell type can significantly influence the outcomes related to the original research question, with some options being more relevant to the human context than others. Primary human cells are generally regarded as the gold standard, particularly in liver toxicology, as they retain characteristics of the original tissue depending on the model's configuration (Zhao 2023). When possible, primary human cells should be the preferred choice for obtaining human-relevant responses at a population level, as they can partially account for inter-donor variability (Vilas-Boas et al. 2021). However, primary cells are often expensive, have limited availability, and are subject to ethical restrictions. Additionally, they typically originate from patient biopsies, which carry inherent risks, especially since these procedures are usually performed under the suspicion of disease.

Recently, significant efforts have been made to develop stem cell-based models, such as organoids, using tissue-resident stem cells from patient biopsies or induced pluripotent stem cells (iPSCs). iPSCs are created by reprogramming adult somatic cells through the overexpression of the so-called Yamanaka factors (Lynch et al. 2019). This approach offers an abundant source of cells with differentiation potential, enabling the recreation of various cell types from the same initial source. However, these models have yet to fully replicate the maturity and cellular diversity found in most adult tissues (Liu et al. 2017). Therefore, further optimization of stem cell-based, human-relevant in vitro models is essential.

In many cases, immortalized cell lines are the most convenient option, providing an unlimited source of human-derived material and expressing key features that make them suitable for quickly addressing important mechanistic questions in a simplified setup. These immortalized cell lines can also be combined to enhance model complexity and more closely resemble the in vivo conditions of human tissues. It is crucial to note, however, that most available cell lines are derived from tumor tissues, which may not always be ideal, as metabolic pathways can be upregulated and cell communication pathways downregulated compared to primary tumor cells or normal cells (Ertel et al. 2006). Fortunately, an increasing number of cell lines derived from normal human tissues are becoming available, presenting a viable alternative.

It is important to recognize that both stem cells and immortalized cell lines do not account for inter-individual variability in human responses. While this limitation allows for more definitive observations regarding a specific genotype or phenotype represented by a particular cell line, it fails to capture the inherent variability of the human population.

3.3 Model Complexity

The susceptibility of cells to NMs depends not only on the cell type but also on cell architecture and model complexity (Juarez-Moreno et al. 2022). For decades, in vitro toxicity studies primarily relied on simple 2D monolayers of a single cell type, typically of cancerous origin. While these models are overly simplistic and do not accurately reflect human physiological complexity, they have been instrumental for screening purposes and in enhancing our understanding of the metabolic pathways and molecular mechanisms triggered by chemicals (Faber and McCullough 2018; Gómez-Lechón et al. 2014). Consequently, simpler models may be more suitable for addressing specific molecular questions compared to more complex systems (Table 2.1).

Recently, there has been a great incentive to develop models that more closely resemble human physiology, such as OoCs and organoids, resulting in an abundance of new complex models for toxicological studies (Leung et al. 2022). These models often integrate multiple cell types and physiological cues, such as perfusion and stretch/strain, making them more appropriate for understanding toxicological effects related to interactions among cells, the extracellular matrix, and different organ systems. For example, co-cultures that include immune cells are generally better suited for studying inflammatory responses triggered by chemicals, including NMs. Similarly, an OoC combining two tissues in tandem may be valuable for assessing the indirect adverse effects on a secondary organ or tissue mediated by messenger molecules released by cells directly exposed to NMs. Additionally, ongoing efforts aim to integrate organoid and OoC technologies, allowing organoids to be exposed to physiologically relevant controlled conditions (Zhao et al. 2024).

It is important to note, however, that most of these more complex models currently exhibit lower throughput and reproducibility. This factor should be taken into account when selecting the appropriate in vitro model for specific studies.

Table 2.1 Association between complexity, type of exposure and context of use of in vitro models for nanotoxicology

Model complexity	Single or co-cultures				Self-generated
	2D Monolayers	Polarized Inserts	(m)OoCs	3D Spheroids	Organoids
In vitro exposure	SUB	SUB, ALI, qALI		SUB	
Context of use	Molecular mechanisms	Barrier tox			
	Tox screening	(inter-)tissue-level tox			
	Cell-level tox		(inter-)organ-level tox		

ALI air-liquid interface, *m* multi, *OcO* organ-on-chip, *qALI* quasi-air-liquid-interface, *SUB* submerged, *Tox* toxicology

More information on advanced models for nanotoxicology assessments can be found in Chap. 7 of this book.

4 The Test Method

4.1 The Outcome

The endpoints to be assessed should align with the research question being addressed. In this context, available AOPs in nanotoxicology can be valuable tools for identifying individual KEs involved in the anticipated pathways. The preferred assay should also facilitate the quantitative detection of the perturbation (in this case, a KE) caused by the stressors (positive control and/or the NM under analysis) in the selected in vitro model. Additionally, the chosen test method should align with the context of use for which it was developed (Hirsch and Schildknecht 2019). Consideration should also be given to the ease or difficulty of sample collection from the selected in vitro model, and vice versa. Other criteria to account for include the assay's predictiveness, robustness, readiness, simplicity, and cost (Ruijter et al. 2023).

Oxidative stress, inflammatory response, cytotoxicity and apoptosis induction, as well as genotoxicity, are outcomes well documented to be triggered by various types of NMs and should be prioritized when testing new NMs (Drasler et al. 2017; Ruijter et al. 2023; Vilas-Boas and Vinken 2021). As discussed in Sect. 2.1. and Chap. 3, it is particularly important to check for endotoxin contamination when studying inflammatory responses, as its presence can trigger inflammation independently of the NM, complicating the ability to discriminate effects purely attributable to the NM.

Research can focus on either the bulk response related to the function of the whole cell population or tissue (e.g., cell viability, membrane integrity) or on the mechanisms behind those outcomes (molecular response). For more detailed studies, it may be necessary to knock-down the cellular expression of relevant proteins or to use inhibitors of specific pathways to elucidate the toxicological pathway triggered by the NM. Bulk responses consider the tissue as a unit, where the most prevalent cellular response is perceived as general. In this scenario, distinctive individual cell responses, which may be significant, are often overlooked. If this approach does not adequately address the initial research question, techniques such as flow cytometry or fluorescence microscopy can help distinguish responses from different cell types, though they are limited to a small number of proteins and mRNAs.

Ultimately, single-cell analysis techniques, such as single-cell omics (transcriptomics, genomics, metabolomics, proteomics), may be required to gain a deeper understanding of the role of each cell in a complex tissue model. Importantly, while bulk analysis methods are generally more affordable and easier to perform in standard biochemistry laboratory settings, single-cell techniques tend to be more costly

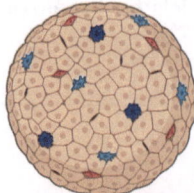

	Bulk response		Single-cell response
	Functional	**Molecular**	
Outcomes or techniques	•Cell viability/death •NP internalization •Cytokine release •ROS production	•Omics •Protein knock-down •Transporter inhibition	•Single-cell omics •Patch-clamp recording •Flow cytometry •Fluorescence microscopy
Advantages	•Tissue-level response		•Highlights cell-to-cell variation •More data from the same cell •Multi-omics available from the same cell •Lower sample amount
	•Speedy •High throughput •Affordable •Simpler data analysis	•Specific cellular targets •More sensitive to sublethal responses •Mechanistic clues	
Limitations	•Cellular heterogeneity is concealed	•Laborious •Intermediate cost •High variability of results	•Challenging sample prep •Onerous data analysis •Labour intensive •Higher cost

Fig. 2.2 Advantages and limitations of analyzing bulk adverse response comparatively to single-cell response in complex in vitro models

and complex to implement, yielding more intricate data, and requiring specific equipment not readily available in every lab (Fig. 2.2).

As a rule of thumb, starting with functional assays and subsequently integrating more single-cell-oriented assays will help determine the necessity for incorporating complex single-cell analysis in subsequent experiments.

4.2 Interference of the NM with Test Methods

It is crucial to exclude potential interferences of the NM with the test method readout to ensure reliable results. This step is fundamental for validating the experiment's outcomes. NMs have been known to interfere with several methods, including the MTT assay, which remains one of the most commonly used assays (Awashra and Młynarz 2023; Kroll et al. 2012; Lebre et al. 2022).

Interference of NPs with test methods can occur in several forms, including:

- Physical interference due to the light-refracting or light-absorbing properties of certain NMs, leading to increased absorbance or decreased fluorescence due to quenching phenomena. If the interference is purely physical, it can generally be addressed by adding a centrifugation step to separate the supernatant for reading, or by subtracting the background absorbance or fluorescence generated by the cells in the presence of the NM (Stone et al. 2009).

- Physical interference from the binding of biomarkers of interest present in the cell culture medium to the NM surface. The increased surface area of NMs provides more docking sites for proteins in the milieu. This is particularly relevant when investigating molecules of the secretome, which are released by cells into the extracellular medium. For instance, it has been documented that NPs can interfere with biomarker detection protocols, such as ELISA, as the cytokines released into the medium may bind nonspecifically to the NP surface (Guadagnini et al. 2015).
- Chemical interference when the NM interacts directly with the detection reagent. Due to their small size and increased surface area, NMs can exhibit heightened reactivity, which may impact assay results. For example, in methods based on cellular redox activity, such as the MTT assay or the generation of ROS, it is essential to assess the direct catalytic effects of NMs beforehand (Awashra and Młynarz 2023). To mitigate this interference, one option is to remove the NM remaining in suspension and wash the cells/tissue before adding the detection reagent. Alternatively, accounting for the signal obtained when mixing the NM with the reagent in the absence of cells may be helpful; however, it is often advisable to employ a different test method based on an alternative assay principle.

To identify potential interference of the NM on the selected assay, the test protocol should be conducted beforehand without cells to detect any reaction attributable solely to the NM. For instance, to assess interference with the method's readout or chemical reactivity with the reagent, various concentrations of the test NM can be mixed with the detection reagent in the absence of cells. These mixtures should be incubated for the expected assay duration, and the resulting assay signal (fluorescence, absorbance, or luminescence) should be measured (Vasimalai et al. 2018). If a signal change is observed compared to blank wells, it indicates that the reagent is affected independently of the presence of cells, confirming interference.

Since interferences are both assay- and NM-dependent, anticipating them without testing is nearly impossible (Karlsson et al. 2015). Therefore, it is highly advisable to perform exploratory tests to exclude such interferences in advance. Table 2.2 lists reported interferences of NMs with some of the most commonly used cytotoxicity assays, their causes, and possible solutions. Guadagnini et al. have reviewed this topic, highlighting other types of NM interferences for different assays (Guadagnini et al. 2015).

4.3 Control Samples and Assay Conditions

A good planning of the assay controls is crucial for obtaining reliable and robust results. Therefore, it is essential to define both positive and negative controls for each adverse outcome or assay. Ideally, these controls should be particulate to match the characteristics of the NM. However, finding nano-specific reference controls remains a challenge, as they must be specific to the endpoint under evaluation and compatible with the selected assay and exposure route (Drasler et al. 2017). In the

Table 2.2 Examples of interferences from nanomaterials with widely used toxicological assays

Methodology	Observed interference		Proposed solution
	Cause	Result interpretation	
MTT reduction WST reduction	NM optical density NM aggregation in cell medium	Falsely ↑ viability	Centrifugation step after cell lysis
	NM direct redox activity	Falsely ↑ or ↓ viability	Choose test not based on redox activity
LDH leakage	NM catalyzes the reaction in the absence of LDH	Falsely ↓ viability	Choose different test
ELISA (cytokine release)	NM adsorb protein of interest	Falsely ↓ cytokine production	Add serum proteins to NM suspension
Comet assay	NM interfere with enzyme activity	Falsely ↓ genotoxicity	Choose different test
	NM induced breaks in naked DNA	Falsely ↑ genotoxicity	
ROS quantification (H_2DCF-DA)	NM direct redox activity	Falsely ↑ or ↓ ROS levels	Choose different test
	NMs quench fluorescence NMs scatter emitted fluorescence	Falsely ↓ ROS levels	Sample centrifugation after cell lysis

Adapted from Lebre et al. (2022)
NM nanomaterial, *MTT* 3-(4,5-dimethylthiazol-2-yl)-2,5-diphenyltetrazolium bromide, *LDH* lactate dehydrogenase, *WST* water-soluble tetrazolium salts, *ELISA* enzyme-linked immunosorbent assay, *ROS* reactive oxygen species, *H2DCF-DA* 2′,7′-dichlorodihydrofluorescein diacetate

absence of a nano reference control, bulk material can serve as a negative control (Wiemann et al. 2016), and at least one conventional chemical should be used as a quality control. Additionally, a non-exposure ("non-treated") control is imperative, where cells are added the same vehicle/dispersant and subjected to the same conditions as those in which the NM is exposed (Drasler et al. 2017). If different time points are being assessed, non-exposure control wells should be included for each time point.

4.4 Confirmation Assay

Given the numerous variables involved in assessing the toxicological profile of a NM in vitro, it is advisable to include a confirmation assay to validate findings. This is particularly crucial if the potential interference of the NM with the assay was not assessed beforehand. A confirmation assay may be based on the same mechanistic principle but, preferably, using a different readout. For instance, metabolic activity assays that measure different metabolic products can be employed. These assays are available with both optical and fluorescent readouts. In metabolic activity testing, such as the MTT reduction assay and resazurin reduction-based assays (e.g., PrestoBlue) decreased signals (absorbance for MTT and fluorescence for resazurin) indicate decreased metabolic activity, suggesting decreased cell viability. An interference with the tests' readouts is expected to affect both absorbance and fluorescence in opposing ways, leading to conflicting cell viability results and raising concerns that factors other than toxicological effects may be influencing the outcomes. Ideally, confirmation tests should be based on different mechanisms (e.g., metabolic function and membrane permeability), and different readouts, to exclude interferences with the mechanistic pathway under analysis.

5 The Experimental Protocol

5.1 Exposure Method

There are various ways to present NMs to cells. Many in vitro experiments have traditionally utilized submerged (SUB) exposure, where cells are fully immersed in the medium containing the test NM. For NMs exposed in SUB conditions, the effective (actual exposure) concentration does not necessarily equal their available (nominal or applied) concentration (see subsection on dosimetry below). Advances in technology and cell model development now allow for exposure at the air-liquid interface (ALI), where the cell layer is in direct contact with air and not covered by cell culture medium. This approach more closely mimics real-life exposure scenarios of organ systems in direct contact with air, such as the airways, skin, or cornea, compared to SUB exposures.

Specific exposure systems, such as the VITROCELL Single Droplet Systems, enable direct exposure of cells at the ALI to various airborne substances, including NMs, brought into suspension. Platforms like the PreciseInhale, developed by Inhalation Sciences, or the PowderX from VITROCELL, allow for exposure to dry aerosols directly applied to the cells under high pressure. These systems eliminate the need to prepare a suspension of the NM before exposure, which can be beneficial for certain NMs that are susceptible to tangling and agglomeration. However, these exposure systems can be costly and may not be readily available in all labs.

As a compromise, the quasi-ALI (qALI) approach, sometimes referred to as pseudo-ALI, can be employed. In this case, the in vitro model is airlifted as in a regular ALI scenario, but the NM is delivered on the cells using a strictly limited amount of medium just to cover the surface (e.g., 50 µL/cm^2). Similarly, the Tecan D300e digital dispenser, primarily developed for chemicals, can be used to deliver NPs <1 µm at a concentration of ≤0.5% using as little as 1.3 µL/cm^2.

It remains challenging to determine the effective exposure concentration when using the systems discussed, as individual materials behave differently. This highlights the importance of a careful characterization and consideration of the test material. Nevertheless, with proper reporting and execution of experiments, all methods discussed should provide reproducible and accurate comparisons between NM studies in vitro.

5.2 Sample Preparation

Defining a stepwise process for preparing NM dispersions is crucial for obtaining reproducible results that lead to robust conclusions. Once the exposure method is selected, several considerations must be made regarding sample preparation to strike a balance between technical requirements and the relevance of the exposure for extrapolating real-life conclusions. After defining the exposure method, carefully consider the following sample preparation steps:

- **Use of Fetal Bovine Serum (FBS) as a Medium Supplement**: FBS provides essential nutrients for cell culture and may also act as a stabilizer/dispersant for some NMs by forming a corona of serum biomolecules (Casals et al. 2010). This protein corona can influence cell-NP interactions, with debates surrounding whether it enhances or decreases NM cytotoxicity (Corbo et al. 2016). On one hand, this biological coating may improve the biocompatibility of the material by blocking surface reactivity or decreasing the NM's surface energy (Vranic et al. 2017). Different amounts of FBS can lead to variations in the protein corona and may trigger distinct uptake mechanisms (Francia et al. 2019). If mimicking blood is the goal, using human plasma instead of serum might be more appropriate, as silica NPs, for example, can adsorb various coagulation factors at their surface (Aliyandi et al. 2021). Conversely, an increased biological component may enhance uptake, potentially raising cytotoxicity, particularly in immune system cells (Corbo et al. 2016). While short exposure periods may be feasible without FBS, it may be necessary for maintaining certain in vitro models in longer-term experiments. Thus, the decision to include FBS should be guided by the characterization of the NPs, the selected exposure method, and the specific needs of the chosen in vitro model. Appropriate controls should be included to rule out any detrimental effects on the biological system arising from the absence of FBS.

  In specific cases, such as respiratory airway exposure, the presence of serum might distance the exposure conditions from real-life scenarios, where NMs would not have a serum-derived corona at the time of exposure (Hsiao and Huang 2013). Consequently, including FBS could lead to an underestimation of some NMs' cytotoxic potential, despite contradictory findings in studies examining NM toxicity in the presence or absence of FBS (Hsiao and Huang 2013; Murugadoss et al. 2020; Vranic et al. 2017). In such cases, a simulated respiratory tract lining fluid, supplemented or not with lung surfactant, may serve as a more suitable dispersion medium (Kumar et al. 2017).

- **Homogenization Techniques**: Some NMs require specific dispersion steps to achieve a homogeneous suspension while balancing the preservation of the NM's properties with the homogeneity and stability of the dispersion. Dispersion steps will vary based on the inherent characteristics of the NMs. For instance, hydrophobic NMs require a pre-wetting step—creating a paste with ultrapure water or a small percentage of ethanol—before sonication to facilitate dispersion in aqueous medium (Hartmann et al. 2015). Direct sonication of NMs in cell culture medium is strongly discouraged, as it may denature proteins and generate ROS from sonolysis. Recently, a milder dispersion protocol for hydrophobic NMs has been described, involving continuous stirring of the NMs in cell culture medium with FBS or bovine serum albumin (Lizonova et al. 2024). In this case, the stability of the NM dispersion is supported by the formation of a protective protein corona. A clearly defined sample preparation protocol will yield reproducible NM suspensions, increasing the reproducibility of results from repeated experiments (Ruijter et al. 2023). Several European projects, such as Nanogenotox and Nanoreg, have tackled this issue and generated dispersion protocols based on sonication, which are freely available for use. Note that acoustic energy and de-agglomeration effects can vary between different sonicator brands and even among the same models. Small procedural differences—such as operator technique, water quality, and temperature—can also influence results. However, sonication might not be suitable when attempting to mimic inhalation conditions or for NMs whose physico-chemical properties (agglomeration, dissolution, etc.) change during the process, as this may impact their toxicity. Other dispersion methods proposed in the literature include the use of detergents or surfactants (e.g., Tween-80, Pluronic), but these have been associated with adverse effects, including potential mutagenic activity (Drasler et al. 2017). Particle suspensions should be stable for at least 30–60 min post-preparation and should be freshly prepared for each experimental repetition, as suggested in the Nanogenotox protocol (Jensen et al. 2009).

- **Sampling from Real-Life Scenarios/Products**: Whenever possible, NMs should be sampled from real-life scenarios or products to enhance the relevance of results related to occupational exposures. More information and practical tips on sampling NMs can be found in (Hyun Lee et al. 2010).
- **Mimicking Biological Processes**: Biological processes should be mimicked whenever feasible. For instance, when investigating the effects of NMs or measuring NM uptake in the intestinal system, it is essential to consider the biotransformations that occur during human digestion before NMs interact with intestinal cells. This includes changes in surface chemistry and biocorona formation. In vitro gastrointestinal digestion protocols, such as INFOGEST 2.0 (Brodkorb et al. 2019), replicate protein digestion using standard laboratory equipment. In this process, materials—whether alone or within a food matrix—undergo sequential steps mimicking the oral, gastric, and intestinal phases of digestion. After digestion is complete, the resulting material can be characterized or applied to in vitro biological systems for toxicity, uptake, or translocation studies.

In conclusion, thorough physico-chemical characterization of the NM, particularly concerning particle size and surface charge, should be considered alongside the selected exposure method to establish and define the sample preparation protocol. Most importantly, to accurately represent real-life exposure scenarios, the properties of the NMs that modulate their cytotoxicity must be preserved in the in vitro setting. For more information on the disadvantages of sonication when preparing NM suspensions, refer to Ruijter et al. (2023).

5.3 Exposure Concentrations

During the innovation process, it is customary—and often expected—to initiate the toxicological assessment of new chemical entities by exposing cells or experimental animals to high concentrations of the chemical. This acute exposure scenario aims to establish the concentrations that elicit positive toxicological responses and to understand the underlying mechanisms of these responses. Despite the importance of acute testing during the innovation phase, most expected exposure scenarios arise from occupational or environmental contexts, where repeated—sometimes continuous—exposure to very low concentrations of the chemical occurs. In this context, the actual concentrations to which humans are exposed in real-life situations are often far below those tested in acute studies, potentially diminishing the relevance of the collected data.

Nevertheless, the disparity between the duration of in vitro tests (typically lasting 24 to 72 hours) and the potential for lifelong, repeated exposure to NMs, may justify testing higher concentrations in the laboratory. Furthermore, the concentrations found in environmental or occupational settings generally exceed those that will reach certain organ systems. For instance, considering the hotspots where inhaled particles interact with the airways—usually at the bifurcations of the respiratory tract (Baláshazy et al. 2003)—particle concentrations per surface area may be in the range of $\mu g/cm^2$ (Alfaro-Moreno et al. 2010). However, if the evaluated particles impact a secondary target by translocating from the lungs into circulation, plausible particle concentrations may be several orders of magnitude lower.

It is crucial, therefore, to begin testing over a wide concentration range in which the NM is stably dispersed (Swartzwelter et al. 2021), subsequently narrowing down the test concentrations based on the observed results. The extrapolation from in vivo to in vitro remains challenging, not only for NM toxicity testing but also for regular chemical toxicity assessments. Nonetheless, unique challenges may arise for NMs due to their distinct properties, including the biological impact of their physicochemical characteristics and their stability in biological fluids.

Importantly, NMs do not exist in isolation within the environment or in consumer products. Thus, the concept of complex mixtures—comprising other NMs or chemical entities—should be considered when attempting to mimic real-life scenarios in toxicological testing. For further insights and methodologies on complex exposure scenarios, refer to Gerloff et al. (2017).

In conclusion, to obtain relevant safety information, test concentrations should be selected based on the specific scenario being investigated. When aiming to replicate real-life conditions, the expected exposure scenario for the organ of interest should be taken into account, along with data from acute toxicological tests to serve as a reference.

5.4 Dose Metrics for In Vitro Nanomaterial Toxicity Testing

Accurate and relevant dose metrics are essential components of in vitro toxicity testing of NMs. In contrast to bulk materials and chemicals, the mass alone does not sufficiently capture the concentration-effect relationship for NMs. Due to the high surface-to-mass ratio of NMs, expressing their concentration solely in terms of mass concentration (e.g., μg/mL) does not provide all the necessary information for the correct interpretation of the results. This is because adding to cells 200 μL instead of 100 μL of the same concentration of a NP suspension results in double NP number and NP's surface area, which can consequently lead to increased toxicity. Therefore, also surface area (e.g., cm^2/mL) and/or particle number concentration (e.g., particles/mL) should be disclosed (DeLoid et al. 2017; OECD 2012) to avoid misinterpretation of results. However, expressing the dose as a function of volume (e.g., cell culture medium) appears rather indirect and makes it difficult to compare to exposures at the ALI, where the concentration is naturally given as a function of surface area. In in vitro toxicity studies, the cell culture surface area has been proposed as the recommended metric for dose expression (Drasler et al. 2017), e.g., particles/cm^2 cell surface.

It is important to make the distinction between the nominal (i.e., the theoretical) and the effect concentration. This is because, in addition to the NM's soluble components, the cells will mostly interact with the NMs in their close proximity. Thus, in a conventional 2D cell culture, with time, NMs sediment on the cells at the bottom of the plate, which results in an increased effective concentration compared to the nominal concentration (Fig. 2.3). Contrastingly, buoyant NMs never or poorly sediment, lessening their interaction with the cells, and possibly leading to an underestimation of their potential toxicity in a conventional cell culture system (Watson et al. 2016). For such NMs, an inverted cell culture system should be preferred to facilitate the contact between the NM and the cells (for a detailed description see (Watson et al. 2016)). The main mechanisms by which NMs reach the cell surface in in vitro assays are diffusion and sedimentation. These processes are strongly influenced by the size and effective density of the NM. Agglomeration affects key properties like particle size and effective density, which in turn influence the fate and transport of particles in suspension, for example during SUB exposure conditions. The sedimentation rate is proportional to the square of a particle's diameter, resulting in a tenfold increase in size leading to a 100-fold increase in sedimentation rate, and a similar change in the delivered dose (DeLoid et al. 2017). Agglomeration and dispersion may vary with each concentration in the test system, resulting in the total surface area to which cells are exposed not being (proportionally) the same at each concentration (OECD 2012). Higher doses of NM often lead to more agglomeration, which reduces the total number of particles and the total surface area of the NM available to the cells (DeLoid et al. 2017). Multiple mathematical models are available to determine the surface-available exposure concentrations (Cheimarios et al. 2022; Hinderliter et al. 2010; Thomas et al. 2018). Furthermore, NMs may exhibit a dynamic behavior over time (e.g., dissolution, aggregation, and sedimentation), leading to time-dependent changes in the

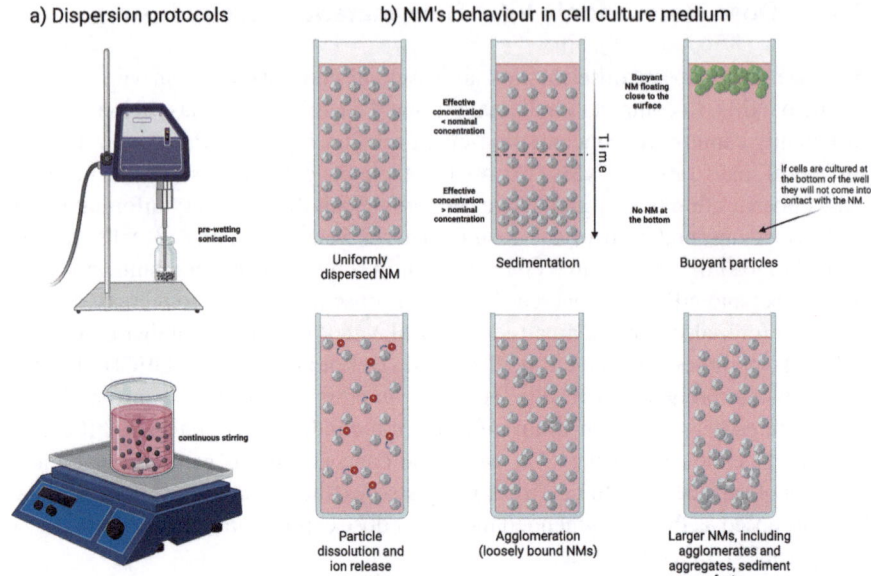

Fig. 2.3 (a) Examples of methods used in the preparation of stable NM suspensions. These include pre-wetting and sonication and, more recently, continuous stirring. (b) The sedimentation of NMs in the medium/dispersion liquid depends on—for single NMs—the density of the medium, the density of the NMs in the medium, and the diameter of the NM. Intrinsically heavier particles, as well as particle agglomerates and/or aggregates, sediment faster. Buoyant NMs, however, will not sediment at all. Certain NMs are expected to release ions which will most likely be available to interact with the cells

delivered dose. For exposure scenarios longer than 24 hours, it is advised that the composition is checked over time (OECD 2012).

6 Other General Considerations

Considering the similarities in testing between conventional chemicals and NMs, some key aspects apply to both contexts. Here we highlight some of those points:

– Apply the GIVIMP whenever possible: the GIVIMP summarizes the guidelines defined by the OECD as good practice to minimize uncertainties and improve predictions when developing or using in vitro methods (OECD 2018). Among many other points, the document provides guidance on SOP preparation, describes the main factors affecting the reliability and relevance of an in vitro method, and advocates for the importance of establishing definite reporting criteria. It is freely accessible via the OECD website, and it is particularly relevant when seeking regulatory acceptance of in vitro models/methods under development. It is a highly recommended go-to tool for all cell culture users, especially for beginners starting their cell culture routines. Another interesting source of

information on this topic, though more focused on increasing the reproducibility of in vitro models and methods, is the paper (Hirsch and Schildknecht 2019).
- Make sure you have all the necessary materials and reagents in advance, before you start the experiment. Set up a list with all the necessary components for your experiment and take a day or two to prepare all the reagents (except NM suspensions) and cell cultures in advance—this will allow you to maximize results/resources.
- Whenever changes to the procedure are needed, make one change at a time to get the best understanding possible of your model/method/result.
- Start simple and then increase the complexity of both the model and the experimental protocol as you refine your technique and deepen your knowledge about the system.
- Favor the reproducibility of your work by:
  - Using cells from reliable sources.
  - Regularly verifying the authenticity/integrity of your cell lines, testing your cell lines for mycoplasma and, if possible, avoiding the use of antibiotics in your cultures. Besides the generation of resistant bacterial species, the long-term use of antibiotics may induce genetic changes that affect cell proliferation, differentiation, survival, and modify drug response (Weiskirchen et al. 2023).
  - Making sure that environmental conditions remain stable during the experiments. Though many times neglected and even known as "silent variables", environmental conditions, such as temperature, relative humidity, and atmospheric carbon dioxide, may play a significant role in the reproducibility of the observed results. When maintained adequately, cell incubators provide a niche of controlled temperature, CO_2, and humidity levels. When removed from the incubator, even for short periods, cells experience different environmental conditions that can slow down or even halt their metabolism and cell proliferation, such as lower temperature and lower CO_2 levels (which disturb medium pH). Concurrently, the physical movement of the medium on the cells generates some level of shear stress, and some of the medium components are sensitive to white light (OECD 2018). These changes are particularly relevant in models and methods requiring many days in culture or prolonged incubation periods. Adequate incubator maintenance and reducing to the minimum possible the amount of time that cells are outside controlled conditions are crucial factors to consider when performing any in vitro studies (Capes-Davis and Freshney 2021).
  - Preparing detailed SOPs including every step of the experimental procedure is fundamental but many times not enough. An additional one-to-one hands-on knowledge transfer is ideal to make sure all the steps are performed similarly by different operators.
  - Using any automation existing in the lab, namely for cell counting, seeding, exposure, and sampling. This will greatly reduce errors associated with the operator. These and other suggestions to improve the reproducibility of

in vitro studies can be found in (Capes-Davis and Freshney 2021; Hirsch and Schildknecht 2019).
- Try to collect as much information as possible from the same in vitro model unit (well, chip, insert, etc.) (Drasler et al. 2017). Multiplexing-compatible assays are a great way to see the bigger picture and provide more power to your assumptions. Nowadays, there are many available non-destructible techniques that can be used in tandem to generate meaningful information about the status of the in vitro cultures and how the tested item affects their health. Supernatants are a great source of information on what is happening in the cell population throughout time in culture, and their analysis in multiplexed systems may be a great alternative to replace destructive methods.

7 Conclusions and Perspectives

While the field of nanotoxicology has benefited from the tradition and knowledge of conventional toxicology, the assessment of the safety of NMs comes with unprecedented challenges, which make a thorough and thoughtful experimental design a real must. The recent investments in the development of new human-relevant advanced in vitro models allowing new routes of exposure open new avenues for recreating closer-to-real-life environments in the lab. Prior investigation and reflection about possible exposure scenarios (route, NM concentrations, etc.) will help define the proper in vitro models and test methods to provide the most relevant data on the effects and mechanisms triggered by NMs.

New and exciting developments in methodological assessments have been observed in recent years, granting access to big data at the cellular population and/or single-cell level. Therefore, many of the conditions seem to be in place to make the most out of both models and methods to advance the state of the art of human-relevant nanotoxicology research.

With these new technologies come new needs for collaborative efforts in harmonization and development of guidelines for best practices. Even though many efforts have already been undertaken in this regard, leading to the generation of valuable standardization protocols, there is substantially more new knowledge that needs to be integrated into these protocols and further disseminated. Consistency in methodologies, especially for characterization, sample preparation, and exposure scenarios, is crucial to enhance the reproducibility and comparability of results across different laboratories.

Particularly, harmonization of exposure methods using generation of aerosols at the ALI would greatly benefit the scientific community. Because the *dose* still *makes the poison*, precise dosimetry must be implemented by means of reliable recording of the actual dose deposited over time and its distribution across the cell layer, together with accurate monitoring of environmental conditions (e.g., humidity and temperature). If widely and openly shared and applied, all these protocols would jointly contribute to advance our understanding of the detrimental effects that NMs may (or may not) pose to human health and the environment.

Acknowledgements This work was funded by the European Union's H2020 project Sinfonia (N.857253) and PHOENIX (N.953110), as well as by Horizon Europe Projects LEARN (N.101057510) and iCare (N.101092971). FC is supported by the Portuguese Foundation for Science and Technology (FCT) through projects UIDP/04378/2020 and UIDB/04378/2020 (UCIBIO), and LA/P/0140/2020 (i4HB).

References

Alfaro-Moreno E, López-Marure R, Montiel-Dávalos A, Symonds P, Osornio-Vargas AR, Rosas I, Clifford Murray J (2007) E-selectin expression in human endothelial cells exposed to PM10: the role of endotoxin and insoluble fraction. Environ Res 103:221–228. https://doi.org/10.1016/j.envres.2006.05.004

Alfaro-Moreno E, Garcia-Cuellar C, De-Vizcaya-Ruiz A, Rojas-Bracho L, Osornio-Vargas A (2010) Cellular mechanisms behind particulate matter air pollution–related health effects. In: Air pollution. CRC Press, pp 249–274. https://doi.org/10.1201/EBK1439809624-c9

Aliyandi A, Reker-Smit C, Bron R, Zuhorn IS, Salvati A (2021) Correlating Corona composition and cell uptake to identify proteins affecting nanoparticle entry into endothelial cells. ACS Biomater Sci Eng 7:5573–5584. https://doi.org/10.1021/acsbiomaterials.1c00804

Awashra M, Młynarz P (2023) The toxicity of nanoparticles and their interaction with cells: an *in vitro* metabolomic perspective. Nanoscale Adv 5:2674–2723. https://doi.org/10.1039/D2NA00534D

Bacher G, Szymanski WW, Kaufman SL, Zöllner P, Blaas D, Allmaier G (2001) Charge-reduced nano electrospray ionization combined with differential mobility analysis of peptides, proteins, glycoproteins, noncovalent protein complexes and viruses. J Mass Spectrom 36:1038–1052. https://doi.org/10.1002/jms.208

Balásházy I, Hofmann W, Heistracher T (2003) Local particle deposition patterns may play a key role in the development of lung cancer. J Appl Physiol 94:1719–1725. https://doi.org/10.1152/japplphysiol.00527.2002

Bell CC, Hendriks DFG, Moro SML, Ellis E, Walsh J, Renblom A, Fredriksson Puigvert L, Dankers ACA, Jacobs F, Snoeys J, Sison-Young RL, Jenkins RE, Nordling Å, Mkrtchian S, Park BK, Kitteringham NR, Goldring CEP, Lauschke VM, Ingelman-Sundberg M (2016) Characterization of primary human hepatocyte spheroids as a model system for drug-induced liver injury, liver function and disease. Sci Rep 6:25187. https://doi.org/10.1038/srep25187

Bergstrand A, Svanberg C, Langton M, Nydén M (2006) Aggregation behavior and size of lipopolysaccharide from Escherichia coli O55:B5. Colloids Surf B Biointerfaces 53:9–14. https://doi.org/10.1016/j.colsurfb.2006.06.007

Beutler B, Rietschel ET (2003) Innate immune sensing and its roots: the story of endotoxin. Nat Rev Immunol 3:169–176. https://doi.org/10.1038/nri1004

Bolden J, Smith K (2017) Application of recombinant factor C reagent for the detection of bacterial endotoxins in pharmaceutical products. PDA J Pharm Sci Technol 71:405–412. https://doi.org/10.5731/pdajpst.2017.007849

Bonner JC, Rice AB, Lindroos PM, O'Brien PO, Dreher KL, Rosas I, Alfaro-Moreno E, Osornio-Vargas AR (1998) Induction of the lung myofibroblast PDGF receptor system by urban ambient particles from Mexico City. Am J Respir Cell Mol Biol 19:672–680. https://doi.org/10.1165/ajrcmb.19.4.3176

Brodkorb A, Egger L, Alminger M, Alvito P, Assunção R, Ballance S, Bohn T, Bourlieu-Lacanal C, Boutrou R, Carrière F, Clemente A, Corredig M, Dupont D, Dufour C, Edwards C, Golding M, Karakaya S, Kirkhus B, Le Feunteun S, Lesmes U, Macierzanka A, Mackie AR, Martins C, Marze S, McClements DJ, Ménard O, Minekus M, Portmann R, Santos CN, Souchon I, Singh RP, Vegarud GE, Wickham MSJ, Weitschies W, Recio I (2019) INFOGEST static in vitro simulation of gastrointestinal food digestion. Nat Protoc 14:991–1014. https://doi.org/10.1038/s41596-018-0119-1

Capes-Davis A, Freshney RI (2021) Freshney's culture of animal cells: a manual of basic technique and specialized applications, 8th edn. Wiley-Blackwell

Casals E, Pfaller T, Duschl A, Oostingh GJ, Puntes V (2010) Time evolution of the nanoparticle protein Corona. ACS Nano 4:3623–3632. https://doi.org/10.1021/nn901372t

Cheimarios N, Pem B, Tsoumanis A, Ilić K, Vrček IV, Melagraki G, Bitounis D, Isigonis P, Dusinska M, Lynch I, Demokritou P, Afantitis A (2022) An in vitro dosimetry tool for the numerical transport modeling of engineered nanomaterials powered by the Enalos RiskGONE cloud platform. Nano 12:3935. https://doi.org/10.3390/nano12223935

Clevers H (2016) Modeling development and disease with organoids. Cell 165:1586–1597. https://doi.org/10.1016/j.cell.2016.05.082

Colas RA, Shinohara M, Dalli J, Chiang N, Serhan CN (2014) Identification and signature profiles for pro-resolving and inflammatory lipid mediators in human tissue. Am J Phys Cell Phys 307:C39–C54. https://doi.org/10.1152/ajpcell.00024.2014

Corbo C, Molinaro R, Parodi A, Toledano Furman NE, Salvatore F, Tasciotti E (2016) The impact of nanoparticle protein Corona on cytotoxicity, immunotoxicity and target drug delivery. Nanomedicine 11:81–100. https://doi.org/10.2217/nnm.15.188

DeLoid GM, Cohen JM, Pyrgiotakis G, Demokritou P (2017) Preparation, characterization, and in vitro dosimetry of dispersed, engineered nanomaterials. Nat Protoc 12:355–371. https://doi.org/10.1038/nprot.2016.172

Deng ZJ, Mortimer G, Schiller T, Musumeci A, Martin D, Minchin RF (2009) Differential plasma protein binding to metal oxide nanoparticles. Nanotechnology 20:455101. https://doi.org/10.1088/0957-4484/20/45/455101

Dobrovolskaia MA, Neun BW, Clogston JD, Ding H, Ljubimova J, McNeil SE (2010) Ambiguities in applying traditional *limulus* Amebocyte lysate tests to quantify endotoxin in nanoparticle formulations. Nanomedicine 5:555–562. https://doi.org/10.2217/nnm.10.29

Drasler B, Sayre P, Steinhäuser KG, Petri-Fink A, Rothen-Rutishauser B (2017) In vitro approaches to assess the hazard of nanomaterials. NanoImpact 8:99–116. https://doi.org/10.1016/j.impact.2017.08.002

Erridge C, Bennett-Guerrero E, Poxton IR (2002) Structure and function of lipopolysaccharides. Microbes Infect 4:837–851. https://doi.org/10.1016/S1286-4579(02)01604-0

Ertel A, Verghese A, Byers SW, Ochs M, Tozeren A (2006) Pathway-specific differences between tumor cell lines and normal and tumor tissue cells. Mol Cancer 5:55. https://doi.org/10.1186/1476-4598-5-55

Faber SC, McCullough SD (2018) Through the looking glass: *In vitro* models for inhalation toxicology and interindividual variability in the airway. Appl In Vitro Toxicol 4:115–128. https://doi.org/10.1089/aivt.2018.0002

FDA (2017) FDA's predictive toxicology roadmap

Francia V, Yang K, Deville S, Reker-Smit C, Nelissen I, Salvati A (2019) Corona composition can affect the mechanisms cells use to internalize nanoparticles. ACS Nano 13:11107–11121. https://doi.org/10.1021/acsnano.9b03824

Gerloff K, Landesmann B, Worth A, Munn S, Palosaari T, Whelan M (2017) The adverse outcome pathway approach in nanotoxicology. Comput Toxicol 1:3–11. https://doi.org/10.1016/j.comtox.2016.07.001

Gómez-Lechón M, Tolosa L, Donato M (2014) Cell-based models to predict human hepatotoxicity of drugs. Rev Toxicol 31:149–156

Gorbet MB, Sefton MV (2005) Endotoxin: the uninvited guest. Biomaterials 26:6811–6817. https://doi.org/10.1016/j.biomaterials.2005.04.063

Griwatz U, Seemayer NH (1995) Tumour necrosis factor-α induction by endotoxin-containing coal mine dusts in cultures of human macrophages and its effects on pneumocyte type II cells. Toxicol In Vitro 9:403–409. https://doi.org/10.1016/0887-2333(95)00026-5

Guadagnini R, Halamoda Kenzaoui B, Walker L, Pojana G, Magdolenova Z, Bilanicova D, Saunders M, Juillerat-Jeanneret L, Marcomini A, Huk A, Dusinska M, Fjellsbø LM, Marano F, Boland S (2015) Toxicity screenings of nanomaterials: challenges due to interference with assay processes and components of classic *in vitro* tests. Nanotoxicology 9:13–24. https://doi.org/10.3109/17435390.2013.829590

Hannon G, Prina-Mello A (2021) Endotoxin contamination of engineered nanomaterials: overcoming the hurdles associated with endotoxin testing. WIREs Nanomed Nanobiotechnol 13. https://doi.org/10.1002/wnan.1738

Hartmann NB, Jensen KA, Baun A, Rasmussen K, Rauscher H, Tantra R, Cupi D, Gilliland D, Pianella F, Riego Sintes JM (2015) Techniques and protocols for dispersing nanoparticle powders in aqueous media—is there a rationale for harmonization? J Toxic Environ Health, Part B 18:299–326. https://doi.org/10.1080/10937404.2015.1074969

Hinderliter PM, Minard KR, Orr G, Chrisler WB, Thrall BD, Pounds JG, Teeguarden JG (2010) ISDD: a computational model of particle sedimentation, diffusion and target cell dosimetry for in vitro toxicity studies. Part Fibre Toxicol 7:36. https://doi.org/10.1186/1743-8977-7-36

Hirsch C, Schildknecht S (2019) In vitro research reproducibility: keeping up high standards. Front Pharmacol 10. https://doi.org/10.3389/fphar.2019.01484

Hsiao I-L, Huang Y-J (2013) Effects of serum on cytotoxicity of nano- and micro-sized ZnO particles. J Nanopart Res 15:1829. https://doi.org/10.1007/s11051-013-1829-5

Huh D, Fujioka H, Tung Y-C, Futai N, Paine R, Grotberg JB, Takayama S (2007) Acoustically detectable cellular-level lung injury induced by fluid mechanical stresses in microfluidic airway systems. Proc Natl Acad Sci 104:18886–18891. https://doi.org/10.1073/pnas.0610868104

Hyun Lee J, Chaul Moon M, Yeob Lee J, Yu J I (2010) Challenges and perspectives of nanoparticle exposure assessment. Toxicol Res

Iavicoli I, Fontana L, Leso V, Bergamaschi A (2013) The effects of nanomaterials as endocrine disruptors. Int J Mol Sci 14:16732–16801. https://doi.org/10.3390/ijms140816732

Jensen KA, Kembouche Y, Christiansen E, Jacobsen NR, Wallin H, Guiot C, Spalla O, Witschger O (2009) Final protocol for producing suitable manufactured nanomaterial exposure media

Juarez-Moreno K, Chávez-García D, Hirata G, Vazquez-Duhalt R (2022) Monolayer (2D) or spheroids (3D) cell cultures for nanotoxicological studies? Comparison of cytotoxicity and cell internalization of nanoparticles. Toxicol In Vitro 85:105461. https://doi.org/10.1016/j.tiv.2022.105461

Karlsson HL, Di Bucchianico S, Collins AR, Dusinska M (2015) Can the comet assay be used reliably to detect nanoparticle-induced genotoxicity? Environ Mol Mutagen 56:82–96. https://doi.org/10.1002/em.21933

Klein SG, Serchi T, Hoffmann L, Blömeke B, Gutleb AC (2013) An improved 3D tetraculture system mimicking the cellular organisation at the alveolar barrier to study the potential toxic effects of particles on the lung. Part Fibre Toxicol 10:31. https://doi.org/10.1186/1743-8977-10-31

Kroll A, Pillukat MH, Hahn D, Schnekenburger J (2012) Interference of engineered nanoparticles with in vitro toxicity assays. Arch Toxicol 86:1123–1136. https://doi.org/10.1007/s00204-012-0837-z

Kumar A, Terakosolphan W, Hassoun M, Vandera K-K, Novicky A, Harvey R, Royall PG, Bicer EM, Eriksson J, Edwards K, Valkenborg D, Nelissen I, Hassall D, Mudway IS, Forbes B (2017) A biocompatible synthetic lung fluid based on human respiratory tract lining fluid composition. Pharm Res 34:2454–2465. https://doi.org/10.1007/s11095-017-2169-4

Lebre F, Chatterjee N, Costa S, Fernández-De-gortari E, Lopes C, Meneses J, Ortiz L, Ribeiro AR, Vilas-Boas V, Alfaro-Moreno E (2022) Nanosafety: an evolving concept to bring the safest possible nanomaterials to society and environment. Nano. https://doi.org/10.3390/nano12111810

Leung CM, de Haan P, Ronaldson-Bouchard K, Kim G-A, Ko J, Rho HS, Chen Z, Habibovic P, Jeon NL, Takayama S, Shuler ML, Vunjak-Novakovic G, Frey O, Verpoorte E, Toh Y-C (2022) A guide to the organ-on-a-chip. Nat Rev Methods Primers 2:33. https://doi.org/10.1038/s43586-022-00118-6

Li J, Chen C, Xia T (2022) Understanding nanomaterial–liver interactions to facilitate the development of safer nanoapplications. Adv Mater 34. https://doi.org/10.1002/adma.202106456

Liu S, Yin N, Faiola F (2017) Prospects and frontiers of stem cell toxicology. Stem Cells Dev 26:1528–1539. https://doi.org/10.1089/scd.2017.0150

Lizonova D, Trivanovic U, Demokritou P, Kelesidis GA (2024) Dispersion and dosimetric challenges of hydrophobic carbon-based nanoparticles in in vitro cellular studies. Nano 14:589. https://doi.org/10.3390/nano14070589

Lu RXZ, Radisic M (2021) Organ-on-a-chip platforms for evaluation of environmental nanoparticle toxicity. Bioact Mater 6:2801–2819. https://doi.org/10.1016/j.bioactmat.2021.01.021

Lynch S, Pridgeon CS, Duckworth CA, Sharma P, Park BK, Goldring CEP (2019) Stem cell models as an *in vitro* model for predictive toxicology. Biochem J 476:1149–1158. https://doi.org/10.1042/BCJ20170780

Malakar A, Kanel SR, Ray C, Snow DD, Nadagouda MN (2021) Nanomaterials in the environment, human exposure pathway, and health effects: a review. Sci Total Environ 759:143470. https://doi.org/10.1016/j.scitotenv.2020.143470

Murugadoss S, Brassinne F, Sebaihi N, Petry J, Cokic SM, Van Landuyt KL, Godderis L, Mast J, Lison D, Hoet PH, van den Brule S (2020) Agglomeration of titanium dioxide nanoparticles increases toxicological responses in vitro and in vivo. Part Fibre Toxicol 17:10. https://doi.org/10.1186/s12989-020-00341-7

OECD (2005) Guidance document on the validation and international acceptance of new or updated test methods for Hazard assessment

OECD (2012) Guidance on sample preparation and dosimetry for the safety testing of manufactured nanomaterials

OECD (2018) Guidance document on good in vitro method practices (GIVIMP). ENV/JM/MONO(2018)19

Olson H, Betton G, Robinson D, Thomas K, Monro A, Kolaja G, Lilly P, Sanders J, Sipes G, Bracken W, Dorato M, Van Deun K, Smith P, Berger B, Heller A (2000) Concordance of the toxicity of pharmaceuticals in humans and in animals. Regul Toxicol Pharmacol 32:56–67. https://doi.org/10.1006/rtph.2000.1399

Raetz CRH, Whitfield C (2002) Lipopolysaccharide endotoxins. Annu Rev Biochem 71:635–700. https://doi.org/10.1146/annurev.biochem.71.110601.135414

Ramos-Godínez MDP, González-Gómez BE, Montiel-Dávalos A, López-Marure R, Alfaro-Moreno E (2013) TiO2 nanoparticles induce endothelial cell activation in a pneumocyte–endothelial co-culture model. Toxicol In Vitro 27:774–781. https://doi.org/10.1016/j.tiv.2012.12.010

Ruijter N, Soeteman-Hernández LG, Carrière M, Boyles M, McLean P, Catalán J, Katsumiti A, Cabellos J, Delpivo C, Sánchez Jiménez A, Candalija A, Rodríguez-Llopis I, Vázquez-Campos S, Cassee FR, Braakhuis H (2023) The state of the art and challenges of in vitro methods for human Hazard assessment of nanomaterials in the context of safe-by-design. Nano 13:472. https://doi.org/10.3390/nano13030472

Savage DT, Hilt JZ, Dziubla TD (2019) In vitro methods for assessing nanoparticle toxicity. Methods Mol Biol 1894:1–29. https://doi.org/10.1007/978-1-4939-8916-4_1

Sharma G, Kodali V, Gaffrey M, Wang W, Minard KR, Karin NJ, Teeguarden JG, Thrall BD (2014) Iron oxide nanoparticle agglomeration influences dose rates and modulates oxidative stress-mediated dose–response profiles *in vitro*. Nanotoxicology 8:663–675. https://doi.org/10.3109/17435390.2013.822115

Stone V, Johnston H, Schins RPF (2009) Development of *in vitro* systems for nanotoxicology: methodological considerations. Crit Rev Toxicol 39:613–626. https://doi.org/10.1080/10408440903120975

Stucki JD, Hobi N, Galimov A, Stucki AO, Schneider-Daum N, Lehr C-M, Huwer H, Frick M, Funke-Chambour M, Geiser T, Guenat OT (2018) Medium throughput breathing human primary cell alveolus-on-chip model. Sci Rep 8:14359. https://doi.org/10.1038/s41598-018-32523-x

Swartzwelter BJ, Mayall C, Alijagic A, Barbero F, Ferrari E, Hernadi S, Michelini S, Navarro Pacheco NI, Prinelli A, Swart E, Auguste M (2021) Cross-species comparisons of nanoparticle interactions with innate immune systems: a methodological review. Nano 11:1528. https://doi.org/10.3390/nano11061528

Thomas DG, Smith JN, Thrall BD, Baer DR, Jolley H, Munusamy P, Kodali V, Demokritou P, Cohen J, Teeguarden JG (2018) ISD3: a particokinetic model for predicting the combined effects of particle sedimentation, diffusion and dissolution on cellular dosimetry for in vitro systems. Part Fibre Toxicol 15:6. https://doi.org/10.1186/s12989-018-0243-7

Ulevitch RJ, Tobias PS (1995) Receptor-dependent mechanisms of cell stimulation by bacterial endotoxin. Annu Rev Immunol 13:437–457. https://doi.org/10.1146/annurev.iy.13.040195.002253

Vasimalai N, Vilas-Boas V, Gallo J, Cerqueira MF, Menéndez-Miranda M, Costa-Fernández JM, Diéguez L, Espiña B, Fernández-Argüelles MT (2018) Green synthesis of fluorescent carbon dots from spices for in vitro imaging and tumour cell growth inhibition. Beilstein J Nanotechnol 9. https://doi.org/10.3762/bjnano.9.51

Vilas-Boas V, Vinken M (2021) Hepatotoxicity induced by nanomaterials: mechanisms and in vitro models. Arch Toxicol 95:27–52. https://doi.org/10.1007/s00204-020-02940-x

Vilas-Boas V, Gijbels E, Leroy K, Pieters A, Baze A, Parmentier C, Vinken M (2021) Primary human hepatocyte spheroids as tools to study the hepatotoxic potential of non-pharmaceutical chemicals. Int J Mol Sci 22:11005. https://doi.org/10.3390/ijms222011005

Vranic S, Gosens I, Jacobsen NR, Jensen KA, Bokkers B, Kermanizadeh A, Stone V, Baeza-Squiban A, Cassee FR, Tran L, Boland S (2017) Impact of serum as a dispersion agent for in vitro and in vivo toxicological assessments of TiO2 nanoparticles. Arch Toxicol 91:353–363. https://doi.org/10.1007/s00204-016-1673-3

Watson CY, DeLoid GM, Pal A, Demokritou P (2016) Buoyant nanoparticles: implications for nano-biointeractions in cellular studies. Small 12:3172–3180. https://doi.org/10.1002/smll.201600314

Weiskirchen S, Schröder SK, Buhl EM, Weiskirchen R (2023) A Beginner's guide to cell culture: practical advice for preventing needless problems. Cells 12:682. https://doi.org/10.3390/cells12050682

Wiemann M, Vennemann A, Sauer UG, Wiench K, Ma-Hock L, Landsiedel R (2016) An in vitro alveolar macrophage assay for predicting the short-term inhalation toxicity of nanomaterials. J Nanobiotechnol 14:16. https://doi.org/10.1186/s12951-016-0164-2

Xuan L, Ju Z, Skonieczna M, Zhou P, Huang R (2023) Nanoparticles-induced potential toxicity on human health: applications, toxicity mechanisms, and evaluation models. MedComm (Beijing) 4. https://doi.org/10.1002/mco2.327

Zhao C (2023) Cell culture: *in vitro* model system and a promising path to *in vivo* applications. J Histotechnol 46:1–4. https://doi.org/10.1080/01478885.2023.2170772

Zhao Y, Landau S, Okhovatian S, Liu C, Lu RXZ, Lai BFL, Wu Q, Kieda J, Cheung K, Rajasekar S, Jozani K, Zhang B, Radisic M (2024) Integrating organoids and organ-on-a-chip devices. Nat Rev Bioeng 2:588–608. https://doi.org/10.1038/s44222-024-00207-z

Open Access This chapter is licensed under the terms of the Creative Commons Attribution-NonCommercial-NoDerivatives 4.0 International License (http://creativecommons.org/licenses/by-nc-nd/4.0/), which permits any noncommercial use, sharing, distribution and reproduction in any medium or format, as long as you give appropriate credit to the original author(s) and the source, provide a link to the Creative Commons license and indicate if you modified the licensed material. You do not have permission under this license to share adapted material derived from this chapter or parts of it.

The images or other third party material in this chapter are included in the chapter's Creative Commons license, unless indicated otherwise in a credit line to the material. If material is not included in the chapter's Creative Commons license and your intended use is not permitted by statutory regulation or exceeds the permitted use, you will need to obtain permission directly from the copyright holder.

Exploring the Interplay Between Particles and the Immune System

3

Thais S. M. Lima and Filipa Lebre

Abstract

In recent years, human exposure to particulate materials, especially nanomaterials (NMs) has become more common and widespread. Nanomaterials are making their way into many aspects of our lives enabling the development of new types of materials for industrial and biomedical applications, leading to the possibility of increased exposure to nanoscale particles for both workers and consumers. Throughout their life cycle (e.g. production; use; disposal), NMs can be released into the environment, potentially causing harmful effects on natural ecosystems, with possible repercussions for both humans and animals health. It is critical to assess the effect of particulate material on short and long-term human health, and this requires a deep understanding of their interplay with the immune system. Immunological assessment of NMs is crucial for determining their safety, and efficacy, how they modulate the immune system, and to prevent undesirable immune reactions. However, assessing the immunomodulatory properties of NMs presents several challenges due to their unique properties that greatly influence their biological interactions and toxicity profile. In this chapter, we will review fundamental concepts about the immune system, how particles can trigger immune responses, as well as some challenges and emerging technologies for assessing the toxicity and immunomodulatory properties of NMs.

T. S. M. Lima
Division of Biological Metrology, National Institute of Metrology Quality and Technology, Rio de Janeiro, Brazil

Postgraduate Program in Metrology and Technology, National Institute of Metrology Quality and Technology, Rio de Janeiro, Brazil

F. Lebre (✉)
Nanosafety Group, International Iberian Nanotechnology Laboratory, Avenida Mestre José Veiga s/n, Braga, Portugal
e-mail: filipa.lebre@inl.int

© The Author(s) 2025
E. Alfaro-Moreno, F. Murphy (eds.), *Nanosafety*,
https://doi.org/10.1007/978-3-031-93871-9_3

Keywords

Immune response · Nanoparticles · Immunomodulation · Inflammatory response · Safety assessment

1 An Introduction to the Immune System

The human immune system's main function is to detect and eliminate pathogens and foreign substances that threaten the host, and two branches of immunity evolved to function together in a complementary fashion to prevent infection. The innate immune system, responds immediately as the body's initial defence mechanism, and in many cases is sufficient to clear the infection; however, if the pathogen is not eliminated from the body, the host will trigger the adaptive immune system. The adaptive immune response is more specific but requires 4–5 days to be activated, during which time the innate system keeps the infection under control until the adaptive system can target the pathogen more effectively.

1.1 Innate Immune System

The human body has evolved barriers capable of protecting against pathogens and foreign substances, including NMs. These anatomical and physiological barriers, such as the skin and mucous membranes, form the first crucial line of defence, imposing physical and chemical obstacles limiting invasion of the host. The skin consists of a multi-layered structure, and its outmost layer, the stratum corneum, creates a tight barrier, is armed with enzymes and antimicrobial peptides and maintains an acidic pH, while mucosal surfaces produce mucus to trap foreign substances (Bulet et al. 2004; Costa et al. 2023) (Fig. 3.1).

The innate system provides a rapid, unspecific defence against pathogens, playing a vital role in the initial recognition and eradication of invaders. Innate cells are one of the first to respond to infection or danger signs, inducing the secretion of many factors including chemokines and cytokines. These signalling molecules orchestrate the immune response by facilitating communication between immune cells, initiating a protective immune response.

Granulocytes are a group of innate cells encompassing neutrophils, eosinophils, and basophils. As the most abundant leukocytes in circulation, neutrophils are generally the first responders to infection sites and foreign objects (Kumar and Sharma 2010). They exhibit a robust phagocytic activity and are highly specialized in killing intracellular pathogens, as the cytoplasm of neutrophils is rich in granules containing antimicrobial peptides (e.g., defensins, cathelicidins) and enzymes (e.g., myeloperoxidase) (Faurschou and Borregaard 2003). Besides directly eliminating pathogens via granule release, neutrophils can also form neutrophil extracellular traps (NETs), which trap and kill extracellular microbes using DNA and antimicrobial proteins, in a process called NETosis (Thiam et al. 2020). Neutrophils are also vital in recruiting other immune cells (e.g. macrophages; dendritic cells) to

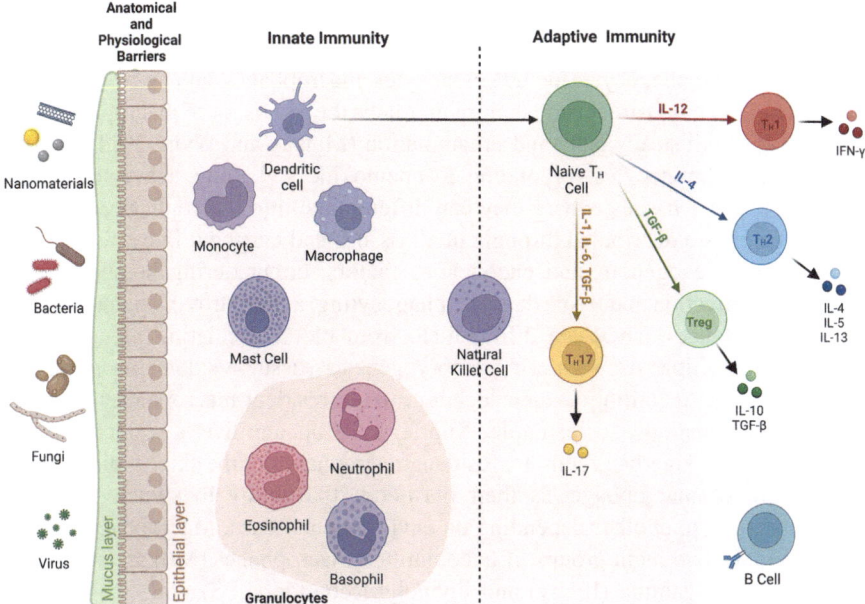

Fig. 3.1 Immune system overview. Anatomical and physiological barriers (e.g. epithelial and mucus layers; enzymes; low pH) are the first defence to prevent the entrance of pathogens (e.g. bacteria; viruses; fungi) or foreign bodies (e.g. micro and nanomaterials) into the host. In response to microbes or NMs that might enter the body, cells of the innate immune system are activated. Cells involved in the innate response include monocytes, macrophages, dendritic cells (DCs), granulocytes (basophils, eosinophils, and neutrophils), and mast cells. Natural killer (NK) cells display features of both innate and adaptive systems. The adaptive system consists of B and T cells. Following activation, naïve T cells can differentiate into distinct phenotypes depending on the cytokines present in the local environment (polarising cytokines) and subsequent expression of transcription factors. In turn, each effector subset secretes a range of specific cytokines that contribute to their function. Created with BioRender

inflammation sites (Bennouna et al. 2003). Eosinophils and basophils are much less abundant than neutrophils, but they play key roles in parasite defence and allergic responses (Bochner 2000; Obata-Ninomiya et al. 2020). Basophils release mediators, such as histamine and heparin, from their granules, promoting vasodilation and increasing vascular permeability, facilitating the recruitment of other immune cells to the infection site, while eosinophils contain granules filled with cytotoxic proteins, including eosinophil cationic protein (ECP) and major basic protein (MBP), which are toxic to parasites. Additionally, eosinophils are capable of secreting pro-inflammatory mediators quickly due to their pre-translated intracellular stock of cytokines (Voehringer et al. 2007). Mast cells are positioned at the level of several host-environment interfaces and they play a crucial role in the immune response against pathogens, due to their ability to quickly release preformed inflammatory mediators (e.g. proteases; histamine; cytokines) from intracellular granules, that can lead to the recruitment of other immune cells (Abraham and St John 2010). Mast cells are also involved in allergic reactions when an allergen binds immunoglobulin E (IgE) antibodies (Amin 2012; Krystel-Whittemore et al. 2016).

Mononuclear phagocytes include monocytes, macrophages, and DCs. Monocytes circulate in the bloodstream and their main role is the provision of macrophages and DCs during both steady state and inflammation (Murray and Wynn 2011). During inflammation, monocytes responding to chemokine signalling, leave the bloodstream and enter tissues, where they can differentiate into macrophages and DCs. Macrophages are distributed throughout all tissues and carry out integral functions responding to exogenous and endogenous factors, either during homeostasis or throughout infection and tissue damage, phagocyting, and destroying pathogens and foreign substances. They can differentiate from blood-circulating monocytes or develop in multiple tissues during embryogenesis (tissue-resident macrophages) (Röszer 2018). According to their location, tissue-resident macrophages can have different designations, for example, Kupffer cells are the liver's resident macrophages, while Langerhans cells are resident macrophages of the skin. A distinguishing feature of macrophages is their plasticity; the ability to switch from one activation state to another depending on environmental cues. Macrophages can be classified into two main groups; if uncommitted macrophages (M0) are stimulated with interferon-gamma (IFN-γ) and lipopolysaccharide (LPS), they will become pro-inflammatory and bactericidal, "classically activated" M1 macrophages, and if exposed to interleukin (IL)-4 and IL-13 they will convert to "alternatively activated" Th2-promoting, anti-inflammatory M2 macrophages (Smith et al. 2016). This dichotomous categorization of macrophages into M1/M2 has become a topic for debate, due to the oversimplification of the broad spectrum of activation states (Murray et al. 2014). Besides their role in the clearance of pathogen and foreign bodies, macrophages are also highly efficient antigen-presenting cells (APCs), displaying processed antigens to T cells (Beutler 2004), and play roles in tissue repair and wound healing by secreting growth factors and remodelling extracellular matrix components (Biswas and Mantovani 2010).

Dendritic cells are a heterogeneous group of cells, first reported in the 70s by Steinman and Cohn (Steinman and Cohn 1973). They have since been described as professional APCs that coordinate the talk between the innate and the adaptive immune response, by priming T cell responses. Due to their essential role in immune surveillance, DCs express a wide array of pattern recognition receptors (PRRs) that recognize pathogen-associated molecular patterns (PAMPs) and danger-associated molecular patterns (DAMPs). Under steady-state conditions, DCs are mostly described as being immature; upon activation, via PRR stimulation, they become mature, a phenotype characterized by upregulation of surface major histocompatibility complex (MHC) and co-stimulatory molecules (e.g. Cluster of Differentiation (CD) 80; CD86; CD40) that promotes DC migration to draining lymph nodes, and present antigen-derived peptides bound to bound to MHC class I or II to the antigen-specific T cells (Collin and Bigley 2018). Furthermore, DCs secrete cytokines (e.g. IL-12; IFN-γ; TGF-β) that impact T cell polarisation, thereby shaping adaptive immunity. In contrast, when DCs are not activated, they fail to upregulate co-stimulatory molecules, and the presentation of self-antigens leads to T cell anergy, potentially resulting in tolerance (Kapsenberg 2003).

Natural Killer (NK) cells are cytotoxic lymphocytes that are primarily implicated in rapid viral and tumour response (Wolf et al. 2023). Upon activation, NK cells secrete perforin that forms pores in the target cell membrane, and granzymes, which penetrate these pores to induce apoptosis (Abel et al. 2018). In addition to directly killing target cells, NK cells also produce cytokines capable of modulating the immune response (Abel et al. 2018). While traditionally recognized as innate cells, NK cells also exhibit some traits of adaptive cells, bridging both systems (Mujal et al. 2021).

Micro and nanoparticles (NPs) can directly modulate the innate immune system in several ways. One of the best-known examples is the use of particulate adjuvants in many licensed vaccine formulations, that can activate APCs (Facciolà et al. 2022; Zhao et al. 2023). It is important to highlight that, even though particles can enhance the body's ability to produce more effective immune responses they may also lead to harmful effects, such as triggering hyper-inflammatory conditions our autoimmune diseases. For instance, exposure to silica particles, especially in occupational settings, can trigger chronic inflammation in the lungs (silicosis) and has been linked to the development of systemic lupus erythematosus and rheumatoid arthritis due to persistent immune activation and inflammatory responses (Pollard 2016).

1.2 Adaptive Immune System

When the innate system is not capable of eliminating an infection, the adaptive system is called into action. While the innate immune system can generate non-specific responses, the adaptive immune response depends on antigen recognition by highly specialized lymphocytes, called T lymphocytes (T cells) and B lymphocytes (B cells). The adaptive system can generate antigen-specific memory cells, which enable the host to respond more rapidly to future infections by the same pathogen, offering long-lasting protective immunity (Marshall et al. 2018) (Fig. 3.2a).

T cells play a crucial role in adaptive immune responses as effector cells, as they are equipped with a T-cell receptor (TCR) that can recognize antigens-bound MHC molecules. $CD4^+$ cells recognise antigens presented in MHC class II, while $CD8^+$ recognise antigens in MHC class I. As previously mentioned, along with antigen-TCR interaction, co-stimulatory molecule engagement, as well as polarising cytokine signalling, are required for a T cell to become activated. Upon activation, these cells differentiate to become effector cells of specialised phenotypes, including T helper 1 (Th1), Th2, Th17, and T regulatory (Treg) cells, each possessing a different immunological function (Fig. 3.1). Naïve $CD4^+$ T cells differentiation into Th1 cells requires the secretion of the cytokines IL-12 from DCs (Seder et al. 1993). Th1 cells primarily secrete IFN-γ, important in mounting a robust cell-mediated immunity against intracellular pathogens. IL-4 secretion is responsible for the polarization of naïve T cells into Th2 cells that promote antibody-mediated humoral immune

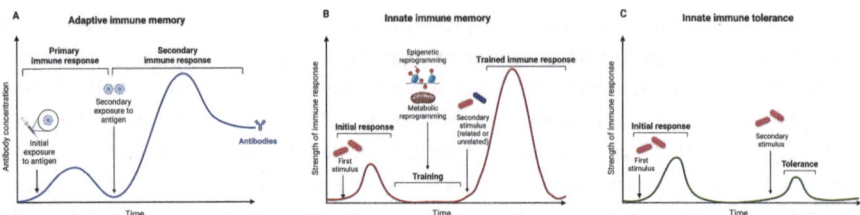

Fig. 3.2 Immunological memory. (**a**) Adaptive immune memory involves antigen recognition following initial exposure, clonal expansion, and the development of long-lasting immunological memory that enables a faster and stronger immune response upon subsequent encounters with the same pathogen (high degree of specificity). It differs from (**b**) innate immune memory, which is characterised by an initial stimulus resulting in epigenetic and metabolic reprogramming giving rise to enhanced responses upon challenge with a second stimulus, and is dependent upon innate cells (e.g. macrophages; monocytes). Priming of innate immune cells can also result in (**c**) tolerance, a state of reduced responsiveness to a secondary stimulus. Created with BioRender

responses. Th2 cells also secrete IL-4, strengthening the Th2 response through a positive feedback loop, as well as secreting IL-5 and IL-13, promoting B cell expansion, and IgG1 and IgE isotype switching (O'Garra 1998; Wan and Flavell 2009; Paul and Zhu 2010). This results in eosinophil infiltration and the degranulation of mast cells that in combination with the production of high levels of antigen-specific antibodies are key to generate protection against extracellular pathogens. Th17 is another T cell subtype abundant at mucosal sites, thus responsible for protection at the mucosal level and homeostasis maintenance (Schnell et al. 2023). They require transforming growth factor-beta (TGF-β) and IL-6 for differentiation, and the phenotype further amplified by IL-1 and IL-23 (Sandquist and Kolls 2018; Schnell et al. 2023) Th17 cells are characterized by the production of IL-17 family members (e.g. IL-17A, IL-17F), which promote local inflammation by inducing the production of cytokine and chemokine, as well as the recruitment of granulocytes. Tregs can be differentiated from naïve CD4$^+$ T cells in the presence of TGF-β. In contrast to T cells described above, Treg cells are immunosuppressive producing cytokines such as IL-10 and TGF-β, essential to maintain immune homeostasis and self-tolerance towards multiple antigens (Vignali et al. 2008). They protect the host from an array of autoimmune and inflammatory pathologies, and their dysfunction can contribute to tumour development and autoimmune diseases (Sakaguchi et al. 2020). CD8$^+$ T cells, also known as cytotoxic T lymphocytes (CTLs), are cytotoxic effectors of the adaptive immune system important to control intracellular infections and tumour. CTLs mediated toxicity through the release of perforin and granzyme, ultimately triggering apoptosis (Chowdhury and Lieberman 2008; Chávez-Galán et al. 2009). Therapeutic vaccines aimed at eliminating viral infections and cancer cells on antigen-specific CD8$^+$ T cells, but the induction of robust CD8$^+$T cell response in vivo is challenging, and this is a field where the use of NPs could prove helpful. Generally speaking, NPs can be specifically tailored to induce and direct adaptive immunity to facilitate the onset of cellular immunity. For instance, immuno-stimulating complexes (ISCOMs) are saponin-based cage-like particles that are approximately 40 nm in size. Studies have shown that they can induce durable antibody responses and robust T cell responses, including CD8+ T cell activation and increased cytokine production (Barr and Mitchell 1996; Stertman et al. 2023). Moreover, polystyrene (PS) and polylactic co-glycolic acid (PLGA) particles in the nanometre range (50–60 nm) were recently found to be optimal at inducing CD8+ responses (Muñoz-Wolf et al. 2023).

The humoral branch of the adaptive system is mediated by B cells. Naïve B cells encounter foreign antigens and recognise them through the B cell receptor (BCR). They can then differentiate into either short- or long-lived plasma cells, responsible for antibody production, or memory B cells that do not secrete antibodies (Amanna and Slifka 2010). Upon activation, short-lived plasma cells mainly produce low-affinity IgM, while long-lived plasma cells primarily secrete antibodies that have undergone isotype switching (Cyster and Allen 2019; Akkaya et al. 2020). These antibodies can be grouped into 5 classes based on the constant region of the heavy chain, namely IgG, IgM, IgA, IgE, and IgD. IgM can be expressed without isotype switching and is therefore the first antibody to be produced. IgD is co-expressed with IgM and they are both involved in complement system activation (Schroeder and Cavacini 2010). The most predominant antibody isotype in the blood is IgG, which can be further divided into 4 groups: IgG1, IgG2, IgG3, and IgG4 (Nimmerjahn and Ravetch 2005). It has a major role in pathogen neutralisation and opsonisation, as is also involved in activating the complement system. IgA is primarily found at mucosal sites such as respiratory and gastrointestinal tracts, where they protect epithelial sites from pathogens. IgE is the least abundant antibody isotype but efficiently sensitises mast cells, leading to their degranulation (Schroeder and Cavacini 2010). It should be highlighted that cytokines secreted by Th cells can influence antibody isotype switching. For example, IL-4 induces B cells to switch isotypes to IgE and IgG1, while TGF-β and IL-10 regulate IgA and IgG2b switching (Kühn et al. 1991; Defrance et al. 1992). Nanoformulations can be used to modulate this response. A study showed that PS NPs in combination with ovalbumin (OVA), were capable of enhancing both antigen-specific B and T cell responses, with smaller particles (≈50 nm) inducing antigen-specific IFN-γ in splenocytes and larger particles (≈100 nm) inducing IL-4 secretion (Mottram et al. 2007).

1.3 Trained Immunity

For years it was accepted that only the adaptive system possessed the capacity to mount a memory response and consequently boost the immune response to a subsequent infection. However, recent findings have challenged this classical view and an increasing amount of studies suggest that cells from the innate immune system can retain memory after initial exposure, generating non-specific protection against secondary infection, a phenomenon termed "trained immunity"(Netea 2013) (Fig. 3.2b). Pioneer studies demonstrated that monocytes derived from healthy volunteers immediately before and 2 weeks post-vaccination with Bacillus Calmette-Guérin (BCG) and challenged with distinct bacterial and fungal microbes, led to increased production of the inflammatory cytokines (IL-1β, TNF-α, IFN-γ), and this effect persisted for at least 3 months post-vaccination (Kleinnijenhuis et al. 2012). Some of the key features that distinguish classical immune memory from innate immune memory are outlined in Fig. 3.2. Unlike adaptive memory, which relies on lymphocytes, specifically B cells and T cells that possess highly specific antigen receptors, the main effector cells involved in innate training include monocytes, macrophages, and NK cells, capable of recognizing a broad range of PAMPS and DMAPs via PRRs. Trained immunity is achieved through epigenetic reprogramming of innate cells, which regulates gene expression and subsequently rewires intracellular signalling and metabolic pathways, increasing the responsiveness to a secondary challenge, and resulting in augmented levels of inflammatory cytokines (Crişan et al. 2016). Adaptive immune memory, involves clonal expansion and differentiation of antigen-specific lymphocytes during an immune response, creating long-lived memory B and T cells. While it takes longer to develop (between days to weeks), it is long-lasting, often persisting for years or even a lifetime. The memory associated with the innate system is generally considered short-term (days to months), however, recent reports support the idea of a long-term effect and connect it to reprograming at the level of myeloid progenitor cells (Kleinnijenhuis et al. 2014; Christ et al. 2018; Giamarellos-Bourboulis et al. 2020). It should be highlighted that while trained immunity can be an important mechanism to give rise to more effective immune responses, it might also have a negative impact generating inadequate responses to future challenges (Netea et al. 2020).

Priming of innate immune cells can also lead to a process called tolerance (Fig. 3.2c). This phenomenon is characterized by diminished responsiveness upon a secondary stimulus and has been reported in numerous cell types including monocytes, macrophages, and DCs (Ochando et al. 2023). Tolerance has been reported for several members of the TLR family, and LPS is the best-characterised substance to induce tolerance (Butcher et al. 2018). Similarly to innate training, LPS-induced tolerance is mainly regulated by epigenetic remodeling (Foster et al. 2007). It was shown that monocytes primed with microbial ligands had their cellular responses affected in a type and concentration-dependent way (Ifrim et al. 2014). Higher doses of PAMPs mainly induced tolerance, which is considered a critical physiological mechanism that prevents excessive inflammation and damage during prolonged or repeated infections (Ifrim et al. 2014; Bauer et al. 2018).

Innate immune memory is well-established in the case of certain pathogens and endogenous molecules, but only recently a connection with NMs has emerged. Exposure to endotoxin-free pristine graphene primed cells of the innate system to enhance their response to a second unrelated stimulus, via epigenetic changes (Lebre et al. 2018a), while gold NPs were able to modulate the BCG-induced memory towards tolerance (Swartzwelter et al. 2020). So far this effect has only been shown with carbon-based materials and gold NPs, but we hypothesize that additional particulate materials can induce reprogramming of innate immune cells. Overall, the ability to reprogram the innate immune response is particularly attractive and can lay the basis for generating new therapeutic strategies.

2   Mechanism of Immune Sensing

Inflammation is a defence mechanism by the innate immune system against stimuli that pose a threat to body homeostasis, in a way to limit tissue damage. Inflammation is mediated by cytokines and chemokines and is characterized by the increase in blood flow to deliver soluble and cellular immune components to the site of insult, causing redness and heat. The symptoms of swelling and pain that accompany inflammation are caused by the accumulation of fluid in the tissues and the release of chemical mediators like histamine and prostaglandin (Rankin 2004; Brannon et al. 2022). Acute inflammation is beneficial to the host, leading to the removal of the original insult, promoting tissue repair, and re-establishing homeostasis. However, if it fails to eliminate the root cause, it can shift to chronic inflammation.

Microbes have evolved to have mechanisms that enable them to evade the host's immune system by limiting their ability to be recognised. However, certain molecular structures expressed by pathogens are conserved and can be detected by immune cells. Nearly 40 years ago, the idea of pathogen recognition receptors was postulated by Charles Janeway Jr. (Janeway 1989) and later known as the "Strange Model". According to this model, the immune system can discriminate between "infectious-nonself" and "noninfectious-self", due to the expression of certain motifs conserved across species, known as pathogen-associated molecular patterns. To date, a range of components from bacteria, viruses, and fungi, have been described as PAMPs (Fig. 3.3). PRRs expressed by cells, serve to monitor intracellular and extracellular compartments, sensing for conserved microbial molecular structures, and initiate a complex signalling cascade leading to inflammation and host defence. PRRs are mainly present in cells of the innate immune defence system, including macrophages, DCs, monocytes, and neutrophils.

The stranger model proposes that APCs are capable of recognising evolutionarily conserved PAMPs, however, it does not explain their ability to recognize and respond to necrotic cell death caused by certain transplants and sterile inflammation (Kono and Rock 2008). To address these exceptions, in 1994 Polly Matzinger proposed the "Danger Model"(Matzinger 1994). This model postulates that the immune system responds specifically to danger-associated molecular patterns rather than those simply recognised as infectious or foreign, and encompasses the idea that the body will tolerate microbes present in the host tissue that does not cause damage, whereas microbes that induce the release of danger molecules will trigger a response and will be eliminated. Similar to PAMPs, DAMPs released into the extracellular space by damaged or necrotic cells, activate APCs, resulting in the activation of signalling pathways mediated by the production of multiple cytokines and chemokines, resulting in a rapid inflammatory response. Examples of intracellular DAMPs include uric acid, adenosine triphosphate (ATP), deoxyribonucleic acid (DNA), and heat shock proteins (HSPs), while extracellular DAMPs include hyaluronic acid, fibronectin, and laminin (Jounai et al. 2012; Zindel and Kubes 2020). Both models complement each other. In addition to PAMPs and DAMPs, the concept of nanoparticle-associated molecular patterns (NAMPs) as a new danger signal in inflammation is emerging (Fadeel 2012; Boraschi et al. 2017).

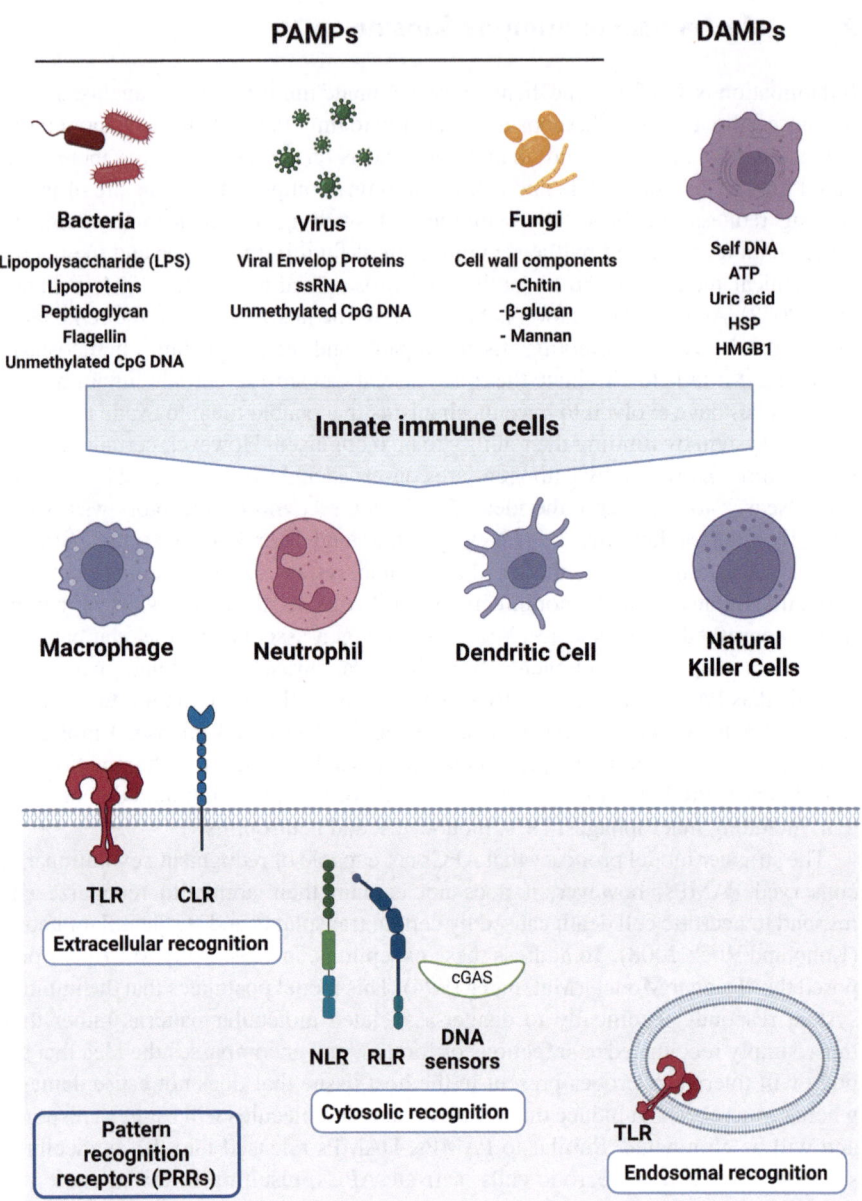

Fig. 3.3 Pathogen recognition. Pattern recognition receptors (PRRs) are capable of recognizing distinct pathogen-associated molecule patterns (PAMPs) on invading pathogens (e.g. bacteria; viruses; fungi) or endogenous danger-associated molecular patterns (DAMPs), binding triggering a signalling cascade, which ultimately results in the expression of inflammatory effector mediators. PRRs are primarily expressed by innate immune cells, both on the cell surface (TLR 1, 2, 4, 5, and 6; CLRs) and intracellularly (TLR 3, 7, 8, and 9; NLRs; RLRs; DNA sensors). Their distributions allow innate cells to sense exogenous and endogenous danger signals. ATP: Adenosine triphosphate; HSP: heat shock protein; HMGB1: high mobility group box 1; TLR: Toll-like receptor; CLR: C-type lectin receptors; RLR: Retinoic acid-inducible gene (RIG)-like receptors; NLR: Nod-like receptors; cGAS: cyclic GMP-AMP synthase. Created with BioRender

PRRs can currently be subdivided into five separate families depending on their cellular localisation and ligand recognition potential: toll-like receptors (TLRs), nucleotide-binding oligomerisation domain (NOD)-like receptors (NLRs), C-type lectin receptors (CLRs), retinoic-acid-inducible gene (RIG)-I-like receptors (RLRs) and cytosolic nucleic acid sensors.

TLRs were the first set of PRRs to be identified. Currently, 10 functional TLRs have been described in humans that can recognize a range of ligands (Akira et al. 2006), and they can be further divided into 2 groups based on their localisation. Extracellular TLRs primarily detect cell wall components from microbes, including lipoproteins (TLR1:2/TLR2:6), LPS of gram-negative bacteria (TLR4), and flagellin (TLR5), while intracellular TLRs mainly recognize nucleic acids from microbes, including double-stranded (ds)ribonucleic acid (RNA) (TLR3), single-stranded (ss) RNA (TLR7) and unmethylated cytidine-phosphate-guanosine (CpG) motifs present in bacterial and viral DNA (TLR9) (Fig. 3.3) (Akira et al. 2006). Following recognition of an agonist, TLRs activate distinct signalling cascades that ultimately induce the transcription of genes involved in inflammation (Fitzgerald and Kagan 2020). Some nanomaterials were shown to engage with TLRs, however, a strong body of evidence suggests that unaccounted contamination with TLR agonists, in particular endotoxins, might be the reason for the reported immunostimulatory effects (Li et al. 2017b).

The NLR family is the most heterogeneous in terms of structure, function, and triggers, from all the PRRs (Sundaram et al. 2024). The NLR family performs cytosolic sensing of PAMPs or DAMPs, responding to a broad range of stimuli and mediating innate immune responses such as the induction of inflammation, autophagy, and cell death. Some NLR members can form macromolecular complexes called inflammasomes that trigger the activation of the inflammatory protease caspase-1 that cleaves the precursors pro-IL-1β and pro-IL-18 into biologically active form IL-1β and IL-18 and induces pyroptosis, a pro-inflammatory form of cell death (Schroder and Tschopp 2010; Broz and Dixit 2016; Zheng et al. 2020).

NOD-like receptor protein 3 (NLRP3) is the most extensively studied inflammasome due to its capacity to be activated by a wide range of stimuli, from microbes and toxins to adenosine triphosphate (ATP), and particulates (e.g. monosodium urate, asbestos, aluminium salts) (Lebre et al. 2016). NLRP3 plays a vital role in initiating an immune response against pathogens and cellular damage; however, inappropriate regulation can lead to damaging consequences, contributing to the pathogenesis of various inflammatory and autoimmune diseases (Alehashemi and Goldbach-Mansky 2020; Li et al. 2020). As a consequence, NLRP3 signalling is tightly regulated involving a two-step process (Kelley et al. 2019; Gritsenko et al. 2020). The priming step (step 1) is initiated by the recognition of PRR ligands that lead to the NF-κB pathway activation, which results in the transcriptional upregulation of NLRP3 and pro-IL-1β. The activation step (step 2) involves the assembly of the NLRP3 inflammasome, mediated by mechanisms including potassium efflux, mitochondrial damage, reactive oxygen species (ROS), and lysosomal damage. NLRP3 oligomerization ultimately leads to protease Caspase-1 activation that processes pro-IL-1β into their active form. Several groups have shown that particulate materials (e.g. asbestos fibres (Dostert et al. 2008); alum (Li et al. 2008); chitosan (Neumann et al. 2014; Lebre et al. 2018b); poly(lactide-co-glycolide) (PLG) and polystyrene particles (Sharp et al. 2009); nano-titanium oxide (Yazdi et al. 2010)) can promote NLRP3 inflammasome activation. Not always NPs contain PAMPs, so DAMPS likely play a key role in the particulate-driven immune response. NMs can induce cell stress, damage, or death, leading to DAMPs release (e.g. extracellular ATP) that trigger an immune response. For instance, the well-known particulate adjuvant alum is now believed to mediate its immune response through the release of DAMPs. Alum can induce necrosis in bone marrow-derived dendritic cells (BMDCs), causing a rapid release of DAMPs (host DNA; uric acid) due to cell swelling and loss of membrane integrity. (Kool et al. 2008; Marichal et al. 2011).

Another PRR family comprises the C-type lectin receptors (CLRs), which can recognise carbohydrate moieties on microbes, such as mannose, fructose, and glucan structures (Brown 2006) (Fig. 3.3). CLRs are mainly predominantly expressed on myeloid cells where they mediate key roles in fungal recognition (Brown 2006; Hardison and Brown 2012). Interestingly, some CLRs modulate TLR signalling but do not induce gene expression by themselves (Robinson et al. 2006). The most studied CLR is the DC-associated C-type lectin-1 (Dectin-1) that recognizes β-glucan. Ligand binding results in a signalling cascade that triggers the secretion of a broad range of pro-inflammatory cytokines (e.g. IL-1β, TNF-α, IL-6) (Rogers et al. 2005). Studies show that particles can engage with CLRs and activate a subsequent response. For instance, one group demonstrated that the Dectin-1 receptor can distinguish between soluble and particulate ligands (Goodridge et al. 2011). Only whole glucan particles induced a Dectin-1-dependent response, whereas soluble β-glucans failed to induce ROS and cytokine secretion. A different group showed that NPs prepared from yeast cell walls were able to activate BMDCs, after engagement with Dectin-1 and modulate the immunosuppressive tumour microenvironment (Xu et al. 2022). Moreover, glucan-based nanoparticles were shown to interact with the Dectin-1 receptor, modulating the immune response of human primary monocytes in a size and chemical composition-dependent way (Jesus et al. 2024), hinting that glucan recognition is vastly dependent upon their physicochemical characteristics.

The RIG-like receptors (RLRs) are a family of PRRs crucial in the response to cytosolic RNA viruses. They are capable of recognising dsRNA viruses or dsRNA, which are generated during the replication of ssRNA viruses (Gerlier and Lyles 2011; Jiang et al. 2023). In line with other PRR signalling pathways, RLR induces NF-κB upregulation of pro-inflammatory genes, and additionally, interferon regulatory factor 3 (IRF3) triggers the expression of antiviral type I interferons (IFNs). (Kato et al. 2005).

Finally, DNA sensors are triggered by the atypical presence of DNA in intracellular compartments, such as endosomes and cytosol (Paludan and Bowie 2013; Yu and Liu 2021). Under homeostatic conditions, host DNA is confined in the nucleus and mitochondria, and therefore the cytosol is generally self-DNA free. A key DNA sensor is cyclic GMP-AMP synthase (cGAS) (Sun et al. 2013). When activated, cGAS triggers a cascade of events that activates the adaptor protein stimulator of

IFN genes (STING), critical in the production of type I IFNs (Zhang et al. 2013; Decout et al. 2021) Some studies highlight the role of NPs in the STING pathway (Luo et al. 2017; Miao et al. 2019).

Overall, it is clear that particles can engage with different PRRs to modulate the immune system, either by resembling PAMPs or by triggering DAMPs, and this interaction depends on factors such as composition, size, shape, surface charge, or the presence of specific surface ligands. Understanding the impact of these factors can help us design systems that either actively engage PRRs to boost immune activation (e.g., in vaccine formulations) or avoid immune detection (to prevent unwanted immune responses).

3 The Challenge of Endotoxin Contamination

Pyrogens are substances that can induce fever. They are broadly divided into 2 classes: endotoxins and non-endotoxin pyrogens (NEPs), and they work by triggering the body's immune system to release endogenous pyrogens (e.g. IL-6, TNF-α, or IL-1-β). Endotoxins are heat-stable lipopolysaccharides, which originate from the outer membrane of Gram-negative bacteria (Gorbet and Sefton 2005) and are the most well-known class of pyrogens. LPS is sensed by TLR-4 mainly expressed in the surface of immune cells (e.g., DCs, macrophages; monocytes), triggering an inflammatory response, and can have a severe impact on human health (Burrell 1994; Copeland et al. 2005). Non-endotoxin pyrogens are contaminants derived from Gram-positive bacteria (e.g. lipoproteins; peptidoglycan; flagellin), viruses (e.g. envelop proteins, RNA), and fungi (cell wall components). The 3 most common testing methods for pyrogen detection are the Rabbit Pyrogen Test (RPT), the *Limulus Amebocyte* Lysate (LAL) assay, and the monocyte activation test (MAT) (Table 3.1). Only the LAL assay is specific to endotoxins, while RPT and MAT can also quantify other NEP.

The Rabbit pyrogen test is the oldest method and monitors the ability of a compound to induce fever in rabbits. Since it involves the use of animals, RPT is becoming less relevant and will be excluded from the European Pharmacopoeia in 2025. Currently, the *Limulus Amebocyte* Lysate assay is the gold standard for the evaluation of endotoxin levels in medical devices and biological products and is based on the enzymatic reaction between endotoxins and a lysate derived from the blood of horseshoe crabs that triggers a clotting cascade. There are 3 versions of the assay: gel-clot, turbidimetric, and chromogenic. The gel-clot method is the least sensitive and only gives qualitative results (presence or absence), while the turbidimetric and chromogenic methods, use either turbidity or colour changes, respectively, to quantify the amount of endotoxin present. This assay has some limitations; for example, some molecules, such as the LPS-binding proteins (LBPs), interact with LPS, preventing the enzymatic reaction and thus interfering with the assay (Dentener et al. 1993; Majerle 2003); beta-glucans are also known to interfere with the assay due to their similarity with endotoxins, leading to false positives. (Cooper et al. 1997) ISO 29701:2010 addresses the possible interference of NPs with the optical readout of

Table 3.1 Common pyrogen testing methods. Comparison between the rabbit pyrogen test (RPT), the Limulus Amebocyte Lysate (LAL) assay, and the monocyte activation test (MAT), based on its principles, advantages, and disadvantages

	Test	Principle	Advantages	Disadvantages
	Rabbit Pyrogen Test (RPT)	*In vivo*: monitors rise in body temperature of rabbits following injection of the test sample	Established physiological standard Detects endotoxins and non-endotoxin pyrogens	Requires the use of animals Less sensitive
	Limulus Amebocyte Lysate (LAL) assay	*In vitro*: detects endotoxins based on the coagulation of blood lysate from horseshoe crabs	Sensitive Can be quantitative Quick results	Restricted to endotoxins Interference by certain substances
	Monocyte activation test (MAT)	*In vitro*: detect the inflammatory response (e.g., IL-1, IL-6, TNF-α) generated by treated human monocytes	Detects endotoxins and non-endotoxin pyrogens No animals required High-throughput	Cytotoxic substances can interfere More complex (requires cell culture)

the assay, as they can affect the light absorption and scattering or impact the colour of the solution (International Organization for Standardization 2010). Additionally, because the LAL assay relies on the interaction between enzymes and LPS, factors such as the presence of denaturing agents, or right pH, can interfere with the test and result in inaccurate outcomes. More recently, the monocyte activation test, an in vitro assay that can detect the inflammatory response of treated human monocytes in response to pyrogenic substances, was recognized as an alternative to the RPT and LAL assays, being included in the European Pharmacopeia in 2010. (Brown 2021) Pyrogens bind to TLRs expressed on the surface of monocytes, leading to the release of quantifiable amounts of pro-inflammatory cytokines by enzyme-linked immunoassay (ELISA). Either human whole blood, peripheral blood mononuclear cells (fresh or cryopreserved), or monocytic cell lines (Mono Mac 6) can be used to assess the release of IL-6 or IL-1-β after exposure and compare it with positive and negative control. The MAT assay can be a good alternative to screen pyrogen contamination in NMs known to interfere with the LAL assay.

All the above assays were developed considering soluble molecules, and as already mentioned, many groups found that NPs can interfere with them (Dobrovolskaia et al. 2009; Li et al. 2017b); thus it is recommended that 2 complementary assays are performed in parallel, with proper controls (e.g., any vehicle used to disperse the NPs, should also be tested). Since endotoxins are the most prevalent and potent exogenous pyrogens, when assessing the effects of NMs on the immune system, it is important to be aware of the potential contamination of the materials during the manufacturing process that can confound the real immunomodulatory effect. For instance, one group showed that carbon-based nanomaterials had a distinct ability to activate macrophages pre and post-endotoxin deactivation. (Lahiani et al. 2017) Likewise, endotoxin-free gold nanoparticles and pristine graphene failed to induce an inflammatory response in vitro. (Li et al. 2017a; Lebre et al. 2018a) Still, the majority of studies fail to report levels of unintentional endotoxin contamination in their formulations, which can jeopardize the interpretation of data related to the observed inflammatory effects. It is important to be aware that LPS release from bacteria happens after the lysis of gram-negative bacteria, which are ubiquitous during NM preparation unless special attention is taken into account while preparing the material. Sterilization methods traditionally used to prevent microbial contamination (autoclaving at 121 °C for 30 min; irradiation) are not effective at removing LPS. Methodologies to eliminate or minimize endotoxin contamination are challenging (e.g., use of high temperatures (250 °C for at least 30 min) or highly acidic or basic solutions), so it is important to address this problem early on material production to have greater confidence on the immunomodulatory data generated.

4 Nanomaterials Characteristics Influence the Immune Response

Although the use of NMs has led to significant advances in many areas, they have also been linked with potential adverse immune effects. When assessing the immunomodulatory effects of NMs it is essential to consider early events, even before they interact with immune cells. Upon contact with biological fluids, NMs often become coated with proteins (e.g. albumin, fibrinogen, apolipoprotein), forming a structure referred to as "protein corona" (Hajipour et al. 2023). This phenomenon confers a new biological identity to the NMs, influencing their interaction with cells. Proteins adsorbed on NMs can also undergo conformational changes and spatial reorganization, which may lead to the exposure of new epitopes, or affect the accessibility of epitopes to antibody recognition, masking the NMs surface. Corona formation is a complex process influenced by the intrinsic properties of NMs, as they can govern the affinity and dynamics of NMs-protein association and dissociation (Hajipour et al. 2023). While protein corona formation has been extensively studied, the relationship between the identity of proteins and the resulting immune responses remains poorly understood. When NMs are coated with complex protein structures, they can act as NAMPs and be recognized by PRRs, which in turn might trigger an inflammatory response, but it can also mitigate NM-related toxicity (Hu et al. 2011). Consequently, protein corona formation can significantly impact cellular interactions and has become a focal point of research interest.

As previously mentioned, NMs characteristics including size, surface area, reactivity, chemical composition, structure, crystallinity, and dissolution rate, can significantly impact the immune response generated (Fig. 3.4) (Pedata et al. 2016). NP size is a critical factor that directly affects the rate of cellular uptake, and toxicity and modulates the immune response. Generally, smaller particles are more easily

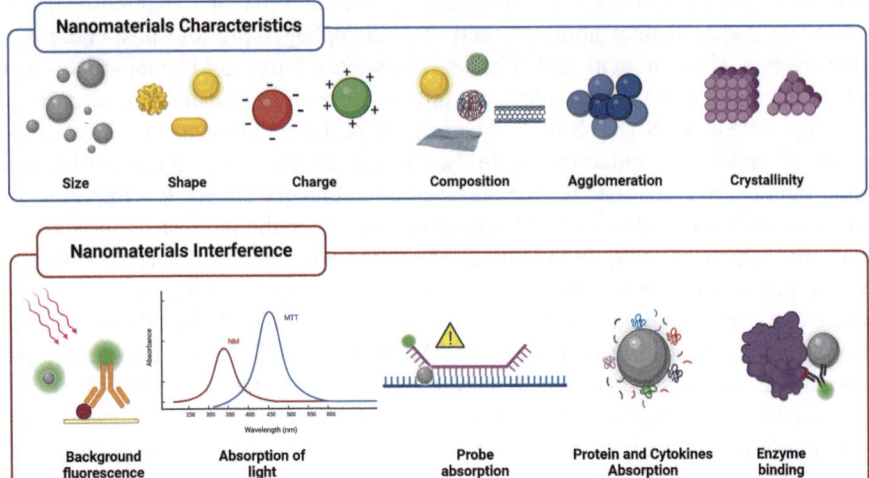

Fig. 3.4 Impact of nanomaterial characteristics on immune response and immunotoxicological tests. The intrinsic characteristics of NMs (e.g. size, crystallinity, surface charge, agglomeration/aggregation state, chemical composition, and shape) directly influence the immunotoxicological responses and testing. The most common interferences are the production of background fluorescence by NMs, adsorption to probes, proteins, cytokines, or test reagents producing erroneous results, absorption of light at the same wavelength as the assay reagents, and binding in enzymatic assays. Created with BioRender

internalized, leading to higher intracellular concentrations and potentially more pronounced biological effects (Bachmann and Jennings 2010; Hoshyar et al. 2016). This occurs due to the exponential increase in surface area relative to volume, which makes NMs more reactive, increasing internalization. Depending on their size, NPs will follow district internalization pathways. In general, NMs smaller than 5 nm will directly translocate across cell membranes, while NPs larger than 25 nm enter through pinocytosis. Macrophages and neutrophils can phagocytose larger particles, increasing their concentration in the blood, liver, and spleen (Wang et al. 2013; Manuja et al. 2021). These differences will impact the amount of NMs that a cell can internalize, as well as the ease of accumulation, localization, and mechanisms of toxicity, as NMs of different sizes can trigger different cell death mechanisms (Pan et al. 2007). Adverse effects related to NM size can also occur indirectly. Larger or clustered NMs, which are not readily internalized, can still exert cytotoxic effects, such as causing DNA damage or impairing critical biological functions (Hoshyar et al. 2016). When a phagocytic cell is unable to fully internalize larger particles, this can lead to frustrated phagocytosis, which can trigger the excessive release of ROS and inflammatory mediators (Baranov et al. 2021). As discussed before, NP size can also dictate the type of immune response generated (Th1 vs Th2)(Muñoz-Wolf et al. 2023) and influence the polarization of macrophages (Jiang et al. 2024).

In addition to size, NMs shape and proportion significantly influence the cellular uptake, clearance, and inflammatory response in the body. For instance, ellipsoidal NMs tend to be more easily phagocytosed than spherical ones (Gratton et al. 2008), while materials with high proportions and rigidity are more resistant to internalization but can rupture the plasma membrane, leading to leakage of intracellular contents (Sharifi et al. 2012). Particles with sharp edges are shown to induce more inflammatory responses (Lebre et al. 2017). In addition to internalization, the shape of NMs can also interfere with the mechanisms underlying clearance. Contrary to their smooth counterparts, NPs with sharp edges could more easily rupture endosomal membranes, preventing lysosome formation, and reducing their excretion rate (Chu et al. 2014). Particles with exceedingly long axes (>20 μm) are not readily phagocytosed, and owing to their accumulation that could result in sustained inflammation and eventually fibrotic changes, impacting immune outcomes (Borm and Kreyling 2004; Hoet et al. 2004).

Surface properties such as charge and hydrophobicity play a crucial role in the interactions of NMs with biological surfaces, cell membranes, and proteins, as well as in cellular uptake. Positive or negative charges are generally associated with increased toxicity compared to neutral ones. This is due to the self-managed equilibrium of neutral NMs. NMs with positive surface charges generally have higher rates of cellular uptake due to strong interactions with negative cell surfaces (Fröhlich 2012). The difference in electrical charges enhances endocytosis as a means of balancing positive charges, which act as a proton pump capable of disrupting the normal function of lysosomes and initiating cell death (Sun et al. 2016). For instance, amino-functionalized polystyrene NPs (positively charged) were able to destabilise the lysosomes, leading to their rupture and triggering NLRP3 inflammasome activation, while carboxyl- (negatively charged) or non-functionalized particles (neutral charge) were unable to do so (Lunov et al. 2011). Furthermore, cationic NMs can interact easily with negatively charged DNA molecules and cause more genotoxic effects compared to anionic NMs (Pedata et al. 2016). Negatively charged NMs, on the other hand, may participate in the destruction of epithelial barriers due to their charge density (Manuja et al. 2021). Altered internalization rates based on surface charge may result in reduced NM circulation and/or increased accumulation within the body.

The type of material can also play a role in the cellular response. For example, when NMs based on metal oxides or alloys are contained within phagolysosomes, they can trigger a "Trojan horse" effect, where toxic metal ions are produced and released, leading to the production of ROS and oxidative stress (Manuja et al. 2021). This means that NMs with high solubility in acidic environments may exhibit greater toxicity when targeted to these organelles. Furthermore, the solvent also influences NM size and dispersion, thus affecting toxicity (Gatoo et al. 2014). The state of agglomeration (particle adhesion by weak forces) or aggregation (particle adhesion by weak forces) of NMs can also impact the immunotoxicological outcomes of NMs on cells (Jiang et al. 2009). The crystal structure of NMs can be a critical determinant of their toxicity (Jiang et al. 2008). NMs with identical physicochemical characteristics but different crystal structures can elicit distinct immunotoxicological effects (Suttiponparnit et al. 2010). Furthermore, the crystal structure of NMs can be altered upon interaction with water or biological fluids (Buzea et al. 2007). The opsonisation process, where NMs are tagged by proteins for immunological recognition, is also influenced by the roughness or smoothness of the NM surface, leading to varied results based on this characteristic (Vianello et al. 2021).

Finally, it is important to underscore that NMs characteristics not only have an impact on the biological responses generated but also create many experimental challenges. Their varied properties may affect the outcomes of numerous immunotoxicity assays, potentially causing misinterpretation of results. Ideally, a systematic approach to analysing the immunotoxicological effects of NMs would involve correlating their adverse effects with their physicochemical properties, which can prove challenging.

5 Methods and Approaches for Immunological Assessment

5.1 Current Methods

Immunotoxicity assessment of NMs is a complex process that involves a range of methodologies used to evaluate important biological endpoints, including cytotoxicity, reactive oxygen species production, and inflammation. Methodologies initially developed for small molecules often require revalidation for NMs, as particles are prone to interfere with the assays leading to inaccurate results, making these assessments inherently more complex (Table 3.2). During immunotoxicological testing, the large reactive surface area of NMs makes them predisposed to interact with analytes and assay reagents, promoting changes in readout signals or intermediate steps (Fig. 3.4).

The impact of NMs exposure on cellular viability can be determined using different endpoints, such as cellular metabolic activity, cell membrane integrity, or mitochondrial activity. From the broad range of assays available to measure cytotoxicity, most of them rely on colorimetric or fluorometric readout, thus NMs have an increased potential to adversely impact the assays, due to their physical, chemical, and biological properties. Many NMs, especially metallic ones like gold, titanium dioxide, and silver NPs, possess intrinsic optical properties that can interfere with colorimetric assays such as MTT, XTT, and WST-1. These assays rely on colour changes that correlates with cell viability, but the absorbance of NMs at similar wavelengths can lead to false positives or negatives (Kroll et al. 2012). For example, gold NPs can absorb light at 570 nm, the same wavelength used in the MTT assay, resulting in an overestimation of cell viability. Furthermore, NMs can interfere with cell viability assays based on fluorometric methods, such as Calcein AM and propidium iodide (PI) assays, or double staining assays with PI and Annexin V (Mello et al. 2020). Quantum dots, for instance, exhibit strong auto-fluorescence that can overlap with the emission spectra of assay reagents, complicating the data

Table 3.2 Potential interferences from nanomaterials

Assay	Reagents	Type of interference	Reference
Cytotoxicity	MTT	Absorption of light and reagents Induction of formazan generation	Monteiro-Riviere et al. (2009), Mello et al. (2020)
	LDH	Light absorption Adsorption to the assay reagents	Kroll et al. (2012), Szymański et al. (2020)
	WST-1	Interference with the conversion of WST-1 to formazan	Kroll et al. (2011), Szymański et al. (2020)
	Live/dead assay	Fluorescence blocking by clustered NMs Reagent adsorption Autofluorescence	Monteiro-Riviere et al. (2009)
	Flow cytometry	Adsorbed NMs can decrease fluorescence Adsorption of NMs on the cell surface	Bohmer et al. (2018)
Apoptosis	Annexin V	Adsorption of the probe Interference with fluorescence	Wilhelmi et al. (2012)
	Caspase assay (Peptide/aminoluciferin)	Interference with recovery of the analyte	Wilhelmi et al. (2012)
	DCFH-DA	NMs background fluorescence	Kroll et al. (2011), Guadagnini et al. (2015)
Oxidative stress	Catalase activity (Red catalase assay)	Changes in dye oxidation	Ong et al. (2014)
Inflammatory response	Cytokine measurement (ELISA)	Cytokine adsorption	Dobrovolskaia et al. (2009), Guadagnini et al. (2015), Abbasi et al. (2021)
	LAL assay	Intrinsic turbidity Interaction with proteins, endotoxin, or reagents	Dobrovolskaia et al. (2010), Mangini et al. (2021)
Protein quantification	Bradford protein assay	Dye reduction	Ong et al. (2014)
	BCA assay	Adsorption to test components	Ong et al. (2014)

MTT 3-[4,5-dimethylthiazol-2-yl]-2,5 diphenyl tetrazolium bromide, *LDH* Lactate dehydrogenase, *WST* water-soluble tetrazolium salt, *NMs* nanomaterials, *DCFH-DA* 2′– 7′- dichlorodihydrofluorescein diacetate, *ELISA* enzyme-linked immunosorbent assay, *LAL Lumulus Amebocyte* lysate, *BCA* bicinchoninic acid protein assay

interpretation. NMs can interact with assay reagents, including dyes and substrates, leading to decreased availability of these reagents for their intended reactions (El Yamani et al. 2024)· (Han et al. 2011). This is particularly problematic in assays like the lactate dehydrogenase (LDH) release assay, where LDH or its substrates can adsorb onto NM surfaces, skewing results (Leudjo Taka et al. 2021). For instance, carbon nanotubes have been shown to adsorb LDH leading to an underestimation of cytotoxicity.

Modulation of the immune response by NMs can be assessed by measuring the cytokine and chemokine release upon immune cell stimulation. The adsorption of proteins to NPs is a common event, thus it is not strange that particles can adsorb cytokines and chemokines, interfering with their quantification. In one study, TiO_2, Fe_3O_4, and SiO_2 NPs were found to adsorb cytokines, including IL-6, IL-8, and GM-CSF (Guadagnini et al. 2015). In a different study, small TiO_2 NPs could specifically bind IL-8 and to a smaller extent IFN-γ, but not TNF-α, IL-1β and IL-6. The authors suggested that the strong interaction between the particles and the chemokine IL-8 (CXCL8) could have clinical implications, as it impaired neutrophil chemotaxis (Batt et al. 2018).

When working with NMs, some mitigation measures can be put in place to avoid biases during immunotoxicological testing. The first approach involves understanding the range of potential interferents, as well as adding appropriate controls for each specific sample. In ELISA, it is recommended to add controls with and without NMs and perform a spike-recovery test with the cytokine of interest (Dobrovolskaia et al. 2009). Another solution is washing or centrifugation to reduce or deplete the amount of NMs present in the sample. In most tests in which NMs are adsorbed to reagents, reducing the concentration of NMs already presents satisfactory results. In addition, it is recommended that a wide range of assays based on different measurement principles be performed with the same sample. Although these measures reduce the risk of errors and biases during the execution of the tests, assay adaptations may not always be sufficient to negligibly minimize the interference of NMs. Robust characterization of NMs and biological fluids, together with accurate methodological approaches and methods adapted to the physicochemical characteristics that may cause interference in immunotoxicological assays, are essential steps to conduct more robust assays with greater translatability to clinical practice. Inadequate characterization will hinder efforts to establish a correlation between NMs properties and the biological effects. Moreover, the data available in the literature often present numerous inconsistencies and the results are highly variable when comparing studies and laboratories. The lack of standardized methodologies and guidelines further complicates the comparison of immunotoxicity assessments across different research groups. While the testing guidelines established by the Organization for Economic Cooperation and Development (OECD) are applicable, certain adaptations may be necessary in how these tests are conducted. Therefore, there is a strong need for the development of novel methods specifically designed for NMs immunotoxicological assessment.

5.2 Emerging Technologies

Due to the interferences that NMs can cause in conventional methods for safety assessment, there is a strong demand for more complex models that can represent an accurate and consistent assessment of the safety of NMs. In recent years, several innovative methodologies focusing on sensitivity, real-time kinetics and integration of multiple parameters have been identified as promising in the investigation of the immunological effects of NMs.

5.3 xCELLigence

It is a novel approach for non-invasive in vitro toxicity assessment. Through this methodology, it is possible to observe all cell growth events in real time (cell growth, proliferation, cell size and morphology) without a label. Thus, the use of xCELLigence can bypass the interference of NMs observed in traditional cytotoxicity methods (Martinez-Serra et al. 2014), in addition to reducing false-positive and false-negative results (Yah and Simate 2020). The methodology is based on the quantification of cell proliferation/viability, morphological changes, and cell fixation through the measurement of electrical impedance. Under normal conditions, the plates used for cell culture (electronic microtiter plates) are permissive to the passage of electric current. However, after cell adhesion, the passage of electric current to be located by the sensors is impeded by the cells. Thus, the differences in electrical impedance values within the electrical circuit are measured and expressed in terms of a cell index (CI), which is directly proportional to the exposure of the sensor electrode over time, reaching a plateau as the cells proliferate and reach 100% confluence (Martinez-Serra et al. 2014). In addition to being considered a sensitive method, real-time monitoring, as opposed to end-point assays, allows continuous data acquisition for multiple studies, allowing assays to be monitored over long periods and time-dependent toxicity tests.

5.4 Genetic Approaches for ROS Detection

Interference in conventional ROS detection tests by fluorescence occurs due to the ability of NMs to scatter or absorb light. To overcome the concerns arising from the detection of fluorescent dyes, new methods based on genetically encoded sensors, primarily based on proteins, that can dynamically monitor ROS with high specificity were developed (Woolley et al. 2013). Transgenic cell lines with fluorescent reporters sensitive to ROS, such as HyPer or roGFP, have been used to quantify intracellular oxidative stress induced by nanoparticle exposure. In addition to functioning as an alternative to fluorescent dyes, theses type of system allows not only the measurement of the magnitude of the response but also the spatial and temporal localization of redox activation peaks within immune cells.

5.5 High-Content Screening (HCS)

High-content screening (HCS) is a multiparametric image analysis method, that enables efficient and reliable simultaneous screening of multiple cellular responses within a single test, reducing the risk of misinterpretation that can occur when investigations focus on a single toxicological endpoint (Huo et al. 2015; Collins et al. 2017). Cell painting is an example of an HCS assay, involving staining different cellular components or processes with various fluorescent dyes, creating detailed images of each cell or cellular process (Bray et al. 2016). Using automated high-throughput fluorescence microscopy, combined with powerful image analysis software, millions of cells can be analyzed in a relatively short amount of time, which can help detect modifications in cellular phenotype and be used to study the immunomodulatory effects induced NMs, as subtle changes in organelle structure or cell shape can be early indicators of immunotoxicity.

5.6 Multiplex Analysis

During immunotoxicological events, cells secrete soluble factors such as cytokines and chemokines. Immunoassays are a powerful tool widely used for the accurate quantification of biomarkers in drug discovery and development, safety assessment, or medical diagnosis. Traditional assays, like ELISA, analyses and quantify a single analyte at a time, which can be very limiting in the context of complex signaling networks. In many circumstances, for instance, levels of cytokines within a pathway relative to each other are significantly more relevant than their presence or absence, or than their absolute levels. This increased recognition of the multifactorial nature of numerous physio-pathological states requires more integrative research approaches. Simultaneous measurement of multiple biomarkers has become an efficient tool for profiling multiple markers in a single sample, thereby minimizing cost, time, and volume. Multiplexing analysis emerges as an alternative to the ELISA, allowing fast analysis of multiple analytes derived from a single experiment, which can be translated into a more robust understanding of the interconnected nature of signaling networks, improving recognition of putative correlations between biomarkers (Collins et al. 2017).

5.7 Atomic Force Microscopy (AFM)

Atomic force microscopy has been identified as a useful tool to generate images that allow various types/sizes of NMs to be localized in different cellular compartments, as well as the analysis of the influence of these NMs on the mechanical properties of the cell (Boyoglu et al. 2013). AFM allows the measurement of adhesion forces, membrane elasticity and endocytosis processes at nanometer resolution. In addition, the technique can be used to map altered cellular topographies after exposure to NMs, providing insights into cellular activation, cytoskeletal reorganization and membrane damage — aspects that often precede immune responses. Thus, it can become a valuable approach in the studies of the internalization pathways and localization of NMs in different study models.

The emerging technologies above mentioned, while promising also have some limitations, such as higher operational costs, complexity, the need for highly skilled labor, and potentially longer execution times. Thus, in silico and machine learning approaches have gained a great deal of attention in recent years (Furxhi et al. 2020). These approaches can assist in screening and modeling models that enable structure-activity relationship analyses to establish a correlation between physicochemical properties and nanotoxicity. Despite this, these studies are still in their infancy and therefore require experimental toxicity tests to be performed to confirm the predicted results (Tirumala et al. 2021).

Overall, understanding the complex interplay between NMs and the immune system is crucial. NMs can have an incredible impact in many fields, particularly in the development of new therapeutic strategies, but it is also important to remember that unintentional exposure to NMs from occupational, environmental, or consumer product sources may lead to unintended immune responses that have significant implications in human and animal health.

References

Abbasi F et al (2021) The synergistic interference effect of silica nanoparticles concentration and the wavelength of ELISA on the colorimetric assay of cell toxicity. Sci Rep 11(1):15133. https://doi.org/10.1038/s41598-021-92419-1

Abel AM et al (2018) Natural killer cells: development, maturation, and clinical utilization. Front Immunol 9. https://doi.org/10.3389/fimmu.2018.01869

Abraham SN, St John AL (2010) Mast cell-orchestrated immunity to pathogens. Nat Rev Immunol 10(6):440–452. https://doi.org/10.1038/nri2782

Akira S, Uematsu S, Takeuchi O (2006) Pathogen recognition and innate immunity. Cell 124(4):783–801. https://doi.org/10.1016/j.cell.2006.02.015

Akkaya M, Kwak K, Pierce SK (2020) B cell memory: building two walls of protection against pathogens. Nat Rev Immunol 20(4):229–238. https://doi.org/10.1038/s41577-019-0244-2

Alehashemi S, Goldbach-Mansky R (2020) Human autoinflammatory diseases mediated by NLRP3-, pyrin-, NLRP1-, and NLRC4-Inflammasome dysregulation updates on diagnosis, treatment, and the respective roles of IL-1 and IL-18. Front Immunol 11. https://doi.org/10.3389/fimmu.2020.01840

Amanna IJ, Slifka MK (2010) Mechanisms that determine plasma cell lifespan and the duration of humoral immunity. Immunol Rev 236(1):125–138. https://doi.org/10.1111/j.1600-065X.2010.00912.x

Amin K (2012) The role of mast cells in allergic inflammation. Respir Med 106(1):9–14. https://doi.org/10.1016/j.rmed.2011.09.007

Bachmann MF, Jennings GT (2010) Vaccine delivery: a matter of size, geometry, kinetics and molecular patterns. Nat Rev Immunol 10(11):787–796. https://doi.org/10.1038/nri2868

Baranov MV et al (2021) Modulation of immune responses by particle size and shape. Front Immunol 11. https://doi.org/10.3389/fimmu.2020.607945

Barr IG, Mitchell GF (1996) ISCOMs (immunostimulating complexes): the first decade. Immunol Cell Biol 74(1):8–25. https://doi.org/10.1038/icb.1996.2

Batt J et al (2018) TiO2 nanoparticles can selectively bind CXCL8 impacting on neutrophil chemotaxis. Eur Cells Mater 35:13–24. https://doi.org/10.22203/eCM.v035a02

Bauer M et al (2018) Remembering pathogen dose: long-term adaptation in innate immunity. Trends Immunol 39(6):438–445. https://doi.org/10.1016/j.it.2018.04.001

Bennouna S et al (2003) Cross-talk in the innate immune system: neutrophils instruct recruitment and activation of dendritic cells during microbial infection. J Immunol (Baltimore, Md. : 1950) 171(11):6052–6058. https://doi.org/10.4049/jimmunol.171.11.6052

Beutler B (2004) Innate immunity: an overview. Mol Immunol 40(12):845–859. https://doi.org/10.1016/j.molimm.2003.10.005

Biswas SK, Mantovani A (2010) Macrophage plasticity and interaction with lymphocyte subsets: cancer as a paradigm. Nat Immunol 11(10):889–896. https://doi.org/10.1038/ni.1937

Bochner BS (2000) Systemic activation of basophils and eosinophils: markers and consequences. J Allergy Clin Immunol 106(5):S292–S302. https://doi.org/10.1067/mai.2000.110164

Bohmer N et al (2018) Interference of engineered nanomaterials in flow cytometry: a case study. Colloids Surf B Biointerf 172:635–645. https://doi.org/10.1016/j.colsurfb.2018.09.021

Boraschi D et al (2017) Nanoparticles and innate immunity: new perspectives on host defence. Semin Immunol 34:33–51. https://doi.org/10.1016/j.smim.2017.08.013

Borm PJA, Kreyling W (2004) Toxicological hazards of inhaled nanoparticles—potential implications for drug delivery. J Nanosci Nanotechnol 4(5):521–531. https://doi.org/10.1166/jnn.2004.081

Boyoglu C et al (2013) Microscopic studies of various sizes of gold nanoparticles and their cellular localizations. ISRN Nanotechnol 2013:1–13. https://doi.org/10.1155/2013/123838

Brannon ER et al (2022) Polymeric particle-based therapies for acute inflammatory diseases. Nat Rev Mater 7(10):796–813. https://doi.org/10.1038/s41578-022-00458-5

Bray M-A et al (2016) Cell painting, a high-content image-based assay for morphological profiling using multiplexed fluorescent dyes. Nat Protoc 11(9):1757–1774. https://doi.org/10.1038/nprot.2016.105

Brown GD (2006) Dectin-1: a signalling non-TLR pattern-recognition receptor. Nat Rev Immunol 6(1):33–43. https://doi.org/10.1038/nri1745

Brown J (2021) Using the monocyte activation test as a stand-alone release test for medical devices. ALTEX:151–156. https://doi.org/10.14573/altex.2012021

Broz P, Dixit VM (2016) Inflammasomes: mechanism of assembly, regulation and signalling. Nat Rev Immunol 16(7):407–420. https://doi.org/10.1038/nri.2016.58

Bulet P, Stöcklin R, Menin L (2004) Anti-microbial peptides: from invertebrates to vertebrates'. Immunol Rev 198(1):169–184. https://doi.org/10.1111/j.0105-2896.2004.0124.x

Burrell R (1994) Human responses to bacterial endotoxin. Circ Shock 43(3):137–153. http://www.ncbi.nlm.nih.gov/pubmed/7850934

Butcher SK et al (2018) Toll-like receptors drive specific patterns of tolerance and training on Restimulation of macrophages. Front Immunol 9. https://doi.org/10.3389/fimmu.2018.00933

Buzea C, Pacheco II, Robbie K (2007) Nanomaterials and nanoparticles: sources and toxicity. Biointerphases 2(4):MR17–MR71. https://doi.org/10.1116/1.2815690

Chávez-Galán L et al (2009) Cell death mechanisms induced by cytotoxic lymphocytes. Cell Mol Immunol 6(1):15–25. https://doi.org/10.1038/cmi.2009.3

Chowdhury D, Lieberman J (2008) Death by a thousand cuts: granzyme pathways of programmed cell death. Annu Rev Immunol 26:389–420. https://doi.org/10.1146/annurev.immunol.26.021607.090404

Christ A et al (2018) Western diet triggers NLRP3-dependent innate immune reprogramming. Cell 172(1–2, 162):–175.e14. https://doi.org/10.1016/j.cell.2017.12.013

Chu Z et al (2014) Unambiguous observation of shape effects on cellular fate of nanoparticles. Sci Rep 4(1):4495. https://doi.org/10.1038/srep04495

Collin M, Bigley V (2018) Human dendritic cell subsets: an update. Immunology 154(1):3–20. https://doi.org/10.1111/imm.12888

Collins AR et al (2017) High throughput toxicity screening and intracellular detection of nanomaterials. WIREs Nanomed Nanobiotechnol 9(1). https://doi.org/10.1002/wnan.1413

Cooper JF, Weary ME, Jordan FT (1997) The impact of non-endotoxin LAL-reactive materials on limulus amebocyte lysate analyses. PDA J Pharm Sci Technol 51(1):2–6. http://www.ncbi.nlm.nih.gov/pubmed/9099058

Copeland S et al (2005) Acute inflammatory response to endotoxin in mice and humans. Clin Vaccine Immunol 12(1):60–67. https://doi.org/10.1128/CDLI.12.1.60-67.2005

Costa S et al (2023) Microfluidic-based skin-on-chip systems for safety assessment of nanomaterials. Trends Biotechnol 41(10):1282–1298. https://doi.org/10.1016/j.tibtech.2023.05.009

Crişan TO, Netea MG, Joosten LAB (2016) Innate immune memory: Implications for host responses to damage-associated molecular patterns. Eur J Immunol 46(4):817–828. https://doi.org/10.1002/eji.201545497

Cyster JG, Allen CDC (2019) B cell responses: cell interaction dynamics and decisions. Cell 177(3):524–540. https://doi.org/10.1016/j.cell.2019.03.016

Decout A et al (2021) The cGAS-STING pathway as a therapeutic target in inflammatory diseases. Nat Rev Immunol 21(9):548–569. https://doi.org/10.1038/s41577-021-00524-z

Defrance T et al (1992) Interleukin 10 and transforming growth factor beta cooperate to induce anti-CD40-activated naive human B cells to secrete immunoglobulin a. J Exp Med 175(3):671–682. https://doi.org/10.1084/jem.175.3.671

Dentener MA et al (1993) Antagonistic effects of lipopolysaccharide binding protein and bactericidal/permeability-increasing protein on lipopolysaccharide-induced cytokine release by mononuclear phagocytes. Competition for binding to lipopolysaccharide. J Immunol (Baltimore, Md. : 1950) 151(8):4258–4265. http://www.ncbi.nlm.nih.gov/pubmed/8409400

Dobrovolskaia MA, Germolec DR, Weaver JL (2009) Evaluation of nanoparticle immunotoxicity. Nat Nanotechnol 4(7):411–414. https://doi.org/10.1038/nnano.2009.175

Dobrovolskaia MA et al (2010) Ambiguities in applying traditional Limulus amebocyte lysate tests to quantify endotoxin in nanoparticle formulations. Nanomedicine 5(4):555–562. https://doi.org/10.2217/nnm.10.29

Dostert C et al (2008) Innate immune activation through Nalp3 Inflammasome sensing of Asbestos and silica. Science 320(5876):674–677. https://doi.org/10.1126/science.1156995

El Yamani N et al (2024) Hazard assessment of nanomaterials using in vitro toxicity assays: guidance on potential assay interferences and mitigating actions to avoid biased results. Nano Today 55:102215. https://doi.org/10.1016/j.nantod.2024.102215

Facciolà A et al (2022) An overview of vaccine adjuvants: current evidence and future perspectives. Vaccine 10(5). https://doi.org/10.3390/vaccines10050819

Fadeel B (2012) Clear and present danger? Engineered nanoparticles and the immune system. Swiss Medical Weekly [Preprint]. https://doi.org/10.4414/smw.2012.13609

Faurschou M, Borregaard N (2003) Neutrophil granules and secretory vesicles in inflammation. Microbes Infect 5(14):1317–1327. https://doi.org/10.1016/j.micinf.2003.09.008

Fitzgerald KA, Kagan JC (2020) Toll-like receptors and the control of immunity. Cell 180(6):1044–1066. https://doi.org/10.1016/j.cell.2020.02.041

Foster SL, Hargreaves DC, Medzhitov R (2007) Gene-specific control of inflammation by TLR-induced chromatin modifications. Nature 447(7147):972–978. https://doi.org/10.1038/nature05836

Fröhlich E (2012) The role of surface charge in cellular uptake and cytotoxicity of medical nanoparticles. Int J Nanomedicine:5577. https://doi.org/10.2147/IJN.S36111

Furxhi I et al (2020) Nanotoxicology data for *in silico* tools: a literature review. Nanotoxicology 14(5):612–637. https://doi.org/10.1080/17435390.2020.1729439

Gatoo MA et al (2014) Physicochemical properties of nanomaterials: implication in associated toxic manifestations. Biomed Res Int 2014:1–8. https://doi.org/10.1155/2014/498420

Gerlier D, Lyles DS (2011) Interplay between innate immunity and negative-strand RNA viruses: towards a rational model. Microbiol Mol Biol Rev: MMBR 75(3):468–490., second page of table of contents. https://doi.org/10.1128/MMBR.00007-11

Giamarellos-Bourboulis EJ et al (2020) Activate: randomized clinical trial of BCG vaccination against infection in the elderly. Cell 183(2):315–323.e9. https://doi.org/10.1016/j.cell.2020.08.051

Goodridge HS et al (2011) Activation of the innate immune receptor Dectin-1 upon formation of a "phagocytic synapse". Nature 472(7344):471–475. https://doi.org/10.1038/nature10071

Gorbet MB, Sefton MV (2005) Endotoxin: the uninvited guest. Biomaterials 26(34):6811–6817. https://doi.org/10.1016/j.biomaterials.2005.04.063

Gratton SEA et al (2008) The effect of particle design on cellular internalization pathways. Proc Natl Acad Sci 105(33):11613–11618. https://doi.org/10.1073/pnas.0801763105

Gritsenko A et al (2020) Mechanisms of NLRP3 priming in inflammaging and age related diseases. Cytokine Growth Factor Rev 55:15–25. https://doi.org/10.1016/j.cytogfr.2020.08.003

Guadagnini R et al (2015) Toxicity screenings of nanomaterials: challenges due to interference with assay processes and components of classic in vitro tests. Nanotoxicology 9 Suppl 1:13–24. https://doi.org/10.3109/17435390.2013.829590

Hajipour MJ et al (2023) An overview of nanoparticle protein Corona literature. Small 19(36). https://doi.org/10.1002/smll.202301838

Han X et al (2011) Validation of an LDH assay for assessing nanoparticle toxicity. Toxicology 287(1–3):99–104. https://doi.org/10.1016/j.tox.2011.06.011

Hardison SE, Brown GD (2012) C-type lectin receptors orchestrate antifungal immunity. Nat Immunol 13(9):817–822. https://doi.org/10.1038/ni.2369

Hoet PH, Brüske-Hohlfeld I, Salata OV (2004) Nanoparticles—known and unknown health risks. J Nanobiotechnol 2(1):12. https://doi.org/10.1186/1477-3155-2-12

Hoshyar N et al (2016) The effect of nanoparticle size on in vivo pharmacokinetics and cellular interaction. Nanomedicine 11(6):673–692. https://doi.org/10.2217/nnm.16.5

Hu W et al (2011) Protein corona-mediated mitigation of cytotoxicity of graphene oxide. ACS Nano 5(5):3693–3700. https://doi.org/10.1021/nn200021j

Huo L et al (2015) High-content screening for assessing nanomaterial toxicity. J Nanosci Nanotechnol 15(2):1143–1149. https://doi.org/10.1166/jnn.2015.9032

Ifrim DC et al (2014) Trained immunity or tolerance: opposing functional programs induced in human monocytes after engagement of various pattern recognition receptors. Clin Vaccine Immunol. Edited by C.J. Papasian 21(4):534–545. https://doi.org/10.1128/CVI.00688-13

International Organization for Standardization (2010) Nanotechnologies—Endotoxin test on nanomaterial samples for in vitro systems—Limulus amebocyte lysate (LAL) test. ISO 29701:2010 [Preprint]. https://www.iso.org/standard/45640.html

Janeway CA (1989) Approaching the asymptote? Evolution and revolution in immunology. Cold Spring Harb Symp Quant Biol 54:1–13. https://doi.org/10.1101/SQB.1989.054.01.003

Jesus S et al (2024) Exploring the immunomodulatory properties of glucan particles in human primary cells. Int J Pharm 655:123996. https://doi.org/10.1016/j.ijpharm.2024.123996

Jiang J et al (2008) Does nanoparticle activity depend upon size and crystal phase? Nanotoxicology 2(1):33–42. https://doi.org/10.1080/17435390701882478

Jiang J, Oberdörster G, Biswas P (2009) Characterization of size, surface charge, and agglomeration state of nanoparticle dispersions for toxicological studies. J Nanopart Res 11(1):77–89. https://doi.org/10.1007/s11051-008-9446-4

Jiang Y et al (2023) Exploiting RIG-I-like receptor pathway for cancer immunotherapy. J Hematol Oncol 16(1):8. https://doi.org/10.1186/s13045-023-01405-9

Jiang W et al (2024) Polystyrene nanoplastics of different particle sizes regulate the polarization of pro-inflammatory macrophages. Sci Rep 14(1):16329. https://doi.org/10.1038/s41598-024-67289-y

Jounai N et al (2012) Recognition of damage-associated molecular patterns related to nucleic acids during inflammation and vaccination. Front Cell Infect Microbiol 2:168. https://doi.org/10.3389/fcimb.2012.00168

Kapsenberg ML (2003) Dendritic-cell control of pathogen-driven T-cell polarization. Nat Rev Immunol 3(12):984–993. https://doi.org/10.1038/nri1246

Kato H et al (2005) Cell type-specific involvement of RIG-I in antiviral response. Immunity 23(1):19–28. https://doi.org/10.1016/j.immuni.2005.04.010

Kelley N et al (2019) The NLRP3 Inflammasome: an overview of mechanisms of activation and regulation. Int J Mol Sci 20(13):3328. https://doi.org/10.3390/ijms20133328

Kleinnijenhuis J et al (2012) Bacille Calmette-Guérin induces NOD2-dependent nonspecific protection from reinfection via epigenetic reprogramming of monocytes. Proc Natl Acad Sci 109(43):17537–17542. https://doi.org/10.1073/pnas.1202870109

Kleinnijenhuis J et al (2014) Long-lasting effects of BCG vaccination on both heterologous Th1/Th17 responses and innate trained immunity. J Innate Immun 6(2):152–158. https://doi.org/10.1159/000355628

Kono H, Rock KL (2008) How dying cells alert the immune system to danger. Nat Rev Immunol 8(4):279–289. https://doi.org/10.1038/nri2215

Kool M et al (2008) Alum adjuvant boosts adaptive immunity by inducing uric acid and activating inflammatory dendritic cells. J Exp Med 205(4):869–882. https://doi.org/10.1084/jem.20071087

Kroll A et al (2011) Cytotoxicity screening of 23 engineered nanomaterials using a test matrix of ten cell lines and three different assays. Part Fibre Toxicol 8:9. https://doi.org/10.1186/1743-8977-8-9

Kroll A et al (2012) Interference of engineered nanoparticles with in vitro toxicity assays. Arch Toxicol 86(7):1123–1136. https://doi.org/10.1007/s00204-012-0837-z

Krystel-Whittemore M, Dileepan KN, Wood JG (2016) Mast cell: a multi-functional master cell. Front Immunol 6. https://doi.org/10.3389/fimmu.2015.00620

Kühn R, Rajewsky K, Müller W (1991) Generation and analysis of interleukin-4 deficient mice. Science (New York, N.Y.) 254(5032):707–710. https://doi.org/10.1126/science.1948049

Kumar V, Sharma A (2010) Neutrophils: Cinderella of innate immune system. Int Immunopharmacol 10(11):1325–1334. https://doi.org/10.1016/j.intimp.2010.08.012

Lahiani MH et al (2017) Graphene and carbon nanotubes activate different cell surface receptors on macrophages before and after deactivation of endotoxins. J Appl Toxicol 37(11):1305–1316. https://doi.org/10.1002/jat.3477

Lebre F, Hearnden CH, Lavelle EC (2016) Modulation of immune responses by particulate materials. Adv Mater 28(27):5525–5541. https://doi.org/10.1002/adma.201505395

Lebre F et al (2017) The shape and size of hydroxyapatite particles dictate inflammatory responses following implantation. Sci Rep 7(1):2922. https://doi.org/10.1038/s41598-017-03086-0

Lebre F et al (2018a) Exfoliation in endotoxin-free albumin generates pristine graphene with reduced inflammatory properties. Adv Biosyst 2(12):284. https://doi.org/10.1002/adbi.201800102

Lebre F et al (2018b) Mechanistic study of the adjuvant effect of chitosan-aluminum nanoparticles. Int J Pharm 552(1–2):7–15. https://doi.org/10.1016/j.ijpharm.2018.09.044

Leudjo Taka A et al (2021) A review on conventional and advanced methods for Nanotoxicology evaluation of engineered nanomaterials. Molecules 26(21):6536. https://doi.org/10.3390/molecules26216536

Li H et al (2008) Cutting edge: Inflammasome activation by alum and Alum's adjuvant effect are mediated by NLRP3. J Immunol 181(1):17–21. https://doi.org/10.4049/jimmunol.181.1.17

Li Y et al (2017a) Bacterial endotoxin (lipopolysaccharide) binds to the surface of gold nanoparticles, interferes with biocorona formation and induces human monocyte inflammatory activation. Nanotoxicology 11(9–10):1157–1175. https://doi.org/10.1080/17435390.2017.1401142

Li Y, Fujita M, Boraschi D (2017b) Endotoxin contamination in nanomaterials leads to the misinterpretation of Immunosafety results. Front Immunol 8. https://doi.org/10.3389/fimmu.2017.00472

Li Z, Guo J, Bi L (2020) Role of the NLRP3 inflammasome in autoimmune diseases. Biomed Pharmacother 130:110542. https://doi.org/10.1016/j.biopha.2020.110542

Lunov O et al (2011) Amino-functionalized polystyrene nanoparticles activate the NLRP3 Inflammasome in human macrophages. ACS Nano 5(12):9648–9657. https://doi.org/10.1021/nn203596e

Luo M et al (2017) A STING-activating nanovaccine for cancer immunotherapy. Nat Nanotechnol 12(7):648–654. https://doi.org/10.1038/nnano.2017.52

Majerle A (2003) Enhancement of antibacterial and lipopolysaccharide binding activities of a human lactoferrin peptide fragment by the addition of acyl chain. J Antimicrob Chemother 51(5):1159–1165. https://doi.org/10.1093/jac/dkg219

Mangini M et al (2021) Interaction of nanoparticles with endotoxin importance in nanosafety testing and exploitation for endotoxin binding. Nanotoxicology 15(4):558–576. https://doi.org/10.1080/17435390.2021.1898690

Manuja A et al (2021) Metal/metal oxide nanoparticles: toxicity concerns associated with their physical state and remediation for biomedical applications. Toxicol Rep 8:1970–1978. https://doi.org/10.1016/j.toxrep.2021.11.020

Marichal T et al (2011) DNA released from dying host cells mediates aluminum adjuvant activity. Nat Med 17(8):996–1002. https://doi.org/10.1038/nm.2403

Marshall JS et al (2018) An introduction to immunology and immunopathology. Allergy Asthma Clin Immunol 14(Suppl 2):49. https://doi.org/10.1186/s13223-018-0278-1

Martinez-Serra J et al (2014) xCELLigence system for real-time label-free monitoring of growth and viability of cell lines from hematological malignancies. Onco Targets Ther:985. https://doi.org/10.2147/OTT.S62887

Matzinger P (1994) Tolerance, danger, and the extended family. Annu Rev Immunol 12(1):991–1045. https://doi.org/10.1146/annurev.iy.12.040194.005015

Mello DF et al (2020) Caveats to the use of MTT, neutral red, Hoechst and Resazurin to measure silver nanoparticle cytotoxicity. Chem Biol Interact 315:108868. https://doi.org/10.1016/j.cbi.2019.108868

Miao L et al (2019) Delivery of mRNA vaccines with heterocyclic lipids increases anti-tumor efficacy by STING-mediated immune cell activation. Nat Biotechnol 37(10):1174–1185. https://doi.org/10.1038/s41587-019-0247-3

Monteiro-Riviere NA, Inman AO, Zhang LW (2009) Limitations and relative utility of screening assays to assess engineered nanoparticle toxicity in a human cell line. Toxicol Appl Pharmacol 234(2):222–235. https://doi.org/10.1016/j.taap.2008.09.030

Mottram PL et al (2007) Type 1 and 2 immunity following vaccination is influenced by nanoparticle size: formulation of a model vaccine for respiratory syncytial virus. Mol Pharm 4(1):73–84. https://doi.org/10.1021/mp060096p

Mujal AM, Delconte RB, Sun JC (2021) Natural killer cells: from innate to adaptive features. Annu Rev Immunol 39(1):417–447. https://doi.org/10.1146/annurev-immunol-101819-074948

Muñoz-Wolf N et al (2023) Non-canonical inflammasome activation mediates the adjuvanticity of nanoparticles. Cell Rep Med 4(1):100899. https://doi.org/10.1016/j.xcrm.2022.100899

Murray PJ, Wynn TA (2011) Protective and pathogenic functions of macrophage subsets. Nat Rev Immunol 11(11):723–737. https://doi.org/10.1038/nri3073

Murray PJ et al (2014) Macrophage activation and polarization: nomenclature and experimental guidelines. Immunity 41(1):14–20. https://doi.org/10.1016/j.immuni.2014.06.008

Netea MG (2013) Training innate immunity: the changing concept of immunological memory in innate host defence. Eur J Clin Investig 43(8):881–884. https://doi.org/10.1111/eci.12132

Netea MG et al (2020) Defining trained immunity and its role in health and disease. Nat Rev Immunol 20(6):375–388. https://doi.org/10.1038/s41577-020-0285-6

Neumann S et al (2014) Activation of the NLRP3 inflammasome is not a feature of all particulate vaccine adjuvants. Immunol Cell Biol 92(6):535–542. https://doi.org/10.1038/icb.2014.21

Nimmerjahn F, Ravetch JV (2005) Divergent immunoglobulin g subclass activity through selective Fc receptor binding. Science (New York, N.Y.) 310(5753):1510–1512. https://doi.org/10.1126/science.1118948

O'Garra A (1998) Cytokines induce the development of functionally heterogeneous T helper cell subsets. Immunity 8(3):275–283. https://doi.org/10.1016/s1074-7613(00)80533-6

Obata-Ninomiya K, Domeier PP, Ziegler SF (2020) Basophils and eosinophils in nematode infections. Front Immunol 11. https://doi.org/10.3389/fimmu.2020.583824

Ochando J et al (2023) Trained immunity — basic concepts and contributions to immunopathology. Nat Rev Nephrol 19(1):23–37. https://doi.org/10.1038/s41581-022-00633-5

Ong KJ et al (2014) Widespread nanoparticle-assay interference: implications for nanotoxicity testing. PLoS One 9(3):e90650. https://doi.org/10.1371/journal.pone.0090650

Paludan SR, Bowie AG (2013) Immune sensing of DNA. Immunity 38(5):870–880. https://doi.org/10.1016/j.immuni.2013.05.004

Pan Y et al (2007) Size-dependent cytotoxicity of gold nanoparticles. Small 3(11):1941–1949. https://doi.org/10.1002/smll.200700378

Paul WE, Zhu J (2010) How are T(H)2-type immune responses initiated and amplified? Nat Rev Immunol 10(4):225–235. https://doi.org/10.1038/nri2735

Pedata P et al (2016) Immunotoxicological impact of occupational and environmental nanoparticles exposure: the influence of physical, chemical, and combined characteristics of the particles. Int J Immunopathol Pharmacol 29(3):343–353. https://doi.org/10.1177/0394632015608933

Pollard KM (2016) Silica, silicosis, and autoimmunity. Front Immunol 7:97. https://doi.org/10.3389/fimmu.2016.00097

Rankin JA (2004) Biological mediators of acute inflammation. AACN Clin Issues Adv Pract Acute Crit Care 15(1):3–17. https://doi.org/10.1097/00044067-200401000-00002

Robinson MJ et al (2006) Myeloid C-type lectins in innate immunity. Nat Immunol 7(12):1258–1265. https://doi.org/10.1038/ni1417

Rogers NC et al (2005) Syk-dependent cytokine induction by Dectin-1 reveals a novel pattern recognition pathway for C type lectins. Immunity 22(4):507–517. https://doi.org/10.1016/j.immuni.2005.03.004

Röszer T (2018) Understanding the biology of self-renewing macrophages. Cells 7(8). https://doi.org/10.3390/cells7080103

Sakaguchi S et al (2020) Regulatory T cells and human disease. Annu Rev Immunol 38:541–566. https://doi.org/10.1146/annurev-immunol-042718-041717

Sandquist I, Kolls J (2018) Update on regulation and effector functions of Th17 cells. F1000Research 7:205. https://doi.org/10.12688/f1000research.13020.1

Schnell A, Littman DR, Kuchroo VK (2023) TH17 cell heterogeneity and its role in tissue inflammation. Nat Immunol 24(1):19–29. https://doi.org/10.1038/s41590-022-01387-9

Schroder K, Tschopp J (2010) The Inflammasomes. Cell 140(6):821–832. https://doi.org/10.1016/j.cell.2010.01.040

Schroeder HW, Cavacini L (2010) Structure and function of immunoglobulins. J Allergy Clin Immunol 125(2):S41–S52. https://doi.org/10.1016/j.jaci.2009.09.046

Seder RA et al (1993) Interleukin 12 acts directly on CD4+ T cells to enhance priming for interferon gamma production and diminishes interleukin 4 inhibition of such priming. Proc Natl Acad Sci USA 90(21):10188–10192. https://doi.org/10.1073/pnas.90.21.10188

Sharifi S et al (2012) Toxicity of nanomaterials. Chem Soc Rev 41(6):2323–2343. https://doi.org/10.1039/C1CS15188F

Sharp FA et al (2009) Uptake of particulate vaccine adjuvants by dendritic cells activates the NALP3 inflammasome. Proc Natl Acad Sci 106(3):870–875. https://doi.org/10.1073/pnas.0804897106

Smith TD et al (2016) Regulation of macrophage polarization and plasticity by complex activation signals. Integr Biol 8(9):946–955. https://doi.org/10.1039/c6ib00105j

Steinman RM, Cohn ZA (1973) Identification of a novel cell type in peripheral lymphoid organs of mice. I. Morphology, quantitation, tissue distribution. J Exp Med 137(5):1142–1162. https://doi.org/10.1084/jem.137.5.1142

Stertman L et al (2023) The matrix-M™ adjuvant: a critical component of vaccines for the 21st century. Hum Vaccin Immunother 19(1):2189885. https://doi.org/10.1080/21645515.2023.2189885

Sun L et al (2013) Cyclic GMP-AMP synthase is a cytosolic DNA sensor that activates the type I interferon pathway. Science (New York, N.Y.) 339(6121):786–791. https://doi.org/10.1126/science.1232458

Sun B, Zhou G, Zhang H (2016) Synthesis, functionalization, and applications of morphology-controllable silica-based nanostructures: a review. Prog Solid State Chem 44(1):1–19. https://doi.org/10.1016/j.progsolidstchem.2016.01.001

Sundaram B et al (2024) The NLR family of innate immune and cell death sensors. Immunity 57(4):674–699. https://doi.org/10.1016/j.immuni.2024.03.012

Suttiponparnit K et al (2010) Role of surface area, primary particle size, and crystal phase on titanium dioxide nanoparticle dispersion properties. Nanoscale Res Lett 6(1):27. https://doi.org/10.1007/s11671-010-9772-1

Swartzwelter BJ et al (2020) Gold nanoparticles modulate BCG-induced innate immune memory in human monocytes by shifting the memory response towards tolerance. Cells 9(2):284. https://doi.org/10.3390/cells9020284

Szymański T et al (2020) Carbon nanotubes interference with luminescence-based assays. Materials (Basel, Switzerland) 13(19). https://doi.org/10.3390/ma13194270

Thiam HR et al (2020) Cellular mechanisms of NETosis. Annu Rev Cell Dev Biol 36(1):191–218. https://doi.org/10.1146/annurev-cellbio-020520-111016

Tirumala MG et al (2021) Novel methods and approaches for safety evaluation of nanoparticle formulations: a focus towards in vitro models and adverse outcome pathways. Front Pharmacol 12. https://doi.org/10.3389/fphar.2021.612659

Vianello F, Cecconello A, Magro M (2021) Toward the specificity of bare nanomaterial surfaces for protein Corona formation. Int J Mol Sci 22(14). https://doi.org/10.3390/ijms22147625

Vignali DAA, Collison LW, Workman CJ (2008) How regulatory T cells work. Nat Rev Immunol 8(7):523–532. https://doi.org/10.1038/nri2343

Voehringer D, van Rooijen N, Locksley RM (2007) Eosinophils develop in distinct stages and are recruited to peripheral sites by alternatively activated macrophages. J Leukoc Biol 81(6):1434–1444. https://doi.org/10.1189/jlb.1106686

Wan YY, Flavell RA (2009) How diverse--CD4 effector T cells and their functions. J Mol Cell Biol 1(1):20–36. https://doi.org/10.1093/jmcb/mjp001

Wang X, Reece SP, Brown JM (2013) Immunotoxicological impact of engineered nanomaterial exposure: mechanisms of immune cell modulation. Toxicol Mech Methods 23(3):168–177. https://doi.org/10.3109/15376516.2012.757686

Wilhelmi V et al (2012) Evaluation of apoptosis induced by nanoparticles and fine particles in RAW 264.7 macrophages: facts and artefacts. Toxicol In Vitro 26(2):323–334. https://doi.org/10.1016/j.tiv.2011.12.006

Wolf NK, Kissiov DU, Raulet DH (2023) Roles of natural killer cells in immunity to cancer, and applications to immunotherapy. Nat Rev Immunol 23(2):90–105. https://doi.org/10.1038/s41577-022-00732-1

Woolley JF, Stanicka J, Cotter TG (2013) Recent advances in reactive oxygen species measurement in biological systems. Trends Biochem Sci 38(11):556–565. https://doi.org/10.1016/j.tibs.2013.08.009

Xu J et al (2022) Yeast-derived nanoparticles remodel the immunosuppressive microenvironment in tumor and tumor-draining lymph nodes to suppress tumor growth. Nat Commun 13(1):110. https://doi.org/10.1038/s41467-021-27750-2

Yah CS, Simate GS (2020) Engineered nanoparticle bio-conjugates toxicity screening: the xCELLigence cells viability impact. BioImpacts 10(3):195–203. https://doi.org/10.34172/bi.2020.24

Yazdi AS et al (2010) Nanoparticles activate the NLR pyrin domain containing 3 (Nlrp3) inflammasome and cause pulmonary inflammation through release of IL-1α and IL-1β. Proc Natl Acad Sci 107(45):19449–19454. https://doi.org/10.1073/pnas.1008155107

Yu L, Liu P (2021) Cytosolic DNA sensing by cGAS: regulation, function, and human diseases. Signal Transduct Target Ther 6(1):170. https://doi.org/10.1038/s41392-021-00554-y

Zhang X et al (2013) Cyclic GMP-AMP containing mixed phosphodiester linkages is an endogenous high-affinity ligand for STING. Mol Cell 51(2):226–235. https://doi.org/10.1016/j.molcel.2013.05.022

Zhao T et al (2023) Vaccine adjuvants: mechanisms and platforms. Signal Transduct Target Ther 8(1):283. https://doi.org/10.1038/s41392-023-01557-7

Zheng D, Liwinski T, Elinav E (2020) Inflammasome activation and regulation: toward a better understanding of complex mechanisms. Cell Discov 6(1):36. https://doi.org/10.1038/s41421-020-0167-x

Zindel J, Kubes P (2020) DAMPs, PAMPs, and LAMPs in immunity and sterile inflammation. Annu Rev Pathol 15(1):493–518. https://doi.org/10.1146/annurev-pathmechdis-012419-032847

Open Access This chapter is licensed under the terms of the Creative Commons Attribution-NonCommercial-NoDerivatives 4.0 International License (http://creativecommons.org/licenses/by-nc-nd/4.0/), which permits any noncommercial use, sharing, distribution and reproduction in any medium or format, as long as you give appropriate credit to the original author(s) and the source, provide a link to the Creative Commons license and indicate if you modified the licensed material. You do not have permission under this license to share adapted material derived from this chapter or parts of it.

The images or other third party material in this chapter are included in the chapter's Creative Commons license, unless indicated otherwise in a credit line to the material. If material is not included in the chapter's Creative Commons license and your intended use is not permitted by statutory regulation or exceeds the permitted use, you will need to obtain permission directly from the copyright holder.

Nanoparticle Mediated Epigenetic and Immunological Modulation

4

Ayse Basak Engin and Atilla Engin

Abstract

Successful delivery of gene-editing tools using nano-systems is dependent on the ability of nanoparticles (NPs) to pass through the cellular membrane. After internalization by phagocytes, NPs are taken up in vesicular phagolysosomes. Phagosomal leakage, redox imbalance and ionic movements induced by toxic NPs result in inflammation. Although their effects in the cytoplasm are limited to targeting circular ribonucleic acids, they have drastic impact on the epigenome. While NPs induce immunotoxicity by affecting immune-specific signaling pathways, they exhibit genotoxic action. NP-induced immunotoxicity and genotoxicity partly share the same metabolic pathways. Following the internalization to the nucleus, NPs participate in chromatin remodeling, thus, epigenetic landscape changes by inducing DNA methylation changes, modifications of histone proteins, and noncoding RNA expression. The metabolic and epigenetic reprogramming involving the development of innate memory are strictly linked. DNA methyltransferases (DNMTs) can induce transcriptional repression by methylating cytosines (m5C) in Cystidine-Guanine dinucleotide-rich sequences. DNMTs can bind and be regulated by long noncoding RNAs. Epigenetic mechanisms play vital roles not only in the activation, differentiation and effector function(s) of immune cells, but also in cancer initiation and progression via epigenetic alterations related to the multiple oncogenic or tumor suppressor gene pathways.

A. B. Engin (✉)
Faculty of Pharmacy, Department of Toxicology, Gazi University, Ankara, Turkey

A. Engin
Faculty of Medicine, Department of General Surgery, Gazi University, Ankara, Turkey

© The Author(s) 2025
E. Alfaro-Moreno, F. Murphy (eds.), *Nanosafety*,
https://doi.org/10.1007/978-3-031-93871-9_4

Keywords

Nanoparticle · DNA methylation · Epigenetic toxicity · Immunotoxicity · Non-coding RNA · Chromatin remodeling · Methyltransferase · Histone modification · Cytosine-phosphate-Guanine dinucleotide · Tumor suppressor gene

1 Introduction

Epigenetics term includes DNA methylation, histone tail modifications, and microRNA (miRNA) mediated mechanisms, which are all able to shape the chromatin structure and/or gene expression levels, without altering the primary DNA sequence (Stoccoro et al. 2013). The transcription response is mainly due to changes in the genomic distribution of histone modifications that can modulate the activity of enhancers. Alteration of the epigenetic landscape is a key mechanism in defining the gene expression program changes resulting in nanotoxicity (Gamberoni et al. 2023). An increasing number of epimutagen nanoparticles (NPs), have been shown to have drastic impact on the epigenome by inducing changes in DNA methylation, modifications of histone proteins, affecting chromatin structure and miRNA expression (Arita and Costa 2009; Cheng et al. 2012). In recent decades, *in vitro* and *in vivo* studies have revealed that nanomaterials (NMs) have cytotoxicity and genotoxicity. However, there are still very limited studies investigating epigenetic mechanisms of NM-induced toxicity, such as changes in DNA methylation, histone modification and miRNA expression (Shyamasundar et al. 2015; Wong et al. 2017). On the one hand, nanosized compounds cause global DNA methylation changes due to disruption of the expression of genes involved in DNA methylation. On the other hand, they play a role in cancer development by causing changes in the gene-specific methylation of tumor suppressor genes as well as inflammatory and DNA repair genes (Stoccoro et al. 2013). While NMs can induce immunotoxic effects by causing death of immune cells or by affecting immune-specific signaling pathways, genotoxic action can be either directly damaging DNA or indirectly resulting in DNA base oxidation. Ultimately, the immunotoxic and genotoxic adverse effects of NPs occur by partly sharing the same metabolic pathways (Dusinska et al. 2017).

Since the misregulation of cellular epigenetic mechanisms or their dynamic response to specific cellular conditions and environmental stimuli may be detrimental to health, the biological effects and toxicity of NMs and NPs have been extensively investigated in the last few years (Ameh and Sayes 2019; Missaoui et al. 2018; Schulte et al. 2019; Sierra et al. 2016). However, there are many unanswered questions as well as limited and inconclusive data regarding their epigenomic and immunological toxic effects (Shyamasundar et al. 2015; Sierra et al. 2016). A growing body of evidence indicates that epigenetic and immunological alterations play a role in the onset of several human diseases, that's why in this chapter cellular uptake, immunotoxicity and genotoxicity mechanisms, as well as novel epigenetic toxicity endpoints will be discussed.

2 Intracellular Fate of Nanoparticles

Successful delivery of gene-editing tools using nano-systems is dependent on the ability of NPs to pass through the cellular membrane, move through the cytoplasm, and cross the nuclear envelope to enter the nucleus (He et al. 2016; Li et al. 2015a; Pozzi et al. 2014; Zheng et al. 2011). Cellular internalization of NPs also depends on their particle size and NPs smaller than 100 nm in size show 27-fold higher gene transfection than those larger than 100 nm (Prabha et al. 2002). Following being taken up by the immune cells, fate of NPs depends on their biodegradability. Biodegradable NMs are digested and, thus, their elimination is facilitated, but non-degradable NMs can be stored in the cells for long periods of time (Dobrovolskaia et al. 2008). Therefore, NPs can be divided into two groups: those that are easily degradable or those that are resistant to degradation. The first group includes nano-liposomes and polymeric nano-arrangements (Kamaly et al. 2012; Li et al. 2015b; Xie et al. 2014). Increasing the circulation time of drug nanocarriers and enhancing the biodegradation process of NMs cause inflammatory responses and toxicity after inadvertent exposures (Sacchetti et al. 2013; Shvedova et al. 2012). Biodegradable NPs are becoming increasingly popular as therapeutics because of the increasing bioavailability, more stability, and solubility properties (DeMarino et al. 2017). The second group of NMs are more persistent and can either display prolonged lifetime at the sites of their entry or migrate to distant locations (Yang et al. 2008). Thus, non-degradable materials, either may continue to be internalized into the adjacent cells (Kagan et al. 2010; Sharifi et al. 2012), or could be taken up by the lymphatic drainage and be eliminated subsequently, after prolonged circulation in the physiological environment.

NPs are engulfed by cells through different pathways depending on their size and surface charge. While the particles with larger size generally enter into the cells by micropinocytosis, the nanoscale products most commonly enter through the endocytosis (Steinman et al. 1983). Indeed, small-sized particles, around 60 nm could conveniently enter the cell by the caveolin-dependent endocytosis, whereas the NPs, which are larger than 120 nm could be internalized through clathrin-dependent/independent endocytosis (Geng et al. 2007).

The internalized NPs are encapsulated in the endosome-based liposome packets, and then they pass to the lysosomes with a highly acidic microenvironment, where they undergo enzymolysis and hydrolysis or discharge the inert nanocontainers as unmodified into the cytoplasm through lysosomal escape pathways (Steinman et al. 1983). These NPs after being released into the cytoplasm, continue to interact with other specific intracellular organelles. They react with the components in the acidic environment of endocytosis/lysosome, activate electron transport chain or produce reactive oxygen species (ROS) in mitochondria, or Golgi apparatus and regulate DNA-control in nucleus or mitochondria (Liu et al. 2020). The endosomal/lysosomal escape mechanism of NPs occurs in four main ways: (1) Formation of membrane apertures in the endosomes or lysosomes, (2) Proton sponge effect, (3) Fusion with the membrane, and (4) Photochemical disruption of endosomal or lysosomal membranes (Liu et al. 2020). The ultra-small positively charged NPs have a higher affinity toward specific organelles, such as mitochondria and nucleus, they are more easily transported to these organelles. The entry of ultrasmall NPs into the cell nucleus is critically dependent on the size of the NPs (Huo et al. 2014). The internalization of NPs toward the nucleus is facilitated by two predominant pathways. Firstly, the composite structure of phospholipids or similar particles interact with the nuclear membrane, facilitate the fusion of the outer layer of phospholipids with the nuclear membrane, and then the internalized NPs are released into the nucleus (Godbey et al. 1999). Thereby, the nano systems are usually modified with specific ligands that can activate nuclear receptors and enable their internalization (Tanaka et al. 2017). Another way of nuclear internalization is the passing of nano-constructs through the nuclear pores or occupying nucleoplasm for transferring nucleoprotein. However, the nuclear diffusion of NPs with functional diameters between 9 and 40 nm is largely limited (Qiu et al. 2015). Ultimately, the internalized NPs in the nucleus predominantly interact with the intranuclear constituents, resulting in different types of mutations and severe DNA damage, the block of DNA cycle, and the activation of the apoptosis signaling pathway (Papageorgiou et al. 2007).

3 Mechanisms of Epigenetic Modifications Induced by Nanomaterials and Nanoparticles

Epigenetic modifications, such as DNA methylation, post-translational modifications of histones, and regulation of non-coding RNAs (ncRNAs) form innate memory (Italiani et al. 2020). Therefore, epigenetic alterations are valuable indicators of NMs and NPs related toxicity and are also potential translational biomarkers for detecting adverse effects of NMs in humans. Both amorphous silica and crystalline silica transforming NPs induce strong cellular myelocytomatosis oncogene (c-Myc) expression in the early stage of cell transformation and this correlates with enrichment in RNA polymerase II as well as histone active marks on its promoter (Seidel et al. 2017). Thus, the increased proliferation that is observed in NM-exposed cells indicates a greater potential to pass the genetic damage to daughter cells (Sargent et al. 2012). Exposure to NMs and NPs results in aberrant expression of protein-coding and protein-non-coding genes. Unfortunately, it is thought that attempt to link the observed changes in DNA methylation and histone modifications to altered functioning of chromatin-modifying proteins are inconclusive (Pogribna and Hammons 2021). However, the impact of some NP's exposure on DNA methylation via histone modification in multiple cell types, as well as deregulated expression of miRNAs indicates the potential involvement of this epigenetic mechanism in the toxicity of NPs (Pogribna et al. 2022; Stoccoro et al. 2013). On the other hand, the alterations in gene-specific methylation and miRNA expression are indicators of inflammatory response to NM exposure as biomarker of both exposure and disease development (Brown et al. 2016). While miRNAs not only control gene expression at the post-transcriptional level, they also directly activate the epigenetic mechanism through a regulatory loop (Iorio et al. 2010). The human genome contains approximately 28 million Cytosine-phosphate-Guanine dinucleotide (CpG) sites, about 60% of which are methylated at the 5' position of the cytosine. Methylation of relatively CpG-rich promoters causes strong transcriptional repression. There is a significant relationship between CpG density and methylation density (Edwards et al. 2010). Current evidence indicates that the primary biological functions of DNA methylation are heritable transcriptional repression of retrotransposons, and the selective exposure of promoters of cellular genes to transcription factors (Edwards et al. 2017). DNA methylation, a covalent modification of cytosine residues in DNA, play role as a major component of the cellular epigenetic regulatory mechanism. It is initiated by DNA methyltransferases, and it develops via either replication-dependent or replication-independent fashion. Latter type of methylation is removed by cytosine via a base excision DNA repair mechanism (Ooi et al. 2009; Pogribna and Hammons 2021; Wu and Zhang 2017). In the genome-wide distribution of histone modifications and compared them with the transcriptome after exposing to NPs indicate that alteration of the epigenetic landscape is a key mechanism in defining the gene expression program changes resulting in nanotoxicity (Gamberoni et al. 2023). Although extensive experimental *in vitro* and *in vivo* studies were made on DNA methylation alterations caused by exposure time to NMs and NPs, there is limited evidence on the engineered NMs and NPs related DNA methylation in humans. Even in a short-term exposure to some NPs, which is inversely associated with oxidative DNA damage, a significant decrease in global DNA methylation is detected (Liou et al. 2017; Pilger and Rüdiger 2006). In fact, epigenetic changes induced by NPs can vary depending on their characteristics such as size, shape, chemical structure, surface functionalization, exposure dose, short-term or long-term exposure, and the type of cells (Italiani et al. 2020). In contrast to short term exposures, long-term occupational exposure to some nanocomposite materials causes substantial changes in CpG site methylation (Rossnerova et al. 2020). Oxidative DNA damage results in dynamic changes in DNA methylation, thereby, simultaneously DNA demethylation and new methylation sites generation are induced (Jiang et al. 2020). Eventually, induction of oxidative stress and inflammation affect the integrity of chromatin and lead to changes in DNA methylation and histone modifications. Although not all NPs cause DNA damage through oxidative stress formation uniformly in the same cells (Wan et al. 2012), oxidative stress-induced DNA lesions, which are produced by NMs, and NPs, inhibit methylation capacity of DNA methyltransferases in some cases, and lead to global DNA hypomethylation (Valinluck et al. 2004) (Fig. 4.1). DNA methylation changes in human monocyte-like cells (THP-1) after incubation with carbon nanotubes (CNTs). It has been observed that CNT induced gene-specific differential methylation, with promoter hypomethylation evident for a thousand different genes, compared with the control samples by assessing methylation of single CpG sites (Öner et al. 2017).

Some inflammation-mediated halogenated cytosine damage products can potentially interfere with normal epigenetic control and alter DNA-protein interactions critical for gene regulation and the hereditary transmission of methylation patterns. NP can induce both global DNA methylation and selective methylation at the level of specific genes (Valinluck and Sowers 2007). Alterations in the gene-specific

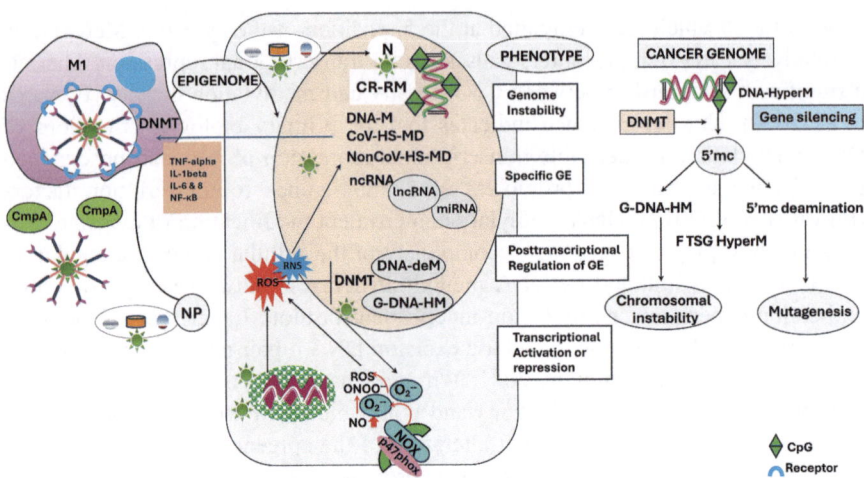

Fig. 4.1 Naked nanomaterials (NMs) which dynamically interact with biomolecules encounter in circulation, form "protein bio-corona". Rapid recruitment of macrophages is triggered by the activation of the complement system by NPs. NPs reprogram macrophages to respond differentially to a second stimulation via changing their epigenetic profiles. Histone-modifying enzymes (covalent histone modification; CoV-HS-MD) have been identified, that control macrophage activation and polarization to M1 or M2 phenotypes. DNA methylation, a covalent modification of cytosine residues in DNA, functions as a major component of the cellular epigenetic regulatory mechanism. Methylation of relatively CpG-rich promoters causes strong transcriptional repression. Induction of oxidative stress and inflammation affect the integrity of chromatin and lead to changes in DNA methylation and histone modifications. Oxidative stress inhibits methylation capacity of DNA methyltransferases. Exposure to NPs causes activation of the inflammatory response that, in turn, may cause DNA hypermethylation and histone modification changes. Global DNA hypomethylation (G-DNA-HM) is associated with increased genomic instability. NM-induced variations in gene expression of non-coding RNAs (ncRNAs) participates in aberrant expression pattern potentially leading to pathological development. Micro RNAs (miRNAs) not only control gene expression at the post-transcriptional level, but also directly activate the epigenetic mechanism through a regulatory loop. Although epigenetic markers modify chromatin structure and regulate expression of downstream genes, aberrant epigenetic modifications are common events in human disease including tumorigenesis and autoimmunity (*Abbreviations: 5'mc:* 5-methylcytosine, *CmpA:* Complement activation, *CoV-HS-MD:* Covalent histone modification, *CpG:* Cytosine-phosphate-Guanine dinucleotide, *CR-RM:* Chromatin remodeling, *DNA-deM:* DNA demethylation, *DNA-HyperM:* DNA Hypermethylation, *DNA-M:* DNA methylation, *DNMT:* DNA methyltransferase, *F TSG HyperM:* Focal tumor suppressor gene hypermethylation, *G-DNA-HM*: Global DNA hypomethylation, *GE:* Gene expression, *IL-1beta:* Interleukin-1beta, *IL-6:* Interleukin 6, *IL-8:* Interleukin 8, *lncRNA:* long noncoding RNAs, *M1:* M1 macrophage, *miRNA:* Micro ribonucleic acid, N: Nucleus, *ncRNAs:* Non-coding RNA, *NF-κB:* Nuclear factor kappa B, *NO:* Nitric oxide, *NonCoV-HS-MD:* Noncovalent histone modification, *NOX:* Reduced nicotinamide adenine dinucleotide phosphate oxidases, *NP:* Nanoparticle, *RNS:* Reactive nitrogen species, *ROS:* Reactive oxygen species, *TNF-alpha:* Tumor necrosis factor alpha)

methylation correspond with an inflammatory response to exposure of NPs (Brown et al. 2016). Activation of inflammasomes can occur through the several mechanisms, such as rupture of lysosomes, direct recognition of NMs by toll-like receptors (Boraschi et al. 2017a). Neutrophils or polymorphonuclear granulocytes play a key role in NM-induced inflammation (Bhattacharya et al. 2013). Activation of mast cells can lead to production of histamines and other substances causing inflammation. Another factor suspected to contribute to the recent dramatic increase in incidence of allergies, lung diseases and asthma is inhalation of ultrafine particles from environmental pollution (Heinrich 2011). The exposure to different NPs creates distinct transcription responses, which are the result of distinct changes of the epigenetic landscape, thus enhancer regions emerge as the genetic elements that are most involved in mediating nanotoxicity-related epigenetic effects (Gamberoni et al. 2023). In addition to alterations in DNA methylation induced by NMs and NPs, disruption of normal patterns of histone modifications is a different epigenetic response. Indeed, as epigenetic regulatory mechanism, histone modifications are the posttranslational covalent modifications of the amino-terminal tails of histone proteins. Like methylation of DNA, histone modifications are dynamic process tightly controlled by the balance between histone phosphorylases, acetyltransferases, methyltransferases and histone phosphatases, deacetylases, demethylases. While first group oversees chemical histone modifications, second group is responsible for removal of chemical modifications (Pogribna and Hammons 2021; Torres and Fujimori 2015). Futhermore, while the epigenetic reprogramming involved in the development of innate memory encompasses mainly methylation and acetylation of histones and the activity of long noncoding RNAs (lncRNA), the metabolic reprogramming involves in changes to glycolysis, tricarboxylic acid cycle, glutaminolysis, and cholesterol metabolism. The metabolic reprogramming and epigenetic reprogramming are strictly linked (Italiani et al. 2020). Changes in the levels of intracellular metabolites produced from metabolic pathways reorganize the functionality of enzymes responsible of 'writing', 'erasing' or 'reading' histone and DNA modifications that alter the epigenetic landscape of the innate immune cells (Domínguez-Andrés et al. 2020). The contribution of ncRNAs, that is, short (miRNA) and long (lncRNA) RNA, to innate memory of immunity have remained largely unexplored or have only recently started to be investigated (Fanucchi and Mhlanga 2019; Fok et al. 2018; Seeley et al. 2018). miRNAs not only control gene expression at the post-transcriptional level, but also directly activate the epigenetic mechanism through a regulatory loop (Iorio et al. 2010), whereas lncRNAs can regulate gene transcription by influencing the access of transcription-regulating proteins to their target genes. Recently, it has been proposed that the enhancer RNAs (eRNAs), play a role in memory of immune responses. The transcripts that arise from enhancers have been shown to play a significant role in the regulation of chromosomal looping and tissue-specific target gene transcription (Lai et al. 2013). In fact, eRNAs derive from the transcription of enhancers, are cell-specific, and are involved in the regulation of chromosomal looping and in the transcription of tissue-specific genes/contacts between target genes and enhancers. eRNAs interact with "cyclic adenosine monophosphate (cAMP) response element-binding protein" (CBP), which stimulates catalytic histone acetyltransferase activity at specific genomic loci. Thus, bi-directional eRNAs contribute to the chromatin structure at active enhancers, which, in turn, is required for regulation of target genes (Bose et al. 2017).

4 Epigenetic Mechanism Regulating miRNA Biogenesis

Small ncRNA (less than 200 nucleotides) sequences are known as miRNA and they have not only been identified as critical regulators of gene expression, but also as assistant in multiple biological and physiological processes (Lee et al. 2019). Epigenetic mechanisms include DNA methylation/demethylation, post-translational changes in histones, together with miRNAs (Lee et al. 2019). The canonical pathway is the major pathway through which the majority of miRNAs are processed. After transcription, primary miRNAs (pri-miRNAs) are processed into precursor-miRNAs (pre-miRNAs) by a microprocessor complex. Primary transcripts are successively cleaved by two RNase III enzymes, DROSHA in the nucleus and DICER in the cytoplasm, to produce approximately 70 nucleotide (nt) hairpin (long) precursor miRNAs and 22 nt long mature miRNAs, respectively (Van Wynsberghe et al. 2011). Microprocessor complex includes RNA binding protein DiGeorge syndrome critical region 8 (DGCR8) and a ribonuclease III enzyme DROSHA, which cleaves the pri-miRNA duplex to form an overhang at 3′ of pre-miRNA of 70 nt (Lee et al. 2003, 2019). Once processed, they are exported to the cytoplasm via exportin 5 (XPO5)/guanosine triphosphate (GTP)-bound Ran (Ran GTP) complex (Bohnsack et al. 2004; Yi et al. 2003). After the pre-miRNA is exported to the cytoplasm to be processed by the ribonuclease DICER, it produces a miRNA duplex that would be processed by the RNA-induced silencing complex (RISC). DICER, RNase III endonuclease along with double-stranded RNA (dsRNA)-binding protein transactivation response RNA binding protein (TRBP), cleaves the pre-miRNA to form a 22 nt mature miRNA complex which has a guide strand and a passenger strand (Ketting et al. 2001). The passenger and guide strands are selected based on various factors, including thermodynamic stability. They both are loaded into argonaute proteins (AGO1–4), where the passenger strand is subsequently degraded (Kobayashi and Tomari 2016). Studies on miRNA mediated gene regulation are predominantly based on gene silencing via translational repression and messenger RNA (mRNA) degradation. miRNA induced gene silencing is performed by miRNA-induced silencing complex (miRISC) which consists of the AGO protein and the guide strand. They bind to the specific sequence at the 3′ untranslated region (UTR) mRNA response element (MRE) of their target mRNA. A full complementary of miRNA:MRE leads to mRNA slicing while most of the miRNA:MREs are partially complementary leading to translational inhibition and mRNA decay (Guo et al. 2010). miRNA has also been shown to bind to the 5′ UTR and other coding regions leading to gene silencing (Zhang et al. 2018). However, various research has shown the ability of miRNA to induce transcription as well as translation (Dharap et al. 2013; Truesdell et al. 2012). Some of the miRNAs share the same pathways leading to toxicity, almost all the altered miRNAs in various toxicants are different from each other, showing an increased specificity of these miRNAs. Even the similar miRNAs in different toxicant exposures have differed targets interacting with varied signaling pathways (Balasubramanian et al. 2020). Naked miRNAs are susceptible to rapid systemic degradation, however, encapsulation of miRNAs into NPs can overcome these delivery challenges. Extracellular miRNAs travel in vesicles and exosomes bound to AGO proteins to avoid the loss of their stability (Figueroa et al. 2022). Exosomes are natural NMs that are known as key mediators in cell-cell communication and facilitate the transfer of genetic and biochemical information between distant cells. Structurally, they are composed of lipids, proteins, and several types of RNAs which enable these vesicles to serve as important disease biomarkers. Exosomes possess high bioavailability, biological stability, targeting specificity, low toxicity, and immune characteristics (Pi et al. 2021). Exosomes contain both mRNA and miRNA, which can be delivered to another cell, and can be functional in this new location (Valadi et al. 2007). The engineered exosome-based chemotherapeutics and miRNA (miR)-21 inhibitor oligonucleotide loaded exosome co-delivery system could efficiently facilitate cellular uptake and significantly down-regulate miR-21 expression in chemotherapeutic resistant colorectal cancer cell line. The downregulation of miR-21 induces cell cycle arrest, reduces tumor proliferation, increases apoptosis of tumor cells (Liang et al. 2020).

Membrane-dependent signaling pathways have also been shown to be responsible for cellular effects of NMs. Epigenetic regulation by NMs affects the NP-protein interactions on cell signaling pathways, and the induction of various cell death modalities (Hussain et al. 2014). NMs toxicity induces generation of oxidative stress, which leads to the dysregulation in antioxidant response, inflammation and other cell death mechanisms. These are regulated by epigenetic modulation of miR-NAs, which targets mRNAs and cause translational repression or degradation. miR-NAs are transcribed by RNA polymerase II/III, either from the intron regions of the protein-coding genes (intragenic) or independently with their own (intergenic) promoters (de Rie et al. 2017; Lee et al. 2003). Exposure to engineered NMs (ENM) can also trigger epigenetic alterations, which are dependent on ENM dose and physicochemical properties including size, shape and surface chemistry, as well as on the cell/organism sensitivity. Some DNA methylation patterns are retained as a form of epigenetic memory (Moreira et al. 2021).

5 Mechanisms of Nanoparticles Mediated Immunomodulation

When a NM enters an organism, its fate can occur in four ways; it accumulates and becomes toxic, becomes available at the biological level and transforms, and interacts with cells or macromolecules within the cell (Accomasso et al. 2018). In this context, NPs interact with both innate and adaptive immune cells, and can stimulate and/or suppress the immune response (Gustafson et al. 2015). NMs provoke considerable inflammation due to their physicochemical properties. How NPs potentially influence the immune system is examined by NP-mediated hemolysis, NP-induced platelet aggregation, and the ability of the NP to activate the complement system (Dobrovolskaia et al. 2008). The compatibility of NMs with the immune system is largely determined by their surface chemistry (Dobrovolskaia and McNeil 2007). Thus, NP-cell membrane interactions may affect the localization of the NPs, their intracellular trafficking, compartmentalization into various parts of the organism, and cellular retention (Adjei et al. 2014). The innate immune system is the organism's first line of defense, and innate cells primarily interact with NPs administered to the body. In this context, there are two options; innate cells either do not response or trigger an innate/inflammatory reaction against the NPs. If innate immune system cells respond, clearance of NPs from the body occurs through complement activation and dynamic corona formation on NPs. NPs also are internalized by dendritic cells (DCs), which are the key antigen-presenting cells of the immune system (Boraschi et al. 2017b). Naked NMs dynamically interact with encountered biomolecules and form the "protein bio-corona" (Neagu et al. 2017). The major characteristics of NPs that influence bio-corona composition are material, size, shape, curvature, solubility, surface functionalization, surface charge and the route of administration to the body, whereas bio-corona compositions characterize the discrete immune patterns (Neagu et al. 2017). Particles bearing cationic or anionic surface charges have been shown to be more attractive to phagocytes than neutral particles of the same size. A hydrophilic surface with a low negative charge reduces protein adsorption and uptake of NPs (Zahr et al. 2006). Direct immunostimulatory effect arise due to NP-mediated production of antigens following tissue penetration and access to the lymphatics (Reddy et al. 2007). NPs act as danger signals because they represent pathogen-associated molecular patterns (PAMPs) and damaged tissues release damage-associated molecular patterns (DAMPs). These molecular signatures are perceived by pattern recognition receptors (PRRs), including innate immunity toll-like receptors. The activation of PRRs triggers inflammation and alerts the adaptive immune system to an impending danger. Upon recognition of NMs, DCs can trigger direct inflammatory response, therewithal activated T cells can likewise trigger adaptive immunity (Neagu et al. 2017). After internalization by phagocytes, NPs are taken up in vesicular phagolysosomes. Intracellular phagosomal leakage, redox unbalance and ionic movements induced by toxic particles result in pro-interleukin 1 beta (pro-IL-1β) expression, inflammasome complex engagement, caspase-1 activation, pro-IL-1β cleavage, biologically active IL-1β release and finally inflammatory cell death. It is described as pyroptosis (Rabolli et al. 2016). Complement system can be activated through three pathways; classical, lectin, or alternative, NP–protein corona complexes seem to initiate the classical pathway (Hulander et al. 2009). The alternative pathway activation provides the complement component 3b (C3b) deposition that adds 25% more mass to the NP surface (Andersson et al. 2005). Whereas the classical pathway of the complement system is initiated when the head group of complement component 1q (C1q) binds to IgG molecules that are present on the surface. C1q binding promotes a damaging effect of excessive complement activation (Salvador-Morales et al. 2006). Classical complement activation is markedly reduced on the nanostructured hydrophilic surface compared with smoother ones. The ability of human IgG to activate the complement system is also significantly reduced by the surface of the nanostructure. In this manner, the physicochemical surface properties of NPs could mediate protein corona formation, and thus control the activation of complement pathways (Hulander et al. 2011). Rapid recruitment of macrophages is triggered by the activation of the complement system. NPs reprogram macrophages to respond differentially to a second stimulation via changing their epigenetic profiles. NPs interact with immunocompetent cells and trigger to release the soluble mediators such as cytokines, chemokines and immunomodulatory molecules that regulate the immune response (Di Gioacchino et al. 2011; Hussain et al. 2012; Zolnik et al. 2010). Macrophages are the "first responders" to NPs (Wang et al. 2012). NPs can recruit macrophages, activate them and are taken up by these immune cells. NP uptake is enhanced in M2-polarized primary human monocyte-derived macrophages compared to the M1 cells. M2 polarization promotes the NP uptake in the monocytic THP-1 cell line (Hoppstädter et al. 2015). Activation of the pro-inflammatory signaling cascade results in the polarization of macrophages from M2 (anti-inflammatory) to M1 (pro-inflammatory) type (Pal et al. 2016). At lower doses of metal NPs, the Th1 or M1 response is counteracted by Th2 or M2 response, resulting in insignificant oxidative damage. However, with increasing dose of NPs, the M1 response increases over the M2 response resulting in significant tissue damage (Kumar et al. 2016). Adaptive immunity can be directed against the NP itself, even subsequent to the formation of a protein corona. The protein corona, when interacting with NPs, can stimulate the adaptive immune process if the complex created represents a danger signal, which in turn activates the nuclear factor kappa-light-chain-enhancer of activated B cells (NF-κB/B cell) signaling pathway (Corbo et al. 2016; Deng et al. 2011). Variations in serum protein adsorption correlate with differences in the mechanism and efficiency of NP uptake by macrophages (Deng et al. 2012; Walkey et al. 2012).

Three main epigenetic mechanisms have been identified, including miRNAs, DNA methylation and histone modification, which may contribute to the altered cellular signaling and signature gene expression during M1 and M2 polarization (Saradna et al. 2018). DNA methylation occurs via DNA methyltransferases (DNMT) which adds methyl groups to 5′CpG dinucleotides leading to condensation of chromatin and prevention of transcription (Wu and Zhang 2010). DNA methylation has been associated with asthma since altered DNA methylation status may cause differential gene expression of cytokines (e.g., IL-4, interferon-gamma; IFN-γ) (Kwon et al. 2008) and transcription factor (e.g., forkhead box P3; FOXP3) (Runyon et al. 2012). DNMT 3a and b are differentially expressed in M1 and M2 macrophages and play a critical role in gene silencing (Kittan et al. 2013). DNMT3b is crucial in regulating macrophage polarization through epigenetic mechanisms.

Histone modification refers to the process in which histones may undergo divergent epigenetic changes, including methylation, acetylation, phosphorylation, ubiquitylation and small ubiquitin-related modifier (SUMO)ylation, which could function as epigenetic markers of chromatin state linked with either transcriptional activation or repression (Van den Bossche et al. 2014; Zhou et al. 2017). Distinct histone-modifying enzymes have been identified that control macrophage activation and polarization to M1 or M2 phenotypes (e.g., methyltransferases, demethylases, acetyltransferases and deacetylases) (Kittan et al. 2013). Histone methylations in several pro-inflammatory genes are also associated with macrophage polarization (e.g., tumor necrosis factor alpha; TNF-α, IL-6) (Chen et al. 2014; Xu et al. 2015). Thus, targeting histone methylations of genes required for M1 or M2 polarization could be beneficial as a therapy for multiple inflammatory diseases.

In contrast, histone acetylation is the most studied histone modification where acetylation of lysine residues is achieved by histone acetyltransferases (HATs), but deacetylation occurs by histone deacetylases (HDACs). Imbalance of HATs/HDACs may cause differential gene expression leading to different diseases. HDAC3 is required for the activation of hundreds of, mainly signal transducer and activator of transcription (STAT)-1-dependent, inflammatory genes in M1 macrophages (Chen et al. 2012). Macrophages lacking HDAC3 show an M2-line phenotype in the absence of external stimuli and are hyper-responsive to IL-4, suggesting that HDAC3 may promote M1 and inhibit M2 polarization (Mullican et al. 2011). Throughout the macrophage genome, HDAC3 deacetylates histone tails at regulatory regions, leading to repression of many IL-4-regulated genes, which are characteristics of M2 macrophages. Pharmacological blockade of HDAC3 functions could be of benefit in the treatment of inflammatory diseases (Saradna et al. 2018).

The interactions of the extracellular matrix (ECM) with NPs depend on the morphological characteristics of intercellular matrix and on the physical characteristics of the NPs. NPs-dependent signaling may seriously impair the ratio between ECM metalloproteinases and tissue inhibitors of matrix proteases which results in pathological turn-over of the ECM and facilitation of inflammatory processes (Engin et al. 2017). The immunostimulatory activity of nanoscale materials occurs with diverse mechanisms: (1) They have a direct immunostimulatory effect on components of the immune system, such as antigen presenting cells (APCs), B cells or T cells. B cell receptor (BCR) co-aggregation, triggering and activation due to spatial organization of the antigens on the particle surface; (2) they deliver compounds that result in immunostimulation and secretion of cytokines are enhanced or they trigger both mechanisms at the same time. Enhanced delivery of antigens and adjuvants might result in apoptosis or necrosis, which enhances vaccine immunogenicity (Smith et al. 2013).

Moreover, NMs either might have a direct immunosuppressive effect on components of the immune response, including APCs, B cells or T cells; or might deliver compounds that result in immunosuppression; or some of them can use both mechanisms at the same time (Smith et al. 2013). Induction of immune tolerance by NPs can be considered as a form of desirable immunosuppression (Zolnik et al. 2010). Direct immunosuppressive effects include the upregulation of transforming growth factor-β (TGFβ), which results in increased cyclooxygenase 2 (COX2) activity, prostaglandin E2 (PGE2) and IL-10, and decreased B cell and T cell activity. Direct immunosuppressive effect may lead to apoptosis. The release of immunosuppressant molecules leads to a decreased response to IL-2, the downregulation of NF-κB and the upregulation of FOXP3, which results in increased regulatory T cell (Treg) activity (Smith et al. 2013). In the last decades, engineered NPs have found wide application in medicine, as diagnostic and therapeutic tools. On the one hand, they were used to avoid unwanted immune system activation, which could contribute to chronic inflammation, autoimmunity, or allergy. On the other hand, exploiting the ability of NP to interact with innate immunity, they were used toward an intentional enhancement or suppression of immune reactions to treat a range of disease (Italiani et al. 2020).

6   Nanoparticle-Mediated Cytoplasmic Immune Response

Circular RNAs (circRNAs) are covalently closed loop structures that lack free 3′ and 5′ ends (Chen and Yang 2015). CircRNAs can be categorized as exonic (ecircRNA), exon-intron (EIcircRNA), or intronic (ciRNA) circRNAs. The majority of circRNAs are ecircRNAs, which are predominantly located in the cytoplasm (Jeck et al. 2013). If NPs cannot enter the nucleus, their effects are limited to targeting circRNAs in the cytoplasm. However, this is not a major issue since most circRNAs exert their functions in the cytoplasm (Jeck et al. 2013). Lipid NPs (LNPs) efficiently deliver circRNA to tumors and a fourfold increase in circRNA transfection occurs in lung cancer. LNPs loaded with circRNA encoding IL-12, induce a robust immune response in cancer tissue, leading to marked tumor regression. The immunological profile of treated tumors indicates LNPs modulate the tumor microenvironment favorably with the substantial increments in cluster of differentiation (CD)45+ leukocytes and CD8+ T cells (Xu et al. 2024). During the industrial and biomedical utilization, exposure to the silver (Ag) NPs or graphene oxide silver (GO-Ag) NPs increases the generation of ROS, causes DNA damage and alters the expression of whole transcriptome including mRNA, miRNA, transfer RNA (tRNA), lncRNA, circRNA. 57 circRNAs, 75 lncRNAs, and 444 mRNAs are upregulated while 35 circRNAs, 21 lncRNAs, and 186 mRNAs are downregulated. These differentially expressed genes mainly involve in the transcriptional mis-regulation and induce toxicity through oxidative damage (Yuan et al. 2023). Canonical small interfering RNA (siRNA) duplexes are potent activators of the mammalian innate immune system. siRNA-based drugs represent an undesirable side effect due to the considerable toxicities associated with excessive cytokine release in humans (Robbins et al. 2009). Gold NPs (AuNPs) covered with polyethylene glycol (PEG) are more durable because PEG prevents them from aggregation and thus the passage of these NPs through different biological barriers is facilitated. PEGylated AuNPs as potential carriers for therapeutical siRNA duplexes directed towards the apolipoprotein E gene (ApoE), which is related to the onset of Alzheimer's disease may be a genetic risk factor (Okła et al. 2023). Different amount of PEG on the particle surface may change its characteristics and mode of action. Although larger molecular weight PEG decreases cellular internalization, it improves cytoplasmic bioavailability due to increased intracellular unpackaging and endosomal release (Miteva et al. 2015). Introducing synthetic siRNAs duplexes into the cytoplasm results in specific degradation of complementary mRNA via a process called RNA interference (RNAi). Using NPs or exosomes as delivery systems for these molecules can improve their stability, intracellular entry, and immunogenicity (Kulkarni et al. 2019). CircRNAs are typically overexpressed using expression plasmids and knocked down using RNAi-based strategies. RNAi molecules have many limitations, including their instability, lack of cell-specificity, low intracellular entry, immune system activation, and other off-target effects (Singh et al. 2011).

Direct damage to immune cells by NMs leads to apoptosis and necrosis, while interactions of NMs with the immune response itself can change immune-specific signaling pathways, resulting in changes in immune cell function measured by expression of surface markers, cytokine production, cell differentiation and immune activation (Hartung and Corsini 2013). The key elements in the identification of NM-induced immunotoxicity are duration and deregulation of the inflammatory response. Due to their small size, NMs may escape the particle-clearing defensive mechanisms (e.g. phagocytosis), and so many of them do not trigger a direct inflammatory response (Kunzmann et al. 2011). The increased aggregation that tends to occur on contact with the biological environment may cause effective clearing or sequestering by immune cell which tends to recognize preferentially larger particles (>0.5 μm) (Geiser 2010). The reactions of self-proteins with NMs and their persistence in the organism can cause autoimmune reactions. Another factor contributing to autoimmunity is the defective clearance of apoptotic cells by scavenger phagocytes (Muñoz et al. 2010).

7 Nano-Immunotherapy

Epigenetic mechanisms play vital roles not only in cancer initiation and progression, but also in the activation, differentiation and effector function(s) of immune cells. Epigenomic dynamics in immune cells not only influence immune cell fate and functionality but also affect the immunogenicity of cancer cells. Epigenetic alterations contribute to carcinogenesis by impacting multiple oncogenic vs. tumor suppressor gene pathways (Dai et al. 2021). Some epigenetic changes occur early in development, preceding the onset of tumor development (Licchesi et al. 2008; Okamoto et al. 1997; Simpson et al. 2004). Indeed, a recent study showed that tissue environment-induced epigenetic programming initiates tumorigenesis (Alonso-Curbelo et al. 2021), with Feinberg and others proposing an epigenetic progenitor origin for human cancer. Cancer epigenetics involve three types of genes: 'epigenetic mediators', corresponding to the tumor progenitor genes as suggested earlier; 'epigenetic modifiers' of the mediators, which are frequently mutated in cancer; and 'epigenetic modulators' upstream of the modifiers, which are responsive to changes in the cellular environment and often linked to the nuclear architecture (Feinberg et al. 2006, 2016). Although epigenetic markers modify chromatin structure and regulate expression of downstream genes, aberrant epigenetic modifications are common events in human disease including tumorigenesis and autoimmunity. Modulators of gene expression, miRNAs have been characterized as tumor suppressors or oncogenes in cancer. Distinct miRNAs are directly regulated by DNA methylation and histone modifications at their promoters. Moreover, miRNAs themselves are key participants in regulating the chromatin modifying machinery (Saito et al. 2014). B lymphocytes are generally recognized as the essential component of humoral immunity and also a regulator of innate immunity. Epigenetic factors, such as DNA methylation, histone modification, and non-coding RNA, play critical roles in establishing B cell lineage-specific gene expression profiles to define and sustain B cell identity and function (Bao and Cao 2016). Autophagy plays a direct role in eliminating invading pathogens by phagocytic processes (Kwon and Song 2018), as well as microtubule-associated protein 1A/1B-light chain 3 (LC3)-associated phagocytosis (LAP) and sequestosome-like receptor recruitment. LC3 is conjugated to phagosome membranes using a portion of the canonical autophagy machinery. LAP may regulate immune function, perhaps through the metabolic reprogramming and polarization of macrophages (Heckmann et al. 2017). Autophagy also limits excessive inflammation during pathogen control by: removing residual microbial debris, known to activate the inflammasome pathway; digesting dysfunctional mitochondria, which typically mediate production of ROS; or through direct removal of inflammasome complexes (Takahama et al. 2018). Regulation is multifactorial and includes a number of epigenetic pathways which can involve modification of DNA-binding histones to induce autophagy-related mRNA synthesis or miRNA and decapping-associated mRNA degradation which results in autophagy suppression (Hargarten and Williamson 2018).

In nanomedicine, innate immune system-NPs interaction is a crucial issue for two reasons: (1) to avoid an immune reaction against NP that could provoke damage in the body; (2) to avoid immune surveillance and clearance of NPs, which are used as diagnostic or therapeutic tools, by the immune cells. NPs can be used to modulate innate memory responses and act as immunosuppressors in diseases with exacerbated immune response/inflammation or as immunostimulants in diseases with reduced/compromised immune responsiveness (Dobrovolskaia et al. 2016; Italiani et al. 2020; Muhammad et al. 2020). Thus, nano-immunotherapy can be applied via three different ways; (1) to target cancer cells, (2) to target the tumor immune microenvironment, and (3) to target the peripheral immune system. When targeting cancer cells, nanodrugs typically aim to induce immunogenic cell death, thereby triggering the release of tumor antigens and DAMPs. Nanodrugs targeting the tumor immune microenvironment potentiate cancer immunotherapy by inhibiting immunosuppressive cells, such as increasing the cytotoxic T cells, as well as by reducing the expression of immunosuppressive molecules. Furthermore, immunomodulatory nanodrugs can restore the functions of the peripheral immune system via potentiating antigen presentation and by engineering T cells (Shi and Lammers 2019). Chitosan hydrogel loaded with lipid-immune regulatory factor 5 (IRF5) mRNA/C-C chemokine ligand 5 (CCL5) siRNA (LPR) NP complexes (LPR@CHG) reprogram the antitumoral immune niche. The LPR@CHG hydrogel upregulates IRF5 and downregulates CCL5 secretion, which cause to a significant increase in M1 phenotype macrophages. Thus, tumor growth is controlled by effective M1 phenotype macrophage that initiate T cell-mediated immune responses (Gao et al. 2022). Tumor exosome-based NPs are a promising and effective drug delivery platform (Yong et al. 2019). A biomimetic NP is created by hybridizing exosomes and liposomes via ultrasound, membrane fusion, and extrusion. On the one hand, miR497 effectively overcome drug resistance in cancer cells (Wang et al. 2021). On the other hand, a transmembrane protein, CD47 is highly expressed in tumor-derived exosomes (Pan et al. 2021). It is observed that hybrid NPs formed by the fusion of CD47-expressing tumor exosomes containing miR497 and triptolide, which suppresses the growth of chemotherapy-resistant cell lines, possess a superior killing effect on cisplatin-resistant cell lines. They are efficiently taken up by tumor cells, thus significantly enhance tumor cell apoptosis (Li et al. 2022). In addition, triptolide modulate the polarization of M2 macrophages to M1 macrophages to assist in the reversal of tumor resistance (Jiang et al. 2021). Gene delivery NP platform enhances the interactions between tumor cells and cytotoxic lymphocytes via reprogramming the tumor microenvironment. Thereby, intratumoral injection of NPs encapsulating mRNA encoding immunostimulatory agents has great translational potential as an immunotherapeutic. The reprogramming NP gel synergizes with immune checkpoint blockade to induce tumor regression and clearance in addition to resistance to tumor rechallenge at a distant site (Neshat et al. 2023).

8 Nanoparticle-Mediated Cell Death

Different NPs contribute to cell death through diverse mechanisms. NP-mitochondria interaction generates enormous levels of ROS. The enhanced levels of intracellular ROS affect not only various intracellular organelles, such as mitochondria and lysosomes, but also show a significant impact on the nuclei. NPs can activate the reduced nicotinamide adenine dinucleotide phosphate (NADPH) oxidases and thus NADPH oxidases-derived ROS is generated, the nuclear translocation is triggered (Zhang et al. 2015), afterward the B-cell lymphoma 2 (Bcl-2)/ Bcl-2–associated X protein (Bax) ratio is reduced, as well as mitochondrial caspase-3-9 is stimulated. The NP induced oxidative attack on DNA is a predominant cause of DNA damage, which results in intracellular toxicity such as deletion or addition of base pairs (Liu et al. 2020; Schins 2002). Plasmolysis, cytosolic and nuclear condensation, as well as subcellular organelle swelling, due to NP-mediated oxidative stress, ultimately leads to cell death (Chen et al. 2024). With the regulation of autophagy-related genes, cell autophagy refers to the process of cells using lysosomes to degrade damaged organelles as well as macromolecules, which are essential for maintaining the stability of cells. Both excessive and low degrees of autophagy may contribute to cell damage, causing autophagic, apoptotic, necrotic, or other kinds of cell death (Chen et al. 2023). Intraperitoneally injected NPs cause apoptotic cell death due to downregulation of autophagy (Lee et al. 2013). DNA methylation was an intimate participant in the process of cell autophagy (Shu et al. 2023). The promotion of cell autophagy in triple-negative breast cancer is associated with DNA hypermethylation due to an increase in DNMT1 (Wong 2021). DNMT1 hypermethylation may enhance the level of autophagy (Yang et al. 2018).

9 A Novel Endpoint in Toxicity Testing of Nanomaterials—Epigenetics

Epigenetics refers to heritable changes in gene expression that occur without alterations in DNA sequence. The epigenetic landscape depends on the interplay among three basic mechanisms—DNA methylation, histone modifications and RNA-mediated post-transcriptional regulation. Indirectly, NMs can influence DNA methylation by their pro-oxidative properties. Oxidatively damaged DNA can affect the ability of DNA methyltransferases, a family of enzymes that catalyze the transfer of a methyl group to DNA, to interact with DNA (Simko et al. 2011). Down- or up-regulation of epigenetic modifiers (enzymes that catalyze epigenetic modifications) were repeatedly found after NM-exposure (Choudhury et al. 2017; Ma et al. 2016; Patil et al. 2016; Qian et al. 2015; Sule et al. 2008). These changes can cause large-scale disruptions of DNA methylation and histone modification patterns. However, there is still a lack of information about NM-induced modifications to histone proteins (Choi et al. 2008). Several NMs exhibit epigenetic effects in terms of deregulation of miRNA expression profiles (Alinovi et al. 2017; Balansky et al. 2013; Chew et al. 2012; Eom et al. 2014; Li et al. 2011; Sun et al. 2015). Because of the scanty of human epidemiological data, epigenetic toxicity and potential health risk of NPs remains controversial. Development of new testing methods will provide the opportunity to distinguish between adverse health effects of NM exposure, and adaptive changes. It is recommended that diverse modes of action are involved in direct immunotoxicity and a set of pathways or genes, rather than one single gene, should be used to screen compounds for immunotoxicity and genotoxicity (Hartung and Corsini 2013; Shao et al. 2013). As immunotoxicity and genotoxicity share, partially, the same adverse outcome pathways, toxicogenomics is a novel approach to clarifying immunotoxicity and genotoxicity using many genes to assess adverse outcome pathways (Hochstenbach et al. 2010).

10   Conclusion

Epigenetic alterations are potential translational biomarkers to evaluate the adverse effects of NMs and NPs in humans. In fact, the innate immune system is a protective mechanism based on epigenetic reprogramming that controls innate/inflammatory responses to repeated stimuli, limiting the side effects of tissue damage and providing effective protection. However, it is fact that various nanoparticles participate in cell death through various mechanisms, primarily by increasing intracellular oxidative stress. It is presumed that engineered NPs may modify innate memory based on their ability to reconvert epigenetic modulation of gene expression. It is not fully elucidated whether NPs exposure directly affects the epigenome or whether the observed exposure-associated changes are due to cell toxicity (Pogribna and Hammons 2021). Future well-designed and well-controlled studies are needed to better understand the mechanisms and processes involved in epigenotoxic, and immunotoxic alterations, which are induced by NPs.

References

Accomasso L, Cristallini C, Giachino C (2018) Risk assessment and risk minimization in Nanomedicine: a need for predictive, alternative, and 3Rs strategies. Front Pharmacol 9:228. https://doi.org/10.3389/fphar.2018.00228

Adjei IM, Sharma B, Labhasetwar V (2014) Nanoparticles: cellular uptake and cytotoxicity. Adv Exp Med Biol 811:73–91. https://doi.org/10.1007/978-94-017-8739-0_5

Alinovi R, Goldoni M, Pinelli S, Ravanetti F, Galetti M, Pelosi G, De Palma G, Apostoli P, Cacchioli A, Mutti A, Mozzoni P (2017) Titanium dioxide aggregating nanoparticles induce autophagy and under-expression of microRNA 21 and 30a in A549 cell line: a comparative study with cobalt(II, III) oxide nanoparticles. Toxicol In Vitro 42:76–85. https://doi.org/10.1016/j.tiv.2017.04.007

Alonso-Curbelo D, Ho Y-J, Burdziak C, Maag JLV, Morris JP, Chandwani R, Chen H-A, Tsanov KM, Barriga FM, Luan W, Tasdemir N, Livshits G, Azizi E, Chun J, Wilkinson JE, Mazutis L, Leach SD, Koche R, Pe'er D, Lowe SW (2021) A gene-environment-induced epigenetic program initiates tumorigenesis. Nature 590:642–648. https://doi.org/10.1038/s41586-020-03147-x

Ameh T, Sayes CM (2019) The potential exposure and hazards of copper nanoparticles: a review. Environ Toxicol Pharmacol 71:103220. https://doi.org/10.1016/j.etap.2019.103220

Andersson J, Ekdahl KN, Lambris JD, Nilsson B (2005) Binding of C3 fragments on top of adsorbed plasma proteins during complement activation on a model biomaterial surface. Biomaterials 26:1477–1485. https://doi.org/10.1016/j.biomaterials.2004.05.011

Arita A, Costa M (2009) Epigenetics in metal carcinogenesis: nickel, arsenic, chromium and cadmium. Metallomics 1:222–228. https://doi.org/10.1039/b903049b

Balansky R, Longobardi M, Ganchev G, Iltcheva M, Nedyalkov N, Atanasov P, Toshkova R, De Flora S, Izzotti A (2013) Transplacental clastogenic and epigenetic effects of gold nanoparticles in mice. Mutat Res 751–752:42–48. https://doi.org/10.1016/j.mrfmmm.2013.08.006

Balasubramanian S, Gunasekaran K, Sasidharan S, Jeyamanickavel Mathan V, Perumal E (2020) MicroRNAs and xenobiotic toxicity: an overview. Toxicol Rep 7:583–595. https://doi.org/10.1016/j.toxrep.2020.04.010

Bao Y, Cao X (2016) Epigenetic control of B cell development and B-cell-related immune disorders. Clin Rev Allergy Immunol 50:301–311. https://doi.org/10.1007/s12016-015-8494-7

Bhattacharya K, Andón FT, El-Sayed R, Fadeel B (2013) Mechanisms of carbon nanotube-induced toxicity: focus on pulmonary inflammation. Adv Drug Deliv Rev 65:2087–2097. https://doi.org/10.1016/j.addr.2013.05.012

Bohnsack MT, Czaplinski K, Gorlich D (2004) Exportin 5 is a RanGTP-dependent dsRNA-binding protein that mediates nuclear export of pre-miRNAs. RNA 10:185–191. https://doi.org/10.1261/rna.5167604

Boraschi D, Fadeel B, Duschl A (2017a) Chapter 13: Immune system adverse effects of engineered nanomaterials. In: Adverse effects of engineered nanomaterials exposure, toxicology, and impact on human health. Elsevier, pp 313–337

Boraschi D, Italiani P, Palomba R, Decuzzi P, Duschl A, Fadeel B, Moghimi SM (2017b) Nanoparticles and innate immunity: new perspectives on host defence. Semin Immunol 34:33–51. https://doi.org/10.1016/j.smim.2017.08.013

Bose DA, Donahue G, Reinberg D, Shiekhattar R, Bonasio R, Berger SL (2017) RNA binding to CBP stimulates histone acetylation and transcription. Cell 168:135–149.e22. https://doi.org/10.1016/j.cell.2016.12.020

Brown TA, Lee JW, Holian A, Porter V, Fredriksen H, Kim M, Cho YH (2016) Alterations in DNA methylation corresponding with lung inflammation and as a biomarker for disease development after MWCNT exposure. Nanotoxicology 10:453–461. https://doi.org/10.3109/17435390.2015.1078852

Chen L-L, Yang L (2015) Regulation of circRNA biogenesis. RNA Biol 12:381–388. https://doi.org/10.1080/15476286.2015.1020271

Chen X, Barozzi I, Termanini A, Prosperini E, Recchiuti A, Dalli J, Mietton F, Matteoli G, Hiebert S, Natoli G (2012) Requirement for the histone deacetylase Hdac3 for the inflammatory gene expression program in macrophages. Proc Natl Acad Sci USA 109:E2865–E2874. https://doi.org/10.1073/pnas.1121131109

Chen C-H, Wang C-Z, Wang Y-H, Liao W-T, Chen Y-J, Kuo C-H, Kuo H-F, Hung C-H (2014) Effects of low-level laser therapy on M1-related cytokine expression in monocytes via histone modification. Mediat Inflamm 2014:625048. https://doi.org/10.1155/2014/625048

Chen J, Zheng D, Cai Z, Zhong B, Zhang H, Pan Z, Ling X, Han Y, Meng J, Li H, Chen X, Zhang H, Liu L (2023) Increased DNMT1 involvement in the activation of LO2 cell death induced by silver nanoparticles via promoting TFEB-dependent autophagy. Toxics 11:751. https://doi.org/10.3390/toxics11090751

Chen Z, Feng Y, Guo Z, Han M, Yan X (2024) Zinc oxide nanoparticles alleviate cadmium toxicity and promote tolerance by modulating programmed cell death in alfalfa (Medicago sativa L.). J Hazard Mater 469:133917. https://doi.org/10.1016/j.jhazmat.2024.133917

Cheng T-F, Choudhuri S, Muldoon-Jacobs K (2012) Epigenetic targets of some toxicologically relevant metals: a review of the literature. J Appl Toxicol 32:643–653. https://doi.org/10.1002/jat.2717

Chew W-S, Poh K-W, Siddiqi NJ, Alhomida AS, Yu LE, Ong W-Y (2012) Short- and long-term changes in blood miRNA levels after nanogold injection in rats—potential biomarkers of nanoparticle exposure. Biomarkers 17:750–757. https://doi.org/10.3109/1354750X.2012.727030

Choi AO, Brown SE, Szyf M, Maysinger D (2008) Quantum dot-induced epigenetic and genotoxic changes in human breast cancer cells. J Mol Med (Berl) 86:291–302. https://doi.org/10.1007/s00109-007-0274-2

Choudhury SR, Ordaz J, Lo C-L, Damayanti NP, Zhou F, Irudayaraj J (2017) From the cover: zinc oxide nanoparticles-induced reactive oxygen species promotes multimodal Cyto- and epigenetic toxicity. Toxicol Sci 156:261–274. https://doi.org/10.1093/toxsci/kfw252

Corbo C, Molinaro R, Parodi A, Toledano Furman NE, Salvatore F, Tasciotti E (2016) The impact of nanoparticle protein corona on cytotoxicity, immunotoxicity and target drug delivery. Nanomedicine (Lond) 11:81–100. https://doi.org/10.2217/nnm.15.188

Dai E, Zhu Z, Wahed S, Qu Z, Storkus WJ, Guo ZS (2021) Epigenetic modulation of antitumor immunity for improved cancer immunotherapy. Mol Cancer 20:171. https://doi.org/10.1186/s12943-021-01464-x

de Rie D, Abugessaisa I, Alam T, Arner E, Arner P, Ashoor H, Åström G, Babina M, Bertin N, Burroughs AM, Carlisle AJ, Daub CO, Detmar M, Deviatiiarov R, Fort A, Gebhard C, Goldowitz D, Guhl S, Ha TJ, Harshbarger J, Hasegawa A, Hashimoto K, Herlyn M, Heutink P, Hitchens KJ, Hon CC, Huang E, Ishizu Y, Kai C, Kasukawa T, Klinken P, Lassmann T, Lecellier C-H, Lee W, Lizio M, Makeev V, Mathelier A, Medvedeva YA, Mejhert N, Mungall CJ, Noma S, Ohshima M, Okada-Hatakeyama M, Persson H, Rizzu P, Roudnicky F, Sætrom P, Sato H, Severin J, Shin JW, Swoboda RK, Tarui H, Toyoda H, Vitting-Seerup K, Winteringham L, Yamaguchi Y, Yasuzawa K, Yoneda M, Yumoto N, Zabierowski S, Zhang PG, Wells CA, Summers KM, Kawaji H, Sandelin A, Rehli M, FANTOM Consortium, Hayashizaki Y, Carninci P, Forrest ARR, de Hoon MJL (2017) An integrated expression atlas of miRNAs and their promoters in human and mouse. Nat Biotechnol 35:872–878. https://doi.org/10.1038/nbt.3947

DeMarino C, Schwab A, Pleet M, Mathiesen A, Friedman J, El-Hage N, Kashanchi F (2017) Biodegradable nanoparticles for delivery of therapeutics in CNS infection. J Neuroimmune Pharmacol 12:31–50. https://doi.org/10.1007/s11481-016-9692-7

Deng ZJ, Liang M, Monteiro M, Toth I, Minchin RF (2011) Nanoparticle-induced unfolding of fibrinogen promotes mac-1 receptor activation and inflammation. Nat Nanotechnol 6:39–44. https://doi.org/10.1038/nnano.2010.250

Deng ZJ, Liang M, Toth I, Monteiro MJ, Minchin RF (2012) Molecular interaction of poly(acrylic acid) gold nanoparticles with human fibrinogen. ACS Nano 6:8962–8969. https://doi.org/10.1021/nn3029953

Dharap A, Pokrzywa C, Murali S, Pandi G, Vemuganti R (2013) MicroRNA miR-324-3p induces promoter-mediated expression of RelA gene. PLoS One 8:e79467. https://doi.org/10.1371/journal.pone.0079467

Di Gioacchino M, Petrarca C, Lazzarin F, Di Giampaolo L, Sabbioni E, Boscolo P, Mariani-Costantini R, Bernardini G (2011) Immunotoxicity of nanoparticles. Int J Immunopathol Pharmacol 24:65S–71S

Dobrovolskaia MA, McNeil SE (2007) Immunological properties of engineered nanomaterials. Nat Nanotechnol 2:469–478. https://doi.org/10.1038/nnano.2007.223

Dobrovolskaia MA, Aggarwal P, Hall JB, McNeil SE (2008) Preclinical studies to understand nanoparticle interaction with the immune system and its potential effects on nanoparticle biodistribution. Mol Pharm 5:487–495. https://doi.org/10.1021/mp800032f

Dobrovolskaia MA, Shurin M, Shvedova AA (2016) Current understanding of interactions between nanoparticles and the immune system. Toxicol Appl Pharmacol 299:78–89. https://doi.org/10.1016/j.taap.2015.12.022

Domínguez-Andrés J, Fanucchi S, Joosten LAB, Mhlanga MM, Netea MG (2020) Advances in understanding molecular regulation of innate immune memory. Curr Opin Cell Biol 63:68–75. https://doi.org/10.1016/j.ceb.2019.12.006

Dusinska M, Tulinska J, El Yamani N, Kuricova M, Liskova A, Rollerova E, Rundén-Pran E, Smolkova B (2017) Immunotoxicity, genotoxicity and epigenetic toxicity of nanomaterials: new strategies for toxicity testing? Food Chem Toxicol 109:797–811. https://doi.org/10.1016/j.fct.2017.08.030

Edwards JR, O'Donnell AH, Rollins RA, Peckham HE, Lee C, Milekic MH, Chanrion B, Fu Y, Su T, Hibshoosh H, Gingrich JA, Haghighi F, Nutter R, Bestor TH (2010) Chromatin and sequence features that define the fine and gross structure of genomic methylation patterns. Genome Res 20:972–980. https://doi.org/10.1101/gr.101535.109

Edwards JR, Yarychkivska O, Boulard M, Bestor TH (2017) DNA methylation and DNA methyltransferases. Epigenetics Chromatin 10:23. https://doi.org/10.1186/s13072-017-0130-8

Engin AB, Nikitovic D, Neagu M, Henrich-Noack P, Docea AO, Shtilman MI, Golokhvast K, Tsatsakis AM (2017) Mechanistic understanding of nanoparticles' interactions with extracellular matrix: the cell and immune system. Part Fibre Toxicol 14:22. https://doi.org/10.1186/s12989-017-0199-z

Eom H-J, Chatterjee N, Lee J, Choi J (2014) Integrated mRNA and micro RNA profiling reveals epigenetic mechanism of differential sensitivity of Jurkat T cells to AgNPs and ag ions. Toxicol Lett 229:311–318. https://doi.org/10.1016/j.toxlet.2014.05.019

Fanucchi S, Mhlanga MM (2019) Lnc-ing trained immunity to chromatin architecture. Front Cell Dev Biol 7:2. https://doi.org/10.3389/fcell.2019.00002

Feinberg AP, Ohlsson R, Henikoff S (2006) The epigenetic progenitor origin of human cancer. Nat Rev Genet 7:21–33. https://doi.org/10.1038/nrg1748

Feinberg AP, Koldobskiy MA, Göndör A (2016) Epigenetic modulators, modifiers and mediators in cancer aetiology and progression. Nat Rev Genet 17:284–299. https://doi.org/10.1038/nrg.2016.13

Figueroa EG, Caballero-Román A, Ticó JR, Miñarro M, Nardi-Ricart A, González-Candia A (2022) miRNA nanoencapsulation to regulate the programming of the blood-brain barrier permeability by hypoxia. Curr Res Pharmacol Drug Discov 3:100129. https://doi.org/10.1016/j.crphar.2022.100129

Fok ET, Davignon L, Fanucchi S, Mhlanga MM (2018) The lncRNA connection between cellular metabolism and epigenetics in trained immunity. Front Immunol 9:3184. https://doi.org/10.3389/fimmu.2018.03184

Gamberoni F, Borgese M, Pagiatakis C, Armenia I, Grazù V, Gornati R, Serio S, Papait R, Bernardini G (2023) Iron oxide nanoparticles with and without cobalt functionalization provoke changes in the transcription profile via epigenetic modulation of enhancer activity. Nano Lett 23:9151–9159. https://doi.org/10.1021/acs.nanolett.3c01967

Gao C, Cheng K, Li Y, Gong R, Zhao X, Nie G, Ren H (2022) Injectable immunotherapeutic hydrogel containing RNA-loaded lipid nanoparticles reshapes tumor microenvironment for pancreatic cancer therapy. Nano Lett 22:8801–8809. https://doi.org/10.1021/acs.nanolett.2c01994

Geiser M (2010) Update on macrophage clearance of inhaled micro- and nanoparticles. J Aerosol Med Pulm Drug Deliv 23:207–217. https://doi.org/10.1089/jamp.2009.0797

Geng Y, Dalhaimer P, Cai S, Tsai R, Tewari M, Minko T, Discher DE (2007) Shape effects of filaments versus spherical particles in flow and drug delivery. Nat Nanotechnol 2:249–255. https://doi.org/10.1038/nnano.2007.70

Godbey WT, Wu KK, Mikos AG (1999) Tracking the intracellular path of poly(ethylenimine)/DNA complexes for gene delivery. Proc Natl Acad Sci USA 96:5177–5181. https://doi.org/10.1073/pnas.96.9.5177

Guo H, Ingolia NT, Weissman JS, Bartel DP (2010) Mammalian microRNAs predominantly act to decrease target mRNA levels. Nature 466:835–840. https://doi.org/10.1038/nature09267

Gustafson HH, Holt-Casper D, Grainger DW, Ghandehari H (2015) Nanoparticle uptake: the phagocyte problem. Nano Today 10:487–510. https://doi.org/10.1016/j.nantod.2015.06.006

Hargarten JC, Williamson PR (2018) Epigenetic regulation of autophagy: a path to the control of autoimmunity. Front Immunol 9:1864. https://doi.org/10.3389/fimmu.2018.01864

Hartung T, Corsini E (2013) Immunotoxicology: challenges in the 21st century and in vitro opportunities. ALTEX 30:411–426. https://doi.org/10.14573/altex.2013.4.411

He Y, Zhou J, Ma S, Nie Y, Yue D, Jiang Q, Wali ARM, Tang JZ, Gu Z (2016) Multi-responsive "turn-on" Nanocarriers for efficient site-specific gene delivery in vitro and in vivo. Adv Healthc Mater 5:2799–2812. https://doi.org/10.1002/adhm.201600710

Heckmann BL, Boada-Romero E, Cunha LD, Magne J, Green DR (2017) LC3-associated phagocytosis and inflammation. J Mol Biol 429:3561–3576. https://doi.org/10.1016/j.jmb.2017.08.012

Heinrich J (2011) Influence of indoor factors in dwellings on the development of childhood asthma. Int J Hyg Environ Health 214:1–25. https://doi.org/10.1016/j.ijheh.2010.08.009

Hochstenbach K, van Leeuwen DM, Gmuender H, Stølevik SB, Nygaard UC, Løvik M, Granum B, Namork E, van Delft JHM, van Loveren H (2010) Transcriptomic profile indicative of immunotoxic exposure: in vitro studies in peripheral blood mononuclear cells. Toxicol Sci 118:19–30. https://doi.org/10.1093/toxsci/kfq239

Hoppstädter J, Seif M, Dembek A, Cavelius C, Huwer H, Kraegeloh A, Kiemer AK (2015) M2 polarization enhances silica nanoparticle uptake by macrophages. Front Pharmacol 6:55. https://doi.org/10.3389/fphar.2015.00055

Hulander M, Hong J, Andersson M, Gervén F, Ohrlander M, Tengvall P, Elwing H (2009) Blood interactions with noble metals: coagulation and immune complement activation. ACS Appl Mater Interfaces 1:1053–1062. https://doi.org/10.1021/am900028e

Hulander M, Lundgren A, Berglin M, Ohrlander M, Lausmaa J, Elwing H (2011) Immune complement activation is attenuated by surface nanotopography. Int J Nanomedicine 6:2653–2666. https://doi.org/10.2147/IJN.S24578

Huo S, Jin S, Ma X, Xue X, Yang K, Kumar A, Wang PC, Zhang J, Hu Z, Liang X-J (2014) Ultrasmall gold nanoparticles as carriers for nucleus-based gene therapy due to size-dependent nuclear entry. ACS Nano 8:5852–5862. https://doi.org/10.1021/nn5008572

Hussain S, Vanoirbeek JAJ, Hoet PHM (2012) Interactions of nanomaterials with the immune system. Wiley Interdiscip Rev Nanomed Nanobiotechnol 4:169–183. https://doi.org/10.1002/wnan.166

Hussain S, Garantziotis S, Rodrigues-Lima F, Dupret J-M, Baeza-Squiban A, Boland S (2014) Intracellular signal modulation by nanomaterials. Adv Exp Med Biol 811:111–134. https://doi.org/10.1007/978-94-017-8739-0_7

Iorio MV, Piovan C, Croce CM (2010) Interplay between microRNAs and the epigenetic machinery: an intricate network. Biochim Biophys Acta 1799:694–701. https://doi.org/10.1016/j.bbagrm.2010.05.005

Italiani P, Della Camera G, Boraschi D (2020) Induction of innate immune memory by engineered nanoparticles in monocytes/macrophages: from hypothesis to reality. Front Immunol 11:566309. https://doi.org/10.3389/fimmu.2020.566309

Jeck WR, Sorrentino JA, Wang K, Slevin MK, Burd CE, Liu J, Marzluff WF, Sharpless NE (2013) Circular RNAs are abundant, conserved, and associated with ALU repeats. RNA 19:141–157. https://doi.org/10.1261/rna.035667.112

Jiang Z, Lai Y, Beaver JM, Tsegay PS, Zhao M-L, Horton JK, Zamora M, Rein HL, Miralles F, Shaver M, Hutcheson JD, Agoulnik I, Wilson SH, Liu Y (2020) Oxidative DNA damage modulates DNA methylation pattern in human breast cancer 1 (BRCA1) gene via the crosstalk between DNA polymerase β and a de novo DNA Methyltransferase. Cells 9:225. https://doi.org/10.3390/cells9010225

Jiang X, Cao G, Gao G, Wang W, Zhao J, Gao C (2021) Triptolide decreases tumor-associated macrophages infiltration and M2 polarization to remodel colon cancer immune microenvironment via inhibiting tumor-derived CXCL12. J Cell Physiol 236:193–204. https://doi.org/10.1002/jcp.29833

Kagan VE, Konduru NV, Feng W, Allen BL, Conroy J, Volkov Y, Vlasova II, Belikova NA, Yanamala N, Kapralov A, Tyurina YY, Shi J, Kisin ER, Murray AR, Franks J, Stolz D, Gou P, Klein-Seetharaman J, Fadeel B, Star A, Shvedova AA (2010) Carbon nanotubes degraded by neutrophil myeloperoxidase induce less pulmonary inflammation. Nat Nanotechnol 5:354–359. https://doi.org/10.1038/nnano.2010.44

Kamaly N, Xiao Z, Valencia PM, Radovic-Moreno AF, Farokhzad OC (2012) Targeted polymeric therapeutic nanoparticles: design, development and clinical translation. Chem Soc Rev 41:2971–3010. https://doi.org/10.1039/c2cs15344k

Ketting RF, Fischer SE, Bernstein E, Sijen T, Hannon GJ, Plasterk RH (2001) Dicer functions in RNA interference and in synthesis of small RNA involved in developmental timing in C. Elegans. Genes Dev 15:2654–2659. https://doi.org/10.1101/gad.927801

Kittan NA, Allen RM, Dhaliwal A, Cavassani KA, Schaller M, Gallagher KA, Carson WF, Mukherjee S, Grembecka J, Cierpicki T, Jarai G, Westwick J, Kunkel SL, Hogaboam CM (2013) Cytokine induced phenotypic and epigenetic signatures are key to establishing specific macrophage phenotypes. PLoS One 8:e78045. https://doi.org/10.1371/journal.pone.0078045

Kobayashi H, Tomari Y (2016) RISC assembly: coordination between small RNAs and Argonaute proteins. Biochim Biophys Acta 1859:71–81. https://doi.org/10.1016/j.bbagrm.2015.08.007

Kulkarni JA, Witzigmann D, Chen S, Cullis PR, van der Meel R (2019) Lipid nanoparticle Technology for Clinical Translation of siRNA therapeutics. Acc Chem Res 52:2435–2444. https://doi.org/10.1021/acs.accounts.9b00368

Kumar S, Meena R, Paulraj R (2016) Role of macrophage (M1 and M2) in titanium-dioxide nanoparticle-induced oxidative stress and inflammatory response in rat. Appl Biochem Biotechnol 180:1257–1275. https://doi.org/10.1007/s12010-016-2165-x

Kunzmann A, Andersson B, Vogt C, Feliu N, Ye F, Gabrielsson S, Toprak MS, Buerki-Thurnherr T, Laurent S, Vahter M, Krug H, Muhammed M, Scheynius A, Fadeel B (2011) Efficient internalization of silica-coated iron oxide nanoparticles of different sizes by primary human macrophages and dendritic cells. Toxicol Appl Pharmacol 253:81–93. https://doi.org/10.1016/j.taap.2011.03.011

Kwon DH, Song HK (2018) A structural view of Xenophagy, a battle between host and microbes. Mol Cells 41:27–34. https://doi.org/10.14348/molcells.2018.2274

Kwon N-H, Kim J-S, Lee J-Y, Oh M-J, Choi D-C (2008) DNA methylation and the expression of IL-4 and IFN-gamma promoter genes in patients with bronchial asthma. J Clin Immunol 28:139–146. https://doi.org/10.1007/s10875-007-9148-1

Lai F, Orom UA, Cesaroni M, Beringer M, Taatjes DJ, Blobel GA, Shiekhattar R (2013) Activating RNAs associate with mediator to enhance chromatin architecture and transcription. Nature 494:497–501. https://doi.org/10.1038/nature11884

Lee Y, Ahn C, Han J, Choi H, Kim J, Yim J, Lee J, Provost P, Rådmark O, Kim S, Kim VN (2003) The nuclear RNase III Drosha initiates microRNA processing. Nature 425:415–419. https://doi.org/10.1038/nature01957

Lee T-Y, Liu M-S, Huang L-J, Lue S-I, Lin L-C, Kwan A-L, Yang R-C (2013) Bioenergetic failure correlates with autophagy and apoptosis in rat liver following silver nanoparticle intraperitoneal administration. Part Fibre Toxicol 10:40. https://doi.org/10.1186/1743-8977-10-40

Lee SWL, Paoletti C, Campisi M, Osaki T, Adriani G, Kamm RD, Mattu C, Chiono V (2019) MicroRNA delivery through nanoparticles. J Control Release 313:80–95. https://doi.org/10.1016/j.jconrel.2019.10.007

Li S, Wang Y, Wang H, Bai Y, Liang G, Wang Y, Huang N, Xiao Z (2011) MicroRNAs as participants in cytotoxicity of CdTe quantum dots in NIH/3T3 cells. Biomaterials 32:3807–3814. https://doi.org/10.1016/j.biomaterials.2011.01.074

Li L-M, Ruan G-X, HuangFu M-Y, Chen Z-L, Liu H-N, Li L-X, Hu Y-L, Han M, Davidson G, Levkin PA, Gao J-Q (2015a) ScreenFect a: an efficient and low toxic liposome for gene delivery to mesenchymal stem cells. Int J Pharm 488:1–11. https://doi.org/10.1016/j.ijpharm.2015.04.050

Li C, Zhang J, Zu Y-J, Nie S-F, Cao J, Wang Q, Nie S-P, Deng Z-Y, Xie M-Y, Wang S (2015b) Biocompatible and biodegradable nanoparticles for enhancement of anti-cancer activities of phytochemicals. Chin J Nat Med 13:641–652. https://doi.org/10.1016/S1875-5364(15)30061-3

Li L, He D, Guo Q, Zhang Z, Ru D, Wang L, Gong K, Liu F, Duan Y, Li H (2022) Exosome-liposome hybrid nanoparticle codelivery of TP and miR497 conspicuously overcomes chemoresistant ovarian cancer. J Nanobiotechnol 20:50. https://doi.org/10.1186/s12951-022-01264-5

Liang G, Zhu Y, Ali DJ, Tian T, Xu H, Si K, Sun B, Chen B, Xiao Z (2020) Engineered exosomes for targeted co-delivery of miR-21 inhibitor and chemotherapeutics to reverse drug resistance in colon cancer. J Nanobiotechnol 18:10. https://doi.org/10.1186/s12951-019-0563-2

Licchesi JDF, Westra WH, Hooker CM, Machida EO, Baylin SB, Herman JG (2008) Epigenetic alteration of Wnt pathway antagonists in progressive glandular neoplasia of the lung. Carcinogenesis 29:895–904. https://doi.org/10.1093/carcin/bgn017

Liou S-H, Wu W-T, Liao H-Y, Chen C-Y, Tsai C-Y, Jung W-T, Lee H-L (2017) Global DNA methylation and oxidative stress biomarkers in workers exposed to metal oxide nanoparticles. J Hazard Mater 331:329–335. https://doi.org/10.1016/j.jhazmat.2017.02.042

Liu C-G, Han Y-H, Kankala RK, Wang S-B, Chen A-Z (2020) Subcellular performance of nanoparticles in cancer therapy. Int J Nanomedicine 15:675–704. https://doi.org/10.2147/IJN.S226186

Ma Y, Fu H, Zhang C, Cheng S, Gao J, Wang Z, Jin W, Conde J, Cui D (2016) Chiral antioxidant-based gold nanoclusters reprogram DNA epigenetic patterns. Sci Rep 6:33436. https://doi.org/10.1038/srep33436

Missaoui WN, Arnold RD, Cummings BS (2018) Toxicological status of nanoparticles: what we know and what we don't know. Chem Biol Interact 295:1–12. https://doi.org/10.1016/j.cbi.2018.07.015

Miteva M, Kirkbride KC, Kilchrist KV, Werfel TA, Li H, Nelson CE, Gupta MK, Giorgio TD, Duvall CL (2015) Tuning PEGylation of mixed micelles to overcome intracellular and systemic siRNA delivery barriers. Biomaterials 38:97–107. https://doi.org/10.1016/j.biomaterials.2014.10.036

Moreira L, Costa C, Pires J, Teixeira JP, Fraga S (2021) How can exposure to engineered nanomaterials influence our epigenetic code? A review of the mechanisms and molecular targets. Mutat Res Rev Mutat Res 788:108385. https://doi.org/10.1016/j.mrrev.2021.108385

Muhammad Q, Jang Y, Kang SH, Moon J, Kim WJ, Park H (2020) Modulation of immune responses with nanoparticles and reduction of their immunotoxicity. Biomater Sci 8:1490–1501. https://doi.org/10.1039/c9bm01643k

Mullican SE, Gaddis CA, Alenghat T, Nair MG, Giacomin PR, Everett LJ, Feng D, Steger DJ, Schug J, Artis D, Lazar MA (2011) Histone deacetylase 3 is an epigenomic brake in macrophage alternative activation. Genes Dev 25:2480–2488. https://doi.org/10.1101/gad.175950.111

Muñoz LE, Lauber K, Schiller M, Manfredi AA, Herrmann M (2010) The role of defective clearance of apoptotic cells in systemic autoimmunity. Nat Rev Rheumatol 6:280–289. https://doi.org/10.1038/nrrheum.2010.46

Neagu M, Piperigkou Z, Karamanou K, Engin AB, Docea AO, Constantin C, Negrei C, Nikitovic D, Tsatsakis A (2017) Protein bio-corona: critical issue in immune nanotoxicology. Arch Toxicol 91:1031–1048. https://doi.org/10.1007/s00204-016-1797-5

Neshat SY, Chan CHR, Harris J, Zmily OM, Est-Witte S, Karlsson J, Shannon SR, Jain M, Doloff JC, Green JJ, Tzeng SY (2023) Polymeric nanoparticle gel for intracellular mRNA delivery and immunological reprogramming of tumors. Biomaterials 300:122185. https://doi.org/10.1016/j.biomaterials.2023.122185

Okamoto K, Morison IM, Taniguchi T, Reeve AE (1997) Epigenetic changes at the insulin-like growth factor II/H19 locus in developing kidney is an early event in Wilms tumorigenesis. Proc Natl Acad Sci USA 94:5367–5371. https://doi.org/10.1073/pnas.94.10.5367

Okła E, Białecki P, Kędzierska M, Pędziwiatr-Werbicka E, Miłowska K, Takvor S, Gómez R, de la Mata FJ, Bryszewska M, Ionov M (2023) Pegylated gold nanoparticles conjugated with siRNA: complexes formation and cytotoxicity. Int J Mol Sci 24:6638. https://doi.org/10.3390/ijms24076638

Öner D, Moisse M, Ghosh M, Duca RC, Poels K, Luyts K, Putzeys E, Cokic SM, Van Landuyt K, Vanoirbeek J, Lambrechts D, Godderis L, Hoet PHM (2017) Epigenetic effects of carbon nanotubes in human monocytic cells. Mutagenesis 32:181–191. https://doi.org/10.1093/mutage/gew053

Ooi SKT, O'Donnell AH, Bestor TH (2009) Mammalian cytosine methylation at a glance. J Cell Sci 122:2787–2791. https://doi.org/10.1242/jcs.015123

Pal R, Chakraborty B, Nath A, Singh LM, Ali M, Rahman DS, Ghosh SK, Basu A, Bhattacharya S, Baral R, Sengupta M (2016) Noble metal nanoparticle-induced oxidative stress modulates tumor associated macrophages (TAMs) from an M2 to M1 phenotype: an in vitro approach. Int Immunopharmacol 38:332–341. https://doi.org/10.1016/j.intimp.2016.06.006

Pan S, Zhang Y, Huang M, Deng Z, Zhang A, Pei L, Wang L, Zhao W, Ma L, Zhang Q, Cui D (2021) Urinary exosomes-based engineered Nanovectors for Homologously targeted chemo-Chemodynamic prostate cancer therapy via abrogating EGFR/AKT/NF-kB/IkB signaling. Biomaterials 275:120946. https://doi.org/10.1016/j.biomaterials.2021.120946

Papageorgiou I, Yin Z, Ladon D, Baird D, Lewis AC, Sood A, Newson R, Learmonth ID, Case CP (2007) Genotoxic effects of particles of surgical cobalt chrome alloy on human cells of different age in vitro. Mutat Res 619:45–58. https://doi.org/10.1016/j.mrfmmm.2007.01.008

Patil NA, Gade WN, Deobagkar DD (2016) Epigenetic modulation upon exposure of lung fibroblasts to TiO2 and ZnO nanoparticles: alterations in DNA methylation. Int J Nanomedicine 11:4509–4519. https://doi.org/10.2147/IJN.S110390

Pi Y-N, Xia B-R, Jin M-Z, Jin W-L, Lou G (2021) Exosomes: powerful weapon for cancer nano-immunoengineering. Biochem Pharmacol 186:114487. https://doi.org/10.1016/j.bcp.2021.114487

Pilger A, Rüdiger HW (2006) 8-Hydroxy-2′-deoxyguanosine as a marker of oxidative DNA damage related to occupational and environmental exposures. Int Arch Occup Environ Health 80:1–15. https://doi.org/10.1007/s00420-006-0106-7

Pogribna M, Hammons G (2021) Epigenetic effects of nanomaterials and nanoparticles. J Nanobiotechnol 19:2. https://doi.org/10.1186/s12951-020-00740-0

Pogribna M, Word B, Lyn-Cook B, Hammons G (2022) Effect of titanium dioxide nanoparticles on histone modifications and histone modifying enzymes expression in human cell lines. Nanotoxicology 16:409–424. https://doi.org/10.1080/17435390.2022.2085206

Pozzi D, Marchini C, Cardarelli F, Salomone F, Coppola S, Montani M, Zabaleta ME, Digman MA, Gratton E, Colapicchioni V, Caracciolo G (2014) Mechanistic evaluation of the transfection barriers involved in lipid-mediated gene delivery: interplay between nanostructure and composition. Biochim Biophys Acta 1838:957–967. https://doi.org/10.1016/j.bbamem.2013.11.014

Prabha S, Zhou W-Z, Panyam J, Labhasetwar V (2002) Size-dependency of nanoparticle-mediated gene transfection: studies with fractionated nanoparticles. Int J Pharm 244:105–115. https://doi.org/10.1016/s0378-5173(02)00315-0

Qian Y, Zhang J, Hu Q, Xu M, Chen Y, Hu G, Zhao M, Liu S (2015) Silver nanoparticle-induced hemoglobin decrease involves alteration of histone 3 methylation status. Biomaterials 70:12–22. https://doi.org/10.1016/j.biomaterials.2015.08.015

Qiu L, Chen T, Öçsoy I, Yasun E, Wu C, Zhu G, You M, Han D, Jiang J, Yu R, Tan W (2015) A cell-targeted, size-photocontrollable, nuclear-uptake nanodrug delivery system for drug-resistant cancer therapy. Nano Lett 15:457–463. https://doi.org/10.1021/nl503777s

Rabolli V, Lison D, Huaux F (2016) The complex cascade of cellular events governing inflammasome activation and IL-1β processing in response to inhaled particles. Part Fibre Toxicol 13:40. https://doi.org/10.1186/s12989-016-0150-8

Reddy ST, van der Vlies AJ, Simeoni E, Angeli V, Randolph GJ, O'Neil CP, Lee LK, Swartz MA, Hubbell JA (2007) Exploiting lymphatic transport and complement activation in nanoparticle vaccines. Nat Biotechnol 25:1159–1164. https://doi.org/10.1038/nbt1332

Robbins M, Judge A, MacLachlan I (2009) siRNA and innate immunity. Oligonucleotides 19:89–102. https://doi.org/10.1089/oli.2009.0180

Rossnerova A, Honkova K, Pelclova D, Zdimal V, Hubacek JA, Chvojkova I, Vrbova K, Rossner P, Topinka J, Vlckova S, Fenclova Z, Lischkova L, Klusackova P, Schwarz J, Ondracek J, Ondrackova L, Kostejn M, Klema J, Dvorackova S (2020) DNA methylation profiles in a group of workers occupationally exposed to nanoparticles. Int J Mol Sci 21:2420. https://doi.org/10.3390/ijms21072420

Runyon RS, Cachola LM, Rajeshuni N, Hunter T, Garcia M, Ahn R, Lurmann F, Krasnow R, Jack LM, Miller RL, Swan GE, Kohli A, Jacobson AC, Nadeau KC (2012) Asthma discordance in twins is linked to epigenetic modifications of T cells. PLoS One 7:e48796. https://doi.org/10.1371/journal.pone.0048796

Sacchetti C, Motamedchaboki K, Magrini A, Palmieri G, Mattei M, Bernardini S, Rosato N, Bottini N, Bottini M (2013) Surface polyethylene glycol conformation influences the protein corona of polyethylene glycol-modified single-walled carbon nanotubes: potential implications on biological performance. ACS Nano 7:1974–1989. https://doi.org/10.1021/nn400409h

Saito Y, Saito H, Liang G, Friedman JM (2014) Epigenetic alterations and microRNA misexpression in cancer and autoimmune diseases: a critical review. Clin Rev Allergy Immunol 47:128–135. https://doi.org/10.1007/s12016-013-8401-z

Salvador-Morales C, Flahaut E, Sim E, Sloan J, Green MLH, Sim RB (2006) Complement activation and protein adsorption by carbon nanotubes. Mol Immunol 43:193–201. https://doi.org/10.1016/j.molimm.2005.02.006

Saradna A, Do DC, Kumar S, Fu Q-L, Gao P (2018) Macrophage polarization and allergic asthma. Transl Res 191:1–14. https://doi.org/10.1016/j.trsl.2017.09.002

Sargent LM, Hubbs AF, Young S-H, Kashon ML, Dinu CZ, Salisbury JL, Benkovic SA, Lowry DT, Murray AR, Kisin ER, Siegrist KJ, Battelli L, Mastovich J, Sturgeon JL, Bunker KL, Shvedova AA, Reynolds SH (2012) Single-walled carbon nanotube-induced mitotic disruption. Mutat Res 745:28–37. https://doi.org/10.1016/j.mrgentox.2011.11.017

Schins RPF (2002) Mechanisms of genotoxicity of particles and fibers. Inhal Toxicol 14:57–78. https://doi.org/10.1080/089583701753338631

Schulte PA, Leso V, Niang M, Iavicoli I (2019) Current state of knowledge on the health effects of engineered nanomaterials in workers: a systematic review of human studies and epidemiological investigations. Scand J Work Environ Health 45:217–238. https://doi.org/10.5271/sjweh.3800

Seeley JJ, Baker RG, Mohamed G, Bruns T, Hayden MS, Deshmukh SD, Freedberg DE, Ghosh S (2018) Induction of innate immune memory via microRNA targeting of chromatin remodelling factors. Nature 559:114–119. https://doi.org/10.1038/s41586-018-0253-5

Seidel C, Kirsch A, Fontana C, Visvikis A, Remy A, Gaté L, Darne C, Guichard Y (2017) Epigenetic changes in the early stage of silica-induced cell transformation. Nanotoxicology 11:923–935. https://doi.org/10.1080/17435390.2017.1382599

Shao J, Katika MR, Schmeits PCJ, Hendriksen PJM, van Loveren H, Peijnenburg AACM, Volger OL (2013) Toxicogenomics-based identification of mechanisms for direct immunotoxicity. Toxicol Sci 135:328–346. https://doi.org/10.1093/toxsci/kft151

Sharifi S, Behzadi S, Laurent S, Forrest ML, Stroeve P, Mahmoudi M (2012) Toxicity of nanomaterials. Chem Soc Rev 41:2323–2343. https://doi.org/10.1039/c1cs15188f

Shi Y, Lammers T (2019) Combining Nanomedicine and immunotherapy. Acc Chem Res 52:1543–1554. https://doi.org/10.1021/acs.accounts.9b00148

Shu F, Xiao H, Li Q-N, Ren X-S, Liu Z-G, Hu B-W, Wang H-S, Wang H, Jiang G-M (2023) Epigenetic and post-translational modifications in autophagy: biological functions and therapeutic targets. Signal Transduct Target Ther 8:32. https://doi.org/10.1038/s41392-022-01300-8

Shvedova AA, Pietroiusti A, Fadeel B, Kagan VE (2012) Mechanisms of carbon nanotube-induced toxicity: focus on oxidative stress. Toxicol Appl Pharmacol 261:121–133. https://doi.org/10.1016/j.taap.2012.03.023

Shyamasundar S, Ng CT, Yung LYL, Dheen ST, Bay BH (2015) Epigenetic mechanisms in nanomaterial-induced toxicity. Epigenomics 7:395–411. https://doi.org/10.2217/epi.15.3

Sierra MI, Valdés A, Fernández AF, Torrecillas R, Fraga MF (2016) The effect of exposure to nanoparticles and nanomaterials on the mammalian epigenome. Int J Nanomedicine 11:6297–6306. https://doi.org/10.2147/IJN.S120104

Simko M, Gazsó A, Fiedeler U, Nentwich M (2011) Nanoparticles, free radicals and oxidative stress

Simpson DJ, McNicol AM, Murray DC, Bahar A, Turner HE, Wass JAH, Esiri MM, Clayton RN, Farrell WE (2004) Molecular pathology shows p16 methylation in nonadenomatous pituitaries from patients with Cushing's disease. Clin Cancer Res 10:1780–1788. https://doi.org/10.1158/1078-0432.ccr-1127-3

Singh S, Narang AS, Mahato RI (2011) Subcellular fate and off-target effects of siRNA, shRNA, and miRNA. Pharm Res 28:2996–3015. https://doi.org/10.1007/s11095-011-0608-1

Smith DM, Simon JK, Baker JR (2013) Applications of nanotechnology for immunology. Nat Rev Immunol 13:592–605. https://doi.org/10.1038/nri3488

Steinman RM, Mellman IS, Muller WA, Cohn ZA (1983) Endocytosis and the recycling of plasma membrane. J Cell Biol 96:1–27. https://doi.org/10.1083/jcb.96.1.1

Stoccoro A, Karlsson HL, Coppedè F, Migliore L (2013) Epigenetic effects of nano-sized materials. Toxicology 313:3–14. https://doi.org/10.1016/j.tox.2012.12.002

Sule N, Singh R, Srivastava DK (2008) Alternative modes of binding of recombinant human histone deacetylase 8 to colloidal gold nanoparticles. J Biomed Nanotechnol 4:463–468. https://doi.org/10.1166/jbn.2008.011

Sun B, Liu R, Ye N, Xiao Z-D (2015) Comprehensive evaluation of microRNA expression profiling reveals the neural signaling specific cytotoxicity of superparamagnetic iron oxide nanoparticles (SPIONs) through N-methyl-D-aspartate receptor. PLoS One 10:e0121671. https://doi.org/10.1371/journal.pone.0121671

Takahama M, Akira S, Saitoh T (2018) Autophagy limits activation of the inflammasomes. Immunol Rev 281:62–73. https://doi.org/10.1111/imr.12613

Tanaka N, Aoyama T, Kimura S, Gonzalez FJ (2017) Targeting nuclear receptors for the treatment of fatty liver disease. Pharmacol Ther 179:142–157. https://doi.org/10.1016/j.pharmthera.2017.05.011

Torres IO, Fujimori DG (2015) Functional coupling between writers, erasers and readers of histone and DNA methylation. Curr Opin Struct Biol 35:68–75. https://doi.org/10.1016/j.sbi.2015.09.007

Truesdell SS, Mortensen RD, Seo M, Schroeder JC, Lee JH, LeTonqueze O, Vasudevan S (2012) MicroRNA-mediated mRNA translation activation in quiescent cells and oocytes involves recruitment of a nuclear microRNP. Sci Rep 2:842. https://doi.org/10.1038/srep00842

Valadi H, Ekström K, Bossios A, Sjöstrand M, Lee JJ, Lötvall JO (2007) Exosome-mediated transfer of mRNAs and microRNAs is a novel mechanism of genetic exchange between cells. Nat Cell Biol 9:654–659. https://doi.org/10.1038/ncb1596

Valinluck V, Sowers LC (2007) Inflammation-mediated cytosine damage: a mechanistic link between inflammation and the epigenetic alterations in human cancers. Cancer Res 67:5583–5586. https://doi.org/10.1158/0008-5472.CAN-07-0846

Valinluck V, Tsai H-H, Rogstad DK, Burdzy A, Bird A, Sowers LC (2004) Oxidative damage to methyl-CpG sequences inhibits the binding of the methyl-CpG binding domain (MBD) of methyl-CpG binding protein 2 (MeCP2). Nucleic Acids Res 32:4100–4108. https://doi.org/10.1093/nar/gkh739

Van den Bossche J, Neele AE, Hoeksema MA, de Winther MPJ (2014) Macrophage polarization: the epigenetic point of view. Curr Opin Lipidol 25:367–373. https://doi.org/10.1097/MOL.0000000000000109

Van Wynsberghe PM, Chan S-P, Slack FJ, Pasquinelli AE (2011) Analysis of microRNA expression and function. Methods Cell Biol 106:219–252. https://doi.org/10.1016/B978-0-12-544172-8.00008-6

Walkey CD, Olsen JB, Guo H, Emili A, Chan WCW (2012) Nanoparticle size and surface chemistry determine serum protein adsorption and macrophage uptake. J Am Chem Soc 134:2139–2147. https://doi.org/10.1021/ja2084338

Wan R, Mo Y, Feng L, Chien S, Tollerud DJ, Zhang Q (2012) DNA damage caused by metal nanoparticles: involvement of oxidative stress and activation of ATM. Chem Res Toxicol 25:1402–1411. https://doi.org/10.1021/tx200513t

Wang H, Wu L, Reinhard BM (2012) Scavenger receptor mediated endocytosis of silver nanoparticles into J774A.1 macrophages is heterogeneous. ACS Nano 6:7122–7132. https://doi.org/10.1021/nn302186n

Wang S, Xu J, Guo Y, Cai Y, Ren X, Zhu W, Geng M, Meng L, Jiang C, Lu S (2021) MicroRNA-497 reduction and increase of its family member MicroRNA-424 Lead to dysregulation of multiple inflammation related genes in synovial fibroblasts with rheumatoid arthritis. Front Immunol 12:619392. https://doi.org/10.3389/fimmu.2021.619392

Wong KK (2021) DNMT1: a key drug target in triple-negative breast cancer. Semin Cancer Biol 72:198–213. https://doi.org/10.1016/j.semcancer.2020.05.010

Wong BSE, Hu Q, Baeg GH (2017) Epigenetic modulations in nanoparticle-mediated toxicity. Food Chem Toxicol 109:746–752. https://doi.org/10.1016/j.fct.2017.07.006

Wu SC, Zhang Y (2010) Active DNA demethylation: many roads lead to Rome. Nat Rev Mol Cell Biol 11:607–620. https://doi.org/10.1038/nrm2950

Wu X, Zhang Y (2017) TET-mediated active DNA demethylation: mechanism, function and beyond. Nat Rev Genet 18:517–534. https://doi.org/10.1038/nrg.2017.33

Xie S, Tao Y, Pan Y, Qu W, Cheng G, Huang L, Chen D, Wang X, Liu Z, Yuan Z (2014) Biodegradable nanoparticles for intracellular delivery of antimicrobial agents. J Control Release 187:101–117. https://doi.org/10.1016/j.jconrel.2014.05.034

Xu G, Liu G, Xiong S, Liu H, Chen X, Zheng B (2015) The histone methyltransferase Smyd2 is a negative regulator of macrophage activation by suppressing interleukin 6 (IL-6) and tumor necrosis factor α (TNF-α) production. J Biol Chem 290:5414–5423. https://doi.org/10.1074/jbc.M114.610345

Xu S, Xu Y, Solek NC, Chen J, Gong F, Varley AJ, Golubovic A, Pan A, Dong S, Zheng G, Li B (2024) Tumor-tailored Ionizable lipid nanoparticles facilitate IL-12 circular RNA delivery for enhanced lung cancer immunotherapy. Adv Mater 36:e2400307. https://doi.org/10.1002/adma.202400307

Yang S-T, Wang X, Jia G, Gu Y, Wang T, Nie H, Ge C, Wang H, Liu Y (2008) Long-term accumulation and low toxicity of single-walled carbon nanotubes in intravenously exposed mice. Toxicol Lett 181:182–189. https://doi.org/10.1016/j.toxlet.2008.07.020

Yang A, Jiao Y, Yang S, Deng M, Yang X, Mao C, Sun Y, Ding N, Li N, Zhang M, Jin S, Zhang H, Jiang Y (2018) Homocysteine activates autophagy by inhibition of CFTR expression via interaction between DNA methylation and H3K27me3 in mouse liver. Cell Death Dis 9:169. https://doi.org/10.1038/s41419-017-0216-z

Yi R, Qin Y, Macara IG, Cullen BR (2003) Exportin-5 mediates the nuclear export of premicroRNAs and short hairpin RNAs. Genes Dev 17:3011–3016. https://doi.org/10.1101/gad.1158803

Yong T, Zhang X, Bie N, Zhang H, Zhang X, Li F, Hakeem A, Hu J, Gan L, Santos HA, Yang X (2019) Tumor exosome-based nanoparticles are efficient drug carriers for chemotherapy. Nat Commun 10:3838. https://doi.org/10.1038/s41467-019-11718-4

Yuan Y-G, Zhang Y-X, Liu S-Z, Reza AMMT, Wang J-L, Li L, Cai H-Q, Zhong P, Kong I-K (2023) Multiple RNA profiling reveal epigenetic toxicity effects of oxidative stress by graphene oxide silver nanoparticles in-vitro. Int J Nanomedicine 18:2855–2871. https://doi.org/10.2147/IJN.S373161

Zahr AS, Davis CA, Pishko MV (2006) Macrophage uptake of core-shell nanoparticles surface modified with poly(ethylene glycol). Langmuir 22:8178–8185. https://doi.org/10.1021/la060951b

Zhang Z, Liu C, Bai J, Wu C, Xiao Y, Li Y, Zheng J, Yang R, Tan W (2015) Silver nanoparticle gated, mesoporous silica coated gold nanorods (AuNR@MS@AgNPs): low premature release and multifunctional cancer theranostic platform. ACS Appl Mater Interfaces 7:6211–6219. https://doi.org/10.1021/acsami.5b00368

Zhang J, Zhou W, Liu Y, Liu T, Li C, Wang L (2018) Oncogenic role of microRNA-532-5p in human colorectal cancer via targeting of the 5'UTR of RUNX3. Oncol Lett 15:7215–7220. https://doi.org/10.3892/ol.2018.8217

Zheng Y, Song X, He G, Cai Z, Zhou Y, Yu B, Xu J, Wei Y, Hou S (2011) Receptor-mediated gene delivery by folate-poly(ethylene glycol)-grafted-trimethyl chitosan in vitro. J Drug Target 19:647–656. https://doi.org/10.3109/1061186X.2010.525650

Zhou D, Yang K, Chen L, Zhang W, Xu Z, Zuo J, Jiang H, Luan J (2017) Promising landscape for regulating macrophage polarization: epigenetic viewpoint. Oncotarget 8:57693–57706. https://doi.org/10.18632/oncotarget.17027

Zolnik BS, González-Fernández A, Sadrieh N, Dobrovolskaia MA (2010) Nanoparticles and the immune system. Endocrinology 151:458–465. https://doi.org/10.1210/en.2009-1082

Open Access This chapter is licensed under the terms of the Creative Commons Attribution-NonCommercial-NoDerivatives 4.0 International License (http://creativecommons.org/licenses/by-nc-nd/4.0/), which permits any noncommercial use, sharing, distribution and reproduction in any medium or format, as long as you give appropriate credit to the original author(s) and the source, provide a link to the Creative Commons license and indicate if you modified the licensed material. You do not have permission under this license to share adapted material derived from this chapter or parts of it.

The images or other third party material in this chapter are included in the chapter's Creative Commons license, unless indicated otherwise in a credit line to the material. If material is not included in the chapter's Creative Commons license and your intended use is not permitted by statutory regulation or exceeds the permitted use, you will need to obtain permission directly from the copyright holder.

Engineered Nanomaterials in Rodent Models of Allergic Lung Disease

5

Logan J. Tisch, Ryan D. Bartone, and James C. Bonner

Abstract

A variety of engineered nanomaterials (ENMs) have been demonstrated to exacerbate allergen-induced lung disease in rodents or cause allergic lung disease in the absence of any allergen. On the other hand, some studies show that ENMs suppress allergen-induced lung disease. Moreover, there are some apparent conflicting reports of exacerbation versus suppression by the same type of ENM. This chapter serves to overview the literature on ENM-allergen interactions that result in exacerbation or suppression of allergic lung disease in rodents. Potential mechanisms are also discussed, including molecular targets revealed by transgenic mouse studies and the role of the allergen corona in mediating the exacerbation of allergic lung disease by ENMs.

Keywords

Asthma · Allergy · Lung · Nanoparticles · Nanotubes · Nanomaterials

1 Introduction

Engineered nanomaterials (ENMs) represent an emerging source of particles that have the potential to initiate or exacerbate allergic lung disease in animal models. Therefore, ENMs could pose a risk for the exacerbation of allergic asthma in humans. Asthma is a chronic inflammatory airway disease that afflicts million of individuals worldwide (Holgate 1999). The pathogenesis of asthma is characterized by eosinophilic inflammation, mucus hypersecretion, and airway smooth muscle

L. J. Tisch · R. D. Bartone · J. C. Bonner (✉)
Toxicology Program, Department of Biological Sciences, North Carolina State University, Raleigh, NC, USA
e-mail: jcbonner@ncsu.edu

© The Author(s) 2025
E. Alfaro-Moreno, F. Murphy (eds.), *Nanosafety*,
https://doi.org/10.1007/978-3-031-93871-9_5

cell hyperplasia and hypertrophy, as well as acute physiological episodes of airway hyperresponsiveness (AHR) (i.e., asthma attacks), all of which lead to airway obstruction (Holgate 1999; Ozdoganoglu and Songu 2012). Allergens such as those derived from house dust mites, cockroaches, mold or pollen are a primary cause of asthma (Kim et al. 2010). Patients with asthma are clinically identified as having allergic or nonallergic disease; the former being more prevalent than the latter (Kim et al. 2010). In the classic paradigm of allergic asthma, antigen-presenting dendritic cells in the lung take up and process inhaled allergens to promote the transformation of naïve T helper lymphocytes to T_H2 lymphocytes (Holgate 1999; Kim et al. 2010). T_H2 lymphocytes then produce a variety of chemokines and cytokines to promote the recruitment and activation of other immune cells. For example, interleukin (IL)-13 and IL-4 produced by T_H2 cells activate B cells, which produce antigen-specific IgE that subsequently binds to allergen and Fc receptors on mast cells, crosslinking them and causing degranulation and release of cytokines, leukotrienes and histamine that contributes to inflammation. In addition to T and B cells, other cell types in the lung play important roles in asthma. For example, innate immune cells such as macrophages, eosinophils and neutrophils have multifaceted roles in promoting the inflammatory environment in airways by releasing cytokines, proteases, and arachidonic acid metabolites (Holgate 1999). Additionally, epithelial cells are activated by inhaled allergens and participate in the initial steps of allergen sensitization by producing chemokines that attract dendritic cells. Moreover, smooth muscle cells surrounding the airways undergo hypertrophy and hyperplasia that contribute to increased AHR during episodes of bronchospasm, while fibroblasts residing beneath the airway epithelium contribute to collagen deposition and airway fibrosis in chronic asthma (Royce et al. 2012).

While allergens are the principal initiating agents that cause asthma, various environmental factors can exacerbate asthma, including viral infections, cigarette smoke, ozone, diesel exhaust particles and ambient air pollution particles (Wark and Gibson 2006; Bernstein et al. 2004). With regards to air pollution particulate matter (PM), epidemiological studies show that ultrafine PM (≤ 0.1 μm) is most significantly associated with asthma (Baldacci et al. 2015). Little is known about the impact of engineered nanomaterials (ENMs) on asthma exacerbation in individuals after occupational or environmental exposure. However, studies with rodents suggest that ENMs could have significant effects on allergic asthma in humans. These animal studies will be reviewed in this chapter.

ENMs are intentionally designed advanced materials with at least one dimension less than or equal to 100 nm and possess unique physicochemical characteristics from their non-nanoscale counterparts of the same composition (Auffan et al. 2009). ENMs are increasingly used in various applications, including electronics, engineering, and medicine (Bonner 2010). The diversity of ENMs includes metal and metal oxide nanoparticles (NPs) (TiO_2, ZnO, Au, NiO), silica NPs, and carbon NPs such as fullerenes, carbon nanofibers and carbon nanotubes (CNTs). ENMs may interact with biological systems in unique ways to produce toxicity since their chemical or physical properties differ substantially from bulk materials of the same composition (Nel et al. 2006). The focus of this chapter is to summarize the current knowledge on ENMs concerning asthma, focusing on the exacerbation of allergen-induced lung disease by ENMs, as well as the initiation of allergic lung disease by ENMs in the absence of allergens.

2   ENM-Induced Exacerbation or Suppression of Allergic Lung Disease

A wide spectrum of ENMs with diverse physicochemical characteristics increase molecular, cellular or pathophysiological aspects of allergen-induced lung disease in mice, while other ENMs decrease aspects of allergen-induced lung disease. Even within a specific type of ENM there are conflicting reports that show either exacerbation or suppression of allergen-induced lung disease and this may be due to a variety of experimental factors including dosing methodology, timing of ENM vs allergen exposure, and rodent species/strain, just to name a few.

2.1 TiO_2 Nanoparticles

Several different mouse models of asthma have been utilized to study the effects titanium dioxide (TiO_2) NPs on the immune response in allergic lung disease. The ovalbumin (OVA) mouse model has been utilized wherein mice are sensitized and then challenged by OVA exposure to produce allergic airway disease (Rossi et al. 2010; Harfoush et al. 2020; Jonasson et al. 2013; Mishra et al. 2016; Kim et al. 2017). Rossi et al. exposed healthy (unsensitized) and OVA-sensitized (asthmatic) mice to nanosized and fine TiO_2 via inhalation for 2 h a day, 3 days a week, for 4 weeks at a concentration of 10 mg/m^3. Surprisingly, exposure to either type of TiO_2 decreased asthma endpoints such as eosinophil numbers, airway mucous production and AHR (Rossi et al. 2010). Harfoush and coworkers exposed female BALB/c mice with or without allergic lung disease induced by OVA sensitization to TiO_2 NPs via intranasal aspiration and showed greater mucous cell metaplasia and T_H2 cytokines (IL-4, IL-5, IL-13) after OVA/TiO_2 exposure compared to OVA alone (Harfoush et al. 2020). Two other studies highlighted the importance of timing of TiO_2 exposure in eliciting an allergic response. Jonasson and colleagues showed that a single inhalation exposure of TiO_2 aggravated AHR and eosinophilia in mice when given prior to OVA sensitization, whereas repeated exposures to TiO_2 during OVA sensitization diminished airway eosinophilia and airway hyperresponsiveness (Jonasson et al. 2013). Mishra and coworkers delivered TiO_2 NPs to mice during OVA sensitization and reported augmented AHR along with elevated T_H2 cytokines (IL-4, IL-13, IL-5) and NF-κB-dependent *Socs3* expression (Mishra et al. 2016). Both studies showed that TiO_2 NPs exacerbated the inflammatory responses in the lungs of sensitized allergic animals. Additionally, Kim and colleagues also reported that TiO_2 NPs exacerbated AHR and airway inflammation in OVA-sensitized/challenged mice where these responses were accompanied by increased production of pro-inflammatory IL-1β, expression of NOD-like receptor pyrin domain-containing

3NLRP3 inflammasome and pro-caspase-1, leading to the production of active caspase-1 in the lung (Kim et al. 2017). Genetic susceptibility also affects the immune response to TiO_2 NPs during OVA sensitization in rats. Gustafsson and coworkers reported that dark agouti (DA) rats and brown Norwegian (BN) rats had increased neutrophils and lymphocytes in the lung following TiO_2 inhalation, but after OVA sensitization and TiO_2 exposure DA rats had significantly increased AHR whereas BN rats did not (Gustafsson et al. 2014). Furthermore, Hussain and colleagues used a toluene diisocyanate (TDI) mouse model of asthma and reported that TiO_2 NPs delivered by oropharyngeal aspiration exacerbated TDI-induced lung inflammation and AHR (Hussain et al. 2011).

2.2 Silica Nanoparticles

The toxicity of silica NPs (SNPs) has been evaluated in the OVA rodent model of asthma. Han et al. demonstrated that silicon dioxide (SiO_2) NPs delivered to the lungs of rats by intratracheal instillation exacerbated lung inflammation and altered the T_H2-T_H1 balance by increasing IL-4 production while decreasing IFN-γ production during OVA-sensitized rats (Han et al. 2011). Yoshida and colleagues compared two SNPs (30 nm and 70 nm) with micro-sized silica particles (300 nm and 1000 nm) delivered to the lungs of mice by intranasal instillation during OVA sensitization (Yoshida et al. 2011). They reported that the smaller 30 nm particles were the most bioactive, causing enhanced IgE and IL-4 production compared to the 70, 300 and 1000 nm particles. Yang and coworkers reported that 20 nm nano-SiO_2 particles delivered by intranasal aspiration in an OVA asthmatic mouse model caused an enhancement of allergic airway inflammation as manifested by OVA-specific serum IgE, AHR, lung inflammation, mucous cell metaplasia, cytokine expression, mast cell activation, and histamine secretion, whereas other NPs tested (Al_2O_3, TiO_2, Fe_2O_3) did not produce this effect (Yang et al. 2022). Ko and colleagues also reported that exposure to SiO_2 NPs delivered by intranasal instillation in OVA-induced allergic mice increased AHR, airway inflammation, and mucous secretion concurrent with increased protein expression levels of thioredoxin-interacting protein (TXNIP), NLRP3 inflammasome, and IL-1β compared OVA-induced allergic mice that received no SiO_2 NPs (Ko et al. 2020). Both studies suggested that SiO_2 NPs in air pollution could be a risk for asthma exacerbation and human health. Polyethylene glycol-coated (PEGylated) SNPs have been shown to enhance OVA-induced eosinophilia, mucous cell metaplasia and T_H2-T_H17 expression in BALB/c mice (Brandenberger et al. 2013). This study also investigated tracheobronchial lymph node cell activation by assessing CD69+ cells by flow cytometry and showed that alveolar macrophages and dendritic cells had increased activation in mice treated with silica NP/OVA compared to OVA alone. Other studies have compared the differences between spherical (S)-SNPs, mesoporous (M)-SNPs (i.e., containing pores between 2 and 50 nm in diameter) and PEGylated (P)-SNPs in OVA-treated mice (Park et al. 2015; Han et al. 2016). These studies showed that all three types of silica NPs increased T_H2 cytokines (IL-5, IL-13), the T_H1 cytokine IFN-γ and IL-1β over

OVA alone. M-SNPs showed the most severe airway inflammation directly (without OVA sensitization) or in combination with OVA sensitization, while P-SNPs induced less inflammation.

2.3 Ag Nanoparticles

Silver (Ag) NPs have potent antimicrobial properties and have been reported to modulate inflammatory signaling in mouse models of asthma. For example, Ag NPs reduced OVA-induced allergic inflammation in mice, decreasing total bronchoalveolar lavage fluid (BALF) cell counts, IL-4 and IL-13 levels, and *Muc5ac* expression (Jang et al. 2012). Moreover, this study showed that Ag NPs decreased VEGF levels in the lungs of mice like that observed with the VEGF inhibitor SU5614, suggesting that attenuation of VEGF signaling by Ag NPs could be involved in the reduction of OVA-induced allergic lung disease. A study by Park and colleagues also found that Ag NPs suppressed allergic lung inflammation in mice as evidenced by decreased T_H2 cytokines (IL-13, IL-4, IL-5), reduced NF-κB levels, and attenuation of AHR (Park et al. 2010). Conversely another study by Chuang et al. found that Ag NPs delivered by inhalation exposure to mice increased IgE and IL-13 levels during OVA sensitization (Chuang et al. 2013). While some of these studies demonstrate suppression of allergic responses by Ag NPs, it is noteworthy that Ag NPs have been shown to increase non-allergic pro-inflammatory endpoints like neutrophilia and circulating levels of TNFα in the absence of any allergen challenge (Silva et al. 2016; Holland et al. 2015).

2.4 Au Nanoparticles

Gold (Au) NPs have also been investigated in mouse models of asthma. For example, the study by Hussain et al. discussed above showed that Au NPs were more pro-inflammatory than TiO_2 NPs in the TDI-induced mouse model of asthma and caused more AHR compared to TiO_2 NPs (Hussain et al. 2011). Baretto and coworkers reported that Au NPs decreased OVA-induced allergic lung disease in mice as evidenced by decreased lung inflammation, suppressed mucus production and reduced cytokine levels (Barreto et al. 2015). Omlor et al. showed that PEGylated and citrated Au NPs attenuated OVA-induced inflammation, AHR, total BALF cell counts and eosinophil numbers (Omlor et al. 2017). Additionally, this study investigated the extrapulmonary uptake of Au NPs and revealed that asthmatic mice had a greater number of Au NPs deposited in the spleen compared to normal mice. Radauer-Preiml and colleagues demonstrated that Au NPs adsorb common allergens including those found in house dust mite extract (e.g., the cysteine protease der p 1) to form a corona that enhanced protease activity and increased basophil activation using an *in vitro* assay (Radauer-Preiml et al. 2015). These findings suggested that co-exposures of allergens with NPs could exacerbate allergic lung disease through corona formation. The allergen corona is discussed in more detail later in this chapter. Finally, it has also been reported that more than 10% of eosinophils in the lungs of OVA-exposed mice contain Au NPs, showing that inflammatory cells in the lung other than macrophages can contribute to Au NP uptake (Geiser et al. 2014).

2.5 Fe, Zn, Cu, and Ni Nanoparticles

In addition to TiO_2, SiO_2, Ag and Au, other types of metal or metal oxide have been investigated in rodent models of allergic asthma, including iron oxide (Fe_2O_3), zinc (Zn), copper (Cu) and nickel (Ni). The intratracheal exposure of mice to Fe_2O_3 NPs has been reported to inhibit OVA-induced allergic lung inflammation, with nano-sized particles decreasing eosinophil cell counts and OVA-specific IgE levels, while larger submicron Fe_2O_3 particles had no effect (Ban et al. 2013). Gustafsson and coworkers showed that Fe_2O_3 (hematite) NPs increased numbers of neutrophils, eosinophils, and lymphocytes in non-sensitized mice, yet decreased total immune cell numbers in the lungs and lymph nodes of OVA-sensitized mice, possibly due to reactive oxygen species (ROS) production by the Fe_2O_3 NPs (Gustafsson et al. 2015). Zinc oxide (ZnO) NPs delivered to the lungs of mice by oropharyngeal aspiration caused more enhanced eosinophilia upon co-exposure with OVA, an effect determined to be due to the Zn NPs rather than Zn ions (Huang et al. 2015). Copper oxide (CuO) NPs exacerbated numerous endpoints in the OVA mouse model of asthma including AHR, inflammatory cell counts, cytokines, IgE and ROS (Park et al. 2016). Glista-Baker and colleagues found that Ni NPs exacerbated allergic lung inflammation in mice lacking the T-bet transcription factor, a transgenic mouse model of asthma susceptibility (Glista-Baker et al. 2014).

2.6 Carbon Black Nanoparticles

Carbon black NPs (CBNPs) can be intentionally manufactured or produced through combustion processes. Koike and colleagues showed that OVA-sensitized mice exposed to 14 nm CBNPs had increased numbers of dendritic cells, macrophages, and B cells, as determined by cell surface markers, while larger 56 nm CBNPs did not elicit any change from OVA alone, indicating the importance of particle size (Koike et al. 2008). The adjuvant effect of CBNPs is supported by another study from Kroker and coworkers which found CBNPs delivered to the lungs of mice by oropharyngeal aspiration during OVA sensitization increased inflammatory cell numbers in the lungs as well as CD8$^+$ T cells, CD4$^+$ T cells and B cells in the lymph nodes (Kroker et al. 2015). These adjuvant effects were probably not due to direct particle action on antigen-presenting cells, as *in vitro* assays using bone marrow-derived dendritic cells only yielded dendritic cell activation with cell free-BALF from CBNP-exposed mice. Lefebvre and colleagues demonstrated that splenic leukocytes sensitized by OVA peptides with CBNPs had enhanced expression of T_H2-associated genes encoding the cytokines IL-13, IL-4 and IL-10, indicating that CBNPs increase T cell activation *in vitro* (Lefebvre et al. 2014).

2.7 Graphene Oxide

Graphene oxide (GO), widely used in industrial applications, is another carbonaceous ENM that has been studied in the OVA mouse model of asthma. Shurin and coworkers reported that GO delivered to the lungs of mice by oropharyngeal aspiration during OVA sensitization increased AHR and airway remodeling (mucous cell metaplasia and smooth muscle cell hypertrophy), yet decreased eosinophil accumulation in the lungs and reduced T_H2 cytokines (IL-4, IL-13, and IL-5) (Shurin et al. 2014). Moreover, this study showed that GO exposure increased macrophage production of mammalian chitinases (CHI3L1 and AMCase), both of which are associated with allergic asthma.

2.8 Carbon Nanotubes

There is an abundance of literature on the impact of carbon nanotubes (CNTs) in mouse models of asthma. Ryman-Rasmussen and colleagues were the first to show that inhalation of multi-walled CNTs (MWCNTs) enhanced the development of airway fibrosis with increased PDGF-AA expression in the OVA mouse model of asthma (Ryman-Rasmussen et al. 2009). Inoue and coworkers showed that intratracheal instillation of MWCNTs enhanced eosinophilic inflammation and mucous cell metaplasia mice during OVA-induced allergic lung disease (Inoue et al. 2009). This same study also showed that MWCNTs increased the proliferation of OVA-specific T cells *in vitro*, suggesting a possible mechanism of asthma exacerbation. In another study, Inoue and colleagues reported that the intratracheal instillation of single-walled (SW)CNT in mice could activate antigen-presenting cells to promote the severity of allergic lung disease during OVA sensitization (Inoue et al. 2010). Delivery of MWCNTs to the lungs of mice by oropharyngeal aspiration during OVA sensitization or administration of SWCNT by intratracheal instillation in rats during OVA sensitization exacerbated AHR (Mizutani et al. 2012; Li et al. 2015). Carbon nanofibers (CNF) have also been reported to exacerbate aspects of OVA-induced allergic lung disease in mice (increased eosinophilic inflammation in the lungs and enhanced serum IgE) when delivered by intranasal aspiration, yet CNFs were not as potent an adjuvant when compared to either SWCNTs or MWCNTs (Nygaard et al. 2013). These differences could be due to the CNTs characteristic of being both longer and thinner compared to CNFs.

The more clinically relevant house dust mite allergen (HDM) model of asthma has also been used to investigate CNT exacerbation of asthma. Like the OVA mouse model of asthma, markers of T_H2 inflammation are increased in mice co-exposed to HDM extract and MWCNTs compared to HDM extract alone. For example, Ronzani and coworkers showed that levels of total IgG1 and HDM-specific IgG1, influx of macrophages, eosinophils and neutrophils, production of collagen, TGF-β1, and mucus, as well as levels of IL-13, eotaxin, and TARC, were dose-dependently increased in mice exposed to HDM and MWCNT compared to HDM alone (Ronzani et al. 2014). Similarly, Ihrie and colleagues reported that co-exposure to MWCNTs exacerbated HDM-induced including eosinophilic lung inflammation, airway fibrosis, mucous cell metaplasia, and serum IgE levels (Ihrie et al. 2021). Moreover, in that study all endpoints of HDM-induced allergic lung disease exacerbated by MWCNTs were ablated in STAT-6 knockout mice (discussed in greater detail under 'transgenic models of allergic lung disease'). In addition to co-exposure scenarios, pre- and post-exposure scenarios to MWCNTs and HDM extract have also been investigated. Shipkowski and coworkers exposed mice repeatedly to HDM extract by intranasal aspiration over 3 weeks prior to a single oropharyngeal aspiration of MWCNTs (Shipkowski et al. 2015). In that study, MWCNT exposure after HDM sensitization increased total cell numbers in BALF and exacerbated airway fibrosis, but decreased numbers of neutrophils and IL-1β in BALF along with reduced pro-caspase-1 in lung tissue. Also, while MWCNTs exacerbated HDM-induced allergic lung disease in mice, IL-1β in the lung was suppressed by co-exposure to HDM extract *in vivo*. Moreover, this corresponded to suppression of inflammasome activation and IL-1β production by T_H2 cytokines IL-4 and IL-13 in human THP-1 monocytes *in vitro* (Shipkowski et al. 2015). In another study by Ihrie and colleagues, B6C33F1/N mice were exposed to three doses of MWCNTs by inhalation for 30 days, then received repeated exposure to HDM extract by intranasal aspiration over a 3-week period (Ihrie et al. 2019). In that study, MWCNTs inhibited HDM-induced serum IgE levels, IL-13 protein in BALF, and airway mucus production. However, the highest dose of MWCNTs in combination with HDM extract post-exposure increased airway fibrosis in the lungs of mice compared to MWCNTs or HDM extract alone. Collectively, these studies show that co-, pre-, or post-exposure to MWCNTs in the context of HDM sensitization/challenge can exacerbate airway fibrosis in the lungs of mice, yet co-exposures elicit the strongest allergic lung disease.

In contrast to OVA and HDM mouse models of asthma, MWCNTs delivered by inhalation exposure to brown Norway rats after sensitization and challenge with trimellitic anhydride (TMA), a chemically induced model of allergic lung disease, resulted in decreased serum IgE levels and BALF lymphocytes (Staal et al. 2014). In that study, MWCNTs did not aggravate the acute allergic reaction induced by TMA, yet large MWCNT agglomerates were found in granulomas in the lungs of allergic rats suggesting decreased clearance of MWCNTs.

2.9 Polymer Nanoparticles

Polymer NPs may be intentionally produced as a ENM or nanoplastics can be derived from the breakdown of plastics in the environment. Hardy and colleagues discovered that glycine-coated polystyrene NPs inhibited airway mucus production, serum IgE and T_H2 cytokines in the lung-draining lymph nodes during OVA-induced allergic lung disease in mice (Hardy et al. 2012). The authors reported that polystyrene NPs decreased the number of migratory dendritic cells in the lymph nodes and inhibited dendritic cell activation in the lung. Enright and colleagues investigated the extrapulmonary transport of ^{64}Cu-labeled 100 nm polystyrene NPs during OVA sensitization in mice (Enright et al. 2013). They reported that asthmatic mice, when compared to control mice, had significantly less lung retention of polystyrene NPs, which were subsequently found in the liver, bladder and gastrointestinal tract. These results indicate that asthma may cause a predisposition for greater extrapulmonary toxicity of NPs. The increasing prevalence of micro- and nanoplastics (MNPs) in air pollution could present a new risk for the exacerbation of asthma. Although there is a well-established connection between ultrafine air pollution particles (i.e., nanoparticles) and asthma exacerbation, the possible role of MNPs in asthma has not yet been studied intensively (Vasse and Melgert 2024). However, a study by Lu and coworkers showed that microplastic spheres purchased from a commercial source exacerbated HDM-induced airway mucus production but not eosinophilia in mice when delivered by intranasal aspiration (Lu et al. 2021).

In summary, a variety of ENMs exacerbate or suppress allergen-induced lung disease in mice, but do not cause allergic lung disease in the absence of an allergen. Most of the ENMs discussed above that exacerbate allergen-induced lung disease in mice are summarized below in Table 5.1.

3   ENMs as a Direct Cause of Allergic Lung Disease

Certain types of ENMs can directly cause allergic lung disease in mice in the absence of allergen exposure. These include certain types of metal or metal oxide nanoparticles and certain types of rigid MWCNTs.

3.1 TiO_2 Nanoparticles

TiO_2 NPs have been reported to directly cause several characteristics of allergic asthma, both *in vivo* and *in vitro*. A single intratracheal administration of TiO_2 NPs delivered to the lungs of rats increased BALF eosinophils and neutrophils, the number of airway cells expressing *Muc5ac*, PAS-positive airway cells, and IL-13 expressing cells (Ahn et al. 2005). *In vivo* findings in mice are supported by *in vitro* experiments wherein human bronchial epithelial cells treated with TiO_2 NPs show a dose-dependent increase in mucus secretion, which was dependent on Ca^{2+} influx into the cell (Chen et al. 2011). TiO_2 NPs also promote T cell proliferation directly *in vitro* and TiO_2-treated dendritic cells display an enhanced capacity to stimulate $CD4^+$ T cell proliferation (Schanen et al. 2013). TiO_2 NPs have also been shown to induce AHR in rats along with increased expression of neurotrophins, which regulate the responsiveness of airway sensory neurons (Scuri et al. 2010). Interestingly, these effects were only seen in newborn and weanling rats but not adults, which suggests that children are more susceptible to TiO_2 NP-induced airway hyperreactivity.

Table 5.1 Effects of ENMs on the exacerbation or suppression of allergen-induced lung disease in rodents

ENM	Allergen	Species	ENM exposure scenario	Allergic lung responses	Reference
TiO$_2$	OVA	Mouse	Inhalation during allergen sensitization	Decreased eosinophilia, AHR, mucus production compared to OVA	Rossi et al. (2010)
	OVA	Mouse	Intranasal aspiration after sensitization	Increased mucous cell metaplasia and T$_H$2 cytokines compared to OVA	Harfoush et al. (2020)
	OVA	Mouse	Inhalation: Single dose prior to or repeated doses during sensitization	Increased AHR & eosinophilia by single dose before OVA; decreased after repeated doses during OVA	Jonasson et al. (2013)
	OVA	Mouse	Intraperitoneal injection during sensitization	Increased AHR and mixed T$_H$2/T$_H$1-dependent response	Mishra et al. (2016)
	OVA	Mouse	Inhalation during sensitization	Increased AHR, inflammation, IL-1β/IL-18 compared to OVA	Kim et al. (2017)
	OVA	Rat	Inhalation before and during sensitization	Increased AHR in OVA sensitized DA rats, but not BN rats	Gustafsson et al. (2014)
	TDI	Mouse	Oropharyngeal aspiration during sensitization	Increased AHR and lung inflammation in TDI-sensitized mice	Hussain et al. (2011)
SiO$_2$	OVA	Rat	Intratracheal instillation during sensitization	Increased lung inflammation and IL-4 production compared to OVA	Han et al. (2011)
	OVA	Mouse	Intranasal aspiration to three sizes during sensitization	Increased IgE & IL-4 with smaller NPs compared to OVA sensitization	Yoshida et al. (2011)
	OVA	Mouse	Intranasal aspiration during sensitization	Increased AHR, mucus, IgE, mast cell activation compared to OVA	Yang et al. (2022), Ko et al. (2020)
	OVA	Mouse	Intranasal aspiration during sensitization	Increased eosinophils, mucus, IgE, T$_H$2 cytokines compared to OVA	Brandenberger et al. (2013)
	OVA	Mouse	Intranasal aspiration of during sensitization	Increased inflammation, mucus, T$_H$2 cytokines compare to OVA	Park et al. (2015), Han et al. (2016)
Ag	OVA	Mouse	Nebulization during sensitization	Decreased AHR, T$_H$2 cytokines, VEGF compared to OVA	Jang et al. (2012), Park et al. (2010)

(continued)

Table 5.1 (continued)

ENM	Allergen	Species	ENM exposure scenario	Allergic lung responses	Reference
	OVA	Mouse	Inhalation during sensitization	Increased eosinophilia and TH2 cytokines compared to OVA	Chuang et al. (2013)
Au	TDI	Mouse	Oropharyngeal aspiration during sensitization	Increased lung inflammation and AHR compared to TDI	Hussain et al. (2011)
	OVA	Mouse	Inhalation during sensitization	Decreased lung inflammation and mucus production compared to OVA	Chuang et al. (2013)
Fe_2O_3	OVA	Mouse	Intratracheal instillation during sensitization	Decreased eosinophilia and IgE compared to OVA alone	Ban et al. (2013), Gustafsson et al. (2015)
ZnO	OVA	Mouse	Oropharyngeal aspiration during sensitization	Increased eosinophilia and T_H2 cytokines alone or compared to OVA	Huang et al. (2015)
CuO	OVA	Mouse	Intranasal aspiration during sensitization	Increased AHR, lung inflammation and IgE compared to OVA	Park et al. (2016)
CB	OVA	Mouse	Intratracheal instillation during sensitization	Increased dendritic cells, T and B cells compared to OVA alone	Koike et al. (2008), Kroker et al. (2015)
GO	OVA	Mouse	Intratracheal instillation during sensitization	Increased AHR and mucus, but decreased eosinophils and T_H2 cytokines	Shurin et al. (2014)
MWCNT	OVA	Mouse	Inhalation after sensitization	Increased airway fibrosis and fibrotic cytokines compared to OVA	Ryman-Rasmussen et al. (2009)
	OVA	Mouses	Oropharyngeal aspiration during sensitization	Increased eosinophils, neutrophils and mucus compared to OVA	Inoue et al. (2009)
	OVA	Mouse	Intranasal aspiration during sensitization	Increased AHR, mucus, TH2 cytokines compared to allergen	Mizutani et al. (2012)
	HDM	Mouse	Intranasal aspiration during sensitization	Increased eosinophilia, mucus, TH2 cytokines compared to allergen	Ronzani et al. (2014), Ihrie et al. (2021)
	HDM	Mouse	Oropharyngeal aspiration after sensitization	Increased lung inflammation & airway fibrosis, but decreased IL-1β	Shipkowski et al. (2015)
	HDM	Mouse	Inhalation exposure before sensitization	Decreased mucus cell metaplasia, IL-13 and IgE compared to allergen	Ihrie et al. (2019)
	TMA	Rat	Inhalation exposure during sensitization	Decreased serum IgE and BALF lymphocytes compared to allergen	Staal et al. (2014)

(continued)

Table 5.1 (continued)

ENM	Allergen	Species	ENM exposure scenario	Allergic lung responses	Reference
SWCNT	OVA	Mouse	Intratracheal instillation during sensitization	Increased mucus production and allergen-induced lung inflammation	Inoue et al. (2009)
	OVA	Rat	Intratracheal instillation during sensitization	Increased AHR mucus, IgE, T_H2 cytokines compared to OVA	Li et al. (2015)

3.2 Zn, Ni, Cu, and Ag Nanoparticles

Zn, Ni and Cu NPs have been reported to recruit eosinophils to the lungs of rats without any allergen co-exposure (Cho et al. 2012; Lee et al. 2015). Zn NPs delivered to the lungs of mice increased eosinophilic inflammation in the absence of any allergen and increased BALF levels of the T_H2 cytokines IL-4, IL-13, and IL-5 (Huang et al. 2015). In addition to eosinophil recruitment in the lungs of mice, Cu NPs also stimulate *Muc5ac* production in bronchial epithelial cells *in vitro*, the latter of which is MAPK-dependent (Ko et al. 2016). Ni NP-induced eosinophil recruitment to the lungs of rats was associated with increased eotaxin production by alveolar macrophages (Lee et al. 2015). Ag NPs promote allergic lung inflammation in mice when administered with LPS, which was dependent on CD4$^+$ T cells and resulted in increased IL-17A (Hirai et al. 2016).

3.3 Carbon Nanotubes

As discussed above, CNTs have been shown to exacerbate allergen-induced lung disease in rodents. However, there is strong evidence that some types of CNTs can elicit allergic lung responses directly in the absence of any allergen. SWCNTs or MWCNTs alone can cause AHR in mice (Hsieh et al. 2012; Beamer et al. 2012). However, while SWCNTs induced a dose-dependent increase in AHR, BALF cell counts in these mice did not show eosinophilia usually associated with allergy but instead showed increased neutrophil numbers (Hsieh et al. 2012). In contrast, MWCNTs caused eosinophilic lung inflammation in addition to neutrophilia and AHR (Beamer et al. 2012). The AHR and eosinophilia observed in this study were found to be dependent on both IL-13 and IL-33, but independent of T and B cells, indicating that MWCNTs modulate the innate immune response to elicit allergy-associated pathologies (Beamer et al. 2012). Park and colleagues reported that intratracheal instillation of MWCNTs in mice increased the proportion of B cells in the blood and increased serum IgE levels while decreasing natural killer cells and T cells (Park et al. 2009). The physicochemical properties of different types of CNTs, namely the rigidity of the tube structure, are strong determinants of allergic lung responses (Rydman et al. 2014; Duke et al. 2017). For example, Rydman and colleagues found that rigid, rod-like MWCNTs are generally more toxic than flexible MWCNTs and cause AHR, mucus secretion, eosinophilia, and T_H2 cytokine release (IL-4, IL-13) (Rydman et al. 2014). These findings are supported by those from Duke and coworkers who compared rod-like and the more flexible, tangled MWCNTs and demonstrated that rod-like but not tangled MWCNTs induced IL-4, airway mucus production, and elevated serum IgE levels (Duke et al. 2017).

4 The Allergen Corona as a Mechanism for ENM Exacerbation of Allergic Lung Disease

The exacerbation of allergen-induced lung disease in rodents from MWCNTs (and potentially other nanoparticles) can be accredited to the formation of a house dust mite (HDM)-bound allergen corona. The term 'biocorona' has become a well-explored topic in the field of nanotoxicology and has been extensively reviewed (Runa et al. 2018; Mahmoudi et al. 2023; Kobos and Shannahan 2020; Lynch et al. 2007; Monopoli et al. 2012; Ke et al. 2017). Originally coined as the adsorption of endogenous biological macromolecules to the surface of nanoparticles, this term has also grown to include exogenous material. The phenomena of protein adsorption to solid surfaces stem from the Vroman effect. According to the Vroman effect, the adsorption of proteins and other macromolecules is a dynamic process where those with higher diffusion rates reach the solid surface and attach (Vroman 1962; Jung et al. 2003). The formation of the corona depends on the nanoparticle size and physicochemical surface characteristics. For example, Shannahan and colleagues compared four types of silver nanoparticles with sizes 20 and 110 nm bound to a subset of 11 proteins (Shannahan et al. 2013). Among the types, the two nanoparticles that were 110 nm in length bound twice as many proteins as the 20 nm nanoparticles. In the same study, the authors discovered that the smaller nanoparticles were bound to hydrophobic proteins compared to the larger-sized particles.

Among these exogenous macromolecules, it has been discovered that introducing MWCNTs to the common allergen HDM results in the formation of an 'allergen corona'. The allergen corona differs from the concentration of allergic proteins found in HDM extract via the selective adsorption of specific molecules to the MWCNTs. For example, two of the most well-known allergic proteins from HDM extract, Der p 1 and Der p 2, were compared following corona formation to MWCNTs via SDS-PAGE and LC-MS/MS (Dominguez et al. 2024). While the concentration of the cysteine protease Der p 1 was reduced on the surface of MWCNTs (~1.8% in HDM extract compared to 0.17% in the HDM-MWCNT corona), the concentration of the lipid binding protein Der p 2 was increased fourfold (~13% in HDM extract compared to ~50% in the HDM-MWCNT corona) (Dominguez et al. 2024).

The inhalation of these unique HDM-MWCNT coronas in rodent models demonstrated the exacerbation of allergic lung disease as previously seen in co-exposures to HDM extract and MWCNTs in mice (Bartone et al. 2024). Mice exposed to MWCNTs containing HDM allergen corona showed significant increases in

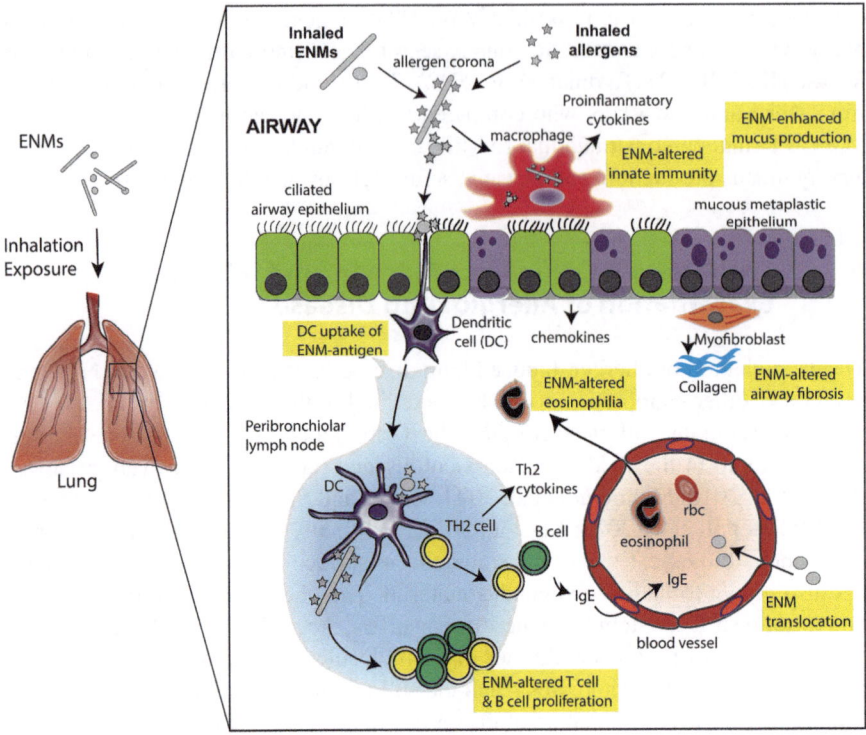

Fig. 5.1 Interaction of ENMs and allergens to form an allergen corona that could mediate innate and adaptive responses involved in the exacerbation of allergic lung disease

eosinophilia, mucous cell metaplasia, airway fibrosis, STAT-6 phosphorylation and arginase-1 compared to pristine MWCNTs (Bartone et al. 2024). All these hallmarks of a T_H2 microenvironment indicated an allergic lung environment.

The mechanism(s) through which ENMs, including MWCNTs, exacerbate allergic lung disease via the allergen corona remain to be elucidated. Possible mechanisms include delivering and presenting allergens to innate professional phagocytes such as macrophages or delivery to antigen-presenting cells (e.g., dendritic cells) leading to adaptive immune responses. The allergen corona on ENMs could play a contributory role in modulating both innate and adaptive immune responses. Some of these potential mechanisms and associated cellular and pathological outcomes are illustrated in Fig. 5.1.

5 Transgenic Mouse Models of ENM-Induced Allergic Lung Disease

Gene targeting to induce genetic modifications in animals is crucial for understanding the regulation and function of specific genes in ENM-induced allergic lung disease studies. Transgenic mouse models are invaluable for identifying novel therapeutic targets to alleviate symptoms associated with allergic lung disease. A limited number of transgenic mouse models have been used to investigate the exacerbation of allergen-induced lung disease by ENMs, including *Stat6*, *Stat1*, *Par2* and *Cox2*. These genes encode transcription factors (STAT6, STAT1), the cell-surface receptor PAR2, and the enzyme COX-2, respectively.

5.1 STAT6

The signal transducer and activator of transcription 6 (STAT6) transgenic mouse model has been utilized in exploring the multifaceted role of STAT6 in T_H2-type inflammation during allergic lung disease. Ihrie and colleagues used STAT6 knockout (KO) mice to investigate the involvement of this transcription factor in the exacerbation of HDM allergen-induced allergic airway disease by MWCNTs (Ihrie et al. 2021). In that study, wildtype (WT) and STAT6 KO mice were exposed via intranasal aspiration to either vehicle, HDM extract, MWCNTs, or a combination of HDM extract and MWCNTs. MWCNTs significantly exacerbated HDM-induced eosinophilic lung inflammation, mucous cell metaplasia, and serum IgE levels. Interestingly, these effects were significantly reduced in STAT6 KO mice, highlighting the critical role of STAT6 in MWCNT-induced exacerbation of allergic lung disease (Ihrie et al. 2021). These findings provided new mechanistic insights, emphasizing the necessity of STAT6 in the exacerbation of allergic lung disease by MWCNTs.

5.2 STAT1

Signal transducer and activator of transcription 1 (STAT-1) transgenic mice have been invaluable in elucidating the mechanisms underlying ENM-induced allergic lung disease. Thompson and colleagues utilized STAT-1 KO mice to explore the role of this transcription factor in allergen-induced airway remodeling and its exacerbation by MWCNTs (Thompson et al. 2015). OVA-sensitized WT and STAT-1 KO mice were exposed to tangled MWCNTs via oropharyngeal aspiration. Notably, STAT-1 KO mice exhibited increased levels of pro-fibrogenic cytokines (TNF-α and osteopontin) in BALF compared to WT mice following treatment with OVA alone or in combination with MWCNTs (Thompson et al. 2015). Additionally, OVA-sensitized STAT-1 KO mice also displayed increased eosinophilia, goblet cell hyperplasia, and airway fibrosis compared to WT mice. MWCNTs further increased these histopathological features in the STAT-1 KO mice. Moreover, lung fibroblasts isolated from the lungs of STAT-1 KO mice produced significantly higher levels of collagen mRNA and protein than those from WT mice (Thompson et al. 2015). These findings demonstrated a protective role for STAT1 in allergen-induced inflammation and exacerbation of airway remodeling by MWCNTs. Duke et al. also used STAT-1 KO mice to investigate the direct allergic and profibrotic lung responses to rod-like, rigid MWCNTs and found that lung injury, airway mucus production and airway fibrosis were greater in STAT-1 KO mice (Duke et al. 2017). The evidence from studies with STAT-1 KO mice shows that STAT-1 suppresses allergic lung disease, in contrast to STAT-6, which promotes allergic lung disease that is enhanced by MWCNT exposure.

5.3 PAR-2

The protease-activated receptor 2 (PAR2) is a G-protein-coupled receptor (GPCR) expressed by a variety of cells (i.e. macrophages, epithelial cells, and fibroblasts) found in the lungs (Cocks and Moffatt 2001). PAR2 has been implicated in the pathogenesis of various respiratory diseases, such as chronic obstructive pulmonary disease (COPD), pulmonary fibrosis, and asthma (Heuberger and Schuepbach 2019; Rayees et al. 2020). Lee and colleagues employed a PAR2 KO mouse model to investigate the role of PAR2 in the exacerbation of HDM-induced allergic lung disease by MWCNTs (Lee et al. 2023). This study showed that in both WT and PAR2 KO mice, co-exposure to MWCNTs and HDM synergistically increased lung inflammation, evidenced by increased BALF cellularity, predominantly eosinophils. In addition, both WT and PAR2 KO mice exhibited similar increases in lung *Ccl-11* mRNA levels. However, PAR2 KO mice displayed significantly reduced airway fibrosis, accompanied by lower protein and mRNA levels of the pro-fibrotic mediator arginase 1 (Arg-1) (Lee et al. 2023). These findings suggest that PAR2 predominantly mediates airway fibrogenesis rather than eosinophilic lung inflammation.

5.4 COX-2

Cyclooxygenase-2 (COX-2) is an enzyme recognized as a protective factor in asthma, primarily due to its role generating prostaglandin E_2 (PGE_2), which has demonstrated protective and anti-inflammatory effects in the airways (Trudeau et al. 2006). In a study by Sayers and colleagues, WT and COX-2-deficient ($COX-2^{-/-}$) mice were sensitized to OVA and then treated with a single dose of MWCNTs to assess the role of COX-2 in the exacerbation of allergen-induced airway remodeling by MWCNTs (Sayers et al. 2013). That study showed that MWCNTs significantly increased OVA-induced lung inflammation and mucus-cell metaplasia in $COX-2^{-/-}$ mice compared with WT mice. However, after exposure to allergens and MWCNTs, airway fibrosis did not differ between WT and $COX-2^{-/-}$ mice (Sayers et al. 2013). Additionally, MWCNTs significantly amplified allergen-induced cytokine responses associated with T_H2 (IL-13 and IL-5), T_H1 (CXCL10), and T_H17 (IL-17A) pathways in $COX-2^{-/-}$ mice but not in WT mice. These findings suggest that COX-2 deficiency exacerbates allergen-induced airway inflammation and mucus-cell metaplasia in response to MWCNTs and is linked to the activation of a mixed immune response.

In summary, transgenic mouse models show that STAT6 and PAR2 mediate exacerbation of allergen-induced lung disease, while STAT1 and COX-2 serve

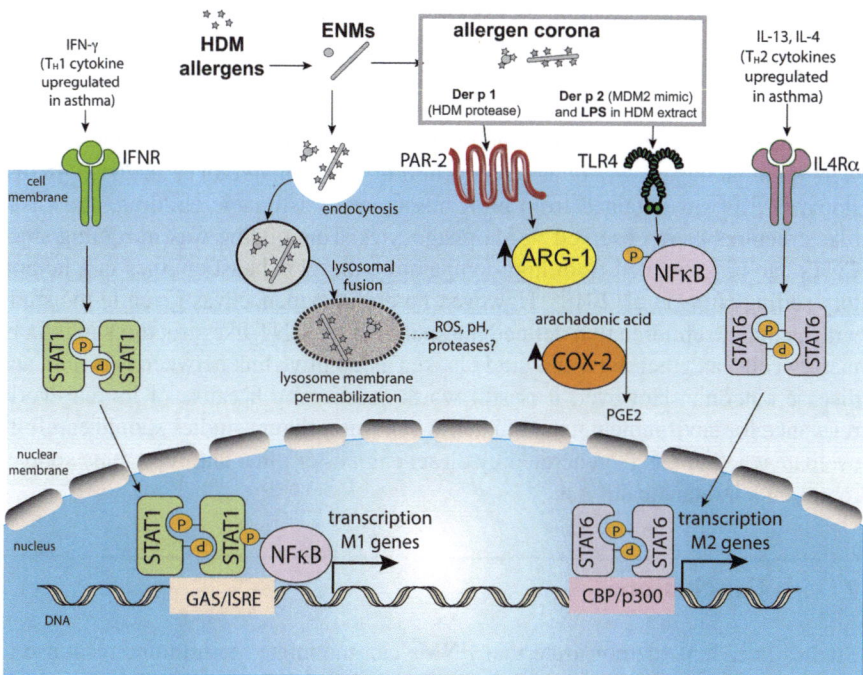

Fig. 5.2 Selected molecular targets involved in the exacerbation of allergen-induced lung disease by ENMs. Emphasis is given to the hypothesis that the allergen corona mediates some of these effects, using house dust mite (HDM) allergens as a specific example. The roles of STAT1, STAT6, PAR2 and COX-2 have been investigated using transgenic mouse models

protective roles in suppressing aspects of allergen-induced lung disease exacerbation by ENMs. All appear to be important in modulating the exacerbation of allergen-induced lung disease in mice by ENMs. The potential interactions of ENMs with these molecular targets are illustrated in Fig. 5.2.

6 Human Evidence for ENM-Induced Asthma

While there is no direct evidence that ENMs cause asthma in humans, the findings of numerous studies with rodents suggest that ENMs may have the potential to cause or exacerbate asthma. A study by Wu and colleagues used fractional exhaled nitric oxide (FENO) measurement as an endpoint to investigate the respiratory function of workers in a ENM handling plant (Wu et al. 2014). FENO is elevated in patients with asthma and FENO measurement has been recognized as a diagnostic tool for airway inflammation. Associations were found between the risk level of NP exposure and FENO, with the strongest association being between TiO_2 NPs and FENO. This study indicated that ENMs, especially TiO_2 NPs, have the potential to exacerbate asthma. A case report by Journeay and Goldman revealed that a chemist working with Ni NPs on an open workbench experienced throat congestion, AHR, and allergic skin reactions (Journeay and Goldman 2014). The patient received bronchodilators, which improved forced expiratory volume, and the patient also tested positive in an allergy test specific for nickel. A study by Kolosnjaj-Tabi and colleagues reported CNTs, presumably generated from diesel automobile engines, were found in the airways of asthmatic children (Kolosnjaj-Tabi et al. 2015). In that study, BALF was obtained from bronchoscopies of asthmatic children, and CNT-like structures were observed inside phagocytes. This finding was intriguing since CNTs can be generated from high-temperature diesel exhaust engines in a laboratory setting (Jung et al. 2013). However, no direct evidence was given in the study with asthmatic children that defined the source of the CNT-like structures in inflammatory cells, and therefore, this study lacked a definitive link between exposure and disease outcome. However, it is still worth mentioning because of its real-world relevance for environmental nanomaterial exposure. Future studies should carefully evaluate whether CNTs generated by diesel engines or other anthropogenic sources represent a human health risk.

7   Conclusion

Studies in rodents demonstrate that ENMs can modulate the immune response to allergens to either exacerbate or attenuate the disease. Moreover, ENMs have been shown to impact various cellular and molecular targets in allergic airway disease including serum IgE, AHR, BALF cell counts, and pro-inflammatory cytokines. The effects that specific ENMs elicit depend on their size, shape, composition, and coating. Because ENMs are so varied it will be necessary in the future to continue to assess their risk to individuals with lung diseases. Progress is being made towards a more mechanistic understanding of ENM-induced allergic lung disease. However, much is still unknown. Additionally, there is a need for more epidemiological studies that focus on workplace exposure to ENMs and their respiratory health effects in humans. It would be advantageous for more studies to look at lower, occupationally relevant doses of ENMs and how exposure affects asthma induced by more relevant, common allergens like those derived from house dust mites.

Acknowledgments The authors would like to acknowledge the following funding sources: National Institute of Environmental Health Sciences (NIEHS) grant R01ES032443 and NIEHS grant P30ES025128. RDB and LJT were partially supported by NIEHS Training Grant T32ES007046.

References

Ahn M-H, Kang C-M, Park C-S, Park S-J, Rhim T, Yoon P-O, Chang HS, Kim S-H, Kyono H, Kim KC (2005) Titanium dioxide particle—induced goblet cell hyperplasia: association with mast cells and IL-13. Respir Res 6:34

Auffan M, Rose J, Bottero J-Y, Lowry GV, Jolivet J-P, Wiesner MR (2009) Towards a definition of inorganic nanoparticles from an environmental, health and safety perspective. Nat Nanotechnol 4:634–641

Baldacci S, Maio S, Cerrai S, Sarno G, Baïz N, Simoni M, Annesi-Maesano I, Viegi G (2015) Allergy and asthma: effects of the exposure to particulate matter and biological allergens. Respir Med 109:1089–1104

Ban M, Langonné I, Huguet N, Guichard Y, Goutet M (2013) Iron oxide particles modulate the ovalbumin-induced Th2 immune response in mice. Toxicol Lett 216:31–39

Barreto E, Serra MF, Dos Santos RV, Dos Santos CE, Hickman J, Cotias AC, Pão CR, Trindade SG, Schmidt V, Giacomelli C, Carvalho VF, Rodrigues E, Silva PM, Cordeiro RS, Martins MA (2015) Local administration of gold nanoparticles prevents pivotal pathological changes in murine models of atopic asthma. J Biomed Nanotechnol 11:1038–1050

Bartone RD, Tisch LJ, Dominguez J, Payne CK, Bonner JC (2024) House dust mite proteins adsorb on multi-walled carbon nanotubes forming an allergen corona that intensifies allergic lung disease in mice. ACS Nano 18(38):26215–26232

Beamer CA, Girtsman TA, Seaver BP, Finsaas KJ, Migliaccio CT, Perry VK, Rottman JB, Smith DE, Holian A (2012) IL-33 mediates multi-walled carbon nanotube (MWCNT)-induced airway hyper-reactivity via the mobilization of innate helper cells in the lung. Nanotoxicology 7:1070–1081

Bernstein JA, Alexis N, Barnes C, Bernstein IL, Nel A, Peden D, Diaz-Sanchez D, Tarlo SM, Williams PB (2004) Health effects of air pollution. J Allergy Clin Immunol 114:1116–1123

Bonner JC (2010) Nanoparticles as a potential cause of pleural and interstitial lung disease. Proc Am Thorac Soc 7(2):138–141

Brandenberger C, Rowley NL, Jackson-Humbles DN, Zhang Q, Bramble LA, Lewandowski RP, Wagner JG, Chen W, Kaplan BL, Kaminski NE, Baker GL, Worden RM, Harkema JR (2013) Engineered silica nanoparticles act as adjuvants to enhance allergic airway disease in mice. Part Fibre Toxicol 10:26

Chen EYT, Garnica M, Wang Y-C, Chen C-S, Chin W-C (2011) Mucin secretion induced by titanium dioxide nanoparticles. PLoS One 6:e16198

Cho W-S, Duffin R, Poland CA, Duschl A, Oostingh GJ, MacNee W, Bradley M, Megson IL, Donaldson K (2012) Differential pro-inflammatory effects of metal oxide nanoparticles and their soluble ions in vitro and in vivo ; zinc and copper nanoparticles, but not their ions, recruit eosinophils to the lungs. Nanotoxicology 6:22–35

Chuang H-C, Hsiao T-C, Wu C-K, Chang H-H, Lee C-H, Chang C-C, Cheng T-J (2013) Allergenicity and toxicology of inhaled silver nanoparticles in allergen-provocation mice models. Int J Nanomedicine 8:4495

Cocks TM, Moffatt JD (2001) Protease-activated receptor-2 (PAR2) in the airways. Pulm Pharmacol Ther 14(3):183–191

Dominguez J, Holmes SK, Bartone RD, Tisch LJ, Tighe RM, Bonner JC, Payne CK (2024) House dust mite extract forms a der p 2 corona on multi-walled carbon nanotubes: implications for allergic airway disease. Environ Sci Nano 11(1):324–335

Duke KS, Taylor-Just AJ, Ihrie MD, Shipkowski KA, Thompson EA, Dandley EC, Parsons GN, Bonner JC (2017) STAT1-dependent and -independent pulmonary allergic and fibrogenic responses in mice after exposure to tangled versus rod-like multi-walled carbon nanotubes. Part Fibre Toxicol 14:26

Enright HA, Bratt JM, Bluhm AP, Kenyon NJ, Louie AY (2013) Tracking retention and transport of ultrafine polystyrene in an asthmatic mouse model using positron emission tomography. Exp Lung Res 39:304–313

Geiser M, Wigge C, Conrad ML, Eigeldinger-Berthou S, Künzi L, Garn H, Renz H, Mall MA (2014) Nanoparticle uptake by airway phagocytes after fungal spore challenge in murine allergic asthma and chronic bronchitis. BMC Pulm Med 14:116

Glista-Baker EE, Taylor AJ, Sayers BC, Thompson EA, Bonner JC (2014) Nickel nanoparticles cause exaggerated lung and airway remodeling in mice lacking the T-box transcription factor, TBX21 (T-bet). Part Fibre Toxicol 11:7

Gustafsson Å, Jonasson S, Sandström T, Lorentzen JC, Bucht A (2014) Genetic variation influences immune responses in sensitive rats following exposure to TiO2 nanoparticles. Toxicology 326:74–85

Gustafsson Å, Bergström U, Ågren L, Österlund L, Sandström T, Bucht A (2015) Differential cellular responses in healthy mice and in mice with established airway inflammation when exposed to hematite nanoparticles. Toxicol Appl Pharmacol 288:1–11

Han B, Guo J, Abrahaley T, Qin L, Wang L, Zheng Y, Li B, Liu D, Yao H, Yang J, Li C, Xi Z, Yang X (2011) Adverse effect of nano-silicon dioxide on lung function of rats with or without ovalbumin immunization. PLoS One 6:e17236

Han H, Park YH, Park HJ, Lee K, Um K, Park J-W, Lee J-H (2016) Toxic and adjuvant effects of silica nanoparticles on ovalbumin-induced allergic airway inflammation in mice. Respir Res 17:60

Hardy CL, LeMasurier JS, Belz GT, Scalzo-Inguanti K, Yao J, Xiang SD, Kanellakis P, Bobik A, Strickland DH, Rolland JM, O'Hehir RE, Plebanski M (2012) Inert 50-nm polystyrene nanoparticles that modify pulmonary dendritic cell function and inhibit allergic airway inflammation. J Immunol 188:1431–1441

Harfoush SA, Hannig M, Le DD, Heck S, Leitner M, Omlor AJ, Tavernaro I, Kraegeloh A, Kautenburger R, Kickelbick G, Beilhack M, Nguyen J, Sester M, Bals R, Dinh QT (2020) High-dose intranasal application of titanium dioxide nanoparticles induces the systemic uptakes and allergic airway inflammation in asthmatic mice. Respir Res 21:168

Heuberger DM, Schuepbach RA (2019) Protease-activated receptors (PARs): mechanisms of action and potential therapeutic modulators in PAR-driven inflammatory diseases. Thromb J 17(1):4

Hirai T, Yoshioka Y, Izumi N, Ichihashi K, Handa T, Nishijima N, Uemura E, Sagami K, Takahashi H, Yamaguchi M, Nagano K, Mukai Y, Kamada H, Tsunoda S, Ishii KJ, Higashisaka K, Tsutsumi Y (2016) Metal nanoparticles in the presence of lipopolysaccharides trigger the onset of metal allergy in mice. Nat Nanotechnol 11:808–816

Holgate ST (1999) The epidemic of allergy and asthma. Nature 402:B2–B4

Holland NA, Becak DP, Shannahan JH, Brown JM, Carratt SA, Van Winkle LS, Pinkerton KE, Wang CM, Munusamy P, Baer DR, Sumner SJ, Fennell TR, Lust RM, Wingard CJ (2015) Cardiac ischemia reperfusion injury following instillation of 20 nm citrate-capped Nanosilver. J Nanomed Nanotechnol 6(Suppl 6):006

Hsieh W-Y, Chou C-C, Ho C-C, Yu S-L, Chen H-Y, Chou H-YE, Chen JJW, Chen H-W, Yang P-C (2012) Single-walled carbon nanotubes induce airway hyper-reactivity and parenchymal injury in mice. Am J Respir Cell Mol Biol 46:257–267

Huang K-L, Lee Y-H, Chen H-I, Liao H-S, Chiang B-L, Cheng T-J (2015) Zinc oxide nanoparticles induce eosinophilic airway inflammation in mice. J Hazard Mater 297:304–312

Hussain S, Vanoirbeek JAJ, Luyts K, De Vooght V, Verbeken E, Thomassen LC, Martens JA, Dinsdale D, Boland S, Marano F, Nemery B, Hoet PH (2011) Lung exposure to nanoparticles modulates an asthmatic response in a mouse model. Eur Respir J 37:299–309

Ihrie MD, Taylor-Just AJ, Walker NJ, Stout MD, Gupta A, Richey JS, Hayden BK, Baker GL, Sparrow BR, Duke KS, Bonner JC (2019) Inhalation exposure to multi-walled carbon nanotubes alters the pulmonary allergic response of mice to house dust mite allergen. Inhal Toxicol 31(5):192–202

Ihrie MD, Duke KS, Shipkowski KA, You DJ, Lee HY, Taylor-Just AJ, Bonner JC (2021) STAT6-dependent exacerbation of house dust mite-induced allergic airway disease in mice by multi-walled carbon nanotubes. NanoImpact 22:100309

Inoue K, Koike E, Yanagisawa R, Hirano S, Nishikawa M, Takano H (2009) Effects of multi-walled carbon nanotubes on a murine allergic airway inflammation model. Toxicol Appl Pharmacol 237:306–316

Inoue K, Yanagisawa R, Koike E, Nishikawa M, Takano H (2010) Repeated pulmonary exposure to single-walled carbon nanotubes exacerbates allergic inflammation of the airway: possible role of oxidative stress. Free Radic Biol Med 48:924–934

Jang S, Park JW, Jang JW, Cha HR, Jung SY, Lee J, Jung SS, Kim JO, Kim S, Lee CS, Park H (2012) Silver nanoparticles modify VEGF signaling pathway and mucus hypersecretion in allergic airway inflammation. Int J Nanomedicine 7:1329–1343

Jonasson S, Gustafsson Å, Koch B, Bucht A (2013) Inhalation exposure of nano-scaled titanium dioxide (TiO$_2$) particles alters the inflammatory responses in asthmatic mice. Inhal Toxicol 25:179–191

Journeay WS, Goldman RH (2014) Occupational handling of nickel nanoparticles: a case report. Am J Ind Med 57:1073–1076

Jung S-Y, Lim SM, Albertorio F (2003) The Vroman effect: a molecular level description of fibrinogen displacement. J Am Chem Soc 125:12782–12786

Jung HS, Miller A, Park K, Kittelson DB (2013) Carbon nanotubes among diesel exhaust particles: real samples or contaminants? J Air Waste Manage Assoc 63:1199–1204

Ke PC, Lin S, Parak WJ, Davis TP, Caruso F (2017) A decade of the protein corona. ACS Nano 11:11773–11776

Kim HY, DeKruyff RH, Umetsu DT (2010) The many paths to asthma: phenotype shaped by innate and adaptive immunity. Nat Immunol 11:577–584

Kim B-G, Lee P-H, Lee S-H, Park M-K, Jang A-S (2017) Effect of TiO2 nanoparticles on Inflammasome-mediated airway inflammation and responsiveness. Allergy Asthma Immunol Res 9:257

Ko J-W, Park J-W, Shin N-R, Kim JH, Cho YK, Shin DH, Kim JC, Lee IC, Oh SR, Ahn KS, Shin IS (2016) Copper oxide nanoparticle induces inflammatory response and mucus production via MAPK signaling in human bronchial epithelial cells. Environ Toxicol Pharmacol 43:21–26

Ko J-W, Shin N-R, Je-Oh L, Jung T-Y, Moon C, Kim T-W, Choi J, Shin I-S, Heo J-D, Kim J-C (2020) Silica dioxide nanoparticles aggravate airway inflammation in an asthmatic mouse model via NLRP3 inflammasome activation. Regul Toxicol Pharmacol 112:104618

Kobos L, Shannahan J (2020) Biocorona-induced modifications in engineered nanomaterial–cellular Interactions Impacting biomedical applications. Wiley Interdiscip Rev Nanomed Nanobiotechnol 12:e1608

Koike E, Takano H, Inoue K-I, Yanagisawa R, Sakurai M, Aoyagi H, Shinohara R, Kobayashi T (2008) Pulmonary exposure to carbon black nanoparticles increases the number of antigen-presenting cells in murine lung. Int J Immunopathol Pharmacol 21:35–42

Kolosnjaj-Tabi J, Just J, Hartman KB, Laoudi Y, Boudjemaa S, Alloyeau D, Szwarc H, Wilson LJ, Moussa F (2015) Anthropogenic carbon nanotubes found in the airways of Parisian children. EBioMedicine 2:1697–1704

Kroker M, Sydlik U, Autengruber A, Cavelius C, Weighardt H, Kraegeloh A, Unfried K (2015) Preventing carbon nanoparticle-induced lung inflammation reduces antigen-specific sensitization and subsequent allergic reactions in a mouse model. Part Fibre Toxicol 12:20

Lee S, Hwang S-H, Jeong J, Han Y, Kim SH, Lee DK, Lee HS, Chung ST, Jeong J, Roh C, Huh YS, Cho WS (2015) Nickel oxide nanoparticles can recruit eosinophils in the lungs of rats by the direct release of intracellular eotaxin. Part Fibre Toxicol 13:30

Lee HY, You DJ, Taylor-Just AJ, Tisch LJ, Bartone RD, Atkins HM, Ralph LM, Antoniak S, Bonner JC (2023) Role of the protease-activated receptor-2 (PAR2) in the exacerbation of house dust mite-induced murine allergic lung disease by multi-walled carbon nanotubes. Part Fibre Toxicol 20(1):32

Lefebvre DE, Pearce B, Fine JH, Chomyshyn E, Ross N, Halappanavar S, Tayabali AF, Curran I, Bondy GS (2014) In vitro enhancement of mouse T helper 2 cell sensitization to ovalbumin allergen by carbon black nanoparticles. Toxicol Sci 138:322–332

Li J, Li L, Chen H, Chang Q, Liu X, Wu Y, Wei C, Li R, Kwan JKC, Yeung KL, Xi Z, Lu Z, Yang X (2015) Application of vitamin E to antagonize SWCNTs-induced exacerbation of allergic asthma. Sci Rep 4:4275

Lu K, Lai KP, Stoeger T, Ji S, Lin Z, Lin X, Chan TF, Fang JK-H, Lo M, Gao L, Qiu C, Chen S, Chen G, Li L, Wang L (2021) Detrimental effects of microplastic exposure on normal and asthmatic pulmonary physiology. J Hazard Mat 416:126069

Lynch I, Cedervall T, Lundqvist M, Cabaleiro-Lago C, Linse S, Dawson KA (2007) The nanoparticle–protein complex as a biological entity; a complex fluids and surface science challenge for the 21st century. Adv Colloid Interf Sci 134:167–174

Mahmoudi M, Landry MP, Moore A, Coreas R (2023) The protein corona from nanomedicine to environmental science. Nat Rev Mater 8:422–438

Mishra V, Baranwal V, Mishra RK, Sharma S, Paul B, Pandey AC (2016) Titanium dioxide nanoparticles augment allergic airway inflammation and Socs3 expression via NF-κB pathway in murine model of asthma. Biomaterials 92:90–102

Mizutani N, Nabe T, Yoshino S (2012) Exposure to multiwalled carbon nanotubes and allergen promotes early- and late-phase increases in airway resistance in mice. Biol Pharm Bull 35:2133–2140

Monopoli MP, Åberg C, Salvati A, Dawson KA (2012) Biomolecular coronas provide the biological identity of nanosized materials. Nat Nanotechnol 7:779–786

Nel A, Xia T, Mädler L, Li N (2006) Toxic potential of materials at the nanolevel. Science 311:622–627

Nygaard UC, Samuelsen M, Marioara CD, Løvik M (2013) Carbon nanofibers have IgE adjuvant capacity but are less potent than nanotubes in promoting allergic airway responses. Biomed Res Int 2013:1–12

Omlor AJ, Le DD, Schlicker J, Hannig M, Ewen R, Heck S, Herr C, Kraegeloh A, Hein C, Kautenburger R, Kickelback G, Bals R, Nguyen J, Thai Dinh Q (2017) Local effects on airway inflammation and systemic uptake of 5 nm PEGylated and citrated gold nanoparticles in asthmatic mice. Small 13:1603070

Ozdoganoglu T, Songu M (2012) The burden of allergic rhinitis and asthma. Ther Adv Respir Dis 6:11–23

Park E-J, Cho W-S, Jeong J, Yi J, Choi K, Park K (2009) Pro-inflammatory and potential allergic responses resulting from B cell activation in mice treated with multi-walled carbon nanotubes by intratracheal instillation. Toxicology 259:113–121

Park HS, Kim KH, Jang S, Park JW, Cha HR, Lee JE, Kim JO, Kim SY, Lee CS, Kim JP, Jung SS (2010) Attenuation of allergic airway inflammation and hyperresponsiveness in a murine model of asthma by silver nanoparticles. Int J Nanomedicine 5:505–515

Park HJ, Sohn JH, Kim Y-J, Park YH, Han H, Park KH, Lee K, Choi H, Um K, Choi IH, Park JW, Lee JH (2015) Acute exposure to silica nanoparticles aggravate airway inflammation: different effects according to surface characteristics. Exp Mol Med 47:e173

Park J-W, Lee I-C, Shin N-R, Jeon C-M, Kwon O-K, Ko J-W, Kim J-C, Oh S-R, Shin I-S, Ahn K-S (2016) Copper oxide nanoparticles aggravate airway inflammation and mucus production in asthmatic mice via MAPK signaling. Nanotoxicology 10:445–452

Radauer-Preiml I, Andosch A, Hawranek T, Luetz-Meindl U, Wiederstein M, Horejs-Hoeck J, Himly M, Boyles M, Duschl A (2015) Nanoparticle-allergen interactions mediate human allergic responses: protein corona characterization and cellular responses. Part Fibre Toxicol 13:3

Rayees S, Rochford I, Joshi JC, Joshi B, Banerjee S, Mehta D (2020) Macrophage TLR4 and PAR2 signaling: role in regulating vascular inflammatory injury and repair. Front Immunol 11:2091

Ronzani C, Casset A, Pons F (2014) Exposure to multi-walled carbon nanotubes results in aggravation of airway inflammation and remodeling and in increased production of epithelium-derived innate cytokines in a mouse model of asthma. Arch Toxicol 88:489–499

Rossi EM, Pylkkänen L, Koivisto AJ, Nykäsenoja H, Wolff H, Savolainen K, Alenius H (2010) Inhalation exposure to nanosized and fine TiO2 particles inhibits features of allergic asthma in a murine model. Part Fibre Toxicol 7:35. https://doi.org/10.1186/1743-8977-7-35

Royce SG, Cheng V, Samuel CS, Tang MLK (2012) The regulation of fibrosis in airway remodeling in asthma. Mol Cell Endocrinol 351:167–175

Runa S, Hussey M, Payne CK (2018) Nanoparticle–cell interactions: relevance for public health. J Phys Chem B 122:1009–1016

Rydman EM, Ilves M, Koivisto AJ, Kinaret PA, Fortino V, Savinko TS, Lehto MT, Pulkkinen V, Vippola M, Hämeri KJ, Matikainen S, Wolff H, Savolainen KM, Greco D, Alenius H (2014)

Inhalation of rod-like carbon nanotubes causes unconventional allergic airway inflammation. Part Fibre Toxicol 11:48

Ryman-Rasmussen JP, Tewksbury EW, Moss OR, Cesta MF, Wong BA, Bonner JC (2009) Inhaled multiwalled carbon nanotubes potentiate airway fibrosis in murine allergic asthma. Am J Respir Cell Mol Biol 40:349–358

Sayers BC, Taylor AJ, Glista-Baker EE, Shipley-Phillips JK, Dackor RT, Edin ML, Lih FB, Tomer KB, Zeldin DC, Langenbach R, Bonner JC (2013) Role of Cyclooxygenase-2 in exacerbation of allergen-induced airway remodeling by multiwalled carbon nanotubes. Am J Respir Cell Mol Biol 49:525–535

Schanen BC, Das S, Reilly CM, Warren WL, Self WT, Seal S, Drake DR (2013) Immunomodulation and T helper TH1/TH2 response polarization by CeO2 and TiO2 nanoparticles. PLoS One 8:e62816

Scuri M, Chen BT, Castranova V, Reynolds JS, Johnson VJ, Samsell L, Walton C, Piedimonte G (2010) Effects of titanium dioxide nanoparticle exposure on neuroimmune responses in rat airways. J Toxicol Environ Health Part A 73:1353–1369

Shannahan JH, Lai X, Ke PC, Podilla R, Brown JM, Witzmann FA (2013) Silver nanoparticle protein corona composition in cell culture media. PLoS One 8:e74001

Shipkowski KA, Taylor AJ, Thompson EA, Glista-Baker EE, Sayers BC, Messenger ZJ, Bauer RN, Jaspers I, Bonner JC (2015) An allergic lung microenvironment suppresses carbon nanotube-induced inflammasome activation via STAT6-dependent inhibition of Caspase-1. PLoS One 10:e0128888

Shurin MR, Yanamala N, Kisin ER, Tkach AV, Shurin GV, Murray AR, Leonard HD, Reynolds JS, Gutkin DW, Star A, Fadeel B, Savolainen K, Kagan VE, Shvedova AA (2014) Graphene oxide attenuates Th2-type immune responses, but augments airway remodeling and hyperresponsiveness in a murine model of asthma. ACS Nano 8:5585–5599

Silva RM, Anderson DS, Peake J, Edwards PC, Patchin ES, Guo T, Gordon T, Chen LC, Sun X, Van Winkle LS, Pinkerton KE (2016) Aerosolized silver nanoparticles in the rat lung and pulmonary responses over time. Toxicol Pathol 44:673–686

Staal YCM, van Triel JJ, Maarschalkerweerd TVP, Arts JHE, Duistermaat E, Muijser H, van de Sandt JJM, Kuper CF (2014) Inhaled multiwalled carbon nanotubes modulate the immune response of Trimellitic anhydride–induced chemical respiratory allergy in Brown Norway rats. Toxicol Pathol 42:1130–1142

Thompson EA, Sayers BC, Glista-Baker EE, Shipkowski KA, Ihrie MD, Duke KS, Taylor AJ, Bonner JC (2015) Role of signal transducer and activator of transcription 1 in murine allergen–induced airway remodeling and exacerbation by carbon nanotubes. Am J Respir Cell Mol Biol 53:625–636

Trudeau J, Hu H, Chibana K, Chu HW, Westcott JY, Wenzel SE (2006) Selective down-regulation of prostaglandin E2–related pathways by the Th2 cytokine IL-13. J Allergy Clin Immunol 117:1446–1454

Vasse GF, Melgert BN (2024) Microplastic and plastic pollution: impact on respiratory disease and health. Eur Respir Rev 33:230226

Vroman L (1962) Effect of absorbed proteins on the wettability of hydrophilic and hydrophobic solids. Nature 196:476–477

Wark PB, Gibson PG (2006) (2006) asthma exacerbations. 3: pathogenesis. Thorax 61:909–915

Wu W-T, Liao H-Y, Chung Y-T, Li W-F, Tsou T-C, Li L-A, Lin M-H, Ho J-J, Wu T-N, Liou S-H (2014) Effect of nanoparticles exposure on fractional exhaled nitric oxide (FENO) in workers exposed to nanomaterials. Int J Mol Sci 15:878–894

Yang Y-S, Cao M-D, Wang A, Liu Q-M, Zhu D-X, Zou Y, Ma L-L, Luo M, Shao Y, Xu D-D, Wei J-F, Sun J-L (2022) Nano-silica particles synergistically IgE-mediated mast cell activation exacerbating allergic inffammation in mice. Front Immunol 13:911300

Yoshida T, Yoshioka Y, Fujimura M, Yamashita K, Higashisaka K, Morishita Y, Kayamuro H, Nabeshi H, Nagano K, Abe Y, Kamada H, Tsunoda S, Itoh N, Yoshikawa T, Tsutsumi Y (2011) Promotion of allergic immune responses by intranasally-administered nanosilica particles in mice. Nanoscale Res Lett 6:195

Open Access This chapter is licensed under the terms of the Creative Commons Attribution-NonCommercial-NoDerivatives 4.0 International License (http://creativecommons.org/licenses/by-nc-nd/4.0/), which permits any noncommercial use, sharing, distribution and reproduction in any medium or format, as long as you give appropriate credit to the original author(s) and the source, provide a link to the Creative Commons license and indicate if you modified the licensed material. You do not have permission under this license to share adapted material derived from this chapter or parts of it.

The images or other third party material in this chapter are included in the chapter's Creative Commons license, unless indicated otherwise in a credit line to the material. If material is not included in the chapter's Creative Commons license and your intended use is not permitted by statutory regulation or exceeds the permitted use, you will need to obtain permission directly from the copyright holder.

Part II

Advanced In Vitro Models and Organ-on-a-Chip Used in Nanotoxicology

Advanced *In Vitro* Airway Models for NM Hazard Assessment

Katie McAllister, Ernesto Alfaro-Moreno, and Fiona Murphy

Abstract

There is an urgent need to identify hazards posed by novel nanomaterials (NMs) however, new models are required to streamline testing approaches, increase our understanding of mechanism of toxicity and incorporate Safer-by-Design (SbD) concepts into NM development. Considering one of the main routes of exposure to NMs is via inhalation, the development of *in vitro* airway models has been a priority area of research in recent years. Over the last 20 years new models and methods have ranged from simple submerged co-cultures of prominent lung cell types to the development of organoids and organ-on-a-chip approaches that include the complex cellular interactions. Current and future efforts to construct more advanced models aim to better mimic the dynamic physiological features of the lung tissue including mechanosignalling from tissue stretching and the incorporation of fluid flow. In this chapter, we will discuss the development and application of advanced *in vitro* models for assessing NM inhalation hazard and the potential impact for promoting SbD in NM development.

Keywords

Airway advanced models · In vitro lung · Safe by design

K. McAllister · F. Murphy (✉)
Strathclyde Institute of Pharmacy and Biomedical Sciences, University of Strathclyde, Glasgow, UK
e-mail: f.murphy@strath.ac.uk

E. Alfaro-Moreno
Nanosafety Group, International Iberian Nanotechnology Laboratory, Braga, Portugal

© The Author(s) 2025
E. Alfaro-Moreno, F. Murphy (eds.), *Nanosafety*,
https://doi.org/10.1007/978-3-031-93871-9_6

1 Introduction

1.1 Exposure by Inhalation

Inhalation of nanomaterials is probably one of the most studied aspects in the evaluation of nanomaterials (NMs) (Lebre et al. 2022). Considering that from bronchi to the alveoli, there is a large variety of cellular types ranging from different populations of epithelial cells (ciliated, pneumocytes type I and II), macrophages, fibroblast, smooth muscle, mastocytes, endothelial cells, just to mention the main subtypes (Tomashefski and Farver 2008), the use and development of simple and complex *in vitro* model may be considered as a far cry from reality, but nonetheless, the use of these type of models have been shown to be a relevant approach to understand the mechanisms of observations made in whole organism approaches (Miller and Spence 2017). On top of these considerations, we must include the challenge related to relevant exposure systems, mimicking real life scenarios. Different approaches from submerged conditions (Alfaro-Moreno et al. 2008), air-liquid-interface (ALI) (Upadhyay and Palmberg 2018), quasi ALI (Franken et al. 2023), including physiological characteristics such as tissue movement (Sengupta et al. 2023) and/or vascularisation (Huh et al. 2011), makes the respiratory tract a very complex system to address *in vitro* (Fig. 6.1).

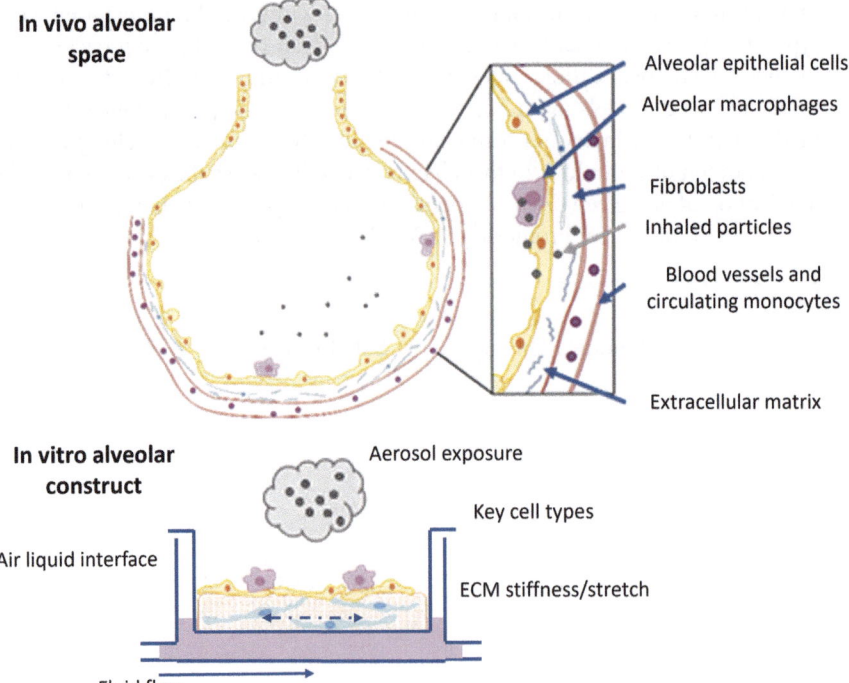

Fig. 6.1 Schematic diagram highlighting the critical features of the alveolar space which should be included in development of physiologically relevant in vitro alveolar constructs

1.2   IATA and Tiered Testing Strategies

There is an urgent need to identify hazards posed by novel NMs however, the vast number of different NMs in production and under development precludes the complete hazard assessment of each individual substance according to current regulatory standards which require long term *in vivo* studies (Moné et al. 2020). New models are essential to streamline testing approaches, increase our understanding of mechanism of toxicity and incorporate Safer-by-Design (SbD) concepts into NM development. This has stimulated efforts to design alternative strategies to streamline the risk assessment process (Fischer et al. 2020). One such approach gaining prominence is the development and use of Integrated Approaches to Testing and Assessment (IATAs) (Sakuratani et al. 2018). Structured as decision trees, IATAs lead the user through an efficient process to identify the key information required to support hazard assessment and risk decision making with regards NMs under investigation (Stone et al. 2020). IATAs, delineated by the primary routes of NM exposure, have successfully been used as the basis of similarity assessment to support grouping of NMs which pose a similar inhalation, ingestion, or dermal hazard (Di Cristo et al. 2022; Murphy et al. 2021; Di Cristo et al. 2021; Braakhuis et al. 2021) and also incorporated into risk decision making framework for the injection of nanobiomedicines (Powell et al. 2022). Recognizing the need to reduce the burden of animal testing in toxicology, the IATAs follow a tiered structure and prioritize data generation utilizing *in silico* and *in vitro* models to categorize hazard before progressing to *in vivo* systems in limited, high priority cases (Fig. 6.2).

Simple 2D cell monocultures are the predominant models used in the Tier 1 safety assessment of inhaled particles to address human health concerns. These models are developed by culturing a single cell line in a tissue culture plate to mimic the epithelial layer of the lung. The different compartments of the lung are lined with distinct epithelial cell types each with specific features tailored to the requirements of the local environment. The epithelial cells of the bronchus and bronchioles are lined with cuboidal epithelial cells which produce surfactant protein and mucins and are covered with cilia which enable the clearance of particles from the lungs forming an important defence mechanism. Alveoli are composed of two distinct types of epithelial cells; Type one alveolar epithelial cells (ATI) are large, thin cells critical for the air–blood barrier function of lungs and gas exchange, whereas type 2 alveolar epithelial cells (ATII) are less numerous but highly metabolically active cells and responsible for surfactant production, ion transport, and proliferation and differentiation to ATI cells after injury.

Primary epithelial cells are difficult to isolate from the lungs therefore the majority of *in vitro* studies rely on lung epithelial cell lines. A number of cell lines are commonly used in Tier 1 nanotoxicology studies which have been derived from different compartments of the lung and therefore recapitulate different physiological features and functionalities. A549 adenocarcinoma line is the most widely used small airway epithelial cell line. A549 cells are similar to ATII cells as they possess lamellar bodies and microvilli, and express surfactant proteins (SP-A, C, D), and mucins (MUC1). However, A549 cells do not form intact barriers. Calu-3 cells

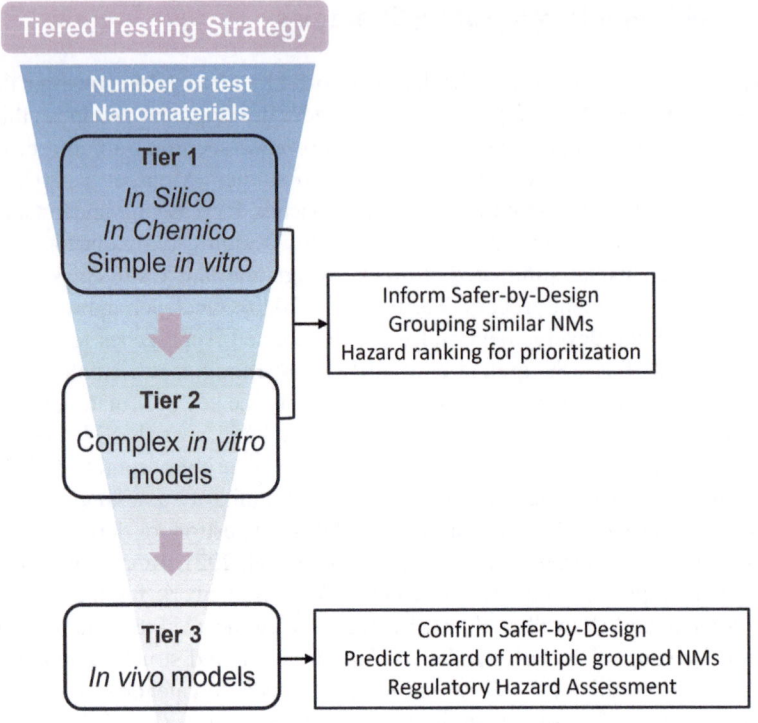

Fig. 6.2 Schematic illustration of tiered testing strategy to reduce number of NMs that require *in vivo* testing. Tier 1 and Tier 2 testing may be used to screen large number of NMs to inform Safer-by-Design innovation, to provide evidence of similarity to support grouping NMs and for comparative hazard ranking to identify NMs of most concern. Select NMs will contribute to Tier 3 *in vivo* testing to confirm predicted hazard and evidence mitigation of hazard by Safer-by-Design interventions, provide *in vivo* data for regulatory hazard assessment for targeted high priority NMs and candidate NMs selected to represent a larger group

exhibit bronchial epithelial phenotype, form mixed ciliated and secretory cell populations, and express the cystic fibrosis transmembrane conductance regulator (CFTR), proSP-C and mucin (Kreft et al. 2015). They form tight monolayers and are widely used to study barrier integrity in response to toxic insult and infection. BEAS-2B cells possess both airway and bronchial epithelial cell properties, expressing vimentin, collagen I, E-cadherin, ICAM-1, VCAM-1 (Stewart et al. 2012). The human bronchial epithelial cell line (16HBE) express CFTR and are commonly used for disease modelling however this cell line is used less frequently to assess NM toxicity (Callaghan et al. 2020). hAELVi cells are a primary cell line which closely resemble that of ATI cells. They have been shown to form tight barriers expressing high levels of ZO-1 and occludens, as well as similar protein expression levels when compared to isolated human ATI-like cells in primary culture (hAEpC)

(Kuehn et al. 2016). The choice of cell line may be dependent on the desire to replicate a specific region of the lung, e.g., A549 cells to replicate alveolar space, or a feature of lung physiology, e.g., barrier formation. A number of new lung epithelial cells have recently come to market but have not yet been tested extensively in applications such as nanotoxicity testing (Sengupta et al. 2023; Carius et al. 2023).

Monocultures exposed to NM test materials can be used to identify biologically reactive NMs providing useful information on the acute toxicity of NMs which may be used to identify inherent drivers of toxicity, inform mechanism of action and for hazard ranking between NMs. For example, a study by Rosario et al. compared the toxicity of silver NPs of differing sizes using the lung alveolar epithelial cell line A549 monoculture model, measuring cell viability and DNA damage with the data suggesting small size and, by extension high surface area, is a critical physicochemical mediator of toxicity (Rosário et al. 2018). Copper oxide (CuO) NPs have also been tested on A549 monocultures revealing that CuO NPs increase oxidative stress by inducing increased levels of reactive oxidative species (ROS) as well as causing genotoxicity through CuO NP induced micronuclei (Akhtar et al. 2016). The mechanism of toxicity appears to be as a result of dissolution of CuO in the biological media to highly reactive Cu^{2+} ions. Endpoints are generally limited to cytotoxicity, oxidative stress, pro-inflammatory responses and direct genotoxicity addressing several hypotheses regarding how NMs induce adverse cellular effects: (i) via induction of oxidative stress paradigm which then leads to pro-inflammatory effects (Donaldson et al. 2013), (ii) through genotoxicity (Schins and Knaapen 2007), and (iii) via NM dissolution, i.e. release of potentially toxic ions and/or other constituents (Bergin and Witzmann 2013; Sabella et al. 2014). These models, although simple, enable us to predict toxicity outcomes to highly reactive NMs in an ethical, cost effective and easily accessible way.

Although simple *in vitro* monocultures may be sufficient to predict acute toxicity outcomes to highly reactive NMs (Wiemann et al. 2016; Landsiedel et al. 2014), these systems fail to replicate the complexity of the lung tissue. Identifying particles with the potential to cause complex and chronic diseases after exposure via inhalation, such as fibrosis and lung cancer, requires an understanding of how interactions between particles and target cells stimulate pro-inflammatory and pro-fibrotic changes leading to pathology, but also how multiple cell types present locally in the target tissue and within systemic systems contribute to the disease microenvironment, as crucial drivers of disease. Co-culture models using two or more cell lines have shown that interactions between cells may lead to altered cellular responses to NM exposure when compared to monocultures. Chortarea et al. (2015) (Chortarea et al. 2015) reported greater resistance of Calu-3 cells to toxic insult when cultured in co-culture models compared to monocultures of Calu-3 alone. Similarly, Müller et al. (2010), (Müller et al. 2010) have previously presented an alternative immune response to NMs when triple cell co-cultures consisting of A549 human epithelial lung cells, human monocyte-derived macrophages and monocyte-derived dendritic cells exposed to combustion-derived NPs (diesel exhaust particles) and to manufactured NPs (titanium dioxide and single-walled carbon nanotubes) were compared to single cell cultures. Interestingly, responses in the triple cell co-culture showed decreased total antioxidant capacity and IL-8 concentrations whereas the TNFα concentrations were higher than the expected values calculated from the monocultures. The above studies exemplify how monoculture models may identify acute toxicity but may not be suitable to understand the complex mechanism underpinning the immunomodulatory effects of NM exposure (Hofer et al. 2022).

The interactions between different cell types and the structural components of the lung tissue response to toxic agents cannot be replicated in simple 2D cell monoculture models (Langhans 2018) e.g., the higher stiffness of the tissue culture dishes compared to the lung parenchyma, i.e. 10^6 kPa vs. 1–15 kPa, significantly influences cellular phenotype (Yanagihara et al. 2020). Lee et al. recently evaluated the potential for a multicell airway tissue model constructed by sequential layering of endothelial cells, collagen impregnated with fibroblasts and Calu-3 lung epithelial cells to assess NM toxicity in comparison to Calu-3 cell monocultures (Lee et al. 2025). Although both models demonstrated an acute and persistent induction of inflammatory cytokine release, inflammatory response of Calu-3 2D cell monoculture was induced by the lowest exposure concentrations whereas higher exposure concentrations were required to stimulate release of measurable levels of cytokine release in the airway tissue model. Furthermore, acute cytotoxicity to ZnO NMs was measured in Calu-3 monoculture starting at 24 hours whereas apoptotic cells in the Calu-3 cells of the airway tissue model appeared only at 14 days post exposure. Differences in time course and magnitude of response to NM exposure detected by models from each Tier could have a significant impact on interpretation of hazard potential with simple Tier 1 models potentially over-representing acute toxicity while lacking the required sensitivity and longevity to detect more subtle immunomodulatory changes.

The conditions employed in the growth of cells in culture are critical to the characteristics of the resultant cell layer. *A wide variety of studies have been performed comparing the culture of lung cells in Air-Liquid Interface (ALI) and in submerged conditions, revealing that cells displayed phenotypic differences* (Bhowmick and Gappa-Fahlenkamp 2016; Kreft et al. 2015; Ghio et al. 2013*)*. A549 cells showed increased expression of alveolar epithelial specific markers such as aquaporin-5 in ALI cultured cells compared to submerged (Wu et al. 2017). Similarly, when Calu-3 cells cultured at the ALI were compared to cells cultured under submerged conditions greater electrical resistance values and barrier formation were observed (Grainger et al. 2006). A wealth of data now exists demonstrating how lung epithelial cells cultured in ALI conditions mature to form a functional and differentiated lung epithelial barrier, better replicating the *in vivo* situation and therefore should be the preferred culture conditions for NM toxicity testing.

In vivo models are important for the evaluation of particle deposition efficiency, tissue distribution and biopersistence and to study the long-term local and systemic pathological responses to inhaled substances. As no Tier 2 *in vitro* airway model has yet undergone the rigorous testing and validation process required to build support for widespread adoption, animal models remain heavily relied upon to confirm whether inhaled particles pose a hazard to human health (Murphy et al. 2022; Murphy et al. 2021; Braakhuis et al. 2021). However, although they can recapitulate key pathological changes in some lung diseases fundamental differences in anatomy and physiology between humans and rodents present critical points of departure which may severely impact interpretation of results. The combination of differences in anatomy, physiology and host immune responses contributes to the varied sensitivity to inhaled toxicants between species (Proudfoot et al. 2011), which presents difficulties for predicting human hazard from *in vivo* data.

1.3 Requirements of a Tier 2 NAM for Predicting Inhalation Toxicity of NM

According to Nossa et al. (Nossa et al. 2021), the specifications for an optimal "physiologically relevant" engineered human *in vitro* alveolus/airway model include (1) the use of human-derived cells that comprise the native alveolar barrier and include an epithelial layer of simple squamous epithelium; a layer of endothelial cells of the capillary wall; and the basement membrane between the two, (2) exposure of cells to the air-liquid interface (ALI) to mimic the *in vivo* microenvironment where the epithelial lung cells are in contact with humid air on one side and blood on the other and (3) incorporation of a substrate for growing cells with properties similar to native tissue in terms of chemical composition and biomechanical behaviour. Additional advantageous mechanical features would be a fluidic system that reproduces the blood flow through the alveolar capillaries (mean velocity$\approx$1 mm/s, flow rate 2–5 mL/min in an adult) to provide adequate oxygenation and nutrients to the cell, as well as physiological shear stress to endothelial cells (around 1.5 Pa), and the ability to subject the substrate to mechanical cyclic stretching to reproduce alveolar barrier motion during breathing.

1.4 Multicell Models

Regarding the construction of the different models, recent progress has been made in the development of multicell 3D models which mimic the cellular constituents of tissues and organs (Summarised in Table 6.1). However, it is not feasible to replicate the full complexity of the lung tissue, therefore decisions are made prioritising the selection of certain cells. From the *in vitro* airway models developed to date, differences in cell selection exist with some approaches focusing on the combination of epithelial and immune cells whereas other approaches prioritize the inclusion of lung epithelial and structural stromal cells including fibroblasts and endothelial cells.

Majority of models include immune cells reflecting the importance of the macrophages as a primary defence mechanism against inhaled foreign substances which directly interact with inhaled substances and propagate the subsequent inflammatory response (Wiemann et al. 2016). Macrophage and dendritic cells reside throughout the lung tissue, in the alveolar space and underlying the epithelial layer, poised to encounter foreign material, infections or tissue damage. Tissue resident

Table 6.1 Multicell airway models used in particle hazard assessment

Cell combinations	Particle treatment	Endpoints assessed	Reference
Alveolar epithelial cells (A549) Macrophages (differentiated THP-1) Mast cells (HMC-1) Endothelial cells (EA.hy926)	PM_{10} collected from the industrial region of the Metropolitan Zone of Mexico City	Pro-inflammatory cytokine release	Alfaro-Moreno et al. (2008)
Alveolar epithelial cells (A549) Monocyte-derived dendritic cells (MDDCs) Macrophages (MDMs)	Multi-walled carbon nanotubes (MWCNTs)	Cytotoxicity (LDH release) Cell morphology Glutathione concentration Pro-inflammatory cytokine release	Chortarea et al. (2015)
Macrophages (differentiated THP-1) Bronchial epithelial cells (Beas-2B)	Tungsten carbide–cobalt nanoparticles	Pro-inflammatory cytokine release Macrophage polarization assay	Armstead and Li (2016)
Bronchial epithelial cells (16HBE) Vascular endothelial cells (HUVEC) Macrophages (differentiated THP-1)	Silver nanoparticles	Trans epithelial electrical resistance (TEER) ROS production Pro-inflammatory cytokine release	Braakhuis et al. (2016)
Alveolar epithelial cells (A549) Endothelial cells (EA.hy926)	Nano Zerovalent iron (ZVI) particles	Cell viability TEER Pro-inflammatory cytokine release	Sun et al. (2016)
Alveolar epithelial lentivirus immortalized cells (hAELVi) Macrophages (differentiated THP-1)	Silver nanoparticles Starch nanoparticles	Cell viability TEER Permeability assay Immunofluorescence for cell markers	Kletting et al. (2018)
Alveolar epithelial cells (A549) Macrophages (differntiated THP-1)	PM_1 from the combustion of three different wood logs (birch, beech and spruce) in a modern masonry heater PM_1 emissions from the combustion of spruce pellets in a fully automated pellet boiler	Cytotoxicity Cell metabolic activity ROS production Pro-inflammatory cytokine release Genotoxicity COMET assay	Kasurinen et al. (2018)

(continued)

Table 6.1 (continued)

Cell combinations	Particle treatment	Endpoints assessed	Reference
Alveolar epithelial cells (A549) Macrophages (differentiated THP-1) Endothelial cells (EA.hy926)	Ambient $PM_{2.5}$ collecting from Shanghai city in China	TEER Cell viability Cytotoxicity Pro-inflammatory cytokine release Expression levels of adhesion related genes	Wang et al. (2019)
Bronchial epithelial cells (Calu-3) Endothelial cells (EA.hy926) Macrophages (differentiated THP-1)	Silver nanoparticles coated with tannic acid	Cytotoxicity Cell viability and proliferation Pro-inflammatory cytokine release	Zhang et al. (2019)
Alveolar epithelial cells (A549) Macrophages (differentiated THP-1) Fibroblast cells (MRC-5)	Quartz (min-U-Sil, DQ12) MWCNT (Mitsui-7, Nanocyl)	Cytotoxicity (LDH release) Cell morphology Glutathione (GSH) content Pro-inflammatory cytokine release Pro-fibrotic cytokine release	Barosova et al. (2020)
Alveolar epithelial cells (A549) Macrophages (differentiated THP-1) Endothelial cells (EA.hy926)	Citrate-capped gold nanoparticles 15% silver on silica nanoparticles Copper oxide nanoparticles	Cell viability ROS production Cytotoxicity (LDH release) Interleukin (IL)-8 concentration	Wang et al. (2020)
Alveolar epithelial cells (A549) Leukemia cells (U937)	10 nm citrate coated silver nanoparticles	Cellular viability Cytotoxicity ROS production Heat shock protein 70 expression Pro-inflammatory cytokines release	Braun et al. (2021)
Bronchial epithelial cells (16HBE) Monocytes (THP-1) Lung microvascular endothelial cells (HLMVEC)	Graphene nanoparticles Graphene oxide nanoparticles	TEER and dextran permeability Immunofluorescene for cell markers and tight junction proteins Cell viability (WST-1) Cytotoxicity (LDH release) Phosphproteomics	Van Den Broucke et al. (2021)
Alveolar epithelial cells (A549) Monocytes (THP-1) Endothelial cells (EA.hy926)	Silver nanomaterials	Fluorescein barrier integrity Cellular viability Genotoxicity COMET assay	Camassa et al. (2022)

(continued)

Table 6.1 (continued)

Cell combinations	Particle treatment	Endpoints assessed	Reference
Bronchial epithelial cells (Beas-2B) Leukemia cells (U937)	Fine particulate matter ($PM_{2.5}$) Nicotine-derived nitrosamine ketone (NNK) Benzo(a)pyrene diol epoxide (BPDE)	Cell viability EdU proliferation assay Apoptosis assay Genotoxicity COMET assay Pro-inflammatory cytokine release	Zhou et al. (2022)
Alveolar epithelial cells (AXiAECs) Bronchial epithelial cells (Calu-3) Human primary alveolar type II cells (AXhAEpCs) Macrophages (differentiated THP-1) Lung microvascular endothelial cells (HLMVECs)	Titanium dioxide nanoparticles Zinc oxide nanoparticles Polyhexamethylene guanidine (PHMG) Fluticasone (FL)	Immunofluorescence staining for cell markers and tight junction proteins. Transbarrier electrical resistance (TER) Cytotoxicity (LDH release) ROS production	Sengupta et al. (2023)
Bronchial epithelial cells (Calu-3) Macrophages (differentiated THP-1)	Lipopolysaccharide Quartz Titanium dioxide nanoparticles	Particle deposition efficiency TEER Cell viability Pro-inflammatory cytokine release	Braakhuis et al. (2023)

immune cells, as well as recruited cells—monocytes and neutrophils—actively regulate homeostasis and provide immediate protection from foreign pathogens and particulates. Inclusion of immune cells into models will allow the immunomodulatory effects of NMs to be explored *in vitro*. Primary monocyte cells isolated from human blood are frequently incorporated into *in vitro* lung models to recapitulate the immune cell niche. However as primary human blood cells are not always readily available the human monocyte cell line, THP-1 is also commonly incorporated into *in vitro* lung models as a surrogate immune cell-line.

Fibroblasts are the major producers of collagen and play an active role in the dynamic organization of collagen network structures in the basement membrane. The inclusion of fibroblasts could therefore support the integrity and longevity of an *in vitro* airway model. Activation of fibroblasts to myofibroblast, key players in the development of pulmonary fibrosis can propagate the inflammatory response; fibroblast activation may occur through direct interaction with toxic substances or alternatively in response to signals released by other cells influencing the tissue microenvironment (Ma et al. 2017; Li et al. 2021).

Endothelial cells are key components of the alveolar tissue and facilitate gas exchange with the circulatory system. As endothelial cells can respond to both direct interaction with NMs or indirect signals from other NM-exposed cells their inclusion in *in vitro* models is critical to faithfully replicate the alveolar tissue. *In vivo* studies have suggested endothelial cells place an important role in the amplification of acute phase responses, a feature of human disease such as metal fume fever.

1.5 Extracellular Matrix

From our review of the literature the majority of complex *in vitro* models do not contain an ECM component, with cells directly grown on tissue culture plastics and polymer membrane inserts. A limited number of studies have reported the inclusion of collagen or Matrigel to provide ECM component to the models (Table 6.2). The ECM of the lungs is composed of a complex 3D network of proteins, glycoproteins, and polysaccharides. Once regarded as a predominantly structural framework, it is now well recognized that the ECM incorporates a diverse array of bioactive molecules with important functional roles (Bissell 2007). Cell-ECM interactions can direct cell survival, adhesion, migration, proliferation, and differentiation. Collagen, which constitute the main structural element of the ECM of the lung (Onursal et al. 2021), providing tensile strength, regulating cell adhesion, supporting chemotaxis and migration, and directing tissue development (Rozario and Desimone 2010) was included in this model to account for the potential influence of the ECM-cell signalling on lung epithelial cell sensitivity to toxic insult. Although there are some examples of alternative models which have coated tissue culture plastic with biomimetic ECM components.

Kang et al. (2022) (Kang et al. 2022) described an artificially constructed airway tissue model used in hazard assessment of large dust particles which utilised the ECM component to perform as a tissue scaffold supporting the growth of interstitial fibroblasts as occurs naturally in the lung tissue. This model incorporates an endothelial cell layer below the ECM scaffold and an epithelial layer composed of Calu-3 cells on the surface. The airway tissue model was recently been employed in hazard assessment of NM (Lee et al. 2025, in press). Cells in the construct remained viable over the 2 week experimental timecourse and successfully differentiated between high and low reactivity NMs; ZnO NM induced expression of inflammatory mediators, release of pro-inflammatory cytokines and apoptosis of the Calu-3 epithelial cells, whereas no changes were observed after exposure to $BaSO_4$ NMs.

1.6 Barriers to Widespread Adoption

Transferability is key to the wider adoption and route to validation of novel alternative models. A primary limitation which may present a barrier to wider adoption of more advance models includes the common requirement for technical expertise and specialist equipment for the construction of complex, multicomponent and dynamic systems. Simple static systems may have lower barriers to wider adoption but as highlighted above at the expense of other aspects of physiological relevance. Increasing recognition of the need to include mechanical components such as the

Table 6.2 Models including extracellular matrix components

Cell types	ECM	Particle treatment	Timeframe	Endpoints assessed	Reference
Human primary alveolar epithelial cells (HPAEpi cells) Human vascular endothelial cells (HUVEC)	Matrigel	Titanium dioxide nanoparticles Zinc oxide nanoparticles	24 hours	TEER and dextran permeability Cadherin expression Intracellular ROS production Apoptosis	Zhang et al. (2018)
Rat primary brain microvascular endothelial cells Rat primary microglia	Collagen IV and fibronectin-coated plates	Ultrafine diesel exhaust particles	24 hours	Lactate dehydrogenase (LDH) leakage ROS productioncell metabolic activity Production of 27 inflammatory markers.	Aquino et al. (2021)
Mouse monocytic macrophages (RAW264.7) human alveolar epithelial cells (A549)human fibroblast cells (MRC-5)	Matrigel	Silica nanoparticles	24 hours	Cell morphologycell migrationpro-inflammatory protein expression Epithelial-mesenchymal transition	Yang et al. (2023)
Human bronchial epithelial cells (Calu-3) Human fibroblast cells (MRC-5) Human lung endothelial cells (HULEC)	Collagen IV	Zinc oxide nanoparticles Barium sulfide nanoparticles	14 days	TEER Tissue morphology Gene expression changes Pro-inflammatory cytokine release Apoptosis	Lee et al. (2025, In press)

incorporation of microfluidics to mimic blood flow and mechanical stress to replicate the breathing action of the lung, to continue to build physiological relevance, emphasizes the need for an interdisciplinary approach to model design and application to prevent the additional technical requirements of these models acting as barriers to wider adoption in future (Table 6.3).

1.7 Incorporation of NAM in Hazard Assessment Strategies

The successful adoption of an *in vitro* skin model for testing irritancy and skin sensitization of substances has greatly reduced the need to test substances included in household products and cosmetics using *in vivo* models (Schäfer-Korting et al. 2008). However, for regulatory acceptance of *in vitro* models, the robustness, reproducibility, and relevance need to be demonstrated and no *in vitro* airway tissue

Table 6.3 Summary of review of complex *in vitro* models used for NM hazard assessment

Cell types	Variety of cell types included in models
	Majority of models rely on cell lines
	Most common combinations include epithelial cells, fibroblasts and macrophages
Extracellular matrix	Majority of models do not include ECM
Model set-up	Majority of models co-culture cells supported by Transwell insert
	Majority of models culture epithelial cells at ALI
	Exposure via cloud system is commonly used
	Recent focus on incorporation of dynamic flow
Particle treatment	Wide variety of particle types have been tested
Timeframe	Majority of studies report acute 24-48 hours exposures.
	Few models report repeat exposures or longer-term responses
Endpoints assessed	Majority of studies report common toxicity endpoints of cytotoxicity, cell viability/metabolic activity, pro-inflammatory cytokine release, TEER and barrier integrity, oxidative stress and genotoxicity
Comparison to simple *in vitro*	Majority of studies include direct comparison to simpler models such as mono-cell culture
	No consistent trend of increased or decreased sensitivity of complex models to particle exposure compared to simple system
Predictivity of *in vivo* response	Few studies report direct comparison to *in vivo* responses to the same material
	Some studies refer directly to *in vivo* responses elicited by the same test materials
	Most studies refer to *in vivo* responses to similar test materials e.g., same chemical composition reported in the literature
	Most models replicate some pathological changes observed *in vivo* however evidence of increased predictivity over simple system or increased physiological relevance of complex model remains limited
Transferability	Most models are reliant on widely available reagents such as commercially available cell lines and Transwell inserts
	Some models require specialized equipment and technical expertise
	Few models have been tested in intralaboratory comparison

model has to date reached this threshold. Relevance of the model or "to what extent does the test correctly measure or predict the biological effect of interest?" is one of the fundamental requirements (Gourmelon and Delrue 2016). Relevance incorporates consideration of the accuracy (concordance) of a test method, biological plausibility and scientific basis of test methods as documented in the scientific literature. Selection of endpoints related to cell viability, cytotoxicity, oxidative stress and pro-inflammatory changes comply with the scientific evidence are considered key to assess human hazard concerns (Ruijter et al. 2023). Additionally, transferability and reliability of the model must be evidenced. This includes demonstrating consistent reproducibility of results expected when the test is repeated, outside of the laboratory that initially developed the test as well as providing an assessment of the extent that a test method can be performed reproducibly within and between laboratories over time, using the same protocol. A recent study (Braakhuis et al. 2023) aimed to evaluate the transferability and reproducibility of a lung model consisting of the human bronchial cell line Calu-3 as a monoculture and, to increase the physiologic relevance of the model, also as a coculture with macrophages (either derived from the THP-1 monocyte cell line or from human blood monocytes). In an inter-laboratory comparison with 7 participating labs, the lung models were exposed to NMs using the VITROCELL® Cloud12 to replicate the ALI. Although results were promising inconsistencies between labs were apparent highlighting the difficulty in aligning protocols and procedures in different lab set-ups. The process of validation is lengthy but will be required to allow hazard data from *in vitro* models to be accepted by regulatory agencies.

Alternative models could however be immediately used for the purpose of Safe(r)-by-Design (SbD) NM development. SbD aims to reduce the human and environmental risk of a substance throughout its entire life cycle by minimizing or eliminating the hazard and/or by reducing exposure (Commission et al. 2017). SbD processes provide a comparative approach, where several versions of an innovative material are compared against each other and against a similar "conventional" material with a well-characterized toxicity profile, including considerations of environmental health and safety concerns alongside functionality. By using SbD, hazard potential is considered from the very early design phases such that the likelihood a new material fails at the market entry stage due to safety issues is drastically decreased. SbD hazard testing aims to identify hazard warnings in the early stages of the innovation process using simple *in vitro* methods. Once a product is designed and produced, the manufacturer should comply with the regulations and perform hazard and risk assessment accordingly. Less stringent hazard assessment is required for initial SbD decision making, however, reliable high/medium-throughput models which balance predictivity, speed and cost are still needed for early-stage screening of multiple NMs to select candidates for further development and investment as the time-to-market is shortened only if SbD decisions are substantiated by evidence that is robust (Kraegeloh et al. 2018). The validity of the SbD process and by extension the biological relevance of methods/models employed within this process is therefore key.

A set of performance criteria has been proposed to evaluate the suitability of *in vitro* methods for hazard assessment of NMs for the purpose of SbD (Ruijter et al. 2023). Criteria were primarily designed to be appropriate to assess the performance and state of readiness of endpoint measurement assays however a number of the criteria could similarly be applied to the evaluation of novel *in vitro* models, such as predictivity of the *in vivo* situation. Prediction accuracy of a model is considered a function of specificity, ability to detect true negatives and sensitivity, ability to detect true positives. It is important to highlight that direct correlation of *in vitro* effects to *in vivo* potency may not be possible due to the simplicity of the endpoints measured in *in vitro* studies. Correlation is not a requirement of SbD but rather the emphasis should be placed on the ability of a model or assay to detect potentially hazardous materials by their ability to trigger known mechanisms of toxicity. For SbD purposes a model should be robust and reliable with evidenced consistency in model construction. Furthermore, the model must be compatible for use with NMs, potential for NM interference in model performance or assessment of endpoint read-outs (as is the case for MTT assay) should be considered and mitigated where possible.

1.8 Adverse Outcome Pathways (AOP)

Physiological relevance of an *in vitro* model should be demonstrated by the ability of the model to respond to toxic insult in a manner which replicates pathological changes observed *in vivo*. Adverse outcome pathways (AOP) are a useful tool to inform selection of relevant endpoints for *in vitro* models based on key events in the pathogenesis of lung disease identified *in vivo* at the cellular, tissue and organ level. The ability to control, tailor and modify *in vitro* model systems may also provide beneficial for the further development of tools such as AOPs. If employed in an iterative process of review and refine, advanced *in vitro* models can contribute to an AOP at mechanistic level by providing specific and targeted information on nano-bio interface interactions and activation/inhibition of key signalling pathways to delineate relationships between key events in a controlled but biologically complex model.

2 Concluding Remarks

The development of advanced *in vitro* lung models for hazard assessment of nanomaterials (NMs) has progressed significantly in recent years, reflecting the growing need for effective toxicity testing in the context of increasing nanomaterial use across various industries. Many complex *in vitro* models have been established, incorporating features such as three-dimensional cell cultures, co-culturing of various cell types, and the use of biomimetic materials to better replicate the lung's microenvironment. Despite this progress, no single model has yet been validated or widely adopted across the field, highlighting the need for a unified approach.

Ongoing optimization efforts focus on enhancing the physiological relevance of these models, particularly in mimicking the dynamic processes of lung physiology, such as airflow, fluid dynamics, and immune responses. By leveraging advances in bioengineering—such as bioprinting, which allows for precise tissue architecture, and microfluidics, which enables the simulation of realistic fluid flows—researchers are striving to create more accurate representations of lung tissue.

However, for these new models to gain traction in regulatory contexts, standardization and harmonization are crucial. Establishing common protocols and benchmarks will promote wider use, facilitate validation, and ensure consistency in results across different laboratories. Ultimately, the integration of these advanced *in vitro* models into regulatory hazard assessments could provide more reliable data on the safety of nanomaterials, helping to protect public health and the environment while fostering innovation in nanotechnology.

References

Akhtar MJ, Kumar S, Alhadlaq HA, Alrokayan SA, Abu-Salah KM, Ahamed M (2016) Dose-dependent genotoxicity of copper oxide nanoparticles stimulated by reactive oxygen species in human lung epithelial cells. Toxicol Ind Health 32:809–821

Alfaro-Moreno E, Nawrot TS, Vanaudenaerde BM, Hoylaerts MF, Vanoirbeek JA, Nemery B, Hoet PHM (2008) Co-cultures of multiple cell types mimic pulmonary cell communication in response to urban PM10. Eur Respir J 32:1184

Aquino GV, Dabi A, Odom GJ, Zhang F, Bruce ED (2021) Evaluating the endothelial-microglial interaction and comprehensive inflammatory marker profiles under acute exposure to ultrafine diesel exhaust particles in vitro. Toxicology 454:152748

Armstead AL, Li B (2016) In vitro inflammatory effects of hard metal (WC-Co) nanoparticle exposure. Int J Nanomedicine 11:6195–6206

Barosova H, Karakocak BB, Septiadi D, Petri-Fink A, Stone V, Rothen-Rutishauser B (2020) An in vitro lung system to assess the proinflammatory Hazard of carbon nanotube aerosols. Int J Mol Sci 21:5335

Bergin IL, Witzmann FA (2013) Nanoparticle toxicity by the gastrointestinal route: evidence and knowledge gaps. Int J Biomed Nanosci Nanotechnol 3

Bhowmick R, Gappa-Fahlenkamp H (2016) Cells and culture systems used to model the small airway epithelium. Lung 194:419–428

Bissell MJ (2007) Architecture is the message: the role of extracellular matrix and 3-D structure in tissue-specific gene expression and breast cancer. Pezcoller Found J 16:2–17

Braakhuis HM, Giannakou C, Peijnenburg WJ, Vermeulen J, van Loveren H, Park MV (2016) Simple in vitro models can predict pulmonary toxicity of silver nanoparticles. Nanotoxicology 10:770–779

Braakhuis HM, Murphy F, Ma-Hock L, Dekkers S, Keller J, Oomen AG, Stone V (2021) An integrated approach to testing and assessment to support grouping and read-across of nanomaterials after inhalation exposure. Appl In Vitro Toxicol 7:112–128

Braakhuis HM, Gremmer ER, Bannuscher A, Drasler B, Keshavan S, Rothen-Rutishauser B, Birk B, Verlohner A, Landsiedel R, Meldrum K, Doak SH, Clift MJD, Erdem JS, Foss OAH, Zienolddiny-Narui S, Serchi T, Moschini E, Weber P, Burla S, Kumar P, Schmid O, Zwart E, Vermeulen JP, Vandebriel RJ (2023) Transferability and reproducibility of exposed air-liquid interface co-culture lung models. NanoImpact 31:100466

Braun NJ, Galaska RM, Jewett ME, Krupa KA (2021) Implementation of a dynamic co-culture model abated silver nanoparticle interactions and nanotoxicological outcomes in vitro. Nanomaterials (Basel):11

Callaghan PJ, Ferrick B, Rybakovsky E, Thomas S, Mullin JM (2020) Epithelial barrier function properties of the 16HBE14o- human bronchial epithelial cell culture model. Biosci Rep 40

Camassa LMA, Elje E, Mariussen E, Longhin EM, Dusinska M, Zienolddiny-Narui S, Rundén-Pran E (2022) Advanced respiratory models for Hazard assessment of nanomaterials-performance of mono-, co- and tricultures. Nanomaterials (Basel) 12

Carius P, Jungmann A, Bechtel M, Grißmer A, Boese A, Gasparoni G, Salhab A, Seipelt R, Urbschat K, Richter C, Meier C, Bojkova D, Cinatl J, Walter J, Schneider-Daum N, Lehr CM (2023) A monoclonal human alveolar epithelial cell line ("Arlo") with pronounced barrier function for studying drug permeability and viral infections. Adv Sci (Weinh) 10:e2207301

Chortarea S, Clift MJD, Vanhecke D, Endes C, Wick P, Petri-Fink A, Rothen-Rutishauser B (2015) Repeated exposure to carbon nanotube-based aerosols does not affect the functional properties of a 3D human epithelial airway model. Nanotoxicology 9:983–993

Commission E, Centre JR, Mech A, Quiros Pesudo L, Rasmussen K, Rauscher H, Riego Sintes J, Olof-Mattsson M, Simko M, Laux P, Bergonzo P, Chevillard S, Clavaguera S, Grall R, Borges T, Micheletti C, Sumrein A, Amenta V, Barberio G, Buttol P, Castelli S, Scalbi S, Stockmann-Juvala H, Walser T, Suarez B, Johansson G, Van Tongeren M, Sanchez Jimenez A, Gouveia H, Vital N, Alessandrelli M, Polci M, Di Prospero Fanghella P, Fito C, Vázquez-Campos S,

Delpivo C, Carlander D, Dusinska M, Atluri R, Dekkers S, Wijnhoven S, Sips A, Oomen A, Bleeker E, Noorlander C, Bekker C, Booth A, Sergent J, Prina-Mello A, Hoehener K, Lehmann H, Ekokoski E, Leinonen R, Einola J, Crutzen H, Jantunen P, Gottardo S (2017) NANoREG framework for the safety assessment of nanomaterials. Publications Office of the European Union

Di Cristo L, Oomen AG, Dekkers S, Moore C, Rocchia W, Murphy F, Johnston HJ, Janer G, Haase A, Stone V, Sabella S (2021) Grouping hypotheses and an integrated approach to testing and assessment of nanomaterials following oral ingestion. Nanomaterials (Basel) 11

Di Cristo L, Janer G, Dekkers S, Boyles M, Giusti A, Keller JG, Wohlleben W, Braakhuis H, Ma-Hock L, Oomen AG, Haase A, Stone V, Murphy F, Johnston HJ, Sabella S (2022) Integrated approaches to testing and assessment for grouping nanomaterials following dermal exposure. Nanotoxicology 16:310–332

Donaldson K, Schinwald A, Murphy F, Cho WS, Duffin R, Tran L, Poland C (2013) The biologically effective dose in inhalation nanotoxicology. Acc Chem Res 46:723–732

Fischer I, Milton C, Wallace H (2020) Toxicity testing is evolving! Toxicol Res 9:67–80

Franken R, Goede H, Shandilya N, Ge C, Kalkman G, Otto M, Van Venrooij B, Van Someren E, Fransman W (2023) 78 Nano exposure quantifier (Neq)—a quantitative tool for assessing exposure in the workplace. Annals Work Exposures Health 67:i57–i57

Ghio AJ, Dailey LA, Soukup JM, Stonehuerner J, Richards JH, Devlin RB (2013) Growth of human bronchial epithelial cells at an air-liquid interface alters the response to particle exposure. Part Fibre Toxicol 10:25

Gourmelon A, Delrue N (2016) Validation in support of internationally harmonised OECD test guidelines for assessing the safety of chemicals. Adv Exp Med Biol 856:9–32

Grainger CI, Greenwell LL, Lockley DJ, Martin GP, Forbes B (2006) Culture of Calu-3 cells at the air interface provides a representative model of the airway epithelial barrier. Pharm Res 23:1482–1490

Hofer S, Hofstätter N, Punz B, Hasenkopf I, Johnson L, Himly M (2022) Immunotoxicity of nanomaterials in health and disease: current challenges and emerging approaches for identifying immune modifiers in susceptible populations. WIREs Nanomed Nanobiotechnol 14:e1804

Huh D, Hamilton GA, Ingber DE (2011) From 3D cell culture to organs-on-chips. Trends Cell Biol 21:745–754

Kang D, Lee H, Jung S (2022) Use of a 3D inkjet-printed model to access dust particle toxicology in the human alveolar barrier. Biotechnol Bioeng 119:3668–3677

Kasurinen S, Happo MS, Rönkkö TJ, Orasche J, Jokiniemi J, Kortelainen M, Tissari J, Zimmermann R, Hirvonen MR, Jalava PI (2018) Differences between co-cultures and monocultures in testing the toxicity of particulate matter derived from log wood and pellet combustion. PLoS One 13:e0192453

Kletting S, Barthold S, Repnik U, Griffiths G, Loretz B, Schneider-Daum N, De Souza Carvalho-Wodarz C, Lehr CM (2018) Co-culture of human alveolar epithelial (hAELVi) and macrophage (Thp-1) cell lines. Altex 35:211–222

Kraegeloh A, Suarez-Merino B, Sluijters T, Micheletti C (2018) Implementation of safe-by-design for nanomaterial development and safe innovation: why we need a comprehensive approach. Nanomaterials (Basel):8

Kreft ME, Jerman UD, Lasič E, Hevir-Kene N, Rižner TL, Peternel L, Kristan K (2015) The characterization of the human cell line Calu-3 under different culture conditions and its use as an optimized in vitro model to investigate bronchial epithelial function. Eur J Pharm Sci 69:1–9

Kuehn A, Kletting S, De Souza Carvalho-Wodarz C, Repnik U, Griffiths G, Fischer U, Meese E, Huwer H, Wirth D, May T, Schneider-Daum N, Lehr CM (2016) Human alveolar epithelial cells expressing tight junctions to model the air-blood barrier. Altex 33:251–260

Landsiedel R, Sauer UG, Ma-Hock L, Schnekenburger J, Wiemann M (2014) Pulmonary toxicity of nanomaterials: a critical comparison of published in vitro assays and in vivo inhalation or instillation studies. Nanomedicine 9:2557–2585

Langhans SA (2018) Three-dimensional in vitro cell culture models in drug discovery and drug repositioning. Front Pharmacol 9

Lebre F, Chatterjee N, Costa S, Fernández-De-Gortari E, Lopes C, Meneses J, Ortiz L, Ribeiro AR, Vilas-Boas V, Alfaro-Moreno E (2022) Nanosafety: an evolving concept to bring the safest possible nanomaterials to society and environment. Nano 12:1810

Li N, Wang L, Shi F, Yang P, Sun K, Zhang J, Yang X, Li X, Shen F, Liu H, Jin Y, Yao S (2021) Silica nanoparticle induces pulmonary fibroblast transdifferentiation via macrophage route: potential mechanism revealed by proteomic analysis. Toxicol In Vitro 76:105220

Ma J, Bishoff B, Mercer RR, Barger M, Schwegler-Berry D, Castranova V (2017) Role of epithelial-mesenchymal transition (EMT) and fibroblast function in cerium oxide nanoparticles-induced lung fibrosis. Toxicol Appl Pharmacol 323:16–25

Miller AJ, Spence JR (2017) In vitro models to study human lung development, disease and homeostasis. Physiology 32:246–260

Moné MJ, Pallocca G, Escher SE, Exner T, Herzler M, Bennekou SH, Kamp H, Kroese ED, Leist M, Steger-Hartmann T, Van De Water B (2020) Setting the stage for next-generation risk assessment with non-animal approaches: the Eu-ToxRisk project experience. Arch Toxicol 94:3581–3592

Müller L, Riediker M, Wick P, Mohr M, Gehr P, Rothen-Rutishauser B (2010) Oxidative stress and inflammation response after nanoparticle exposure: differences between human lung cell monocultures and an advanced three-dimensional model of the human epithelial airways. J R Soc Interface 7:S27–S40

Murphy F, Dekkers S, Braakhuis H, Ma-Hock L, Johnston H, Janer G, Di Cristo L, Sabella S, Jacobsen NR, Oomen AG, Haase A, Fernandes T, Stone V (2021) An integrated approach to testing and assessment of high aspect ratio nanomaterials and its application for grouping based on a common mesothelioma hazard. NanoImpact 22:100314

Murphy F, Jacobsen NR, Di Ianni E, Johnston H, Braakhuis H, Peijnenburg W, Oomen A, Fernandes T, Stone V (2022) Grouping MWCNTs based on their similar potential to cause pulmonary hazard after inhalation: a case-study. Part Fibre Toxicol 19:50

Nossa R, Costa J, Cacopardo L, Ahluwalia A (2021) Breathing in vitro: designs and applications of engineered lung models. J Tissue Eng 12:20417314211008696

Onursal C, Dick E, Angelidis I, Schiller HB, Staab-Weijnitz CA (2021) Collagen biosynthesis, processing, and maturation in lung ageing. Front Med (Lausanne) 8:593874

Powell LG, Gillies S, Fernandes TF, Murphy F, Giubilato E, Cazzagon V, Hristozov D, Pizzol L, Blosi M, Costa AL, Prina-Mello A, Bouwmeester H, Sarimveis H, Janer G, Stone V (2022) Developing integrated approaches for testing and assessment (IATAs) in order to support nanomaterial safety. Nanotoxicology 16:484–499

Proudfoot AG, Mcauley DF, Griffiths MJ, Hind M (2011) Human models of acute lung injury. Dis Model Mech 4:145–153

Rosário F, Hoet P, Nogueira AJA, Santos C, Oliveira H (2018) Differential pulmonary in vitro toxicity of two small-sized polyvinylpyrrolidone-coated silver nanoparticles. J Toxicol Environ Health A 81:675–690

Rozario T, Desimone DW (2010) The extracellular matrix in development and morphogenesis: a dynamic view. Dev Biol 341:126–140

Ruijter N, Soeteman-Hernández LG, Carrière M, Boyles M, Mclean P, Catalán J, Katsumiti A, Cabellos J, Delpivo C, Sánchez Jiménez A, Candalija A, Rodríguez-Llopis I, Vázquez-Campos S, Cassee FR, Braakhuis H (2023) The state of the art and challenges of in vitro methods for human Hazard assessment of nanomaterials in the context of safe-by-design. Nano 13:472

Sabella S, Carney RP, Brunetti V, Malvindi MA, Al-Juffali N, Vecchio G, Janes SM, Bakr OM, Cingolani R, Stellacci F, Pompa PP (2014) A general mechanism for intracellular toxicity of metal-containing nanoparticles. Nanoscale 6:7052–7061

Sakuratani Y, Horie M, Leinala E (2018) Integrated approaches to testing and assessment: OECD activities on the development and use of adverse outcome pathways and case studies. Basic Clin Pharmacol Toxicol 123:20–28

Schäfer-Korting M, Bock U, Diembeck W, Düsing H-J, Gamer A, Haltner-Ukomadu E, Hoffmann C, Kaca M, Kamp H, Kersen S, Kietzmann M, Korting HC, Krächter H-U, Lehr C-M, Liebsch M, Mehling A, Müller-Goymann C, Netzlaff F, Niedorf F, Rübbelke MK, Schäfer U, Schmidt

E, Schreiber S, Spielmann H, Vuia A, Weimer M (2008) The use of reconstructed human epidermis for skin absorption testing: results of the validation study. Altern Lab Anim 36:161–187

Schins RPF, Knaapen AM (2007) Genotoxicity of poorly soluble particles. Inhal Toxicol 19:189–198

Sengupta A, Dorn A, Jamshidi M, Schwob M, Hassan W, De Maddalena LL, Hugi A, Stucki AO, Dorn P, Marti TM, Wisser O, Stucki JD, Krebs T, Hobi N, Guenat OT (2023) A multiplex inhalation platform to model in situ like aerosol delivery in a breathing lung-on-chip. Front Pharmacol 14

Stewart CE, Torr EE, Mohd Jamili NH, Bosquillon C, Sayers I (2012) Evaluation of differentiated human bronchial epithelial cell culture systems for asthma research. J Allergy (Cairo) 2012:943982

Stone V, Gottardo S, Bleeker EA, Braakhuis H, Dekkers S, Fernandes T, Haase A, Hunt N, Hristozov D, Jantunen P (2020) A framework for grouping and read-across of nanomaterials-supporting innovation and risk assessment. Nano Today 35:100941

Sun Z, Yang L, Chen KF, Chen GW, Peng YP, Chen JK, Suo G, Yu J, Wang WC, Lin CH (2016) Nano zerovalent iron particles induce pulmonary and cardiovascular toxicity in an in vitro human co-culture model. Nanotoxicology 10:881–890

Tomashefski JF, Farver CF (2008) Anatomy and histology of the lung. In: Tomashefski JF, Cagle PT, Farver CF, Fraire AE (eds) Dail and Hammar's pulmonary pathology: volume I: nonneoplastic lung disease. Springer, New York

Upadhyay S, Palmberg L (2018) Air-liquid Interface: relevant in vitro models for investigating air pollutant-induced pulmonary toxicity. Toxicol Sci 164:21–30

Van Den Broucke S, Vanoirbeek JAJ, Derua R, Hoet PHM, Ghosh M (2021) Effect of graphene and graphene oxide on airway barrier and differential phosphorylation of proteins in tight and Adherens junction pathways. Nanomaterials (Basel) 11

Wang G, Zhang X, Liu X, Zheng J, Chen R, Kan H (2019) Ambient fine particulate matter induce toxicity in lung epithelial-endothelial co-culture models. Toxicol Lett 301:133–145

Wang Y, Adamcakova-Dodd A, Steines BR, Jing X, Salem AK, Thorne PS (2020) Comparison of in vitro toxicity of aerosolized engineered nanomaterials using air-liquid interface mono-culture and co-culture models. NanoImpact 18

Wiemann M, Vennemann A, Sauer UG, Wiench K, Ma-Hock L, Landsiedel R (2016) An in vitro alveolar macrophage assay for predicting the short-term inhalation toxicity of nanomaterials. J Nanobiotechnol 14:16

Wu J, Wang Y, Liu G, Jia Y, Yang J, Shi J, Dong J, Wei J, Liu X (2017) Characterization of air-liquid interface culture of A549 alveolar epithelial cells. Braz J Med Biol Res 51:e6950

Yanagihara T, Chong SG, Vierhout M, Hirota JA, Ask K, Kolb M (2020) Current models of pulmonary fibrosis for future drug discovery efforts. Expert Opin Drug Discov 15:931–941

Yang X, Zhang J, Xiong M, Yang Y, Yang P, Li N, Shi F, Zhu Y, Guo K, Jin Y (2023) NF-κB pathway affects silica nanoparticle-induced fibrosis via inhibited inflammatory response and epithelial-mesenchymal transition in 3D co-culture. Toxicol Lett 383:141–151

Yunji, Lee K, McAllister H-R, Lee S, Jung F, Murphy (2025) Evaluation of a bioprinted 3D airway tissue model for toxicity testing of nanomaterials; Pathway to integration into a tiered testing strategy for hazard assessment to support safety-by-design Nano Today 61102655. https://doi.org/10.1016/j.nantod.2025.102655

Zhang M, Xu C, Jiang L, Qin J (2018) A 3D human lung-on-a-chip model for nanotoxicity testing. Toxicol Res (Camb) 7:1048–1060

Zhang F, Aquino GV, Dabi A, Bruce ED (2019) Assessing the translocation of silver nanoparticles using an in vitro co-culture model of human airway barrier. Toxicol In Vitro 56:1–9

Zhou J, Zou H, Liu Y, Chen Y, Du Y, Liu J, Huang Z, Liang L, Xie R, Yang Q (2022) Acute cytotoxicity test of PM(2.5), NNK and BPDE in human normal bronchial epithelial cells: A comparison of a co-culture model containing macrophages and a mono-culture model. Toxicol In Vitro 85:105480

Open Access This chapter is licensed under the terms of the Creative Commons Attribution-NonCommercial-NoDerivatives 4.0 International License (http://creativecommons.org/licenses/by-nc-nd/4.0/), which permits any noncommercial use, sharing, distribution and reproduction in any medium or format, as long as you give appropriate credit to the original author(s) and the source, provide a link to the Creative Commons license and indicate if you modified the licensed material. You do not have permission under this license to share adapted material derived from this chapter or parts of it.

The images or other third party material in this chapter are included in the chapter's Creative Commons license, unless indicated otherwise in a credit line to the material. If material is not included in the chapter's Creative Commons license and your intended use is not permitted by statutory regulation or exceeds the permitted use, you will need to obtain permission directly from the copyright holder.

Advanced Skin Models for Nanomaterials Safety Assessment

7

A. R. Ribeiro, S. Costa, S. Nogueira, M. González-Durruthy, H. Colley, N. Oliva, R. De Vecchi, and E. Alfaro-Moreno

Abstract

The human skin acts as a biological shield against prolonged exposure to nanomaterials (NMs) and nanoparticles (NPs) coming from cosmetics, textiles, and environmental pollutants that are known to lead to adverse effects such as oxidative stress, skin irritation, and skin diseases. This chapter reviews the main pollutants that our skin are exposed to daily as well as the advanced *in vitro* skin models used for assessing nanotoxicity. It is widely known that the existing 2D and 3D skin models try to mimic the complexity of skin physiology however they still lack specific skin structures such as vascularization and hair follicles. Skin-on-a-chip (SoC) devices, employing microfluidic technologies, bring the advantage of offering dynamic environments for more realistic evaluations of NMs' safety assessment. In this chapter, we analyze critically how these models could accelerate nanotoxicity testing and support regulatory decisions. Additionally, we also review existing biological assays for skin toxicity as well as the available computational models (*e.g.,* Nano-QSR) that could help in

---

A. R. Ribeiro (✉) · S. Nogueira · M. González-Durruthy · E. Alfaro-Moreno
Nanosafety Group, International Iberian Nanotechnology Laboratory, Braga, Portugal
e-mail: ana.ribeiro@inl.int

S. Costa
Nanosafety Group, International Iberian Nanotechnology Laboratory, Braga, Portugal

BiotechHealth Program, Institute for Biomedical Sciences Abel Salazar of the University of Porto, Porto, Portugal

H. Colley
School of Clinical Dentistry, University of Sheffield, Sheffield, UK

N. Oliva
Department of Bioengineering, Institut Quimic de Sarria, Barcelona, Spain

R. De Vecchi
EPSKIN Brazil, Rio de Janeiro, Brazil

© The Author(s) 2025
E. Alfaro-Moreno, F. Murphy (eds.), *Nanosafety*,
https://doi.org/10.1007/978-3-031-93871-9_7

predicting nanotoxicity taking into consideration the physicochemical properties of NMs. Future research should focus on enhancing skin model complexity and employing computational methods to predict NM behavior, ensuring the safe development of nanomaterials for dermal applications.

Keywords

Nanomaterials · *In vitro* and computational models · Nanotoxicology · Safety assessment

1 Skin a Route of Exposure to NMs

The skin serves as the primary barrier against repeated and persistent stressors, such as nanomaterials (NMs) and nanoparticles (NPs) (Domingues et al. 2022; Gupta et al. 2022; Parrado et al. 2019) Exposure occurs through interaction with NM-enabled consumer products (such as textiles, cosmetics, tattoos, and drug delivery systems), emissions from anthropogenic industrial processes that release NMs into the environment (*e.g.,* pollutants), and occupational exposure, where construction workers, painters, hairdressers, manufacturing, and laboratory personnel are in continuous contact with NMs (Asmatulu et al. 2022; Domingues et al. 2022; Gupta et al. 2022; Omari Shekaftik et al. 2022; Wang et al. 2022).

Regarding consumer exposure, the market offers many cosmetic products containing NPs (see Table 7.1). For example, inorganic NPs (metals and their oxides) are highly stable, hydrophilic, biocompatible, and typically non-toxic. These properties make them frequently employed in sunscreen applications (*e.g.,* titanium dioxide and zinc oxide) due to their effective UV radiation filtering properties (Chauhan and Chauhan 2021). Moreover, silver NPs possess broad antimicrobial activity and are often used in cosmetic products due to their preservative effects, particularly in acne formulations. They are also incorporated into topical medical treatments for burns as an anti-infective agent (Asmatulu et al. 2022; Domingues et al. 2022; Gupta et al. 2022; Omari Shekaftik et al. 2022; Wang et al. 2022). Silicon dioxide NPs are applied in cosmetic formulations and as drug carriers, commonly recognized in rinse-off and leave-on cosmetic products for hair and face (Asmatulu et al. 2022; Domingues et al. 2022; Gupta et al. 2022; Omari Shekaftik et al. 2022; Wang et al. 2022). In recent years, the presence of gold NPs in the cosmetics market has increased significantly, as they stimulate collagen secretion and cell regeneration, providing a glow and preventing skin wrinkling (Mascarenhas-Melo et al. 2023). Organic particles such as micelles, liposomes, and dendrimers are considered non-toxic, and biodegradable, with their shapes (nanospheres or nanocapsules) ideal for transdermal uptake. Carbon-based particles such as fullerenes, carbon fibers, and graphene have been employed in skin applications to enhance the delivery of cosmetic agents (Asmatulu et al. 2022; Domingues et al. 2022; Gupta et al. 2022; Omari Shekaftik et al. 2022; Wang et al. 2022). Besides exposure to consumer products, the skin is also exposed daily to air pollutants (a multi-component mixture that includes particulate matter (metals, organic compounds, inorganic

Table 7.1 Overview of reported NP and NMs present in consumer products and occupational settings

Products occupational settings	NMs/NPs	Applications and properties	Potential effects on the skin
Textiles	Carbon nanofibers	Increase textiles' tensile strength, durability, and thermal stability.	Prolonged exposure might cause skin irritation and/or allergic reactions (Bacakova et al. 2020)
	Nanoclays	Improve textiles resistance to fire, water, and gases.	Potential toxicity (Saleem and Zaidi 2020)
	Ag	Inhibit bacteria, fungi, and other microorganisms' growth.	Prolonged exposure might cause skin irritation, allergic reactions and alter skin microbiome (Koivisto et al. 2024; Melnik et al. 2023)
	Au	Thermal conductivity, optical, and antimicrobial properties, employed in smart textiles that integrate electronic components or sensors.	Prolonged exposure might cause skin sensitivity or allergic reactions (Rosie Broadhead 2023)
	Cu	Textiles for medical or therapeutic applications due to their antimicrobial and anti-inflammatory properties.	Persistent exposure might cause skin irritation or allergic reaction and can disrupt skin microbiota (Broadhead et al. 2021)
	TiO_2	Provide UV protection and durability.	Continued exposure might cause skin irritation, leading to redness and swelling (Rashid et al. 2021)
Cosmetics	TiO_2	Sunscreens, lip balms, and makeup products such as powders and blushes, creams, and lotions to provide UV protection and transparency to the formulations.	Skin penetration in damaged or compromised skin (Lee et al. 2020a)
	ZnO	Sunscreens, makeup products such as powders, ointments, creams, and acne treatment products to provide UV protection, control shine and oiliness on the skin, having antimicrobial and anti-inflammatory properties.	Potential increase oxidative stress markers, which might have implications for skin health and can disrupt the natural microbiota of the skin (Lee et al. 2020a)
	CeO_2	Sunscreens since it provides UV protection, and antioxidant properties.	Can induce oxidative stress and affect cellular health (Ali et al. 2015)
	Al_2O_3	Sunscreens, creams, and exfoliants to improve cosmetics' texture and remove dead skin cells.	High concentrations can cause skin irritation (Dobler et al. 2019)
	MgO	Creams for oily skin, and makeup products such as powders and blushes to control shine and oiliness on the skin, improve cosmetics' texture.	Potential to disrupt the natural balance of skin microbiota (Dobler et al. 2019)
	Au	Skincare products to prevent loss of collagen and elastin and protect from free radicals.	Skin irritation or allergic reactions (Liu et al. 2022)

(continued)

Table 7.1 (continued)

Products occupational settings	NMs/NPs	Applications and properties	Potential effects on the skin
Tattoos	Quantum dots	Tattoo inks to create vibrant colors, and due to their photostability.	Skin penetration with the potential to enter the bloodstream and cause systemic effects (Ryman-Rasmussen et al. 2006)
	Carbon nanotubes	Provide stability of the tattoo ink, reducing fading over time, and create tattoos with fine lines and precision.	Skin inflammation, or allergic reactions (Battistini et al. 2020)
	TiO_2	White pigment in tattoo inks, enhancing the contrast of tattoos, and providing resistance to fading.	Skin inflammatory, or allergic reactions (Battistini et al. 2020)
	ZnO	White pigment in tattoo inks, with antibacterial properties, which can reduce the risk of infection.	Skin irritation, sensitization, and phototoxicity (Jang et al. 2012)
	Ag	Metallic or reflective effects in tattoos.	Skin irritation and allergic reactions such as redness, itching, and swelling (Islam et al. 2016)
Construction workers	Nanoclays	Mechanical improvement of mortars.	Long exposure cause skin irritation, redness, itching, dryness (Ferreira et al. 2023)
	TiO_2	Mortars to confer abrasion resistance and cement hydration, in glass to confer fouling resistance, and in paints to confer UV resistance.	Can induce oxidative stress and affect cellular health (Ferreira et al. 2023; Malte et al. 2020)
	Fe_2O_3	Mortars for electrical conductivity, mechanical improvement, and permeability reduction.	Skin irritation and allergic reactions and induce oxidative stress (Ferreira et al. 2023)
	SiO_2	Mortars to improve abrasion and freeze-thaw resistance, in roads for mechanical improvement.	Prolonged exposure might cause skin irritation, leading to redness, itching, and dryness (Ferreira et al. 2023; Malte et al. 2020)
	$Ca(OH)_2$	Wall paintings, mortars, renders, and plaster due to biocidal activity providing protection.	prolonged exposure can cause severe skin irritation and even chemical burns (Ferreira et al. 2023; Malte et al. 2020)

(continued)

Table 7.1 (continued)

Products occupational settings	NMs/NPs	Applications and properties	Potential effects on the skin
Hairdressers	TiO_2	Hair care products for their UV-blocking properties.	Skin penetration and induction of oxidative stress and inflammatory responses (Malte et al. 2020; Rosen et al. 2015)
	SiO_2	Hair products for their conditioning properties.	Prolonged exposure can lead to skin irritation (Rosen et al. 2015)
	Ag	Hair products for their antimicrobial properties.	Prolonged exposure can cause argyria and other toxic effects (Rosen et al. 2015)
Manufacturing and laboratory personnel	Carbon nanotubes	Production of conductive films and electronic displays.	Skin irritation and inflammation (Hu et al. 2015)
	Au	Diagnostic tests and drug delivery systems.	High concentrations can cause oxidative stress and inflammation (Dhasmana et al. 2017)
	Fe_2O_3	Magnetic resonance imaging, drug delivery, and hyperthermia treatment for cancer.	Skin irritation, inflammation and induce oxidative stress (Dhasmana et al. 2017)
Anthropogenic industrial processes	Ultrafine particles	Emitted from vehicles and industrial combustion processes.	Skin penetration, oxidative stress and prolonged exposure can cause skin aging (Dijkhoff et al. 2020)

carbonaceous material, sulfate, and nitrate), nitrogen oxides, polyaromatic hydrocarbons, nicotine, formaldehyde, and microorganisms) from sources such as wildfires, factory emissions, automobile exhaust, and smoke. These pollutants vary in time and space, and it is known that they induce oxidative stress, impair skin barrier function and ultimately lead to the development of skin diseases (Gu et al. 2024).

Due to its distinctive location, human skin works as a biological shield against NMs, where prolonged and repetitive exposure has been demonstrated to have severe deleterious effects on cutaneous tissue. The human skin is essential for maintaining body temperature, preventing the loss of water, inhibiting the entry of bacteria and xenobiotics, and performing metabolic processes (Larese Filon et al. 2015). Advanced interactions between skin cells and the microbiota establish protective skin mechanisms, where a mechanosensory system continuously detects and reacts to different external stimuli, including NMs (Pelikh et al. 2021; Shapira et al. 2022). The skin is divided into hypodermis, dermis, and epidermis where the outermost layer of the epidermis, called stratum corneum (SC), is responsible for the skin barrier function (see Fig. 7.1) (Gupta et al. 2022; Larese Filon et al. 2015; Sanches et al. 2020). The deepest subcutaneous layer, known as the hypodermis, is made up of loose, fatty connective tissues that support the dermis. The dermis layer, which supports the epidermis and skin appendages, is the skin's vascularized elastic

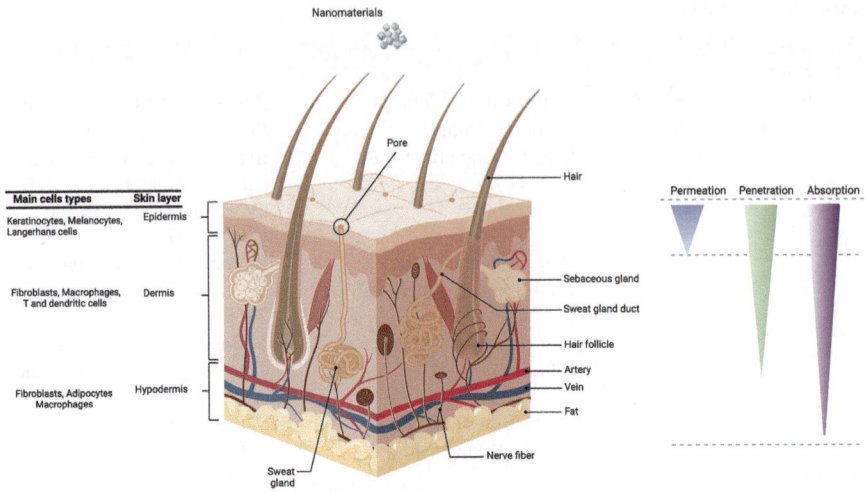

Fig. 7.1 Schematic diagram of the three fundamental skin structural layers and the main cell types found within each layer. (Constructed in Biorender)

connective tissue. Rich in fibroblasts, collagen fibres, elastin, and proteoglycans that form an extracellular matrix (ECM) where immune cells (mast cells, lymphocytes, and macrophages) adhere to (Gupta et al. 2022; Larese Filon et al. 2015; Sanches et al. 2020). The dermis also includes sympathetic fibres, sweat glands, and blood vessels that fund the dermis sensory system. Three cellular types make up the epidermis: melanocytes, Langerhans cells (LC, a type of specialized immune cell), and undifferentiated keratinocytes. As they go towards the outer layers, keratinocytes go through a differentiation process that involves morphological and biochemical alterations that drive cells from an undifferentiated, proliferative state toward metabolic inactivity. The epidermis is innervated by subepidermal nerve bundles and intraepidermal nerve fibres. The distinct qualities of the human skin barrier are established by the skin lipid matrix, which is made up of ceramides, cholesterol, and fatty acids arranged in an orthorhombic lattice (Gupta et al. 2022; Larese Filon et al. 2015; Sanches et al. 2020). Resuming the cutaneous microenvironment is constituted by microbial, immunological, chemical, and physical barriers that preserve skin homeostasis (Parrado et al. 2019).

When considering transdermal absorption of NPs, it can be defined as their ability to reach the circulatory system through penetration across various skin layers, typically occurring via passive diffusion following Fick's Law. Consequently, NPs can become systemically available and accumulate in other organs (*e.g.*, liver and spleen) (Alkilani et al. 2015). Additionally, skin permeation refers to the diffusion of NMs into a specific skin layer, whereas skin penetration involves diffusion into deeper layers, as illustrated in Fig. 7.1 (Alkilani et al. 2015). The *in vitro* and *in vivo data* on the potential dermal absorption and/or penetration of NPs exhibit controversial results. Although several studies report contrary findings, NPs penetration in both healthy and compromised skin (*e.g.,* scarring, sunburn, and depilated skin) has been demonstrated within the scientific community (Sanches et al. 2020). The main pathways of NPs penetration are intracellular, intercellular, and transappendageal. The intercellular route is the most prevalent pathway, wherein NPs cross the stratum corneum by diffusion among the cells. For this pathway, the size, as well as the mechanical properties of NPs, need to be taken into consideration, as they need to have the right flexibility. Rigid particles, such as metal NPs, have been found to scarcely penetrate the SC via the intercellular route due to their lack of flexibility which hinders their diffusion between the cells. On the other side, the intracellular pathway is challenging as the NPs must overcome both the lipophilic (cell membrane and the lipid matrix) and lipophobic structures (inside the cells) within the skin cells (Asmatulu et al. 2022; Domingues et al. 2022; Gupta et al. 2022; Larese Filon et al. 2016; Niska et al. 2018; Salvioni et al. 2021; Tordesillas et al. 2018). NPs pass through both the lipid bilayer and the cytoplasm of keratinocytes diffusing through the corneocytes of the SC (Asmatulu et al. 2022; Domingues et al. 2022; Gupta et al. 2022; Larese Filon et al. 2016; Niska et al. 2018; Salvioni et al. 2021; Tordesillas et al. 2018). Nanocarriers with a definite degree of amphiphilicity may be worthy candidates to avoid this difficulty. In contrast, skin appendages such as hair follicles (HF), sebaceous glands, and sweat glands have been identified as potential alternative entry routes. First, NPs tend to accumulate in HFs, where they may persist for several days or weeks (Sun et al. 2018; Yuan et al. 2019). Second, due to the presence of the SC in the lower regions of HFs, there is an increased likelihood that NPs will penetrate the surrounding tissues or be taken up by specialized immune cells, such as macrophages, Langerhans cells (LCs), and dermal dendritic cells. Once activated, these cells can either function locally by clearing NPs and secreting specialized immune mediators or migrate to lymph nodes to initiate an adaptive immune response (Sun et al. 2018; Yuan et al. 2019). The HF interface with lymph and blood streams through capillary vessels is the quickest and most efficient mechanism for NPs to enter the systemic circulation. A high density of Langerhans cells around hair follicles, were capable of internalize NPs of various sizes, whereas the transport across the epidermis was restricted to 40 nm particles. For rigid NPs, the most projected path for penetrating the skin is via the follicular route. It is important to stress that the transappendageal pathway is limited, as HF and glandular ducts make up only 0.1% and 0.01% of the total surface area of the skin, respectively (Asmatulu et al. 2022; Domingues et al. 2022; Gupta et al. 2022; Larese Filon et al. 2016; Niska et al. 2018; Salvioni et al. 2021; Tordesillas et al. 2018).

NPs penetration appears to be limited to the SC due to its robust barrier formed by dermal and robust tight junctions. However, increasing evidence suggests that NPs can penetrate healthy skin, although the precise mechanisms remain unclear. The ability of NPs to penetrate the skin is influenced by the nature of formulation (containing solvents or detergents), NP's physicochemical properties such as size, shape, rigidity, and surface charge as well as skin characteristics, including follicular density, thickness, metabolism, and structural variations across different anatomical regions (Asmatulu et al. 2022; Domingues et al. 2022; Gupta et al. 2022; Larese Filon et al. 2016; Niska et al. 2018; Salvioni et al. 2021; Tordesillas et al.

2018). Additionally, aged or damaged skin due to factors such as sunburn, depilation, scaling, scarring, skin hydration, or any skin disease exhibits increased susceptibility to NPs penetration, allowing larger particles (>45 nm) to permeate more easily (Gupta et al. 2022; Palmer et al. 2019).

2 The Evolution of *In Vitro* Skin Models for Nanotoxicological Studies

The safety assessment of NPs for skin application starts with hazard identification, where the toxicological profile is undertaken via several tests (*in vivo, in vitro*), and clinical and epidemiological studies. With all this information some parameters such as no observed adverse effect level (NOAEL) and no observed effect level (NOEL) can be measured to study the exposure—toxic response, exposure assessment, and ultimately the risk characterization (Coimbra et al. 2022). Different types of safety studies can be employed to ensure the safety of consumers. The effect of NPs on the skin and their underlying mechanisms can be explored by employing *in vitro* 2D and 3D models that mimic skin physiology using genuine or synthetic skins and *in vivo* models (mice, pigs, rabbits, guinea pigs, and human explants) (Melnik et al. 2023).

Nonetheless, it is widely reported that non-clinical *in vivo* investigations of cosmetics frequently fail to translate to human clinical trials due to species-specific physiological characteristics (*e.g.*, skin thickness, cell populations, HF density, and immunology). Human skin explants are far more physiologically relevant since they contain the entire skin architecture as well as most skin cell types. However, they can be greatly influenced by individual variables (age and living patterns) limiting reproducibility, but they carry ethical concerns and availability issues due to the dependence on discarded tissue from the plastic surgery of healthy individuals (Melnik et al. 2023).

Driven by ethical concerns and societal ambition to reduce animal experimentation (prohibited by the European Cosmetic legislation) reliable and animal-free systems that try to emulate human skin have been employed. The passage of the FDA Modernization Act 2.0, as well as growing confidence in more complicated human-based *in vitro* models, is paving the way for more human-relevant research into nanotoxicity processes (see Fig. 7.2).

2.1 Two-Dimensional Models

Two-dimensional models of the dermis and epidermis have expanded the knowledge of how skin cells behave and respond to a variety of stimuli, including NPs. The most usual 2D skin models share a combination of key features such as human- or animal-derived fibroblasts, epidermal keratinocytes, and a basal supportive substrate. *In vitro*, studies demonstrate the importance of ECM proteins such as collagen fibres in maintaining dermal fibroblast integrity (Dijkhoff et al. 2020; Wistner et al.

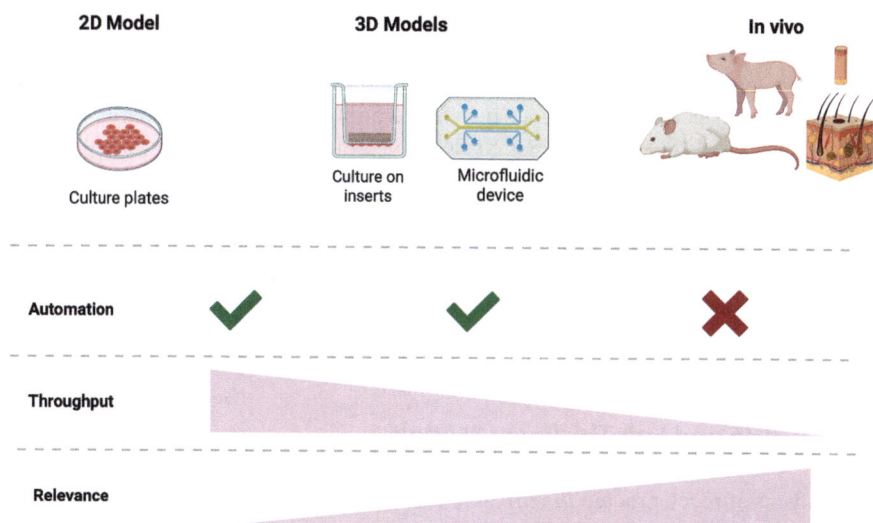

Fig. 7.2 Available skin models their possible automation and throughput relevance. (Constructed in Biorender)

2023). When skin cells are isolated from their dynamic environment and grown in stiff and static Petri dishes, they demonstrate low predictive power over clinical outcomes and as such they are limited largely to cytotoxicity studies but are useless at elucidating efficacy in drug testing (Dijkhoff et al. 2020; Wistner et al. 2023). Human adult low calcium high temperature (HaCaT) cells and recently human telomerase reverse transcriptase (hTERT)-immortalized human keratinocytes cell lines together with murine fibroblasts (3T3) are frequently employed for 2D *in vitro* studies (Alépée et al. 2014; Boukamp et al. 1988). Nevertheless, primary human cells isolated directly from the neonatal foreskin or adult skin of healthy donors are the more biologically relevant. Interestingly it was already described that keratinocytes need to be co-cultured with supporting fibroblasts or fibroblast-conditioned media to produce durable epithelium, suggesting the importance of fibroblasts-keratinocytes communication for epidermal formation, growth, and differentiation. Normal human epidermal keratinocytes (NHEK) or normal human dermal fibroblasts (NHDF) monocultures are frequently isolated and cultured on specific substrates in an optimized cell culture medium and as well as keratinocytes can receive NPs treatment directly in their culture medium. 2D models were employed in the 1980s to predict cytotoxicity (OECD 2010) and phototoxicity (in 3T3 cells; (OECD 2015)). In 2004, OECD guideline 432 was the first *in vitro* test that largely replaced animal hazard analysis. Although the usage of 2D models allows for reproducible, throughput, and low-cost experiments for the basic study of cell growth and survival of epithelial systems, they still fail to mimic skin physiology, lacking a physiological skin barrier with a competent SC layer (Abd et al. 2016; Augustine 2018; Avci et al. 2013; Cui et al. 2010; De Wever et al. 2015; Elaine 1999; Filaire et al. 2022; Guichard et al. 2022; Gupta et al. 2013; Krieg and Aumailley 2011; Kwak et al.

2020; Lee et al. 2020b; Motter Catarino et al. 2022; Pereira et al. 2013; Ramadan and Ting 2016; Venus et al. 2010; Yurchenco' and Schittny 1990; Zeb et al. 2019). For example, it has already reported that the oxidative stress and pro-inflammatory response of silver NPs were higher in a 2D keratinocytes model compared to a 3D epidermal model due to the impaired SC (Chen et al. 2019). Resuming, 2D skin models do not recapitulate the cutaneous environment, they overestimate toxicity, have limited predictive capacity, and have differences in physiology and genetic background (Domingues et al. 2022; Rogal et al. 2022; Schneider et al. 2021; Wu et al. 2020). Thus, these 2D models inspired the development of skin equivalents that employ 3D scaffolds to mimic skin structure and biological complexity (Sun et al. 2006).

2.2 RHE and Full-Thickness Models

The advent of reconstructed *in vitro* skin models marked a significant advancement in dermatological and toxicological research. Moving beyond traditional two-dimensional cell cultures, these models provide a more physiologically relevant platform to study skin biology, and disease pathogenesis, and assess the safety and efficacy of cosmetics, pharmaceuticals, and NPs/NMs (Alépée et al. 2019a, b; Hofmann et al. 2023; Moon et al. 2021; Singh et al. 2024). Reconstructed human skin models, developed as an animal-free alternative to previous models, are obtained from human cells like epidermal keratinocytes and fibroblasts, but cultured in a multilayered format (epidermis and dermis), generated by combining biomolecules such as collagen and human-derived cells. One important aspect regarding the creation of fully differentiated *in vitro* skin models is that they require culture at the air-liquid interface (ALI) (Chen and Schoen 2019). To create an ALI culture, epithelial cells are seeded in compartmentalized culture systems on the top of porous filter supports that keep them physically isolated from the underlying fluid. Typically, porous membranes are employed to divide the layers and replicate structures found in native human skin while facilitating intercellular crosstalk. After the initial attachment and proliferation phase and the creation of a confluent monolayer, the culture media on the apical side is removed. The cells 'interface' with the surrounding air, differentiate, and create an apical microenvironment through transudation and apical secretion. The basal surface of the cells has access to the culture media, including nutrients and other additives, via diffusion through the porous membrane (Chen and Schoen 2019).

The Reconstructed Human Epidermal (RHE) model consists of organized basal, spinous, and granular layers, with a multilayered stratum corneum containing intercellular lamellar lipid layers arranged in patterns, representing the main lipids found *in vivo* (Alépée et al. 2019a, b). RHE is typically constituted of keratinocytes differentiated to form a multilayered epidermis, mimicking the skin's barrier function. This permits researchers to evaluate the potential of substances to induce irritation, corrosion, sensitization, and absorption (see Table 7.2). Considering NPs, RHE models are particularly useful in evaluating their safety, providing insights into NPs

Table 7.2 *In vitro* models used for cosmetic screening employing OECD guidelines

In vitro model	Biological output	OECD	Characteristics
3T3 NRU	Phototoxicity	432	BALB/c 3T3 mouse fibroblasts to measure the concentration-dependent reduction in neutral red uptake by the cells after exposure to a test material (presence or absence of UVA).
SkinEthic™	Skin irritation	439	The biological properties of the RhE model prevent the passage of a material around the stratum corneum to the viable tissue.
KeratinoSens™	Skin sensitization	442 D	Employ immortalised cell line derived from human keratinocytes and measure luciferase gene induction as an indicator of the activity of Nrf2 transcription.
epiCS®	Skin irritation	431	The model consists of organized basal, spinous and granular layers, and a multi-layered stratum corneum containing intercellular lamellar lipid layers.
Human or animals skin	Skin absorption	428	The test system includes the donor chamber, the skin surface rinsing, the skin preparation and the receptor chamber.

penetration, uptake, and potential toxicity to epidermal layers (Moon et al. 2021). For example, EpiDerm™ skin was incubated with iron, aluminium oxide, titanium dioxide, and silver NPs and tested for skin corrosion and irritation based on the OECD TG431 and TG439. Results demonstrated that NPs were non-corrosive and non-irritant and that the *in vitro* model is suitable for NPs safety assessment. EpiSkin™ was also employed to evaluate the skin irritation caused by metallic NPs such as aluminium oxide NPs, TiO_2, and Ag NPs after short and long-term incubations and the epidermal penetration of gold NPs (Filaire et al. 2022; Hao et al. 2017; Kim et al. 2016; Strüver et al. 2017). However, the skin is a complex organ, and

RHE models, while valuable, lack the full complement of cell types and structures present *in vivo*. To address this, full-thickness skin models have been developed, where collagen matrices are widely employed and known to provide fibroblasts with an adequate environment to support ECM protein synthesis and paracrine factor secretion, thus promoting keratinocyte growth, maturation, and formation of stratified epithelium. These advanced models besides incorporating a dermal compartment, often populated with fibroblasts, may include additional cell types such as melanocytes and immune cells (Abdayem et al. 2016; Hofmann et al. 2023). The inclusion of these components allows for a more comprehensive assessment of NPs interactions with the skin, including potential inflammatory responses and immunotoxicity.

Commercially available RhE models already validated are the EpiSkin™ (L'Oreal, France), EpiDerm™ (MatTek Corporation, Massachusetts, USA), SkinEthic™ (SkinEthics, France), and epiCS® (CellSystems, Germany) (Chen et al. 2024). Indeed, some can recapitulate the epidermis, and others already replicate both dermal and epidermal compartments. When considering air pollutants, Phenion^{FT} skin equivalents were employed to evaluate the synergistic effect of ozone and particulate matter (PM). PM was observed to decrease the epidermal thickness and promote a matrix-building phenotype, while ozone was found to alter lipid homeostasis and induce inflammation (Reynolds et al. 2023). Skin barrier dysfunction, dose-dependent inflammatory reaction, and modifications in differentiation protein markers and water transport were observed upon exposure of PM to an in-house RHE (Hieda et al. 2020). All these events could eventually aggravate several skin diseases. Resuming, the use of 3D skin models, as an alternative to animal testing is validated to test chemicals and recommended when testing the cutaneous effects of NPs. However, we must keep in mind this model possesses weaker barrier properties compared to native skin, they lack cellular and biomolecular variety as well as vasculature or adnexal structures. This is still a main issue since their absence impacts skin functionality; however, efforts are underway to develop more skin-relevant models (Filaire et al. 2022). Also, adequate controls are necessary to avoid the interference of NPs on the biological assays.

2.3 Biofabrication of 3D Equivalent Models

Fabricating skin in a lab setting that recapitulates all the structural and functional aspects of a native dermis is a challenging task. Several biofabrication techniques have been applied to the development of the skin (see Fig. 7.3), highlighted next in this section with their respective advantages and disadvantages.

2.3.1 3D Bioprinting

This technique relies on the layer-by-layer deposition of bioinks to create complex 3D structures. There are different bioprinting techniques, each with its own characteristics, resolution, and print speed (Table 7.3). Inkjet Bioprinting deposits bioink onto a substrate using thermal or piezoelectric actuators at high print speed and

Fig. 7.3 Biofabrication techniques employed for the development of 3D *ex vivo* models of skin. (Constructed in Biorender)

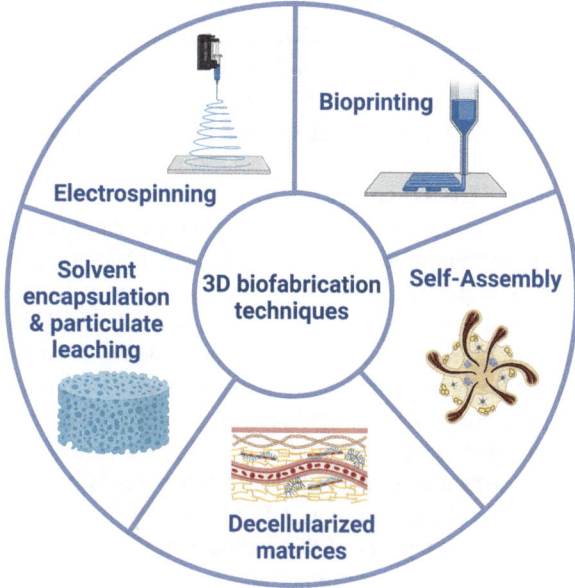

Table 7.3 Types of 3D bioprinting, along with their advantages and disadvantages

Technique	Advantages	Disadvantages
Inkjet bioprinting	High resolution, fast print speed, cost-effective	Nozzle clogging, limited viscosity range, low cell viability
Extrusion bioprinting	Bioink viscosity range, scalable	Lower resolution, slow print speed, low cell viability
Laser-assisted bioprinting	High precision, high cell viability	Technically complex, expensive, slow print speed

resolution. It is therefore suitable for creating detailed patterns and depositing multiple cell types. However, it is prone to nozzle clogging, can only print with low-viscosity bioinks, and the thermal stresses involved cause low cell viability (Murphy and Atala 2014; Nakamura 2005). Extrusion bioprinting uses a continuous flow of bioink extruded through a nozzle but at a lower resolution and print speed than inkjet bioprinting. The shear stresses experienced by cells during extrusion also decrease the viability significantly. However, it is a more versatile technique, as it can use bioinks of a range of viscosities and print larger and more complex structures (Murphy and Atala 2014). Laser-assisted bioprinting uses a laser to transfer a bioink from a donor slide to a substrate, offering excellent resolution and precision with high cell viability. However, the print speed is slow, and it is more expensive and technically demanding than the other techniques (Murphy and Atala 2014).

Overall, 3D bioprinting enables precise control over the placement of cells, biomaterials, and growth factors. Therefore, it can be used to create complex structures that recreate the multiple layers of the skin and incorporate other key structures like hair follicles, blood vessels, sweat glands, etc. It also enables customization for patient-specific models. However, there are still limitations related to maintaining the original mechanical properties of biopolymers (like collagen) after they have been modified into bioinks, as well as the predominant animal origin of bioinks and its implications in regulatory frameworks and ethical considerations.

2.3.2 Electrospinning

Electrospinning is a technique by which nanofibers are created and deposited as scaffolds that mimic the ECM. The nanofibers are generated from both natural and synthetic polymers in a solution that is drawn from a syringe because of a high-voltage electric field, creating fine fibers. Several studies have demonstrated that electrospinning is suitable for creating dermis-like equivalents made of a range of polymers (e.g., collagen, polycaprolactone, or chitosan, amongst many others) that support the attachment and proliferation of cells (Law et al. 2017; Lizarazo-Fonseca et al. 2023; Tamilarasi et al. 2023). The main advantages of electrospinning are the high surface area for cell attachment and high porosity for nutrient and waste exchange of the resulting scaffolds. However, scaffold uniformity and, most importantly, the balancing of the mechanical properties (strength and flexibility) of the scaffolds are the most important challenges (Venugopal and Ramakrishna 2005).

2.3.3 Hydrogel-Based Encapsulation

This technique employs hydrogels, which are three-dimensional hydrophilic polymer networks that can retain large amounts of water, just like the natural ECM in the dermis. Moreover, and very much like the natural ECM, hydrogels present unique viscoelastic properties. They are also highly biocompatible and support cell attachment and growth. The large pores and high water content facilitate nutrient, oxygen, and waste exchange through diffusion. For these reasons, hydrogels are ideal biomaterials to mimic the dermis enabling full-thickness model development (Hoffman 2012; Peppas and Khare 1993).

The polymers forming the hydrogel network can be natural (like collagen, gelatin, or alginate), synthetic (like polyethylene glycol (PEG) or polyvinyl alcohol (PVA)), or a hybrid of both. Cells are suspended in the precursor solution of polymers, which are then mixed with the crosslinkers to form the 3D network. The mechanical properties (like stiffness and elasticity) of hydrogels can be easily controlled by tuning the physicochemical properties of the polymers and the ratios between polymers and crosslinkers. This enables the development of skin models of a range of mechanical properties that can mimic different diseases or processes associated with altered ECM composition of the skin. For example, scarring has been associated with higher stiffness, owing to the excessive deposition of ECM, while ageing is linked to a decrease in elasticity. Hydrogels' tunability enables the development of these types of models using the same base polymers for the scaffolds (Drury and Mooney 2003; Place et al. 2009). On the other hand, there is very little control over the cell distribution in the hydrogel, which is often not uniformly distributed within the scaffold. Moreover, it is challenging to develop complex or hierarchical structures using this technique.

2.3.4 Decellularization

Decellularization, as the name indicates, is a technique by which all cellular components from a donor tissue are removed, ideally preserving the native ECM structure, composition, and mechanical properties. By doing so, decellularization prevents immunogenicity while maintaining the scaffolding for cells to grow into new tissue. There are different methods of decellularization, divided into chemical, physical, and biological (Crapo et al. 2011; Reing et al. 2010). Chemical (detergents), physical (freeze-thaw cycles) and biological (enzymatic) methods have been widely employed to remove all cell types from the tissue, where the next step is recellularizing it, meaning repopulating the decellularized ECM with human cells to restore tissue functionality. Cells can be seeded directly on the scaffold's surface and allowed to migrate and proliferate throughout, or they can be perfused through the scaffold using bioreactors. The latter approach results in a more uniform cell distribution. The main challenges of this biofabrication technique are ensuring complete decellularization, preservation of the ECM structure and properties, and efficient recellularization of the donor tissue scaffold.

2.3.5 Self-Assembly

Self-assembly is a biofabrication technique that leverages the inherent ability of cells to organize into structured tissues. The rationale behind this is to allow cells to follow the natural processes of tissue formation and development, driven by cellular interactions, ECM production, and biochemical and biophysical cues. In this process, fibroblasts produce collagen and other ECM components to form the dermis, while keratinocytes form the epidermal layer. Self-assembly of skin models can include more cell types, like melanocytes to produce pigmentation, and endothelial and smooth muscle cells to form blood vessels (Jakab et al. 2010).

The process of self-assembly starts with a monolayer culture of the cells, followed by 3D cultures in spheroids or organoids, which promotes the self-assembly of the cells into tissue-like structures. The most obvious advantage of self-assembly biofabrication is the reduced need for synthetic scaffolds *that might cause* or degrade unpredictably. Overall, this approach reduces the hurdles for clinical applications. The resulting architecture of the *ex vivo* model is also much more physiologically relevant, compared to scaffold-based methods, and incorporates all the naturally occurring and necessary cell-cell and cell-ECM interactions. However, scaling up this biofabrication technique is very challenging, due to nutrient and oxygen diffusion limitations. Moreover, the control of tissue self-assembly of multiple cell types to produce complex tissue structures like blood vessels is also challenging, as is to incorporate these different cell types in the correct spatial organization to mimic the complex native skin.

There are many biofabrication techniques to build and mimic the complex skin architecture, each with their advantages and disadvantages, and therefore suitable for specific uses and purposes. This means that the "perfect" model does not exist, but rather a collection of models that enable the study of different aspects of the skin. For example, if only the mechanical and physicochemical environment of fibroblasts is being studied, a simple hydrogel encapsulation will suffice, providing the ECM component and an easy platform to alter the properties and assess the effects. However, if the study aims to understand the effects of blood supply to the skin, sebum production, or hair growth on other physiological events, 3D bioprinting or self-assembly techniques that can recapitulate the complex architecture of these skin appendages and structures are key to attaining a relevant model.

2.4 Skin-on-a-Chip Models

One limitation of all the techniques highlighted in this section is that they cannot simulate the dynamic physiological environment of the skin, in particular the fluid flow and mechanical forces experienced by the tissue. Alternatively, microfluidic devices, such as skin-on-a-chip models, provide this missing aspect.

The latest advancements in microfluidics have enabled the development of organ-on-chip (OoC) devices that are established by perfused microfluidic chambers populated by cells that mimic tissue- and/or organ physiology. In OoC devices, the cellular microenvironment can be controlled with high spatiotemporal precision, allowing extracellular cues to guide cells into physiologically accurate configurations. The goal is that these OoC model cells remain viable for extended periods replicating one or a few specific functional properties of organs, such as the barrier function in the case of the skin. They can also simulate the dynamic interactions between cells, the cell-ECM, and the mechanical stresses that cells encounter within tissues (Costa et al. 2023b).

Approaches from tissue engineering are employed in skin-on-chip (SoCs), which involve cell culture on scaffolds, application of physical signals (fluid-dynamic, mechanical, electrical), and microfabrication techniques of culture spaces and channels (Costa et al. 2023b). These features enable SoCs to provide improved consistency of skin barrier function and structure while incorporating only a few cell types. In SoC systems, cells typically grow on the surfaces of microcavities or porous membranes, with continuous perfusion of medium culture supplied through the microchannels. In terms of chip design, two general trends emerge: (1) closed devices, with sealed channels and pump-driven flow; and (2) open devices that resemble well plates and are perfused by gravity-driven flow or use rocking platforms. The medium flows between microfluidic chambers via gravity, a 3D tilting mechanism, or active pumps to mimic blood flow and sustain cell viability by averting the accumulation of metabolic wastes and offering necessary fluid shear stress (Zoio and Oliva 2022). Medium flow, growth factors, and cytokines through these channels accentuate cell differentiation and augment skin tissue longevity, overcoming the limitations of pump-driven systems. Curiously, the pumpless microfluidic technology overcomes challenges, such as air bubble formation, since it provides high controllability and enables complex flow patterns, however, it can complicate device operation and cell seeding. Effective perfusion between the skin tissue and the microfluidic channel is essential for waste removal but also for tissue sustenance that is accomplished through a porous membrane (polydimethylsiloxane, polycarbonate, and polyethylene terephthalate) (Costa et al. 2023b; Zoio and

Oliva 2022). Conversely, open device chips facilitate easy tissue and media retrieval but offer less precise environmental control. To emulate skin multiple layers, the chip design should enable an air-liquid interface so that the stratum corneum layer of the epidermis is exposed to air, while the dermal layer is exposed to culture media. Simultaneously, the field of microfluidics advanced rapidly with the invention of soft lithography and the ability to prototype devices using polydimethylsiloxane (PDMS), a soft silicone-based material (Cho et al. 2024). Regarding chip fabrication, soft lithography with PDMS has been the predominant method. However, mechanical or laser-based structuring of thermoplastics polystyrene (PS), polycarbonate (PC) and polymethyl methacrylate (PMMA) are also used to overcome PDMS limitations, such as absorption of hydrophobic molecules (e.g., culture media components, NPs or tested drugs), limited scalability for industrial production, and challenges with sensor integration. PDMS advantages are the low manufacturing cost, air permeability, ease of handling, efficient sealing process, and ability to form complex micropatterns (Cho et al. 2024). All the referred materials are optically transparent and have good biocompatibility suitable for observing and performing several cell-based studies.

In recent years, numerous SoC models have been presented. These models include (i) devices where skin biopsies or reconstructed skin models are transferred to the chip, it incorporates either patient-derived skin, harvested through biopsies, or post-mortem skin samples (Costa et al. 2023b). One of the first examples is the chip designed by Abaci et al. for placing a human skin equivalent (HSE) to test its viability and maintenance. The HSE was cultured on a porous membrane to enable nutrient diffusion from the channel. The group investigated the transdermal transport of substances and the potential of this device for drug testing applications (Fig. 7.4a) (Abaci et al. 2015). The transferred SoC model more accurately represents the cellular population and interactions of the skin at a microscopic level. This method is also simpler than cultivating skin cells and creating an optimized environment within the device for long-term tissue maintenance and experimental purposes. However, obtaining sufficient skin tissue from individuals can be logistically challenging due to the extensive tissue volume requirements for research. (ii) The second strategy involves the cultivation of human skin cells on the chip the so-called in situ SoC, whereas 2D cell monolayers are cultured on porous membranes. As an example, Wufuer et al. developed an SoC model containing three layers—epidermal, dermal, and vascular—to study inflammation and edema (Wufuer et al. 2016). The layers were represented by keratinocytes, fibroblasts, and human umbilical vein endothelial cells, respectively, and were separated by transparent porous membranes to enable interlayer communication and mimic skin biology. They generated a model of skin inflammation by perfusing the chip with tumor necrosis factor (TNF-α) and measured proinflammatory cytokine levels and tight junction integrity. The efficacy of the drug dexamethasone was evaluated using this inflammation model, demonstrating that the drug could mitigate TNF-α-induced endothelial barrier dysfunction (see Fig. 7.4b). It is important to refer to that in this model the epidermis with its different layers is not achieved since the model doesn't allow culture of cells in ALI. The last approach (iii) works with platforms with perfusable

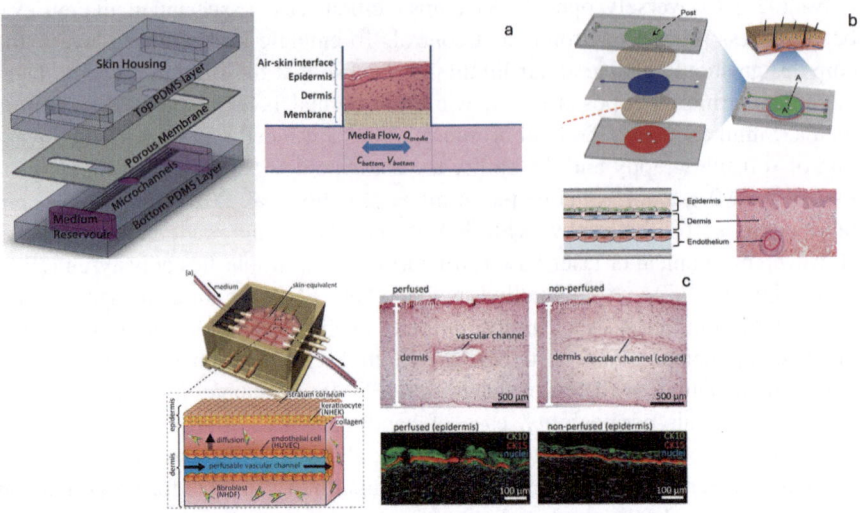

Fig. 7.4 Skin-on-chip examples: (**a**) Pumpless chip with transferred skin designed for testing HSEs viability and maintenance; (**b**) 3D schematic of the skin-on-a-chip system, which comprises three PDMS layers and two PET porous membranes, with a representative histological section of the skin; (**c**) the schematic diagram of the skin chip integrated with the perfusion vascular channel which can be used to test the toxicity of cosmetics. (Copyright permission is conveyed through Copyright Clearance Center, Inc.)

lumens (mainly for vasculogenesis) or patterned microchannels, onto which skin tissue is assembled directly using membranes or custom scaffolds. As an example, Mori et al. fabricated a culture device using 3D templating techniques (Mori et al. 2017). They developed a device with anchoring structures and nylon wires strung across connectors. A collagen structure was fixed into the device, and perfusable vascular channels were created by removing the nylon wires. This approach allowed the recreation of dermal/epidermal within a vascular channel. However, this technique lacks a complete microvascular network in the dermis. Additionally, the contraction of the collagen used for the dermal compartment affected the permeation assay, limiting it to the central portion of the (Fig. 7.4c) (Mori et al. 2017).

Resuming, most available SoC models focus on recapitulating the dermis and epidermis, primarily using cell lines. Some models also incorporate additional relevant cell types, such as vascular cells (human umbilical vein endothelial cells, human primary microvascular endothelial cells) and immune cells (human leukemic monocyte lymphoma cell line). The inclusion of vascularization in the models allows for to study NPs and drug absorption, gaining insight into the transport of intravenously injected drugs/NPs, and the opportunity to construct thick skin models with enhanced deposition of cell-derived ECM components (Rimal et al. 2024). The most reported benefit of SoC is the increased expression of filaggrin and involucrin (role in forming the epidermal skin barrier) and enhanced skin barrier function. A consistent finding is the higher mean transepithelial electrical resistance in

SoC models compared to static controls, confirming a more robust barrier with a lower permeability (Zoio and Oliva 2022).

Concerning toxicological studies, the available SoC models have yet to be investigated for NMs. Given the rising market for NM-containing products and NP-associated pollutants, monitoring toxicity has become a top issue. As a result, advanced SoC models are urgently needed to accelerate NM/NPs safety evaluation and give human-relevant data. The benefit of SoC in nanotoxicology is that tailored microfluidic channels on top of the epidermal layer could provide well-controlled and homogeneous particle exposure. The use of laminar flow profiles can be employed to mitigate the agglomeration or aggregation of NMs (Costa et al. 2023b). Besides that, SoC systems could serve as a platform for independently monitoring the characteristics of the most frequent types of NM-induced skin harm, including skin corrosion, irritation, sensitization, genotoxicity, and phototoxicity (Costa et al. 2023b).

While individual SoC models hold significant potential, coupling multiple tissue compartments in a single microfluidic circuit, known as multi-organ chips (MoCs), brings another layer of complexity. These systems enable systemic safety evaluations and the capture of metabolite toxicity (Sanches et al. 2020; Salvioni et al. 2021; Alépée et al. 2014; Boukamp et al. 1988). MoCs to capture interorgan communication between the gut and the skin as well as the liver and skin have been already developed (Chong et al. 2018). However, MoC use in understanding critical biological mechanisms related to exposure, uptake, translocation, and the effects of NMs on the skin and secondary organs remains absent.

3 Safety Testing of NM Toxicity in *In Vitro* Skin Models

As has been stated from the beginning of this chapter, the skin is a critical route of exposure for NMs and NPs and therefore a key target organ for potential adverse effects. The development and use of *in vitro* skin models have become a central and relevant aspect of the safety assessment of NMs and NPs. With these models, we can obtain valuable information on dermal toxicity while reducing the need for animal testing. When conducting NM toxicity testing in these *in vitro* skin models, several key endpoints are commonly evaluated:

- *Cytotoxicity:*
  Cell viability assays such as the MTT (3-(4,5-Dimethylthiazol-2-yl)-2,5-Diphenyltetrazolium Bromide) or LDH (lactate dehydrogenase) release assay are used to assess the cytotoxic effects of NMs on skin cells. This provides information on the concentrations at which NMs become toxic. The MTT (and similar assays) provide information related to the metabolic activity of the cells, less activity, less viability), while the LDH gives information regarding the liberation of an enzyme that should not be found outside of cells, and only being liberated once a cell is damaged and dead (Kroll et al. 2011).

- *Oxidative stress:*
  NMs can induce oxidative stress in skin cells, leading to inflammation and other toxic effects. Assays to measure reactive oxygen species, glutathione levels, and antioxidant enzyme activity can provide insights into this mechanism of NM toxicity. Recently, it has been shown that measuring the presence of reactive oxygen species in keratinocytes, may alter the redox status of melanocytes when explored in co-culture conditions. Also, *in vitro* models have been useful in evaluating the potential protective effect of some compounds, like reveratrol (Barygina et al. 2019; Shukla et al. 2011; Soeur et al. 2015).
- *Genotoxicity:*
  A main concern related to the use of NM in cosmetics and topical medication is the potential to cause DNA damage. *In vitro,* skin models can be used to assess NM genotoxicity using assays like the comet assay or micronucleus test. An interesting approach has been reported using a commercial reconstructed epidermis model, where the genotoxicity of chemicals has been tested employing the micronucleous assay (Chen et al. 2021; Magdolenova et al. 2014).
- *Skin irritation:*
  Skin irritation is mediated by innate immune responses, involving some cell types and expression of specific biomarkers. Different *in vitro* methods have been described to be useful in the evaluation of TiO_2 as irritant, and even some OECD guidelines have been developed to use *in vitro* approaches instead of animals (Samberg et al. 2010; Sanches et al. 2020).

When designing and conducting NM toxicity studies in *in vitro* skin models, it is important to carefully consider factors such as NM characterization, dosimetry, and the relevance of the model to the intended exposure scenario. Appropriate positive and negative controls, as well as thorough statistical analysis, are also crucial for ensuring the reliability and reproducibility of the results. Overall, *in vitro* skin models have become an invaluable tool for assessing the dermal toxicity of NMs. By providing information on a range of toxicological endpoints, these models can help identify potentially hazardous NMs and guide the development of safer nanomaterials. As the field of nanotoxicology continues to evolve, *in vitro* skin models will likely play an increasingly important role in the safety assessment of NMs (Warheit 2018).

4 Computational Structure-Based Approach for Skin Nanotoxicity Prediction

The two approaches for NPs hazard assessment are experiential toxicology (*in vitro* or *in vivo* biological experiments) and *in silico* approaches (computational studies). In silico toxicology applies computational techniques to analyze, simulate, visualize, and predict NMs/NPs toxicity as well as chemicals and drugs (Costa et al. 2023a; Enoch et al. 2008; Kalantari et al. 2021; Khanna et al. 2015; Thwala et al. 2022; Trott and Olson 2010). To accurately predict particle toxicity, computational models must account for their physicochemical complexity, which requires comprehensive characterization, as well as diverse exposure routes. Several models of different complexity have been developed, they predominantly use statistical and machine learning (ML) algorithms to establish relationships between NPs physicochemical properties and their consequent biological effects. Frequently used ML algorithms include regression, decision trees, support vector machines, artificial neural networks, partial least squares, and principal component analysis. On the other hand, structure-based techniques such as molecular docking coupled with molecular dynamics (MD) is also employed (Forest 2022). Predictive techniques such as Nano-Quantitative-Structure-Toxicity Relationships (Nano-QSTR models) are exclusive to NPs since it requires both previous determination of the physicochemical characteristics of NMs such as size, shape, surface area, and solubility and a proper experimental determination of the relevant skin nanotoxicity output (see Fig. 7.5). Physiologically based pharmacokinetic (PBPK) modelling coupled with molecular dynamics simulations are also known to foster a deeper understanding of NMs behavior, while, grouping and read-across strategies have significantly contributed to clustering, categorization, and classification of the most relevant NM properties linked to skin nanotoxicity, even when limited data are available. It is important to note that, in this computational context, from the methodological point of view the Nano-QSTR stands out for its predictive versatility oriented to multiple skin nanotoxicity outputs. This makes them better aligned with *in vitro* skin strategies compared with their in-silico counterparts (*i.e.*, molecular docking, molecular dynamics, DFT methods). The latter is more focused on answering specific questions at the molecular level from a mechanistic perspective on the interaction of NM and relevant molecular targets for skin nanotoxicity. All cited computational

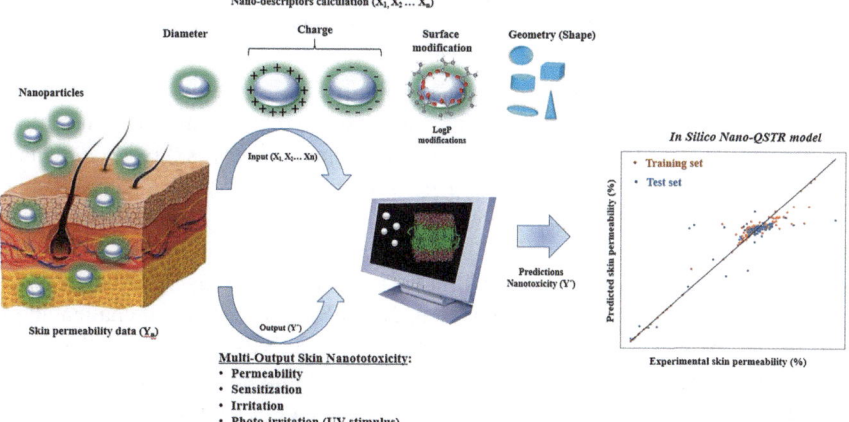

Fig. 7.5 Schematic diagram of a general workflow for in silico prediction of nanoparticle skin permeability (%) in humans by using a Nano-QSTR approach. Herein, nanodescriptors represent the physicochemical properties (e.g., diameter, charge, surface, and shape) of the NPs (from X1 to Xn) while the model output or predicted skin permeability (%) is denoted by the Yn

approaches are known to reduce the dependency on animal testing, saving time and resources as well as allowing a faster screening of larger numbers of NMs.

Recent advancements in computational modelling offer a promising approach for unveiling new insights into nanotoxicological evaluations, guiding risk assessment, and informed decision-making in nano-cosmeceuticals and skin health. By employing advanced in silico algorithms, researchers can anticipate potential adverse effects of NPs on skin permeability, sensitization, photo-induced irritation, as well as general irritation. Regarding skin sensitization of chemicals, several in silico models have been established (Toxtree, PredSkin, OECD's QSAR Toolbox, UL's REACHAcross™, Danish QSAR Database, TIMES-SS, and Lhasa Limited's Derek Nexus). These models utilize machine learning techniques and QSAR models to predict skin sensitization accurately, with some achieving correct classification rates of 70–80% on human data sets (Golden et al. 2021). Dermal permeation and absorption of substances have been also explored by in silico models, where mathematical equations can estimate the permeability coefficient of chemicals across the skin, taking into consideration features like skin anatomy, but also the physicochemical properties of the compounds to predict both local and systemic bioavailability applied to the skin, aiding in formulation risk assessment (Patel et al. 2022).

Quasi-QSAR was already employed to predict human keratinocyte cells (HaCaT) and human bronchial epithelial cells (BEAS-2B) cell viability when exposed to different 20 metal oxide nanomaterials. Hierarchical cluster analysis (HCA) and min-max normalization methods were employed in allocating codes for numerical descriptors (*e.g.,* core size, hydrodynamic size, surface charge, and dose). The established model provided good statistical and predictive performance (Choi et al. 2019). A new in silico model called Computational Indicator of Nanotoxicity exploits free energy analysis coupled with molecular dynamics simulations to evaluate the cytotoxicity of 2D nanomaterials can be relevant for skin toxicity evaluations (Tsukanov et al. 2022).

From the mechanistic point of view, several in silico strategies can be proposed, which to the best of our knowledge remain unexplored and represent a current gap in the field. The modelling of skin and epidermis nanotoxicity is directly influenced by critical NM physicochemical descriptors such as NPs size/diameter and shape which are recognized to have more weight during the design of computational models. For example, we could efficiently model the impact of the surface charge and/or coating of different NPs on the skin toxicodynamics by using structure-based molecular docking approaches (Lian et al. 2008) However, it is still challenging to implement the same modelling strategy for inorganic NPs since the mathematical scoring function to predict the thermodynamics binding affinity between NPs surface and the skin target receptors fails, when we want to model the influence of the different crystallographic planes of NP inorganic surfaces which significantly impact in the skin nanotoxicity (Norioka et al. 2021). To solve this issue, computational simulations based on Density Functional Theory (DFT) could be more successful in describing the influence of surface reactivity-based crystallographic planes for inorganic NPs. For example, DFT can simulate the TiO_2NPs anatase crystallographic facets (101) which are highly stable, and its stability could lead to prolonged persistence in the human skin, potentially inducing oxidative stress and skin inflammation depending on the exposure time. In the case of anatase, the DFT modelling of the (001) plane could be more relevant to explain potential skin cytotoxicity-based interactions because this plane is known for its high reactivity (Kang et al. 2023; McLean and Zhan 2022; Tsukanov et al. 2022; Wilm et al. 2018).

To computationally predict the influence of these structural determinants together with the concentration and exposure time on the skin nanotoxicity; one could follow these methodological steps: (i) collect the existing experimental or theoretical data on NPs properties from public repositories or specific databases and simultaneously perform physicochemical nano descriptors calculation (as model inputs), (ii) data collection on toxicological skin effects (*i.e.,* irritation, photo-irritation, inflammation, sensitization, skin permeability, cytotoxicity, etc.), (iii) divide the dataset into different subsets, typically training (containing the 70% of the total dataset) and test sets (containing the 30% of the total dataset). Step (iv) performs a single or integrated multi-target output as the predictive Nano-QSTR model with machine learning procedure by implementing appropriate predictive algorithms using Python or state-of-the-art visual programming Knime-based workflow pipeline. (v) choose the best predictive model with the most relevant structural nano descriptors for skin nanotoxicity and statistical performance-based metrics (sensitivity and specificity), and (vi) model validation with an ad-hoc external dataset or experimental validation, when possible. This in silico strategy can help identify potential nano risks in skin-related applications.

5 Conclusion Remarks and Future Trends

Currently available skin models specifically developed to test nanomaterials and nanoparticles are scarce, indicating that this field is in its infancy. 2D, 3D static and dynamic models have been developed with technologies allowing precise control of cells placement creating complex skin structures that recreate the multiple layers of the dermis and epidermis and allowing researchers to conduct longer-term studies and gain a deeper understanding of NP's safety assessment. The design of new models for risk assessment purposes should consider: (i) the inclusion of hair follicles since it can support the evaluation of alternative routes of NPs absorption and excretion, (ii) tissue vascularization to allow the study of systemic exposure of dermally absorbed NPs, (iii) the inclusion of immune cells to grant to the model immune competency and (iv) tissue innervation with neural cells to allow the study of sensory reactions upon NMs exposure.

Computational modelling for predicting nanotoxicity has experienced significant advancements, however in the field of skin there still exist a gap of knowledge. From our point of view computational methods offer a promising alternative to traditional animal-based testing with possible high acceptance from the new approach methodologies regulatory context. Boosting machine learning algorithms in conjunction with extensive human data repositories, exhibit considerable promise to predict skin permeability, sensitization, photo-irritation, and irritation induced by NPs/NMs. The continuous refinement and validation of these computational approaches are expected to lead to more precise nanotoxicity predictions and assist in the development of novel NPs with safer profiles for dermatological applications.

Acknowledgments This research was funded by the European Union's H2020 project Sinfonia (N.857253) and LEARN (N.101057510). SbDToolBox, with reference NORTE-01-0145-FEDER-000047, supported by Norte Portugal Regional Operational Programme (NORTE 2020), under the PORTUGAL 2020 Partnership Agreement, through the European Regional Development Fund. This publication is based upon work from COST Action European Network for Skin Engineering and Modeling (NETSKINMODELS), CA21108, supported by COST (European Cooperation in Science and Technology).

References

Abaci HE, Gledhill K, Guo Z, Christiano AM, Shuler ML (2015) Pumpless microfluidic platform for drug testing on human skin equivalents. Lab Chip 15:882–888. https://doi.org/10.1039/b000000x/HHS

Abd E, Yousef SA, Pastore MN, Telaprolu K, Mohammed YH, Namjoshi S, Grice JE, Roberts MS (2016) Skin models for the testing of transdermal drugs. Clin Pharmacol 8:163–176. https://doi.org/10.2147/CPAA.S64788

Abdayem R, Formanek F, Minondo AM, Potter A, Haftek M (2016) Cell surface glycans in the human stratum corneum: distribution and depth-related changes. Exp Dermatol 25:865–871. https://doi.org/10.1111/exd.13070

Alépée N, Bahinski A, Daneshian M, De Wever B, Fritsche E, Goldberg A, Hansmann J, Hartung T, Haycock J, Hogberg HT, Hoelting L, Kelm JM, Kadereit S, Mcvey E, Landsiedel R, Leist M, Lübberstedt M, Noor F, Pellevoisin C, Petersohn D, Pfannenbecker U, Reisinger K, Ramirez T, Rothen-Rutishauser B, Schäfer-Korting M, Zeilinger K, Zurich M-G (2014) State-of-the-art of 3D cultures (organs-on-a-chip) in safety testing and pathophysiology HHS public access. ALTEX 31:441–477. https://doi.org/10.14573/altex.1406111

Alépée N, Adriaens E, Abo T et al (2019a) Development of a defined approach for eye irritation or serious eye damage for neat liquids based on Cosmetics Europe analysis of in vitro RhCE and BCOP test methods. Toxicol in Vitro 59:100–114. https://doi.org/10.1016/j.tiv.2019.04.011

Alépée N, Adriaens E, Abo T et al (2019b) Development of a defined approach for eye irritation or serious eye damage for liquids, neat and in dilution, based on Cosmetics Europe analysis of in vitro STE and BCOP test methods. Toxicol in Vitro 54:147–167. https://doi.org/10.1016/j.tiv.2019.02.019

Ali D, Alarifi S, Alkahtani S, AlKahtane AA, Almalik A (2015) Cerium oxide nanoparticles induce oxidative stress and genotoxicity in human skin melanoma cells. Cell Biochem Biophys 71:1643–1651. https://doi.org/10.1007/s12013-014-0386-6

Alkilani AZ, McCrudden MTC, Donnelly RF (2015) Transdermal drug delivery: innovative pharmaceutical developments based on disruption of the barrier properties of the stratum corneum. Pharmaceutics 7:438–470. https://doi.org/10.3390/pharmaceutics7040438

Asmatulu E, Andalib MN, Subeshan B, Abedin F (2022) Impact of nanomaterials on human health: a review. Environ Chem Lett 20:2509–2529. https://doi.org/10.1007/s10311-022-01430-z

Augustine R (2018) Skin bioprinting: a novel approach for creating artificial skin from synthetic and natural building blocks. Prog Biomater 12:77–92. https://doi.org/10.1007/s40204-018-0087-0

Avci P, Sadasivam M, Gupta A, De Melo WC, Huang YY, Yin R, Chandran R, Kumar R, Otufowora A, Nyame T, Hamblin MR (2013) Animal models of skin disease for drug discovery. Expert Opin Drug Discov 8:331–355. https://doi.org/10.1517/17460441.2013.761202

Bacakova L, Zikmundova M, Pajorova J, Broz A, Filova E, Blanquer A, Matejka R, Stepanovska J, Mikes P, Jencova V, Kuzelova Kostakova E, Sinica A (2020) Nanofibrous scaffolds for skin tissue engineering and wound healing based on synthetic polymers. In: Applications of nanobiotechnology. IntechOpen. https://doi.org/10.5772/intechopen.88744

Barygina V, Becatti M, Lotti T, Moretti S, Taddei N, Fiorillo C (2019) ROS-challenged keratinocytes as a new model for oxidative stress-mediated skin diseases. J Cell Biochem 120:28–36. https://doi.org/10.1002/jcb.27485

Battistini B, Petrucci F, De Angelis I, Failla CM, Bocca B (2020) Quantitative analysis of metals and metal-based nano- and submicron-particles in tattoo inks. Chemosphere 245:125667. https://doi.org/10.1016/j.chemosphere.2019.125667

Boukamp P, Petrussevska RT, Breitkreutz D, Hornung J, Markham A, Fusenig NE (1988) Normal keratinization in a spontaneously immortalized aneuploid human keratinocyte cell line. J Cell Biol 106:761–771. https://doi.org/10.1083/jcb.106.3.761

Broadhead R (2023) Skin and textile interaction and the future of fashion as therapeutics. In: Woolley D, Johnstone F, Sampson E, Chambers P (eds) Wearable objects and curative things. Palgrave Macmillan, Cham/Ghent, pp 189–204

Broadhead R, Craeye L, Callewaert C (2021) The future of functional clothing for an improved skin and textile microbiome relationship. Microorganisms 31:1192. https://doi.org/10.3390/microorganisms9061192

Chauhan A, Chauhan C (2021) Emerging trends of nanotechnology in beauty solutions: a review. In: Materials today: proceedings. Elsevier Ltd, pp 1052–1059. https://doi.org/10.1016/j.matpr.2021.04.378

Chen S, Schoen J (2019) Air-liquid interface cell culture: from airway epithelium to the female reproductive tract. Reprod Domest Anim 54:38–45. https://doi.org/10.1111/rda.13481

Chen L, Wu M, Jiang S, Zhang Y, Li R, Lu Y, Liu L, Wu G, Liu Y, Xie L, Xu L (2019) Skin toxicity assessment of silver nanoparticles in a 3D epidermal model compared to 2D keratinocytes. Int J Nanomedicine 14:9707–9719. https://doi.org/10.2147/IJN.S225451

Chen L, Li N, Liu Y, Faquet B, Alépée N, Ding C, Eilstein J, Zhong L, Peng Z, Ma J, Cai Z, Ouedraogo G (2021) A new 3D model for genotoxicity assessment: EpiSkin™ micronucleus assay. Mutagenesis 36:51–61. https://doi.org/10.1093/mutage/geaa003

Chen S, Wang Q, Ding Y, Li S, Yu R (2024) Progress in the application of 3D skin models for cosmetic assessment. https://doi.org/10.20944/preprints202402.1340.v1

Cho SW, Malick H, Kim SJ, Grattoni A (2024) Advances in skin-on-a-chip technologies for dermatological disease modeling. J Invest Dermatol 144:1707–17015. https://doi.org/10.1016/j.jid.2024.01.031

Choi JS, Trinh TX, Yoon TH, Kim J, Byun HG (2019) Quasi-QSAR for predicting the cell viability of human lung and skin cells exposed to different metal oxide nanomaterials. Chemosphere 217:243–249. https://doi.org/10.1016/j.chemosphere.2018.11.014

Chong LH, Li H, Wetzel I, Cho H, Toh YC (2018) A liver-immune coculture array for predicting systemic drug-induced skin sensitization. Lab Chip 18:3239–3250. https://doi.org/10.1039/c8lc00790j

Coimbra SC, Sousa-Oliveira I, Ferreira-Faria I, Peixoto D, Pereira-Silva M, Mathur A, Pawar KD, Raza F, Mazzola PG, Mascarenhas-Melo F, Veiga F, Paiva-Santos AC (2022) Safety assessment of nanomaterials in cosmetics: focus on dermal and hair dyes products. Cosmetics 9:83. https://doi.org/10.3390/cosmetics9040083

Costa IGF, Ribeiro SRFL, Nascimento LL, Patrocinio AOT, Cardoso VL, Batista FRX, Reis MHM (2023a) Well-dispersed titanium dioxide and silver nanoparticles on external and internal surfaces of asymmetric alumina hollow fibers for enhanced chromium (VI) photoreductions. Environ Sci Pollut Res 30:62508–62521. https://doi.org/10.1007/s11356-023-26528-x

Costa S, Vilas-Boas V, Lebre F, Granjeiro JM, Catarino CM, Moreira Teixeira L, Loskill P, Alfaro-Moreno E, Ribeiro AR (2023b) Microfluidic-based skin-on-chip systems for safety assessment of nanomaterials. Trends Biotechnol 41:1282–1298. https://doi.org/10.1016/j.tibtech.2023.05.009

Crapo PM, Gilbert TW, Badylak SF (2011) An overview of tissue and whole organ decellularization processes. Biomaterials 32:3233–3243. https://doi.org/10.1016/j.biomaterials.2011.01.057

Cui X, Dean D, Ruggeri ZM, Boland T (2010) Cell damage evaluation of thermal inkjet printed Chinese hamster ovary cells. Biotechnol Bioeng 106:963–969. https://doi.org/10.1002/bit.22762

De Wever B, Kurdykowski S, Descargues P (2015) Human skin models for research applications in pharmacology and toxicology: introducing NativeSkin®, the "missing link" bridging cell culture and/or reconstructed skin models and human clinical testing. Appl In Vitro Toxicol 1:26–32. https://doi.org/10.1089/aivt.2014.0010

Dhasmana A, Firdaus S, Singh KP, Raza S, Jamal QMS, Kesari KK, Rahman Q, Lohani M (2017) Nanoparticles: applications, toxicology and safety aspects. Environ Sci Eng:47–70. https://doi.org/10.1007/978-3-319-46248-6_3

Dijkhoff IM, Drasler B, Karakocak BB, Petri-Fink A, Valacchi G, Eeman M, Rothen-Rutishauser B (2020) Impact of airborne particulate matter on skin: a systematic review from epidemiology to in vitro studies. Part Fibre Toxicol 25:35. https://doi.org/10.1186/s12989-020-00366-y

Dobler D, Schmidts T, Wildenhain S, Seewald I, Merzhäuser M, Runkel F (2019) Impact of selected cosmetic ingredients on common microorganisms of healthy human skin. Cosmetics 6:1–13. https://doi.org/10.3390/cosmetics6030045

Domingues C, Santos A, Alvarez-Lorenzo C, Concheiro A, Jarak I, Veiga F, Barbosa I, Dourado M, Figueiras A (2022) Where is nano today and where is it headed? A review of nanomedicine and the dilemma of nanotoxicology. ACS Nano 16:9994–10041. https://doi.org/10.1021/acsnano.2c00128

Drury JL, Mooney DJ (2003) Hydrogels for tissue engineering: scaffold design variables and applications. Biomaterials 24:4337–4351. https://doi.org/10.1016/S0142-9612(03)00340-5

Elaine F (1999) Mini-review epidermal differentiation: the bare essentials. J Cell Biol 111:2807–2814. https://doi.org/10.1083/jcb.111.6.2807

Enoch SJ, Madden JC, Cronin MTD (2008) Identification of mechanisms of toxic action for skin sensitisation using a SMARTS pattern-based approach. SAR QSAR Environ Res 19:555–578. https://doi.org/10.1080/10629360802348985

Ferreira MT, Soldado E, Borsoi G, Mendes MP, Flores-Colen I (2023) Nanomaterials applied in the construction sector: environmental, human health, and economic indicators. Appl Sci 13:12896. https://doi.org/10.3390/app132312896

Filaire E, Nachat-Kappes R, Laporte C, Harmand MF, Simon M, Poinsot C (2022) Alternative in vitro models used in the main safety tests of cosmetic products and new challenges. Int J Cosmet Sci 44:604–613. https://doi.org/10.1111/ics.12803

Forest V (2022) Experimental and computational nanotoxicology—complementary approaches for nanomaterial hazard assessment. Nanomaterials 12:1346. https://doi.org/10.3390/nano12081346

Golden E, Macmillan DS, Dameron G, Kern P, Hartung T, Maertens A (2021) Evaluation of the global performance of eight in silico skin sensitization models using human data. ALTEX 38:33–48. https://doi.org/10.14573/altex.1911261

Gu X, Li Z, Su J (2024) Air pollution and skin diseases: a comprehensive evaluation of the associated mechanism. Ecotoxicol Environ Saf 278:116429. https://doi.org/10.1016/j.ecoenv.2024.116429

Guichard A, Remoué N, Honegger T (2022) In vitro sensitive skin models: review of the standard methods and introduction to a new disruptive technology. Cosmetics 9:67. https://doi.org/10.3390/cosmetics9040067

Gupta S, Bansal R, Gupta S, Jindal N, Jindal A (2013) Nanocarriers and nanoparticles for skin care and dermatological treatments. Indian Dermatol Online J 4:267. https://doi.org/10.4103/2229-5178.120635

Gupta V, Mohapatra S, Mishra H, Farooq U, Kumar K, Ansari MJ, Aldawsari MF, Alalaiwe AS, Mirza MA, Iqbal Z (2022) Nanotechnology in cosmetics and cosmeceuticals—a review of latest advancements. Gels 8:173. https://doi.org/10.3390/gels8030173

Hao F, Jin X, Liu QS, Zhou Q, Jiang G (2017) Epidermal penetration of gold nanoparticles and its underlying mechanism based on human reconstructed 3D Episkin model. ACS Appl Mater Interfaces 9:42577–42588. https://doi.org/10.1021/acsami.7b13700

Hieda DS, da Costa A, Carvalho L, Vaz de Mello B, de Oliveira EA, Romano de Assis S, Wu J, Du-Thumm L, Viana da Silva CL, Roubicek DA, Maria-Engler SS, de Moraes B, Barros S (2020) Air particulate matter induces skin barrier dysfunction and water transport alteration on a reconstructed human epidermis model. J Invest Dermatol 140:2343–2352.e3. https://doi.org/10.1016/j.jid.2020.03.971

Hoffman AS (2012) Hydrogels for biomedical applications. Adv Drug Deliv Rev 64:18–23. https://doi.org/10.1016/j.addr.2012.09.010

Hofmann E, Schwarz A, Fink J, Kamolz LP, Kotzbeck P (2023) Modelling the complexity of human skin in vitro. Biomedicine 11:794. https://doi.org/10.3390/biomedicines11030794

Hu Q, Tuck C, Wildman R, Hague R (2015) Application of nanoparticles in manufacturing. In: Handbook of nanoparticles. Springer, pp 1219–1278. https://doi.org/10.1007/978-3-319-15338-4_55

Islam PS, Chang C, Selmi C, Generali E, Huntley A, Teuber SS, Gershwin ME (2016) Medical complications of tattoos: a comprehensive review. Clin Rev Allergy Immunol 50:273–286. https://doi.org/10.1007/s12016-016-8532-0

Jakab K, Norotte C, Marga F, Murphy K, Vunjak-Novakovic G, Forgacs G (2010) Tissue engineering by self-assembly and bio-printing of living cells. Biofabrication 2:022001. https://doi.org/10.1088/1758-5082/2/2/022001

Jang YS, Lee EY, Park YH, Jeong SH, Lee SG, Kim YR, Kim MK, Son SW (2012) The potential for skin irritation, phototoxicity, and sensitization of ZnO nanoparticles. Mol Cell Toxicol 8:171–177. https://doi.org/10.1007/s13273-012-0021-9

Kalantari L, Tran F, Blaha P (2021) Density functional theory study of metal and metal-oxide nucleation and growth on the anatase TiO_2 (101) surface. Computation 9:125. https://doi.org/10.3390/computation9110125

Kang Y, Kim MG, Lim KM (2023) Machine-learning based prediction models for assessing skin irritation and corrosion potential of liquid chemicals using physicochemical properties by XGBoost. Toxicol Res 39:295–305. https://doi.org/10.1007/s43188-022-00168-8

Khanna P, Ong C, Bay BH, Baeg GH (2015) Nanotoxicity: an interplay of oxidative stress, inflammation and cell death. Nanomaterials 5:1163–1180. https://doi.org/10.3390/nano5031163

Kim H, Choi J, Lee H, Park J, Yoon BI, Jin SM, Park K (2016) Skin corrosion and irritation test of nanoparticles using reconstructed three-dimensional human skin model, EpiDermTM. Toxicol Res 32:311–316. https://doi.org/10.5487/TR.2016.32.4.311

Koivisto AJ, Burrueco-Subirà D, Candalija A, Vázquez-Campos S, Nicosia A, Ravegnani F, Furxhi I, Brigliadori A, Zanoni I, Blosi M, Costa A, Belosi F, Lopez de Ipiña J (2024) Exposure assessment and risks associated with wearing silver nanoparticle-coated textiles. Open Res Eur 4:100. https://doi.org/10.12688/openreseurope.17254.1

Krieg T, Aumailley M (2011) The extracellular matrix of the dermis: flexible structures with dynamic functions. Exp Dermatol 20:689–695. https://doi.org/10.1111/j.1600-0625.2011.01313.x

Kroll A, Dierker C, Rommel C, Hahn D, Wohlleben W, Schulze-Isfort C, Göbbert C, Voetz M, Hardinghaus F, Schnekenburger J (2011) Cytotoxicity screening of 23 engineered nanomaterials using a test matrix of ten cell lines and three different assays. Part Fibre Toxicol 8. https://doi.org/10.1186/1743-8977-8-9

Kwak BS, Jin SP, Kim SJ, Kim EJ, Chung JH, Sung JH (2020) Microfluidic skin chip with vasculature for recapitulating the immune response of the skin tissue. Biotechnol Bioeng 117:1853–1863. https://doi.org/10.1002/bit.27320

Larese Filon F, Mauro M, Adami G, Bovenzi M, Crosera M (2015) Nanoparticles skin absorption: new aspects for a safety profile evaluation. Regul Toxicol Pharmacol 72:310–322. https://doi.org/10.1016/j.yrtph.2015.05.005

Larese Filon F, Bello D, Cherrie JW, Sleeuwenhoek A, Spaan S, Brouwer DH (2016) Occupational dermal exposure to nanoparticles and nano-enabled products: part I—factors affecting skin absorption. Int J Hyg Environ Health 219:536–544. https://doi.org/10.1016/j.ijheh.2016.05.009

Law JX, Liau LL, Saim A, Yang Y, Idrus R (2017) Electrospun collagen nanofibers and their applications in skin tissue engineering. Tissue Eng Regen Med 14:699–718. https://doi.org/10.1007/s13770-017-0075-9

Lee CC, Lin YH, Hou WC, Li MH, Chang JW (2020a) Exposure to ZnO/TiO$_2$ nanoparticles affects health outcomes in cosmetics salesclerks. Int J Environ Res Public Health 17:1–12. https://doi.org/10.3390/ijerph17176088

Lee J, Rabbani CC, Gao H, Steinhart MR, Woodruff BM, Pflum ZE, Kim A, Heller S, Liu Y, Shipchandler TZ, Koehler KR (2020b) Hair-bearing human skin generated entirely from pluripotent stem cells. Nature 582:399–404. https://doi.org/10.1038/s41586-020-2352-3

Lian G, Chen L, Han L (2008) An evaluation of mathematical models for predicting skin permeability. J Pharm Sci 97:584–598. https://doi.org/10.1002/jps.21074

Liu C, Wang Y, Zhang G, Pang X, Yan J, Wu X, Qiu Y, Wang P, Huang H, Wang X, Zhang H (2022) Dermal toxicity influence of gold nanomaterials after embedment in cosmetics. Toxics 10:276. https://doi.org/10.3390/toxics10060276

Lizarazo-Fonseca L, Correa-Araujo L, Prieto-Abello L, Camacho-Rodríguez B, Silva-Cote I (2023) In vitro and in vivo evaluation of electrospun poly (ε-caprolactone)/collagen scaffolds and Wharton's jelly mesenchymal stromal cells (hWJ-MSCs) constructs as potential alternative for skin tissue engineering. Regen Ther 24:11–24. https://doi.org/10.1016/j.reth.2023.05.005

Magdolenova Z, Collins A, Kumar A, Dhawan A, Stone V, Dusinska M (2014) Mechanisms of genotoxicity. A review of in vitro and in vivo studies with engineered nanoparticles. Nanotoxicology 8:233–278. https://doi.org/10.3109/17435390.2013.773464

Malte S, Jeanne J, Johansen D, Rustemeyer T, Elsner P, Maibach HI (2020) Kanerva's occupational dermatology, 3rd edn. Springer

Mascarenhas-Melo F, Mathur A, Murugappan S, Sharma A, Tanwar K, Dua K, Singh SK, Mazzola PG, Yadav DN, Rengan AK, Veiga F, Paiva-Santos AC (2023) Inorganic nanoparticles in dermopharmaceutical and cosmetic products: properties, formulation development, toxicity, and regulatory issues. Eur J Pharm Biopharm 192:25–40. https://doi.org/10.1016/j.ejpb.2023.09.011

McLean K, Zhan W (2022) Mathematical modelling of nanoparticle-mediated topical drug delivery to skin tissue. Int J Pharm 611:121322. https://doi.org/10.1016/j.ijpharm.2021.121322

Melnik AV, Callewaert C, Dorrestein K, Broadhead R, Minich JJ, Ernst M, Humphrey G, Ackermann G, Gathercole R, Aksenov AA, Knight R, Dorrestein PC (2023) The molecular effect of wearing silver-threaded clothing on the human skin. mSystems 8:e00922–e00922. https://doi.org/10.1128/msystems.00922-22

Moon S, Kim DH, Shin JU (2021) In vitro models mimicking immune response in the skin. Yonsei Med J 62:969–980. https://doi.org/10.3349/ymj.2021.62.11.969

Mori N, Morimoto Y, Takeuchi S (2017) Skin integrated with perfusable vascular channels on a chip. Biomaterials 116:48–56. https://doi.org/10.1016/j.biomaterials.2016.11.031

Motter Catarino C, Kaiser K, Baltazar T, Motter Catarino L, Brewer JR, Karande P (2022) Evaluation of native and non-native biomaterials for engineering human skin tissue. Bioeng Transl Med 7. https://doi.org/10.1002/btm2.10297

Murphy SV, Atala A (2014) 3D bioprinting of tissues and organs. Nat Biotechnol 32:773–785. https://doi.org/10.1038/nbt.2958

Nakamura M et al (2005) Biocompatible inkjet printing technique for designed seeding of individual living cells. Tissue engineering 11(11–12):1658–66. https://doi.org/10.1089/ten.2005.11.1658

Niska K, Zielinska E, Radomski MW, Inkielewicz-Stepniak I (2018) Metal nanoparticles in dermatology and cosmetology: interactions with human skin cells. Chem Biol Interact 295:38–51. https://doi.org/10.1016/j.cbi.2017.06.018

Norioka C, Inamoto Y, Hajime C, Kawamura A, Miyata T (2021) A universal method to easily design tough and stretchable hydrogels. NPG Asia Mater 13:34. https://doi.org/10.1038/s41427-021-00302-2

OECD (2010) Guidance Document on Using Cytotoxicity Tests to Estimate Starting Doses for Acute Oral Systemic Toxicity Tests. Series on Testing and Assessment No. 129, ENV/JM/MONO(2010)20. Organisation for Economic Co-operation and Development, Paris

OECD (2015) Test No. 404: Acute Dermal Irritation/Corrosion, OECD Guidelines for the Testing of Chemicals, Section 4. OECD Publishing. https://doi.org/10.1787/9789264242678-en

Omari Shekaftik S, Shirazi FH, Yarahmadi R, Rasouli M, Ashtarinezhad A (2022) Investigating the relationship between occupational exposure to nanomaterials and symptoms of nanotechnology companies' employees. Arch Environ Occup Health 77:209–218. https://doi.org/10.1080/19338244.2020.1863315

Palmer BC, Phelan-Dickenson SJ, Delouise LA (2019) Multi-walled carbon nanotube oxidation dependent keratinocyte cytotoxicity and skin inflammation. Part Fibre Toxicol 16. https://doi.org/10.1186/s12989-018-0285-x

Parrado C, Mercado-Saenz S, Perez-Davo A, Gilaberte Y, Gonzalez S, Juarranz A (2019) Environmental stressors on skin aging. Mechanistic insights. Front Pharmacol 10:759. https://doi.org/10.3389/fphar.2019.00759

Patel N, Clarke JF, Salem F, Abdulla T, Martins F, Arora S, Tsakalozou E, Hodgkinson A, Arjmandi-Tash O, Cristea S, Ghosh P, Alam K, Raney SG, Jamei M, Polak S (2022) Multiphase multi-layer mechanistic dermal absorption (MPML MechDermA) model to predict local and systemic exposure of drug products applied on skin. CPT Pharmacometrics Syst Pharmacol 11:1060–1084. https://doi.org/10.1002/psp4.12814

Pelikh O, Eckert RW, Pinnapireddy SR, Keck CM (2021) Hair follicle targeting with curcumin nanocrystals: influence of the formulation properties on the penetration efficacy. J Control Release 329:598–613. https://doi.org/10.1016/j.jconrel.2020.09.053

Peppas NA, Khare AR (1993) Preparation, structure and diffusional behavior of hydrogels in controlled release. Adv Drug Deliv Rev 11:1–35. https://doi.org/10.1016/0169-409X(93)90025-Y

Pereira RF, Barrias CC, Granja PL, Bartolo PJ (2013) Advanced biofabrication strategies for skin regeneration and repair. Nanomedicine:603–621. https://doi.org/10.2217/nnm.13.50

Place ES, George JH, Williams CK, Stevens MM (2009) Synthetic polymer scaffolds for tissue engineering. Chem Soc Rev 38:1139–1151. https://doi.org/10.1039/b811392k

Ramadan Q, Ting FCW (2016) In vitro micro-physiological immune-competent model of the human skin. Lab Chip 16:1899–1908. https://doi.org/10.1039/c6lc00229c

Rashid MM, Tavčer PF, Tomšič B (2021) Influence of titanium dioxide nanoparticles on human health and the environment. Nano 11:2354. https://doi.org/10.3390/nano11092354

Reing JE, Brown BN, Daly KA, Freund JM, Gilbert TW, Hsiong SX, Huber A, Kullas KE, Tottey S, Wolf MT, Badylak SF (2010) The effects of processing methods upon mechanical and biologic properties of porcine dermal extracellular matrix scaffolds. Biomaterials 31:8626–8633. https://doi.org/10.1016/j.biomaterials.2010.07.083

Reynolds WJ, Eje N, Christensen P, Li WH, Daly SM, Parsa R, Chavan B, Birch-Machin MA (2023) Biological effects of air pollution on the function of human skin equivalents. FASEB Bioadv 5:470–483. https://doi.org/10.1096/fba.2023-00068

Rimal R, Muduli S, Desai P, Marquez AB, Möller M, Platzman I, Spatz J, Singh S (2024) Vascularized 3D human skin models in the forefront of dermatological research. Adv Health Mater. https://doi.org/10.1002/adhm.202303351

Rogal J, Schlünder K, Loskill P (2022) Developer's guide to an organ-on-chip model. ACS Biomater Sci Eng 8:4643–4647. https://doi.org/10.1021/acsbiomaterials.1c01536

Rosen J, Landriscina A, Friedman AJ (2015) Nanotechnology-based cosmetics for hair care. Cosmetics 2:211–224. https://doi.org/10.3390/cosmetics2030211

Ryman-Rasmussen JP, Riviere JE, Monteiro-Riviere NA (2006) Penetration of intact skin by quantum dots with diverse physicochemical properties. Toxicol Sci 91:159–165. https://doi.org/10.1093/toxsci/kfj122

Saleem H, Zaidi SJ (2020) Sustainable use of nanomaterials in textiles and their environmental impact. Materials 13:5134. https://doi.org/10.3390/ma13225134

Salvioni L, Morelli L, Ochoa E, Labra M, Fiandra L, Palugan L, Prosperi D, Colombo M (2021) The emerging role of nanotechnology in skincare. Adv Colloid Interf Sci 293:102437. https://doi.org/10.1016/j.cis.2021.102437

Samberg ME et al (2010) Evaluation of silver nanoparticle toxicity in skin in vivo and keratinocytes in vitro. Environ Health Perspect 118(3):407–413

Sanches PL, Geaquinto LRDO, Cruz R, Schuck DC, Lorencini M, Granjeiro JM, Ribeiro ARL (2020) Toxicity evaluation of TiO_2 nanoparticles on the 3d skin model: a systematic review. Front Bioeng Biotechnol 8:575. https://doi.org/10.3389/fbioe.2020.00575

Schneider MR, Oelgeschlaeger M, Burgdorf T, van Meer P, Theunissen P, Kienhuis AS, Piersma AH, Vandebriel RJ (2021) Applicability of organ-on-chip systems in toxicology and pharmacology. Crit Rev Toxicol 51:540–554. https://doi.org/10.1080/10408444.2021.1953439

Shapira C, Itshak D, Duadi H, Harel Y, Atkins A, Lipovsky A, Lavi R, Lellouche JP, Fixler D (2022) Noninvasive nanodiamond skin permeation profiling using a phase analysis method: ex vivo experiments. ACS Nano 16:15760–15769. https://doi.org/10.1021/acsnano.2c03613

Shukla RK, Sharma V, Pandey AK, Singh S, Sultana S, Dhawan A (2011) ROS-mediated genotoxicity induced by titanium dioxide nanoparticles in human epidermal cells. Toxicol in Vitro 25:231–241. https://doi.org/10.1016/j.tiv.2010.11.008

Singh G, Mishra A, Mathur A, Shastri S, Nizam A, Rizwan A, Dadial AS, Firdous A, Hassan H (2024) Advancement of organ-on-chip towards next-generation medical technology. Biosens Bioelectron X 18:100480. https://doi.org/10.1016/j.biosx.2024.100480

Soeur J, Eilstein J, Léreaux G, Jones C, Marrot L (2015) Skin resistance to oxidative stress induced by resveratrol: from Nrf2 activation to GSH biosynthesis. Free Radic Biol Med 78:213–223. https://doi.org/10.1016/j.freeradbiomed.2014.10.510

Strüver K, Friess W, Hedtrich S (2017) Development of a perfusion platform for dynamic cultivation of in vitro skin models. Skin Pharmacol Physiol 30:180–189. https://doi.org/10.1159/000476071

Sun T, Jackson S, Haycock JW, MacNeil S (2006) Culture of skin cells in 3D rather than 2D improves their ability to survive exposure to cytotoxic agents. J Biotechnol 122:372–381. https://doi.org/10.1016/j.jbiotec.2005.12.021

Sun X, Chang Y, Cheng Y, Feng Y, Zhang H (2018) Band alignment-driven oxidative injury to the skin by anatase/rutile mixed-phase titanium dioxide nanoparticles under sunlight exposure. Toxicol Sci 164:300–312. https://doi.org/10.1093/toxsci/kfy088

Tamilarasi GP, Sabarees G, Manikandan K, Gouthaman S, Alagarsamy V, Solomon VR (2023) Advances in electrospun chitosan nanofiber biomaterials for biomedical applications. Mater Adv 4:3114–3139. https://doi.org/10.1039/d3ma00010a

Thwala MM, Afantitis A, Papadiamantis AG, Tsoumanis A, Melagraki G, Dlamini LN, Ouma CNM, Ramasami P, Harris R, Puzyn T, Sanabria N, Lynch I, Gulumian M (2022) Using the Isalos platform to develop a (Q)SAR model that predicts metal oxide toxicity utilizing facet-based electronic, image analysis-based, and periodic table derived properties as descriptors. Struct Chem 33:527–538. https://doi.org/10.1007/s11224-021-01869-w

Tordesillas L, Lozano-Ojalvo D, Dunkin D, Mondoulet L, Agudo J, Merad M, Sampson HA, Berin MC (2018) PDL2+ CD11b+ dermal dendritic cells capture topical antigen through hair follicles to prime LAP+ Tregs. Nat Commun 9. https://doi.org/10.1038/s41467-018-07716-7

Trott O, Olson AJ (2010) AutoDock Vina: improving the speed and accuracy of docking with a new scoring function, efficient optimization, and multithreading. J Comput Chem 31:455–461. https://doi.org/10.1002/jcc.21334

Tsukanov AA, Turk B, Vasiljeva O, Psakhie SG (2022) Computational indicator approach for assessment of nanotoxicity of two-dimensional nanomaterials. Nano 12(4):650. https://doi.org/10.3390/nano12040650

Venugopal J, Ramakrishna S (2005) Applications of polymer nanofibers in biomedicine and biotechnology. Appl Biochem Biotechnol 125:147–158. https://doi.org/10.1385/abab:125:3:147

Venus M, Waterman J, McNab I (2010) Basic physiology of the skin. Surgery 28:469–472. https://doi.org/10.1016/j.mpsur.2010.07.011

Wang W, Lin Y, Yang H, Ling W, Liu L, Zhang W, Lu D, Liu Q, Jiang G (2022) Internal exposure and distribution of airborne fine particles in the human body: methodology, current understandings, and research needs. Environ Sci Technol 56(11):6857–6869. https://doi.org/10.1021/acs.est.1c07051

Warheit DB (2018) Hazard and risk assessment strategies for nanoparticle exposures: how far have we come in the past 10 years? F1000Res. https://doi.org/10.12688/f1000research.12691.1

Wilm A, Kühnl J, Kirchmair J (2018) Computational approaches for skin sensitization prediction. Crit Rev Toxicol 48:738–760. https://doi.org/10.1080/10408444.2018.1528207

Wistner SC, Rashad L, Slaughter G (2023) Advances in tissue engineering and biofabrication for in vitro skin modeling. Bioprinting 35:e00306. https://doi.org/10.1016/j.bprint.2023.e00306

Wu Q, Liu J, Wang X, Feng L, Wu J, Zhu X, Wen W, Gong X (2020) Organ-on-a-chip: recent breakthroughs and future prospects. Biomed Eng Online 19:9. https://doi.org/10.1186/s12938-020-0752-0

Wufuer M, Lee GH, Hur W, Jeon B, Kim BJ, Choi TH, Lee SH (2016) Skin-on-a-chip model simulating inflammation, edema and drug-based treatment. Sci Rep 6. https://doi.org/10.1038/srep37471

Yuan J, Zhang Q, Liu Z, Du J, Qin W, Lu M, Cui H, Li X, Ding S, Li R (2019) Dermal exposure to nano-TiO_2 induced cardiovascular toxicity through oxidative stress, inflammation and apoptosis. J Toxicol Sci 44:35–45. https://doi.org/10.2131/jts.44.35

Yurchenco PD, Schittny JC (1990) Molecular architecture of basement membranes. FASEB 4:1577–1590. https://doi.org/10.1096/fasebj.4.6.2180767

Zeb A, Arif ST, Malik M, Shah FA, Din FU, Qureshi OS, Lee ES, Lee GY, Kim JK (2019) Potential of nanoparticulate carriers for improved drug delivery via skin. J Pharm Investig 49:485–517. https://doi.org/10.1007/s40005-018-00418-8

Zoio P, Oliva A (2022) Skin-on-a-chip technology: microengineering physiologically relevant in vitro skin models. Pharmaceutics 14:682. https://doi.org/10.3390/pharmaceutics14030682

Open Access This chapter is licensed under the terms of the Creative Commons Attribution-NonCommercial-NoDerivatives 4.0 International License (http://creativecommons.org/licenses/by-nc-nd/4.0/), which permits any noncommercial use, sharing, distribution and reproduction in any medium or format, as long as you give appropriate credit to the original author(s) and the source, provide a link to the Creative Commons license and indicate if you modified the licensed material. You do not have permission under this license to share adapted material derived from this chapter or parts of it.

The images or other third party material in this chapter are included in the chapter's Creative Commons license, unless indicated otherwise in a credit line to the material. If material is not included in the chapter's Creative Commons license and your intended use is not permitted by statutory regulation or exceeds the permitted use, you will need to obtain permission directly from the copyright holder.

Advanced Gut-on-a-Chip *In Vitro* Models for ENMs Safety Assessment

8

M. D. Neto, L. Pastrana, and C. Gonçalves

Abstract

Over the last decades, the use of engineered nanomaterials (ENMs) as delivery systems has been rising, finding applications in different fields, such as pharmaceutical, medical, and food technology. ENMs offer opportunities for targeted delivery, improved stability through the digestive conditions, increased solubility and bioavailability, and sustained release. Therefore, the risk assessment of those materials is essential to ensure their safety, in which *in vitro* models of the human gastrointestinal tract arise as fundamental tools to address this need. This chapter highlights the physiological features that play a significant role in a reliable risk assessment of orally administered ENMs and reviews the current advances and developments in *in vitro* models, pinpointing the opportunities that arise with organ-on-chip technology, organoids and the combination thereof. Moreover, we unveil the lag in using organ-on-a-chip systems to assess the fate and biological interactions of ENMs intended for oral administration despite the significant advances in the field. Finally, we outline that a closer collaboration between bioengineers and clinical toxicologists is fundamental to maximize the availability of human clinical data for *in vitro* model validation, fostering the wide acceptance of microfluidic models in pre-clinical studies.

Keywords

Oral route of administration · *In vitro* digestion · Intestinal transport · Gut-on-a-chip Intestinal organoids

M. D. Neto · L. Pastrana · C. Gonçalves (✉)
International Iberian Nanotechnology Laboratory, Braga, Portugal
e-mail: catarina.goncalves@inl.int

© The Author(s) 2025
E. Alfaro-Moreno, F. Murphy (eds.), *Nanosafety*,
https://doi.org/10.1007/978-3-031-93871-9_8

1 Introduction

The use of engineered nanomaterials (ENMs) has grown exponentially in the last years, finding applications in different industrial sectors. Inevitably, human exposure to nanomaterials is expected to keep increasing in the next years, upon deliberate administration (food, medicines) or incidentally (release into the environment—air, water, soil), calling for rigorous safety testing. Indeed, the potential adverse effects that size-reduced materials, such as micro- and nanoplastics can have in the human health, is growing public concern (Bredeck et al. 2022).

To address these concerns, regulatory agencies have developed guidelines and regulations for the risk assessment of ENMs to ensure their safe use. In the European Union, application of nanotechnology in food products is governed by the European Chemicals Agency (ECHA) and the European Food Safety Authority (EFSA), while in the United States, the responsible agency is the Food and Drug Administration (FDA) (Amenta et al. 2015). The EFSA published a Guidance in 2021 elaborating on the main steps to be followed for the risk assessment of ingested ENMs. The flow chart begins with a thorough physicochemical characterization of the material, followed by an exposure assessment, determining whether the nanomaterial (i) persists in the food/feed matrix, (ii) migrates from a Food Contact Material (FCM) or (iii) persists after digestion in conditions representative of the human gastrointestinal tract. If the nanomaterial persists after digestive conditions, *in vitro* tests should follow to identify and characterize specific hazards, including geno- and cytotoxicity assessment, using cells representative of the human intestine. In case a full degradation of the material occurs, there is no need to follow the guidance for ENMs and guidance for conventional materials can be followed (More et al. 2021; Xavier et al. 2021).

As the intestine follows the mouth, esophagus and stomach, it is important to present to the intestinal cells, the samples as they reach the intestine *in vivo*. While traveling through the gastrointestinal environment, the properties of nanoparticles can be altered, such as their aggregation state, determining biological effects and potential toxicity (Zhou and McClements 2022). However, only a few studies on ENMs have combined *in vitro* digestion (upper digestion) and cell-based approaches for intestinal absorption studies (Silva et al. 2024; Sousa Ribeiro et al. 2023). After testing intestinal absorption, the role of intestinal microbiota must be considered, as non-absorbed particles continue traveling along the intestine having the possibility to be in contact with the microbiota for some time, before being eventually excreted. These interactions might have adverse effects on the microbial community, reducing specific populations or leading to a compositional shift of the microbiota landscape.

The intricate nature of the human gastrointestinal tract makes it a big challenge to mimic, requiring the recreation of very different microenvironments along the way. Cell-based intestinal *in vitro* models vary from mono- to co-cultures of cell lines, organoids and gut-on-a-chip. The complexity integrated into a selected testing strategy may be based on the research question aiming to be addressed (Kämpfer et al. 2020).

Intestinal *in vitro* models have been used in different industrial sectors, such as pharmaceuticals, food, medical devices and others. The pharmaceutical sector employs *in vitro* tests mainly for screening purposes to study pharmacokinetics and identify the toxicological potential of drug candidates. In medicine-related fields, studies on the uptake and targeted delivery of nanotechnology-enabled drugs into and across intestinal tissues are usually performed. In the food sector, intestinal *in vitro* models are mainly used for screening and ranking purposes instead of getting mechanistic information (Bredeck et al. 2022). This chapter addresses how advanced intestinal *in vitro* models can support the risk assessment of nanomaterials, considering oral exposure.

2   Gut Anatomy

Digestion, absorption, secretion and motility are the main functions of the human intestine. In addition, it constitutes a physical and biochemical barrier between the external and the internal environments of the human body. Its wall comprises four distinct layers: (i) mucosa, facing the lumen, which includes the columnar epithelial cell monolayer, the *lamina propria*, and muscularis mucosae, (ii) submucosa, beneath the mucosa, (iii) muscularis externa and (iv) serosa, which is the outmost layer (Fig. 8.1a). Tissue architecture and cell composition vary along the different

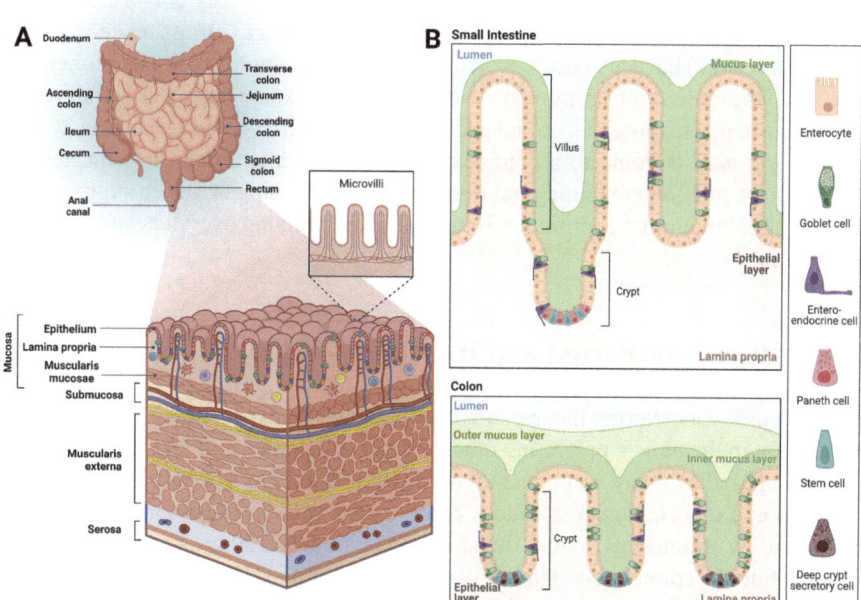

Fig. 8.1 Intestinal anatomical features and organization. (**a**) Segments of the human intestine and structure of the intestinal wall. (**b**) Small intestine and colonic mucosa, highlighting the differences in topographical features, mucus layers and cellular composition. (Created with BioRender.com)

regions of the intestine as they serve distinct physiological functions (Mowat and Agace 2014). The mucosa of the small intestine features finger-like protrusions—villi—projected into the lumen enhancing the intestinal surface area and facilitating absorption of nutrients and other compounds; while they are absent in the mucosa of the colon, which has little to no digestive function and where mainly water and microbiota-derived metabolites are absorbed. The epithelial cell monolayer that lines each villus is composed of different cell types, such as enterocytes (absorptive in nature and representing the most abundant cell type), goblet cells (mucus-producing) and enteroendocrine cells (hormone-producing). In the small intestine, at the base of villi, minute depressions extend into the lamina propria forming the crypts, where stem cells are interspaced between Paneth cells (Crawley et al. 2014). In the colon, deep crypt secretory (DCS) cells located at the colonic crypt base are a critical component of the colonic stem cell niche, sharing some similarities to Paneth and goblet cells (Fig. 8.1b) (McCarthy and Keely 2023).

The intestine is considered the largest immune organ in the human body. Intestinal lymphoid cells are mainly present in the intestinal epithelium, mucosal lamina propria (LP) or organized in the gut-associated lymphoid tissue (GALT), which includes Peyer's patches (PPs), mesenteric lymph nodes (MLNs), and isolated lymphoid follicles. B cells are the major immune cells population in PPs, and T cells make up the highest percentage in MLNs and the LP (Lin et al. 2022). M cells are specialized epithelial cells, lacking microvilli, that can be found in the epithelia above Peyer's patches where the mucus layer is reduced or absent (Fig. 8.2). These cells represent <1% of the total population, however, they play a crucial role in the uptake of microorganisms and antigens. The goblet cell-associated antigen passages (GAPs) have also been associated to antigen sampling from the lumen and deliver to the underlying immune system (Fig. 8.2) (Kulkarni et al. 2020). Pathogenic organisms can use those cells to invade the host (Gustafsson and Johansson 2022).

The gut wall is home to a complex population of microorganisms that play a determinant role in maintaining gut homeostasis. Increasing evidence suggests that the intestinal microbial community is crucial in maintaining overall health (Talapko et al. 2022).

3   Intestinal Barrier and Transport of ENMs

The mucus layer covering the intestinal epithelial surface constitutes the first line of protection in the lumen, protecting mucosa from pathogens or other external agents. It is composed of many components: water (90–95%), electrolytes, lipids (1–2%), proteins and others. However, mucus composition and thickness vary significantly along the gastrointestinal tract. The small intestine has a single unattached mucus layer, while the colon has two layers of mucus (Fig. 8.1b). The outer layer is thinner and looser serving as the natural habitat for commensal bacteria while the inner layer is typically thicker and denser, and impermeable to microorganisms, thus preventing direct interaction between the commensal bacteria and the host epithelium (Pelaseyed et al. 2014). However, the inner mucus layer can be impaired due to

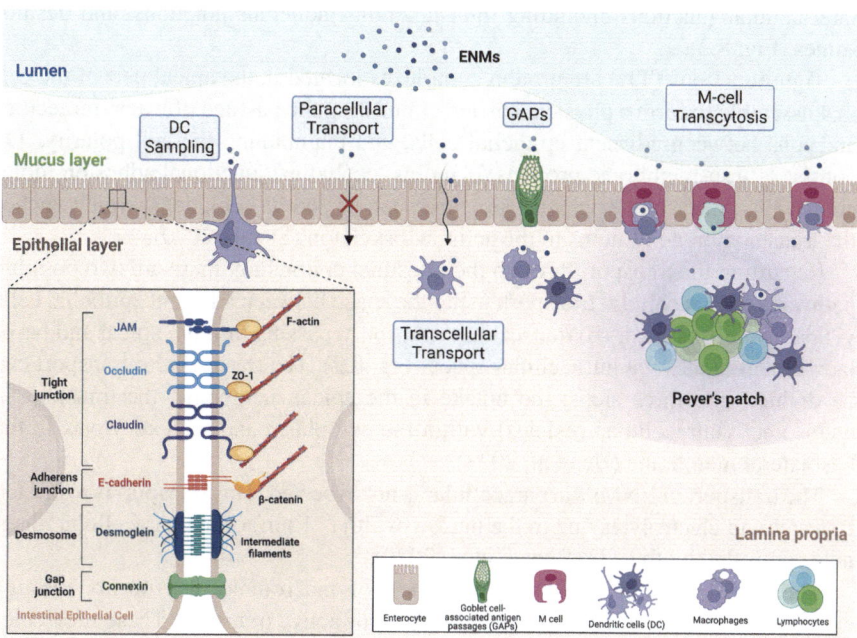

Fig. 8.2 Diagram showing inter-cellular junctional complexes and possible pathways to permeate the intestinal epithelium and mechanisms for ENMs uptake, transport and presentation to the underlying immune system. (Created with BioRender.com)

aberrations in the immune system or due to the action of proteases of certain microorganisms that can dissolve the mucus allowing bacteria penetration. Reaching the epithelium, bacteria activate the immune system and inflammation is triggered (Johansson et al. 2013). Mucins are highly glycosylated proteins, the major (1–5%) structural and functional constituent (Paone and Cani 2020). The overall negative charge of the mucus layer entraps positively charged pathogens or particulates through electrostatic interactions. In addition, the mucins have hydrophobic regions that interact with hydrophobic materials (McCright and Maisel 2020).

Depending on their properties, ENMs can be entrapped in the mucus layer, preventing their interaction with the epithelium and subsequent uptake (Stalder et al. 2022). However, while mucus can represent a barrier to effective oral delivery, it can also enable longer residence time, improving the efficacy of oral drug delivery systems. Biodegradable mucoadhesive nanoparticles have been studied for vaccine delivery due to their inherent immune adjuvant property and ability to protect the antigen from degradation, sustaining its release in the intestine due to the increased residence time (Zhao et al. 2022).

The gut epithelium constitutes the second barrier, which integrity is crucial for maintaining overall health. The impairment of this barrier can lead to the entry of microorganisms and harmful substances into the bloodstream, activating pathological processes (Chelakkot et al. 2018). Intestinal barrier integrity depends on inter-cellular junctions, including tight junctions, adherens junctions, and desmosomes (Fig. 8.2).

Tight junctions (TJs) are protein complexes located in the apical part of the cell that assemble to form a physical barrier, controlling the passage of water, molecules and ions between adjacent epithelial cells, and maintaining the cell polarity. TJs consist of transmembrane proteins (claudins, occludins, junctional adhesion molecule-A) and intracellular plaque proteins, such as zonula occludens, which anchor the transmembrane proteins to the actin cytoskeleton (Suzuki 2020).

Regarding the transport through the intestinal epithelium, there are two possible pathways (i) paracellular transport, using the space between adjacent epithelial cells via passive diffusion or (b) transcellular transport, passing through apical and basolateral membranes via intracellular space (Fig. 8.2). The transcellular transport can be divided into three steps: the uptake in the apical membrane, the transport in endosomes (intracellular vesicles) within the cytoplasm and the exocytosis in the basolateral membrane (He et al. 2023).

The transport of ENMs via paracellular is not expected. This transport is restricted to water and electrolytes due to the narrow width (<1 nm) of the paracellular channels controlled by the TJs (Babadi et al. 2020).

The main route of intestinal absorption of macromolecules occurs through enterocytes via transcellular passive diffusion or active transport. These molecules are subsequently delivered through the portal vein to the liver and finally reaching the systemic circulation. Macromolecules that possess high solubility and permeability exhibit enhanced bioavailability and satisfactory therapeutic efficacy (Shugarts and Benet 2009). Adversely, some are susceptible to the first-pass hepatic metabolism, leading to low bioavailability (Milligan and Saha 2022; Ye et al. 2020).

The lymphatic pathway appears as a promising alternative for oral delivery, as molecules are transported first through the lymphatic vessels and finally to the systemic circulation. This pathway offers two major targets (i) chylomicrons in enterocytes and (ii) M cells. Moreover, the leaky capillaries of the lymphatics allow the transport of macromolecules and particles that have larger sizes (Zhang et al. 2021).

Different drug delivery systems have been engineered to overcome mucosal barriers for immunotherapy and vaccination, targeting lymphoid tissues (lymph nodes, lymphatic vessels, and M cells) (McCright and Maisel 2020). The oral delivery of proteins has been extensively explored as well, representing a big challenge, as the gastrointestinal tract is prepared to digest peptides and proteins (Chen et al. 2022). Moreover, only small molecules and products of protein digestion, such as di- and tri-peptides can cross the epithelium into systemic circulation. Innovative encapsulation techniques of peptides and proteins have improved their stability under harsh digestive conditions, and the use of permeation enhancers in the formulation increases intestinal permeability. Permeation enhancers can either improve transcellular transport by alteration of the structure (permeabilization) of the cell membrane or paracellular transport by rearranging the tight junctions (Whitehead and Mitragotri 2008). The first oral peptide formulation containing a permeation enhancer was approved by the FDA in 2019 (Rybelsus®, Eligen™ SNAC technology, licensed by Novo Nordisk). Nanoparticles themselves have shown the capacity to induce tight junction relaxation, increasing intestinal permeability and enabling the oral delivery of proteins (Chen et al. 2013; Lamson et al. 2020; Pangua et al. 2024).

Recently, research has been focused on the mechanism of transport of ENMs using intestinal *in vitro* models of different complexities and integrating different cell types, such as enterocyte-like cells (Caco-2), mucus-secreting cells (HT29-MTX) and/or follicle-associated epithelial cells (M-cells). Inflamed intestinal models have been explored, as well, to mimic different aspects of the intestinal epithelium (Beloqui et al. 2016). The surface properties of nanoparticles, such as charge, hydrophilicity/hydrophobicity, and the presence of ligands, play a crucial role in determining their interactions with cells (Zheng et al. 2024). The ultimate goal would be to anticipate the ENMs fate across the intestinal barrier based on physicochemical properties.

4   Gut Microbiota

The human small intestine is the site where nutrients are absorbed. Uptake of electrolytes and water occurs further down in the colon, as well as fermentation of undigested foods, such as polysaccharides, proteins and fibers, by colonic microbiota.

The human gut microbiota is a complex, dynamic and balanced ecosystem composed of a multitude of microorganisms that play a crucial role in maintaining gut homeostasis, influencing immunity, metabolism and barrier function (Thursby and Juge 2017). An imbalance in the microorganism community, known as dysbiosis, has been associated with several diseases, such as diabetes, obesity, inflammatory bowel disease (IBD), and even cancer (Toor et al. 2019). The composition of the gut microbiota throughout the longitudinal and cross-sectional axes of the lower gastrointestinal tract (GIT) is known to be spatially segregated, creating distinct microbial habitats influenced by the different physiological environments (Donaldson et al. 2015). Colonic microorganisms are responsible for the final digestion process, through fermentation, producing short-chain fatty acids (SCFA). Acetate, propionate, and butyrate are the main metabolites produced in the colon by bacterial fermentation of dietary fibers and resistant starch. In addition to contributing to intestinal homeostasis and regulation of energy metabolism, growing evidence supports the idea that SCFAs also exert crucial physiological effects on organs, beyond the gut, through their circulation in the blood (Silva et al. 2020; van der Hee and Wells 2021). The effect of microbiome-derived SCFAs on the severity of ulcerative colitis was studied in a human microphysiological system of the gut, liver, and circulating Treg and Th17 cells, using a multi-omics approach. An impact on the immune response, gut barrier and hepatic function was demonstrated (Trapecar et al. 2020). Moreover, microbiota-gut-brain crosstalk has been extensively investigated aiming to unravel the direct and indirect impact of SCFAs on physiology and brain function, supporting microbiota-targeted therapeutics, such as prebiotics, probiotics and diet, for brain disorders (Dalile et al. 2019).

In the last decades, *in vitro* simulators have been developed to mimic the physiological GIT conditions that impact the microbial community and its metabolic activity (Venema and Van Den Abbeele 2013). Both EU and US authorities, EFSA and FDA, respectively, support the use of *in vitro* models to provide evidence of a purpose for food/ingredient (Marzorati et al. 2014). Studying the gut microbiome interaction with human intestinal cells in conventional/static *in vitro* models is only feasible for a short period due to bacterial overgrowth compromising the epithelial monolayer (Taebnia et al. 2023). More sophisticated models, such as organoids and gut-on-a-chip can revolutionize the *in vitro* study of host-microbial interactions (Puschhof et al. 2021). Fluid flow, an inherent property of microfluidic devices, and mechanical forces are key features to enable long-term co-cultures of epithelial intestinal cells and gut microbiota, avoiding bacterial overgrowth, through the continuous delivery of nutrients and removal of waste products, excess of bacteria and dead cells (Kim et al. 2016). Moreover, oxygen gradient within the model has profound impacts on cellular function, highlighting that controlling oxic conditions is crucial to mimic the microbiome and host interactions. The lumen is an anaerobic environment, due to the oxygen consumption by the microbes but also due to oxidative chemical reactions, such as lipid oxidation, while the *lamina propria* is oxygenated, and rich in blood and lymphatic vessels (Valiei et al. 2023).

5 *In Vitro* Models of Upper Digestion

The *in vitro* models to mimic the upper human digestion (mouth, stomach and intestine) can be divided into static, semi-dynamic, or dynamic models. The majority of methodologies available in the literature are based on static conditions. However, there is a big effort in designing new semi-dynamic or dynamic systems that are more representative of human physiological conditions.

Static models consist of a reservoir where the oral, gastric, and intestinal phases are sequentially simulated. In each phase, the sample is incubated for a specific time with simulant fluids (salivary, gastric, or intestinal) using constant physicochemical conditions and enzyme activities throughout each phase. Dynamic models are complex computer-controlled systems, with flow allowing time-resolved addition of simulant fluids and enzymes, gastric emptying and pH and temperature monitoring. Other dynamic features, such as peristalsis can also be included (Dupont et al. 2019).

The static models, while simple, often lack standardization of experimental conditions, including the type and concentration/activity of digestive enzymes, the composition of simulant fluids, pH levels, duration of each digestion phase, and the type and velocity of agitation. These variations, even small, do not allow the comparison of the results among different studies (Xavier and Mariutti 2021). Efforts have been made to standardize conditions. The INFOGEST protocols, a result of a COST Action that gathered an international network (more than 40 countries) to provide protocols as close as possible to what occurs *in vivo*, and simple enough to be applied in different laboratories to different compounds, have been published for static (Brodkorb et al. 2019; Minekus et al. 2014) and semi-dynamic (Mulet-Cabero et al. 2020) conditions. The semi-dynamic models represent an intermediate approach between the static and dynamic ones (Calero et al. 2024). Those models include dynamics only in the gastric phase closely mimicking the *in vivo* conditions: gradual addition of gastric secretions, gradual acidification (from pH 7 to 3) and different gastric emptying, allowing the assessment of the role of these dynamics in the digestion of tested materials. The intestinal phase is kept static (2 h of incubation using constant conditions: pH and enzyme activities), as in the static model. However, there are currently few models offering the finesse required to study nanomaterials at the development stage (Xavier et al. 2023). Dynamic models reported in the literature were built aiming to study digestion of food.

In the last years, the miniaturization trend has led to new systems (microfluidic and millifluidic devices), easier and cheaper to manufacture, that represent promising solutions to rapidly test compounds for their therapeutic efficiency and safety. To date, only two examples of miniaturized digestion simulators have been reported (De Haan et al. 2019; Xavier et al. 2023). Considering the increasing production and use of ENMs in multiple applications, a more elaborated and detailed hazard evaluation is recognized as needed, including upper *in vitro* digestion coupled with informative *in vitro* cell-based studies (cytotoxicity, oxidative stress, inflammation) (Vital et al. 2024). The main challenge when coupling both parts is to avoid the cytotoxicity of simulant digestion fluids toward the cultured intestinal epithelial cells. Different approaches are reported to mitigate the cytotoxicity of the simulant digestion fluids, based on the physical separation of enzymes and/or bile salts (filtration, ultracentrifugation, dialysis) or enzymes inactivation (heat or chemical inhibitors) (Kondrashina et al. 2023). However, there is always the need to dilute the digestion products, at least 10 times, to maintain low cytotoxicity levels (Vital et al. 2024; Xavier et al. 2023). It is worth considering that the initial sample concentration undergoes 1:8 dilution upon *in vitro* digestion (INFOGEST protocol) plus 1:10 dilution for the subsequent assays with cells, to avoid cytotoxicity.

6 Cell-Based *In Vitro* Models of the Intestinal Epithelium

Cell-based intestinal *in vitro* models vary in complexity from simple 2D models using mono- or co-cultures of cell lines to 3D models, such as organoids or gut-on-a-chip systems.

Early 2D *in vitro* models are reductionist but reproducible models usually based on single or multiple cultures of cell lines on a dish or microplate. Later, Transwell® inserts provided the possibility to evaluate barrier function and intestinal permeability, granting access to both apical and basolateral sides of polarized cell monolayers. More recently, the generation of more intricate and physiologically relevant models has revolutionized the field, contributing to different areas ranging from development and disease modelling to compound screening and toxicology. IBD is a disease condition that has been receiving considerable attention regarding *in vitro* models (Macedo et al. 2023).

Organ-on-a-chip systems are powerful tools for assessing ENMs toxicity, filling the gap between the conventional 2D cell culture models and *in vivo* experiments, using animal models. On the one hand, 2D *in vitro* systems fail to accurately reproduce key features, such as villus differentiation, production of mucus, expression of tissue-specific transporters and enzymes, and also hinder the long-term culture of microbiome. On the other hand, animal models are time-consuming and expensive, and raise ethical questions and concerns regarding the recapitulation of human physiology, due to interspecies variability. Indeed, researchers and regulatory entities are planning a future without lab animals, using alternative non-animal models (Moutinho 2023).

The ability to recapitulate the structure and function of the intestine is particularly important for nanotoxicity, as some features play a significant role in a reliable assessment, as previously highlighted. However, the selection of an *in vitro* model may be guided by the research question. Patient-derived intestinal organoids associated to organ-on-a-chip technology represent the most promising *in vitro* model to mimic the native human intestine (Campbell et al. 2021).

6.1 Immortalized Cell Lines

Cytotoxicity, uptake and/or transport of ENMs have been extensively investigated *in vitro* using mainly cancer-derived cells, such as the colon carcinoma cell lines Caco-2 and HT29, which differentiate into enterocyte- and goblet-like cells, respectively.

Mono- (Caco-2) or co-cultures (Caco-2/HT29-MTX) in confluent or differentiated stage, in well-plates or in Transwell® inserts for permeability studies, are extensively reported (Banerjee et al. 2016; Liu et al. 2022; Mahler et al. 2009; Strugari et al. 2019; Yang et al. 2018; Ye et al. 2017). Interestingly, a study found that differentiated mono- and co-culture models behaved very similarly upon exposure to TiO_2 nanoparticles (NPs), with both models revealing an increase in the accumulation of reactive oxygen species (ROS) and downregulation of the expression of enzymes related with cellular redox homoeostasis, while cell viability was not impaired (Dorier et al. 2017). The same authors have later demonstrated that the same NPs caused similar responses on an undifferentiated co-culture model (Dorier et al. 2019). Despite these results, it has been reported that after differentiation, cells become more robust and less sensitive to particle toxicity, thus revealing the importance of choosing the adequate model or experimental setup that displays a physiologically relevant exposure scenario (Becht et al. 2024; Gerloff et al. 2013; Kämpfer et al. 2021; Song et al. 2015).

Another intriguing aspect is the transport of particles being usually low in these models. This low transport across the epithelial barrier is probably due to particle bioaccumulation or deposition on the epithelial surface, as shown by Ribeiro and colleagues (Sousa Ribeiro et al. 2023). Particle internalization will most likely contribute to the generation of intracellular ROS, and accumulation around the nucleus can potentially lead to oxidative damage in DNA (Augustine et al. 2020;

Kusi-Appiah et al. 2017). Despite having proved to be of indisputable value, Caco-2 cells often misrepresent the *in vivo* due to inappropriate expression levels or mutations of transporters and enzymes, resulting in inaccurate biological responses, mainly predicting the intestinal permeability of hydrophilic compounds, particularly substrates of highly expressed intestinal transporters, such as members of the amino acid, nucleoside, and peptide transporter families (Larregieu and Benet 2013).

M-cells, representing a minor population in the follicle-associated epithelium, have been pointed out as responsible for the uptake and transport of macromolecules, microorganisms, and antigens to the underlying mucosal immune system (lymphoid follicles) via transepithelial transport (Kernéis et al. 2000). The absence of mature M cells in mice has been demonstrated to lead to a reduced accumulation of particulate matter in Peyer's Patches (Kanaya et al. 2012). In this context, the presence of M cells is of particular importance when studying the permeability or biological responses of ENMs *in vitro*. Kerneis et al. developed an *in vitro* model based on Caco-2 cells and mouse Peyer's patch lymphocytes, triggering the development of major M-cell-like features. Later, Caco-2 cells were co-cultured with human Raji B-cells, demonstrating an M-cell-like morphology, an altered expression of alkaline phosphatase (down-regulation) and Sialyl Lewis A antigen (up-regulation) and a 40-fold increase in microparticle transport (Gullberg et al. 2000). Intestinal *in vitro* permeability of nanoparticles with different properties has been evaluated in more complex models, including triple cultures (Caco-2/HT29-MTX/Raji B) demonstrating the role of M-cells on nanoparticles translocation (Schimpel et al. 2014).

6.2 Primary Cultures and Organoids

Primary cells are harvested from tissues to be immediately used or cultured *in vitro* for very limited periods. Despite the technical challenges associated with the culture of primary human intestinal epithelial cells, the generation of intestinal organoids represents a significant breakthrough in the *in vitro* primary cell-based models. Organoids can be generated from either intestinal crypts containing adult stem cells, using human biopsy samples (Sato et al. 2011), or pluripotent stem cells, such as embryonic stem cells (ESCs) or induced pluripotent stem cells (iPSCs) (Spence et al. 2011). Irrespective of the source, organoids retain long-term self-renewal properties owing to stable intestinal stem cells interspersed between differentiated cells and give rise to all the intestinal epithelial subtypes (Sato and Clevers n.d.).

Intestinal organoids are miniature versions of epithelia that grow in three dimensions (3D) and rely on the ability of the stem cells to expand seemingly indefinitely. When cultured within a 3D extracellular matrix and culture medium containing Wnt, noggin, R-spondin, and other relevant growth factors, intestinal organoids spontaneously undergo villus-crypt morphological organization and intestinal histogenesis. The organoids derived from an intestinal tissue biopsy of a healthy or diseased individual patient can be grown, frozen, and revived for multiple purposes enabling the creation of biobanks for the easy generation of on-demand patient-specific *in vitro* models. Moreover, organoids capture the specific characteristics of the donor opening up new opportunities for personalized medicine (Clevers 2016). Additionally, organoids are recognized as more physiologically relevant models by reflecting not only key structural features but also functional properties. They are able to perform basic functions, such as molecular absorption, transport and secretion, highly resembling the *in vivo* architecture and function (Zietek et al. 2015). Conversely to immortalized cell lines, organoids show similar expression levels of different transporters and enzymes to that of their *in vivo* counterparts, revealing to be a better alternative when assessing molecular transport (Gunasekara et al. 2018; Speer et al. 2019).

However, there are still substantial technical challenges associated with this technology. Organoids lack other cell types that are found in the living intestine, such as endothelium-lined blood vessels and immune cells. Additionally, organoids are heterogeneous structures (size and shape) and their 3D closed structure requires advanced techniques to access the lumen. A high-throughput organoid microinjection system for cargo delivery and content sampling in the organoid lumen was reported (Williamson et al. 2018). Alternatively, organoids have been cultured as monolayers, after dissociation, in conventional 2D formats, such as cell culture inserts, or even within microfluidic systems, allowing manipulation of both apical and basolateral sides in a compartmentalized manner (Altay et al. 2019).

Organoids have been employed as monolayers in assessing safety of small molecules. Kourula et al. applied enterocyte-enriched human duodenal and colonic organoids cultured as monolayers, as a tool to assess absorption, distribution, metabolism and excretion (ADME) and safety properties of different compounds (Kourula et al. 2023). This study demonstrated that intestinal toxic compounds identified in humans were similar to the ones detected in organoid models, which also displayed metabolic activity for most enzymes expressed in the intestine. Apart from testing small molecules and compounds, organoids have also been used to assess interactions and transport of particulate systems. Regarding the use of NPs as drug delivery systems, a PLGA-PEG functionalized NPs were exploited as a Trojan-horse strategy to improve its delivery across the intestinal epithelial barrier. The NPs were designed not only to protect the drug from the harsh conditions of the GIT, but to also target the intestinal Fc receptor located on the luminal surface of the epithelial cells. This improved its bioavailability and permeability, therefore enhancing its transcellular transport (Pinto et al. 2024).

From a nanosafety assessment perspective, intestinal organoids and their derived monolayers have also been used to evaluate the effects of micro- and nano-plastics on human gut health. An interesting work by Hou and co-workers revealed that ~50 nm sized polystyrene (PS) NPs show a distinct accumulation in organoids, particularly in secretory cells (goblet, Paneth and enteroendocrine cells), and that long-term exposure led to cell apoptosis, ROS production and elicit an inflammatory response (Hou et al. 2022). Moreover, the authors elucidated the underlying mechanism involved in PS particle uptake, showing a critical role played by active endocytosis. A recent work demonstrated the ability to generate organoid-derived monolayers with the inclusion or exclusion of M cells aiming at testing the effects of PS micro- and nano-plastics. A size-, concentration-, and time-dependent internalization of particles by epithelial cells was observed, linked to the secretion of inflammatory cytokines for higher concentrations. The presence of M cells increased the degree of particle translocation. Large-sized particles (sizes of 500 nm and 1 μm) preferentially aggregated in M cells, crossing the epithelial barrier through them. In comparison, smaller-sized particles (100 and 30 nm) were randomly uptaken by enterocytes and secretory cells. With these results, the authors demonstrated the relevance of M cells in sensing, sampling and role in transporting large-sized particles across the intestinal barrier (Chen et al. 2023).

6.3 Advanced Models and Nanosafety Assessment

Gut-on-chips are advanced models consisting of microfabricated channels or chambers lined with living cells cultured under fluid flow recapitulating, to some extent, the functionality of the human intestine. A common design consists of two overlapping channels separated by a semi-permeable membrane that supports mono- or co-cultures of different cell types. Both channels are independently perfused to mimic different physiological compartments. The production of microdevices often relies on soft lithography using elastomeric materials, such as poly (dimethylsiloxane) (PDMS). PDMS offers low cost, ease of fabrication, optical transparency, biocompatibility and gas permeability. However, absorption of small hydrophobic molecules is a major drawback. Different approaches were tested to reduce PDMS absorption (Rodrigues et al. 2022) and non-absorptive materials have been explored as an alternative to PDMS (Campbell et al. 2021).

The complexity of gut-on-chip platforms can vary from mono- or co-cultures of cell lines to organoid-derived monolayers, in combination with microbial cells and other non-epithelial cells, such as immunological, endothelial and fibroblasts, along with mechanical stimulation that emulate the natural compressive and tensile strains, as well as the fluid shear stress (Fig. 8.3) (Apostolou et al. 2021; Beaurivage et al. 2020; Deguchi et al. 2024; Nikolaev et al. 2020; Sontheimer-Phelps et al. 2020; Workman et al. 2018).

The first report of gut-on-chip devices co-culturing human cells (Caco-2 cell line) and living microorganisms (Lactobacillus rhamnosus GG) experiencing intestinal peristalsis-like motions dates back to 2012. Under the mechanical stimulus (cyclic strain), Caco-2 cells polarize, forming a 3D structure that resembles intestinal villi (Kim et al. 2012). Later, a gut-on-a-chip was reported to co-culture multiple commensal microbes with Caco-2 cells. Under exposure to bacterial lipopolysaccharides, human peripheral blood mononuclear cells (PBMCs) released inflammatory molecules that damaged the epithelial monolayer. The effect of gut microbiome, inflammatory cells (PBMCs), and peristalsis-like movements on intestinal bacterial overgrowth and inflammation were analyzed (Kim et al. 2016).

More recently, efforts have been made to recapitulate the intestinal epithelial–endothelial interface. In 2018, Kasendra et al. described a method for fabricating an Intestine-Chip from human intestinal biopsies. The primary epithelial cells were

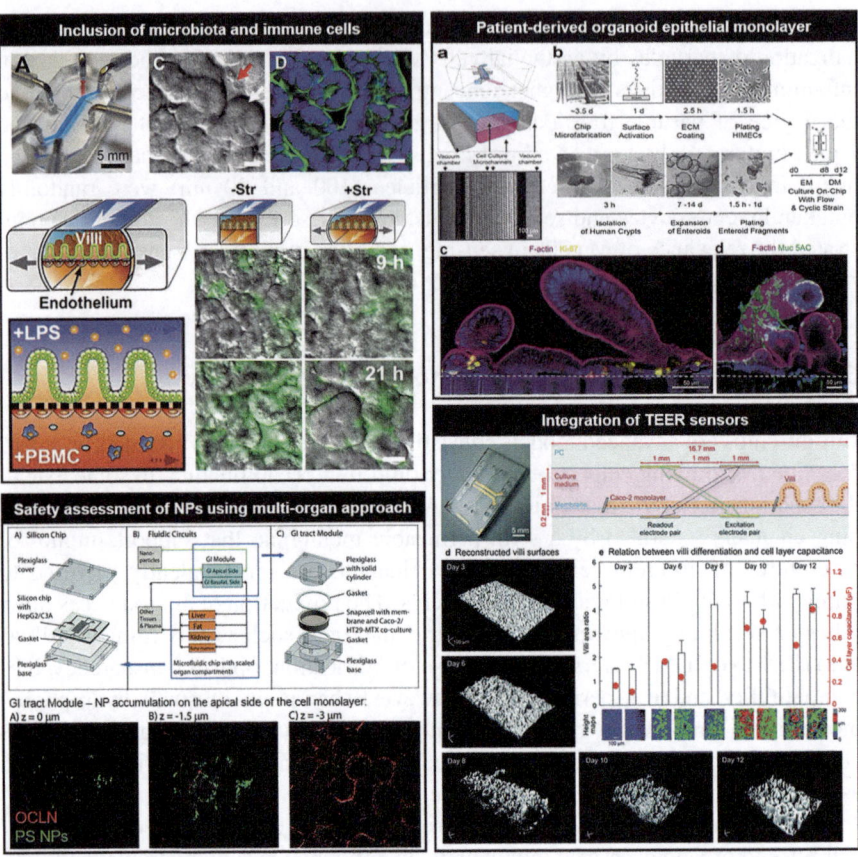

Fig. 8.3 Different set-ups and microfluidic devices established as cell-based intestinal *in vitro* models integrating physiological features that play a significant role in a reliable risk assessment of ENMs, such as the (i) inclusion of microbiota and immune cells within a fully differentiated and polarized Caco-2 monolayer, challenged with LPS and PBMCs. Peristaltic-like motions employed through mechanical deformations revealed overgrowth of the bacterial components (shown as green) when this stimulation ceased. (Reproduced from Kim et al. (2016) without modifications under the terms of the Creative Commons Attribution (CC-BY) Copyright 2016, National Academy of Sciences). (ii) integration of patient-derived cells using organoid-derived monolayers generating a polarized epithelial monolayer with finger-like protrusions and presence of proliferating cells (marked with Ki67, shown in yellow) and specialized (mucus-producing) differentiated cells (marked with Muc5AC, shown in green). (Reproduced from Kasendra et al. (2018) without modifications under the terms of the Creative Commons Attribution (CC-BY) Copyright 2018, Springer Nature). (iii) considering a multi-organ-on-a-chip approach combining human intestinal epithelium, represented by a co-culture of enterocytes (Caco-2) and mucin-producing cells (HT29-MTX), and the liver, represented by HepG2/C3A cells. The device also contained chambers that together represented the liquid portions of all other organs of the human body to test 50 nm carboxylated polystyrene nanoparticles (green) that accumulate at different focal planes (A: $z = 0$ μm, B: $z = -1.5$ μm, and C: $z = -3$ μm) of Caco-2/HT29-MTX co-cultures. Their slightly higher location and grouping in such accumulations suggest that the nanoparticles could reside in or above patches of mucous layers. (Reproduced with permission from Esch et al. (2014) Copyright 2014, Royal Society of Chemistry) and (iv) integration of sensors in a gut-on-a-chip device, where the villus cultured on-chip, increase in number and height over time and the measured epithelial capacitance effectively predicted the villi area ratio and hence the degree of villus differentiation. (Reproduced with permission from van der Helm et al. (2019) Copyright 2019, Royal Society of Chemistry)

first expanded as 3D organoids, and then dissociated, and cultured in a microfluidic device combined in a parallel channel with human intestinal microvascular endothelium cells. The Intestine-Chip integrating both flow and cyclic deformation demonstrated villi-like projections lined by polarized epithelial cells that undergo multi-lineage differentiation (Kasendra et al. 2018).

Another promising research direction is the integration of sensing components into organ-on-a-chip models introducing real-time monitoring capabilities. The capacity of real-time measurement of biochemical changes or morphological evaluation over time will be instrumental for mechanistic understanding of biological responses to ENMs. Monitoring barrier integrity is of utmost importance when studying barrier-forming tissues for permeability purposes. Impedance spectroscopy is a powerful tool and is widely used to monitor barrier permeability and integrity in real-time (Marr et al. 2023; Marrero et al. 2023; van der Helm et al. 2019; Vera et al. 2024).

Nowadays, multiple companies, such as Emulate Inc., MIMETAS and others, commercialize distinct organ-chips. The technology comprises different applications with great potential to revolutionize the next-generation risk assessment of novel drugs/formulations or ingredients in the pharma or food industries, respectively.

Organ-on-chip is an emerging technology in the field of nanosafety with still very few publications addressing nanotoxicity after oral exposure, in a gut-on-chip system (Kang et al. 2021). A liver-gut model was developed to test toxicity of 50 nm carboxylated PS nanoparticles. Co-cultures of Caco-2/HT29-MTX and HepG2/C3A cells were used to represent intestinal epithelium and the liver, respectively. The results showed a significantly higher aspartate aminotransferase (AST), an intracellular enzyme of the liver that indicates liver cell injury, release in the gut-liver chip compared to the gut-chip alone, demonstrating the relevance of inter-organ crosstalk when testing nanoparticle-mediated toxicity (Esch et al. 2014). In a different study, gene expression of Caco-2 cells was analysed upon exposure to TiO_2 (E171) and ZnO (NM110) nanoparticles, referred as the most produced nanomaterials, when cultured under dynamic (gut-on-chip) and static conditions (Transwells®). Whole-genome transcriptome analyses revealed that cells cultured in the gut-on-chip displayed a stronger response when compared with cells cultured in Transwells, suggesting that cells cultured under dynamic conditions are more sensitive and, therefore more suitable for toxicological hazard characterization (Kulthong et al. 2021).

7   Outlook and Future Perspectives

The remarkable progress in the nanotechnology application in different sectors has driven efforts to develop new tools for *in vitro* prediction of ENMs fate under physiological conditions regarding their efficacy and safety. Considering the oral route, the complexity of the human intestine and all the processes associated, appear as a challenge. The small intestine is the main site of absorption. However, before reaching the small intestine, ingested samples are exposed to different environments (pH, enzymes, bile salts) that can completely alter their initial properties, influencing the subsequent biological interactions, therefore it must not be neglected.

Throughout this chapter we highlight the complexity of the human gastrointestinal tract and point out determinant factors that may be considered when testing ENMs. Organ-on-a-chip technology represents the most advanced model providing a promising approach to overcome the limitations of 2D models and reduce the use of animal models. Considering the 3R principle (replacement, reduction, and refinement), the use of *in vitro* models is of extreme importance in the nanosafety research.

It is our belief that future research will focus on the development of next-generation *in vitro* models, combining organoid cultures with tissue engineering approaches and other enabling technologies, such as organ-on-chips and 3D bioprinting, allowing the integration of crucial intestinal architectural and functional features that better mimic the intricate native microenvironment. Those features include crypt-villi domains, the presence of representative epithelial cell types, stromal compartment, immune cells and also microbiota. Moreover, multi-organ platforms, mainly the coupling of liver and kidney systems, will be valuable for a more integrative and comprehensive approach shedding light on inter-organ crosstalk for ADME-Toxicology studies. In the context of ENMs safety assessment, the inclusion of such elements might find relevance when assessing possible interactions with the microbiota, or its by-products, and their fate after encountering the immune system upon absorption by the epithelial layer, as well as the process for their metabolism and subsequent excretion.

It is not only the development of the model systems that is important, but also the data acquisition with integrative analytics which will be significant for retrieving critical information in a real-time and *in situ* manner. The integration of biosensors allows a close and more sensitive monitoring, as well as spatio-temporal resolution of the biological generated data. Omics techniques will continue to be a very powerful tool for a deeper and comprehensive understanding at a spatially-resolved single-cell level.

Despite the significant advances in the field of *in vitro* models, there is still a lag in using organ-on-a-chip systems to assess the fate and biological interactions of ENMs. The slow process is partly attributed to the lack of standardization, throughput and scale-up of manufacturing processes representing a huge challenge when trying to integrate such models in the roadmap of nanosafety. To the best of our knowledge, there are only a few examples of application of gut-on-a-chip technology in the safety assessment of ENMs. We foresee the next step to be the validation of *in vitro* findings, that require close cooperation and collaboration between scientists developing *in vitro* models, clinical toxicologists, and pharmaceutical companies, to maximize the availability of human clinical data that can be used for comparative analysis and validation purposes, in which artificial intelligence and machine learning will have the leading role.

References

Altay G, Larrañaga E, Tosi S, Barriga FM, Batlle E, Fernández-Majada V, Martínez E (2019) Self-organized intestinal epithelial monolayers in crypt and villus-like domains show effective barrier function. Sci Rep 9(1). https://doi.org/10.1038/s41598-019-46497-x

Amenta V, Aschberger K, Arena M, Bouwmeester H, Botelho Moniz F, Brandhoff P, Gottardo S, Marvin HJP, Mech A, Quiros Pesudo L, Rauscher H, Schoonjans R, Vettori MV, Weigel S, Peters RJ (2015) Regulatory aspects of nanotechnology in the agri/feed/food sector in EU and non-EU countries. Regul Toxicol Pharmacol 73(1):463–476. https://doi.org/10.1016/j.yrtph.2015.06.016

Apostolou A, Panchakshari RA, Banerjee A, Manatakis DV, Paraskevopoulou MD, Luc R, Abu-Ali G, Dimitriou A, Lucchesi C, Kulkarni G, Maulana TI, Kasendra M, Kerns JS, Bleck B, Ewart L, Manolakos ES, Hamilton GA, Giallourakis C, Karalis K (2021) A novel microphysiological colon platform to decipher mechanisms driving human intestinal permeability. CMGH 12(5):1719–1741. https://doi.org/10.1016/j.jcmgh.2021.07.004

Augustine R, Hasan A, Primavera R, Wilson RJ, Thakor AS, Kevadiya BD (2020) Cellular uptake and retention of nanoparticles: insights on particle properties and interaction with cellular components. Mater Today Commun 25. Elsevier Ltd. https://doi.org/10.1016/j.mtcomm.2020.101692

Babadi D, Dadashzadeh S, Osouli M, Daryabari MS, Haeri A (2020) Nanoformulation strategies for improving intestinal permeability of drugs: a more precise look at permeability assessment methods and pharmacokinetic properties changes. J Control Release 321:669–709. Elsevier B.V. https://doi.org/10.1016/j.jconrel.2020.02.041

Banerjee A, Qi J, Gogoi R, Wong J, Mitragotri S (2016) Role of nanoparticle size, shape and surface chemistry in oral drug delivery. J Control Release 238:176–185. https://doi.org/10.1016/j.jconrel.2016.07.051

Beaurivage C, Kanapeckaite A, Loomans C, Erdmann KS, Stallen J, Janssen RAJ (2020) Development of a human primary gut-on-a-chip to model inflammatory processes. Sci Rep 10(1). https://doi.org/10.1038/s41598-020-78359-2

Becht JM, Kohlleppel H, Schins RPF, Kämpfer AAM (2024) Effect of butyrate on food-grade titanium dioxide toxicity in different intestinal in vitro models. Chem Res Toxicol. https://doi.org/10.1021/acs.chemrestox.4c00086

Beloqui A, des Rieux A, Préat V (2016) Mechanisms of transport of polymeric and lipidic nanoparticles across the intestinal barrier. Adv Drug Deliv Rev 106:242–255. Elsevier B.V. https://doi.org/10.1016/j.addr.2016.04.014

Bredeck G, Halamoda-Kenzaoui B, Bogni A, Lipsa D, Bremer-Hoffmann S (2022) Tiered testing of micro- and nanoplastics using intestinal in vitro models to support hazard assessments. Environ Int 158. Elsevier Ltd. https://doi.org/10.1016/j.envint.2021.106921

Brodkorb A, Egger L, Alminger M, Alvito P, Assunção R, Ballance S, Bohn T, Bourlieu-Lacanal C, Boutrou R, Carrière F, Clemente A, Corredig M, Dupont D, Dufour C, Edwards C, Golding M, Karakaya S, Kirkhus B, Le Feunteun S et al (2019) INFOGEST static in vitro simulation of gastrointestinal food digestion. Nat Protoc 14(4):991–1014. https://doi.org/10.1038/s41596-018-0119-1

Calero V, Rodrigues PM, Dias T, Ainla A, Vilaça A, Pastrana L, Xavier M, Gonçalves C (2024) A miniaturised semi-dynamic in-vitro model of human digestion. Sci Rep 14(1). https://doi.org/10.1038/s41598-024-54612-w

Campbell SB, Wu Q, Yazbeck J, Liu C, Okhovatian S, Radisic M (2021) Beyond polydimethylsiloxane: alternative materials for fabrication of organ-on-a-chip devices and microphysiological systems. ACS Biomater Sci Eng 7(7):2880–2899. American Chemical Society. https://doi.org/10.1021/acsbiomaterials.0c00640

Chelakkot C, Ghim J, Ryu SH (2018) Mechanisms regulating intestinal barrier integrity and its pathological implications. Exp Mol Med 50(8) Nature Publishing Group. https://doi.org/10.1038/s12276-018-0126-x

Chen MC, Mi FL, Liao ZX, Hsiao CW, Sonaje K, Chung MF, Hsu LW, Sung HW (2013) Recent advances in chitosan-based nanoparticles for oral delivery of macromolecules. Adv Drug Deliv Rev 65(6):865–879. https://doi.org/10.1016/j.addr.2012.10.010

Chen G, Kang W, Li W, Chen S, Gao Y (2022) Oral delivery of protein and peptide drugs: from non-specific formulation approaches to intestinal cell targeting strategies. Theranostics 12(3):1419–1439. Ivyspring International Publisher. https://doi.org/10.7150/thno.61747

Chen Y, Williams AM, Gordon EB, Rudolph SE, Longo BN, Li G, Kaplan DL (2023) Biological effects of polystyrene micro- and nano-plastics on human intestinal organoid-derived epithelial tissue models without and with M cells. Nanomedicine 50. https://doi.org/10.1016/j.nano.2023.102680

Clevers H (2016) Modeling development and disease with organoids. Cell 165(7):1586–1597. Cell Press. https://doi.org/10.1016/j.cell.2016.05.082

Crawley SW, Mooseker MS, Tyska MJ (2014) Shaping the intestinal brush border. J Cell Biol 207(4):441–451. Rockefeller University Press. https://doi.org/10.1083/jcb.201407015

Dalile B, Van Oudenhove L, Vervliet B, Verbeke K (2019) The role of short-chain fatty acids in microbiota–gut–brain communication. Nat Rev Gastroenterol Hepatol 16(8):461–478. Nature Publishing Group. https://doi.org/10.1038/s41575-019-0157-3

De Haan P, Ianovska MA, Mathwig K, Van Lieshout GAA, Triantis V, Bouwmeester H, Verpoorte E (2019) Digestion-on-a-chip: a continuous-flow modular microsystem recreating enzymatic digestion in the gastrointestinal tract. Lab Chip 19(9):1599–1609. https://doi.org/10.1039/c8lc01080c

Deguchi S, Kosugi K, Takeishi N, Watanabe Y, Morimoto S, Negoro R, Yokoi F, Futatsusako H, Nakajima-Koyama M, Iwasaki M, Yamamoto T, Kawaguchi Y, Torisawa Y, Suke, Takayama K (2024) Construction of multilayered small intestine-like tissue by reproducing interstitial flow. Cell Stem Cell. https://doi.org/10.1016/j.stem.2024.06.012

Donaldson GP, Lee SM, Mazmanian SK (2015) Gut biogeography of the bacterial microbiota. Nat Rev Microbiol 14(1):20–32. Nature Publishing Group. https://doi.org/10.1038/nrmicro3552

Dorier M, Béal D, Marie-Desvergne C, Dubosson M, Barreau F, Houdeau E, Herlin-Boime N, Carriere M (2017) Continuous in vitro exposure of intestinal epithelial cells to E171 food additive causes oxidative stress, inducing oxidation of DNA bases but no endoplasmic reticulum stress. Nanotoxicology 11(6):751–761. https://doi.org/10.1080/17435390.2017.1349203

Dorier M, Tisseyre C, Dussert F, Béal D, Arnal ME, Douki T, Valdiglesias V, Laffon B, Fraga S, Brandão F, Herlin-Boime N, Barreau F, Rabilloud T, Carriere M (2019) Toxicological impact of acute exposure to E171 food additive and TiO2 nanoparticles on a co-culture of Caco-2 and HT29-MTX intestinal cells. Mutat Res Genet Toxicol Environ Mutagen 845. https://doi.org/10.1016/j.mrgentox.2018.11.004

Dupont D, Alric M, Blanquet-Diot S, Bornhorst G, Cueva C, Deglaire A, Denis S, Ferrua M, Havenaar R, Lelieveld J, Mackie AR, Marzorati M, Menard O, Minekus M, Miralles B, Recio I, Van den Abbeele P (2019) Can dynamic in vitro digestion systems mimic the physiological reality? Crit Rev Food Sci Nutr 59(10):1546–1562. Taylor and Francis Inc. https://doi.org/10.1080/10408398.2017.1421900

Esch MB, Mahler GJ, Stokol T, Shuler ML (2014) Body-on-a-chip simulation with gastrointestinal tract and liver tissues suggests that ingested nanoparticles have the potential to cause liver injury. Lab Chip 14(16):3081–3092. https://doi.org/10.1039/c4lc00371c

Gerloff K, Pereira DIA, Faria N, Boots AW, Kolling J, Förster I, Albrecht C, Powell JJ, Schins RPF (2013) Influence of simulated gastrointestinal conditions on particle-induced cytotoxicity and interleukin-8 regulation in differentiated and undifferentiated Caco-2 cells. Nanotoxicology 7(4):353–366. https://doi.org/10.3109/17435390.2012.662249

Gullberg E, Leonard M, Karlsson J, Hopkins AM, Brayden D, Baird AW, Artursson P (2000) Expression of specific markers and particle transport in a new human intestinal M-cell model. Biochem Biophys Res Commun 279(3):808–813. https://doi.org/10.1006/bbrc.2000.4038

Gunasekara DB, Speer J, Wang Y, Nguyen DL, Reed MI, Smiddy NM, Parker JS, Fallon JK, Smith PC, Sims CE, Magness ST, Allbritton NL (2018) A monolayer of primary colonic epithelium generated on a scaffold with a gradient of stiffness for drug transport studies. Anal Chem 90(22):13331–13340. https://doi.org/10.1021/acs.analchem.8b02845

Gustafsson JK, Johansson MEV (2022) The role of goblet cells and mucus in intestinal homeostasis. Nat Rev Gastroenterol Hepatol 19(12):785–803. Nature Research. https://doi.org/10.1038/s41575-022-00675-x

He Y, Cheng M, Yang R, Li H, Lu Z, Jin Y, Feng J, Tu L (2023) Research progress on the mechanism of nanoparticles crossing the intestinal epithelial cell membrane. Pharmaceutics 15(7) Multidisciplinary Digital Publishing Institute (MDPI). https://doi.org/10.3390/pharmaceutics15071816

Hou Z, Meng R, Chen G, Lai T, Qing R, Hao S, Deng J, Wang B (2022) Distinct accumulation of nanoplastics in human intestinal organoids. Sci Total Environ 838. https://doi.org/10.1016/j.scitotenv.2022.155811

Johansson MEV, Sjövall H, Hansson GC (2013) The gastrointestinal mucus system in health and disease. Nat Rev Gastroenterol Hepatol 10(6):352–361. https://doi.org/10.1038/nrgastro.2013.35

Kämpfer AAM, Busch M, Schins RPF (2020) Advanced in vitro testing strategies and models of the intestine for nanosafety research. Chem Res Toxicol 33(5):1163–1178. American Chemical Society. https://doi.org/10.1021/acs.chemrestox.0c00079

Kämpfer AAM, Busch M, Büttner V, Bredeck G, Stahlmecke B, Hellack B, Masson I, Sofranko A, Albrecht C, Schins RPF (2021) Model complexity as determining factor for in vitro nanosafety studies: effects of silver and titanium dioxide nanomaterials in intestinal models. Small 17(15). https://doi.org/10.1002/smll.202004223

Kanaya T, Hase K, Takahashi D, Fukuda S, Hoshino K, Sasaki I, Hemmi H, Knoop KA, Kumar N, Sato M, Katsuno T, Yokosuka O, Toyooka K, Nakai K, Sakamoto A, Kitahara Y, Jinnohara T, Mcsorley SJ, Kaisho T et al (2012) The Ets transcription factor Spi-B is essential for the differentiation of intestinal microfold cells. Nat Immunol 13(8):729–736. https://doi.org/10.1038/ni.2352

Kang S, Park SE, Huh DD (2021) Organ-on-a-chip technology for nanoparticle research. Nano Converg 8(1) Korea Nano Technology Research Society. https://doi.org/10.1186/s40580-021-00270-x

Kasendra M, Tovaglieri A, Sontheimer-Phelps A, Jalili-Firoozinezhad S, Bein A, Chalkiadaki A, Scholl W, Zhang C, Rickner H, Richmond CA, Li H, Breault DT, Ingber DE (2018) Development of a primary human small intestine-on-a-chip using biopsy-derived organoids. Sci Rep 8(1). https://doi.org/10.1038/s41598-018-21201-7

Kernéis S, Caliot E, Stubbe H, Bogdanova A, Kraehenbuhl J-P, Pringault E (2000) Molecular studies of the intestinal mucosal barrier physiopathology using cocultures of epithelial and immune cells: a technical update. Microbes Infect 2(9):1119–1124

Kim HJ, Huh D, Hamiltona G, Ingber DE (2012) Human gut-on-achip inhabited by microbial flora that experiences intestinal peristalsis-like motions and flow. Lab Chip, 12, 2165–2174. https://doi.org/10.1039/c2lc40074j

Kim HJ, Li H, Collins JJ, Ingber DE (2016) Contributions of microbiome and mechanical deformation to intestinal bacterial overgrowth and inflammation in a human gut-on-a-chip. Proc Natl Acad Sci USA 113(1):E7–E15. https://doi.org/10.1073/pnas.1522193112

Kondrashina A, Arranz E, Cilla A, Faria MA, Santos-Hernández M, Miralles B, Hashemi N, Rasmussen MK, Young JF, Barberá R, Mamone G, Tomás-Cobos L, Bastiaan-Net S, Corredig M, Giblin L (2023) Coupling in vitro food digestion with in vitro epithelial absorption; recommendations for biocompatibility. Crit Rev Food Sci Nutr. Taylor and Francis Ltd. https://doi.org/10.1080/10408398.2023.2214628

Kourula S, Derksen M, Jardi F, Jonkers S, van Heerden M, Verboven P, Theuns V, Van Asten S, Huybrechts T, Kunze A, Frazer-Mendelewska E, Lai KW, Overmeer R, Roos JL, Vries RGJ, Boj SF, Monshouwer M, Pourfarzad F, Snoeys J (2023) Intestinal organoids as an in vitro platform to characterize disposition, metabolism, and safety profile of small molecules. Eur J Pharm Sci 188. https://doi.org/10.1016/j.ejps.2023.106481

Kulkarni DH, Gustafsson JK, Knoop KA, McDonald KG, Bidani SS, Davis JE, Floyd AN, Hogan SP, Hsieh CS, Newberry RD (2020) Goblet cell associated antigen passages support the induction and maintenance of oral tolerance. Mucosal Immunol 13(2):271–282. https://doi.org/10.1038/s41385-019-0240-7

Kulthong K, Hooiveld GJEJ, Duivenvoorde LPM, Miro Estruch I, Bouwmeester H, van der Zande M (2021) Comparative study of the transcriptomes of Caco-2 cells cultured under dynamic vs. static conditions following exposure to titanium dioxide and zinc oxide nanomaterials. Nanotoxicology 15(9):1233–1252. https://doi.org/10.1080/17435390.2021.2012609

Kusi-Appiah AE, Mastronardi ML, Qian C, Chen KK, Ghazanfari L, Prommapan P, Kübel C, Ozin GA, Lenhert S (2017) Enhanced cellular uptake of size-separated lipophilic silicon nanoparticles. Sci Rep 7. https://doi.org/10.1038/srep43731

Lamson NG, Berger A, Fein KC, Whitehead KA (2020) Anionic nanoparticles enable the oral delivery of proteins by enhancing intestinal permeability. Nat Biomed Eng 4(1):84–96. https://doi.org/10.1038/s41551-019-0465-5

Larregieu CA, Benet LZ (2013) Drug discovery and regulatory considerations for improving in silico and in vitro predictions that use caco-2 as a surrogate for human intestinal permeability measurements. AAPS J 15(2):483–497. https://doi.org/10.1208/s12248-013-9456-8

Lin S, Wu F, Cao Z, Liu J (2022) Advances in nanomedicines for interaction with the intestinal barrier. Adv NanoBiomed Res 2(6) John Wiley and Sons Inc. https://doi.org/10.1002/anbr.202100147

Liu Y, Shen J, Shi J, Gu X, Chen H, Wang X, Wang L, Wang P, Hou X, He Y, Zhu C, Wang Z, Guo T, Guo S, Feng N (2022) Functional polymeric core–shell hybrid nanoparticles overcome intestinal barriers and inhibit breast cancer metastasis. Chem Eng J 427. https://doi.org/10.1016/j.cej.2021.131742

Macedo MH, Dias Neto M, Pastrana L, Gonçalves C, Xavier M (2023) Recent advances in cell-based in vitro models to recreate human intestinal inflammation. Adv Sci 10(31) John Wiley and Sons Inc. https://doi.org/10.1002/advs.202301391

Mahler GJ, Shuler ML, Glahn RP (2009) Characterization of Caco-2 and HT29-MTX cocultures in an in vitro digestion/cell culture model used to predict iron bioavailability. J Nutr Biochem 20(7):494–502. https://doi.org/10.1016/j.jnutbio.2008.05.006

Marr EE, Mulhern TJ, Welch M, Keegan P, Caballero-Franco C, Johnson BG, Kasaian M, Azizgolshani H, Petrie T, Charest J, Wiellette E (2023) A platform to reproducibly evaluate human colon permeability and damage. Sci Rep 13(1). https://doi.org/10.1038/s41598-023-36020-8

Marrero D, Guimera A, Maes L, Villa R, Alvarez M, Illa X (2023) Organ-on-a-chip with integrated semitransparent organic electrodes for barrier function monitoring. Lab Chip 23(7):1825–1834. https://doi.org/10.1039/d2lc01097f

Marzorati M, Vanhoecke B, De Ryck T, Sadaghian Sadabad M, Pinheiro I, Possemiers S, Van Den Abbeele P, Derycke L, Bracke M, Pieters J, Hennebel T, Harmsen HJ, Verstraete W, Van De Wiele T (2014) The HMI™ module: a new tool to study the host-microbiota interaction in the human gastrointestinal tract in vitro. BMC Microbiol 14(1). https://doi.org/10.1186/1471-2180-14-133

McCarthy H, Keely S (2023) Re-evaluating the role of deep crypt secretory cells in intestinal homeostasis. CMGH 15(4):1020–1021. Elsevier Inc. https://doi.org/10.1016/j.jcmgh.2023.01.005

McCright JC, Maisel K (2020) Engineering drug delivery systems to overcome mucosal barriers for immunotherapy and vaccination. Tissue Barriers 8(1) Taylor and Francis Inc. https://doi.org/10.1080/21688370.2019.1695476

Milligan JJ, Saha S (2022) A nanoparticle's journey to the tumor: strategies to overcome first-pass metabolism and their limitations. Cancer 14(7) MDPI. https://doi.org/10.3390/cancers14071741

Minekus M, Alminger M, Alvito P, Ballance S, Bohn T, Bourlieu C, Carrière F, Boutrou R, Corredig M, Dupont D, Dufour C, Egger L, Golding M, Karakaya S, Kirkhus B, Le Feunteun S, Lesmes U, MacIerzanka A, MacKie A et al (2014) A standardised static in vitro digestion method suitable for food-an international consensus. Food Funct 5(6):1113–1124. https://doi.org/10.1039/c3fo60702j

More S, Bampidis V, Benford D, Bragard C, Halldorsson T, Hernández-Jerez A, Hougaard Bennekou S, Koutsoumanis K, Lambré C, Machera K, Naegeli H, Nielsen S, Schlatter J, Schrenk D, Silano V, Turck D, Younes M, Castenmiller J, Chaudhry Q et al (2021) Guidance on risk assessment of nanomaterials to be applied in the food and feed chain: human and animal health. EFSA J 19(8). https://doi.org/10.2903/j.efsa.2021.6768

Moutinho S (2023) Researchers and regulators plan for a future without lab animals. Nat Med. https://doi.org/10.1038/s41591-023-02362-z

Mowat AM, Agace WW (2014) Regional specialization within the intestinal immune system. Nat Rev Immunol 14(10):667–685. Nature Publishing Group. https://doi.org/10.1038/nri3738

Mulet-Cabero AI, Egger L, Portmann R, Ménard O, Marze S, Minekus M, Le Feunteun S, Sarkar A, Grundy MML, Carrière F, Golding M, Dupont D, Recio I, Brodkorb A, Mackie A (2020) A standardised semi-dynamic in vitro digestion method suitable for food-an international consensus. Food Funct 11(2):1702–1720. https://doi.org/10.1039/c9fo01293a

Nikolaev M, Mitrofanova O, Broguiere N, Geraldo S, Dutta D, Tabata Y, Elci B, Brandenberg N, Kolotuev I, Gjorevski N, Clevers H, Lutolf MP (2020) Homeostatic mini-intestines through scaffold-guided organoid morphogenesis. Nature 585(7826). https://doi.org/10.1038/s41586-020-2724-8

Pangua C, Espuelas S, Martínez-Ohárriz MC, Vizmanos JL, Irache JM (2024) Mucus-penetrating and permeation enhancer albumin-based nanoparticles for oral delivery of macromolecules: application to bevacizumab. Drug Deliv Transl Res 14(5):1189–1205. https://doi.org/10.1007/s13346-023-01454-0

Paone P, Cani PD (2020) Mucus barrier, mucins and gut microbiota: the expected slimy partners? Gut 69(12):2232–2243. BMJ Publishing Group. https://doi.org/10.1136/gutjnl-2020-322260

Pelaseyed T, Bergström JH, Gustafsson JK, Ermund A, Birchenough GMH, Schütte A, van der Post S, Svensson F, Rodríguez-Piñeiro AM, Nyström EEL, Wising C, Johansson MEV, Hansson GC (2014) The mucus and mucins of the goblet cells and enterocytes provide the first defense line of the gastrointestinal tract and interact with the immune system. Immunol Rev 260(1):8–20. Blackwell Publishing Ltd. https://doi.org/10.1111/imr.12182

Pinto S, Hosseini M, Buckley ST, Yin W, Garousi J, Gräslund T, van Ijzendoorn S, Santos HA, Sarmento B (2024) Nanoparticles targeting the intestinal Fc receptor enhance intestinal cellular trafficking of semaglutide. J Control Release 366:621–636. https://doi.org/10.1016/j.jconrel.2024.01.015

Puschhof J, Pleguezuelos-Manzano C, Clevers H (2021) Organoids and organs-on-chips: insights into human gut-microbe interactions. Cell Host Microbe 29(6):867–878. Cell Press. https://doi.org/10.1016/j.chom.2021.04.002

Rodrigues PM, Xavier M, Calero V, Pastrana L, Gonçalves C (2022) Partitioning of small hydrophobic molecules into polydimethylsiloxane in microfluidic analytical devices. Micromachines 13(5). https://doi.org/10.3390/mi13050713

Sato T, Clevers H (n.d.) Growing self-organizing mini-guts from a single intestinal stem cell: mechanism and applications. https://www.science.org

Sato T, Stange DE, Ferrante M, Vries RGJ, Van Es JH, Van Den Brink S, Van Houdt WJ, Pronk A, Van Gorp J, Siersema PD, Clevers H (2011) Long-term expansion of epithelial organoids from human colon, adenoma, adenocarcinoma, and Barrett's epithelium. Gastroenterology 141(5):1762–1772. https://doi.org/10.1053/j.gastro.2011.07.050

Schimpel C, Teubl B, Absenger M, Meindl C, Fröhlich E, Leitinger G, Zimmer A, Roblegg E (2014) Development of an advanced intestinal in vitro triple culture permeability model to study transport of nanoparticles. Mol Pharm 11(3):808–818. https://doi.org/10.1021/mp400507g

Shugarts S, Benet LZ (2009) The role of transporters in the pharmacokinetics of orally administered drugs. Pharm Res 26(9):2039–2054. https://doi.org/10.1007/s11095-009-9924-0

Silva YP, Bernardi A, Frozza RL (2020) The role of short-chain fatty acids from gut microbiota in gut-brain communication. Front Endocrinol 11. Frontiers Media S.A. https://doi.org/10.3389/fendo.2020.00025

Silva PM, Neto MD, Cerqueira MA, Rodriguez I, Bourbon AI, Azevedo AG, Pastrana LM, Coimbra MA, Vicente AA, Gonçalves C (2024) Resveratrol-loaded octenyl succinic anhydride modified starch emulsions and hydroxypropyl methylcellulose (HPMC) microparticles: cytotoxicity and antioxidant bioactivity assessment after in vitro digestion. Int J Biol Macromol 259. https://doi.org/10.1016/j.ijbiomac.2024.129288

Song ZM, Chen N, Liu JH, Tang H, Deng X, Xi WS, Han K, Cao A, Liu Y, Wang H (2015) Biological effect of food additive titanium dioxide nanoparticles on intestine: an in vitro study. J Appl Toxicol 35(10):1169–1178. https://doi.org/10.1002/jat.3171

Sontheimer-Phelps A, Chou DB, Tovaglieri A, Ferrante TC, Duckworth T, Fadel C, Frismantas V, Sutherland AD, Jalili-Firoozinezhad S, Kasendra M, Stas E, Weaver JC, Richmond CA, Levy O, Prantil-Baun R, Breault DT, Ingber DE (2020) Human colon-on-a-chip enables continuous in vitro analysis of colon mucus layer accumulation and physiology. CMGH 9(3):507–526. https://doi.org/10.1016/j.jcmgh.2019.11.008

Sousa Ribeiro IR, da Silva RF, Rabelo RS, Marin TM, Bettini J, Cardoso MB (2023) Flowing through gastrointestinal barriers with model nanoparticles: from complex fluids to model human intestinal epithelium permeation. ACS Appl Mater Interfaces 15(30):36025–36035. https://doi.org/10.1021/acsami.3c07048

Speer JE, Gunasekara DB, Wang Y, Fallon JK, Attayek PJ, Smith PC, Sims CE, Allbritton NL (2019) Molecular transport through primary human small intestinal monolayers by culture on a collagen scaffold with a gradient of chemical cross-linking. J Biol Eng 13(1). https://doi.org/10.1186/s13036-019-0165-4

Spence JR, Mayhew CN, Rankin SA, Kuhar MF, Vallance JE, Tolle K, Hoskins EE, Kalinichenko VV, Wells SI, Zorn AM, Shroyer NF, Wells JM (2011) Directed differentiation of human pluripotent stem cells into intestinal tissue in vitro. Nature 470(7332):105–110. https://doi.org/10.1038/nature09691

Stalder T, Zaiter T, El-Basset W, Cornu R, Martin H, Diab-Assaf M, Béduneau A (2022) Interaction and toxicity of ingested nanoparticles on the intestinal barrier. Toxicology 481. Elsevier Ireland Ltd. https://doi.org/10.1016/j.tox.2022.153353

Strugari AFG, Stan MS, Gharbia S, Hermenean A, Dinischiotu A (2019) Characterization of nanoparticle intestinal transport using an in vitro co-culture model. Nanomaterials 9(1). https://doi.org/10.3390/nano9010005

Suzuki T (2020) Regulation of the intestinal barrier by nutrients: the role of tight junctions. Anim Sci J 91(1) Blackwell Publishing. https://doi.org/10.1111/asj.13357

Taebnia N, Römling U, Lauschke VM (2023) In vitro and ex vivo modeling of enteric bacterial infections. Gut Microbes 15(1) Taylor and Francis Ltd. https://doi.org/10.1080/19490976.2022.2158034

Talapko J, Včev A, Meštrović T, Pustijanac E, Jukić M, Škrlec I (2022) Homeostasis and dysbiosis of the intestinal microbiota: comparing hallmarks of a healthy state with changes in inflammatory bowel disease. Microorganisms 10(12) MDPI. https://doi.org/10.3390/microorganisms10122405

Thursby E, Juge N (2017) Introduction to the human gut microbiota. Biochem J 474(11):1823–1836. Portland Press Ltd. https://doi.org/10.1042/BCJ20160510

Toor D, Wasson MK, Kumar P, Karthikeyan G, Kaushik NK, Goel C, Singh S, Kumar A, Prakash H (2019) Dysbiosis disrupts gut immune homeostasis and promotes gastric diseases. Int J Mol Sci 20(10) MDPI AG. https://doi.org/10.3390/ijms20102432

Trapecar M, Communal C, Velazquez J, Maass CA, Huang YJ, Schneider K, Wright CW, Butty V, Eng G, Yilmaz O, Trumper D, Griffith LG (2020) Gut-liver physiomimetics reveal paradoxical modulation of IBD-related inflammation by short-chain fatty acids. Cell Syst 10(3):223–239. e9. https://doi.org/10.1016/j.cels.2020.02.008

Valiei A, Aminian-Dehkordi J, Mofrad MRK (2023) Gut-on-a-chip models for dissecting the gut microbiology and physiology. APL Bioeng 7(1) American Institute of Physics Inc. https://doi.org/10.1063/5.0126541

van der Hee B, Wells JM (2021) Microbial regulation of host physiology by short-chain fatty acids. Trends Microbiol 29(8):700–712. Elsevier Ltd. https://doi.org/10.1016/j.tim.2021.02.001

van der Helm MW, Henry OYF, Bein A, Hamkins-Indik T, Cronce MJ, Leineweber WD, Odijk M, van der Meer AD, Segerink LI, Ingber DE (2019) Non-invasive sensing of transepithelial barrier function and tissue differentiation in organs-on-chips using impedance spectroscopy. Lab Chip 19(3):452–463. https://doi.org/10.1039/c8lc00129d

Venema K, Van Den Abbeele P (2013) Experimental models of the gut microbiome. Best Pract Res Clin Gastroenterol 27(1):115–126. Bailliere Tindall Ltd. https://doi.org/10.1016/j.bpg.2013.03.002

Vera D, García-Díaz M, Torras N, Castillo Ó, Illa X, Villa R, Alvarez M, Martinez E (2024) A 3D bioprinted hydrogel gut-on-chip with integrated electrodes for transepithelial electrical resistance (TEER) measurements. Biofabrication 16(3). https://doi.org/10.1088/1758-5090/ad3aa4

Vital N, Gramacho AC, Silva M, Cardoso M, Alvito P, Kranendonk M, Silva MJ, Louro H (2024) Challenges of the application of in vitro digestion for nanomaterials safety assessment. Food Secur 13(11). https://doi.org/10.3390/foods13111690

Whitehead K, Mitragotri S (2008) Mechanistic analysis of chemical permeation enhancers for oral drug delivery. Pharm Res 25(6):1412–1419. https://doi.org/10.1007/s11095-008-9542-2

Williamson IA, Arnold JW, Samsa LA, Gaynor L, DiSalvo M, Cocchiaro JL, Carroll I, Azcarate-Peril MA, Rawls JF, Allbritton NL, Magness ST (2018) A high-throughput organoid microinjection platform to study gastrointestinal microbiota and luminal physiology. CMGH 6(3):301–319. https://doi.org/10.1016/j.jcmgh.2018.05.004

Workman MJ, Gleeson JP, Troisi EJ, Estrada HQ, Kerns SJ, Hinojosa CD, Hamilton GA, Targan SR, Svendsen CN, Barrett RJ (2018) Enhanced utilization of induced pluripotent stem cell–derived human intestinal organoids using microengineered chips. CMGH 5(4):669–677.e2. https://doi.org/10.1016/j.jcmgh.2017.12.008

Xavier AA, Mariutti LR (2021) Static and semi-dynamic in vitro digestion methods: state of the art and recent achievements towards standardization. Curr Opin Food Sci 41:260–273. Elsevier Ltd. https://doi.org/10.1016/j.cofs.2021.08.002

Xavier M, Parente IA, Rodrigues PM, Cerqueira MA, Pastrana L, Gonçalves C (2021) Safety and fate of nanomaterials in food: the role of in vitro tests. Trends Food Sci Technol 109:593–607. Elsevier Ltd. https://doi.org/10.1016/j.tifs.2021.01.050

Xavier M, Rodrigues PM, Neto MD, Guedes MI, Calero V, Pastrana L, Gonçalves C (2023) From mouth to gut: microfluidic in vitro simulation of human gastro-intestinal digestion and intestinal permeability. Analyst 148(14):3193–3203. https://doi.org/10.1039/d2an02088b

Yang D, Liu D, Qin M, Chen B, Song S, Dai W, Zhang H, Wang X, Wang Y, He B, Tang X, Zhang Q (2018) Intestinal mucin induces more endocytosis but less transcytosis of nanoparticles across enterocytes by triggering nanoclustering and strengthening the retrograde pathway. ACS Appl Mater Interfaces 10(14):11443–11456. https://doi.org/10.1021/acsami.7b19153

Ye D, Bramini M, Hristov DR, Wan S, Salvati A, Åberg C, Dawson KA (2017) Low uptake of silica nanoparticles in Caco-2 intestinal epithelial barriers. Beilstein J Nanotechnol 8(1):1396–1406. https://doi.org/10.3762/bjnano.8.141

Ye JY, Chen ZY, Huang CL, Huang B, Zheng YR, Zhang YF, Lu BY, He L, Liu CS, Long XY (2020) A non-lipolysis nanoemulsion improved oral bioavailability by reducing the first-pass metabolism of raloxifene, and related absorption mechanisms being studied. Int J Nanomedicine 15:6503–6518. https://doi.org/10.2147/IJN.S259993

Zhang Z, Lu Y, Qi J, Wu W (2021) An update on oral drug delivery via intestinal lymphatic transport. Acta Pharm Sin B 11(8):2449–2468. Chinese Academy of Medical Sciences. https://doi.org/10.1016/j.apsb.2020.12.022

Zhao K, Xie Y, Lin X, Xu W (2022) The mucoadhesive nanoparticle-based delivery system in the development of mucosal vaccines. Int J Nanomedicine 17:4579–4598. Dove Medical Press Ltd. https://doi.org/10.2147/IJN.S359118

Zheng Y, Luo S, Xu M, He Q, Xie J, Wu J, Huang Y (2024) Transepithelial transport of nanoparticles in oral drug delivery: from the perspective of surface and holistic property modulation. Acta Pharm Sin B. https://doi.org/10.1016/j.apsb.2024.06.015

Zhou H, McClements DJ (2022) Recent advances in the gastrointestinal fate of organic and inorganic nanoparticles in foods. Nano 12(7) MDPI. https://doi.org/10.3390/nano12071099

Zietek T, Rath E, Haller D, Daniel H (2015) Intestinal organoids for assessing nutrient transport, sensing and incretin secretion. Sci Rep 5. https://doi.org/10.1038/srep16831

Open Access   This chapter is licensed under the terms of the Creative Commons Attribution-NonCommercial-NoDerivatives 4.0 International License (http://creativecommons.org/licenses/by-nc-nd/4.0/), which permits any noncommercial use, sharing, distribution and reproduction in any medium or format, as long as you give appropriate credit to the original author(s) and the source, provide a link to the Creative Commons license and indicate if you modified the licensed material. You do not have permission under this license to share adapted material derived from this chapter or parts of it.

The images or other third party material in this chapter are included in the chapter's Creative Commons license, unless indicated otherwise in a credit line to the material. If material is not included in the chapter's Creative Commons license and your intended use is not permitted by statutory regulation or exceeds the permitted use, you will need to obtain permission directly from the copyright holder.

BBB-on-a-Chip: Microfluidic Tools as an Alternative to *In Vivo* Experiments for Nanosafety Studies

M. C. Lefevre, M. C. Ceccarelli, M. Bernardeschi, and G. Ciofani

Abstract

Nanoparticles (NPs) have attracted increasing interest due to their unique properties. However, their production and use have raised concerns about their unwanted effects on human health, particularly their potential to penetrate the blood-brain barrier (BBB) and their subsequent consequences on the central nervous system. Traditional *in vivo* models have been widely used to study NP interactions with the BBB, but these models often suffer from poor translatability to human physiology. This has driven the development of advanced *in vitro* models, such as BBB-on-chip systems, which offer a promising alternative for studying NP neurotoxicity and evaluating nanosafety. This chapter provides a comprehensive review of BBB-on-chip technology, including its design, biological integration, and application in nanosafety studies. The discussion highlights both the advantages and limitations of these models, such as the challenges in balancing tissue complexity with engineering constraints, the importance of dynamic conditions, and the need for improved sensorization. Despite these challenges, recent advances in the field are promising and BBB-on-chip models offer more and more reliable alternatives to animal testing.

Keywords

Organs-on-chip · Blood-brain barrier · Nanoparticles · Nanosafety

---

M. C. Lefevre (✉) · M. Bernardeschi · G. Ciofani (✉)
Smart Bio-Interfaces, Istituto Italiano di Tecnologia, Pontedera, Italy
e-mail: marie.lefevre@iit.it; gianni.ciofani@iit.it

M. C. Ceccarelli
Smart Bio-Interfaces, Istituto Italiano di Tecnologia, Pontedera, Italy

The Biorobotics Institute, Scuola Superiore Sant'Anna, Pontedera, Italy

Abbreviations

AC	alternative current
BBB	blood-brain barrier
BMEC	brain microvascular endothelial cell
CNS	central nervous system
COC	cyclic olefin copolymer
DC	direct current
EC	endothelial cells
EIS	eletrochemical impedance spectroscopy
FDA	U.S. Food and drug administration
GDNF	glial-derived neurotrophic factor
GelMA	gelatin methacryloyl
GLUT1	glucose transporter 1
HPLC	high-performance liquid chromatography
iPSC	induced pluripotent stem cell
JAM	junctional adhesion molecules
LRP	lipoprotein receptor-related protein
MS	mass spectroscopy
MTT	3-(4,5-dimethylthiazol-2-yl)-2,5-diphenyl-2H-tetrazolium bromide
NP	Nanoparticle
OoC	Oorgan-on-chip
PC	polycarbonate
PCL	polycaprolactone
PDGF	platelet-derived growth factor
PDMS	polydimethylsiloxane
PE	polyester
PMMA	poly(methyl methacrylate)
PS	polystyrene
RNS	reactive nitrogen species
ROS	reactive oxygen specie
SERS	amplified surface Raman scattering
SiN	silicon nitride
SiO_2	silicon dioxide
SLC	solute carrier
SPR	surface plasmon reference
TEER	transendothelial electrical resistance
TGF-β	transforming growth factor- β
TJ	tight junction
VEGF	vascular endothelial growth factor
ZO-1	zonula occludens-1

1 Introduction

In the last decades, nanoparticles (NP) have gained growing interest due to their unique properties, making them promising in many fields. Industry and medicine are taking advantage of their small size and large surface area for various applications, from nanotextiles to facade coatings, water treatment, cosmetics, therapies, and medical imaging (Kumah et al. 2023). As their production increases, so does the unintended release of nanomaterials, increasing the chance of affecting the environment and human health. Indeed, due to their small size, NPs can enter the body through multiple routes thus, affecting diverse organs such as the lungs, heart, and brain (Martínez et al. 2020). These properties are also being exploited for therapeutic applications, as nanomaterials have a high potential to cross biological barriers, making them suitable for gene and drug delivery to organs such as the brain, where standard chemotherapeutic agents are usually restricted (Ealia and Saravanakumar 2017). Studying the effects of frequent exposure and possible accumulation of such materials has therefore become a public health issue.

As the central organ of the nervous system, the brain is very well protected. The uptake of harmful compounds such as bacteria or viruses is severely limited by the blood-brain barrier (BBB), a structure separating the brain from the circulatory system. However, the unique properties of NPs enable their passage through this highly selective membrane, to reach the brain parenchyma, thus raising questions about their potential effects. To date, most studies have relied on *in vivo* experiments to model the BBB and assess its permeability, whether for toxicological studies or, conversely, for therapeutic purposes. However, numerous studies have highlighted the poor translatability to humans, in addition to the fact that animal models are expensive and time-consuming, and raise several ethical concerns. With the FDA no longer requiring animal testing before approval, the development of more complex *in vitro* models such as organs-on-chips (OoC) has received growing interest (Wadman 2023).

BBB-on-chip are microfluidic models aiming to advance the understanding of NPs effect on the brain and evaluate their potential neurotoxic effects on humans. First, an overview of the BBB is provided, explaining how its specific anatomy allows it to protect the brain from pathogens, followed by the dual-edged implications of nanomaterials able to cross this barrier, with potential benefits and associated risks. The main section then focuses on BBB-on-chip models, from their fabrication to the integration of biological components, the implementation of dynamic conditions, and their sensorization. The last part discusses their application to nanosafety studies, emphasizing their advantages and limitations, and concluding with future challenges to obtain a reliable and predictive model capable of replacing animal testing.

2 The Blood-Brain Barrier: Guardian of the Central Nervous System

2.1 BBB Anatomy

The BBB is a physical barrier lining the blood vessels in the brain. A complex network of cells including primarily brain microvascular endothelial cells (BMECs), astrocytes, and pericytes forms this semipermeable membrane, which limits the passive diffusion of molecules, mediates the transport of nutrients, prevents the entry of pathogens and toxins, and regulates the migration of circulating immune cells (Fig. 9.1). Each cell type has a distinct role in forming and maintaining the barrier, and their interactions are crucial for the functionality of the BBB (Pardridge 2005).

BMECs form the inner lining of the brain's capillaries separating blood from the brain parenchyma. In the BBB, endothelial cells (EC) display a unique phenotype characterized by the extensive presence of tight junctions (TJs). The TJs are proteins overlapping occlusions between BMECs, greatly restricting paracellular diffusion of polar substances and preventing harmful substances but also peripheral immune cells from entering the brain. These proteins, including claudins, occludin, and junctional adhesion molecules (JAMs), create a physical barrier that tightly regulates the movement of molecules, further reinforced by the low rates of transcytosis in BMECs, thus limiting the passage of potentially harmful substances (Sweeney et al. 2019).

Pericytes, located at the abluminal surface of ECs within precapillary arterioles, capillaries, and postcapillary venules, play a critical role in the formation, maintenance, and function of the BBB. Their unique position allows them to partially envelop capillaries and communicate directly with ECs through the gap and adherent junctions, significantly impacting EC function and blood flow modulation (Armulik et al. 2010). Pericytes regulate BBB integrity by influencing the development and maintenance of TJs. They secrete signaling molecules such as platelet-derived growth factor (PDGF), which recruits pericytes to develop blood vessels, ensuring proper BBB formation and function. Furthermore, they regulate the expression and localization of TJs proteins, such as occludin and claudin-5, and

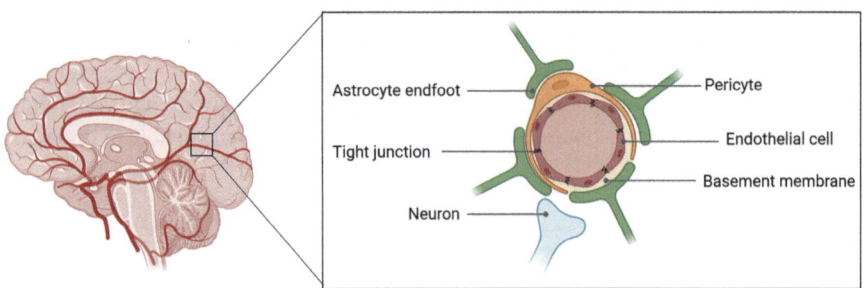

Fig. 9.1 Schematic of the BBB anatomy. (Created with BioRender)

modulate EC transcytosis rates, which are crucial for maintaining BBB permeability (Dohgu et al. 2015; Liu et al. 2012).

Astrocytes, characterized by a star-shaped morphology, surround the ECs with their end-feet processes. They secrete extracellular matrix components such as laminins, which are essential for the structural integrity of the BBB. They secrete signaling molecules, including vascular endothelial growth factor (VEGF), glial-derived neurotrophic factor (GDNF), and transforming growth factor-beta (TGF-β), which promote the establishment of TJs and participate in the maintenance of BBB integrity. Continuous signaling and interaction with other cells in the neurovascular unit, including pericytes and BMECs, ensure the selective permeability of the BBB, which is critical for protecting the brain from neurotoxic substances and maintaining neuronal function. Finally, they regulate blood flow and ion balance, and provide metabolic support to neurons, ensuring the stability of the brain's microenvironment.

The interactions among BMECs, astrocytes, and pericytes are dynamic and reciprocal, ensuring the BBB's adaptability and responsiveness to physiological changes. Glial cells influence the behavior of ECs, enhancing their barrier properties through direct contact and the release of soluble factors. This intricate cellular interplay is essential for the development, function, and maintenance of the BBB. Dysfunction or structural changes of astrocytes and pericytes can contribute to BBB breakdown, which is implicated in various neuropathological conditions. Therefore, they are not only pivotal in the initial formation of the BBB but also in its ongoing function and integrity, underscoring their importance in central nervous system (CNS) health (Abbott et al. 2010; Schiera et al. 2024).

2.2 BBB Function

As explained previously, the BBB has multiple critical functions, primarily protection and regulation, achieved through both physical and biochemical barriers shielding the brain from fluctuations in blood composition that could disrupt neural function (Alahmari 2021). The biochemical aspect of the BBB involves various enzymatic systems and efflux transporters. Enzymes such as monoamine oxidases (Behl et al. 2021) and cytochrome P450 oxidases (Liu et al. 2004) metabolize the potentially harmful substances, while efflux pumps like P-glycoprotein actively transport toxins and xenobiotics back into the bloodstream, thereby reducing their accumulation in the brain (Ahmed Juvale et al. 2022). The BBB controls the selective passage of essential nutrients, ions, and metabolic waste products between the bloodstream and the brain through various transport mechanisms, including carrier-mediated transport, receptor-mediated transcytosis, and efflux pumps (Fig. 9.2) (Daneman and Prat 2015). For instance, glucose is transported by glucose transporter 1 (GLUT1) (Peng and Zeng 2024), and amino acids are transported by various solute carrier (SLC) proteins, ensuring that neurons receive the necessary substrates for energy production and neurotransmitter synthesis (Pizzagalli et al. 2021). Larger molecules (6 kDa) like insulin and transferrin cross the BBB via

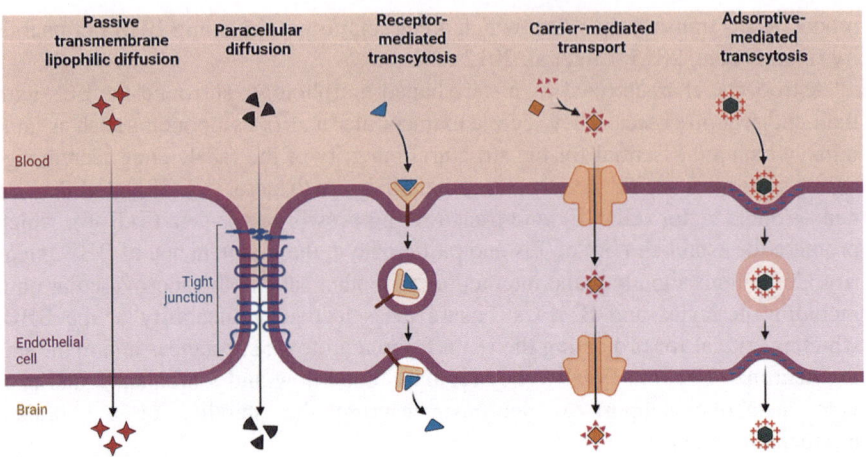

Fig. 9.2 Schematic of different pathways for NPs crossing the BBB. (Created with BioRender)

receptor-mediated transcytosis, involving the binding of these molecules to their specific receptors on the EC surface, followed by endocytosis and transcytosis across the EC to the brain side. Finally, the BBB ensures the control of the ionic composition of the brain's extracellular fluid, through the regulation of potassium and calcium levels, which is vital for proper neuronal function and preventing excitotoxicity.

3 Nanomaterials Crossing the Blood-Brain Barrier: Dual-Edged Implications

Nanomaterials are becoming increasingly important in many fields, particularly in biomedicine. The unique properties of nanomaterials, such as their small size, large surface area, and reactivity, make them highly desirable for numerous applications. However, their increasing use raises significant concerns regarding their potential impact on the environment and human health, especially their interactions with the human brain. Understanding how nanostructures interact with biological systems is crucial to assessing their safety and developing effective risk management strategies (Lebre et al. 2022; Oberdörster et al. 2005).

3.1 Types of Nanostructures

Nanostructures are generally classified into two main groups based on their composition: organic and inorganic. Each category has specific applications in the fields of nanomedicine and the environment, resulting in different exposure scenarios.

Organic NPs include polymeric nanostructures, lipid nanostructures, and protein-based nanostructures. They are typically biodegradable and non-toxic, making them suitable for medical applications, particularly in drug delivery and diagnostics. Environmentally, they are found in biodegradable plastics and agricultural products designed to release pesticides or fertilizers in a controlled manner (Kumar and Lal 2014).

Inorganic NPs include metal- and metal oxide-based nanostructures, as well as NPs made entirely of carbon. Almost all metals can be synthesized into NPs, the most common being aluminum, cadmium, cobalt, copper, gold, iron, lead, silver, and zinc (Ealia and Saravanakumar 2017). Metal oxides offer superior reactivity and efficiency compared to their metallic counterparts, making them ideal candidates for use in nanomedicine applications such as imaging and cancer treatment. In addition, the antimicrobial properties of metals and metal oxides can be exploited in wound dressings and medical devices (Alshammari et al. 2023). Titanium dioxide nanomaterials are widely used in sunscreens and cosmetics for their UV-blocking properties. In the environment, they are prevalent in industrial processes, coatings, paints, and water treatment systems (Rashid et al. 2021). Finally, carbon-based NPs including fullerenes, graphene, and carbon nanotubes, are valued for their unique electrical, mechanical, and thermal properties. For example, carbon nanotubes are promising for drug delivery, imaging, and biosensing due to their ability to penetrate cell membranes (Ealia and Saravanakumar 2017).

Despite their numerous applications in nanomedicine, nanomaterials raise several questions regarding human health due to their ability to cross the BBB, and thus potentially induce neurotoxicity. In addition, the growing interest in nanomaterials is accompanied by an increased risk of unintentional release into the environment, e.g. during the manufacture or disposal of electronic devices, contributing to air and water pollution and thus increasing the risk of unwanted human exposure (Oberdörster et al. 2005).

3.2 Nanostructures Exposure

As nanostructures are dispersed in the environment, they can interfere with the human body in different ways, each carrying unique risks and potential impacts on human health, particularly for the brain. Human exposure to these nanomaterials can happen through a variety of routes, presenting unique challenges and risks (Asmatulu et al. 2022). One common route is ingestion. Nanostructures can be consumed through food, water, and oral medications. Once they enter the digestive system, they encounter several barriers such as the gastric and intestinal mucosa. The ability of these nanostructures to be absorbed through the gut largely depends on their size and surface properties. If they succeed in penetrating these barriers, they can enter systemic circulation, where there is a potential for them to reach the brain (Asmatulu et al. 2022; Kumah et al. 2023). Another significant route of exposure is inhalation, particularly relevant for nanostructures present in industrial and consumer products like sprays, paints, and cosmetics. When inhaled, these particles can lodge in the lungs and may translocate into the bloodstream. From there, they have the potential to reach the brain either through the olfactory bulb or via systemic circulation (Bhattacharjee et al. n.d.; Kumah et al. 2023). Cutaneous exposure represents yet another pathway. Although the penetration of nanostructures through intact skin is generally limited, it can be more pronounced if the skin barrier is compromised, allowing entry into the bloodstream. Finally, injection provides a direct method of introducing nanostructures into the body, a practice commonly utilized in medical applications. Nanostructures used in drug delivery systems or as imaging agents are specifically engineered to navigate the bloodstream and reach target sites, including the brain (Asmatulu et al. 2022).

3.3 Mechanisms of Crossing the Blood-Brain Barrier

Understanding how nanomaterials can cross the BBB is crucial for assessing their impact on brain health, particularly concerning their potential toxicity. Nanostructures can cross the BBB through several mechanisms, depending on their intrinsic properties (Fig. 9.2).

As explained previously, the paracellular pathway can only be exploited by very small nanomaterials, due to the widespread presence of TJs connecting the BMECs (Gomes et al. 2014). Some nanomaterials can bypass this restrictive pathway, by penetrating the brain through the transcellular pathway via transcytosis. This process involves endocytosis on one side of the cell and exocytosis on the other, effectively allowing the NP to pass through the cell (Hersh et al. 2022). Adsorptive-mediated transcytosis can be achieved by positively charged nanostructures that can bind to the negatively charged cell membranes, facilitating their transport across the BBB (Haqqani et al. 2024; Zhang et al. 2022). Finally, nanostructures can be functionalized with ligands that bind to specific receptors on the ECs of the BBB to achieve receptor-mediated transcytosis. This binding triggers endocytosis and subsequent transport into the brain. This method leverages the natural receptor-mediated processes of the BBB, providing a targeted and efficient way to deliver therapeutic agents to the brain (Haqqani et al. 2024).

3.4 Brain Accumulation and Toxicity

The impact of nanomaterial accumulation in the brain is twofold. Therapeutically, nanostructures can improve the delivery of drugs to target sites within the brain, potentially revolutionizing the treatment of neurological disorders. However, there are also significant safety concerns. Once nanostructures cross the BBB, their accumulation in brain tissue can lead to various toxic effects. Many nanostructures can generate reactive oxygen species (ROS), leading to oxidative stress. This stress damages cellular components such as DNA, proteins, and lipids, potentially resulting in significant neuronal damage (Kopatz et al. 2023; Zhang et al. 2022). In addition, nanomaterials can activate microglia, the brain's resident immune cells, triggering inflammatory responses. This activation can exacerbate neuronal damage and contribute to the progression of neuroinflammatory conditions. Prolonged exposure and accumulation of nanostructures in the brain can also lead to genotoxicity causing mutations and chromosomal aberrations and induce neurodegenerative changes (Kopatz et al. 2023; Prüst et al. 2020).

3.5 Challenges for Nanosafety Studies

The increasing use of nanostructures in various industries, coupled with their potential to penetrate biological barriers and reach the brain, underscores the need for comprehensive studies on their safety. While nanostructures offer significant benefits in medicine and other fields, standardized toxicological assessments are necessary to mitigate the risks associated with NP exposure. However, studying the BBB and the effects of NPs in the environment on the CNS presents significant challenges. One of the primary challenges lies in the limitations of existing BBB models. Traditional *in vitro* models such as transwell systems often fail to replicate the complexity and functionality of the *in vivo* barrier, particularly regarding TJs, cell-cell interactions, and the three-dimensional structure of the BBB. Although animal models offer a closer approximation, significant species-specific differences can limit the translatability of findings to human physiology (Pérez-López et al. 2023). The development of realistic BBB models such as microfluidic models is promising to better mimic *in vivo* physiological conditions and obtain more predictive models for nanosafety (Carvalho et al. 2022; Guarino et al. 2023).

4 BBB-on-Chip: A Three-Dimensional Microfluidic Model

4.1 Organs-on-Chip as an Alternative to Traditional Biological Models of BBB

In the last years, OoCs have gained interest in the development of more predictive biological models. These *in vitro* devices are biomimetic systems combining cell biology, engineering, and biomaterial technology to mimic key aspects of human physiology and disease. The OoC are microfluidic devices involving the manipulation of very small volumes in microchannels, which have nonetheless a large surface area and a high mass transfer (Whitesides 2006; Wu et al. 2020). Moreover, working on the microscale allows a better level of control over the microenvironment of biological samples grown inside the chip, as well as direct observation through microscopy techniques.

OoCs join the non-exhaustive list of the array of biological models available for researchers in preclinical studies for the evaluation of human physiology and disease. While *in vivo* experiments on mammal models are the most predictive in terms of biological complexity and three-dimensional structure, they often fail to replicate human physiology due to species differences (Saeidnia et al. 2015). They are also low throughput, expensive, and poorly accessible, despite significant advances in microscopy techniques, in addition to the ethical concerns they are raising. For decades, *in vitro* experiments based on 2D cell culture were routinely used for the study of the human body or the development of new drugs and treatments. However, these models fail to accurately mimic physiological events in the human body due to their lower complexity and multicellularity, as well as the lack of 3D architecture and cell-cell interactions (Urzì et al. 2023). Advances in tissue engineering led to the development of more complex *in vitro* models such as 3D cell cultures or stem cell-derived organoids that are more complex but are still not dynamic systems contrary to physiological conditions. OoCs are bridging models offering the possibility to grow complex cell models in an engineered dynamic microenvironment to ensure more physiological conditions while overcoming the shortcomings of alternative models.

OoCs are highly suitable for mimicking the BBB as they can combine the dynamic behavior of the vasculature on one side and the 3D complex structure of the brain parenchyma on the other side. Indeed, developing *in vitro* BBB models has been a challenge for decades, since 1978, when a study reported that ECs isolated from rat cerebral capillaries can be grown in sterile conditions (Panula et al. 1978). The first major advancement was made in 1983 when Bowman et al. pioneered the use of culture inserts made of a porous membrane, on which primary brain ECs were grown (Bowman et al. 1983). Using this technique, they could obtain for the first time a model containing two fluid compartments, thus allowing permeability assays. In the following years, the model was progressively improved and made more complex by culturing ECs with glial cells such as astrocytes and pericytes, resulting in a strengthened and more predictive barrier (Dehouck et al. 1990; Nakagawa et al. 2007, 2009). Many advances were also made from the biological point of view, starting from isolated rat cells to human models and more recently, with the differentiation of induced pluripotent stem cells (iPSCs) or hematopoietic stem cells into brain-like ECs (Cecchelli et al. 2014; Lippmann et al. 2012). However, it was demonstrated that static models cannot recapitulate BBB physiology, as fluid flow in the blood compartments plays a crucial role in maintaining barrier functions and homeostasis (Li et al. 2005). Moreover, the BBB architecture can be barely reconstituted using cell insert, due to the poor 3D interactions between the cells. BBB-on-chips have several advantages regarding static culture inserts including controlled and limited volumes, three-dimensional cell cultures, the possibility of integrating sensors, and the use of microfluidic pumps for performing experiments in fluidic conditions.

4.2 BBB-on-Chip Devices

The BBB-on-chip devices generally consist of different compartments mimicking cerebral blood vessels and the brain parenchyma. The vascular side, where ECs are cultured, can be operated under dynamic conditions, while the perivascular cells of the brain parenchyma are cultured under static conditions in the second part. These chambers are either physically separated by solid and porous membranes or

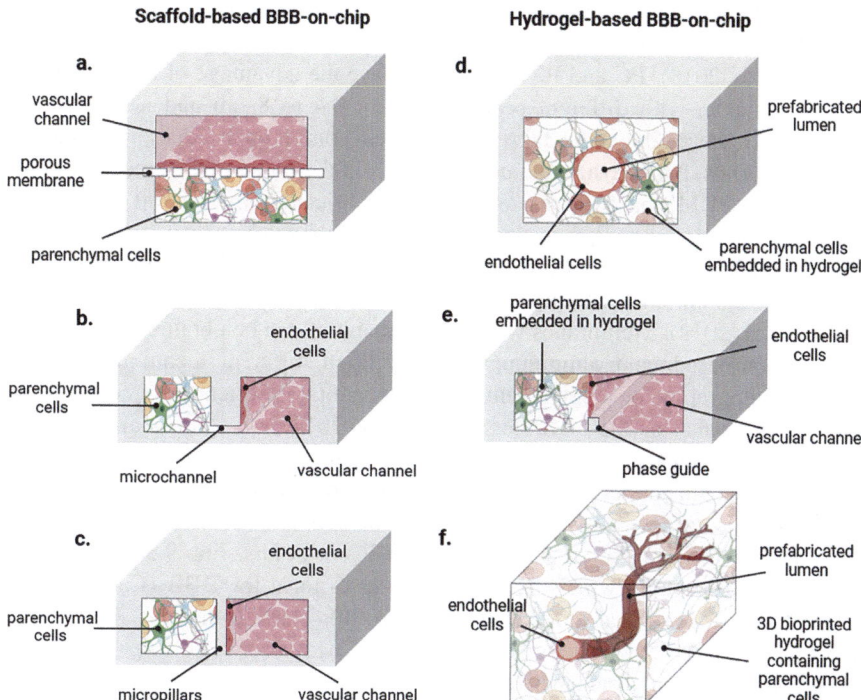

Fig. 9.3 The major types of BBB-on-chip models divided in two categories: scaffold-based devices (**a–c**) and hydrogel-based devices (**d–f**). (**a**) "Sandwich configuration". (**b**) Horizontal configuration with the vascular and the parenchymal compartments connected through microchannels. (**c**) Horizontal configuration with the vascular and the parenchymal compartments separated with micropillars. (**d**) Preformed cylindrical lumen in hydrogel obtained through insertion of acupuncture needles or hydrostatically controlled tunneling of cell culture medium. (**e**) Horizontal configuration obtained with a phase guide restricting the hydrogel in the parenchymal part. (**f**) 3D bioprinted complex BBB model. (Created with BioRender)

scaffolds, or based on hydrogels to allow direct contact between the cells (Fig. 9.3) (Deli et al. 2024; Pérez-López et al. 2023).

4.2.1 Scaffold-Based Models

The physical separation of the different chambers was the first approach developed for BBB-on-chip. The delimitation is placed between the different channels, which are aligned either vertically or horizontally. The "sandwich configuration" consists of an upper and a lower chamber, separated by a porous membrane providing mechanical support for cell growth while allowing cellular communication and molecular diffusion through the pores (Fig. 9.3a). The mechanical and optical properties of the material, the pore size, the porosity, and the membrane thickness are key parameters as they affect transmigration and biochemical communication, but also the strain level that can be applied to the cells or the possibility of microscopy imaging. Most of the studies use polycarbonate (PC), polycaprolactone (PCL), polydimethylsiloxane (PDMS), polyester (PE), silicon nitride (SiN), or silicon dioxide (SiO_2) as a material for porous membranes depending on the application (Chung et al. 2018). PC and PE membranes have the advantage of being commercially available, with different pore sizes from 0.4 to 8 µm and a thickness of 7–10 µm. However, due to the fabrication process combining irradiation and etching of the materials, the porosity remains low to avoid the possibility of merging pores. Conversely, PDMS-based membranes can be fabricated through soft lithography, enabling total control over pore size and spacing in the micrometric scale. However, this material is known to absorb hydrophobic molecules such as drugs but silicon oxide or titanium oxide coating of the material can reduce the absorption of drugs (Wang et al. 2012). Membranes with nanoscale pores can be obtained with silicon- and carbon-based, and aluminum oxide, through different techniques such as e-beam lithography, or nanoimprint for instance, but are often more complex and costly (Agrawal et al. 2010; Mireles and Gaborski 2017; Nasrollahi et al. 2017). The main drawback of the vertical configuration remains the impossibility of imaging simultaneously the upper and the lower chamber, contrary to the horizontal configuration. In this case, two or more different chambers are horizontally aligned and separated by arrays of microchannels (Fig. 9.3b) or pillars (Fig. 9.3c), acting as a physical barrier along which ECs are growing to form the BBB (Campisi et al. 2018; Palma-Florez et al. 2023; Peng et al. 2020). These techniques simplify the fabrication of the device as the barrier is part of the microfluidic chip. There is no need to bond the membrane between the two layers, thus avoiding the risks of leakage or flow disruption. Moreover, the surface area between the different compartments can be finely tuned and adapted to the application and the type of cells used. Finally, these scaffold-based models have the advantage of being robust and versatile in terms of hydrogel and cells, as they are inserted after the fabrication of the chip and thus they are not affected by the fabrication process.

4.2.2 Hydrogel-Based Models

Hydrogel-based BBB-on-chip models allow direct contact and communication between the cells as there is no physical barrier to separate the hydrogel from the vascular microchannel. Several techniques were developed over time, such as the insertion of acupuncture needles, or the hydrostatically controlled tunneling of cell culture medium in the hydrogel before its complete gelation (Fig. 9.3d) (Herland et al. 2016; Partyka et al. 2017). Despite being more physiologically relevant, the cylindrical shape of the vascular channel produced makes it more difficult to sensorize, in addition to the lower reproducibility and throughput compared to the scaffold-based models. To address this problem, two-lanes, and three-lanes horizontal devices were developed using phase guides acting as meniscus pinning barriers to confine the hydrogel and the medium in their respective lanes (Fig. 9.3e) (Wevers et al. 2018). Finally, recent advances in 3D bioprinting have made it possible to develop new techniques for tissue engineering employing hydrogels, especially regarding vascular networks (Fig. 9.3f). For extrusion-based 3D bioprinting, a cell-laden hydrogel precursor or bio-ink is extruded through a nozzle along a printing bed to create stacked layers (Vera et al. 2021). However, the choice of the hydrogel precursor remains critical, as it must possess optimal viscosity and mechanical stability while having a low gelation time. Moreover, the nozzle extrusion can damage embedded cells due to shear stress. Light-based 3D bioprinting is another technique based on photocrosslinkable polymers (Yu et al. 2020b). A light source is used to perform layer-by-layer photopolymerization of the hydrogel. This can be achieved with a focused laser beam inducing spatially localized photocrosslinking (Wang et al. 2018). However, the long printing time can compromise the viability of the embedded cells. Digital light projection stereolithography can speed up the process by allowing full x-y plane photopolymerization per exposure (Xue et al. 2018). In both techniques, photoinitiators and exposure conditions must be carefully optimized to avoid photoablation and cytotoxic effects. Finally, laser-based photopatterning used the reversed phenomenon: hollow microchannels are generated with a focused laser by photodegradation of the hydrogel due to multiphoton absorption (Pradhan et al. 2017). This technique enables high-resolution printing, with diameters as small as 10 µm, but requires a long-term process with expensive and complex equipment, which remains the main limitation for most 3D bioprinting techniques (Arakawa et al. 2017).

4.2.3 BBB-on-Chip Material Selection and Fabrication

The choice of material is crucial for the establishment of relevant BBB-on-chip models and depends on the functionalities and the read-outs required for the application, as well as the microfabrication facilities accessible. For nanosafety studies, critical parameters include biocompatibility and sterilization, gas permeability, optical transparency, elasticity, and molecule adsorption (Leung et al. 2022). The most widely used material is PDMS, a silicone polymer rubber, due to its biocompatibility, gas permeability, and relatively good optical transparency (Berthier et al. 2012). The fabrication of high-resolution complex microstructures and nanostructures can be achieved through soft lithography (replica molding from microfabricated templates) or laser pyrolysis (Leung et al. 2022; Shin et al. 2021). Due to its elasticity, PDMS is also compatible with applications requiring mechanical stimuli (Huh et al. 2010). However, the main drawback of PDMS remains its ability to adsorb and absorb hydrophobic small molecules, limiting the possibilities of drug testing studies (Toepke and Beebe 2006; Wang et al. 2012). Regarding nanosafety applications, the adsorption rate of hydrophobic NPs can be reduced by increasing the hydrophilicity of the surfaces in contact with the biological environment through plasma activation or using a slightly alkaline pH environment, increasing the flow rate, and minimizing surface roughness (Hirama et al. 2021; Simovic and Prestidge 2003). Finally, the partial optical transparency of PDMS regarding conventional inverted microscopes can be overcome by attaching the microfluidic chip to a glass substrate, which provides a fully transparent, inert, and biocompatible bottom support. The chemical bonding of PDMS onto glass can be easily achieved by plasma activation (Borók et al. 2021). Fully glass devices are barely used because of their fragility, the high cost of fabrication, and the complex manufacturing that requires advanced processing facilities. Thermoplastics such as polystyrene (PS), poly(methyl methacrylate) (PMMA), polycarbonate (PC), and cyclic olefin copolymer (COC) are interesting alternatives due to their better optical transparency and lower absorption, but also the possibility of mass production through injection molding (Lee et al. 2018). While this technique enables high-throughput studies, it is not suitable for complex designs, and the materials used are rigid and poorly permeable (Kim et al. 2021). New resins suitable for 3D printing are currently being investigated to generate complex structures with high-throughput complex structures (Homan et al. 2019). However, their use for biological experiments is still limited due to their opacity and/or poor biocompatibility (Leung et al. 2022). Finally, special consideration must be given to sterilization techniques depending on the material chosen. Indeed, PMMA and PC are not suitable for autoclave sterilization due to their lower thermal resistance and opaque materials might limit UV penetration. Ethanol soaking is not compatible with all materials, such as PMMA, which can be dissolved. Gamma irradiation and ethylene oxide treatment are the most compatible techniques for clinical applications (Leung et al. 2022).

3D bioprinting is a relatively recent technique for generating hydrogel microfluidic chips. Regarding BBB-on chip, several hydrogels were investigated to develop vascular networks, in which brain cells can be seeded to generate a BBB model. Extrusion-based bioprinting is mainly based on gelatin, gelatin methacryloyl (GelMA) collagen, methacrylated alginate and methacrylated hyaluronic acid, but they exhibit limitations in terms of structural stability, which can be improved by adding a copolymer such as poly(ethylene glycol) derivatives (Ji et al. 2019; Lee and Cho 2016; Pi et al. 2018; Zhang et al. 2016). Light-based 3D bioprinting requires photocrosslinkable polymers such as poly(ethylene glycol) diacrylate but its use is still debated due to possible cytotoxic effects of the initiator (Grigoryan et al. 2019; Pi et al. 2018). Finally, poly(ethylene glycol), collagen I, and Matrigel are investigated for laser-based photopatterning but the use of this technique remains limited due to its cost and the specific equipment required (Brandenberg and Lutolf 2016). In the case of BBB-on-chip, hydrogels are also used inside the microfluidic chip, in the parenchymal compartment, as a support to brain cells. Cells are mixed with the liquid precursor of the chosen material and loaded in the chip where it forms a gel, providing a scaffold in which cells can grow in 3D. These hydrogels are generally based on laminin, collagen and/or fibrin and can contain extracellular matrix proteins and growth factors to promote cell survival or neuronal differentiation (Uemura et al. 2010).

4.2.4 Biological Elements

Physiological human BBB models are generally composed of ECs lining the vascular compartment and glial cells, mainly astrocytes and pericytes, in the parenchymal part. Indeed, the integration of astrocytes and/or pericytes is essential for promoting BBB properties as they induce the development of a tighter barrier (Dehouck et al. 1990; Nakagawa et al. 2007). Neurons can be included in the case of nanosafety studies focusing on neuronal toxicity for instance. Different approaches are used, the organotypic approach based on the incorporation of primary tissues or engineered tissues, or the bottom-up approach in which cells are seeded in the chip to remodel into a functional neo-tissue, the latter one being the main one for

BBB-on-chip. Primary brain cells are directly isolated from human tissues and immortalized to obtain cell lines without cellular senescence. Immortalized cell lines grow faster and can be used for several passages and thus are easy to use and maintain, contrary to primary cells which are more challenging and can be used for a very limited time as they lose most of their phenotype within three passages when cultured *in vitro* (Verma et al. 2020).

Regarding brain ECs, which have a pivotal role in the development of the BBB-on-chip model, primary cells can better recreate physiological barrier properties than immortalized cell lines, which exhibit weaker paracellular barrier properties due to the immortalization process (Veszelka et al. 2018). However, the availability of primary cells is way more limited, and variations between individual donors must be taken into consideration regarding the reproducibility of the experiments. Brain-like cells differentiated from human stem cells are an interesting alternative because iPSCs are more easily accessible and have better scalability. However, iPSCs require generally 2–3 weeks to be fully differentiated with challenging techniques, and often express limited functionalities compared with primary cells, depending on the cells and techniques used (Diederichs and Tuan 2014; Lippmann et al. 2020). Two main groups obtained through different differentiation protocols are currently defined based on their cellular identity. The first group exhibits a high paracellular barrier tightness but limited vascular character, whereas the second has a definitive vascular character but results in leakier barriers (Nishihara et al. 2020). Nanosafety studies mainly focus on the BBB permeability to NPs and the possible accumulation and toxicity in the brain, therefore the priority should be the formation of a predictive BBB model with strong paracellular barrier properties.

4.2.5 Fluidic Integration

The effective mimicking of functional BBB models is accompanied by the implementation of fluidic conditions similar to the physiological ones. Media perfusion and circulation in the vascular compartment are essential to maintain a concentration gradient for nutrients. Moreover, the shear stress induced by the flow over the ECs plays a non-negligible role in BBB properties, even if the effect on paracellular tightness remains controversial (Li et al. 2005). Increases of TEER are commonly measured but are poorly reproducible between the different studies, and results regarding permeability of the barrier and expression of TJ proteins are contrasted and highly context-dependent (Deli et al. 2024; Griep et al. 2013; Lyu et al. 2021; Santa-Maria et al. 2021; Vatine et al. 2019; Walter et al. 2016).

Different systems can enable media perfusion, including active pumps such as syringe pumps, microvalve-driven actuator pumps, and peristaltic pumps. Other groups also developed pump-free systems driven by gravity, for instance, to minimize system complexity (Ong et al. 2017; Yu et al. 2020a). The choice of the pump depends on several parameters, including the flow type and rate, and the need for recirculation of the fluids for instance. Indeed, one-pass flow allows a constant and stable supply of fresh nutrients, contrary to recirculation which can lead to an accumulation of waste in the circulatory system. On the other hand, in a one-pass configuration, intercellular communication is only possible in the direction of the flow, whereas recirculation allows systemic propagation of chemical signaling. The choice of the perfusion system depends on the need to sample a fraction of the medium from the system for off-line analysis, or simply to refresh the medium. The choice of flow rate must be chosen following the geometry of the microfluidic channels, the type of cells and their sensitivity to shear stress, mass transport regime of soluble factors.

4.3 BBB Sensorization

BBB sensorization is essential, both for ensuring the reliability of the model and for further studies, in this case, nanosafety studies. Different types of analysis can be performed in OoC models, with various space and time resolutions, and read-outs must be carefully chosen depending on the applications. They can be divided into three categories: *in situ*, online and offline analysis (Fig. 9.4). Offline analysis requires sampling of the supernatant, which is further analyzed for analyzing chemicals or biomarkers, with high-performance liquid chromatography (HPLC) coupled with mass spectroscopy (MS) or multiplexed bead-based protein-binding/DNA-binding assays (Fig. 9.4b) (Leung et al. 2022). However, sampling of supernatant requires a specific setup and might disrupt the system, especially in very low-volume systems, and time resolution might be limiting for analyzing short

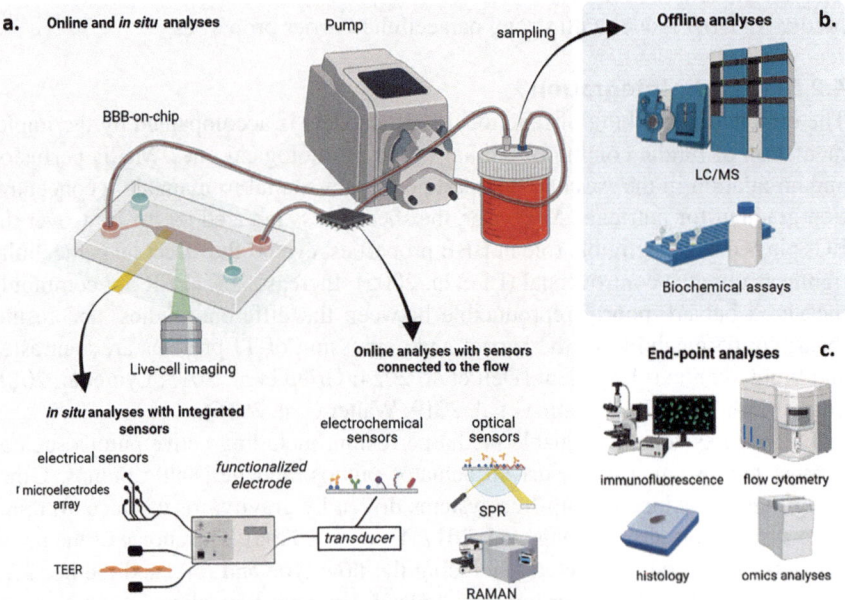

Fig. 9.4 Schematic of the analyses that can be performed in BBB-on-chip models. (**a**) integrated sensors for online and in situ analyses. (**b**) offline analyses. (**c**) end-point analyses. (Created with BioRender)

events or molecules with a short degradation time. The integration of online and *in situ* sensors allows reducing this time delay and providing continuous and real-time measurements, with a better spatial resolution (Fig. 9.4a) (Liang and Yoon 2021). Microscopy is one of the most common techniques as it allows direct visualization inside the chip while being not invasive. Indeed, as explained previously, most OoCs are made of transparent material and can be imaged with epifluorescence and/or confocal microscopes after staining of cells for instance, in order to assess cell viability or expression of specific biomarkers. Progress in microscopy techniques, sample preparation, and image analysis provide high-content imaging of complex 3D structures which is essential for nanosafety studies in such complex microenvironments. The specificities and the cost of equipment remain limiting factors, as well as the need to use fluorescent stains or antibodies, which might affect the biological environment for mid- and long-term studies. The integration of sensors *in situ* such as electrical, electrochemical and optical sensors enables real-time monitoring over time while being minimally invasive (Liang and Yoon 2021). Electrical sensors are extensively used in BBB-on-chip for measuring the TEER, a parameter defining the integrity and the permeability of barrier tissues such as the BBB, whose TEER is comprised between 1500 and 8000 ohm.cm^2 (Wolff et al. 2015). Measurement of TEER in microscale systems can be done with the Ohm's law method or electrochemical impedance spectroscopy (EIS), using either microelectrodes directly integrated into the device during the fabrication, or with external electrodes inserted during the experiment, which are more invasive. The most accurate representation of the TEER can be obtained by combining impedance spectroscopy with fitting algorithms compared to Ohm's law method but is also more complicated to obtain (Bürgel et al. 2016). EIS can also provide information on tissue size and integrity. Additionally, microelectrodes integrated into the parenchymal part of the model can be used to measure the electrical activity of neurons and thus evaluate their potential degeneration due to drugs or NPs (Liu et al. 2022). Electrochemical biosensors are based on the same principle but the working electrode is modified with a biological recognition element, whose binding to a specific target molecule will affect the measured electrical signal (Mir et al. 2022). However, the application of such sensors into BBB-on-chip models remains limited due to space restriction, and up to now, they have been mainly used for oxygen level monitoring (Sticker et al. 2019). Several studies exploited these sensors to measure pH, molecules involved in metabolic activity (glucose, lactate, …) or short-term reactive species such as reactive oxygen species (ROS) which are of interest to monitor neuroinflammation reactions (Kavand et al. 2022; Mir et al. 2022). Optical biosensors are using different features of light (resonance, reflection, polarization, and plasmon effect) making them easily integrable into OoCs, as they do not require direct contact between the transducer and the detector (Cecen et al. 2023). Optical signals can be generated with labeling agents or genetically encoded fluorescent signals and detected with various methods such as colorimetry, fluorescence, luminescence, surface plasmon resonance (SPR), and amplified surface Raman scattering (SERS) (Cecen et al. 2023). Several examples can be found in the literature about BBB-on-chip, for the detection of neurotransmitters and neuromodulators, neural activation, and calcium imaging (Bi et al. 2021; Marino et al. 2023; Shaver et al. 2022; Shemetov et al. 2021). Other types of optical sensors do not require labeling and are based on Interactions between specific analytes and a transducer located *in situ*, modifying light properties. A recent study used for instance a Mach-Zehnder interferometer to investigate the penetration capability of SARS-CoV-2 through the BBB (Petrovszki et al. 2022). Finally, some analyses require end-point assays, such as genomics, proteomics, transcriptomics, or histology (Fig. 9.4c). This might be taken into consideration when designing the BBB-on-chip model, to ensure the possible recovery of cells and tissues for further analyses.

5 Applications to Nanosafety: BBB-on-Chip for Permeability and Toxicity Studies

As explained previously, the increased use of NPs is accompanied by a higher chance of human exposure, which might be detrimental to the CNS. Additionally, innovative nanotherapeutic drugs are developed for several pathologies including brain diseases, therefore there is a crucial need to evaluate whether NPs can cross and disrupt the BBB, and if their potential accumulation in the parenchyma can induce neurotoxic effects. BBB-on-chip models offer a reliable alternative to animal models for human nanosafety studies, especially since a new law, signed by the FDA in 2022, has been enacted to allow animal-free alternatives before performing human trials (Wadman 2023).

5.1 Biochemical Assays

Permeability is one of the most critical parameters when developing BBB-on-chip models for nanosafety studies. Indeed, the effectiveness of the generated barrier must be confirmed to ensure reliable studies and evaluate the potential changes induced by NPs. The widest technique used is based on measuring the barrier's permeability to paracellular tracers of different molecular weights, affected mainly by the presence of TJs between the ECs of the barrier. Radiolabeled tracers enable quantitative studies but cause safety issues for handling and storage, and have a short half-life (Srinivasan et al. 2015). For these reasons, fluorescent-labeled tracers are preferred even if they present lower stability and lower sensitivity to small changes in the barrier (Duffy and Murphy 2001). The tracer is placed in the influent and the concentration of the tracer able to pass the barrier is measured in the effluent to calculate the apparent permeability calculated with the following equation (Bree et al. 1988):

$$P_{app}\left(cm.s^{-1}\right) = \frac{C_t \times V\left(cm^3\right)}{C_0 \times t(s) \times A\left(cm^2\right)}, \qquad (9.1)$$

where P_{app} is the apparent permeability, C_0 is the initial concentration of tracer in the influent, C_t is the concentration of tracer in the effluent at the end-time, V is the volume of effluent recovered, t is the time of the experiment and A is the surface area of the barrier. This technique ensures the generation of an effective barrier, ensuring the reliability of further experiments. Moreover, the same technique can be used for the evaluation of the apparent permeability of the barrier to NPs, performing dosage of NPs in the parenchyma through various analytical techniques.

Several studies evaluated transcytosis of nanocarriers through the BBB by fluorescent labeling of NPs, in the scope of nanotherapies (Fig. 9.5). Fluorescence microscopy or spectroscopy was performed in the compartment mimicking the brain parenchyma to determine the apparent permeability adapting the previously described formula (9.1). As an example, many groups focused on the functionalization of NPs with the angiopep-2 peptide, which targets low-density lipoprotein receptor-related proteins (LRPs) expressed in the BBB and therefore promotes transcytosis of NPs to the brain parenchyma (Vanlandewijck et al. 2018). Penetration of functionalized NPs such as rhodamine-labeled liposomes (Papademetriou et al. 2018), quantum dots (Park et al. 2019), gold nanorods (Fan et al. 2023; Palma-Florez et al. 2023), or cisplatin-loaded PS NPs (Papademetriou et al. 2018; Straehla et al. 2022) were evaluated to emphasize the ability of angiopep-2 functionalization to facilitate the permeability of NPs (Fig. 9.5a). Similar studies performed with apolipoproteins and transferrin were for instance reviewed (Deli et al. 2024). These studies can be performed in any BBB-on-chip configurations, either directly with microscopy for horizontal configurations, or by recovering the content of the parenchymal part for further analyses with fluorescence spectroscopy, flow cytometry, or liquid chromatography. However, the use of tracers can affect the viability of the model and often requires sampling that can be complicated depending on the model. If a hydrogel was used, the matrix must be digested to recover the parenchymal content, therefore ending the experiment.

Regarding potential neurotoxicity, evaluating the effects of NPs on the BBB itself is crucial to avoid further penetration of potential pathogens and toxins that are normally prevented. Imaging of TJs proteins such as Zonula occludens-1 (ZO-1), claudin-5 or occludin is a common method to ensure the integrity of the barrier after NPs exposure but the model needs to be fixated and stained with specific antibodies (Booth and Kim 2012). The cytocompatibility of NPs can also be evaluated on any cellular population, including ECs, by recovering the cells and performing viability assay such as MTT (Lee et al. 2020). In both cases, fixation or recovery of the cells leads to the end of the experiment. In addition, the measurement of specific biomarkers or molecules allows to assess of neurotoxicity directly on cells from the brain parenchyma. As an example, a BBB-on-chip model was developed for ecotoxicology studies regarding the neurotoxic effects of indoor airborne NPs (Li et al. 2020) (Fig. 9.5b). After translocation of NPs through the BBB, pro-inflammatory and oxidative responses were observed, emphasized by a higher level of oxidative stress-relative biomarkers, in addition to abnormal astrocyte proliferation and lower cell viability.

5.2 Electrical Measurements

The integration of electrodes in BBB-on-chip models enables measuring the TEER, which depends mainly on the presence of TJs between the ECs (Srinivasan et al. 2015). The simplest method is based on Ohm's law, in which a direct current (DC) is applied between two electrodes located on each side of the barrier and the ohmic resistance is directly calculated with Ohm's law as a ratio of the voltage and current measured. However, DC can damage cells and the measurement system, compromising further experiments. For this reason, most of the commercially available systems are based on the application of alternative current (AC) signals at a frequency of 12.5 Hz to avoid any charging effects on the electrodes and the cells (Srinivasan et al. 2015). The TEER is then calculated with the following formula:

$$TEER\left(\Omega.cm^2\right) = R_{barrier}\left(\Omega\right) \times A\left(cm^2\right), \tag{9.2}$$

Where $R_{barrier}$ is the cell-specific resistance and A is the surface area of the barrier. Commercially available voltohmmeter uses chopstick electrodes that are difficult to integrate into microfluidic models. Moreover, the quality of the measurement depends on the electrode's positions, affecting the reproducibility of the technique, which can be solved by integrating microelectrodes in the microfluidic system during its fabrication. EIS provides a more reliable TEER value than the one obtained with DC or single-frequency AC measurements (Douville et al. 2010). A small amplitude AC signal is applied with a frequency sweep on the barrier, and the amplitude and the phase of the resulting current are measured. The electrical impedance (Z), which is a ratio of the voltage-time function V(t) and the resulting current-time function I(t), combined with fitting algorithms provides a more reliable TEER value and additionally, the capacitance of the cell layer (Benson et al. 2013). Indeed, the measured impedance spectrum can be analyzed with an equivalent circuit of the barrier to extract the signal due to the paracellular (TEER) and transcellular (capacitance) routes (Lazanas and Prodromidis 2023). Most studies limit the use of electrical measurements to confirm the effective formation of the biological barrier and to ensure the reliability of the model before testing NPs (Ahn et al. 2020; Lee et al. 2020; Palma-Florez et al. 2023). However, including TEER measurement is also a powerful technique to evaluate the potentially detrimental effect of NPs in real time since electrical resistance is known to drop drastically when ECs and TJs are disrupted (Kuzmanov et al. 2016). As an example, the effect of gold nanorods functionalized with angiopep-2 on BBB integrity was evaluated *in situ* with integrated thin-film microelectrodes, emphasizing that this protein enhances NPs permeability through endocytosis, while increasing TEER values at the same time, proving their non-toxic effects and the strengthening of TJs between ECs (Palma-Florez et al. 2023). Apart from TEER measurement, integrated microelectrodes can be used for detecting lower or abnormal electrical activity of neurons (Gramowski et al. 2010). Only a few studies are solely focusing on NPs toxicity as they are often tested as potential therapeutic agents with pathological models representative of glioblastoma or Alzheimer's disease for instance (Palma-Florez et al. 2023; Straehla et al.

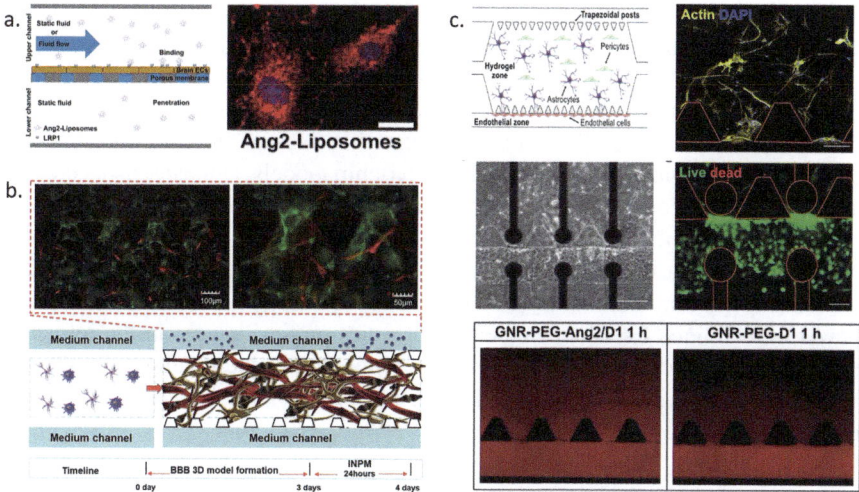

Fig. 9.5 BBB-on-chip models used for the evaluation of permeability and toxicity studies of NPs. (**a**) Fluorescence assessment of permeability of angiopep-decorated rhodamine-labeled NPs in a "sandwich configuration" human BBB-on-chip (Papademetriou et al. 2018). (**b**) Confocal imaging of astrocytes for the evaluation of the toxic effect of indoor airborne NPs on glial cells, performed in a "horizontal" BBB-on-chip (Li et al. 2020). (**c**) Permeability evaluation of functionalized gold nanorods using integrated microelectrodes and fluorescence imaging in a "horizontal" BBB-on-chip (Palma-Florez et al. 2023)

2022) (Fig. 9.5c). However, real-time impedance monitoring systems developed for studying the formation and degeneration of neuronal networks in pathological models of brain-on-chip could be adapted for nanosafety studies (Liu et al. 2022).

5.3 Potential of Electrochemical and Optical Sensors in BBB-on-Chip for Nanosafety

As discussed previously, the working electrodes can be functionalized with biological recognition elements to detect specific substances of interest. Up to now, they are mostly used for neurological investigations, drug toxicology, or pathological models, but offer promising options for nanosafety. Detection of reactive oxygen species (ROS) and reactive nitrogen species (RNS) is important for potential neurotoxicity, as they induce oxidative stress that can damage cellular components such as DNA, proteins, and lipids. However, they are highly reactive and therefore have a short life, requiring highly sensitive sensors (Zhao et al. 2021). Even if such sensors have not been yet integrated into BBB-on-chip models, several technologies developed could be of interest (Marino et al. 2023; Shamkhalichenar 2020). As an example, Dou et al. developed a trimetallic hybrid nanoflower-decorated MoS_2 nanosheet-modified sensor for the real-time and *in situ* subnanomolar level detection of hydrogen peroxide (Dou et al. 2018). Enzymatic biosensors are also investigated, for instance, based on the horseradish peroxidase, but exhibit a lower sensitivity due to a potential inactivation of enzymes over time (Zhao et al. 2021). Glucose and lactate, as markers of metabolic activity, were also quantified in a microfluidic analyzer able to monitor concentrations with a high temporal resolution (Samper et al. 2019). This study, performed in a microdialysate stream *in vivo* is promising for implementation into BBB-on-chip models. Optical sensors are also a valuable option for monitoring biomarkers in situ, as they are already investigated in the scope of microfluidic devices. For instance, Su et al. developed integrated digital immunosensors for the detection of pro-inflammatory cytokines released in the case of neuroinflammation (Su et al. 2023). Finally, SPR is a promising and powerful technique for nanotherapeutics. It relies on the changes in the refractive index at the surface of functionalized electrodes due to the binding of a specific ligand and can be used both for detecting lipid NPs (Schneider et al. 2015) and biomarkers (Liao et al. 2019).

6 Advantages and Limitations of BBB-on-Chip for Nanosafety Studies and Future Challenges

Recent advances in BBB-on-chip models made them an attractive alternative for the development of nanotherapies and related nanosafety studies. However, both device and biological components must be carefully selected depending on the application. Indeed, for the evaluation of neurotoxicity, outputs from the model can be completely different depending on the number and the type of cells included in the chip (Pediaditakis et al. 2022). Moreover, a balance needs to be found between tissue complexity for better physiological relevance, and engineering limitations. The manufacture of complex geometries using 3D bioprinting makes it possible to obtain more predictive models, but it is, therefore, more complicated to operate, image, and sensorize. Moreover, the lack of standardization for manufacturing and operating procedures limits the robustness and throughput of the developed models (Leung et al. 2022).

The implementation of dynamic conditions in the vascular microchannel is also crucial, as numerous studies have shown the importance of shear stress on brain ECs in BBB models, affecting paracellular tightness through TJ proteins, but also signal transduction pathways amongst others (Li et al. 2005). The connection between the microfluidic system and the pumping system can be challenging and often leads to leaks of the formation of bubbles. This can be solved by using a pumpless system, for instance gravity-driven flow obtained by tilting the device to drive fluid from one reservoir to another, but cells experience bidirectional flow that can affect their behavior (Yu et al. 2020a). Pneumatic pressure-driven microfluidic devices can induce circulation without using an active pump, but currently developed systems must be optimized to avoid pressure variability due to the geometry of the microfluidic circuit, as well as potential evaporation of the medium (Narayanamurthy et al. 2020).

Regarding sensorization, TEER measurement is the gold standard to evaluate BBB models, but they are barely comparable between studies due to the variability in measurement systems (Deli et al. 2024). Commercially available devices mostly use chopsticks that are poorly compatible with microfluidic devices contrary to integrated microelectrodes, which, however, need to be carefully optimized in terms of material, size, and placement to provide reliable data. EIS should be preferred over DC or single-frequency AC measurements, but require more complex data analysis. The integration of microelectrode arrays can provide quantitative data on the integrity and permeability of the barrier, as well as the proper functioning of the neuronal network.

While most nanosafety studies performed in BBB-on-chip models are based on imaging techniques, adapting electrochemical sensors to microfluidic systems could significantly improve the throughput of the models regarding neurotoxicity, including less invasive and more sensitive sensors able to detect simultaneously and in real-time several biomarkers. However, this must go hand in hand with real efforts to develop systems capable of simultaneously recording and processing large amounts of data.

7 Conclusion

BBB-on-chip models are very powerful tools for nanosafety studies but they require taking into account several fields of expertise to obtain a relevant biological model that can be properly analyzed. Bioengineering, biology, material sciences, data processing, and physics must be combined to obtain reliable and predictive results, each of which must be adapted and optimized depending on the application. The significant advances observed in all these fields in recent years suggest that BBB-on-chip could become an essential tool for human nanosafety studies, replacing progressively animal models.

Acknowledgements The authors acknowledge the European Health and Digital Executive Agency (iCare, 101092971).

References

Abbott NJ, Patabendige AAK, Dolman DEM, Yusof SR, Begley DJ (2010) Structure and function of the blood-brain barrier. Neurobiol Dis 37:13–25. https://doi.org/10.1016/j.nbd.2009.07.030

Agrawal AA, Nehilla BJ, Reisig KV, Gaborski TR, Fang DZ, Striemer CC, Fauchet PM, McGrath JL (2010) Porous nanocrystalline silicon membranes as highly permeable and molecularly thin substrates for cell culture. Biomaterials 31:5408–5417. https://doi.org/10.1016/j.biomaterials.2010.03.041

Ahmed Juvale II, Abdul Hamid AA, Abd Halim KB, Che Has AT (2022) P-glycoprotein: new insights into structure, physiological function, regulation and alterations in disease. Heliyon 8:e09777. https://doi.org/10.1016/j.heliyon.2022.e09777

Ahn SI, Sei YJ, Park H-J, Kim J, Ryu Y, Choi JJ, Sung H-J, MacDonald TJ, Levey AI, Kim Y (2020) Microengineered human blood–brain barrier platform for understanding nanoparticle transport mechanisms. Nat Commun 11:175. https://doi.org/10.1038/s41467-019-13896-7

Alahmari A (2021) Blood-brain barrier overview: structural and functional correlation. Neural Plast 2021:6564585. https://doi.org/10.1155/2021/6564585

Alshammari BH, Lashin MMA, Adil Mahmood M, Al-Mubaddel FS, Ilyas N, Rahman N, Sohail M, Khan A, Shukhratovich Abdullaev S, Khan R (2023) Organic and inorganic nanomaterials: fabrication, properties and applications. RSC Adv 13:13735–13785. https://doi.org/10.1039/D3RA01421E

Arakawa CK, Badeau BA, Zheng Y, DeForest CA (2017) Multicellular vascularized engineered tissues through user-programmable biomaterial photodegradation. Adv Mater 29:1703156. https://doi.org/10.1002/adma.201703156

Armulik A, Genové G, Mäe M, Nisancioglu MH, Wallgard E, Niaudet C, He L, Norlin J, Lindblom P, Strittmatter K, Johansson BR, Betsholtz C (2010) Pericytes regulate the blood-brain barrier. Nature 468:557–561. https://doi.org/10.1038/nature09522

Asmatulu E, Andalib M, Subeshan B, Abedin F (2022) Impact of nanomaterials on human health: a review. Environ Chem Lett 20. https://doi.org/10.1007/s10311-022-01430-z

Behl T, Kaur D, Sehgal A, Singh S, Sharma N, Zengin G, Andronie-Cioara FL, Toma MM, Bungau S, Bumbu AG (2021) Role of monoamine oxidase activity in Alzheimer's disease: an insight into the therapeutic potential of inhibitors. Molecules 26:3724. https://doi.org/10.3390/molecules26123724

Benson K, Cramer S, Galla H-J (2013) Impedance-based cell monitoring: barrier properties and beyond. Fluids Barriers CNS 10:5. https://doi.org/10.1186/2045-8118-10-5

Berthier E, Young EWK, Beebe D (2012) Engineers are from PDMS-land, biologists are from polystyrenia. Lab Chip 12:1224–1237. https://doi.org/10.1039/C2LC20982A

Bhattacharjee B, Dutta S, Maity T, Dey S, Mondal S, Bhowmick S (n.d.) Impacts of nanofluids and nanomaterials on environment and human health: a review. Nanosci Nanotechnol Asia 13:7–23. https://doi.org/10.2174/2210681213666230601103342

Bi X, Beck C, Gong Y (2021) Genetically encoded fluorescent indicators for imaging brain chemistry. Biosensors 11:116. https://doi.org/10.3390/bios11040116

Booth R, Kim H (2012) Characterization of a microfluidic in vitro model of the blood-brain barrier (μBBB). Lab Chip 12:1784–1792. https://doi.org/10.1039/C2LC40094D

Borók A, Laboda K, Bonyár A (2021) PDMS bonding technologies for microfluidic applications: a review. Biosensors 11:292. https://doi.org/10.3390/bios11080292

Bowman PD, Ennis SR, Rarey KE, Betz AL, Goldstein GW (1983) Brain microvessel endothelial cells in tissue culture: a model for study of blood-brain barrier permeability. Ann Neurol 14:396–402. https://doi.org/10.1002/ana.410140403

Brandenberg N, Lutolf MP (2016) In situ patterning of microfluidic networks in 3D cell-laden hydrogels. Adv Mater 28:7450–7456. https://doi.org/10.1002/adma.201601099

Bürgel SC, Diener L, Frey O, Kim J-Y, Hierlemann A (2016) Automated, multiplexed electrical impedance spectroscopy platform for continuous monitoring of microtissue spheroids. Anal Chem 88:10876–10883. https://doi.org/10.1021/acs.analchem.6b01410

Campisi M, Shin Y, Osaki T, Hajal C, Chiono V, Kamm RD (2018) 3D self-organized microvascular model of the human blood-brain barrier with endothelial cells, pericytes and astrocytes. Biomaterials 180:117–129. https://doi.org/10.1016/j.biomaterials.2018.07.014

Carvalho V, Bañobre-López M, Minas G, Teixeira SFCF, Lima R, Rodrigues RO (2022) The integration of spheroids and organoids into organ-on-a-chip platforms for tumour research: a review. Bioprinting 27:e00224. https://doi.org/10.1016/j.bprint.2022.e00224

Cecchelli R, Aday S, Sevin E, Almeida C, Culot M, Dehouck L, Coisne C, Engelhardt B, Dehouck M-P, Ferreira L (2014) A stable and reproducible human blood-brain barrier model derived from hematopoietic stem cells. PLoS One 9:e99733. https://doi.org/10.1371/journal.pone.0099733

Cecen B, Saygili E, Zare I, Nejati O, Khorsandi D, Zarepour A, Alarcin E, Zarrabi A, Topkaya SN, Yesil-Celiktas O, Mostafavi E, Bal-Öztürk A (2023) Biosensor integrated brain-on-a-chip platforms: progress and prospects in clinical translation. Biosens Bioelectron 225:115100. https://doi.org/10.1016/j.bios.2023.115100

Chung HH, Mireles M, Kwarta BJ, Gaborski TR (2018) Use of porous membranes in tissue barrier and co-culture models. Lab Chip 18:1671–1689. https://doi.org/10.1039/c7lc01248a

Daneman R, Prat A (2015) The blood-brain barrier. Cold Spring Harb Perspect Biol 7:a020412. https://doi.org/10.1101/cshperspect.a020412

Dehouck MP, Méresse S, Delorme P, Fruchart JC, Cecchelli R (1990) An easier, reproducible, and mass-production method to study the blood-brain barrier in vitro. J Neurochem 54:1798–1801. https://doi.org/10.1111/j.1471-4159.1990.tb01236.x

Deli MA, Porkoláb G, Kincses A, Mészáros M, Szecskó A, Kocsis AE, Vigh JP, Valkai S, Veszelka S, Walter FR, Dér A (2024) Lab-on-a-chip models of the blood–brain barrier: evolution, problems, perspectives. Lab Chip 24:1030–1063. https://doi.org/10.1039/D3LC00996C

Diederichs S, Tuan RS (2014) Functional comparison of human-induced pluripotent stem cell-derived mesenchymal cells and bone marrow-derived mesenchymal stromal cells from the same donor. Stem Cells Dev 23:1594–1610. https://doi.org/10.1089/scd.2013.0477

Dohgu S, Takata F, Kataoka Y (2015) Brain pericytes regulate the blood-brain barrier function. Nihon Yakurigaku Zasshi 146:63–65. https://doi.org/10.1254/fpj.146.63

Dou B, Yang J, Yuan R, Xiang Y (2018) Trimetallic hybrid nanoflower-decorated MoS2 nanosheet sensor for direct in situ monitoring of H2O2 secreted from live cancer cells. Anal Chem 90:5945–5950. https://doi.org/10.1021/acs.analchem.8b00894

Douville N, Tung Y-C, Li R, Wang J, El-Sayed M, Takayama S (2010) Fabrication of two-layered channel system with embedded electrodes to measure resistance across epithelial and endothelial barriers. Anal Chem 82:2505–2511. https://doi.org/10.1021/ac9029345

Duffy SL, Murphy JT (2001) Colorimetric assay to quantify macromolecule diffusion across endothelial monolayers. Biotechniques 31:495–496, 498, 500–501. https://doi.org/10.2144/01313st02

Ealia SAM, Saravanakumar MP (2017) A review on the classification, characterisation, synthesis of nanoparticles and their application. IOP Conf Ser: Mater Sci Eng 263:032019. https://doi.org/10.1088/1757-899X/263/3/032019

Fan Y, Xu C, Deng N, Gao Z, Jiang Z, Li X, Zhou Y, Pei H, Li L, Tang B (2023) Understanding drug nanocarrier and blood–brain barrier interaction based on a microfluidic microphysiological model. Lab Chip 23:1935–1944. https://doi.org/10.1039/D2LC01077A

Gomes MJ, das Neves J, Sarmento B (2014) Nanoparticle-based drug delivery to improve the efficacy of antiretroviral therapy in the central nervous system. Int J Nanomedicine 9:1757–1769. https://doi.org/10.2147/IJN.S45886

Gramowski A, Flossdorf J, Bhattacharya K, Jonas L, Lantow M, Rahman Q, Schiffmann D, Weiss DG, Dopp E (2010) Nanoparticles induce changes of the electrical activity of neuronal networks on microelectrode array neurochips. Environ Health Perspect 118:1363–1369. https://doi.org/10.1289/ehp.0901661

Griep LM, Wolbers F, de Wagenaar B, ter Braak PM, Weksler BB, Romero IA, Couraud PO, Vermes I, van der Meer AD, van den Berg A (2013) BBB on chip: microfluidic platform to mechanically and biochemically modulate blood-brain barrier function. Biomed Microdevices 15:145–150. https://doi.org/10.1007/s10544-012-9699-7

Grigoryan B, Paulsen SJ, Corbett DC, Sazer DW, Fortin CL, Zaita AJ, Greenfield PT, Calafat NJ, Gounley JP, Ta AH, Johansson F, Randles A, Rosenkrantz JE, Louis-Rosenberg JD, Galie PA, Stevens KR, Miller JS (2019) Multivascular networks and functional intravascular topologies within biocompatible hydrogels. Science 364:458–464. https://doi.org/10.1126/science.aav9750

Guarino V, Zizzari A, Bianco M, Gigli G, Moroni L, Arima V (2023) Advancements in modelling human blood brain-barrier on a chip. Biofabrication 15. https://doi.org/10.1088/1758-5090/acb571

Haqqani AS, Bélanger K, Stanimirovic DB (2024) Receptor-mediated transcytosis for brain delivery of therapeutics: receptor classes and criteria. Front Drug Deliv 4. https://doi.org/10.3389/fddev.2024.1360302

Herland A, van der Meer AD, FitzGerald EA, Park T-E, Sleeboom JJF, Ingber DE (2016) Distinct contributions of astrocytes and pericytes to neuroinflammation identified in a 3D human blood-brain barrier on a chip. PLoS One 11:e0150360. https://doi.org/10.1371/journal.pone.0150360

Hersh AM, Alomari S, Tyler BM (2022) Crossing the blood-brain barrier: advances in nanoparticle technology for drug delivery in neuro-oncology. Int J Mol Sci 23:4153. https://doi.org/10.3390/ijms23084153

Hirama H, Otahara R, Kano S, Hayase M, Mekaru H (2021) Characterization of nanoparticle adsorption on polydimethylsiloxane-based microchannels. Sensors 21:1978. https://doi.org/10.3390/s21061978

Homan KA, Gupta N, Kroll KT, Kolesky DB, Skylar-Scott M, Miyoshi T, Mau D, Valerius MT, Ferrante T, Bonventre JV, Lewis JA, Morizane R (2019) Flow-enhanced vascularization and maturation of kidney organoids in vitro. Nat Methods 16:255–262. https://doi.org/10.1038/s41592-019-0325-y

Huh D, Matthews BD, Mammoto A, Montoya-Zavala M, Hsin HY, Ingber DE (2010) Reconstituting organ-level lung functions on a chip. Science 328:1662–1668. https://doi.org/10.1126/science.1188302

Ji S, Almeida E, Guvendiren M (2019) 3D bioprinting of complex channels within cell-laden hydrogels. Acta Biomater 95:214–224. https://doi.org/10.1016/j.actbio.2019.02.038

Kavand H, Nasiri R, Herland A (2022) Advanced materials and sensors for microphysiological systems: focus on electronic and electrooptical interfaces. Adv Mater 34:2107876. https://doi.org/10.1002/adma.202107876

Kim S, Ko J, Lee S-R, Park D, Park S, Jeon NL (2021) Anchor-IMPACT: a standardized microfluidic platform for high-throughput antiangiogenic drug screening. Biotechnol Bioeng 118:2524–2535. https://doi.org/10.1002/bit.27765

Kopatz V, Wen K, Kovács T, Keimowitz AS, Pichler V, Widder J, Vethaak AD, Hollóczki O, Kenner L (2023) Micro- and nanoplastics breach the blood–brain barrier (BBB): biomolecular corona's role revealed. Nano 13:1404. https://doi.org/10.3390/nano13081404

Kumah EA, Fopa RD, Harati S, Boadu P, Zohoori FV, Pak T (2023) Human and environmental impacts of nanoparticles: a scoping review of the current literature. BMC Public Health 23:1059. https://doi.org/10.1186/s12889-023-15958-4

Kumar R, Lal S (2014) Synthesis of organic nanoparticles and their applications in drug delivery and food nanotechnology: a review. J Nanomater Mol Nanotechnol 3. https://doi.org/10.4172/2324-8777.1000150

Kuzmanov I, Herrmann AM, Galla H-J, Meuth SG, Wiendl H, Klotz L (2016) An in vitro model of the blood-brain barrier using impedance spectroscopy: a focus on T cell-endothelial cell interaction. J Vis Exp:54592. https://doi.org/10.3791/54592

Lazanas AC, Prodromidis MI (2023) Electrochemical impedance spectroscopy—a tutorial. ACS Meas Sci Au 3:162–193. https://doi.org/10.1021/acsmeasuresciau.2c00070

Lebre F, Chatterjee N, Costa S, Fernández-de-Gortari E, Lopes C, Meneses J, Ortiz L, Ribeiro AR, Vilas-Boas V, Alfaro-Moreno E (2022) Nanosafety: an evolving concept to bring the safest possible nanomaterials to society and environment. Nano 12:1810. https://doi.org/10.3390/nano12111810

Lee H, Cho D-W (2016) One-step fabrication of an organ-on-a-chip with spatial heterogeneity using a 3D bioprinting technology. Lab Chip 16:2618–2625. https://doi.org/10.1039/C6LC00450D

Lee Y, Choi JW, Yu J, Park D, Ha J, Son K, Lee S, Chung M, Kim H-Y, Jeon NL (2018) Microfluidics within a well: an injection-molded plastic array 3D culture platform. Lab Chip 18:2433–2440. https://doi.org/10.1039/C8LC00336J

Lee SWL, Campisi M, Osaki T, Possenti L, Mattu C, Adriani G, Kamm RD, Chiono V (2020) Modeling nanocarrier transport across a 3D in vitro human blood-brain–barrier microvasculature. Adv Healthc Mater 9:1901486. https://doi.org/10.1002/adhm.201901486

Leung CM, de Haan P, Ronaldson-Bouchard K, Kim G-A, Ko J, Rho HS, Chen Z, Habibovic P, Jeon NL, Takayama S, Shuler ML, Vunjak-Novakovic G, Frey O, Verpoorte E, Toh Y-C (2022)

A guide to the organ-on-a-chip. Nat Rev Methods Primers 2:1–29. https://doi.org/10.1038/s43586-022-00118-6

Li Y-SJ, Haga JH, Chien S (2005) Molecular basis of the effects of shear stress on vascular endothelial cells. J Biomech 38:1949–1971. https://doi.org/10.1016/j.jbiomech.2004.09.030

Li Y, Liu Y, Hu C, Chang Q, Deng Q, Yang X, Wu Y (2020) Study of the neurotoxicity of indoor airborne nanoparticles based on a 3D human blood-brain barrier chip. Environ Int 143:105598. https://doi.org/10.1016/j.envint.2020.105598

Liang Y, Yoon J-Y (2021) In situ sensors for blood-brain barrier (BBB) on a chip. Sens Actuators Rep 3:100031. https://doi.org/10.1016/j.snr.2021.100031

Liao Z, Zhang Y, Li Y, Miao Y, Gao S, Lin F, Deng Y, Geng L (2019) Microfluidic chip coupled with optical biosensors for simultaneous detection of multiple analytes: a review. Biosens Bioelectron 126:697–706. https://doi.org/10.1016/j.bios.2018.11.032

Lippmann ES, Azarin SM, Kay JE, Nessler RA, Wilson HK, Al-Ahmad A, Palecek SP, Shusta EV (2012) Derivation of blood-brain barrier endothelial cells from human pluripotent stem cells. Nat Biotechnol 30:783–791. https://doi.org/10.1038/nbt.2247

Lippmann ES, Azarin SM, Palecek SP, Shusta EV (2020) Commentary on human pluripotent stem cell-based blood–brain barrier models. Fluids Barriers CNS 17:64. https://doi.org/10.1186/s12987-020-00222-3

Liu M, Hurn PD, Alkayed NJ (2004) Cytochrome P450 in neurological disease. Curr Drug Metab 5:225–234. https://doi.org/10.2174/1389200043335540

Liu S, Agalliu D, Yu C, Fisher M (2012) The role of pericytes in blood-brain barrier function and stroke. Curr Pharm Des 18:3653–3662. https://doi.org/10.2174/138161212802002706

Liu N-C, Liang C-C, Li Y-CE, Lee I-C (2022) A real-time sensing system for monitoring neural network degeneration in an Alzheimer's disease-on-a-chip model. Pharmaceutics 14:1022. https://doi.org/10.3390/pharmaceutics14051022

Lyu Z, Park J, Kim K-M, Jin H-J, Wu H, Rajadas J, Kim D-H, Steinberg GK, Lee W (2021) A neurovascular-unit-on-a-chip for the evaluation of the restorative potential of stem cell therapies for ischaemic stroke. Nat Biomed Eng 5:847–863. https://doi.org/10.1038/s41551-021-00744-7

Marino A, Battaglini M, Lefevre MC, Ceccarelli MC, Ziaja K, Ciofani G (2023) Sensorization of microfluidic brain-on-a-chip devices: towards a new generation of integrated drug screening systems. TrAC Trends Anal Chem 168:117319. https://doi.org/10.1016/j.trac.2023.117319

Martínez G, Merinero M, Pérez-Aranda M, Pérez-Soriano EM, Ortiz T, Begines B, Alcudia A (2020) Environmental impact of nanoparticles' application as an emerging technology: a review. Materials (Basel) 14:166. https://doi.org/10.3390/ma14010166

Mir M, Palma-Florez S, Lagunas A, López-Martínez MJ, Samitier J (2022) Biosensors Integration in blood–brain barrier-on-a-chip: emerging platform for monitoring neurodegenerative diseases. ACS Sens 7:1237–1247. https://doi.org/10.1021/acssensors.2c00333

Mireles M, Gaborski TR (2017) Fabrication techniques enabling ultrathin nanostructured membranes for separations. Electrophoresis 38:2374–2388. https://doi.org/10.1002/elps.201700114

Nakagawa S, Deli MA, Nakao S, Honda M, Hayashi K, Nakaoke R, Kataoka Y, Niwa M (2007) Pericytes from brain microvessels strengthen the barrier integrity in primary cultures of rat brain endothelial cells. Cell Mol Neurobiol 27:687–694. https://doi.org/10.1007/s10571-007-9195-4

Nakagawa S, Deli MA, Kawaguchi H, Shimizudani T, Shimono T, Kittel A, Tanaka K, Niwa M (2009) A new blood-brain barrier model using primary rat brain endothelial cells, pericytes and astrocytes. Neurochem Int 54:253–263. https://doi.org/10.1016/j.neuint.2008.12.002

Narayanamurthy V, Jeroish ZE, Bhuvaneshwari KS, Bayat P, Premkumar R, Samsuri F, Yusoff MM (2020) Advances in passively driven microfluidics and lab-on-chip devices: a comprehensive literature review and patent analysis. RSC Adv 10:11652–11680. https://doi.org/10.1039/D0RA00263A

Nasrollahi S, Banerjee S, Qayum B, Banerjee P, Pathak A (2017) Nanoscale matrix topography influences microscale cell motility through adhesions, actin organization, and cell shape. ACS Biomater Sci Eng 3:2980–2986. https://doi.org/10.1021/acsbiomaterials.6b00554

Nishihara H, Gastfriend BD, Soldati S, Perriot S, Mathias A, Sano Y, Shimizu F, Gosselet F, Kanda T, Palecek SP, Du Pasquier R, Shusta EV, Engelhardt B (2020) Advancing human induced pluripotent stem cell-derived blood-brain barrier models for studying immune cell interactions. FASEB J 34:16693–16715. https://doi.org/10.1096/fj.202001507RR

Oberdörster G, Oberdörster E, Oberdörster J (2005) Nanotoxicology: an emerging discipline evolving from studies of ultrafine particles. Environ Health Perspect 113:823–839. https://doi.org/10.1289/ehp.7339

Ong LJY, Chong LH, Jin L, Singh PK, Lee PS, Yu H, Ananthanarayanan A, Leo HL, Toh Y-C (2017) A pump-free microfluidic 3D perfusion platform for the efficient differentiation of human hepatocyte-like cells. Biotechnol Bioeng 114:2360–2370. https://doi.org/10.1002/bit.26341

Palma-Florez S, López-Canosa A, Moralez-Zavala F, Castaño O, Kogan MJ, Samitier J, Lagunas A, Mir M (2023) BBB-on-a-chip with integrated micro-TEER for permeability evaluation of multi-functionalized gold nanorods against Alzheimer's disease. J Nanobiotechnol 21:115. https://doi.org/10.1186/s12951-023-01798-2

Panula P, Joó F, Rechardt L (1978) Evidence for the presence of viable endothelial cells in cultures derived from dissociated rat brain. Experientia 34:95–97. https://doi.org/10.1007/BF01921925

Papademetriou I, Vedula E, Charest J, Porter T (2018) Effect of flow on targeting and penetration of angiopep-decorated nanoparticles in a microfluidic model blood-brain barrier. PLoS One 13:e0205158. https://doi.org/10.1371/journal.pone.0205158

Pardridge WM (2005) The blood-brain barrier: bottleneck in brain drug development. NeuroRx 2:3–14

Park T-E, Mustafaoglu N, Herland A, Hasselkus R, Mannix R, FitzGerald EA, Prantil-Baun R, Watters A, Henry O, Benz M, Sanchez H, McCrea HJ, Goumnerova LC, Song HW, Palecek SP, Shusta E, Ingber DE (2019) Hypoxia-enhanced blood-brain barrier chip recapitulates human barrier function and shuttling of drugs and antibodies. Nat Commun 10:2621. https://doi.org/10.1038/s41467-019-10588-0

Partyka PP, Godsey GA, Galie JR, Kosciuk MC, Acharya NK, Nagele RG, Galie PA (2017) Mechanical stress regulates transport in a compliant 3D model of the blood-brain barrier. Biomaterials 115:30–39. https://doi.org/10.1016/j.biomaterials.2016.11.012

Pediaditakis I, Kodella KR, Manatakis DV, Le CY, Barthakur S, Sorets A, Gravanis A, Ewart L, Rubin LL, Manolakos ES, Hinojosa CD, Karalis K (2022) A microengineered brain-chip to model neuroinflammation in humans. iScience 25:104813. https://doi.org/10.1016/j.isci.2022.104813

Peng Q, Zeng W (2024) The protective role of endothelial GLUT1 in ischemic stroke. Brain Behav 14:e3536. https://doi.org/10.1002/brb3.3536

Peng B, Tong Z, Tong WY, Pasic PJ, Oddo A, Dai Y, Luo M, Frescene J, Welch NG, Easton CD, Thissen H, Voelcker NH (2020) In situ surface modification of microfluidic blood–brain-barriers for improved screening of small molecules and nanoparticles. ACS Appl Mater Interfaces 12:56753–56766. https://doi.org/10.1021/acsami.0c17102

Pérez-López A, Torres-Suárez AI, Martín-Sabroso C, Aparicio-Blanco J (2023) An overview of *in vitro* 3D models of the blood-brain barrier as a tool to predict the *in vivo* permeability of nanomedicines. Adv Drug Deliv Rev 196:114816. https://doi.org/10.1016/j.addr.2023.114816

Petrovszki D, Walter FR, Vigh JP, Kocsis A, Valkai S, Deli MA, Dér A (2022) Penetration of the SARS-CoV-2 spike protein across the blood–brain barrier, as revealed by a combination of a human cell culture model system and optical biosensing. Biomedicine 10:188. https://doi.org/10.3390/biomedicines10010188

Pi Q, Maharjan S, Yan X, Liu X, Singh B, van Genderen AM, Robledo-Padilla F, Parra-Saldivar R, Hu N, Jia W, Xu C, Kang J, Hassan S, Cheng H, Hou X, Khademhosseini A, Zhang YS (2018) Digitally tunable microfluidic bioprinting of multilayered cannular tissues. Adv Mater 30:1706913. https://doi.org/10.1002/adma.201706913

Pizzagalli MD, Bensimon A, Superti-Furga G (2021) A guide to plasma membrane solute carrier proteins. FEBS J 288:2784–2835. https://doi.org/10.1111/febs.15731

Pradhan S, Keller KA, Sperduto JL, Slater JH (2017) Fundamentals of laser-based hydrogel degradation and applications in cell and tissue engineering. Adv Healthc Mater 6:1700681. https://doi.org/10.1002/adhm.201700681

Prüst M, Meijer J, Westerink RHS (2020) The plastic brain: neurotoxicity of micro- and nanoplastics. Part Fibre Toxicol 17:24. https://doi.org/10.1186/s12989-020-00358-y

Rashid MM, Forte Tavčer P, Tomšič B (2021) Influence of titanium dioxide nanoparticles on human health and the environment. Nanomaterials 11:2354. https://doi.org/10.3390/nano11092354

Saeidnia S, Manayi A, Abdollahi M (2015) From in vitro experiments to in vivo and clinical studies; pros and cons. Curr Drug Discov Technol 12:218–224. https://doi.org/10.2174/1570163813666160114093140

Samper IC, Gowers SAN, Rogers ML, Murray D-SRK, Jewell SL, Pahl C, Strong AJ, Boutelle MG (2019) 3D printed microfluidic device for online detection of neurochemical changes with high temporal resolution in human brain microdialysate. Lab Chip 19:2038–2048. https://doi.org/10.1039/C9LC00044E

Santa-Maria AR, Walter FR, Figueiredo R, Kincses A, Vigh JP, Heymans M, Culot M, Winter P, Gosselet F, Dér A, Deli MA (2021) Flow induces barrier and glycocalyx-related genes and negative surface charge in a lab-on-a-chip human blood-brain barrier model. J Cereb Blood Flow Metab 41:2201–2215. https://doi.org/10.1177/0271678X21992638

Schiera G, Di Liegro CM, Schirò G, Sorbello G, Di Liegro I (2024) Involvement of astrocytes in the formation, maintenance, and function of the blood–brain barrier. Cells 13:150. https://doi.org/10.3390/cells13020150

Schneider CS, Perez JG, Cheng E, Zhang C, Mastorakos P, Hanes J, Winkles JA, Woodworth GF, Kim AJ (2015) Minimizing the non-specific binding of nanoparticles to the brain enables active targeting of Fn14-positive glioblastoma cells. Biomaterials 42:42–51. https://doi.org/10.1016/j.biomaterials.2014.11.054

Shamkhalichenar H (2020) Review—non-enzymatic hydrogen peroxide electrochemical sensors based on reduced graphene oxide. J Electrochem Soc 167(3):037531

Shaver A, Mahlum JD, Scida K, Johnston ML, Aller Pellitero M, Wu Y, Carr GV, Arroyo-Currás N (2022) Optimization of vancomycin aptamer sequence length increases the sensitivity of electrochemical, aptamer-based sensors in vivo. ACS Sens 7:3895–3905. https://doi.org/10.1021/acssensors.2c01910

Shemetov AA, Monakhov MV, Zhang Q, Canton-Josh JE, Kumar M, Chen M, Matlashov ME, Li X, Yang W, Nie L, Shcherbakova DM, Kozorovitskiy Y, Yao J, Ji N, Verkhusha VV (2021) A near-infrared genetically encoded calcium indicator for in vivo imaging. Nat Biotechnol 39:368–377. https://doi.org/10.1038/s41587-020-0710-1

Shin J, Ko J, Jeong S, Won P, Lee Y, Kim J, Hong S, Jeon NL, Ko SH (2021) Monolithic digital patterning of polydimethylsiloxane with successive laser pyrolysis. Nat Mater 20:100–107. https://doi.org/10.1038/s41563-020-0769-6

Simovic S, Prestidge CA (2003) Adsorption of hydrophobic silica nanoparticles at the PDMS droplet–water interface. Langmuir 19:8364–8370. https://doi.org/10.1021/la0347197

Srinivasan B, Kolli AR, Esch MB, Abaci HE, Shuler ML, Hickman JJ (2015) TEER measurement techniques for in vitro barrier model systems. SLAS technology, special issue: microengineered cell- and tissue-based assays for drug screening and toxicology applications (Part 1 of 2) 20:107–126. https://doi.org/10.1177/2211068214561025

Sticker D, Rothbauer M, Ehgartner J, Steininger C, Liske O, Liska R, Neuhaus W, Mayr T, Haraldsson T, Kutter JP, Ertl P (2019) Oxygen management at the microscale: a functional biochip material with long-lasting and tunable oxygen scavenging properties for cell culture applications. ACS Appl Mater Interfaces 11:9730–9739. https://doi.org/10.1021/acsami.8b19641

Straehla JP, Hajal C, Safford HC, Offeddu GS, Boehnke N, Dacoba TG, Wyckoff J, Kamm RD, Hammond PT (2022) A predictive microfluidic model of human glioblastoma to assess trafficking of blood–brain barrier-penetrant nanoparticles. Proc Natl Acad Sci USA 119:e2118697119. https://doi.org/10.1073/pnas.2118697119

Su S-H, Song Y, Stephens A, Situ M, McCloskey MC, McGrath JL, Andjelkovic AV, Singer BH, Kurabayashi K (2023) A tissue chip with integrated digital immunosensors: in situ brain endothelial barrier cytokine secretion monitoring. Biosens Bioelectron 224:115030. https://doi.org/10.1016/j.bios.2022.115030

Sweeney MD, Zhao Z, Montagne A, Nelson AR, Zlokovic BV (2019) Blood-brain barrier: from physiology to disease and back. Physiol Rev 99:21–78. https://doi.org/10.1152/physrev.00050.2017

Toepke MW, Beebe DJ (2006) PDMS absorption of small molecules and consequences in microfluidic applications. Lab Chip 6:1484–1486. https://doi.org/10.1039/B612140C

Uemura M, Refaat MM, Shinoyama M, Hayashi H, Hashimoto N, Takahashi J (2010) Matrigel supports survival and neuronal differentiation of grafted embryonic stem cell-derived neural precursor cells. J Neurosci Res 88:542–551. https://doi.org/10.1002/jnr.22223

Urzì O, Gasparro R, Costanzo E, De Luca A, Giavaresi G, Fontana S, Alessandro R (2023) Three-dimensional cell cultures: the bridge between in vitro and in vivo models. Int J Mol Sci 24:12046. https://doi.org/10.3390/ijms241512046

van Bree JB, de Boer AG, Danhof M, Ginsel LA, Breimer DD (1988) Characterization of an "in vitro" blood-brain barrier: effects of molecular size and lipophilicity on cerebrovascular endothelial transport rates of drugs. J Pharmacol Exp Ther 247:1233–1239

Vanlandewijck M, He L, Mäe MA, Andrae J, Ando K, Del Gaudio F, Nahar K, Lebouvier T, Laviña B, Gouveia L, Sun Y, Raschperger E, Räsänen M, Zarb Y, Mochizuki N, Keller A, Lendahl U, Betsholtz C (2018) A molecular atlas of cell types and zonation in the brain vasculature. Nature 554:475–480. https://doi.org/10.1038/nature25739

Vatine GD, Barrile R, Workman MJ, Sances S, Barriga BK, Rahnama M, Barthakur S, Kasendra M, Lucchesi C, Kerns J, Wen N, Spivia WR, Chen Z, Van Eyk J, Svendsen CN (2019) Human iPSC-derived blood-brain barrier chips enable disease modeling and personalized medicine applications. Cell Stem Cell 24:995–1005.e6. https://doi.org/10.1016/j.stem.2019.05.011

Vera D, García-Díaz M, Torras N, Álvarez M, Villa R, Martinez E (2021) Engineering tissue barrier models on hydrogel microfluidic platforms. ACS Appl Mater Interfaces 13:13920–13933. https://doi.org/10.1021/acsami.0c21573

Verma A, Verma M, Singh A (2020) Animal tissue culture principles and applications. Anim Biotechnol:269–293. https://doi.org/10.1016/B978-0-12-811710-1.00012-4

Veszelka S, Tóth A, Walter FR, Tóth AE, Gróf I, Mészáros M, Bocsik A, Hellinger É, Vastag M, Rákhely G, Deli MA (2018) Comparison of a rat primary cell-based blood-brain barrier model with epithelial and brain endothelial cell lines: gene expression and drug transport. Front Mol Neurosci 11:166. https://doi.org/10.3389/fnmol.2018.00166

Wadman M (2023) FDA no longer has to require animal testing for new drugs. Science 379:127–128. https://doi.org/10.1126/science.adg6276

Walter FR, Valkai S, Kincses A, Petneházi A, Czeller T, Veszelka S, Ormos P, Deli MA, Dér A (2016) A versatile lab-on-a-chip tool for modeling biological barriers. Sensors Actuators B Chem 222:1209–1219. https://doi.org/10.1016/j.snb.2015.07.110

Wang JD, Douville NJ, Takayama S, ElSayed M (2012) Quantitative analysis of molecular absorption into PDMS microfluidic channels. Ann Biomed Eng 40:1862–1873. https://doi.org/10.1007/s10439-012-0562-z

Wang Z, Jin X, Tian Z, Menard F, Holzman JF, Kim K (2018) A novel, well-resolved direct laser bioprinting system for rapid cell encapsulation and microwell fabrication. Adv Healthc Mater 7:1701249. https://doi.org/10.1002/adhm.201701249

Wevers NR, Kasi DG, Gray T, Wilschut KJ, Smith B, van Vught R, Shimizu F, Sano Y, Kanda T, Marsh G, Trietsch SJ, Vulto P, Lanz HL, Obermeier B (2018) A perfused human blood-brain barrier on-a-chip for high-throughput assessment of barrier function and antibody transport. Fluids Barriers CNS 15:23. https://doi.org/10.1186/s12987-018-0108-3

Whitesides GM (2006) The origins and the future of microfluidics. Nature 442:368–373. https://doi.org/10.1038/nature05058

Wolff A, Antfolk M, Brodin B, Tenje M (2015) In vitro blood–brain barrier models—an overview of established models and new microfluidic approaches. J Pharm Sci 104:2727–2746. https://doi.org/10.1002/jps.24329

Wu Q, Liu J, Wang X, Feng L, Wu J, Zhu X, Wen W, Gong X (2020) Organ-on-a-chip: recent breakthroughs and future prospects. Biomed Eng Online 19:9. https://doi.org/10.1186/s12938-020-0752-0

Xue D, Wang Y, Zhang J, Mei D, Wang Y, Chen S (2018) Projection-based 3D printing of cell patterning scaffolds with multiscale channels. ACS Appl Mater Interfaces 10:19428–19435. https://doi.org/10.1021/acsami.8b03867

Yu F, Kumar NDS, Foo LC, Ng SH, Hunziker W, Choudhury D (2020a) A pump-free tricellular blood-brain barrier on-a-chip model to understand barrier property and evaluate drug response. Biotechnol Bioeng 117:1127–1136. https://doi.org/10.1002/bit.27260

Yu C, Schimelman J, Wang P, Miller KL, Ma X, You S, Guan J, Sun B, Zhu W, Chen S (2020b) Photopolymerizable biomaterials and light-based 3D printing strategies for biomedical applications. Chem Rev 120:10695–10743. https://doi.org/10.1021/acs.chemrev.9b00810

Zhang YS, Davoudi F, Walch P, Manbachi A, Luo X, Dell'Erba V, Miri AK, Albadawi H, Arneri A, Li X, Wang X, Dokmeci MR, Khademhosseini A, Oklu R (2016) Bioprinted thrombosis-on-a-chip. Lab Chip 16:4097–4105. https://doi.org/10.1039/C6LC00380J

Zhang N, Xiong G, Liu Z (2022) Toxicity of metal-based nanoparticles: challenges in the nano era. Front Bioeng Biotechnol 10. https://doi.org/10.3389/fbioe.2022.1001572

Zhao S, Zang G, Zhang Y, Liu H, Wang N, Cai S, Durkan C, Xie G, Wang G (2021) Recent advances of electrochemical sensors for detecting and monitoring ROS/RNS. Biosens Bioelectron 179:113052. https://doi.org/10.1016/j.bios.2021.113052

Open Access This chapter is licensed under the terms of the Creative Commons Attribution-NonCommercial-NoDerivatives 4.0 International License (http://creativecommons.org/licenses/by-nc-nd/4.0/), which permits any noncommercial use, sharing, distribution and reproduction in any medium or format, as long as you give appropriate credit to the original author(s) and the source, provide a link to the Creative Commons license and indicate if you modified the licensed material. You do not have permission under this license to share adapted material derived from this chapter or parts of it.

The images or other third party material in this chapter are included in the chapter's Creative Commons license, unless indicated otherwise in a credit line to the material. If material is not included in the chapter's Creative Commons license and your intended use is not permitted by statutory regulation or exceeds the permitted use, you will need to obtain permission directly from the copyright holder.

Part III
In Vitro-In Vivo Bridging Models

Planarians as an Alternative Standard Invertebrate Model to Assess the Eco-safety of Nanoparticles/Nanomaterials

10

M. Bernardeschi, M. C. Lefevre, M. C. Ceccarelli, A. Salvetti, and G. Ciofani

Abstract

With their ability to repair and/or regenerate any damaged or missing part of their body, freshwater planarians have gained more and more attention over the years for their potential application as bio-indicators of environmental and chemical stress. These free-living flatworms can be easily cultured under laboratory conditions, and they can be used to study regeneration, developmental biology, and behavior (feeding and locomotor activity) after exposure under strictly controlled conditions. In this chapter, after a general overview, we want to summarize some key factors that make them suitable for assessing the safety of emerging contaminants such as nanoparticles and nanomaterials. These substances, due to their peculiar and remarkable characteristics, are exponentially employed in many different fields of the production chain. So, it is fundamental to deeply investigate the biological mechanisms that occur after exposure, from molecular to population level, and freshwater planarians seem to be perfect to fulfill this goal.

Keywords

Planarians · Nanoparticles · Ecotoxicology · Toxicity · Genotoxicity · Neurobehavior

---

M. Bernardeschi (✉) · M. C. Lefevre · G. Ciofani (✉)
Smart Bio-Interfaces, Istituto Italiano di Tecnologia, Pontedera, Italy
e-mail: margherita.bernardeschi@iit.it; gianni.ciofani@iit.it

M. C. Ceccarelli
Smart Bio-Interfaces, Istituto Italiano di Tecnologia, Pontedera, Italy

The Biorobotics Institute, Scuola Superiore Sant'Anna, Pontedera, Italy

A. Salvetti
Dipartimento di Medicina Clinica e Sperimentale, Università di Pisa, Pisa, Italy

© The Author(s) 2025
E. Alfaro-Moreno, F. Murphy (eds.), *Nanosafety*,
https://doi.org/10.1007/978-3-031-93871-9_10

Abbreviations

AChE	Acetylcholinesterase
DNA	Deoxyribonucleic acid
DSB	Double-stranded break
dUTP	Deoxyuridine triphosphate
NP	Nanoparticle
NM	Nanomaterial
PCR	Polymerase chain reaction
TUNEL	Terminal deoxynucleotide transferase dUTP nick end labeling
TCDD	2,3,7,8-tetrachlorodibenzo-p-dioxin
ROS	Reactive Oxygen Specie
RNS	Reactive Nitrogen Specie
SCGE	Single Cell Gel Electrophoresis
SSB	Single-stranded break

1 Introduction

1.1 Planarians: A General Overview of Classification, General Anatomy and Lifestyle

"Freshwater planarians" is a common name referring to numerous free-living organisms belonging to the traditional taxon "Turbellaria" in Platyhelminthes, especially those placed in the order Tricladida (Ehlers 1986; Noreña et al. 2015; Schockaert et al. 2008). Planarians are triploblastic, acoelomate, and bilaterally symmetrical metazoans whose body consists of tissues derived from all three basic germ layers (endoderm, mesoderm, and ectoderm). It presents just one opening, which leads to a three-branched digestive system, but no anus (Fig. 10.1a) (Rompolas et al. 2009). The planarian central nervous system consists of a cephalic ganglion (brain) in the head and two ventral nerve cords that are connected by multiple commissures.

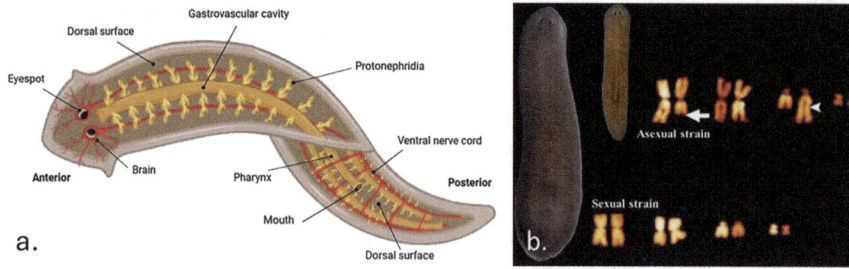

Fig. 10.1 Planarian's anatomy and classification: (**a**) schematic of planarian anatomy; (**b**) planarian Schmidtea mediterranea: sexual (left) and asexual (right) biotypes with their corresponding diploid karyotypes modified from Guo et al. (2018)

These structures can be further subdivided into specific functional and molecular regions, demonstrating their molecular complexity (Cebrià et al. 2002; Fraguas et al. 2012). The planarian nervous system shares key characteristics with the vertebrate nervous system, including all key neurotransmitters ("New Worm on the Block: Planarians in (Neuro)Toxicology—Ireland—2022—Current Protocols—Wiley Online Library" n.d.). Specialized receptors, with the most prominent being a set of eyespots in the head region, are used to effectively respond to environmental stimuli. Planarians have no apparent circulatory or respiratory systems; however, they do possess muscles, which line the inner wall of the body in longitudinal, diagonal, and circular orientations (Fig. 10.1a).

Planarians are ubiquitous organisms, found in nature at all latitudes and at different altitudes. As an example, background freshwater planarians of the genus *Dugesia* are distributed in a major part of the Old World and Australia, although until recently only very few species were known from China (Song et al. 2020). Planarians can be found in freshwater rivers, streams, or ponds, but they have colonized marine waters and lands too. They usually reside on the underside of rocks, leaves, and other objects in shallow waters as well as in aquatic vegetation (i.e. algae or aquatic plants) (Schockaert et al. 2008).

In addition to geographical diversity, planarians show remarkable variability in their position in the trophic chain. There are omnivorous detritivore species that feed on decomposing biological material, both of animal and plant origin, as well as carnivorous predatory species that prey on other invertebrates (Boll et al. 2023). This ability to adapt to different food sources and habitats makes them extremely versatile and resilient organisms.

Considering their preferred habitat, as well as their diet, freshwater planarians can reproduce both asexually and sexually (Fig. 10.1b). Sexual worms have hermaphroditic reproductive organs. In contrast, asexual worms regenerate lost body parts after fission without developing reproductive organs (Pearse et al. 1987). As for many other animal species, changes in environmental conditions can induce a switch from asexual to sexual reproduction. It is well established that turbellarian species, and among them freshwater planarians, can post-embryonically produce germ line cells from pluripotent stem cells called neoblasts, which enables some of them to switch between an asexual and a sexual state in response to environmental changes (Nakagawa et al. 2018). When asexual worms switch to a sexual state because of environmental stimuli (Curtis 1902; Jenkins 1967; Kenk 1937; "Proceedings of the United States National Museum" n.d.), which can be both because of a natural or anthropogenic stimulus, they differentiate hermaphroditic reproductive organs from the neoblasts (Lange 1968; Newmark and Sánchez Alvarado 2000; Orii et al. 2005; Saló and Baguñà 1985, 2002; Sanchez-Hernandez 2006; Shibata et al. 2012).

The strains of freshwater planarians most commonly used for research and education are summarized in Table 10.1: *Schmidtea mediterranea*, *Schmidtea polychroa*, and *Dugesia japonica* (Sluys and Riutort 2018), which in addition to excellent regenerative abilities, are easy to culture in the lab. In recent decades, *S. mediterranea* has emerged as the species of choice for modern molecular biology research,

Table 10.1 Most common strains of planarians used in research, and their applications

Planarian strain	Reproduction	Applications	References
S. mediterranea	Asexual and sexual	Aging, ciliary assembly and motility, chemical toxicity, teratogenicity and tumorigenicity, neurotoxicity, carcinogenicity, regeneration	Gentile et al. (2011), Hagstrom et al. (2016), King and Patel-King (2016), Ofoegbu et al. (2019), Oviedo et al. (2008a), Sheiman and Kreshchenko (2015) and Stevens et al. (2015)
S. polychroa	Asexual and sexual	Toxicity, regeneration, embryonic development	Mazzitelli et al. (2018), Monjo and Romero (2015) and Póti et al. (2024)
D. japonica	Asexual and sexual	Chemical toxicity, teratogenicity and tumorigenicity, neurotoxicity, neurodegenerative diseases	Hagstrom et al. (2016), Inden et al. (2004), Nishimura et al. (2011), Wu et al. (2014), Yuan et al. (2018) and Zhang et al. (2014)

due to its diploid chromosomes and the availability of both asexual and sexual strains (Newmark and Alvarado 2002). Other common species used are the blackish *Planaria maculata* and *Girardia dorotocephala*. They have been divided into categories considering geographical distribution, position in the food chain, lifestyle, and their way of reproduction.

1.2 Regeneration and Wound Healing: Insight from a Molecular and Cellular Point of View

Planarians are mostly known for their ability to regenerate any missing body region after an injury, a process that usually takes from 1 to 2 weeks. As anticipated in the previous paragraph, pluripotent stem cells called neoblasts are the source of their amazing power or adaptation and regeneration. Neoblasts are widely distributed throughout the body, consisting of 25–30% of the body cells in the whole worms (Baguñà 2012). Planarian neoblasts are heterogenous cells with a high nucleo-cytoplasmic ratio, with a cell diameter of about 5–10 μm. Also, a commonly recognized characteristic of planarian stem cells is that the cytoplasm contains more chromatoid bodies, which enable the neoblasts to respond quickly to external stimuli (Rossi et al. 2008). Upon injury, 6 and 48 h post-amputation, these cells can respond with two peaks of mitotic cell numbers. The first peak of neoblast proliferation occurs at 6 h, involving a wide range of neoblasts distributed throughout the body and being responsible for initiating the wound healing and regeneration process. The second peak occurs specifically following major injuries (amputation) and is located near the wound site (Ge et al. 2022).

Briefly, stem cells do migrate to the site of the injury, and then differentiate into progenitor cells able to build a transparent tissue called blastema (Fig. 10.2). Positional Control Genes (PCGs) are then expressed from the blastema itself, to

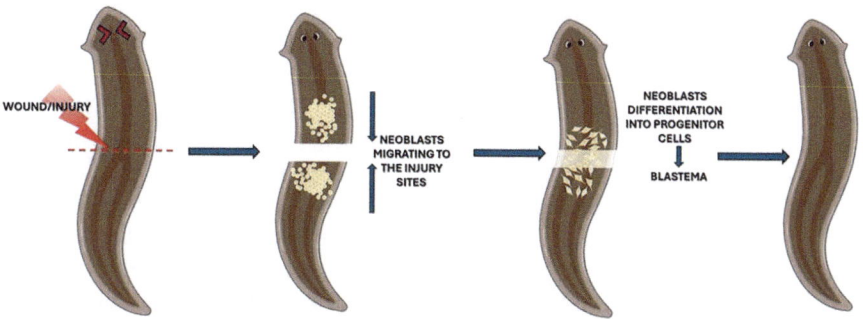

Fig. 10.2 Exemplary diagram of the regeneration process of a freshwater planarian following amputation. (Created with Biorender)

determine the polarity of the "rebuilding" process, i.e. the regeneration of the appropriate body-missing fragment. Biological processes (such as cell cycle, apoptosis, and cell proliferation) regulate the migration, oriented differentiation, and self-renewal of neoblast during planarian regeneration, together with various signaling pathways and molecular mechanisms (Ge et al. 2022).

Due to their being significantly larger than nematodes, planarian regeneration can be easily observed without specialized equipment. This feature, among the other previously cited, has made planarians popular for chemical exposure studies to investigate different endpoints, such as neurodevelopment (Ireland and Collins 2022).

2 Planarians as Invertebrate Model Species

As previously mentioned, freshwater planarians possess some unique biological characteristics that make them attractive as experimental models. The primitive form of the central nervous system and notable capability to regenerate tissues are the two most remarkable features (Wu and Li 2018), and this extraordinary plasticity and regenerative capacity, together with their sensitivity to toxins, provide unique opportunities for investigating the reorganization of the different tissues and organs after injury (Sarnat and Netsky 1985), as well as the effects of toxins in general. For years, freshwater planarians have been implied in studies related to many different fields (Table 10.1): aging (Oviedo et al. 2008b); pharmacology/drug abuse (Pagán 2017; Pagán et al. 2009; Rawls et al. 2009); human diseases (Lemieux and Warren 2012; Prokai-Tatrai et al. 2012); ciliary assembly and motility (King and Patel-King 2016); chemical toxicity, teratogenicity, and tumorigenicity (Hagstrom et al. 2016); carcinogenicity (Stevens et al. 2018), stem cell biology and regenerative medicine (Gentile et al. 2011; Saló and Agata 2012; Sheiman and Kreshchenko 2015; Simanov et al. 2012) and neurotoxicology (Hagstrom et al. 2015, 2016).

The planarians evolved before the divergence of the phylogenetic line leading to vertebrates (Sarnat and Netsky 1985), so even if their anatomical structure is quite simple, they share many of the same neurotransmitters and neuronal populations with the mammalian brain (Hagstrom et al. 2016). Such a similarity, together with their low-cost maintenance and high rate of reproduction, makes these flatworms the best candidates for studies concerning neuro-regeneration (Hagstrom et al. 2016). Indeed, one of the most challenging fields is the study of the effects of chemicals on the brain and, more in general, on the nervous system. To fully understand how certain toxins act, and interact, at the neurological level it is fundamental to explore their effects on a living organism (Ireland and Collins 2022). Since traditional mammalian testing is too time- and cost-intensive to keep up with the large number of environmental chemicals needing assessment, the use of this sentinel species might be a useful way to understand the effects of chemicals on nervous system development and function on a system level (Ireland and Collins 2022). Currently, the goal of the researchers devoted to this field of science is to find alternative ways to avoid experimentation on upper vertebrates, such as frogs and fishes, relying on the so-called "non-animal models" (Ireland and Collins 2022). Planarians, together with other species such as the earthworm *Caenorhabditis elegans* and the embryos and larvae of the zebrafish *Danio rerio*, are believed to be very good candidates to fulfill this target.

2.1 Planarians' Maintenance in Laboratory Conditions

Planarians are easy to maintain and propagate, have inexpensive needs, and are reasonably macroscopic (1 mm to 1 cm in length). These characteristics make them excellent organisms to use in both complex academic research and hands-on teaching laboratories (Dean and Duncan 2020). It is crucial to keep a stable, healthy population of animals in a consistent environment to avoid inter-animal variability and modifier effects that can mask true phenotypes from experimental perturbation (Oviedo et al. 2008b). For this purpose, the variables required for maintenance must be strictly controlled for each strain. Although different species usually require different approaches for maintenance, mostly in terms of water composition and food typology, many of the procedures (such as feeding strategy, food supply, and environmental condition setup) can be applied to other species.

Planarians can be directly collected in nature, since they usually colonize ponds and streams (under rocks or attached to other surfaces), or (for some strains) ordered from commercial sources (Table 10.2). To the best of our knowledge, there are no commercial suppliers for either *S. mediterranea* or *D. japonica*. An alternative way to get some specimens and install a colony under laboratory conditions is to obtain them from research laboratories working with the desired species (Table 10.3).

Table 10.2 Commercial sources of planarians

Company name	Company website
Carolina Biological Supply Company	https://www.carolina.com/
Southern Biological	https://www.southernbiological.com/
Home Science Tools	https://www.homesciencetools.com/
Blades Biological Ltd	https://blades-bio.co.uk/

Table 10.3 Some of worldwide laboratories working with planarians

Principal investigator	Lab link
Gianni Ciofani	https://www.iit.it/web/smart-bio-interfaces
Takashi Gojobori	https://www.kaust.edu.sa/en/study/faculty/takashi-gojobori
Michael Levin	https://www.drmichaellevin.org/publications/planaria.html
Nico Michiels	https://insidescientific.com/profile/nico-k-michiels/
Phillip Newmark	https://crb.wisc.edu/staff/newmark-phillip/
Robert Raffa	https://neumentum.com/leadership/
Peter Reddien	https://reddienlab.wi.mit.edu/
Leonardo Rossi	https://unimap.unipi.it/cercapersone/dettaglio.php?ri=4233
Teresa Adell	http://www.ub.edu/planariabcn/people/teresa-adell
Alejandro Sánchez Alvarado	https://planaria.stowers.org/
Eva-Maria Schoetz	https://schoetzlab.ucsd.edu/
Josien van Wolfswinkel	https://vanwolfswinkellab.org/?page_id=509
Ricardo Zayas	https://biology.sdsu.edu/faculty/zayas.html
Nestor Oviedo	https://sites.ucmerced.edu/oviedolab
Christian Petersen	https://sites.northwestern.edu/petersenlab/
Jason Pellettieri	https://www.pellettierilab.com/projects
Jochen Rink	https://www.mpinat.mpg.de/rink
Francesc Cebrià	http://www.ub.edu/planariabcn/people/francesc-cebria
Aziz Aboobaker	https://www.biology.ox.ac.uk/people/aziz-aboobaker
Marta Riutort	https://www.researchgate.net/lab/Marta-Riutort-Lab
Kerstin Bartscherer	https://www.hubrecht.eu/research/bartscherer-tissue-regeneration/
Jordi Solana	https://lsi.exeter.ac.uk/groups/solana-group/
Cristina González-Estévez	https://www.leibniz-fli.de/research/former-research-groups/gonzalez-estevez/
Karen Smeets	https://www.uhasselt.be/en/who-is-who/detail/karen-smeets
Luca Gentile	www.ibmt.fraunhofer.de
Omri Wurtzel	https://www.wurtzellab.sites.tau.ac.il/
Juliette Azimzadeh	https://www.insb.cnrs.fr/fr/personne/juliette-azimzadeh
Prasada Abnave	https://www.prasadabnavelab.com/

Updated and modified from Oviedo et al. (2008b)

As anticipated, the maintenance of the colonies under laboratory conditions mainly focuses on two points: water composition and food supply. The aqueous medium in which each species lives differs slightly. The natural habitat (i.e. streams and ponds) is usually characterized by different salt concentrations, and to have healthy colonies the goal should be to reproduce as faithfully as possible in these conditions. In each case, the medium should be freshly prepared (e.g., weekly) because the pH tends to drift, and salts may precipitate with time. Planarians can be kept long-term in any food-safe, Bisphenol A-free, water-tight container with a lid and a relatively flat bottom. Containers commonly used are plastic/glass food-storage boxes and small fish aquaria with an ideal volume of at least 2 liters (Oviedo et al. 2008b). For short periods, or during experimental investigations, also petri dishes or wells can be used. The worms must be constantly submerged in artificial freshwater (18–22 °C, pH ranging from 6.9 to 8.1) (Dean and Duncan 2020), avoiding keeping the colonies under direct light. Planarians are nocturnal, and for this *S. mediterranea* and *D. japonica* are best maintained in dark environments (although they are exposed to light during feeding and cleaning). A refrigerated incubator or a

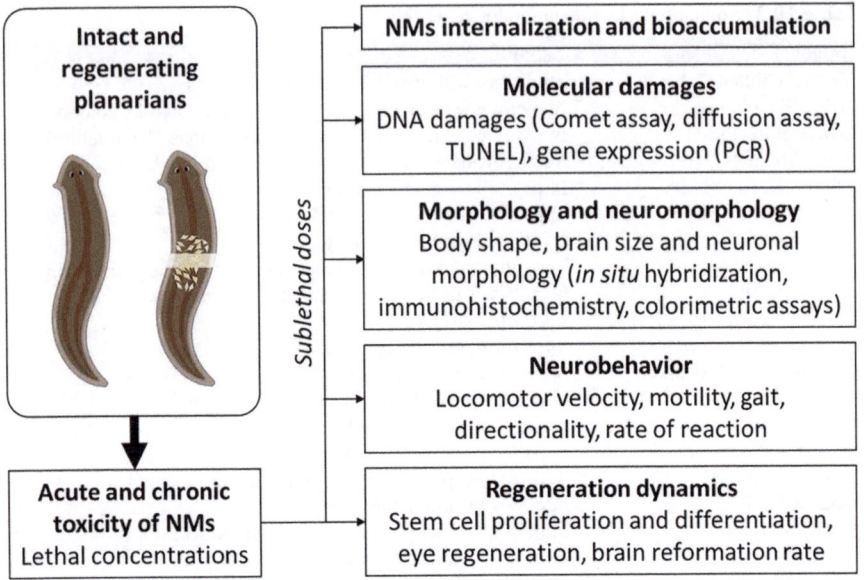

Fig. 10.3 Overview of the different assays available for evaluating toxicity in planarians at different levels of organization

room with a controlled temperature and light (Fig. 10.3b) are the preferable solutions for storing the animals. Worms kept in dark cabinets, exposed to light only during feeding and cleaning operations, develop an increased aversion to light. This must be considered when planning experiments (especially behavioral ones). Higher temperatures (~25 °C) are acceptable to planarians, but a warmer environment encourages bacterial growth, likely to result in infections that can cause particular concern in experimental animals that have been cut. Indeed, regenerating worms may have higher mortality rates when kept in warmer environments (Oviedo et al. 2008b).

As mentioned above, planarians are resilient animals that can easily adapt to prolonged periods with relatively still water. Considering this, healthy colonies should be grown in water changed every 2–5 days, but this duration can safely be extended for 1 or 2 weeks in case of experimental need. A typical indicator of the wellness of the population is the smell, as healthy cultures should be virtually odorless. To avoid algae growth or bacterial infection, the bottom and the walls of the tanks should be regularly cleaned by simply smearing them gently with the fingertips. Planarians are very sensitive to chemicals, so the use of any kind of detergent is strictly forbidden.

Freshwater planarians used in research are typically maintained on a diet of fresh organic chicken/calf/beef liver or boiled egg yolk in the laboratory. These diets are inexpensive, relatively easy to prepare, and promote rapid growth. The liver can be given directly cut into small pieces (approximately 2 cm), after the removal of all visible veins and connective tissue as well as any fatty inclusions, or after a procedure of mincing (Oviedo et al. 2008b). In both cases it must be stored in refrigerated conditions after the preparation, at −20 °C in the first case and at −80 °C in the second, defrosting it just right before feeding the animals.

2.2 Studies in Ecotoxicology

Freshwater planarians can serve not only as alternative models for chemical toxicity screening in laboratories but also as potential bioindicators for the quality of freshwater environments (Wu and Li 2018). They can be used as sentinel species able to detect the overall quality of the aquatic environment, due mostly to anthropogenic activities. As an example, global warming-induced environmental changes have resulted in salinity fluctuations that markedly impact aquatic organisms. It was assessed that the reproduction rate was altered in *Dugesia bursagrossa* specimens, with an increase in the asexual one and a decrease in the sexual ones (Harrath et al. 2024). Planarians also serve as promising bio-indicators for freshwater ecosystems as they undergo bioaccumulation, have sensitive functional and morphological endpoints, and play different roles (predator, prey) in the trophic aquatic system (Leynen et al. 2024). As planarians are aquatic, they can also be easily exposed under laboratory conditions to chemical solutions, which are absorbed through the skin and/or the pharynx (Balestrini et al. 2014; Kapu and Schaeffer 1991). The relatively high sensitivity of invertebrates to environmental chemicals makes them promising screening tools for predicting the acute/chronic toxicities of pollutants to mammals (Calleja et al. 1994; Neuhauser et al. 1985, 1986). As the basic knowledge of the biology and physiology of invertebrate species is sufficient to assess the effects of a given chemical and the differences between invertebrates and vertebrates, replacing vertebrate animals with invertebrates in toxicity testing can likely yield more insightful results (Lagadic and Caquet 1998). Among the most common fields where planarians might be employed as sentinel species, the assessment of aquatic environment pollution is at the top of the row. For years, it was highlighted how the sensitivity of these animals makes it possible to discriminate the presence of chemicals (both inorganic and organic) by analyzing parameters such as survival rate, feeding rate, and locomotion.

2.3 Neurotoxicity Studies

Similarly, planarians are insightful models for toxicity studies involving chemicals. In this scope, several techniques were developed to evaluate neurotoxicity and developmental neurotoxicity, thus offering similarly great potential regarding nanosafety studies.

In terms of locomotion for instance, exposure to certain chemicals induced a change from "gliding", a ciliary-driven motion used normally by planarians, to "peristalsis" or "scrunching", respectively driven by musculature and asymmetric body shape oscillations (Cochet-Escartin et al. 2015). Modifications of motility, gait, directionality, rate of reaction, and body shape were also demonstrated as proof of neurotoxicity (Hagstrom et al. 2016; Inoue et al. 2015; Ireland et al. 2020). The development of automated techniques based on algorithms combined with multi-well screening provides quantitative data on motility and shape descriptors in a more reliable and high throughput way (Ireland et al. 2022; Werner et al. 2014; Zhang et al. 2019). Machine learning is also increasingly used for evaluating eye regeneration as a marker of toxicity (Zhang et al. 2019).

Apart from mobility and morphological readouts, more specific tests were developed to evaluate the neurotoxicity of chemicals in planarians. Changes in neuronal morphology or brain size due to chemical exposure can be evaluated via *in situ* hybridization or immunohistochemistry with specific neuronal markers, as done for evaluating pesticide toxicity (Hagstrom et al. 2015). Cellular damages caused by exposure to metals such as copper and cadmium were assessed using terminal deoxynucleotide transferase deoxyuridine triphosphate (dUTP) nick end labeling (TUNEL) staining and Comet assay, respectively linked with cell death and deoxyribonucleic acid (DNA) damages (Majid et al. 2022). Pesticides and environmental pollutants effects were also extensively investigated via colorimetric assays, through the quantification of neuronal enzymes such as acetylcholinesterase (AChE) (Ireland et al. 2022), or enzymes related to oxidative stress (Zhang et al. 2014). Gene expression of target enzymes can be studied via quantitative polymerase chain reaction (PCR) to investigate toxicity pathways after exposure to berberine (Balestrini et al. 2014) or diazinon and physostigmine (Hagstrom et al. 2018), which are inhibitors of AChE.

3 Nanoparticles and Nanomaterials: New Challenges for Health and Environment

Nanotechnology has been defined as research and development at the atomic, molecular or macromolecular scales. Nanoparticles (NPs), which are the building blocks for nanotechnology, comprehend all those particles with at least one dimension <100 nm (Biswas and Wu 2005). They are extensively utilized in commercial products and industries owing to their tiny size and unique properties. During the last years, the ability to synthesize and manipulate such materials was developed, creating more complex structures to cover a broader range of applications. Nanomaterials (NMs) are defined as materials of which 50% or more of the constituent particles have one or more external diameters in the size range of 1–100 nm (Leynen 2021). As the particle size decreases, the number of atoms on the particle surface increases. This leads to an increased chemical reactivity, making these materials promising tools for many applications, despite potentially increasing their toxicity (Leynen et al. 2024). Due to their size and surface area, NM exhibit unique chemical, physical, electrical, and mechanical properties that are more pronounced compared to bulk material (Joudeh and Linke 2022). In addition, the same nanomaterial can occur in different shapes or have different surface charges and modifications, properties that affect particle uptake and toxicity. NMs are very promising, the possibilities given by nanoscience and technology are intensive (George et al. 2023), and their rapid development promotes their applications in various fields (Xie et al. 2023). As anticipated, the peculiar properties of NPs, and consequently of NMs, might not just possess a positive lapel. On the contrary, the high reactivity correlated with the high surface/volume ratio can result in negative effects on the biota, from cellular to organism level. The effects, also, become unpredictable when the particles/materials are released into the environment since they can interact with other chemicals potentially enhancing their toxicity or increasing their own. As an example, nanosized titanium dioxide has been demonstrated to be responsible for a Trojan-horse effect by increasing 2,3,7,8-tetrachlorodibenzo-p-dioxin (TCDD) bioaccumulation, associated with synergistic or antagonistic effects depending on experimental conditions, as well as cell/tissue and parameter measured (Canesi et al. 2014; Nigro et al. 2015).

NPs and NMs can be present in the aqueous phase or can aggregate and precipitate due to their physicochemical properties, making it important to assess their interaction with different compartments of the ecosystem, choosing different endpoints to have a spectrum as broad as possible of their potential toxicological effects. Nanotoxicity patterns can be derived from the biological responses elicited by different NMs, in *in vitro* and *in vivo* experiments. The few toxicity tests that are described in the literature were performed, in most cases, *in vitro*, whereas the number of *in vivo* tests is still quite limited, which can be explained by the cost differences and by the early stage of development of NMs. The use of laboratory animals for *in vivo* tests is a matter of debate, and the search for new, robust and adequate methods to determine the ecotoxicity of NMs without resorting to traditional laboratory animal testing, is a great challenge (Kustov et al. 2014).

The choice of the biological target for nanotoxicity tests is governed, first, by the field of NPs and NMs applications. For applications in the industrial and technology sectors, the utmost concern is the environment, so marine and/or freshwater organisms are considered to be priority targets because they are the first point of contact with waste NMs released into the environment (Kustov et al. 2014). The aquatic compartment is the main receptacle for "nano-waste". The sources are multiple, going from wastewater deriving from productive processes to the simple discharge of customer products such as cosmetics, pharmaceuticals and food. Taking this into account, together with the need for *in vivo* trials on alternative animal models, it appears clear that freshwater flatworms might be good sentinel species for assessing NPs and NMs potential toxicity. Hydrobionts in general are considered to be priority targets since they are the first point of contact with waste NMs released into the environment. Such organisms are at the lower levels in the trophic chain, therefore, their use to measure the nanotoxicity is quite indicative and predictive of potential environmental toxicity (Canesi et al. 2014; Kustov et al. 2014; Nigro et al. 2015). An overview of the different techniques developed is depicted in Fig. 10.3.

Both in the water phase and the sediments, NPs will mix and interact with other environmental pollutants, such as organic xeno-biotics and metals, that may interact in many ways (i.e. additively, synergistically or antagonistically) to induce responses at different levels of biological organization in aquatic organisms. Developing organisms and tissues are often more susceptible to environmental pollutants, including nanoparticles. Stem cells, which undergo rapid cell division, are highly vulnerable to environmental chemicals, impairing their differentiation and development (Pearson and Alvarado 2010). As a benthic animal in the water body, planarians use their motile cilia to glide over sediment surfaces (Rompolas et al. 2009) and interact with the deposited NPs and eventually NMs, which may make them an excellent model for investigating the effects of NPs on aquatic animals (Xie et al. 2023). Planarians have been used for assessing the nanotoxicity of silver NPs (Leynen et al. 2024), micro- (Cesarini et al. 2023; Gambino et al. 2020; Gao et al. 2022) and nanoplastics (Cesarini et al. 2023; Gambino et al. 2020; Gao et al. 2022), iron oxide NPs (Gentile 2019), silver sulfide NPs (Silva et al. 2022), titanium dioxide NPs (Leynen et al. 2024) and ceria NPs (Salvetti et al. 2020; Xie et al. 2023). This research mainly focuses on assessing the regenerative potential of the animals after exposure (alone or in co-exposure with other stressors) to the different types of NPs. In some cases, it was highlighted as a protective effect of the materials. In some others, a toxic outcome was underlined. Due to this uncertainty, it is fundamental to apply a broad spectrum of biomarkers to assess the nano-safety of the materials employed.

3.1 Nanoparticles and Nanomaterials-Induced DNA Damage

As previously stated, the toxicity of NPs and NMs can affect many different levels of organization, from population to molecular ones. Among this latter category, DNA damage can be considered as one of the most critical outcomes, since DNA damage is the central mechanism through which changes in the genome occur (Fig. 10.4).

DNA integrity is a fundamental requirement for preventing mutagenic and carcinogenic events, which can verify if DNA damage is not properly repaired (Bernardeschi et al. 2021). The genome undergoes continuously different types of DNA lesions (abasic sites, mismatches, interstrand crosslinks, or single-stranded (SSB) and double-stranded breaks (DSB)), and consequently, cells have evolved specialized DNA damage response mechanisms to sustain genome integrity (Fig. 10.4) (Carusillo and Mussolino 2020). Spontaneous errors during the replication of DNA might occur and, similarly, Reactive Oxygen Species (ROS) or Reactive Nitrogen Species (RNS) are generally produced as byproducts of multiple

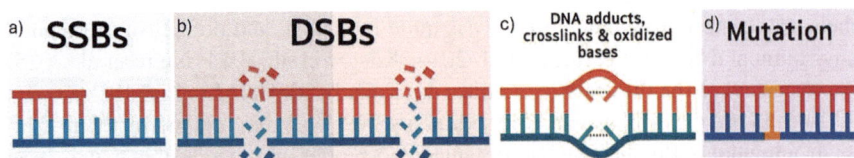

Fig. 10.4 DNA damages: (**a**) single-strand break (SSBs), (**b**) double-strand break (DSBs), (**c**) bulk adducts crosslinks and oxidized bases, (**d**) mutation. (Adapted from Moon et al. 2023)

physiological activities in diverse subcellular sites (Cadet and Wagner 2013). These naturally occurring mechanisms usually are well regulated by the cell, but the repair pathways can be altered by environmental stressors leading to a failure in the repairing processes needed for the survival of the cell. Among the numerous biological responses proposed in the last decades, those based on variations at the molecular and cellular levels represent the earliest signals of environmental disturbance and have been increasingly applied in ecotoxicological investigations (Guidi et al. 2010).

NPs, due to their small size and their peculiar physic-chemical properties, have been found to provoke DNA damage both directly (by interfering with the molecule) and indirectly (for example, through the generation of ROS). So, it is mandatory to clarify as much as possible not just *in vitro* but also *in vivo* the safety of the nanosized materials used in the manufacturing processes.

The Comet assay is used to microscopically detect primary repairable, and therefore reversible, DNA lesions (DNA SSBs, DSBs, alkali labile sites) at the level of a single cell, providing an instantaneous picture of DNA primary damage (Bernardeschi et al. 2021; Collins et al. 2023; Guidi et al. 2010; Tice et al. 2000). It is a simple, reliable technique that can be applied to any kind of nucleated cell, with just slight modifications according to the specimens investigated. This assay is also known as Single Cell Gel Electrophoresis (SCGE), and as the name suggests, its peculiarity relies on the fact that the single, naked nuclei undergo electrophoresis typically performed under alkaline conditions (the buffer pH is generally equal or above 13, but it can be lowered depending on the type of damage to be highlighted). When analyzed, these nuclei may appear comet-shaped, with the length and/or the intensity of the fluorescence of the comet "tail" directly proportional to the number of DNA fragments (Fig. 10.5). With a slight modification, i.e. the omission of the electrophoresis run, with the same protocol, it is possible to assess also the number of apoptotic nuclei following the exposure. The so-called "diffusion assay" (or "halo assay") allows to detect the level of DNA damage in terms of cell death, which cannot be identified with the Comet assay due to the very small molecular weight of the fragments that typically characterize this process (Salvetti et al. 2015, 2020).

Planarians have been proven to be suitable organisms also for this type of *in vivo* investigation (Gajski et al. 2019), at least in its alkaline version, since apart from

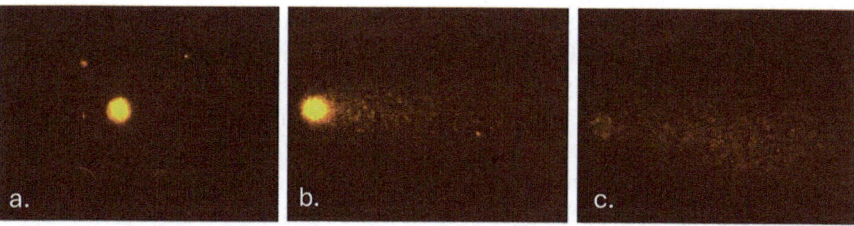

Fig. 10.5 Representative images of nuclei showing different levels of DNA migrated in the tail. (**a**) Undamaged nucleus, with intact chromatin; (**b, c**) "Comet-shaped" nuclei, showing the presence of DNA primary damage. The longer the tail, the higher the extent of DNA fragmentation

being easy and cheap to maintain, they can be acute or chronically exposed to selected NMs and then dissociated to single cells (Leynen et al. 2024; Salvetti et al. 2015, 2020). Controversial results have been obtained, depending on the experimental conditions and the type of NPs or NMs used. For example, some particles have been proven to exert protective action against physical stressors, such as infrared radiation. Indeed, a reduction of nuclei with undamaged DNA was observed after irradiation of animals treated with ceria nanoparticles (Salvetti et al. 2020). Also, boron nitride nanotubes were proven to be harmless to DNA molecule (Salvetti et al. 2015), while titanium dioxide caused DNA fragmentation after chronic exposure (Androschuk 2013). These results, which are just a part of those available online, prove that planarians are both versatile and sensitive since they can highlight properly the effect of different types of materials.

3.2 NP and NM Internalization

One of the main objectives of *in vitro* and *in vivo* studies concerning NPs and/or NMs should be to assess their uptake and, eventually, their distribution inside the target. This type of investigation can give fundamental information about the behavior of the particles once inside the biota and about the main routes of exposure. To assess the safety of NPs and NMs it appears fundamental to explore this process since the internalization of these materials can translate into metabolic issues at different levels. Understanding the exposure, toxicokinetics and bioaccumulation potential of NPs has become crucial for assessing their environmental risk (Petersen et al. 2019). The potential bioaccumulation of NPs and NMs has been widely investigated (Handy et al. 2018; Kuehr et al. 2021; Petersen et al. 2019; Wang and Liu 2022), being the majority of the studies conducted under standard laboratory conditions (Kuehr et al. 2021). Such studies can provide important mechanistic information on the NP uptake by individual organisms but do not reflect the complexity of environmental systems (Colman et al. 2014; Lead et al. 2018).

Since planarians undergo bioaccumulation, have sensitive functional and morphological endpoints, and play different roles (predator, prey) in the trophic aquatic system (Wu and Li 2018), they can serve as bio-indicators of particle intake in freshwater ecosystems. The assumption of particles can happen through food or by dermal contact, depending on the type of material (Salvetti et al. 2021; Silva et al. 2022). In laboratory exposure, NPs and/or NMs can also be directly injected into the specimens (Salvetti et al. 2015). Once inside the organism, regardless of the way of assumption, NPs are mostly accumulated in the gut tissues of organisms. The passage through the epithelial surface, and its consequent transportation via the circulatory system, is not immediate (Mao et al. 2016a, b). Once ingested, they remain in the digestive tracts of organisms for days up to weeks, allowing their transfer to the food web (Lammel et al. 2019; Mei et al. 2021; Zhao et al. 2017). Biomagnification amplifies the concentration of the chemicals at upper trophic levels, and NPs are no exception. For this reason, understanding the mechanisms that stand beyond this process is mandatory, although the potential adverse effects of NPs are hard to predict. It appears fundamental, with this background, to integrate quantitative studies to produce a more comprehensive and objective evaluation of the differential biological responses triggered by NPs exposure (Wang and Liu 2022).

3.3 Neurobehavioral Effects

There are several reasons to consider the experimental analysis of planarian behavior. The most important is that, although the evolutionary lineages of planaria and humans diverged long ago, the neurobiology of a planarian is surprisingly similar to that of vertebrates, and it is considered one of the first organisms to have a "true brain" (Pagan and Pagan 2014). Also, taking into consideration the unique ability to regenerate that characterizes these animals, it may be possible to train a planarian to engage in a target behavior, then dissect the planarian and determine if the head or tail segment retained prior training (Deochand et al. 2018). To properly evaluate changes in the behavior of planarians, a clear understanding of the physiological spectrum of movements that characterize these animals is required. Typically, at least under laboratory conditions, planarians do prefer to avoid moving in open spaces, showing a preference for the walls of the container they are hosted in (Akiyama et al. 2015). For this reason, tracking procedures should be performed in Petri dishes with a maximum of 1 cm of shallow depth. In these conditions, the upward lifting of the head is minimized, together with the fact that this action is usually made in response to food or obstacles encountered (Talbot and Schötz 2011). When a planarian is placed in a novel environment, there is an initial exploratory phase. After 5–30 min without any kind of stimuli, planarians mostly remain motionless, with the head "scrunched" into the body. This behavior can also be induced by physically touching the planarian's head and presumably confers some protection to the brain region (Deochand et al. 2018). Another typical behavior is that after dissection, both head and tail segments typically stay close to the concave curves of the outer walls of containers. In this position, they are less likely to engage in head-turning unless out in the open (Akiyama et al. 2015). One of the protocols mostly used when assessing planarians' (neuro)behavior is aimed to measure the locomotor velocity of the animals, after placing them individually into a clear plastic Petri dish containing planarians' media at 19 °C and located over graph paper with grid lines spaced 0.5 cm apart. Locomotor velocity was quantified as the number of grid lines crossed or recrossed by each planarian per minute over a 10-min observation (Raffa et al. 2001). Nowadays, automated systems have been developed to track the movements and speed of animals. Since typically a higher number of specimens can be scored, these devices can collect more reliable and affordable data (Liu and Ella 2024).

The analysis of planarians' behavior can be a useful tool for understanding the potential toxicity of chemicals, both classical and emerging. NPs and NMs are no exception. A recent study found that Ag NPs can affect the locomotion of freshwater planarians *S. mediterranea* by inducing neurotoxicity and disorder in neurodevelopment (Leynen et al. 2019). Other researchers found that planarians exposed to ZnO

NPs exhibited obvious behavioral changes and hyperkinesia, in some cases together with alterations of body morphology ("https://www.thaiscience.info/Journals/Article/NUSJ/10991464.pdf" n.d.). As another example, it was found that n-CeO$_2$ and Ce^{3+} can reduce the mobility of regenerating planarians, with a consequent locomotive inhibition which be attributed to a neurotoxic effect exerted by the NPs (Xie et al. 2023). In other cases, on the contrary, it was found that the exposure to NPs (particularly, nanoplastics) did not affect the behavior of the planarians at all (Cesarini et al. 2023). These data show how planarians are a useful tool to evaluate the neuro-safety of the NPs and NMs, being able to discriminate between one type of material and the other.

4 Conclusion

Due to their extraordinary regeneration capabilities, freshwater planarians have recently gained in popularity for neurotoxicity studies as an alternative to standard animal models. Their use remains limited in the context of nanosafety, even though this model can be used to assess the effects of NM exposure on the nervous system. Indeed, they represent an easy and quick way of assessing toxicity and provide direct information on induced neurobehavioral and developmental effects. However, their use is still limited by a lack of robustness and standardization in terms of species selected, maintenance conditions, exposure to NMs, screening techniques and data analysis. Improvements in these areas could greatly improve the cost-effectiveness and reliability of the model, thus paving the way for more complex studies, for instance on pathological models as planarians are good candidates for tissue transplantations (Rojo-Laguna et al. 2019).

Acknowledgments The authors acknowledge the European Health and Digital Executive Agency (iCare, 101092971). This work is part of the "Technologies for Sustainability" Flagship Program of the Italian Institute of Technology.

References

Akiyama Y, Agata K, Inoue T (2015) Spontaneous behaviors and wall-curvature lead to apparent wall preference in planarian. PLoS One 10:e0142214. https://doi.org/10.1371/journal.pone.0142214

Androschuk A (2013) The genotoxic effects of titanium dioxide nanoparticles on Dugesia dorotocephala genomic DNA [WWW Document]. ERA. https://doi.org/10.7939/r3-pas8-0e62

Baguñà J (2012) The planarian neoblast: the rambling history of its origin and some current black boxes. Int J Dev Biol 56:19–37. https://doi.org/10.1387/ijdb.113463jb

Balestrini L, Isolani ME, Pietra D, Borghini A, Bianucci AM, Deri P, Batistoni R (2014) Berberine exposure triggers developmental effects on planarian regeneration. Sci Rep 4:4914. https://doi.org/10.1038/srep04914

Bernardeschi M, Guidi P, Palumbo M, Genovese M, Alfè M, Gargiulo V, Lucchesi P, Scarcelli V, Falleni A, Bergami E, Freyria FS, Bonelli B, Corsi I, Frenzilli G (2021) Suitability of nanoparticles to face benzo(a)pyrene-induced genetic and chromosomal damage in M. galloprovincialis. An In Vitro Approach Nanomater 11:1309. https://doi.org/10.3390/nano11051309

Biswas P, Wu C-Y (2005) Nanoparticles and the environment. J Air Waste Manage Assoc 55:708–746. https://doi.org/10.1080/10473289.2005.10464656

Boll PK, Rossi I, Amaral SVD, Leal-Zanchet AM (2023) Regeneration in a neotropical land planarian (Platyhelminthes, Tricladida). Neotrop Biol Conserv 18:163–176. https://doi.org/10.3897/neotropical.18.e103357

Cadet J, Wagner JR (2013) DNA base damage by reactive oxygen species, oxidizing agents, and UV radiation. Cold Spring Harb Perspect Biol 5:a012559. https://doi.org/10.1101/cshperspect.a012559

Calleja MC, Persoone G, Geladi P (1994) Comparative acute toxicity of the first 50 multicentre evaluation of in vitro cytotoxicity chemicals to aquatic non-vertebrates. Arch Environ Contam Toxicol 26:69–78. https://doi.org/10.1007/BF00212796

Canesi L, Frenzilli G, Balbi T, Bernardeschi M, Ciacci C, Corsolini S, Della Torre C, Fabbri R, Faleri C, Focardi S, Guidi P, Kočan A, Marcomini A, Mariottini M, Nigro M, Pozo-Gallardo K, Rocco L, Scarcelli V, Smerilli A, Corsi I (2014) Interactive effects of n-TiO2 and 2,3,7,8-TCDD on the marine bivalve Mytilus galloprovincialis. Aquat Toxicol. Proceedings from the 17th international symposium on pollutant responses in marine organisms (PRIMO17) 153:53–65. https://doi.org/10.1016/j.aquatox.2013.11.002

Carusillo A, Mussolino C (2020) DNA damage: from threat to treatment. Cells 9:1665. https://doi.org/10.3390/cells9071665

Cebrià F, Kudome T, Nakazawa M, Mineta K, Ikeo K, Gojobori T, Agata K (2002) The expression of neural-specific genes reveals the structural and molecular complexity of the planarian central nervous system. Mech Dev 116:199–204. https://doi.org/10.1016/S0925-4773(02)00134-X

Cesarini G, Coppola F, Campos D, Venditti I, Battocchio C, Di Giulio A, Muzzi M, Pestana JLT, Scalici M (2023) Nanoplastic exposure inhibits feeding and delays regeneration in a freshwater planarian. Environ Pollut 332:121959. https://doi.org/10.1016/j.envpol.2023.121959

Cochet-Escartin O, Mickolajczyk KJ, Collins E-MS (2015) Scrunching: a novel escape gait in planarians. Phys Biol 12:056010. https://doi.org/10.1088/1478-3975/12/5/056010

Collins A, Møller P, Gajski G, Vodenková S, Abdulwahed A, Anderson D, Bankoglu EE, Bonassi S, Boutet-Robinet E, Brunborg G, Chao C, Cooke MS, Costa C, Costa S, Dhawan A, de Lapuente J, Bo CD, Dubus J, Dusinska M, Duthie SJ, Yamani NE, Engelward B, Gaivão I, Giovannelli L, Godschalk R, Guilherme S, Gutzkow KB, Habas K, Hernández A, Herrero O, Isidori M, Jha AN, Knasmüller S, Kooter IM, Koppen G, Kruszewski M, Ladeira C, Laffon B, Larramendy M, Hégarat LL, Lewies A, Lewinska A, Liwszyc GE, de Cerain AL, Manjanatha M, Marcos R, Milić M, de Andrade VM, Moretti M, Muruzabal D, Novak M, Oliveira R, Olsen A-K, Owiti N, Pacheco M, Pandey AK, Pfuhler S, Pourrut B, Reisinger K, Rojas E, Rundén-Pran E, Sanz-Serrano J, Shaposhnikov S, Sipinen V, Smeets K, Stopper H, Teixeira JP, Valdiglesias V, Valverde M, van Acker F, van Schooten F-J, Vasquez M, Wentzel JF, Wnuk M, Wouters A, Žegura B, Zikmund T, Langie SAS, Azqueta A (2023) Measuring DNA modifications with the comet assay: a compendium of protocols. Nat Protoc 18:929–989. https://doi.org/10.1038/s41596-022-00754-y

Colman BP, Espinasse B, Richardson CJ, Matson CW, Lowry GV, Hunt DE, Wiesner MR, Bernhardt ES (2014) Emerging contaminant or an old toxin in disguise? Silver nanoparticle impacts on ecosystems. Environ Sci Technol 48:5229–5236. https://doi.org/10.1021/es405454v

Curtis WC (1902) The life history, the normal fission and the reproductive organs of Planaria maculata. [Boston] Society [of Natural History] from the Gurdon Saltonstall Fund

Dean MRP, Duncan EM (2020) Laboratory maintenance and propagation of freshwater planarians. Curr Protoc Microbiol 59:e120. https://doi.org/10.1002/cpmc.120

Deochand N, Costello MS, Deochand ME (2018) Behavioral research with planaria. Perspect Behav Sci 41:447–464. https://doi.org/10.1007/s40614-018-00176-w

Ehlers U (1986) Comments on a phylogenetic system of the Platyhelminthes. In: Tyler S (ed) Advances in the biology of turbellarians and related Platyhelminthes. Springer Netherlands, Dordrecht, pp 1–12. https://doi.org/10.1007/978-94-009-4810-5_1

Fraguas S, Barberán S, Ibarra B, Stöger L, Cebri F (2012) Regeneration of neuronal cell types in Schmidtea mediterranea: an immunohistochemical and expression study. Int J Dev Biol 56:143–153. https://doi.org/10.1387/ijdb.113428sf

Gajski G, Žegura B, Ladeira C, Pourrut B, Del Bo C, Novak M, Sramkova M, Milić M, Gutzkow KB, Costa S, Dusinska M, Brunborg G, Collins A (2019) The comet assay in animal models: from bugs to whales—(part 1 invertebrates). Mutat Res Mutat Res 779:82–113. https://doi.org/10.1016/j.mrrev.2019.02.003

Gambino G, Falleni A, Nigro M, Salvetti A, Cecchettini A, Ippolito C, Guidi P, Rossi L (2020) Dynamics of interaction and effects of microplastics on planarian tissue regeneration and cellular homeostasis. Aquat Toxicol 218:105354. https://doi.org/10.1016/j.aquatox.2019.105354

Gao T, Sun B, Xu Z, Chen Q, Yang M, Wan Q, Song L, Chen G, Jing C, Zeng EY, Yang G (2022) Exposure to polystyrene microplastics reduces regeneration and growth in planarians. J Hazard Mater 432:128673. https://doi.org/10.1016/j.jhazmat.2022.128673

Ge X-Y, Han X, Zhao Y-L, Cui G-S, Yang Y-G (2022) An insight into planarian regeneration. Cell Prolif 55:e13276. https://doi.org/10.1111/cpr.13276

Gentile L (2019) Assessment of iron oxide nanoparticle ecotoxicity on regeneration and homeostasis in the replacement model system Schmidtea mediterranea. ALTEX 583–596. https://doi.org/10.14573/altex.1902061

Gentile L, Cebrià F, Bartscherer K (2011) The planarian flatworm: an in vivo model for stem cell biology and nervous system regeneration. Dis Model Mech 4:12–19. https://doi.org/10.1242/dmm.006692

George J, Palanisamy K, Kulandaivel S, Saravanan P, Peedika DM, Rajagopalan K, Kaliyannan MK (2023) Impact of cerium oxide nanoparticles on survivability and reproduction of earthworm Eudrilus eugeniae and its compost quality. BioNanoScience 13:1911–1921. https://doi.org/10.1007/s12668-023-01173-3

Guidi P, Frenzilli G, Benedetti M, Bernardeschi M, Falleni A, Fattorini D, Regoli F, Scarcelli V, Nigro M (2010) Antioxidant, genotoxic and lysosomal biomarkers in the freshwater bivalve (Unio pictorum) transplanted in a metal polluted river basin. Aquat Toxicol 100:75–83. https://doi.org/10.1016/j.aquatox.2010.07.009

Guo L, Accorsi A, He S, Guerrero-Hernández C, Sivagnanam S, McKinney S, Gibson M, Sánchez Alvarado A (2018) An adaptable chromosome preparation methodology for use in invertebrate research organisms. BMC Biol 16:25. https://doi.org/10.1186/s12915-018-0497-4

Hagstrom D, Cochet-Escartin O, Zhang S, Khuu C, Collins E-MS (2015) Freshwater planarians as an alternative animal model for neurotoxicology. Toxicol Sci 147:270–285. https://doi.org/10.1093/toxsci/kfv129

Hagstrom D, Cochet-Escartin O, Collins E-MS (2016) Planarian brain regeneration as a model system for developmental neurotoxicology. Regeneration 3:65–77. https://doi.org/10.1002/reg2.52

Hagstrom D, Zhang S, Ho A, Tsai ES, Radić Z, Jahromi A, Kaj KJ, He Y, Taylor P, Collins E-MS (2018) Planarian cholinesterase: molecular and functional characterization of an evolutionarily ancient enzyme to study organophosphorus pesticide toxicity. Arch Toxicol 92:1161–1176. https://doi.org/10.1007/s00204-017-2130-7

Handy RD, Ahtiainen J, Navas JM, Goss G, Bleeker EAJ, von der Kammer F (2018) Proposal for a tiered dietary bioaccumulation testing strategy for engineered nanomaterials using fish. Environ Sci Nano 5:2030–2046. https://doi.org/10.1039/C7EN01139C

Harrath AH, Aldahmash W, Mansour L, Elfaki K, Alwasel S (2024) Salinity-induced modulations in sexual and asexual reproduction in the freshwater planarian Dugesia bursagrossa (nomen nudum species): insights from microtubular cytoskeleton and oxidative stress marker analyses. J King Saud Univ Sci 36:103147. https://doi.org/10.1016/j.jksus.2024.103147

https://www.thaiscience.info/Journals/Article/NUSJ/10991464.pdf [WWW Document] (n.d.). https://www.thaiscience.info/Journals/Article/NUSJ/10991464.pdf. Accessed 12 Sept 2024

Inden M, Kitamura Y, Taniguchi T, Watanabe K, Agata K (2004) Parkinsonian model of planarian, an invertebrate flatworm. In: International congress series, the senescence-accelerated mouse (SAM): an animal model of senescence. Proceedings of the 2nd international conference on senescence: the SAM model 1260, pp 291–295. https://doi.org/10.1016/S0531-5131(03)01574-7

Inoue T, Hoshino H, Yamashita T, Shimoyama S, Agata K (2015) Planarian shows decision-making behavior in response to multiple stimuli by integrative brain function. Zool Lett 1:7. https://doi.org/10.1186/s40851-014-0010-z

Ireland D, Collins E-MS (2022) New worm on the block: planarians in (neuro)toxicology. Curr Protoc 2:e637. https://doi.org/10.1002/cpz1.637

Ireland D, Bochenek V, Chaiken D, Rabeler C, Onoe S, Soni A, Collins E-MS (2020) Dugesia japonica is the best suited of three planarian species for high-throughput toxicology screening. Chemosphere 253:126718. https://doi.org/10.1016/j.chemosphere.2020.126718

Ireland D, Zhang S, Bochenek V, Hsieh J-H, Rabeler C, Meyer Z, Collins E-MS (2022) Differences in neurotoxic outcomes of organophosphorus pesticides revealed via multi-dimensional screening in adult and regenerating planarians. Front Toxicol 4. https://doi.org/10.3389/ftox.2022.948455

Jenkins MM (1967) Aspects of planarian biology and behavior. In: Corning WC, Ratner SC (eds) Chemistry of learning: invertebrate research. Springer US, Boston, pp 116–143. https://doi.org/10.1007/978-1-4899-6565-3_9

Joudeh N, Linke D (2022) Nanoparticle classification, physicochemical properties, characterization, and applications: a comprehensive review for biologists. J Nanobiotechnol 20:262. https://doi.org/10.1186/s12951-022-01477-8

Kapu MM, Schaeffer DJ (1991) Planarians in toxicology. Responses of asexual Dugesia dorotocephala to selected metals. Bull Environ Contam Toxicol 47:302–307. https://doi.org/10.1007/BF01688656

Kenk R (1937) Sexual and asexual reproduction in Euplanaria tigrina (girard). Biol Bull. https://doi.org/10.2307/1537589

King SM, Patel-King RS (2016) Planaria as a model system for the analysis of ciliary assembly and motility. In: Satir P, Christensen ST (eds) Cilia: methods and protocols. Springer, New York, pp 245–254. https://doi.org/10.1007/978-1-4939-3789-9_16

Kuehr S, Kaegi R, Maletzki D, Schlechtriem C (2021) Testing the bioaccumulation potential of manufactured nanomaterials in the freshwater amphipod Hyalella azteca. Chemosphere 263:127961. https://doi.org/10.1016/j.chemosphere.2020.127961

Kustov L, Tiras K, Al-Abed S, Golovina N, Ananyan M (2014) Estimation of the toxicity of silver nanoparticles by using planarian flatworms. Altern Lab Anim 42:51–58. https://doi.org/10.1177/026119291404200108

Lagadic L, Caquet T (1998) Invertebrates in testing of environmental chemicals: are they alternatives? Environ Health Perspect 106:593–611. https://doi.org/10.1289/ehp.98106593

Lammel T, Thit A, Mouneyrac C, Baun A, Sturve J, Selck H (2019) Trophic transfer of CuO NPs and dissolved Cu from sediment to worms to fish—a proof-of-concept study. Environ Sci Nano 6:1140–1155. https://doi.org/10.1039/C9EN00093C

Lange CS (1968) Studies on the cellular basis of radiation lethality: I. The pattern of mortality in the whole-body irradiated planarian (Tricladida, Paludicola). Int J Radiat Biol Relat Stud Phys Chem Med 13:511–530. https://doi.org/10.1080/09553006814550581

Lead JR, Batley GE, Alvarez PJJ, Croteau M-N, Handy RD, McLaughlin MJ, Judy JD, Schirmer K (2018) Nanomaterials in the environment: behavior, fate, bioavailability, and effects—an updated review. Environ Toxicol Chem 37:2029–2063. https://doi.org/10.1002/etc.4147

Lemieux H, Warren BE (2012) An animal model to study human muscular diseases involving mitochondrial oxidative phosphorylation. J Bioenerg Biomembr 44:503–512. https://doi.org/10.1007/s10863-012-9451-2

Leynen N (2021) Does Size Matter? A Study on the in Vivo Effects of Inorganic and Organic Nanoparticles in Schmidtea Mediterranea: Doctoral Dissertation (Doctoral dissertation)

Leynen N, Van Belleghem FGAJ, Wouters A, Bove H, Ploem J-P, Thijssen E, Langie SAS, Carleer R, Ameloot M, Artois T, Smeets K (2019) In vivo toxicity assessment of silver nanoparticles in homeostatic versus regenerating planarians. Nanotoxicology 13:476–491. https://doi.org/10.1080/17435390.2018.1553252

Leynen N, Tytgat JS, Bijnens K, Jaenen V, Verleysen E, Artois T, Van Belleghem F, Saenen ND, Smeets K (2024) Assessing the in vivo toxicity of titanium dioxide nanoparticles in Schmidtea mediterranea: uptake pathways and (neuro)developmental outcomes. Aquat Toxicol 270:106895. https://doi.org/10.1016/j.aquatox.2024.106895

Liu Y, Ella L (2024) PlanariaScan: development of a video-based monitoring system on planaria learning and memory under various stressors. In: 2024 IEEE 3rd international conference on computing and machine intelligence (ICMI). Presented at the 2024 IEEE 3rd international conference on computing and machine intelligence (ICMI), pp 1–6. https://doi.org/10.1109/ICMI60790.2024.10585963

Majid S, Van Belleghem F, Ploem J-P, Wouters A, Blust R, Smeets K (2022) Interactive toxicity of copper and cadmium in regenerating and adult planarians. Chemosphere 297:133819. https://doi.org/10.1016/j.chemosphere.2022.133819

Mao L, Hu M, Pan B, Xie Y, Petersen EJ (2016a) Biodistribution and toxicity of radio-labeled few layer graphene in mice after intratracheal instillation. Part Fibre Toxicol 13:7. https://doi.org/10.1186/s12989-016-0120-1

Mao L, Liu C, Lu K, Su Y, Gu C, Huang Q, Petersen EJ (2016b) Exposure of few layer graphene to Limnodrilus hoffmeisteri modifies the graphene and changes its bioaccumulation by other organisms. Carbon 109:566–574. https://doi.org/10.1016/j.carbon.2016.08.037

Mazzitelli J-Y, Budzinski H, Cachot J, Geffard O, Marty P, Chiffre A, François A, Bonnafe E, Geret F (2018) Evaluation of psychiatric hospital wastewater toxicity: what is its impact on aquatic organisms? Environ Sci Pollut Res 25:26090–26102. https://doi.org/10.1007/s11356-018-2501-5

Mei N, Hedberg J, Ekvall MT, Kelpsiene E, Hansson L-A, Cedervall T, Blomberg E, Odnevall I (2021) Transfer of cobalt nanoparticles in a simplified food web: from algae to zooplankton to fish. Appl Nanosci 2:184–205. https://doi.org/10.3390/applnano2030014

Monjo F, Romero R (2015) Embryonic development of the nervous system in the planarian Schmidtea polychroa. Dev Biol 397:305–319. https://doi.org/10.1016/j.ydbio.2014.10.021

Moon J, Kitty I, Renata K, Qin S, Zhao F, Kim W (2023) DNA damage and its role in cancer therapeutics. Int J Mol Sci 24:4741. https://doi.org/10.3390/ijms24054741

Nakagawa H, Sekii K, Maezawa T, Kitamura M, Miyashita S, Abukawa M, Matsumoto M, Kobayashi K (2018) A comprehensive comparison of sex-inducing activity in asexual worms of the planarian Dugesia ryukyuensis: the crucial sex-inducing substance appears to be present in yolk glands in Tricladida. Zool Lett 4:14. https://doi.org/10.1186/s40851-018-0096-9

Neuhauser EF, Loehr RC, Milligan DL, Malecki MR (1985) Toxicity of metals to the earthworm Eisenia fetida. Biol Fertil Soils 1:149–152. https://doi.org/10.1007/BF00301782

Neuhauser EF, Durkin PR, Malecki MR, Anatra M (1986) Comparative toxicity of ten organic chemicals to four earthworm species. Comp Biochem Physiol C 83:197–200. https://doi.org/10.1016/0742-8413(86)90036-8

New Worm on the Block: Planarians in (Neuro)Toxicology—Ireland—2022—Current Protocols—Wiley Online Library [WWW Document] (n.d.). https://currentprotocols.onlinelibrary.wiley.com/doi/abs/10.1002/cpz1.637. Accessed 30 July 2024

Newmark PA, Alvarado AS (2002) Not your father's planarian: a classic model enters the era of functional genomics. Nat Rev Genet 3:210–219. https://doi.org/10.1038/nrg759

Newmark PA, Sánchez Alvarado A (2000) Bromodeoxyuridine specifically labels the regenerative stem cells of planarians. Dev Biol 220:142–153. https://doi.org/10.1006/dbio.2000.9645

Nigro M, Bernardeschi M, Costagliola D, Della Torre C, Frenzilli G, Guidi P, Lucchesi P, Mottola F, Santonastaso M, Scarcelli V, Monaci F, Corsi I, Stingo V, Rocco L (2015) n-TiO2 and CdCl2 co-exposure to titanium dioxide nanoparticles and cadmium: genomic, DNA and chromosomal damage evaluation in the marine fish European sea bass (Dicentrarchus labrax). Aquat Toxicol 168:72–77. https://doi.org/10.1016/j.aquatox.2015.09.013

Nishimura K, Inoue T, Yoshimoto K, Taniguchi T, Kitamura Y, Agata K (2011) Regeneration of dopaminergic neurons after 6-hydroxydopamine-induced lesion in planarian brain. J Neurochem 119:1217–1231. https://doi.org/10.1111/j.1471-4159.2011.07518.x

Noreña C, Damborenea C, Brusa F (2015) Phylum Platyhelminthes. In: Thorp and Covich's freshwater invertebrates. Elsevier, pp 181–203. https://doi.org/10.1016/B978-0-12-385026-3.00010-3

Ofoegbu PU, Campos D, Soares AMVM, Pestana JLT (2019) Combined effects of NaCl and fluoxetine on the freshwater planarian, Schmidtea mediterranea (Platyhelminthes: Dugesiidae). Environ Sci Pollut Res 26:11326–11335. https://doi.org/10.1007/s11356-019-04532-4

Orii H, Sakurai T, Watanabe K (2005) Distribution of the stem cells (neoblasts) in the planarian Dugesia japonica. Dev Genes Evol 215:143–157. https://doi.org/10.1007/s00427-004-0460-y

Oviedo NJ, Nicolas CL, Adams DS, Levin M (2008a) Planarians: a versatile and powerful model system for molecular studies of regeneration, adult stem cell regulation, aging, and behavior. CSH Protoc 2008:pdb.emo101. https://doi.org/10.1101/pdb.emo101

Oviedo NJ, Nicolas CL, Adams DS, Levin M (2008b) Establishing and maintaining a colony of planarians. Cold Spring Harb Protoc 2008:pdb.prot5053. https://doi.org/10.1101/pdb.prot5053

Pagán O (2017) Planaria: an animal model that integrates development, regeneration and pharmacology. Int J Dev Biol 61:519–529. https://doi.org/10.1387/ijdb.160328op

Pagan OR, Pagan OR (2014) The first brain: the neuroscience of planarians. Oxford University Press, Oxford/New York

Pagán OR, Rowlands AL, Fattore AL, Coudron T, Urban KR, Bidja AH, Eterović VA (2009) A cembranoid from tobacco prevents the expression of nicotine-induced withdrawal behavior in planarian worms. Eur J Pharmacol 615:118–124. https://doi.org/10.1016/j.ejphar.2009.05.022

Pearse V, Pearse J, Buchsbaum M, Buchsbaum R (1987) Flatworm body plan: bilateral symmetry, three layers of cells, organ-system level of construction. Regeneration. In: Living invertebrates. The Boxwood Press, Pacific Grove, pp 204–221

Pearson BJ, Alvarado AS (2010) A planarian p53 homolog regulates proliferation and self-renewal in adult stem cell lineages. Development 137:213–221. https://doi.org/10.1242/dev.044297

Petersen EJ, Mortimer M, Burgess RM, Handy R, Hanna S, Ho KT, Johnson M, Loureiro S, Selck H, Scott-Fordsmand JJ, Spurgeon D, Unrine J, van den Brink NW, Wang Y, White J, Holden P (2019) Strategies for robust and accurate experimental approaches to quantify nanomaterial bioaccumulation across a broad range of organisms. Environ Sci Nano 6:1619–1656. https://doi.org/10.1039/C8EN01378K

Póti Á, Szüts D, Vermezovic J (2024) Mutational profile of the regenerative process and de novo genome assembly of the planarian Schmidtea polychroa. Nucleic Acids Res 52:1779–1792. https://doi.org/10.1093/nar/gkad1250

Prokai-Tatrai K, Szarka S, Nguyen V, Sahyouni F, Walker C, White S, Talamantes T, Prokai L (2012) "All in the mind"? Brain-targeting chemical delivery system of 17β-estradiol (Estredox) produces significant uterotrophic side effect. Pharm Anal Acta Suppl 7:10.4172/2153-2435. S7–002. https://doi.org/10.4172/2153-2435.S7-002

Raffa RB, Holland LJ, Schulingkamp RJ (2001) Quantitative assessment of dopamine D2 antagonist activity using invertebrate (Planaria) locomotion as a functional endpoint. J Pharmacol Toxicol Methods 45:223–226. https://doi.org/10.1016/S1056-8719(01)00152-6

Rawls SM, Thomas T, Adeola M, Patil T, Raymondi N, Poles A, Loo M, Raffa RB (2009) Topiramate antagonizes NMDA- and AMPA-induced seizure-like activity in planarians. Pharmacol Biochem Behav 93:363–367. https://doi.org/10.1016/j.pbb.2009.05.005

Rojo-Laguna JI, Garcia-Cabot S, Saló E (2019) Tissue transplantation in planarians: a useful tool for molecular analysis of pattern formation. Semin Cell Dev Biol 87:116–124. https://doi.org/10.1016/j.semcdb.2018.05.022

Rompolas P, Patel-King RS, King SM (2009) Chapter 4—Schmidtea mediterranea: a model system for analysis of motile cilia. In: King SM, Pazour GJ (eds) Methods in cell biology, methods in cell biology. Academic, pp 81–98. https://doi.org/10.1016/S0091-679X(08)93004-1

Rossi L, Salvetti A, Batistoni R, Deri P, Gremigni V (2008) Planarians, a tale of stem cells. Cell Mol Life Sci CMLS 65:16–23. https://doi.org/10.1007/s00018-007-7426-y

Saló E, Agata K (2012) Planarian regeneration: a classic topic claiming new attention. Int J Dev Biol 56:1–4. https://doi.org/10.1387/ijdb.123495es

Saló E, Baguñà J (1985) Cell movement in intact and regenerating planarians. Quantitation using chromosomal, nuclear and cytoplasmic markers. Development 89:57–70. https://doi.org/10.1242/dev.89.1.57

Saló E, Baguñà J (2002) Regeneration in planarians and other worms: new findings, new tools, and new perspectives. J Exp Zool 292:528–539. https://doi.org/10.1002/jez.90001

Salvetti A, Rossi L, Iacopetti P, Li X, Nitti S, Pellegrino T, Mattoli V, Golberg D, Ciofani G (2015) In vivo biocompatibility of boron nitride nanotubes: effects on stem cell biology and tissue regeneration in planarians. Nanomedicine 10:1911–1922. https://doi.org/10.2217/nnm.15.46

Salvetti A, Gambino G, Rossi L, De Pasquale D, Pucci C, Linsalata S, Degl'Innocenti A, Nitti S, Prato M, Ippolito C, Ciofani G (2020) Stem cell and tissue regeneration analysis in low-dose irradiated planarians treated with cerium oxide nanoparticles. Mater Sci Eng C 115:111113. https://doi.org/10.1016/j.msec.2020.111113

Salvetti A, Degl'Innocenti A, Gambino G, van Loon JJWA, Ippolito C, Ghelardoni S, Ghigo E, Leoncino L, Prato M, Rossi L, Ciofani G (2021) Artificially altered gravity elicits cell homeostasis imbalance in planarian worms, and cerium oxide nanoparticles counteract this effect. J Biomed Mater Res A 109:2322–2333. https://doi.org/10.1002/jbm.a.37215

Sanchez-Hernandez JC (2006) Earthworm biomarkers in ecological risk assessment. In: Ware GW, Whitacre DM, Albert LA, de Voogt P, Gerba CP, Hutzinger O, Knaak JB, Mayer FL, Morgan DP, Park DL, Tjeerdema RS, Yang RSH, Gunther FA (eds) Reviews of environmental contamination and toxicology: continuation of residue reviews. Springer, New York, pp 85–126. https://doi.org/10.1007/978-0-387-32964-2_3

Sarnat HB, Netsky MG (1985) The brain of the planarian as the ancestor of the human brain. Can J Neurol Sci 12:296–302. https://doi.org/10.1017/S031716710003537X

Schockaert ER, Hooge M, Sluys R, Schilling S, Tyler S, Artois T (2008) Global diversity of free living flatworms (Platyhelminthes, "Turbellaria") in freshwater. In: Balian EV, Lévêque C, Segers H, Martens K (eds) Freshwater animal diversity assessment. Springer Netherlands, Dordrecht, pp 41–48. https://doi.org/10.1007/978-1-4020-8259-7_5

Sheiman IM, Kreshchenko ND (2015) Regeneration of planarians: experimental object. Russ J Dev Biol 46:1–9. https://doi.org/10.1134/S1062360415010075

Shibata N, Hayashi T, Fukumura R, Fujii J, Kudome-Takamatsu T, Nishimura O, Sano S, Son F, Suzuki N, Araki R, Abe M, Agata K (2012) Comprehensive gene expression analyses in pluripotent stem cells of a planarian, Dugesia japonica. Int J Dev Biol 56:93–102. https://doi.org/10.1387/ijdb.113434ns

Silva PV, Pinheiro C, Morgado RG, Verweij RA, van Gestel CAM, Loureiro S (2022) Bioaccumulation but no biomagnification of silver sulfide nanoparticles in freshwater snails and planarians. Sci Total Environ 808:151956. https://doi.org/10.1016/j.scitotenv.2021.151956

Simanov D, Mellaart-Straver I, Sormacheva I, Berezikov E (2012) The flatworm Macrostomum lignano is a powerful model organism for ion channel and stem cell research. Stem Cells Int 2012:167265. https://doi.org/10.1155/2012/167265

Sluys R, Riutort M (2018) Planarian diversity and phylogeny. In: Rink JC (ed) Planarian regeneration: methods and protocols. Springer, New York, pp 1–56. https://doi.org/10.1007/978-1-4939-7802-1_1

Song X-Y, Li W-X, Sluys R, Huang S-X, Li S-F, Wang A-T (2020) A new species of Dugesia (Platyhelminthes, Tricladida, Dugesiidae) from China, with an account on the histochemical structure of its major nervous system. Zoosyst Evol 96:431–447. https://doi.org/10.3897/zse.96.52484

Stevens A-S, Pirotte N, Plusquin M, Willems M, Neyens T, Artois T, Smeets K (2015) Toxicity profiles and solvent–toxicant interference in the planarian Schmidtea mediterranea after dimethylsulfoxide (DMSO) exposure. J Appl Toxicol 35:319–326. https://doi.org/10.1002/jat.3011

Stevens A-S, Wouters A, Ploem J-P, Pirotte N, Van Roten A, Willems M, Hellings N, Franken C, Koppen G, Artois T, Plusquin M, Smeets K (2018) Planarians customize their stem cell responses following genotoxic stress as a function of exposure time and regenerative state. Toxicol Sci 162:251–263. https://doi.org/10.1093/toxsci/kfx247

Talbot J, Schötz E-M (2011) Quantitative characterization of planarian wild-type behavior as a platform for screening locomotion phenotypes. J Exp Biol 214:1063–1067. https://doi.org/10.1242/jeb.052290

Tice RR, Agurell E, Anderson D, Burlinson B, Hartmann A, Kobayashi H, Miyamae Y, Rojas E, Ryu J-C, Sasaki YF (2000) Single cell gel/comet assay: guidelines for in vitro and in vivo genetic toxicology testing. Environ Mol Mutagen 35:206–221. https://doi.org/10.1002/(SICI)1098-2280(2000)35:3<206::AID-EM8>3.0.CO;2-J

Wang T, Liu W (2022) Emerging investigator series: metal nanoparticles in freshwater: transformation, bioavailability and effects on invertebrates. Environ Sci Nano 9:2237–2263. https://doi.org/10.1039/D2EN00052K

Werner S, Rink JC, Riedel-Kruse IH, Friedrich BM (2014) Shape mode analysis exposes movement patterns in biology: flagella and flatworms as case studies. PLoS One 9:e113083. https://doi.org/10.1371/journal.pone.0113083

Wu J-P, Li M-H (2018) The use of freshwater planarians in environmental toxicology studies: advantages and potential. Ecotoxicol Environ Saf 161:45–56. https://doi.org/10.1016/j.ecoenv.2018.05.057

Wu J-P, Lee H-L, Li M-H (2014) Cadmium neurotoxicity to a freshwater planarian. Arch Environ Contam Toxicol 67:639–650. https://doi.org/10.1007/s00244-014-0056-0

Xie C, Li X, Hei L, Chen Y, Dong Y, Zhang S, Ma S, Xu J, Pang Q, Lynch I, Guo Z, Zhang P (2023) Toxicity of ceria nanoparticles to the regeneration of freshwater planarian Dugesia japonica: the role of biotransformation. Sci Total Environ 857:159590. https://doi.org/10.1016/j.scitotenv.2022.159590

Yuan Z, Shao X, Miao Z, Zhao B, Zheng Z, Zhang J (2018) Perfluorooctane sulfonate induced neurotoxicity responses associated with neural genes expression, neurotransmitter levels and acetylcholinesterase activity in planarians Dugesia japonica. Chemosphere 206:150–156. https://doi.org/10.1016/j.chemosphere.2018.05.011

Zhang X, Zhang B, Yi H, Zhao B (2014) Mortality and antioxidant responses in the planarian (Dugesia japonica) after exposure to copper. Toxicol Ind Health 30:123–131. https://doi.org/10.1177/0748233712452600

Zhang S, Hagstrom D, Hayes P, Graham A, Collins E-MS (2019) Multi-behavioral endpoint testing of an 87-chemical compound library in freshwater planarians. Toxicol Sci 167:26–44. https://doi.org/10.1093/toxsci/kfy145

Zhao X, Yu M, Xu D, Liu A, Hou X, Hao F, Long Y, Zhou Q, Jiang G (2017) Distribution, bioaccumulation, trophic transfer, and influences of CeO2 nanoparticles in a constructed aquatic food web. Environ Sci Technol 51:5205–5214. https://doi.org/10.1021/acs.est.6b05875

Open Access This chapter is licensed under the terms of the Creative Commons Attribution-NonCommercial-NoDerivatives 4.0 International License (http://creativecommons.org/licenses/by-nc-nd/4.0/), which permits any noncommercial use, sharing, distribution and reproduction in any medium or format, as long as you give appropriate credit to the original author(s) and the source, provide a link to the Creative Commons license and indicate if you modified the licensed material. You do not have permission under this license to share adapted material derived from this chapter or parts of it.

The images or other third party material in this chapter are included in the chapter's Creative Commons license, unless indicated otherwise in a credit line to the material. If material is not included in the chapter's Creative Commons license and your intended use is not permitted by statutory regulation or exceeds the permitted use, you will need to obtain permission directly from the copyright holder.

Caenorhabditis elegans: A Bridging Model to Assess the Safety of Nanomaterials

11

Nivedita Chatterjee

Abstract

This chapter focuses on the role of *Caenorhabditis elegans* as a bridging model in nanotoxicology and nanosafety research. With its simple multicellular structure, well-characterized genetics, low maintenance costs, short life cycle, and suitability for high-throughput screening, *C. elegans* is effective for evaluating nanoparticle toxicity across various exposure scenarios, including acute and chronic treatments. The chapter examines key physiological endpoints—such as survival rates, growth, reproduction, and behavior—and employs mutant and transgenic strains alongside advanced omics technologies to investigate the molecular pathways affected by nanoparticle exposure, particularly oxidative stress, genotoxicity, and neurotoxicity. By integrating multi-endpoint assessments and behavioral investigations, *C. elegans* provides valuable insights into the safety and potential risks of nanomaterials, contributing to a broader understanding of nanotoxicology in alignment with the 'One Health' framework.

Keywords

Caenorhabditis elegans (*C. elegans*) · Engineered nanomaterial · Nanotoxicology

1 Introduction

The soil-dwelling, non-parasitic nematode *Caenorhabditis elegans* (*C. elegans*) has been a foundational model organism in biological research since the 1970s when it was first was first proposed as a model organism by Sydney Brenner in 1965 and employed to study the genetic regulation of development (Tejeda-Benitez and

N. Chatterjee (✉)
NanoSafety Group, International Iberian Nanotechnology Laboratory, Braga, Portugal
e-mail: nivedita.chatterjee@inl.int

Olivero-Verbel 2016; Brenner 2009; Avila et al. 2011). Its popularity has grown significantly due to numerous advantages. These include its small size (approximately 1 mm in length for adults), rapid life cycle (about 3 days at 20 °C to reach adulthood), short lifespan (around 2.5 weeks), self-fertilization capability, large brood size (over 300 offspring per hermaphrodite), and ease of genetic manipulation (Leung et al. 2008). The requirements for maintaining *C. elegans* in the lab are minimal—ambient temperature, humidity, oxygen, and a bacterial food source—making it a cost-effective and accessible model for research (Avila et al. 2011).

1.1 Ecology and Natural Environment

The nematode *C. elegans* thrives in environments rich in decaying organic matter, such as rotting fruits and compost heaps, where it primarily feeds on bacteria. Its population follows a "boom-and-bust" dynamic, increasing when food is abundant. When resources are scarce, it enters the dauer stage, a dormant phase that enables survival under harsh conditions and dispersal to new environments. *C. elegans* demonstrates adaptability and plays a crucial role in nutrient cycling across various habitats, including soil and decomposing plant material (Frézal and Félix 2015).

1.2 Anatomy and Tissues

C. elegans features a simple yet high differentiated anatomical structure. Adult hermaphrodites consist of 959 somatic cells, while males possess 1031. Despite its simplicity, *C. elegans* develops specialized tissues including muscle, hypodermis, intestine, gonads, glands, an excretory system, and a nervous system composed of 302 neurons and their synapses (Sulston 1983; Avila et al. 2011) (Fig. 11.1).

1.2.1 Epidermis
The epidermis consists of a single layer of hypodermal cells, covered by a protective cuticle.

1.2.2 Muscles
Body wall muscles are arranged into four quadrants, enabling the nematode's characteristic sinusoidal movement.

1.2.3 Digestive System
This system includes a pharynx, intestine, and anus, ensuring efficient nutrient absorption.

1.2.4 Nervous System
The hermaphrodite's nervous system contains 302 neurons and 56 glial cells, while the male has 381 neurons. The complete neural network (connectome) has been fully mapped, making *C. elegans* an excellent model for studying neural function, development, and degeneration.

Anatomy of *Caenorhabditis elegans*

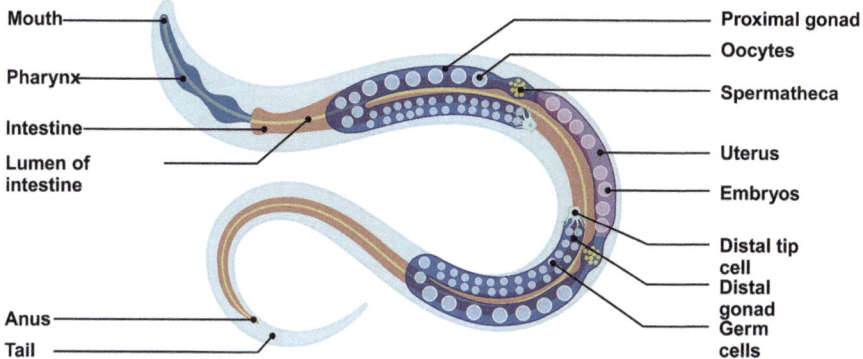

Fig. 11.1 Anatomy of adult *C. elegans* hermaphrodite (schematic). (Figure created with BioRender.com)

1.2.5 Reproductive System

Hermaphrodites have a bilobed gonad, each lobe containing an ovary, oviduct, and spermatheca, while males have a single-lobed gonad with a vas deferens leading to the cloaca.

1.3 Development and Reproduction

Mature oocytes pass through the spermatheca, where they are fertilized by sperm from either the hermaphrodite or a male. The resulting zygote forms a tough chitinous shell and vitelline membrane, rendering it impermeable to most solutes. Eggs are typically retained in the uterus through the first few cleavages before being laid around the time of gastrulation, approximately 3 h after fertilization. During embryogenesis, cell division, organogenesis, and morphogenesis occur, resulting in the first-stage larva. Post-embryonic development sees continuous growth, with somatic cell nuclei increasing from 558 in the first-stage larva to 959 in adult hermaphrodites (Avila et al. 2011; Ferreira et al. 2014).

Larval development proceeds through four stages (L1-L4), with significant cellular differentiation occurring during each phase (Avila et al. 2011; Tejeda-Benitez and Olivero-Verbel 2016). For example, certain proteins such as Cu^{2+}/Zn^{2+} superoxide dismutase and aspartyl proteinase are highly expressed in the L1 stage but decrease as the nematode matures (Mádi et al. 2003). By the L4 stage, gonadogenesis is complete, enabling reproductive capability. The entire life cycle, from egg to reproductive adult, takes just 3.5 days at 20 °C. Under optimal conditions, the lifespan of wild-type *C. elegans* is about 2.5 weeks (~18 days). In response to food scarcity or high population density, an alternative dauer stage can form at the L2/L3

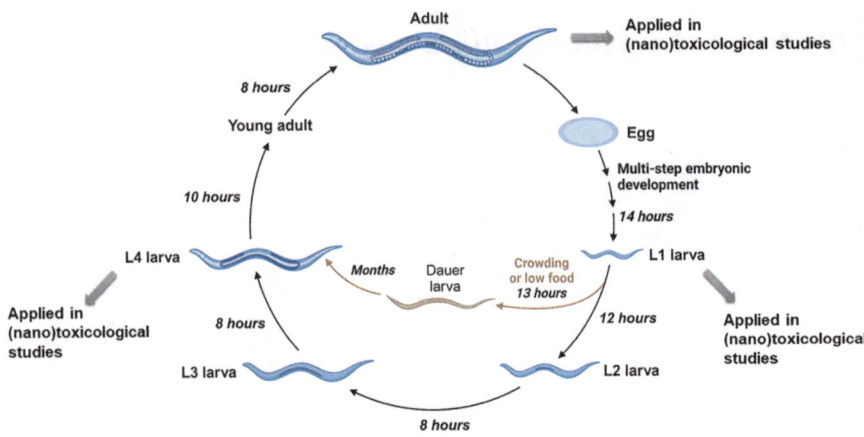

Fig. 11.2 Life cycle of *C. elegans* (schematic representation): relevance of different developmental stages to nanotoxicological study applications. (Created with BioRender.com)

molt. Dauers are resistant to desiccation and can survive up to 3 months without developing further (Avila et al. 2011) (Fig. 11.2).

1.4 Reproduction

C. elegans exists as either hermaphrodite or male. Hermaphrodites can self-fertilize, producing only hermaphrodite offspring, while cross-fertilization between hermaphrodites and males produces both sexes in equal proportions. This unique reproductive strategy is particularly useful for genetic studies. Hermaphrodites possess a bilobed gonad, while males have a single-lobed gonad that connects with the cloaca near the tail. Males also have specialized structures in their tail for mating, including 18 sensory rays and spicules that assist with sperm transfer during copulation (Tejeda-Benitez and Olivero-Verbel 2016; Avila et al. 2011; Ferreira et al. 2014).

1.5 Genome and Genetic Manipulation

The *C. elegans* genome, one of the first multicellular organisms to be fully sequenced, consists of approximately 100 million base pairs and 20,000 genes spread across six chromosomes. This wealth of genetic information is accessible through databases such as WormBase. Various genetic techniques, including mutagenesis, transgenesis, and RNA interference (RNAi), are employed to study *C. elegans*. Knockout mutant libraries and genetic manipulation tools, such as GFP-tagging, have been particularly valuable for in vivo studies of cells and molecular pathways (Avila et al. 2011; Chalfie et al. 1994).

1.6 *C. elegans* as a Model in Biology

C. elegans has been instrumental in advancing biological research since its adoption as a model organism. Key discoveries include the genetic mechanisms behind the development, apoptosis, and neural function, with landmark achievements like the complete mapping of its cell lineage, the sequencing of its genome, and the discovery of RNA interference (RNAi). Its rapid life cycle, transparent body, and self-fertilization simplify genetic studies and cellular observations, while its simple maintenance makes it a cost-effective research tool. Despite its biological simplicity compared to higher organisms, *C. elegans* continues to provide profound insights into fundamental biological processes, driving breakthroughs in science and medicine (Tejeda-Benitez and Olivero-Verbel 2016; Avila et al. 2011).

2 *C. elegans*—The Bridging Model Organism and Its Applications in Toxicity Research

C. elegans became a preferred model for toxicity studies in the late 1990s, owing to its low maintenance cost, short life cycle, and suitability for high-throughput screening (Helmcke et al. 2010; Avila et al. 2011). Unlike isolated cell cultures, *C. elegans* provides a complete multicellular organism to assess whole-system responses to toxicants. It possesses functional nervous, digestive, and reproductive systems, offering insights into the holistic impact of toxin exposure. Its fully sequenced genome allows for easy manipulation through RNA interference (RNAi) and mutagenesis, and researchers can access thousands of transgenic and mutant strains from the Caenorhabditis Genetics Center (Ferreira et al. 2014).

Toxicity assays typically test endpoints such as growth, reproduction, feeding, and movement (Wu et al. 2019; Avila et al. 2011; Tejeda-Benitez and Olivero-Verbel 2016). Growth and reproduction are often more sensitive indicators than lethality for many toxicants like polycyclic aromatic hydrocarbons (Sese et al. 2009). *C. elegans*, especially transgenic strains, is extensively utilized as a bioindicator in ecotoxicology, with a focus on sublethal conditions (Anbalagan et al. 2013; Lagido et al. 2009). Its application is significant across both terrestrial and aquatic environments (Ellegaard-Jensen et al. 2012; Kuhn et al. 2021). Toxicant exposure can be conducted on solid agar plates or in liquid media, providing flexibility in experimental design and distinct advantages for toxicology assays. The transparent nature of the worm's cuticle eliminates the need for dissection, allowing researchers to directly observe a wide range of endpoints and simplifying toxicity assessments. In essence, *C. elegans* enables researchers to collect data on a whole living organism using a methodology often similar to that of cell line monocultures (Ferreira et al. 2014).

To evaluate the toxic effects of chemicals, researchers use various bioassays with *C. elegans*. Typically, young adult worms are exposed to different concentrations of the test substance in a liquid medium. The absence of food during these acute exposures allows for a focused assessment of the chemical's impact. For long-term

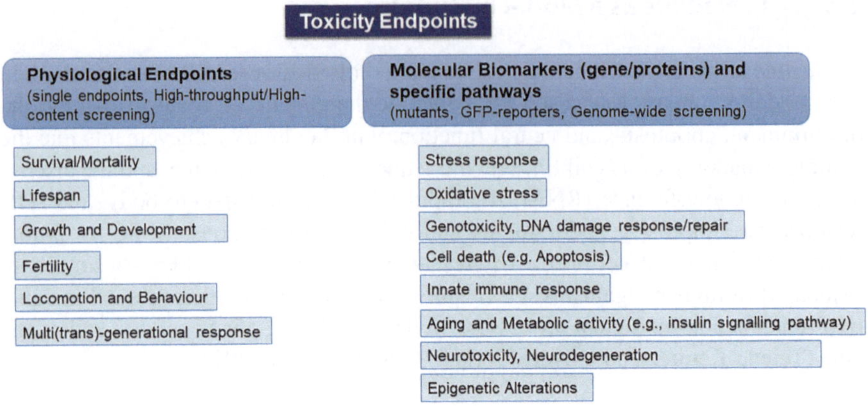

Toxicity Endpoints

Physiological Endpoints
(single endpoints, High-throughput/High-content screening)
- Survival/Mortality
- Lifespan
- Growth and Development
- Fertility
- Locomotion and Behaviour
- Multi(trans)-generational response

Molecular Biomarkers (gene/proteins) and specific pathways
(mutants, GFP-reporters, Genome-wide screening)
- Stress response
- Oxidative stress
- Genotoxicity, DNA damage response/repair
- Cell death (e.g. Apoptosis)
- Innate immune response
- Aging and Metabolic activity (e.g., insulin signalling pathway)
- Neurotoxicity, Neurodegeneration
- Epigenetic Alterations

Fig. 11.3 Key endpoints in toxicological assessments with the *C. elegans* model

studies, L1 larvae are exposed to the chemical in the presence of a food source such as *E. coli* OP50. Toxicity endpoints in *C. elegans* encompass a wide range of biological responses, including lethality, growth rate, locomotion, and reproductive capacity. To gain deeper insights into the mechanisms of toxicity, molecular markers like those for oxidative stress, gene expression, DNA damage, or green fluorescent protein (GFP) expression can be employed (Tejeda-Benitez and Olivero-Verbel 2016; Wu et al. 2019). A classification of commonly used endpoints in *C. elegans* toxicity research is presented in Fig. 11.3. This comprehensive approach enables the identification of sensitive endpoints and the characterization of the toxicant's mode of action.

2.1 Adaptability to High-Throughput, Automated Behaviour System, and Genome-Wide Toxicity Screening

C. elegans is an ideal model organism for high-throughput screening due to its adaptability to both aquatic and terrestrial environments, prolific reproduction, and short life cycles. These features enable the analysis of toxicant effects through various methods (Helmcke et al. 2010). Additionally, multi-endpoint, high-content screening platforms have been developed and applied in various toxicity fields (Wu et al. 2022; Jung et al. 2015). Automated tools such as the Biosort (Union Biometrica, Inc.) and COPAS biosorter can analyze parameters like length, motion, fluorescence, and reproductive endpoints in 96-well plates (Shin et al. 2019; Helmcke et al. 2010). Furthermore, microfluidic devices and robotic systems improve the precision of worm manipulation and immobilization for imaging and microsurgery (Hulme et al. 2007; Mondal et al. 2016; Rohde et al. 2007).

Computer-based assays also offer automated readouts for assessing toxicant impacts on behaviors like thrashing, fluorescence, and developmental endpoints such as egg-laying, dauer formation, and lifespan in wild-type, mutant, and transgenic worms (Buckingham and Sattelle 2009; Rohde et al. 2007; Leung et al. 2011; Rahman et al. 2020).

Genome-wide screens for molecular contributors to toxicity, using methods such as microarray, RNA sequencing, RNAi screening, and transgenic approaches, have identified genes involved in toxicant responses (McElwee et al. 2013; Chatterjee et al. 2017; Kim et al. 2017a, 2020a, b). Thus, *C. elegans* offers efficient, high-throughput capabilities for studying toxicant effects, supported by advanced genetic tools and automated technologies.

The *C. elegans* model holds promise for connecting in vivo and in vitro approaches (Kaletta and Hengartner 2006; Chakravarty 2022). It addresses the challenges of mammalian models by providing a more affordable, efficient, and ethically favorable alternative. Additionally, *C. elegans* features a fully sequenced genome, the availability of transgenic knock-out mutants, and compatibility with high-throughput automation techniques. Despite its evolutionary distance from humans, *C. elegans* shares many conserved metabolic pathways and gene homologs with humans, enabling in-depth analysis of these shared mechanisms. This makes *C. elegans* a key model for bridging both in vitro and in vivo systems, as well as for advancing research on human and environmental health, aligning with the 'One Health' framework (von Mikecz 2022).

3 *C. elegans* in the Field of Nanotoxicology and Nanosafety

The toxicological potential of engineered nanoparticles (ENPs) has become a growing concern due to their significant release into the environment, positioning them among the group of emerging contaminants. Despite their widespread application in medical and clinical settings, the interactions between these nanomaterials and biological systems are not yet fully understood. This nano-bio interaction knowledge gap has prompted extensive studies using various biological models, including *C. elegans* (Table 11.1). Utilizing *C. elegans* allows researchers to explore the fate and toxicity of NPs within a multicellular organism. *C. elegans* proves to be a valuable model for assessing NP toxicity across different exposure scenarios, including acute, prolonged, and chronic treatments through oral ingestion, topical application, or microinjection. The model supports the evaluation of numerous endpoints, such as physiological effects like average body length and brood size, which indicate developmental and reproductive health. Furthermore, *C. elegans* facilitates the study of molecular mechanisms by employing biological markers like gene expression, green fluorescent protein (GFP) reporters, and application of specific mutant strains which provide insights into the specific pathways and mechanisms affected by NP exposure (Tejeda-Benitez and Olivero-Verbel 2016; Wu et al. 2019; Gonzalez-Moragas et al. 2015a). The efforts are in line with high throughput screening for several nanomaterials with various doses to target several physiological endpoints (Jung et al. 2015).

Table 11.1 Assessment of nanosfaety of manufactured nanomaterials in the bridging model, *C. elegans*, with various toxicological endpoints

Nanomaterials Name	*C. elegans* (Stage)	Exposure Conc.	Time	Media (food '+')	Endpoints	References
Silver nanoparticles (AgNP)	Young adult (3 days old)	0.05, 0.1, and 0.5 mg/L	24 h and 72 h	K-media	Lethality, growth, reproduction, gene expressions (microarray)	Roh et al. (2009)
Citrate-coated (CIT10), PVP-coated smaller size range (PVPS) and larger size range (PVPL)	L1	0.5, 5, and 50 mg/L	3 days	K-media (+)	Growth, uptake	Meyer et al. (2010)
Silver nanoparticles (AgNP)	Young adult (3 days old)	0.05 mg/L	24 h	EPA water	Lethality, ROS formation, gene/protein expressions	Eom et al. (2013)
Citrate-coated silver nanoparticles (cAgNPs)				NGM (?)	Lethality and reproduction	Kim et al. (2012)
Commercial nano-silver (AgNP1 and PVP coated AgNP28)	L2	0.5–10 mg/L (AgNP1) 0.6–3 mg/L (AgNP28)	24, 48, 72 and 96 h	K-media (+/−)	Lethality	Ellegaard-Jensen et al. (2012)
Bare (AgNP) and two PVP-coated (PVP8 and PVP38) silver nanoparticles	Young adult (3 days old)	0.025, 0.05, 0.075 μg/mL	24 h	EPA water	Lethality, ROS formation, oxidative DNA damage (8OHdG), mitochondrial membrane potential	Ahn et al. (2014)
Citrate-coated (CIT7), small and large PVP -coated (PVPS, PVPL, PVP8 and PVP38). Gum Arabic-coated (GA5 and GA22)	L1	20–46 μM of total silver	3 days	K-media and EPA water (+)	Growth (in COPAS biosort)	Yang et al. (2012)

Material	Stage	Concentration	Time	Medium	Endpoints	Reference
Silver nanoparticles (AgNP)	Young adult (3 days old)	0.025 mg/L (LC10) and 0.05 mg/L (LC50)	24 h	EPA water	Lethality, oxidative DNA damage (8OHdG), gene expression	Chatterjee et al. (2014a)
Citrate (cit-), PVP, PEG, BPEI-coated silver nanoparticles and silver plates	L1 and adult	0–100 µg/mL	24 h and 3 days	C. elegans habitation reagent (CeHR), water, and organic non-fat cows' milk (CeHRM)	Lethality, growth (in COPAS biosort), locomotion, gene expression (microarray), uptake	Hunt et al. (2014)
Silver nanoparticles (as 1 mg/mL in citrate buffer)	L1, L2 and adult	0–50 µg/mL	3 days	CeHRM	Growth (in COPAS biosort), morphology, oxidative DNA damage (several lesions GC/MS/MS) detected in and uptake	Hunt et al. (2013)
Citrate-coated AgNPs (CIT-AgNPs)	L1 and adult	0.1–1.5 mg-Ag/L	24 h and 3 days	EPA water (+)	Lethality, growth	Maurer et al. (2016)
Ag-MNP and sulfidized Ag-MNPs (sAg-MNPs)	L1 and L3	No food: 50–1500 µg/L, with food: 1000–10,000	24 h, 48 h and 50 h	Low chloride MHRW medium	Lethality, growth, reproduction	Starnes et al. (2015)
Silver nanoparticles (NM300K)	L1	0.2–10 mg/L	20 h and 96 h	M9 media (+)	Reproduction, feeding	Kleiven et al. (2018)
PVP-coated silver nanoparticles	L1	0.01–10 µg/mL	24, 48 and 72 h	Semi-fluid nematode growth gelrite media (Dengg and van Meel) (+)	Reproduction, lifespan, growth, ROS formation, mitochondrial membrane potential, apoptosis	Luo et al. (2017)
Silver nanoparticles using PVP as dispersant	L4	0.005–1 mg/L	6 h and 24 h	Multi-well and microfluidic chip-based chamber (+)	Development, growth, protein expression (reporter)	Kim et al. (2017b)

(continued)

Table 11.1 (continued)

Nanomaterials Name	C. elegans (Stage) Stage	Exposure Conc.	Time	Media (food '+')	Endpoints	References
(PVP)-coated (Ag-PVP) and sulfidized (Ag$_2$S) silver nanoparticles	L1	0.75–24 mg Ag/L (Ag-PVP) and 7.5–240 mg Ag/L (Ag$_2$S)	24 h and multigenerational	NGM (multigenerational maintenance) and exposure to simulated soil pore water (SSPW) (+)	Reproduction, lifespan, growth, multi-generation	Schultz et al. (2016)
Silver nanoparticles (AgNP)	L1	1–25 mg/mL	24 h and multigenerational	NGM (+) OP50 E. coli was treated with AgNP for 12 h and used to fed worms	Uptake & accumulation, reproduction, lifespan	Luo et al. (2016)
Silver nanoparticles (AgNP)	Egg, L4	0.0625–4 mg Ag/L	48 h and 72 h	NGM and simulated soil pore water (SSPW) (+)	Reproduction and growth	Tyne et al. (2015)
Oleic acid coated-silver nanoparticles	L4	1–100 mg Ag/L	24 h	NGM (+)	Reproduction, lifespan, growth, locomotion behaviour	Contreras et al. (2014)
Aluminium nanoparticle (Al$_2$O$_3$-NP)	L1, L4 and young adult		3 days, 48 h and 24 h	K-media	Lethality, stress response (reporter strain), intestinal autofluorescence	Wu et al. (2011)
Aluminium nanoparticle (Al$_2$O$_3$-NP)	Young adults	0.01–23.1 mg/L	6 h, 24 h and 10 days	K-media and NGM (+)	Locomotion, gene expression, stress response (reporter strain), oxidative stress (ROS, superoxide dismutase (Soddu et al.) activity, carbonylated proteins)	Li et al. (2012)
Gold nanoparticles (AuNP)	L4	100 μg/mL	24 h	NGM (+)	Lethality, reproduction, gene expression	Gonzalez-Moragas et al. (2017a)

Material	Stage	Concentration	Time	Medium	Endpoints	Reference
Citrate-coated gold nanoparticles (AuNP)	L1, L4	1, 10 and 100 µg/mL	24 h, 3 days	S-media, NGM (+)	Growth, behaviour (feeding, pharyngeal pumping, chemotaxis), morphology of neurons, calcium transit, gene expressions (RNA-seq)	Wang et al. (2023b)
Bare gold (Bare-Au) and different proportions of 11-mercaptoundecanoic acid (MUA)-coated (MUA/Au-0.5 MUA/Au-1 MUA/Au-3)	Eggs	0.1 mg/mL	72 h	S-media (+)	Uptake, growth, locomotion (thrashing), reproduction, primary neuron visualization (isolated and cultured), gene expressions (microarray)	Hu et al. (2018b)
Gold nanocolloids (AuNP)	L1	0, 5, 25, and 50×10^{10} particles/mL (mixed with *E. coli* OP50)	48 h	NGM (+)	Survival, reproduction (number of offspring and reproductive system abnormalities), lipofuscin accumulation, multigenerational effects	Moon et al. (2017)
Gold nanocolloids (AuNP)	L1	5, 25, and 50×10^{10} particles/mL (mixed with *E. coli* OP50)	12 h	NGM (+)	Survival, reproduction, transgenerational effects	Kim et al. (2013)
Gold nanoparticles (AuNP)	L3	2.5, 5.5, 7, 15, and 30 mg/L	12 h	K-media	Gene expressions (microarray ad qPCR) and pathway analysis	Tsyusko et al. (2012)
Two boron nitride nanospheres BN-NS and BN-800-2	L1, L4	1, 10, 100, and 500 µg/mL	24, 48 & 72 h	K-media (+)	Growth, lifespan, reproduction, locomotion, oxidative stress (gene expression, ROS formation), GABA neuron (reporter)	Wang et al. (2017)

(continued)

Table 11.1 (continued)

Nanomaterials	C. elegans (Stage)	Exposure		Media (food '+')	Endpoints	References
Name	Stage	Conc.	Time			
CdTe quantum dots (QDs)	L1	2.5–20 mg/L	72 h	NGM (+)	Autofluorescence, locomotion, GFP reporters, gene expression	Wu et al. (2016a)
CdTe quantum dots (QDs)	L1, L2	1–10 μM	12 h, 24 h	NGM (−)	Development, lifespan, autofluorescence	Wang et al. (2016)
CdTe quantum dots (QDs)	L4	400–1600 μg/mL	24 h, 72 h	NGM (+)	Locomotion, behaviour (learning & memory), gene expression, oxidative stress	Wu et al. (2015)
CdTe quantum dots (QDs)	L1, L4	10 nM	24 h	NGM (−)	Autophagy, lifespan	Zhou et al. (2015)
CdTe quantum dots (QDs)	L1	0.1–1 μg/L	3.5, 5 days	NGM (+)	Behaviour (feeding, defecation), development, reproduction, neurodegeneration, oxidative stress	Zhao et al. (2015)
Cerium oxide (CeO$_2$) nanoparticles	L4	0.17–17.21 μg/mL	24 h	NGM (+)	Growth, reproduction, oxidative stress	Rogers et al. (2015)
Polymer-cationic (diethylaminoethyl dextran; DEAE), anionic (carboxymethyl dextran; CM), and non-ionic (dextran; DEX) polymers-coated cerium oxide nanomaterials	L1, L2, L3	0–5000 mg Ce/L	24 h, 48 h	NGM (±)	Oxidative stress, metabolism, GFP reporters	Arndt et al. (2017)

Nanoparticle	Stage	Concentration	Time	Medium	Endpoints	Reference
Citrate- and bovine serum albumine-coated of superparamagnetic iron oxide nanoparticles (C-SPIONs and BSA-SPIONs)	L4	0–500 μg Fe/mL	24 h	Milli-Q water (−)	Lethality, reproduction	Gonzalez-Moragas et al. (2015b)
Zero-valent iron nanoparticles (Fe0 NPs)	L1, L4	5–500 mg/L	48 h, 65 h	NGM (+)	Bioaccumulation, chemical analysis, bioassays (fertility, locomotion, development) biodynamics modelling (TBTK) dose–response-based TD modelling	Yang et al. (2017)
Carboxymethyl cellulose (CMC)-stabilized nZVI, nanoscale iron oxide (nFe_3O_4),	L4	5–100 mg/L	48 h	NGM (+)	Reproduction, oxidative stress	Yang et al. (2016)
Copper oxide nanoparticles (CuONP)	L1	3.8–15.9 mg cu/L	96 h	NGM (+)	Growth, feeding, neurodegeneration, stress response	Mashock et al. (2016)
Copper oxide nanoparticles (CuONP)	L4	150 mg/L	48 h	K-media	Growth, reproduction, transgenerational assays	Wei et al. (2020)
Nickel nanoparticles (NiNP)	L1, L4	1.0, 2.5, and 5.0 μg/cm^2	24 h, 48 h	NGM (+)	Reproduction (fertility, brood size, egg laying rate, spermatogenesis), oxidative stress (ROS formation, SOD, CAT, and GSH level), apoptosis-related gene expressions	Kong et al. (2017)
Silica-nanoparticles (SiO_2)	L4	2.5 mg/mL	24 h	NGM (+)	Behaviour, bag of worm (BOW), protein (aggregome) expression, neurodegeneration (polyQ aggregates, reporter)	Scharf et al. (2016)

(continued)

Table 11.1 (continued)

Nanomaterials	C. elegans (Stage)	Exposure				
Name	Stage	Conc.	Time	Media (food '+')	Endpoints	References
Four types mesoporous silica particles (bare and functionalized with hydrolyzed starch—M0, M1, N0 and N1)	L1	0.5, 5.0, 50 µg/mL	2 days, 21 days	NGM (+)	Uptake, lifespan, locomotion, reproduction, oxidative stress (defence to H_2O_2 exposure)	Acosta et al. (2018)
Silica nanoparticles (SiNPs)	Young adult	20–500 mg/L	24 h, 48 h, 72 h	K-media (±)	Lethality, reproduction, ROS formation, genome-wide transcriptional changes (microarray), endocytosis (inhibitor, gene expressions)	Eom and Choi (2019)
Amorphous silica nanoparticles (SiNPs)	L4	100 µg/mL	24 h	K-media (+)	Genome-wide transcriptional changes (microarray of mRNA and miRNA)	Liang et al. (2020a)
Mesoporous silica nanoparticles (mSiNPs)	L1, L4	3, 30, 300 and 3000 µg/L	24 h, 48 h	K-media (±)	Intestinal autofluorescence, behaviour (shrinking, foraging, locomotion, pharyngeal pumping, defecation), GABA neurodegeneration (reporter, gene expression), ROS formation	Liang et al. (2020b)
Silica nanoparticles (SiO2 NPs)	L4	0.25–1 mg/mL	24 h	K-media (+)	Lethality, growth, reproduction, locomotion, apoptosis, oxidative stress (ROS formation, GST, MDA level), gene expressions	Zhang et al. (2020)

Fluorescent silicon nanoparticles (SiNPs)	L1, L2, L4, young adult	50 mg/mL	4 h, 24 h, 3 days	NGM (+)	Uptake, growth, lifespan, reproduction, endocytic sorting, stress response, endoplasmic reticulum and mitochondrial stress (reporter), gene expressions (immune response, apoptosis, hypoxia, metal detoxification and aging)	Wang et al. (2022)
TiO2: anatase and rutile	Egg, L1, L4	100–500 mg/mL	72 h	S-media (+)	Uptake and internalization, growth and development, locomotion, reproduction, neuronal function (isolated and cultured), genome-wide transcriptional changes (microarray)	Hu et al. (2018a)
TiO$_2$	Young adult	2–10 mg/L (±UV)	24 h, 72 h	K-media (±)	Lethality, reproduction, ROS formation, genome-wide transcriptional changes (RNA-seq), global metabolomics, gene expressions	Kim et al. (2017a)
TiO$_2$	L4	0.01–1 mg/L	2 h	Distilled water (−) followed by NGM (+)	Pharyngeal pumping, reproduction (egg laying and larval growth), development	Iannarelli et al. (2016)
TiO$_2$	L4	1–500 mg/L	24 h, 72 h	NGM (±)	Locomotion, growth, neurodegeneration, genome-wide transcriptional changes (microarray)	Rocheleau et al. (2015)
TiO$_2$	L4	7.7–38.5 µg/mL	24 h	NGM (−)	Lethality, oxidative stress, global metabolomics	Ratnasekhar et al. (2015)

(continued)

Table 11.1 (continued)

Nanomaterials Name	C. elegans (Stage) Stage	Exposure Conc.	Time	Media (food '+')	Endpoints	References
TiO$_2$	L1, L4	1–100 µg/L, 10–100 mg/L	24 h, 72 h		Lethality, reproduction, oxidative stress, gene expression intestinal autofluorescence, oxidative stress, defecation cycle	Zhao et al. (2014)
TiO$_2$	L1	0.0001–1 µg/L	72 h	NGM (+)	Lethality, reproduction, lifespan, development, growth, oxidative stress, intestinal autofluorescence, behaviour (locomotion, pharyngeal pumping, defecation)	Wu et al. (2014c)
TiO$_2$	L1	1–100 mg/L	96 h	NGM (+)	Lethality, reproduction, gene expression, oxidative stress	Angelstorf et al. (2014)
TiO$_2$	Young adult	20 µg/L, 25 mg/L	24 h	K-media (+)	Lethality, growth, reproduction, locomotion, intestinal autofluorescence, ROS formation, gene expression	Rui et al. (2013)
ZnO	L1	0.614–614 µM	72 h	NGM (+)	Reproduction, locomotion	O'Donnell et al. (2017)
ZnO	L4		24 h, 72 h	Milli-Q water (±)	Growth, reproduction, locomotion, gene expression, fertility, oxidative stress	Khare et al. (2015)
ZnO	L4	0.1–2.0 g/L	24 h	NGM (−)	Lethality, oxidative stress	Gupta et al. (2015)
ZnO	L1	5–50 mg/L	48 h, 72 h	NGM (+)	Growth, reproduction, lifespan, ROS formation, gene expression, bacterial growth inhibition	Polak et al. (2014)
TiO$_2$ and ZnO	L1, L4	20–200 mg/mL (TiO$_2$) 0.125–0.8 mg/mL (ZnO)	24 h, 72 h	NGM (±)	Lethality, oxidative stress (ROS formation, antioxidant recovery)	Sonane et al. (2017)

AgNP, ZnO and CeO$_2$	L1/L2	1, 10, 100 μg/mL AgNP; 20, 80, 160 μg/mL for ZnO and CeO$_2$	~40 days (total life)	S-media (+)	Lifespan, behaviour, bag of worm (BOW), neurodegeneration (reporter)	Piechulek and von Mikecz (2018)
Inorganic nanoparticles (SiO$_2$, TiO$_2$, and ZnO)	L4	0.05 mg/mL	~22 days (total life)	NGM (+)	Life span	Ma et al. (2018)
Silver nanoparticles, multiwalled carbon nanotubes (MWCNT) and polyamidoamine dendrimers (PAMAM)	L1	10^{10} particles/mL	24 h, 48 h, 72 h	NGM	Lethality, growth, genome-wide transcriptional changes (microarray)	Walczynska et al. (2018)
Graphene oxide (GO) and reduced graphene oxide (rGO)	Young adult	20–100 mg/L	24 h, 48 h, 72 h	K-media (±)	Lethality, reproduction, gene expression, fluorescence in reporter strains	Chatterjee et al. (2017)
Graphene oxide (GO)	L1	1–100 mg/L	72 h	K-media (+)	Distribution and translocation, behaviour (locomotion, defecation), ROS formation, fluorescence in reporter strains, fat content (lipid and triglyceride level)	Zhi et al. (2016)
Graphene oxide (GO)	L1	1–100 mg/L	72 h	K-media (+)	Lethality, locomotion, oxidative stress	Zhao et al. (2016c)
Graphene oxide (GO)	L1	1–100 mg/L	72 h	K-media (+)	Cell apoptosis, fertility, reproduction, DNA damage, gene expression	Zhao et al. (2016a)
PEG modified graphene oxide (GO-PEG)	L1	1, 10, and 100 mg/L	72 h	K-media (+)	Distribution and translocation, biological permeability, gene expressions	Zhao et al. (2018)
Graphene oxide (GO)	L1	1, 10, and 100 mg/L	72 h	K-media (+)	Genome-wide functional analysis of long-noncoding RNA	Wu et al. (2016b)

(continued)

Table 11.1 (continued)

Nanomaterials	C. elegans (Stage)	Exposure		Media (food '+')	Endpoints	References
Name	Stage	Conc.	Time			
Graphene oxide (GO) and PEG modified graphene oxide (GO-PEG)	L1, young adult	0.001–1 mg/L	11 days, 8 days	K-media (+)	Lethality, behaviour (locomotion, defecation), intestinal auto-fluorescence, ROS formation, immune response (OP50 accumulation, related gene expression), AVL and DVB neurons visualization	Wu et al. (2014a)
Graphene oxide (GO)	L1	100 mg/L	3.5 days	K-media (+)	Distribution and translocation, lifespan, reproduction, effects on targeted organs, behaviour (locomotion, defecation), intestinal permeability (lipid level and triglyceride content); oxidative stress (ROS formation, related mutants)	Wu et al. (2014b)
Graphene oxide (GO)	L1	0.1–100 mg/L	3.5 days	K-media (+)	Distribution and translocation, lifespan, locomotion, autofluorescence, ROS formation, global micro-RNA expression (SOLiD sequencing)	Wu et al. (2014d)
Graphite (G), graphite oxide nanoplatelets (GO) and graphene quantum dots (GQDs)	L1	1–100 mg/L	6 days	K-media (+)	Distribution, lethality, behaviour (locomotion), head thrashing, body bending, pharyngeal pumping, neuronal analysis (reporter strains)	Li et al. (2017)
Graphene oxide (GO)	L1, young adult	10–100 mg/L	24 h, 72 h	K-media (+)	Locomotion, ROS formation, gene expression, GFP reporters	Ren et al. (2017)

Material	Stage	Concentration	Duration	Medium	Endpoints	Reference
Graphene oxide (GO)	L4	0.1, 1, 10 mg/L	24 h	K-media (+)	Distribution-translocation, lipid level, lipofuscin	Ren et al. (2018)
Graphene oxide (GO)	L1	10–100 mg/L	72 h	K-media (+)	Lethality, lifespan, autofluorescence, ROS formation, locomotion, lipid content, gene expression	Qu et al. (2017)
Multiwalled carbon nanotubes (MWCNT)	L1	1 mg/L	3.5 days	K-media (+)	Distribution and translocation, lethality, lifespan, locomotion, GFP reporters	Zhuang et al. (2016)
Multiwalled carbon nanotubes (MWCNT)	L1	1 mg/L	72 h	K-media (+)	Global gene (mRNA, miRNA) expressions (RNA-seq), mRNA-miRNA network, reproduction, locomotion, ROS formation, lipid content	Zhao et al. (2016b)
Pristine and hydroxylated (OH)-MWCNTs	L4	1–500 mg/L	4 h, 24 h, 48 h, 72 h	K-media (±)	Lethality, reproduction, feeding, oxidative stress, global gene expressions (microarray) and proteins (proteomics)	Eom et al. (2015)
Pristine and carboxylated single-walled carbon nanotubes (SWCNTs–COOH)	L1, L4	0.001–1000 µg/L	24 h, 48 h	K-media	Lethality, growth, lifespan, reproduction, ROS generation, locomotion, gene expression	Lu et al. (2022)
Pristine and amide-modified single-walled carbon nanotubes (a-SWCNTs)	L1	1–500 µg/mL	48 h, 72 h	NGM (+)	Lifespan, growth, reproduction, feeding, endocytosis, global transcriptomics (microarray), oxygen consumption rate, DaF-16 GFP reporter	Chen et al. (2013)
Fluorescent single-walled carbon nanotubes	Adult	0.1–300 mg/L	24 h, 8 day	NGM (−)	Viability, imaging	Hendler-Neumark et al. (2021)

PVP polyvinylpyrrolidone, *NGM* Nematode growth medium, *ROS* Reactive oxygen species, *BPEI* Branched polyethyleneimine (BPEI), *PEG* Polyethylene glycol

3.1 General Physiological Endpoints Assessment

3.1.1 Survival/Mortality

In *C. elegans*, assessing survival rates is a primary approach for understanding the toxicity of nanoparticles (NPs). Mortality is usually determined by constructing concentration-response curves, which reflect how different doses of NPs influence the death rate of the nematodes (Gonzalez-Moragas et al. 2015a; Tejeda-Benitez and Olivero-Verbel 2016). A critical aspect of this evaluation is distinguishing between lethality and paralysis. While death is indicated by the complete cessation of movement and physiological activity, paralysis refers to immobility where nematodes still exhibit basic life functions, such as normal pharyngeal pumping (Wang 2018). This distinction is crucial to avoid overestimating the toxic effects of NPs based on immobility alone.

3.1.2 Growth and Development

Growth and developmental outcomes in *C. elegans* provide important sublethal endpoints for assessing NP toxicity. One key indicator of developmental progress is the body length of nematodes, as exposure to NPs can delay their growth, especially at early stages like the first and second larval stages (Hu et al. 2018a, b). The inhibition of growth is often associated with disruptions in key biological processes, such as the endocytic process, which plays a significant role in mitigating NP-induced stress. For example, studies suggest that normal lysosomal function is vital for nematode growth under stress from silver nanoparticles (AgNPs) (Maurer et al. 2016). Additionally, NP toxicity may reduce the availability of food or reduces the food sensation which can further hinder growth by limiting nutrient intake (Meyer et al. 2010; Wang et al. 2023b).

3.1.3 Reproduction

Reproductive health is one of the most sensitive indicators of NP toxicity in *C. elegans*, often affected at lower concentrations than those that impair survival or movement. Reproductive toxicity is measured by comparing the reproductive capabilities of NP-exposed nematodes to a control group, focusing on factors such as the number of offspring, brood size, and rate of egg laying (Kong et al. 2017; Zhao et al. 2016a). A decline in reproductive output, often reflected in reduced brood size or increased sterility, is a common observation following NP exposure. In particular, nanoparticles like ZnO and graphene oxide (GO) have been found to induce damage in the gonads of nematodes through mechanisms such as germline apoptosis and cell cycle arrest, which are triggered by DNA damage (Zhao et al. 2016a; O'Donnell et al. 2017). Furthermore, NP-induced damage may not be confined to a single generation; reproductive abnormalities can be passed down to future generations. For instance, after exposure to AuNPs, the F2 generation exhibited significant reproductive system abnormalities, though these effects gradually diminished by the F4 generation, suggesting an adaptive response across generations (Kim et al. 2013). Similar multi-generational studies' impact underscores the need to consider long-term reproductive effects in NP toxicity studies (Contreras et al. 2013; Moon et al. 2017).

3.1.4 Behavioural Alterations

Behavioral changes in response to environmental stressors, including chemicals and pollutants, have long been recognized as critical indicators of organismal and ecological health. Various environmental contaminants can adversely affect an organism's behavior, influencing key activities such as feeding, locomotion, reproduction, and cognitive functions. These behavioral disruptions can cascade into broader ecological consequences, affecting species interactions, predator-prey dynamics, and ecosystem balance. However, behavioral studies have been underrepresented in regulatory ecotoxicology, primarily due to a lack of standardized methods for assessing these effects (Ford et al. 2021). The growing understanding of how environmental stressors alter behavior has emphasized the need for including behavioral metrics in risk assessments to better capture the full scope of toxicity.

When it comes to nanoparticle (NP) exposure, behavioral toxicity has been extensively studied in *C. elegans*, a key model organism. Nanomaterials such as Al_2O_3NPs, CdTe QDs, oleic acid-coated AgNP impair both locomotion and learning abilities in *C. elegans*, indicating neurotoxic effects (Contreras et al. 2014; Wu et al. 2015; Li et al. 2012). Feeding behavior is also disrupted, with nanoparticles like CdTe QDs and Zein-NPs altering pharyngeal pumping speed, RMEs motor neurons, and defecation cycles, which can lead to increased fat storage (Zhao et al. 2015; Lucio et al. 2017). Moreover, chronic exposure to graphene-based NPs causes a significant reduction in crawling distance, mean speed, and bending reversal frequency, all of which indicate a loss of motor coordination and balance (Li et al. 2017).

3.2 Mechanistic Endpoints Evaluations

3.2.1 Application of Mutants and Transgenic *C. elegans* Strains

The use of reverse genetics allows precise manipulation of gene activity in *C. elegans*, enabling researchers to target any gene in the organism. Tools like small interfering RNAs (siRNAs) are valuable for studying the function of single genes in *C. elegans*. Additionally, the extensive library of transgenic, mutant, and reporter strains from the *C. elegans* consortium offers a valuable resource for studying nanoparticle toxicity. Researchers can use these strains to explore molecular pathways, cellular responses, and genetic variations, providing insights into toxicity mechanisms. Various phenotypic effects, including survival, growth, reproduction, and lifespan changes, have been examined in both wild-type and mutant strains in nanotoxicology studies (Rogers et al. 2015; Wang et al. 2017; Qu et al. 2018; Chatterjee et al. 2017). Moreover, transgenic *C. elegans* strains that replicate human molecular disease mechanisms, which are difficult to study in other models, are utilized to assess the toxic effects of nanoparticles (NPs) in organisms affected by chronic conditions like neurodegenerative diseases. For instance, Soria et al. demonstrated that silver nanoparticles (AgNPs) had a more severe impact on movement and oxidative stress in *C. elegans* strains mimicking Alzheimer's disease than in wild-type strains (Soria et al. 2015).

3.2.2 OMICS Platforms for Gene Expression and Toxicity Pathways

Transcriptomics is a powerful tool to study large-scale gene expression changes in *C. elegans* exposed to NPs. Various studies have shown that exposure to NPs affects genes involved in oxidative stress, metal detoxification, DNA damage, endocytosis, and intestinal integrity, with the extent of these effects dependent on the concentration and exposure duration (Starnes et al. 2019; Tsyusko et al. 2012; Hunt et al. 2014; Rocheleau et al. 2015; Gonzalez-Moragas et al. 2017b). Integrating transcriptomics data with proteomics and metabolomics provides a comprehensive understanding of how NPs influence biological processes, contributing to the development of adverse outcomes relevant to risk assessment (Eom et al. 2015; Ratnasekhar et al. 2015). The combination of multiple OMICS techniques allows a more detailed mapping of NP-induced biological changes at various levels of organization, enabling researchers to link molecular alterations with functional outcomes in *C. elegans*.

3.2.3 Oxidative Stress, Innate Immunity, and Signalling Pathway Alterations

Oxidative stress is considered a key mechanism through which NPs cause toxicity in *C. elegans*. The accumulation of reactive oxygen species (ROS) in NP-treated nematodes has been linked to adverse outcomes such as reduced lifespan, impaired growth, and reproductive damage, in a dose- and time-dependent manner (Wu et al. 2012a, b; Ahn et al. 2014; Eom et al. 2013; Lim et al. 2012; Yu et al. 2011; Li et al. 2012). Excessive ROS generation can lead to functional defects even in organs that do not retain NPs, such as reduced locomotion and reproductive issues. Interestingly, pre-treatment with antioxidants like ascorbate or N-acetyl-l-cysteine (NAC) can mitigate these effects (Wu et al. 2013; Lim et al. 2012; Li et al. 2012).

Several signalling pathways, including mitochondrial complex I and MAPK pathways, have been identified as critical regulators in controlling NP-induced oxidative stress and toxicity (Lim et al. 2012; Li et al. 2020; Teng et al. 2024; Eom et al. 2013). Additionally, genes like sod-3, gst-4, and hsp-16, which are associated with stress responses, have been highlighted as sensitive markers for NP toxicity (Li et al. 2012; Zhao et al. 2015; Rui et al. 2013; Wu et al. 2014b).

The exposure of *C. elegans* to nanoparticles (NPs) leads to significant alterations in multiple signalling pathways, which are essential for understanding the mechanisms of NP-induced toxicity. One key pathway is the Wnt signaling pathway, where ligands like CWN-1, CWN-2, and LIN-44 regulate NP toxicity by controlling NP accumulation, with mutations in these genes either increasing resistance or susceptibility (Zhi et al. 2016; Chatterjee et al. 2017). Similarly, the insulin/IGF-1 pathway, particularly through the DAF-2/DAF-16 axis, is involved in longevity and stress resistance, with miRNAs such as mir-355 modulating NP toxicity via insulin signalling (Zhao et al. 2016b). Additionally, the TGF-β pathway is implicated in reproductive toxicity, where disruption by NPs, such as titanium dioxide, causes damage to reproductive capacity and developmental processes (Kim et al. 2017a).

The MAPK signalling pathway is also critical in stress responses. In particular, the p38 MAPK-SKN-1/Nrf cascade is involved in the innate immune response, offering protection against oxidative stress induced by NPs like graphene oxide (Zhao et al. 2016c). Chronic GO exposure impairs immune function by causing the accumulation of pathogenic microbes like OP50 in the intestine, which disrupts innate immunity. However, surface modification, such as PEG, reduces this toxicity (Wu et al. 2014a). GO also activates the p38 MAPK pathway, with PMK-1 playing a key protective role, while amino-functionalized GO shows less immunotoxicity, highlighting the importance of nanoparticle modification (Rive et al. 2019). AgNPs trigger oxidative stress and activate PMK-1, leading to immune defence responses (Lim et al. 2012). The ERK signalling pathway is also involved in regulating GO toxicity, working in synergy with p38 MAPK to control immune responses (Qu et al. 2017). Additionally, ZnO-NPs suppress innate immunity regulated by SKN-1/Nrf and the p38 MAPK signalling pathway, decreasing survival during infection and downregulating key immune genes (Li et al. 2020).

3.2.4 Neurotoxicity and Neurodegeneration

Nearly all behavioural endpoints in *C. elegans*—such as locomotion, body bending, feeding, defecation, pharyngeal pumping, egg-laying, sensory perception, learning, and memory—are controlled by the nervous system and achieved through muscle contractions. Exposure to nanoparticles (NPs) has been shown to disrupt these behaviours. For instance, a reduction in feeding and defecation behaviours is often linked to NP-induced stress and alterations in pharyngeal pumping and defecation cycles (Wu et al. 2015). CdTe quantum dots (QDs), graphene-based nanomaterials, and copper oxide nanoparticles have been found to cause significant damage to dopamine and glutamatergic neurons in *C. elegans*, leading to abnormal feeding behaviour developmental deficits, neurodegeneration, and abnormalities in the neural network (Zhao et al. 2015; Mashock et al. 2016; Li et al. 2017). Nanoparticles like silver (AgNPs) have been shown to impair a range of neuronal systems, including dopaminergic, GABAergic, and cholinergic neurons, affecting locomotion and sensory perception. The severity of these effects depends on both the dose and duration of exposure (Zhang et al. 2021). Additionally, hybrid nanoparticles such as Fe_3O_4@Ag-NPs have been linked to neurotoxicity by disrupting cholinergic neurons and inducing oxidative stress, leading to behavioural impairments and apoptosis in *C. elegans* (Silva et al. 2023). Graphene oxide (GO) NPs also exhibit considerable neurotoxicity. GO exposure causes damage to AFD sensory neurons, reduces neurotransmitter levels such as dopamine, GABA, and tyramine, and leads to altered locomotion behaviors like reduced speed and coordination (Kim et al. 2020a). Silica (SiO_2) nanoparticles have also shown neurotoxic effects, particularly in disrupting serotonergic neurotransmission. These impairments are associated with neuromuscular defects, notably affecting the egg-laying apparatus in *C. elegans*, which can be mitigated by anti-amyloid compounds (Scharf et al. 2016). This indicates that SiO_2 NPs can interfere with reproductive and muscular systems, compounding their neurotoxic effects. Additionally, exposure to titanium dioxide (TiO_2) NPs has been linked to neuron damage and impaired locomotion, further highlighting the broad toxicological impact of various nanomaterials (Hu et al. 2018a).

3.2.5 Genotoxicity, Mutation, DNA Damage Response, and Apoptosis

Genotoxicity in *C. elegans* can be assessed through several established techniques. Methods like qPCR measure DNA damage by detecting how lesions inhibit polymerase progression, with the extent of damage indicated by the length of the PCR products (Leung et al. 2010). The comet assay has been used to evaluate the genotoxicity of environmental pollutants (Imanikia et al. 2016). Additionally, transgenic strains like hus-1::GFP are utilized to visualize DNA double-strand breaks, where fluorescent foci in gonadal germ cells indicate the extent of damage, allowing precise quantification (Wang et al. 2014; Hofmann et al. 2002). These methods provide a clear understanding of the DNA damage response, involving checkpoint activation that leads to either cell cycle arrest or repair or, in severe cases, apoptosis (Gartner et al. 2004; Craig et al. 2012).

Nanoparticles, particularly silver nanoparticles (AgNPs), graphene oxide (GO), and zinc oxide nanoparticles (ZnO NPs), have been shown to induce significant genotoxicity in *C. elegans*. Smaller, uncoated AgNPs, for example, cause oxidative stress that leads to mitochondrial membrane damage and oxidative DNA damage, such as 8-OHdG lesions (Ahn et al. 2014). This oxidative DNA damage triggers the activation of DNA repair mechanisms, such as DNA glycosylases like NTH-1, which specifically repair oxidative lesions. PMK-1, a p38 MAPK homolog, also plays a protective role in mitigating AgNP-induced DNA damage through repair pathways (Chatterjee et al. 2014a). Similarly, GO nanoparticles activate key components of the apoptosis pathway, such as cep-1 (a homolog of p53), egl-1, ced-4, and ced-3, which either arrest the cell cycle or induce apoptosis when DNA damage becomes too severe to repair, highlighting the role of these pathways in maintaining genomic integrity (Zhao et al. 2016a). Surface modifications, such as coating GO nanoparticles with bovine serum albumin (BSA), have been shown to reduce the activation of DNA damage checkpoints and apoptosis-related genes, thus lowering toxicity (Sivaselvam et al. 2020). Furthermore, prolonged exposure to AgNPs over multiple generations has been linked to the accumulation of DNA damage, insufficient repair activation, and the inheritance of reproductive and developmental defects (Wamucho et al. 2019). Similarly, ZnO NPs disrupt germ cell development, triggering apoptosis through DNA damage checkpoints and causing chromosomal deletions, which impair reproductive capacity (Wang et al. 2023a).

3.2.6 Epigenetic Biomarkers

Emerging research suggests that epigenetic mechanisms, particularly microRNAs (miRNAs), play crucial roles in mediating protective or harmful responses in *C. elegans* exposed to NPs. For example, prolonged exposure to graphene oxide (GO) was shown to significantly affect miRNA-regulated biological processes like development, reproduction, and cell cycle regulation (Wu et al. 2014d). Certain miRNAs, such as mir-259 and mir-360, have been identified as key players in protecting against NP-induced oxidative stress and DNA damage in nematodes (Zhuang et al. 2016; Zhao et al. 2016a). Furthermore, miRNA-mRNA interaction networks, including the regulation of mir-355 with the DAF-2/insulin receptor, have been linked to the modulation of NP toxicity in *C. elegans* (Zhao et al. 2016b). Similarly, long non-coding RNAs (lncRNAs) have been implicated in controlling NP toxicity, further emphasizing the importance of epigenetic regulation in the organism's response to environmental stressors (Wu et al. 2016b).

3.3 Factors Affecting the Nano-Bio Interaction in *C. elegans*

3.3.1 Exposure

When assessing the toxicity of nanoparticles, it is crucial to consider both exposure concentration and duration. Researchers often use exposure ranges from non-toxic to threshold levels to establish dose-effect relationships. However, creating a precise dose tolerance curve for a specific nanoparticle is challenging due to variations in study conditions. Lower-order developmental stages, such as L1 larvae, are typically more sensitive than later stages, such as young adults. Moreover, longer exposure times generally lead to more severe effects compared to shorter ones, though hormesis effects observed in short-term exposures may diminish with prolonged exposure (Tyne et al. 2015). Additionally, intermittent exposure can sometimes produce more pronounced effects than continuous exposure, highlighting the importance of considering both exposure time and historical exposure in toxicity evaluations (Moon et al. 2017).

Nanoparticles often exhibit unstable behaviour in liquid media, such as K-medium and S-medium, where they can aggregate to sizes over 100 times their original dimensions and precipitate, thus reducing the effective exposure dose to organisms. Additionally, some metal nanoparticles in liquid media may partially dissolve or release ions due to hydration kinetics, complicating toxicity assessments. While the release of metallic ions is believed to contribute to observed toxicity, it remains unclear whether the effects are due to the particles themselves or the ions. Researchers suggest that simulated soil pore water (SSPW) provides a more realistic testing environment for metal nanoparticle toxicity in *C. elegans* due to its low ionic strength and organic content, which stabilize the nanoparticles (Tyne et al. 2013). In contrast, applying nanomaterials to whole NGM agar media can affect the effective exposure dose because the worms interact only with the solid surface of the NGM, not with the entire medium. Mixing nanoparticles with viable E. coli OP50, used as food, can alter nanoparticle transformation and toxicity evaluations. Applying a mixture of deactivated E. coli and selected nanomaterials spread over the surface of solid NGM plates as a lawn provides a more reliable exposure medium by minimizing biotransformation and enhancing nano-bio interactions, as demonstrated for diesel exhaust particles (Chatterjee et al. 2024). Additionally, semi-fluid nematode growth gelrite medium (Dengg and van Meel 2004) is suitable for nanoparticle toxicity evaluation compared to standard nematode growth medium (NGM) and K-medium, with Ag-NPs demonstrating stability in NGG without increased dissolution of Ag ions over time (Luo et al. 2017). Therefore, the choice of exposure medium—liquid, solid, or bacterial suspension—plays a crucial role in determining the effective concentration and toxicity of nanoparticles, underscoring the need for standardized testing protocols.

3.3.2 Physiochemical Properties of Nanomaterials

The physicochemical properties of nanomaterials, such as size, shape, surface modification, and charge, significantly influence their toxicity and biological interactions. These properties can affect how nanomaterials are absorbed, distributed, and accumulated within organisms, ultimately impacting their potential health risks and environmental effects.

Size

The correlation between nanoparticle size and toxicity is significant, with smaller nanoparticles generally causing more severe effects in *C. elegans* compared to larger ones. Smaller particles can penetrate more easily, leading to increased toxicity (Khare et al. 2015; Roh et al. 2010), possibly through mechanisms such as alterations in metabolic pathways (Ratnasekhar et al. 2015) or the formation of aggregates that limit food availability (Luo et al. 2016). However, the same study suggests that larger particles may accumulate more within the body, potentially causing long-term effects such as reduced lifespan, while impaired reproductive capacity was observed with smaller particle exposure (Contreras et al. 2014). The impact of nanoparticle size on toxicity is complex and may depend on factors such as agglomeration state and particle-specific effects (Jung et al. 2015).

Coating and Surface Modification

Surface modifications and coatings can significantly influence the toxicity of nanoparticles. Sulfidized silver nanoparticles (AgNPs), for example, exhibit reduced toxicity compared to uncoated AgNPs due to decreased solubility and limited silver ion release, which lowers their bioavailability and particle-specific toxicity (Starnes et al. 2015). Similarly, citrate coatings on AgNPs reduce silver ion availability, although they are less effective than BSA coatings (Yang et al. 2012; Hunt et al. 2014; Meyer et al. 2010). CdTe quantum dots (QDs) with ZnS coatings, unlike bare CdTe QDs, did not translocate into motor neurons, thereby avoiding neurotoxicity (Zhao et al. 2015). Surface modifications such as hydroxylation, carboxylation, and amination have also reduced the reproductive toxicity of multi-walled carbon nanotubes (MWCNTs), especially carboxylation, which might facilitate the elimination of functionalized MWCNTs than the pristine one (Chatterjee et al. 2014b). Additionally, PEG modification, commonly used in nanoparticles, effectively mitigates the negative effects of graphene oxide (GO) on both primary and secondary target organs (Wu et al. 2016b). However, some coatings, like gum arabic, can increase nanoparticle toxicity, while others, such as polyvinylpyrrolidone (PVP), show conflicting results, with studies reporting both higher and lower toxicity compared to uncoated nanoparticles (Bone et al. 2015; Yang et al. 2012; Ellegaard-Jensen et al. 2012; Ahn et al. 2014).

Charge

Positively charged nanoparticles tend to be more toxic to *C. elegans* and accumulate more than neutral or negatively charged particles. This increased toxicity and bioaccumulation are observed in most cases, highlighting the importance of particle charge in toxicity assessments (Collin et al. 2014; Arndt et al. 2017).

Shape

The shape of nanoparticles can influence their toxic effects in *C. elegans*. For example, different shapes of TiO_2 nanoparticles exhibit varying effects on pharyngeal function, reproduction, and larval growth (Iannarelli et al. 2016). Anatase-TiO_2 had a stronger impact on metabolic pathways compared to rutile, while rutile-TiO_2 influenced developmental processes more significantly (Rocheleau et al. 2015). Silver nanocubes generally show lower toxicity compared to quasi-spherical silver nanoparticles and silver nanowires, indicating that shape engineering can optimize nanoparticle properties while minimizing adverse effects (Gorka et al. 2015). Additionally, the crystalline structure could explain the differences in agglomeration behaviour observed in the intestine, which in turn influenced the reproductive toxicity of the TiO_2 material (Angelstorf et al. 2014).

3.3.3 Other Factors

Environmental factors such as UV irradiation can enhance the toxicity of metal oxide nanoparticles like ZnO and TiO_2 through mechanisms such as photocatalytic ROS generation and photo-enhanced dissolution (Ma et al. 2011, 2014; Lee and An 2013). Moreover, the stability and toxicity of nanoparticles are influenced by dissolved organic matter and the physiological properties of the test organism, such as pH and biomolecular interactions within the intestinal lumen (Gonzalez-Moragas et al. 2017a). Variations in toxicity may also result from differences in material formulation, nematode life stage, and testing procedures (Ma et al. 2013).

4 Conclusion and Perspectives

C. elegans has proven to be an effective and versatile model in nanotoxicology studies, particularly for initial biological screenings of nanoparticles (NPs). Its small size, low cost, and short lifespan facilitate large-scale, long-term toxicity assessments under controlled conditions, making it ideal for chronic exposure studies (Leung et al. 2008). Additionally, the transparency of *C. elegans* enables straightforward observation of NPs at both molecular and cellular levels, especially when using transgenic strains that express fluorescent markers (Scharf et al. 2013). Advances such as microfluidic chip platforms further enhance its utility, offering a high-throughput, on-site system for rapidly assessing NP uptake and toxicity while reducing labor and time requirements (Mondal et al. 2016; Rohde et al. 2007). These features, along with the nematode's genetic tractability and the conservation of many molecular pathways with humans (Kaletta and Hengartner 2006; Markaki and Tavernarakis 2020), make *C. elegans* a robust platform for nanotoxicology research (Wu et al. 2019).

Nevertheless, *C. elegans* has inherent limitations when used in nanotoxicology studies, particularly in comparison to mammalian models. For example, it lacks key mammalian organs such as the heart, kidneys, bones, and eyes, rendering it unsuitable for evaluating NP toxicity in these organ-specific systems. Additionally, the absence of a circulatory system restricts its ability to mimic intravenous NP exposure scenarios (Tejeda-Benitez and Olivero-Verbel 2016).

Despite these limitations, *C. elegans* continues to excel as a bridging model between ecological and human health risk assessments, aligning well with the 3R principles (Replacement, Reduction, and Refinement) and New Approach Methodologies (NAMs). By connecting in vitro and in vivo assessments, it supports more ethical, cost-effective, and efficient toxicity testing. Its capacity to evaluate a range of endpoints—including lethality, growth, reproduction, fertility, and locomotion—makes it invaluable for early-stage evaluations of nanomaterials. Furthermore, as a fully sequenced organism with high genetic tractability, *C. elegans* offers the added benefit of creating transgenic strains to study gene expression changes in response to toxicants and nanomaterials. This capability allows researchers to gain mechanistic insights into gene regulation and biochemical pathways affected by pollutants, toxicants, and nanoparticles. By observing direct molecular responses, such as changes in gene expression, *C. elegans* helps uncover the biological mechanisms underlying toxicity at various levels—from single-cell interactions to whole-organism responses. Consequently, *C. elegans* remains a highly effective and versatile model for advancing nanotoxicology, providing critical data that can enhance the safety and regulation of emerging nanomaterials.

Acknowledgments This study was funded by the European Union's H2020 projects, Sinfonia (N.857253), LEARN (N.101057510), iCare (N.101092971) and SbDToolBox, with reference NORTE-01-0145-FEDER-000047, supported by Norte Portugal Regional Operational Programme (NORTE 2020), under the PORTUGAL 2020 Partnership Agreement, through the European Regional Development Fund.

Declaration of Competing Interest None.

References

Acosta C, Barat JM, Martínez-Máñez R, Sancenón F, Llopis S, González N, Genovés S, Ramón D, Martorell P (2018) Toxicological assessment of mesoporous silica particles in the nematode Caenorhabditis elegans. Environ Res 166:61–70. https://doi.org/10.1016/j.envres.2018.05.018

Ahn JM, Eom HJ, Yang X, Meyer JN, Choi J (2014) Comparative toxicity of silver nanoparticles on oxidative stress and DNA damage in the nematode Caenorhabditis elegans. Chemosphere 108:343–352. https://doi.org/10.1016/j.chemosphere.2014.01.078

Anbalagan C, Lafayette I, Antoniou-Kourounioti M, Gutierrez C, Martin JR, Chowdhuri DK, De Pomerai DI (2013) Use of transgenic GFP reporter strains of the nematode Caenorhabditis elegans to investigate the patterns of stress responses induced by pesticides and by organic extracts from agricultural soils. Ecotoxicology 22:72–85. https://doi.org/10.1007/s10646-012-1004-2

Angelstorf JS, Ahlf W, Von Der Kammer F, Heise S (2014) Impact of particle size and light exposure on the effects of TiO2 nanoparticles on Caenorhabditis elegans. Environ Toxicol Chem 33:2288–2296. https://doi.org/10.1002/etc.2674

Arndt DA, Oostveen EK, Triplett J, Butterfield DA, Tsyusko OV, Collin B, Starnes DL, Cai J, Klein JB, Nass R, Unrine JM (2017) The role of charge in the toxicity of polymer-coated cerium oxide nanomaterials to Caenorhabditis elegans. Comp Biochem Physiol C Toxicol Pharmacol 201:1–10. https://doi.org/10.1016/j.cbpc.2017.08.009

Avila DS, Adams MR, Chakraborty S, Aschner M (2011) Chapter 16—Caenorhabditis elegans as a model to assess reproductive and developmental toxicity. In: Gupta RC (ed) Reproductive and developmental toxicology. Academic, San Diego. https://doi.org/10.1016/B978-0-12-382032-7.10016-5

Bone AJ, Matson CW, Colman BP, Yang X, Meyer JN, Di Giulio RT (2015) Silver nanoparticle toxicity to Atlantic killifish (Fundulus heteroclitus) and Caenorhabditis elegans: a comparison of mesocosm, microcosm, and conventional laboratory studies. Environ Toxicol Chem 34:275–282. https://doi.org/10.1002/etc.2806

Brenner S (2009) In the beginning was the worm. Genetics 182:413–415. https://doi.org/10.1534/genetics.109.104976

Buckingham SD, Sattelle DB (2009) Fast, automated measurement of nematode swimming (thrashing) without morphometry. BMC Neurosci 10:84. https://doi.org/10.1186/1471-2202-10-84

Chakravarty B (2022) The evolving role of the Caenorhabditis elegans model as a tool to advance studies in nutrition and health. Nutr Res 106:47–59. https://doi.org/10.1016/j.nutres.2022.05.006

Chalfie M, Tu Y, Euskirchen G, Ward WW, Prasher DC (1994) Green fluorescent protein as a marker for gene expression. Science 263:802–805. https://doi.org/10.1126/science.8303295

Chatterjee N, Eom HJ, Choi J (2014a) Effects of silver nanoparticles on oxidative DNA damage-repair as a function of p38 MAPK status: a comparative approach using human Jurkat T cells and the nematode Caenorhabditis elegans. Environ Mol Mutagen 55:122–133. https://doi.org/10.1002/em.21844

Chatterjee N, Yang J, Kim HM, Jo E, Kim PJ, Choi K, Choi J (2014b) Potential toxicity of differential functionalized multiwalled carbon nanotubes (MWCNT) in human cell line (BEAS2B) and Caenorhabditis elegans. J Toxicol Environ Health A 77:1399–1408. https://doi.org/10.1080/15287394.2014.951756

Chatterjee N, Kim Y, Yang J, Roca CP, Joo SW, Choi J (2017) A systems toxicology approach reveals the Wnt-MAPK crosstalk pathway mediated reproductive failure in Caenorhabditis elegans exposed to graphene oxide (GO) but not to reduced graphene oxide (RGO). Nanotoxicology 11:76–86. https://doi.org/10.1080/17435390.2016.1267273

Chatterjee N, González-Durruthy M, Costa MD, Ribeiro AR, Vilas-Boas V, Vilasboas-Campos D, Maciel P, Alfaro-Moreno E (2024) Differential impact of diesel exhaust particles on glutamatergic and dopaminergic neurons in Caenorhabditis elegans: a neurodegenerative perspective. Environ Int 186:108597. https://doi.org/10.1016/j.envint.2024.108597

Chen PH, Hsiao KM, Chou CC (2013) Molecular characterization of toxicity mechanism of single-walled carbon nanotubes. Biomaterials 34:5661–5669. https://doi.org/10.1016/j.biomaterials.2013.03.093

Collin B, Oostveen E, Tsyusko OV, Unrine JM (2014) Influence of natural organic matter and surface charge on the toxicity and bioaccumulation of functionalized ceria nanoparticles in Caenorhabditis elegans. Environ Sci Technol 48:1280–1289. https://doi.org/10.1021/es404503c

Contreras EQ, Cho M, Zhu H, Puppala HL, Escalera G, Zhong W, Colvin VL (2013) Toxicity of quantum dots and cadmium salt to Caenorhabditis elegans after multigenerational exposure. Environ Sci Technol 47:1148–1154. https://doi.org/10.1021/es3036785

Contreras EQ, Puppala HL, Escalera G, Zhong W, Colvin VL (2014) Size-dependent impacts of silver nanoparticles on the lifespan, fertility, growth, and locomotion of Caenorhabditis elegans. Environ Toxicol Chem 33:2716–2723. https://doi.org/10.1002/etc.2705

Craig AL, Moser SC, Bailly AP, Gartner A (2012) Methods for studying the DNA damage response in the Caenorhabditis elegans germ line. In: Rothman JH, Singson A (eds) Methods in cell biology. Academic. https://doi.org/10.1016/B978-0-12-394620-1.00011-4

Dengg M, Van Meel JC (2004) Caenorhabditis elegans as model system for rapid toxicity assessment of pharmaceutical compounds. J Pharmacol Toxicol Methods 50:209–214. https://doi.org/10.1016/j.vascn.2004.04.002

Ellegaard-Jensen L, Jensen KA, Johansen A (2012) Nano-silver induces dose-response effects on the nematode Caenorhabditis elegans. Ecotoxicol Environ Saf 80:216–223. https://doi.org/10.1016/j.ecoenv.2012.03.003

Eom HJ, Choi J (2019) Clathrin-mediated endocytosis is involved in uptake and toxicity of silica nanoparticles in Caenorhabditis elegans. Chem Biol Interact 311:108774. https://doi.org/10.1016/j.cbi.2019.108774

Eom HJ, Ahn JM, Kim Y, Choi J (2013) Hypoxia inducible factor-1 (HIF-1)–flavin containing monooxygenase-2 (FMO-2) signaling acts in silver nanoparticles and silver ion toxicity in the nematode, Caenorhabditis elegans. Toxicol Appl Pharmacol 270:106–113. https://doi.org/10.1016/j.taap.2013.03.028

Eom HJ, Roca CP, Roh JY, Chatterjee N, Jeong JS, Shim I, Kim HM, Kim PJ, Choi K, Giralt F, Choi J (2015) A systems toxicology approach on the mechanism of uptake and toxicity of MWCNT in Caenorhabditis elegans. Chem Biol Interact 239:153–163. https://doi.org/10.1016/j.cbi.2015.06.031

Ferreira DW, Chen Y, Allard P (2014) Using the alternative model C. elegans in reproductive and developmental toxicology studies. In: Faqi AS (ed) Developmental and reproductive toxicology. Springer New York, New York. https://doi.org/10.1007/7653_2014_27

Ford AT, Ågerstrand M, Brooks BW, Allen J, Bertram MG, Brodin T, Dang Z, Duquesne S, Sahm R, Hoffmann F, Hollert H, Jacob S, Klüver N, Lazorchak JM, Ledesma M, Melvin SD, Mohr S, Padilla S, Pyle GG, Scholz S, Saaristo M, Smit E, Steevens JA, Van Den Berg S, Kloas W, Wong BBM, Ziegler M, Maack G (2021) The role of behavioral ecotoxicology in environmental protection. Environ Sci Technol 55:5620–5628. https://doi.org/10.1021/acs.est.0c06493

Frézal L, Félix MA (2015) C. elegans outside the Petri dish. elife 4:e05849. https://doi.org/10.7554/eLife.05849

Gartner A, Macqueen AJ, Villeneuve AM (2004) Methods for analyzing checkpoint responses in Caenorhabditis elegans. Methods Mol Biol 280:257–274. https://doi.org/10.1385/1-59259-788-2:257

Gonzalez-Moragas L, Roig A, Laromaine A (2015a) C. elegans as a tool for in vivo nanoparticle assessment. Adv Colloid Interf Sci 219:10–26. https://doi.org/10.1016/j.cis.2015.02.001

Gonzalez-Moragas L, Yu SM, Carenza E, Laromaine A, Roig A (2015b) Protective effects of bovine serum albumin on superparamagnetic iron oxide nanoparticles evaluated in the nematode Caenorhabditis elegans. ACS Biomater Sci Eng 1:1129–1138. https://doi.org/10.1021/acsbiomaterials.5b00253

Gonzalez-Moragas L, Berto P, Vilches C, Quidant R, Kolovou A, Santarella-Mellwig R, Schwab Y, Stürzenbaum S, Roig A, Laromaine A (2017a) In vivo testing of gold nanoparticles using the Caenorhabditis elegans model organism. Acta Biomater 53:598–609. https://doi.org/10.1016/j.actbio.2017.01.080

Gonzalez-Moragas L, Yu S-M, Benseny-Cases N, Stürzenbaum S, Roig A, Laromaine A (2017b) Toxicogenomics of iron oxide nanoparticles in the nematode C. elegans. Nanotoxicology 11:647–657. https://doi.org/10.1080/17435390.2017.1342011

Gorka DE, Osterberg JS, Gwin CA, Colman BP, Meyer JN, Bernhardt ES, Gunsch CK, Di Giulio RT, Liu J (2015) Reducing environmental toxicity of silver nanoparticles through shape control. Environ Sci Technol 49:10093–10098. https://doi.org/10.1021/acs.est.5b01711

Gupta S, Kushwah T, Vishwakarma A, Yadav S (2015) Optimization of ZnO-NPs to investigate their safe application by assessing their effect on soil nematode Caenorhabditis elegans. Nanoscale Res Lett 10:303. https://doi.org/10.1186/s11671-015-1010-4

Helmcke KJ, Avila DS, Aschner M (2010) Utility of Caenorhabditis elegans in high throughput neurotoxicological research. Neurotoxicol Teratol 32:62–67. https://doi.org/10.1016/j.ntt.2008.11.005

Hendler-Neumark A, Wulf V, Bisker G (2021) In vivo imaging of fluorescent single-walled carbon nanotubes within C. elegans nematodes in the near-infrared window. Mater Today Bio 12:100175. https://doi.org/10.1016/j.mtbio.2021.100175

Hofmann ER, Milstein S, Boulton SJ, Ye M, Hofmann JJ, Stergiou L, Gartner A, Vidal M, Hengartner MO (2002) Caenorhabditis elegans hus-1 is a DNA damage checkpoint protein required for genome stability and egl-1-mediated apoptosis. Curr Biol 12:1908–1918. https://doi.org/10.1016/s0960-9822(02)01262-9

Hu CC, Wu GH, Hua TE, Wagner OI, Yen TJ (2018a) Uptake of TiO2 nanoparticles into C. elegans neurons negatively affects axonal growth and worm locomotion behavior. ACS Appl Mater Interfaces 10:8485–8495. https://doi.org/10.1021/acsami.7b18818

Hu CC, Wu GH, Lai SF, Muthaiyan Shanmugam M, Hwu Y, Wagner OI, Yen TJ (2018b) Toxic effects of size-tunable gold nanoparticles on Caenorhabditis elegans development and gene regulation. Sci Rep 8:15245. https://doi.org/10.1038/s41598-018-33585-7

Hulme SE, Shevkoplyas SS, Apfeld J, Fontana W, Whitesides GM (2007) A microfabricated array of clamps for immobilizing and imaging C. elegans. Lab Chip 7:1515–1523. https://doi.org/10.1039/B707861G

Hunt PR, Marquis BJ, Tyner KM, Conklin S, Olejnik N, Nelson BC, Sprando RL (2013) Nanosilver suppresses growth and induces oxidative damage to DNA in Caenorhabditis elegans. J Appl Toxicol 33:1131–1142. https://doi.org/10.1002/jat.2872

Hunt PR, Keltner Z, Gao X, Oldenburg SJ, Bushana P, Olejnik N, Sprando RL (2014) Bioactivity of nanosilver in Caenorhabditis elegans: effects of size, coat, and shape. Toxicol Rep 1:923–944. https://doi.org/10.1016/j.toxrep.2014.10.020

Iannarelli L, Giovannozzi AM, Morelli F, Viscotti F, Bigini P, Maurino V, Spoto G, Martra G, Ortel E, Hodoroaba VD, Rossi AM, Diomede L (2016) Shape engineered TiO2 nanoparticles in Caenorhabditis elegans: a Raman imaging-based approach to assist tissue-specific toxicological studies. RSC Adv 6:70501–70509. https://doi.org/10.1039/C6RA09686G

Imanikia S, Galea F, Nagy E, Phillips DH, Stürzenbaum SR, Arlt VM (2016) The application of the comet assay to assess the genotoxicity of environmental pollutants in the nematode Caenorhabditis elegans. Environ Toxicol Pharmacol 45:356–361. https://doi.org/10.1016/j.etap.2016.06.020

Jung SK, Qu X, Aleman-Meza B, Wang T, Riepe C, Liu Z, Li Q, Zhong W (2015) Multi-endpoint, high-throughput study of nanomaterial toxicity in Caenorhabditis elegans. Environ Sci Technol 49:2477–2485. https://doi.org/10.1021/es5056462

Kaletta T, Hengartner MO (2006) Finding function in novel targets: C. elegans as a model organism. Nat Rev Drug Discov 5:387–398. https://doi.org/10.1038/nrd2031

Khare P, Sonane M, Nagar Y, Moin N, Ali S, Gupta KC, Satish A (2015) Size dependent toxicity of zinc oxide nanoparticles in soil nematode Caenorhabditis elegans. Nanotoxicology 9:423–432. https://doi.org/10.3109/17435390.2014.940403

Kim SW, Nam SH, An YJ (2012) Interaction of silver nanoparticles with biological surfaces of Caenorhabditis elegans. Ecotoxicol Environ Saf 77:64–70. https://doi.org/10.1016/j.ecoenv.2011.10.023

Kim SW, Kwak JI, An YJ (2013) Multigenerational study of gold nanoparticles in Caenorhabditis elegans: transgenerational effect of maternal exposure. Environ Sci Technol 47:5393–5399. https://doi.org/10.1021/es304511z

Kim H, Jeong J, Chatterjee N, Roca CP, Yoon D, Kim S, Kim Y, Choi J (2017a) Jak/Stat and TGF-β activation as potential adverse outcome pathway of TiO2 nanoparticles phototoxicity in Caenorhabditis elegans. Sci Rep 7:17833. https://doi.org/10.1038/s41598-017-17495-8

Kim JH, Lee SH, Cha YJ, Hong SJ, Chung SK, Park TH, Choi SS (2017b) C. elegans-on-a-chip for in situ and in vivo Ag nanoparticles' uptake and toxicity assay. Sci Rep 7:40225. https://doi.org/10.1038/srep40225

Kim M, Eom HJ, Choi I, Hong J, Choi J (2020a) Graphene oxide-induced neurotoxicity on neurotransmitters, AFD neurons and locomotive behavior in Caenorhabditis elegans. Neurotoxicology 77:30–39. https://doi.org/10.1016/j.neuro.2019.12.011

Kim Y, Jeong J, Lee S, Choi I, Choi J (2020b) Identification of adverse outcome pathway related to high-density polyethylene microplastics exposure: Caenorhabditis elegans transcription factor RNAi screening and zebrafish study. J Hazard Mater 388:121725. https://doi.org/10.1016/j.jhazmat.2019.121725

Kleiven M, Rossbach LM, Gallego-Urrea JA, Brede DA, Oughton DH, Coutris C (2018) Characterizing the behavior, uptake, and toxicity of NM300K silver nanoparticles in Caenorhabditis elegans. Environ Toxicol Chem 37:1799–1810

Kong L, Gao X, Zhu J, Zhang T, Xue Y, Tang M (2017) Reproductive toxicity induced by nickel nanoparticles in Caenorhabditis elegans. Environ Toxicol 32:1530–1538. https://doi.org/10.1002/tox.22373

Kuhn EC, Jacques MT, Teixeira D, Meyer S, Gralha T, Roehrs R, Camargo S, Schwerdtle T, Bornhorst J, Ávila DS (2021) Ecotoxicological assessment of Uruguay River and affluents pre- and post-pesticides' application using Caenorhabditis elegans for biomonitoring. Environ Sci Pollut Res Int 28:21730–21741. https://doi.org/10.1007/s11356-020-11986-4

Lagido C, McLaggan D, Flett A, Pettitt J, Glover LA (2009) Rapid sublethal toxicity assessment using bioluminescent *Caenorhabditis elegans*, a novel whole-animal metabolic biosensor. Toxicol Sci 109(1):88–95. https://doi.org/10.1093/toxsci/kfp058

Lee WM, An YJ (2013) Effects of zinc oxide and titanium dioxide nanoparticles on green algae under visible, UVA, and UVB irradiations: no evidence of enhanced algal toxicity under UV pre-irradiation. Chemosphere 91(4):536–544. https://doi.org/10.1016/j.chemosphere.2012.12.033

Leung MC, Williams PL, Benedetto A, Au C, Helmcke KJ, Aschner M, Meyer JN (2008) *Caenorhabditis elegans*: an emerging model in biomedical and environmental toxicology. Toxicol Sci 106(1):5–28. https://doi.org/10.1093/toxsci/kfn121

Leung MC, Goldstone JV, Boyd WA, Freedman JH, Meyer JN (2010) *Caenorhabditis elegans* generates biologically relevant levels of genotoxic metabolites from aflatoxin B1 but not benzo[a]pyrene in vivo. Toxicol Sci 118(2):444–453. https://doi.org/10.1093/toxsci/kfq295

Leung CK, Deonarine A, Strange K, Choe KP (2011) High-throughput screening and biosensing with fluorescent *C. elegans* strains. J Vis Exp 51:e2745. https://doi.org/10.3791/2745

Li Y, Yu S, Wu Q, Tang M, Pu Y, Wang D (2012) Chronic Al2O3-nanoparticle exposure causes neurotoxic effects on locomotion behaviors by inducing severe ROS production and disruption of ROS defense mechanisms in nematode *Caenorhabditis elegans*. J Hazard Mater 219–220:221–230. https://doi.org/10.1016/j.jhazmat.2012.03.083

Li P, Xu T, Wu S, Lei L, He D (2017) Chronic exposure to graphene-based nanomaterials induces behavioral deficits and neural damage in *Caenorhabditis elegans*. J Appl Toxicol 37(10):1140–1150. https://doi.org/10.1002/jat.3468

Li SW, Huang CW, Liao VH (2020) Early-life long-term exposure to ZnO nanoparticles suppresses innate immunity regulated by SKN-1/Nrf and the p38 MAPK signaling pathway in *Caenorhabditis elegans*. Environ Pollut 256:113382. https://doi.org/10.1016/j.envpol.2019.113382

Liang S, Duan J, Hu H, Zhang J, Gao S, Jing H, Li G, Sun Z (2020a) Comprehensive analysis of SiNPs on the genome-wide transcriptional changes in *Caenorhabditis elegans*. Int J Nanomedicine 15:5227–5237. https://doi.org/10.2147/ijn.S251269

Liang X, Wang Y, Cheng J, Ji Q, Wang Y, Wu T, Tang M (2020b) Mesoporous silica nanoparticles at predicted environmentally relevant concentrations cause impairments in GABAergic motor neurons of nematode *Caenorhabditis elegans*. Chem Res Toxicol 33(6):1665–1676. https://doi.org/10.1021/acs.chemrestox.9b00477

Lim D, Roh JY, Eom HJ, Choi JY, Hyun J, Choi J (2012) Oxidative stress-related PMK-1 p38 MAPK activation as a mechanism for toxicity of silver nanoparticles to reproduction in the nematode *Caenorhabditis elegans*. Environ Toxicol Chem 31(3):585–592. https://doi.org/10.1002/etc.1706

Lu JH, Hou WC, Tsai MH, Chang YT, Chao HR (2022) The impact of background-level carboxylated single-walled carbon nanotubes (SWCNTs-COOH) on induced toxicity in *Caenorhabditis elegans* and human cells. Int J Environ Res Public Health 19(12):7525. https://doi.org/10.3390/ijerph19031218

Lucio D, Martínez-Ohárriz MC, Jaras G, Aranaz P, González-Navarro CJ, Radulescu A, Irache JM (2017) Optimization and evaluation of zein nanoparticles to improve the oral delivery of glibenclamide: in vivo study using *C. elegans*. Eur J Pharm Biopharm 121:104–112. https://doi.org/10.1016/j.ejpb.2017.09.018

Luo X, Xu S, Yang Y, Li L, Chen S, Xu A, Wu L (2016) Insights into the ecotoxicity of silver nanoparticles transferred from *Escherichia coli* to *Caenorhabditis elegans*. Sci Rep 6:36465. https://doi.org/10.1038/srep36465

Luo X, Xu S, Yang Y, Zhang Y, Wang S, Chen S, Xu A, Wu L (2017) A novel method for assessing the toxicity of silver nanoparticles in *Caenorhabditis elegans*. Chemosphere 168:648–657. https://doi.org/10.1016/j.chemosphere.2016.11.011

Ma H, Kabengi NJ, Bertsch PM, Unrine JM, Glenn TC, Williams PL (2011) Comparative phototoxicity of nanoparticulate and bulk ZnO to a free-living nematode *Caenorhabditis elegans*: the importance of illumination mode and primary particle size. Environ Pollut 159(5):1473–1480. https://doi.org/10.1016/j.envpol.2011.03.013

Ma H, Williams PL, Diamond SA (2013) Ecotoxicity of manufactured ZnO nanoparticles—a review. Environ Pollut 172:76–85. https://doi.org/10.1016/j.envpol.2012.08.011

Ma H, Wallis LK, Diamond S, Li S, Cañas-Carrell J, Parra A (2014) Impact of solar UV radiation on toxicity of ZnO nanoparticles through photocatalytic reactive oxygen species (ROS) generation and photo-induced dissolution. Environ Pollut 193:165–172. https://doi.org/10.1016/j.envpol.2014.06.027

Ma Z, Garrido-Maestu A, Lee C, Chon J, Jeong D, Yue Y, Sung K, Park Y, Jeong KC (2018) Comprehensive in vitro and in vivo risk assessments of chitosan microparticles using human epithelial cells and *Caenorhabditis elegans*. J Hazard Mater 341:248–256. https://doi.org/10.1016/j.jhazmat.2017.07.071

Mádi A, Mikkat S, Ringel B, Thiesen HJ, Glocker MO (2003) Profiling stage-dependent changes of protein expression in *Caenorhabditis elegans* by mass spectrometric proteome analysis leads to the identification of stage-specific marker proteins. Electrophoresis 24(10):1809–1817. https://doi.org/10.1002/elps.200305390

Markaki M, Tavernarakis N (2020) *Caenorhabditis elegans* as a model system for human diseases. Curr Opin Biotechnol 63:118–125. https://doi.org/10.1016/j.copbio.2019.12.011

Mashock MJ, Zanon T, Kappell AD, Petrella LN, Andersen EC, Hristova KR (2016) Copper oxide nanoparticles impact several toxicological endpoints and cause neurodegeneration in *Caenorhabditis elegans*. PLoS One 11(1):e0167613. https://doi.org/10.1371/journal.pone.0167613

Maurer LL, Yang X, Schindler AJ, Taggart RK, Jiang C, Hsu-Kim H, Sherwood DR, Meyer JN (2016) Intracellular trafficking pathways in silver nanoparticle uptake and toxicity in *Caenorhabditis elegans*. Nanotoxicology 10(8):831–835. https://doi.org/10.3109/17435390.2015.1110759

McElwee MK, Ho LA, Chou JW, Smith MV, Freedman JH (2013) Comparative toxicogenomic responses of mercuric and methyl-mercury. BMC Genomics 14:698. https://doi.org/10.1186/1471-2164-14-698

Meyer JN, Lord CA, Yang XY, Turner EA, Badireddy AR, Marinakos SM, Chilkoti A, Wiesner MR, Auffan M (2010) Intracellular uptake and associated toxicity of silver nanoparticles in *Caenorhabditis elegans*. Aquat Toxicol 100(2):140–150. https://doi.org/10.1016/j.aquatox.2010.07.016

Mondal S, Hegarty E, Martin C, Gökçe SK, Ghorashian N, Ben-Yakar A (2016) Large-scale microfluidics providing high-resolution and high-throughput screening of Caenorhabditis elegans poly-glutamine aggregation model. Nat Commun 7:13023. https://doi.org/10.1038/ncomms13023

Moon J, Kwak JI, Kim SW, An YJ (2017) Multigenerational effects of gold nanoparticles in Caenorhabditis elegans: continuous versus intermittent exposures. Environ Pollut 220:46–52. https://doi.org/10.1016/j.envpol.2016.09.021

O'Donnell B, Huo L, Polli JR, Qiu L, Collier DN, Zhang B, Pan X (2017) From the cover: ZnO nanoparticles enhanced germ cell apoptosis in Caenorhabditis elegans, in comparison with ZnCl2. Toxicol Sci 156:336–343. https://doi.org/10.1093/toxsci/kfw258

Piechulek A, Von Mikecz A (2018) Life span-resolved nanotoxicology enables identification of age-associated neuromuscular vulnerabilities in the nematode Caenorhabditis elegans. Environ Pollut 233:1095–1103. https://doi.org/10.1016/j.envpol.2017.10.012

Polak N, Read DS, Jurkschat K, Matzke M, Kelly FJ, Spurgeon DJ, Stürzenbaum SR (2014) Metalloproteins and phytochelatin synthase may confer protection against zinc oxide nanoparticle induced toxicity in Caenorhabditis elegans. Comp Biochem Physiol C: Toxicol Pharmacol 160:75–85. https://doi.org/10.1016/j.cbpc.2013.12.001

Qu M, Li Y, Wu Q, Xia Y, Wang D (2017) Neuronal ERK signaling in response to graphene oxide in nematode Caenorhabditis elegans. Nanotoxicology 11:520–533. https://doi.org/10.1080/17435390.2017.1315190

Qu M, Xu K, Li Y, Wong G, Wang D (2018) Using ACS-22 mutant Caenorhabditis elegans to detect the toxicity of nanopolystyrene particles. Sci Total Environ 643:119–126. https://doi.org/10.1016/j.scitotenv.2018.06.173

Rahman M, Edwards H, Birze N, Gabrilska R, Rumbaugh KP, Blawzdziewicz J, Szewczyk NJ, Driscoll M, Vanapalli SA (2020) Nemalife chip: a micropillar-based microfluidic culture device optimized for aging studies in crawling C. elegans. Sci Rep 10:16190. https://doi.org/10.1038/s41598-020-73002-6

Ratnasekhar C, Sonane M, Satish A, Mudiam MK (2015) Metabolomics reveals the perturbations in the metabolome of Caenorhabditis elegans exposed to titanium dioxide nanoparticles. Nanotoxicology 9:994–1004. https://doi.org/10.3109/17435390.2014.993345

Ren M, Zhao L, Lv X, Wang D (2017) Antimicrobial proteins in the response to graphene oxide in Caenorhabditis elegans. Nanotoxicology 11:578–590. https://doi.org/10.1080/17435390.2017.1329954

Ren M, Zhao L, Ding X, Krasteva N, Rui Q, Wang D (2018) Developmental basis for intestinal barrier against the toxicity of graphene oxide. Part Fibre Toxicol 15:26. https://doi.org/10.1186/s12989-018-0262-4

Rive C, Reina G, Wagle P, Treossi E, Palermo V, Bianco A, Delogu LG, Rieckher M, Schumacher B (2019) Improved biocompatibility of amino-functionalized graphene oxide in Caenorhabditis elegans. Small 15:e1902699. https://doi.org/10.1002/smll.201902699

Rocheleau S, Arbour M, Elias M, Sunahara GI, Masson L (2015) Toxicogenomic effects of nano- and bulk-TiO2 particles in the soil nematode Caenorhabditis elegans. Nanotoxicology 9:502–512. https://doi.org/10.3109/17435390.2014.948941

Rogers S, Rice KM, Manne ND, Shokuhfar T, He K, Selvaraj V, Blough ER (2015) Cerium oxide nanoparticle aggregates affect stress response and function in Caenorhabditis elegans. SAGE Open Med 3:2050312115575387. https://doi.org/10.1177/2050312115575387

Roh JY, Sim SJ, Yi J, Park K, Chung KH, Ryu DY, Choi J (2009) Ecotoxicity of silver nanoparticles on the soil nematode Caenorhabditis elegans using functional ecotoxicogenomics. Environ Sci Technol 43:3933–3940. https://doi.org/10.1021/es803477u

Roh JY, Park YK, Park K, Choi J (2010) Ecotoxicological investigation of CeO2 and TiO2 nanoparticles on the soil nematode Caenorhabditis elegans using gene expression, growth, fertility, and survival as endpoints. Environ Toxicol Pharmacol 29:167–172. https://doi.org/10.1016/j.etap.2009.12.003

Rohde CB, Zeng F, Gonzalez-Rubio R, Angel M, Yanik MF (2007) Microfluidic system for on-chip high-throughput whole-animal sorting and screening at subcellular resolution. Proc Natl Acad Sci 104:13891–13895. https://doi.org/10.1073/pnas.0706513104

Rui Q, Zhao Y, Wu Q, Tang M, Wang D (2013) Biosafety assessment of titanium dioxide nanoparticles in acutely exposed nematode Caenorhabditis elegans with mutations of genes required for oxidative stress or stress response. Chemosphere 93:2289–2296. https://doi.org/10.1016/j.chemosphere.2013.08.007

Scharf A, Piechulek A, Von Mikecz A (2013) Effect of nanoparticles on the biochemical and behavioral aging phenotype of the nematode Caenorhabditis elegans. ACS Nano 7:10695–10703. https://doi.org/10.1021/nn403443r

Scharf A, Gührs KH, Von Mikecz A (2016) Anti-amyloid compounds protect from silica nanoparticle-induced neurotoxicity in the nematode C. elegans. Nanotoxicology 10:426–435. https://doi.org/10.3109/17435390.2015.1073399

Schultz CL, Wamucho A, Tsyusko OV, Unrine JM, Crossley A, Svendsen C, Spurgeon DJ (2016) Multigenerational exposure to silver ions and silver nanoparticles reveals heightened sensitivity and epigenetic memory in Caenorhabditis elegans. Proc R Soc B Biol Sci 283:20152911. https://doi.org/10.1098/rspb.2015.2911

Sese BT, Grant A, Reid BJ (2009) Toxicity of polycyclic aromatic hydrocarbons to the nematode Caenorhabditis elegans. J Toxicol Environ Health A 72:1168–1180. https://doi.org/10.1080/15287390903091814

Shin N, Cuenca L, Karthikraj R, Kannan K, Colaiácovo MP (2019) Assessing effects of germline exposure to environmental toxicants by high-throughput screening in C. elegans. PLoS Genet 15:e1007975. https://doi.org/10.1371/journal.pgen.1007975

Silva AC, Dos Santos AGR, Pieretti JC, Rolim WR, Seabra AB, Ávila DS (2023) Iron oxide/silver hybrid nanoparticles impair the cholinergic system and cause reprotoxicity in Caenorhabditis elegans. Food Chem Toxicol 179:113945. https://doi.org/10.1016/j.fct.2023.113945

Sivaselvam S, Mohankumar A, Thiruppathi G, Sundararaj P, Viswanathan C, Ponpandian N (2020) Engineering the surface of graphene oxide with bovine serum albumin for improved biocompatibility in Caenorhabditis elegans. Nanoscale Adv 2:5219–5230. https://doi.org/10.1039/d0na00574f

Sonane M, Moin N, Satish A (2017) The role of antioxidants in attenuation of Caenorhabditis elegans lethality on exposure to TiO2 and ZnO nanoparticles. Chemosphere 187:240–247. https://doi.org/10.1016/j.chemosphere.2017.08.080

Soria C, Coccini T, De Simone U, Marchese L, Zorzoli I, Giorgetti S, Raimondi S, Mangione PP, Ramat S, Bellotti V, Manzo L, Stoppini M (2015) Enhanced toxicity of silver nanoparticles in transgenic Caenorhabditis elegans expressing amyloidogenic proteins. Amyloid 22:221–228. https://doi.org/10.3109/13506129.2015.1077216

Starnes DL, Unrine JM, Starnes CP, Collin BE, Oostveen EK, Ma R, Lowry GV, Bertsch PM, Tsyusko OV (2015) Impact of sulfidation on the bioavailability and toxicity of silver nanoparticles to Caenorhabditis elegans. Environ Pollut 196:239–246. https://doi.org/10.1016/j.envpol.2014.10.009

Starnes D, Unrine J, Chen C, Lichtenberg S, Starnes C, Svendsen C, Kille P, Morgan J, Baddar ZE, Spear A, Bertsch P, Chen KC, Tsyusko O (2019) Toxicogenomic responses of Caenorhabditis elegans to pristine and transformed zinc oxide nanoparticles. Environ Pollut 247:917–926. https://doi.org/10.1016/j.envpol.2019.01.077

Sulston JE (1983) Neuronal cell lineages in the nematode Caenorhabditis elegans. Cold Spring Harb Symp Quant Biol 48(Pt 2):443–452. https://doi.org/10.1101/sqb.1983.048.01.049

Tejeda-Benitez L, Olivero-Verbel J (2016) Caenorhabditis elegans, a biological model for research in toxicology. In: De Voogt WP (ed) Reviews of environmental contamination and toxicology 237. Springer, Cham. https://doi.org/10.1007/978-3-319-23573-8_1

Teng J, Yu T, Yan F (2024) GABA attenuates neurotoxicity of zinc oxide nanoparticles due to oxidative stress via DAF-16/FOXO and SKN-1/Nrf2 pathways. Sci Total Environ 934:173214. https://doi.org/10.1016/j.scitotenv.2024.173214

Tsyusko OV, Unrine JM, Spurgeon D, Blalock E, Starnes D, Tseng M, Joice G, Bertsch PM (2012) Toxicogenomic responses of the model organism Caenorhabditis elegans to gold nanoparticles. Environ Sci Technol 46:4115–4124. https://doi.org/10.1021/es2033108

Tyne W, Lofts S, Spurgeon DJ, Jurkschat K, Svendsen C (2013) A new medium for Caenorhabditis elegans toxicology and nanotoxicology studies designed to better reflect natural soil solution conditions. Environ Toxicol Chem 32:1711–1717. https://doi.org/10.1002/etc.2247

Tyne W, Little S, Spurgeon DJ, Svendsen C (2015) Hormesis depends upon the life-stage and duration of exposure: examples for a pesticide and a nanomaterial. Ecotoxicol Environ Saf 120:117–123. https://doi.org/10.1016/j.ecoenv.2015.05.024

Von Mikecz A (2022) Exposome, molecular pathways and one health: the invertebrate Caenorhabditis elegans. Int J Mol Sci 23:9084. https://doi.org/10.3390/ijms23169084

Walczynska M, Jakubowski W, Wasiak T, Kadziola K, Bartoszek N, Kotarba S, Siatkowska M, Komorowski P, Walkowiak B (2018) Toxicity of silver nanoparticles, multiwalled carbon nanotubes, and dendrimers assessed with multicellular organism Caenorhabditis elegans. Toxicol Mech Methods 28:432–439. https://doi.org/10.1080/15376516.2018.1449277

Wamucho A, Unrine JM, Kieran TJ, Glenn TC, Schultz CL, Farman M, Svendsen C, Spurgeon DJ, Tsyusko OV (2019) Genomic mutations after multigenerational exposure of Caenorhabditis elegans to pristine and sulfidized silver nanoparticles. Environ Pollut 254:113078. https://doi.org/10.1016/j.envpol.2019.113078

Wang D (2018) Endpoints for toxicity assessment of nanomaterials. In: Nanotoxicology in Caenorhabditis elegans. Springer Singapore, Singapore. https://doi.org/10.1007/978-981-13-0233-6_2

Wang Y, Wang S, Luo X, Yang Y, Jian F, Wang X, Xie L (2014) The roles of DNA damage-dependent signals and MAPK cascades in tributyltin-induced germline apoptosis in Caenorhabditis elegans. Chemosphere 108:231–238. https://doi.org/10.1016/j.chemosphere.2014.01.045

Wang Q, Zhou Y, Song B, Zhong Y, Wu S, Cui R, Cong H, Su Y, Zhang H, He Y (2016) Linking subcellular disturbance to physiological behavior and toxicity induced by quantum dots in Caenorhabditis elegans. Small 12:3143–3154. https://doi.org/10.1002/smll.201600766

Wang N, Wang H, Tang C, Lei S, Shen W, Wang C, Wang G, Wang Z, Wang L (2017) Toxicity evaluation of boron nitride nanospheres and water-soluble boron nitride in Caenorhabditis elegans. Int J Nanomedicine 12:5941–5957. https://doi.org/10.2147/ijn.S130960

Wang Q, Zhu Y, Song B, Fu R, Zhou Y (2022) The in vivo toxicity assessments of water-dispersed fluorescent silicon nanoparticles in Caenorhabditis elegans. Int J Environ Res Public Health 19:4101. https://doi.org/10.3390/ijerph19074101

Wang M, Feng Y, Cao Z, Yu N, Wang J, Wang X, Kang D, Su M, Hu J, Du H (2023a) Multiple generation exposure to ZnO nanoparticles induces loss of genomic integrity in Caenorhabditis elegans. Ecotoxicol Environ Saf 249:114383. https://doi.org/10.1016/j.ecoenv.2022.114383

Wang M, Zhang Z, Sun N, Yang B, Mo J, Wang D, Su M, Hu J, Wang M, Wang L (2023b) Gold nanoparticles reduce food sensation in Caenorhabditis elegans via the voltage-gated channel EGL-19. Int J Nanomedicine 18:1659–1676. https://doi.org/10.2147/ijn.S394666

Wei CC, Yen PL, Chaikritsadakarn A, Huang CW, Chang CH, Liao VHC (2020) Parental CuO nanoparticles exposure results in transgenerational toxicity in Caenorhabditis elegans associated with possible epigenetic regulation. Ecotoxicol Environ Saf 203:111001. https://doi.org/10.1016/j.ecoenv.2020.111001

Wu S, Lu J, Rui Q, Yu S, Cai T, Wang D (2011) Aluminum nanoparticle exposure in L1 larvae results in more severe lethality toxicity than in L4 larvae or young adults by strengthening the formation of stress response and intestinal lipofuscin accumulation in nematodes. Environ Toxicol Pharmacol 31:179–188. https://doi.org/10.1016/j.etap.2010.10.005

Wu Q, Li Y, Tang M, Wang D (2012a) Evaluation of environmental safety concentrations of DMSA coated Fe2O3-NPs using different assay systems in nematode Caenorhabditis elegans. PLoS One 7:e43729. https://doi.org/10.1371/journal.pone.0043729

Wu Q, Wang W, Li Y, Li Y, Ye B, Tang M, Wang D (2012b) Small sizes of TiO2-NPs exhibit adverse effects at predicted environmental relevant concentrations on nematodes in a modified chronic toxicity assay system. J Hazard Mater 243:161–168. https://doi.org/10.1016/j.jhazmat.2012.10.013

Wu Q, Nouara A, Li Y, Zhang M, Wang W, Tang M, Ye B, Ding J, Wang D (2013) Comparison of toxicities from three metal oxide nanoparticles at environmentally relevant concentrations in nematode Caenorhabditis elegans. Chemosphere 90:1123–1131. https://doi.org/10.1016/j.chemosphere.2012.09.019

Wu Q, Zhao Y, Fang J, Wang D (2014a) Immune response is required for the control of in vivo translocation and chronic toxicity of graphene oxide. Nanoscale 6:5894–5906. https://doi.org/10.1039/C4NR00699B

Wu Q, Zhao Y, Li Y, Wang D (2014b) Molecular signals regulating translocation and toxicity of graphene oxide in the nematode Caenorhabditis elegans. Nanoscale 6:11204–11212. https://doi.org/10.1039/c4nr02688h

Wu Q, Zhao Y, Li Y, Wang D (2014c) Susceptible genes regulate the adverse effects of TiO2-NPs at predicted environmental relevant concentrations on nematode Caenorhabditis elegans. Nanomedicine 10:1263–1271. https://doi.org/10.1016/j.nano.2014.03.010

Wu Q, Zhao Y, Zhao G, Wang D (2014d) MicroRNAs control of in vivo toxicity from graphene oxide in Caenorhabditis elegans. Nanomedicine 10:1401–1410. https://doi.org/10.1016/j.nano.2014.04.005

Wu T, He K, Zhan Q, Ang S, Ying J, Zhang S, Zhang T, Xue Y, Tang M (2015) MPA-capped CdTe quantum dots exposure causes neurotoxic effects in nematode Caenorhabditis elegans by affecting the transporters and receptors of glutamate, serotonin and dopamine at the genetic level, or by increasing ROS, or both. Nanoscale 7:20460–20473. https://doi.org/10.1039/C5NR05914C

Wu Q, Zhi L, Qu Y, Wang D (2016a) Quantum dots increased fat storage in intestine of Caenorhabditis elegans by influencing molecular basis for fatty acid metabolism. Nanomedicine 12:1175–1184. https://doi.org/10.1016/j.nano.2016.01.016

Wu Q, Zhou X, Han X, Zhuo Y, Zhu S, Zhao Y, Wang D (2016b) Genome-wide identification and functional analysis of long noncoding RNAs involved in the response to graphene oxide. Biomaterials 102:277–291. https://doi.org/10.1016/j.biomaterials.2016.06.041

Wu T, Xu H, Liang X, Tang M (2019) Caenorhabditis elegans as a complete model organism for biosafety assessments of nanoparticles. Chemosphere 221:708–726. https://doi.org/10.1016/j.chemosphere.2019.01.021

Wu J, Gao Y, Xi J, You X, Zhang X, Zhang X, Cao Y, Liu P, Chen X, Luan Y (2022) A high-throughput microplate toxicity screening platform based on Caenorhabditis elegans. Ecotoxicol Environ Saf 245:114089. https://doi.org/10.1016/j.ecoenv.2022.114089

Yang X, Gondikas AP, Marinakos SM, Auffan M, Liu J, Hsu-Kim H, Meyer JN (2012) Mechanism of silver nanoparticle toxicity is dependent on dissolved silver and surface coating in Caenorhabditis elegans. Environ Sci Technol 46:1119–1127. https://doi.org/10.1021/es202417t

Yang YF, Chen PJ, Liao VH (2016) Nanoscale zerovalent iron (Nzvi) at environmentally relevant concentrations induced multigenerational reproductive toxicity in Caenorhabditis elegans. Chemosphere 150:615–623. https://doi.org/10.1016/j.chemosphere.2016.01.068

Yang YF, Lin YJ, Liao CM (2017) Toxicity-based toxicokinetic/toxicodynamic assessment of bioaccumulation and nanotoxicity of zerovalent iron nanoparticles in Caenorhabditis elegans. Int J Nanomedicine 12:4607–4621. https://doi.org/10.2147/ijn.S138790

Yu S, Rui Q, Cai T, Wu Q, Li Y, Wang D (2011) Close association of intestinal autofluorescence with the formation of severe oxidative damage in intestine of nematodes chronically exposed to Al(2)O(3)-nanoparticle. Environ Toxicol Pharmacol 32:233–241. https://doi.org/10.1016/j.etap.2011.05.008

Zhang F, You X, Zhu T, Gao S, Wang Y, Wang R, Yu H, Qian B (2020) Silica nanoparticles enhance germ cell apoptosis by inducing reactive oxygen species (Ros) formation in Caenorhabditis elegans. J Toxicol Sci 45:117–129. https://doi.org/10.2131/jts.45.117

Zhang W, Li W, Li J, Chang X, Niu S, Wu T, Kong L, Zhang T, Tang M, Xue Y (2021) Neurobehavior and neuron damage following prolonged exposure of silver nanoparticles with/without polyvinylpyrrolidone coating in Caenorhabditis elegans. J Appl Toxicol 41:2055–2067. https://doi.org/10.1002/jat.4197

Zhao Y, Wu Q, Tang M, Wang D (2014) The in vivo underlying mechanism for recovery response formation in nano-titanium dioxide exposed Caenorhabditis elegans after transfer to the normal condition. Nanomedicine 10:89–98. https://doi.org/10.1016/j.nano.2013.07.004

Zhao Y, Wang X, Wu Q, Li Y, Wang D (2015) Translocation and neurotoxicity of CdTe quantum dots in RMEs motor neurons in nematode Caenorhabditis elegans. J Hazard Mater 283:480–489. https://doi.org/10.1016/j.jhazmat.2014.09.063

Zhao Y, Wu Q, Wang D (2016a) An epigenetic signal encoded protection mechanism is activated by graphene oxide to inhibit its induced reproductive toxicity in Caenorhabditis elegans. Biomaterials 79:15–24. https://doi.org/10.1016/j.biomaterials.2015.11.052

Zhao Y, Yang J, Wang D (2016b) A microRNA-mediated insulin signaling pathway regulates the toxicity of multi-walled carbon nanotubes in nematode Caenorhabditis elegans. Sci Rep 6:23234. https://doi.org/10.1038/srep23234

Zhao Y, Zhi L, Wu Q, Yu Y, Sun Q, Wang D (2016c) P38 MAPK-SKN-1/Nrf signaling cascade is required for intestinal barrier against graphene oxide toxicity in Caenorhabditis elegans. Nanotoxicology 10:1469–1479. https://doi.org/10.1080/17435390.2016.1235738

Zhao L, Kong J, Krasteva N, Wang D (2018) Deficit in the epidermal barrier induces toxicity and translocation of PEG modified graphene oxide in nematodes. Toxicol Res (Camb) 7:1061–1070. https://doi.org/10.1039/c8tx00136g

Zhi L, Ren M, Qu M, Zhang H, Wang D (2016) Wnt ligands differentially regulate toxicity and translocation of graphene oxide through different mechanisms in Caenorhabditis elegans. Sci Rep 6:39261. https://doi.org/10.1038/srep39261

Zhou Y, Wang Q, Song B, Wu S, Su Y, Zhang H, He Y (2015) A real-time documentation and mechanistic investigation of quantum dots-induced autophagy in live Caenorhabditis elegans. Biomaterials 72:38–48. https://doi.org/10.1016/j.biomaterials.2015.08.044

Zhuang Z, Li M, Liu H, Luo L, Gu W, Wu Q, Wang D (2016) Function of RSKS-1-AAK-2-DAF-16 signaling cascade in enhancing toxicity of multi-walled carbon nanotubes can be suppressed by mir-259 activation in Caenorhabditis elegans. Sci Rep 6:32409. https://doi.org/10.1038/srep32409

Open Access This chapter is licensed under the terms of the Creative Commons Attribution-NonCommercial-NoDerivatives 4.0 International License (http://creativecommons.org/licenses/by-nc-nd/4.0/), which permits any noncommercial use, sharing, distribution and reproduction in any medium or format, as long as you give appropriate credit to the original author(s) and the source, provide a link to the Creative Commons license and indicate if you modified the licensed material. You do not have permission under this license to share adapted material derived from this chapter or parts of it.

The images or other third party material in this chapter are included in the chapter's Creative Commons license, unless indicated otherwise in a credit line to the material. If material is not included in the chapter's Creative Commons license and your intended use is not permitted by statutory regulation or exceeds the permitted use, you will need to obtain permission directly from the copyright holder.

Galleria Mellonella as a Potential Bridging Model for Nanotoxicology

12

Aaron Curtis, Kevin Kavanagh, and Fiona Murphy

Abstract

The use of Galleria mellonella, or the greater wax moth larva, as a model organism for toxicology assessments of nanomaterials offers a promising bridge between *in vitro* and *in vivo* studies. Traditional *in vitro* models often fail to replicate complex biological responses, while *in vivo* studies are costly and raise ethical concerns. The shared similarities of insect and mammalian innate immune systems has enabled the use of G. mellonella as an effective infection model to study host-pathogen interactions without the ethical issues associated with vertebrates. Recent studies indicate that G. mellonella can effectively mimic human immune responses, making it suitable for assessing the biocompatibility and toxicity of various nanomaterials, including carbon nanotubes and metal nanoparticles. Its immunological characteristics make it particularly valuable for immunotoxicity studies, helping to evaluate the impact of nanomaterials on immune responses. This innovative approach not only streamlines the assessment process but also aligns with the principles of the 3Rs (Replacement, Reduction, Refinement) in toxicology research. As research in nanotechnology progresses, further development of the G. mellonella model could play a pivotal role in ensuring the safe development and application of nanomaterials and in enhancing our understanding of host-pathogen dynamics and immunotoxicity.

Keywords

Invertebrate models · 3Rs · Immunotoxicity · Nanomaterial safety

---

A. Curtis · K. Kavanagh
Maynooth University, Maynooth, Ireland

F. Murphy (✉)
Strathclyde Institute of Pharmacy and Biomedical Sciences, University of Strathclyde, Glasgow, UK
e-mail: f.murphy@strath.ac.uk

© The Author(s) 2025
E. Alfaro-Moreno, F. Murphy (eds.), *Nanosafety*,
https://doi.org/10.1007/978-3-031-93871-9_12

1 Introduction

Vertebrates and insects diverged approximately 700 million years ago but many of the systems and process that sustain life have remained similar despite the passage of time (Mantica et al. 2024), for example the immune system of insects shares many similarities with the innate immune system of mammals. As a result, a variety of insects (e.g. Drosophila melanogaster, Manduca sexta, Bombyx mori, Galleria mellonella) are now widely used in biomedical research to model disease processes, to assess the virulence of pathogens or measure the *in vivo* toxicity and efficacy of antimicrobial agents (Curtis et al. 2022). The results generated using insects show a strong correlation with those obtained from conventional mammalian testing (Jander et al. 2000; Brennan et al. 2002). Conservation of fundamental biological processes across species and genera enables invertebrate models to act as surrogate whole organism systems to study developmental biology, tissue homeostasis, microorganism pathogenesis and immune responses.

Characterising the insect immune response to pathogens highlights similarities with the innate immune response of mammals. Haemocytes (insect immune cells) phagocytose pathogens, degranulate and produce superoxide in an identical manner to mammalian neutrophils (Bergin et al. 2005; Renwick et al. 2007). The pathogen recognition receptors in G. mellonella larvae show many structural and functional similarities to those in mammals and allow the immune system to rapidly and effectively identify pathogens and mount a protective response (Wang et al. 2019). Insects lack an adaptive immune response, which is only found in the jawed vertebrates, but many insect species demonstrate an analogous system called immune priming where exposure to a sublethal inoculum of a pathogen elicits an enhanced immune response that protects against a subsequent potentially lethal infection (Sheehan et al. 2020a). Immune priming can also be triggered by administration of certain antimicrobials (Kelly and Kavanagh 2011) or microbial cell wall material (Mowlds et al. 2008). Because of their small size and ease of use, insects have become a popular choice for assessing the virulence of pathogens or for determining the toxicity and *in vivo* antimicrobial efficacy of antimicrobial agents prior to time and resource intensive murine testing.

2 Galleria Mellonella as an In Vivo Model for Infection Studies

Taxonomically, Galleria mellonella (the greater wax moth) is classified as a member of the order Lepidoptera, family Pyralidae, and subfamily Galleriinae. G. mellonella is a pest species which reside in honeybee hives. The destructive properties of G. mellonella to honeybee colonies is considered to cause a significant worldwide economic cost, especially in Africa and Asia (Kwadha et al. 2017). Galleria mellonella is a holometabolous insect, i.e. its life cycle has four developmental stages: egg, larva, pupa, and adult (Fig. 12.1).

Fig. 12.1 Galleria mellonella larvae, final instar stage

Larvae of G. mellonella are now frequently used in biomedical research and have numerous advantages over murine testing including low cost, easy to house, high throughput model for generation of rapid results. G. mellonella are also not subject to the legal/ethical restrictions associated with mammalian testing (Curtis et al. 2022) (Fig. 12.2). G. mellonella larvae can be infected with a pathogen orally by force feeding, or by injection directly into the haemocoel through one of the prolegs (Ramarao et al. 2012). In some cases, larvae can be infected by application of a pathogen to the cuticular surface. The response of the larva to infection can be monitored by assessing the time to death, observing the extent of melanisation, measuring the proliferation of the pathogen *in vivo*, characterising changes in the population of circulating haemocytes or by quantifying the alterations in the expression of larval immune genes or proteins (Genç et al. 2024; Fuchs et al. 2010; Champion et al. 2018).

G. mellonella larvae can be employed to characterise the pathogen—host interactions that lead to the pathologies evident in cases of human infection. Infection of G. mellonella larvae with Listeria monocytogenes, for example, results in the formation of melanised nodules on the brain which are similar in structure to the nodules that form in the human brain during L. monocytogenes infection (Mukherjee et al. 2013). Larvae infected with the pulmonary fungal pathogen Aspergillus fumigatus show the development of granulomas which are also evident in case of human infection (Sheehan et al. 2018a). The melanised grains evident during

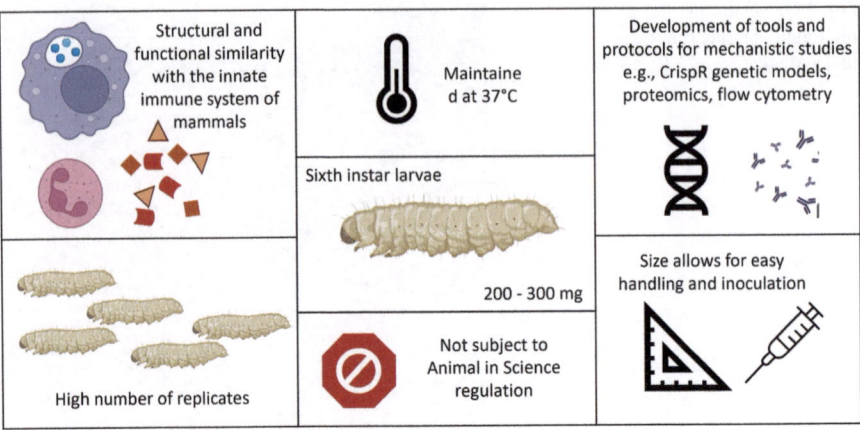

Fig. 12.2 Advantages of G. mellonella as a replacement model. (Adapted from Pereira et al. 2020)

infection of G. mellonella larvae with the pathogenic fungus Madurella mycotomatis show strong similarities to those that occur during human infection (Sheehan et al. 2020b). Using this knowledge it is possible to develop and optimise strategies to assess the development of diseases-associated pathologies in Galleria larvae before assessing them in mammalian model systems.

G. mellonella larvae are also used to assess the *in vivo* efficacy and toxicity of novel antimicrobial compounds and to help identify promising candidates before murine evaluation is conducted. In this application it is wise to first assess the effect of the antimicrobial agents on larval viability and to establish if they induce an immune response which might account for all or part of the *in vivo* antimicrobial effect of the compounds (Kelly and Kavanagh 2011). One way to determine this is to quantify the haemocyte density in larvae following administration with the compound. A significant increase in haemocyte density is indicative of an immune response which can also be characterised by assessing changes in gene expression or by utilising proteomic techniques to determine changes in the relative abundance of antimicrobial peptides.

3   Insects as In Vivo Toxicity Models

Conventional *in vivo* assays to assess the toxicity of novel therapeutics or food additives has relied heavily upon the use of a variety of vertebrate systems including mice, rats, guinea pigs and birds. The use of mammals for studying toxicity of compounds has contributed enormously to our knowledge due to the homology between mammalian genomes and the similarities between the cell biology and physiology of model species and our own. However, despite these advances, there is recognition that the number of mammals used in such tests must be reduced to the minimum and that alternative, ethically acceptable and scientifically valid systems must be developed (Burden et al. 2021; Hartung 2009).

The insect and mammalian gastrointestinal tracts share similar tissue, anatomical and physiological functions. The midgut surface of insects secretes a peritrophic membrane that consists of glycoproteins and chitinous fibrils. One proposed function of peritrophic matrices is to prevent microorganisms in the ingested food from entering the haemocoel (Orihel 1975). The microvilli in the midgut of G. mellonella contain microbes that resemble those found in the intestinal microvilli of mammals (Mukherjee et al. 2013). The insect fat body plays a very important role in detoxification, endocrinology, reproduction and nourishment of insects and acts as a homologue to the mammalian liver. The fat body is an organ that functions in drug metabolism, similar to the liver in mammals (Büyükgüzel et al. 2013). A number of cytochrome P450 and sulfo-, glutathione- or glucose-conjugation enzymes which are involved in drug detoxification are present in the fat body (Ahmad and Hopkins 1993; Luque et al. 2002; Matsumoto et al. 2003; Yamamoto et al. 2010; Niwa et al. 2011). However, knowledge of the drug metabolism pathways in insects, including G. mellonella, is limited (Hamamoto et al. 2009).

A positive correlation between toxicity of a range of food preservative in G. mellonella larvae and in rats ($R^2 = 0.65$, $p = 0.01$) and with LD50 in cultured cells ($R^2 = 0.72$, $p = 0.007$) was established (Maguire et al. 2016). G. mellonella larvae were also utilised to study the metabolism of the widely used food preservative, potassium nitrate. The results showed that administration of potassium nitrate to larvae resulted in an increased haemocyte density and an increase in the abundance of several proteins associated with nitrate metabolism (Maguire et al. 2017a). Similarly force feeding of caffeine to larvae resulted in the production of the metabolites (theobromine and theophylline) also found in mammals and a transient reduction in movement (Maguire et al. 2017b).

4 Galleria Mellonella Use in Nanotoxicology

G. mellonella larvae are gaining momentum as an alterative model for the hazard assessment of particles and nanomaterials (NMs). A review of the literature identified 24 studies between 2018 and 2024 which exposed Galleria larvae to different NMs to measure toxicity or assess biocompatibility.

The majority of the studies were concerned with the potential environmental toxicity of NMs using G. mellonella to as an indicator species to predict the ecotoxicological effects of NMs released in the environment. The primary route of exposure for these studies was through diet from first to last instar stage and toxicity was assessed based on larval survival, developmental changes or effects on the immune system. Exposure to TiO_2 NMs in diet lead to developmental delay and a reduced lifespan of the adult moths, increased activity of antioxidant enzymes in the haemolymph and decreased number of total haemocytes (Zorlu et al. 2018; Sugeçti et al. 2021; Tuncsoy and Mese 2021). Similarly, CuO NMs accumulated in the larval mid gut and fat body and altered the activity of antioxidant enzymes such as catalase, superoxide dismutase and acetylcholinesterase in a tissue-specific manner (Sezer Tuncsoy et al. 2019). In a follow up study significant changes in the haemocyte populations were also reported in larvae fed a CuO NM-containing diet (Tunçsoy et al. 2021). The immunomodulatory effects of a NM-containing diet were also assessed where Al_2O_3 NM caused a reduction in encapsulation efficiency, melanisation and phenoloxidase (PO) activity in response to further challenge with polystyrene beads, whereas a diet spiked with ZnO nanorods had the opposite effect (Eskin and Nurullahoğlu 2023; Demirtürk et al. 2024). Force-feeding has also been used as an alternative mode of exposure to assess more acute effects of NMs (24–48 h) on haemocyte populations and antioxidant capacity (Eskin et al. 2019, 2022; Eskin 2022; Eskin and Nurullahoğlu 2022). Both ZnO nanorods and SiO_2 NPs caused haemocyte cytotoxicity (Eskin and Nurullahoğlu 2023; Eskin 2022) and changes in haemocyte cell populations (Eskin and Nurullahoğlu 2022) whereas copper phosphate nanoflowers caused increased haemocyte number and increased antioxidant enzyme activity after 24 h (Eskin et al. 2022).

Given the similarities between the G. mellonella and the mammalian innate immune systems, G. mellonella larvae present an attractive alternative to rodent testing for human toxicity studies. However, only a handful of studies published to date have utilized G. mellonella to assess human hazard potential of NMs in lieu of *in vivo* models. Moya-Anderico et al. compared acute toxicity effects caused by silver, selenium, and functionalized gold NP using an array of indicators of toxicity within the larvae including survival, haemocyte proliferation, NP distribution, behavioral changes, and histological alterations. Toxicity was measured at timepoints up to 72 h after exposure to NM via injection into the lower left pro-leg. SeNPs were the most toxic to the larvae (LD50: 89 mg/kg), AgNPs resulted in intermediate level of toxicity (LD50 939 mg/kg) while AuNPs were the least toxic (2023 mg/kg) (Moya-Andérico et al. 2021). The survival results replicated the hazard ranking determined in a murine model demonstrating the ability of the G. mellonella larvae to differentiate between high, intermediate, and low toxicity NMs. Further analysis of NM-mediated tissue and cell responses may be indicative of the underlying immune response and mechanisms of toxicity. The impact of NM size was assessed by Tuncsoy and Tuncsoy (2023) who compared the biodistribution of Al_2O_3 micro- and nanoparticles after injection and subsequent immune responses. Nano-sized Al_2O_3 preferentially accumulated in the larval fat body whereas micro-sized particles were found in abundance in the larval midgut. Antioxidant enzyme activity was found to be increased in a tissue-specific manner mirroring particle accumulation (Tuncsoy and Tuncsoy 2023). Understanding the biodistribution and biopersistence of particles after exposure will be critical to understand and interpret subsequent immunological changes and toxic responses.

The G. mellonella model has also been used to support safety and biocompatibility assessment of NMs designed for use as drug carriers. Polymeric nanocapsules and mesoporous silica are promising novel modes of drug delivery which have both been tested using G. mellonella as a pre-clinical screening model (Cé et al. 2020; Carvalho et al. 2022). Both materials were judged as non-toxic to G. mellonella and therefore progressed to further investigation into their potential utility in drug delivery.

The potential anti-microbial properties of NMs themselves are being exploited to develop novel therapeutics. Nanomedicines are being developed to provide an alternative option to antibiotics for a range of difficult to treat infections. G. mellonella larvae are commonly used for studying microbial virulence and for screening new antimicrobial agents testing both efficacy and toxicity in a physiologically relevant model system and can therefore readily be applied to the safety assessment of nanomedicines. Interestingly co-exposure of G. mellonella to pathogens and nanomedicine NMs has allowed the interplay between the host, pathogen and NM to be examined e.g., AgNPs have been shown to protect G. mellonella larvae from Pseudomonas aeruginosa infection by directly killing the bacteria but also indirectly by preventing an excessive and destructive immunological response against the pathogen (Thomaz et al. 2020). Conversely ZnO nanorods, administered to G. mellonella prophylactically, significantly prolonged survival of Candida albicans-infected larvae by priming the immune system; exposure to ZnO nanorods increased haemocytes density, enhanced phagocytosis and activated PO activity in larva allowing an efficient immune response to be activated against challenge with C. albicans (Xu et al. 2021).

Since the US Food and Drug Administration (FDA) Modernization Act 2.0 was signed into law in 2022 the FDA no longer requires toxicity data obtained from animal studies to make a judgement on the relative safety of a new drug or substance (Wadman 2023). Specifically, the FDA Modernization Act 2.0 "allows for alternatives to animal testing for purposes of drug and biological product applications," which, while it does not ban animal testing outright, may significantly decrease it (Han 2023). Although this Act allows for acceptance of hazard data from *in vitro* studies, a general lack of confidence in the robustness of data derived from artificial microphysiological systems may in reality hinder the acceptance of safety data solely based on *in vitro* systems. This opens the potential for bridging models such as G. mellonella which can better replicate the dynamic responses of a whole organism system to exogenous agents to act as a promising alternative to build a robust hazard assessment profile of a new substance and simultaneously address ethical concerns.

The use of G. mellonella larvae to meet the need for more rapid and streamlined hazard assessment of ever-increasing numbers of NM currently remains under-exploited. To realise the potential of the G. mellonella model for nanotoxicology studies the boundaries of use and model limitations must first be well-defined. Key questions to be answered include whether immune responses measurable in the larvae are indictive of NM hazard posed to human populations and whether the model is sufficiently sensitive and selective to differentiate between different mechanism of NM toxicity.

5   Galleria Mellonella and NM Immunotoxicity

Unique physicochemical properties such as size, surface coating/charge and shape can dictate toxicological and pathological consequences following exposure to NMs. A number of mechanisms through which NMs can mediate toxic effects have been identified, including (i) dissolution to reactive ions in extracellular biological fluid (Naz et al. 2020), (ii) dissolution in the intracellular environment and delivery of toxic ions via Trojan horse pathway (Sabella et al. 2014), (iii) surface reactivity and interactions at nano-bio interface, including NM protein corona activation of pathogen recognition receptors (Pavan et al. 2022), (iv) failure to phagocytose high aspect ratio fibres leading lysosomal disruption, NALP3 inflammasome activation, release of damage-associated molecular patterns (DAMPs) and frustrated phagocytosis (Donaldson et al. 2011), and (v) particle overload via accumulation of biopersistent NM in phagolysosome of immune cells resulting in reduced motility and inhibition of clearance (Driscoll and Borm 2020). Cellular consequences indicative of toxicity which result from the interactions between NMs and the biological environment include oxidative stress, surface membrane injury, lysosomal disruption, inflammasomes activation, autophagy activation/blockade, mitochondria damage, proinflammatory responses and DNA damage. The determination of equivalent responses which are readily measurable in G. mellonella after exposure to different NMs will be crucial to support the wider adoption of this model for NMs toxicity testing. Below we briefly summarise how immune responses of G. mellonella larva to infection may prove useful indicators of NM toxicity.

Reactive NMs are highly cytotoxic to cells *in vitro* and *in vivo* and are likely to cause similar detrimental damage to cells in the larval haemolymph. High levels of NM-mediated cytotoxicity can overwhelm the larval immune system leading to larval death, therefore reactive NMs may be readily identified through simple survival studies. Melanization is an important defence mechanism of the G. mellonella immune system which involves synthesizing and depositing melanin to encapsulate pathogens at the site of a wound. The process is initiated when soluble pathogen recognition receptors engage with target surfaces, triggering the serine protease cascade that cleaves pro-phenoloxidase (ProPO) to phenoloxidase (PO) (Cerenius et al. 2008). PO oxidizes the phenolic compounds into quinones that are then metabolized into melanin, resulting in black spots forming in infected larvae (Kopácek et al. 1995). The melanization pathway participates in the antimicrobial activity against bacteria or fungi (Hoffmann et al. 1996; Pereira et al. 2015) and could potentially be triggered by interactions at the nano-bio interface. Melanization can be used as a visual indicator of larval health providing a broad measure of progressive immune responses over the time course of exposure. G. mellonella survival assays are routinely used to validate the results of a high throughput mutant screens for virulence factor expression in pathogenic microorganisms. This approach could be adopted for high throughput screening of different forms of NMs to identify physiochemical drivers of toxicity or to measure similarity between nanoforms.

Although limited in the level of mechanistic insight such studies can provide, gross observation of larval melanisation and survival are simple to conduct and may prove useful for initial screening of NM panels, hazard banding and informing dose-setting to refine subsequent *in vivo* studies (Fig. 12.3).

A common theme arising from many toxicological studies of NM exposure is their immunomodulatory effects leading to concerns that exposure to NMs may contribute to the pathogenesis of chronic inflammatory diseases such as fibrosis and

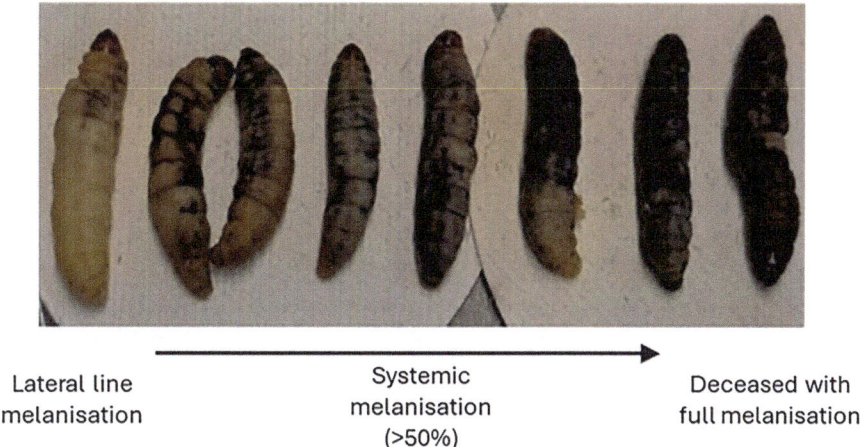

Fig. 12.3 Progressive melanisation. Degree of melanisation is clearly visible which can be indicative of immune response and health status of the larvae

tumour formation. Understanding how physicochemical interactions between different NMs and the body triggers adverse immunological responses remains a challenge but is of critical importance for hazard assessment of new materials. Due to the dynamic, temporal and multicomponent nature of the immune response to foreign particles *in vitro* models have proven inadequate to differentiate between aberrant immune activation, immune modulation or immune responses which may be tolerated (Hofer et al. 2022). Exposure of *in vitro* monocultures of immune cells to reactive NMs may result in activation of pro-inflammatory signaling and cell death. However this response may not be a sign of immunotoxicity but of a normal homeostatic immune reaction if, *in vivo*, exposure to the same NM triggered an appropriate immune response sufficient to neutralise the threat (Boraschi et al. 2020).

Many parallels exist between invertebrate and mammalian immune responses which could be exploited in the use of G. mellonella as a model for assessing immunotoxicity of NMs (Fig. 12.4). Haemocytes are the primary immune cells of G. mellonella. In general, insects possess a number of inter-related haemocyte populations each with particular morphological, histological and functional features (Gupta 1985). Prohaemocytes are progenitor cells, having the property to differentiate into several cell types (Browne et al. 2013), whereas plasmatocytes and granular cells are the predominant immune cells, and are key members of cellular immunity due to their role in phagocytosis, nodule formation and encapsulation (Tojo et al. 2000; Wu et al. 2016). In mammals, neutrophils, macrophages and dendritic cells phagocytose foreign pathogens which is mediated by opsonization and recognition. When a NM is introduced into a biological system proteins and other biomolecules present in the biological milieu adsorb to the surface and create a "biomolecular corona" (Zhang et al. 2024). Opsonization with complement components can occur upon adsorption of complement proteins in the NM biomolecular corona. During opsonization, complement (C3a), antibodies (IgG/M) or soluble C-type lectins (such as

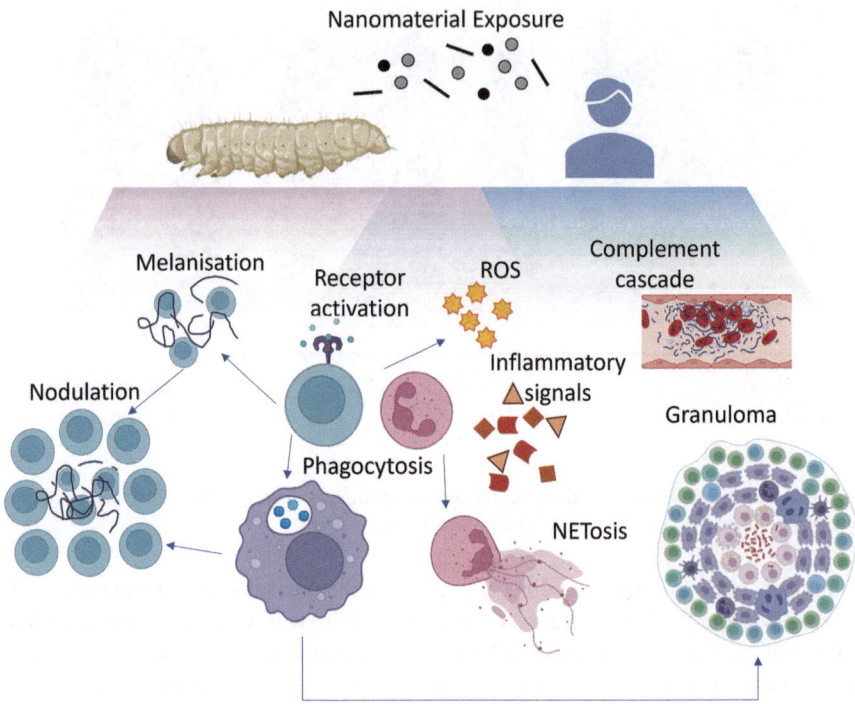

Fig. 12.4 Similarities between G. mellonella and rodent responses to foreign pathogens which may be exploited for hazard assessment of nanomaterials. (Adapted from Smith and Casadevall 2021)

SP-D) bind to the foreign object. Complement receptor (CR3), FcR receptor and mannose receptors on the surface of the immune cells bind to the opsonins, respectively, which triggers phagocytosis. In insects, thioester-containing proteins (TEPs) are hypothesised to play a C3-like role, and multibinding protein (MBP) and hemolin are known to act as opsonins to facilitate phagocytosis although the corresponding haemocyte cell surface receptors have not yet been identified.

In both mammals and insects, phagocytosis is a key defence mechanism to contain and kill pathogens. Upon phagocytosis, pathogens and foreign objects reside within the phagosome, which merges with lysosomes to form the phagolysosome where pathogens are subject to acid degradation, reactive oxygen species (ROS) and acid hydrolases in both mammals and insects. Complete phagocytosis provokes killing of pathogen by mechanisms including NADPH oxidase dependent production of ROS. Homologues to a number of proteins essential for superoxide production in human neutrophils have been identified in G. mellonella larvae. Bergin et al. (2005) propose insect haemocytes phagocytose and kill bacterial and fungal cells by a mechanism similar to the NADPH oxidase dependent production of superoxide described in human immune neutrophils (Bergin et al. 2005).

If a pathogen cannot be killed or successfully contained within the phagolysosome both mammalian and insect immune cells trigger processes to neutralise pathogens by forming large structures known as granulomas and nodules, respectively. These contain phagocytic immune cells, fibrotic networks and other immune factors. Granulomas are organized aggregates of macrophages which form in response to persistent particulate infectious and non-infectious stimuli that individual macrophages cannot eradicate (Pagán and Ramakrishnan 2018). Granulomas evolved as protective responses to destroy or sequester particles but can become pathogenic if the inflammatory response persists. As granulomas mature, they may undergo structural changes, such as becoming fibrotic or developing areas of necrosis, causing excessive tissue damage. Granuloma formation is a key pathological feature of tuberculosis infection and has also been described as a chronic response to exposure to pathogenic, biopersistent particles such as silica and asbestos fibres (Pagán and Ramakrishnan 2018). Therefore, granuloma formation is a pathogenic response of concern in the hazard assessment of biopersistent NMs. Nodulation in insects is a comparable nonspecific response involving plasmatocytes and granulocytes (analogous to neutrophils and macrophages) triggered to isolate foreign material. Nodulation begins with the attachment of granulocytes to an aggregate of microbes or infected cells. Degranulation of granulocytes releases a plasmatocyte spreading peptide, which aggregates plasmatocytes leading to the formation of granuloma-like structures (Kavanagh and Reeves 2004). G. mellonella nodules are associated with melanisation and often have dark structures within them. Histological analysis of G. mellonella infected with a highly virulent strain of mycobacterium tuberculosis (MTB) revealed the formation of nodules composed of haemocytes surrounding a central mass of bacterial aggregates throughout the larval body which closely resemble early-stage granulomas formed by innate immune cells in human lungs (Asai et al. 2022). In another disease context G. mellonella have been shown to faithfully recapitulate the characteristic grain formation observed in mycetoma, a chronic and granulomatous infection primarily associated with the fungal pathogen Madurella mycetomatis (Sheehan et al. 2020b). Therefore G. mellonella represents a promising model system to assess the chronic inflammatory effects and enduring tissue changes which may be triggered in response to exposure to biopersistent NMs.

When activated by innate immune stimuli, neutrophils may initiate NETosis to release neutrophil extracellular traps (NETs), consisting of 15–17 nm chromatin strands decorated with different antimicrobial proteins and peptides including myeloperoxidase (MPO) and neutrophil elastase (NE) (Brinkmann et al. 2004). Several kinds of NMs such as gold, silver, cationic lipid and polystyrene nanoparticles, nanodiamonds, and graphene oxide (GO) platelets were found to trigger NETosis in human and murine neutrophils. G. mellonella haemocytes have been shown to release extracellular traps after bacterial stimulation *in vivo* and *ex vivo* (Chen and Keddie 2021) further supporting the relevance of G. mellonella for assessing NM immunomodulation.

6   Immune Recognition and Humoral Responses

The NM biocorona is composed by self-proteins and other self-molecules which can mask the NM surface and prevent immune recognition. However, the interaction with the NM surface can alter the 3D structure and folding of proteins, thereby making them signals for non-inflammatory elimination by phagocytes or triggers for a defensive innate inflammatory response (Borgognoni et al. 2015). Mammalian immune cells and insect haemocytes share a number of comparable receptors and signalling pathways to sense foreign pathogens and orchestrate the downstream immune response.

Toll-like receptors (TLRs) are a group of type I transmembrane receptors that play a role in innate humoral immunity in both insects and mammals. As Toll and TLRs are conserved throughout evolution, they can be found in mammals, invertebrates and plants. Upon activation of Toll and Toll-like receptors a series of similar signalling pathways is initiated in insects and mammals which lead to activation of homologous transcription factors; NF-κB in mammals and Dorsal and Dif in insects (Sheehan et al. 2018b). The insect IMD pathway displays some similarities to the mammalian tumour necrosis factor-α (TNF-α) pathway as well as the Toll-like pathway. Both the TNF-α and IMD pathways ultimately result in the production of the homologous transcription factors NF-κB and Relish, respectively. Activation of these signalling pathways upon sensing infection results in the production of antimicrobial peptides and inflammatory cytokines (Sheehan et al. 2018b).

Direct interaction between NM biocorona and immune cells can activate pattern recognition receptors (PRR) on the cell membrane such as Toll-like receptors or intracellular Nod-like receptors triggering downstream signalling pathways and inflammasomes involved in propagating an inflammatory response (Danielsen et al. 2021). Upon activation of pro-inflammatory signalling pathways a wide array of soluble molecules are produced that either directly exert defensive functions, e.g., antimicrobial peptides (AMP), or communicate with surrounding tissues for mounting a defensive response e.g., inflammatory cytokines. AMPs are a major component of the invertebrate defensive system. They are small cationic, amphipathic molecules effective against Gram-positive and Gram-negative bacteria, yeasts, fungi, and some protozoa and enveloped viruses and are conserved throughout different groups of species. Although the production of pro-inflammatory cytokines has been less well characterised in G. mellonella a potential analogue of the mammalian TNF- family, has been isolated from G. mellonella, namely Gallysin 2 (Wittwer et al. 1999). Furthermore, granular cells from G. mellonella larvae were shown to strongly react with both anti-human IL-1α and TNFα primary antibodies, whereas plasmatocytes stained to a lesser extent. Fluorescence microscopy studies revealed the presence of a protein similar to human IFN gamma in the culture of G. mellonella haemocytes which was confirmed by flow cytometry using anti-IFN gamma monoclonal antibodies (Wrońska et al. 2022). To better understand the role played by G. mellonella analogues to mammalian inflammatory mediators requires further mechanistic study however the detection of AMP and soluble inflammatory mediators in G. mellonella haemolymph may act as a useful measurable endpoint to indicate immune activation in response to NM exposure.

NMs which are designed for use as nanomedicines i.e., as platforms for drug delivery, will come into direct contact with blood cells, endothelial cells and plasma proteins, where they can affect the hemostatic balance leading to adverse cardiovascular effects including thrombus formation. It is of critical importance to conduct comprehensive hemocompatibility studies to detect NM interference with the coagulation cascade and activation of cellular components of the blood system to facilitate the translation of nanomedicines into clinical application. Perturbation of the coagulation system by specific NMs has raised concerns exposure may result in serious life-threatening conditions such as deep vein thrombosis (de la Harpe et al. 2019).

As in vertebrates, blood clotting or coagulation in G. mellonella is a complex process designed to prevent blood loss and seal wounds. Clotting in G. mellonella occurs with the participation of haemocytes, functioning in similar fashion to platelets in mammals, which become highly adhesive and aggregate when stimulated. Galleria clotting also involves the participation of soluble factors, including transglutaminase, lipophorin and apolipoproteins, which are all part of a complex cascade (Dushay 2009). Insect transglutaminases are homologous to human clotting factor XIIIa (Sheehan et al. 2018b). Similarities between the blood-clotting cascades suggests the potential to utilise G. mellonella as a suitable intact circulatory system within which the interactions between NMs and mediators of blood clotting can be identified.

Intravenous injection of nanomedicines has also been shown to trigger hypersensitivity reactions. Although the mechanism may differ from case-to-case activation of the complement (C) system is commonly implicated as a causative or contributing factor (Szebeni 2014). Activation of the complement system by intravenously administered NMs is therefore of critical concern regarding the hemocompatibility of NMs for use as nanomedicines. The development of standardised and validated *in vitro* and *in vivo* tests for the quantitation and prediction of hypersensitivity reactions by nanomedicine is an urgent unmet need in regulatory toxicology. The insect proPO system which controls melanisation in response to pathogen sensing displays similarities to the complement system of vertebrates. In both the complement system of mammals and the proPO system of insects, there is production of cytotoxic and opsonic components (Cerenius et al. 2010). Furthermore, there is some similarity between the sequences of insect proPO and the mammalian complement proteins C3 and C4 (Shokal and Eleftherianos 2017). G. mellonella may provide a relevant model system for preliminary assessment of the propensity of NM to activate the complement system and the subsequent immunomodulatory consequences within the context of a dynamic whole organism system.

Particle phagocytosis, complement activation, oxidative burst and nodulation are all immune responses which may be measured in G. mellonella upon exposure to NMs and indicative of a potential to trigger comparable immune responses in humans. Further investigation into how the G. mellonella immune system detects and interactions with different NM may help illuminate NM mechanisms of toxicity in a human system and identify further measurable endpoints in the Galleria system indicative of the potential of specific NM to activate immune responses in humans.

7 Protocol Standardization

To enable the widespread uptake of the G. mellonella model for nanotoxicology studies and support acceptance of hazard data derived from the use of the model, protocols for NM exposure and assessment of tissue responses require optimization, standardization and validation.

The procurement of larva from different sources ranging from scientific research labs to commercial bait shops can introduce significant variability between labs. G. mellonella breeding colonies are not genetically identical, various rearing procedures and diets are used, and there is no standard, well-characterised G. mellonella type strain or detailed information on the diversity of different lab stocks worldwide (Champion et al. 2018).

Regardless of larva source stringent selection criteria should be followed when setting up experimental groups to ensure intra-lab reproducibility and reduce inter-lab variability. Selected larvae should fall within a narrow weight range and larvae with any visible sign of melanisation should be discarded, as this may be a sign of underlying inflammation.

Many studies describe injection into the left rear proleg of G. mellonella for systemic exposure directly into the larval haemolymph, however parameters such as method of injection, needle size, exposure volumes may differ from lab to lab. Absolute haemocyte numbers are rarely comparable between studies due to the inherent differences in haemolymph extraction, washing and isolation of cells and method of counting. To fully appreciate the impact of methodological differences would require direct comparison of protocols and strategies to minimize points of departure which would be best facilitated through multi-partner, interlaboratory studies. Such multi-partner studies could then progress to consensus-driven and evidence-based development of standardized protocols.

Different populations of haemocytes have been reported in multiple studies however there is no general consensus on the number of distinct cell types, proportions and functions of each haemocyte population and methods for cell typing remain inconsistent. Recent advances in protocol development for haemocyte phenotyping using flow cytometry are promising (Campbell et al. 2024). The publication of protocols outlining procedures for the intracellular detection of cytokines by flow cytometry and the isolation and *ex vivo* culture of G. mellonella haemocytes will further advance the characterization of the native populations and changes observed when stimulated (Wrońska et al. 2022; Admella and Torrents 2022). As a contribution to the promotion of standardized protocols for G. mellonella use below we have provided a detailed description of a number of G. mellonella techniques which are applicable to NM toxicity studies.

In the coming years, increased interest in the G. mellonella model will lead to the further development of new protocols and tools to further extend the potential of G. mellonella as a replacement model. Further use of G. mellonella for different applications will benefit from and contribute to advances in G. mellonella genetics, imaging, metabolomics, proteomics, and transcriptomic methodologies, alongside the development and accessibility of reagents to quantify immune markers.

Application of CrispR technology will allow for the development of genetically modified G. mellonella lines (Pearce et al. 2024), which could be designed to further our understanding of G. mellonella immune responses, NM mechanism of toxicity and mode of action of novel nanomedicines. G. mellonella larvae are emerging as a valuable and accessible *in vivo* model capable of exploring complex biological questions relating to the toxicity of NMs, illuminating mechanisms of actions that may not be readily identified in *in vitro* models. In this way G. mellonella presents an attractive option to act as a bridging model between simple *in vitro* based acute toxicity studies and time- and resource-intensive *in vivo* models to streamline the hazard assessment of novel NMs and advanced materials.

8   Protocols

8.1   Larval Selection

Sixth instar larvae ready for inoculation should be maintained at 15 °C with woodchips as this can delay pupation and prolong the window in which larvae are suitable for toxicity and infection studies.

1. Prior to inoculation larvae should be removed from feed substrate and allowed to acclimatize at room temperature for 1–2 h prior to inoculation.
2. Larvae weighing between 200 and 300 mg are most widely utilised in studies.
3. Examine the larvae for any signs of melanisation as it can serve as a marker of low-lying preexisting infection.
4. The movement of the larva is also an important exclusion criterion as it can provide insight into any neurological or physical damage to the individual.
5. Healthy larvae tend to be firm to the touch while some larvae are softer, and the tissue gives way under gentle pressure and this factor often correlated with melanisation and larval viability.
6. Selected larvae should be stored in clean petri dishes with either wood chips or a sheet of filter paper and used within a 2-h window of selection for optimal results.

8.2   Larval Inoculation via Intra-Haemocoel Injection

1. Larvae should be held with sufficient pressure between the thumb and index finger of the nondominant hand.
2. The needle should be inserted under the last left proleg ensuring the needle remains parallel to the surface of the larva as insertion at an angle increases the damage at the wound site and increases the risk of puncturing the midgut. Hamilton or insulin syringes with a needle size of approximately 26G are most suitable.

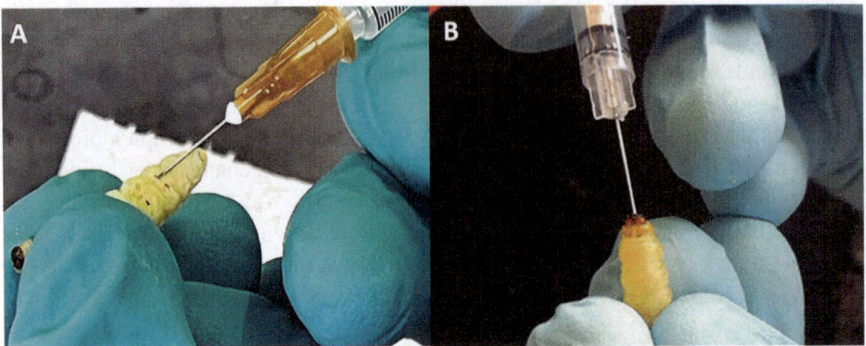

Fig. 12.5 (a) Larval inoculation via intra-haemocoel injection. (b) Inoculation via force feeding

3. The needle should be inserted bevel side up so the sharpest point of the syringe initially pierces the cuticle and should only be inserted to the point where the bevel of the needle is within the larva (Fig. 12.5a).
4. Due to the low volume of haemolymph in an individual larva the dose should not exceed 20 μl and to avoid osmotic stress phosphate buffer saline or other suitable Salines should be used for experiments.
5. Once the dose has been administered the larva should be gentle eased off the needle.
6. The larvae can then be kept in an incubator at 37 °C and larval response can be measured at 24-h intervals.
7. Larvae may also be force fed if required using a blunt needle (Fig. 12.5b).

8.3 Haemolymph Extraction

1. The larva is held firmly between the thumb and index fingers of the nondominant hand by the tail end.
2. A small incision is made using a 25G microlance at the lower left thoracic leg while holding the larva above a collection tube on ice. The incision should be just deep enough to draw haemolymph to the surface and once haemolymph is flowing additional application of gentle pressure rolling upward from the non-dominant hand can increase the yield of haemolymph extracted (Fig. 12.6). If fat body, visualised as white clumps is observed to be exiting the incision site, cease the extraction.
3. Extracted haemolymph begins to clot and melanise once expose to oxygen and should be utilised as soon as possible after extraction. The addition of β-mercaptoethanol and N-phenylthiourea can inhibit melanisation. Diluting the haemolymph after extraction can also inhibit the affect.

Fig. 12.6 Isolation of haemolymph by piercing lower left thoracic leg

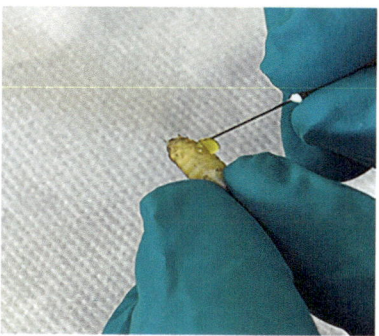

8.4 Haemocyte Enumeration

1. Dilute the haemolymph 1 in 5 with saline and gently mixed using a micropipette.
2. Place 10 µl of the dilute haemolymph at the interface between the edge of the coverslip and the haemocytometer.
3. Locate the counting grid under the microscope and count the haemocytes either in the whole grid or the corners and centre box (Fig. 12.7).
4. Haemocyte density is calculated based on the dilution factor and the grids counted to get a total concentration per ml haemolymph.

8.5 Assessing Haemocyte Killing Activity

1. Haemocytes can be harvested by centrifugation at 5000× g for 8 min. The cell free haemolymph can be used to opsonise a pathogen of interest e.g. *Candida albicans*.
2. The remaining haemocytes can be gently resuspended in sterile saline avoiding rapid pipetting, as this may lyse the cells.
3. The haemocytes can be enumerated as described above and kept on ice while the pathogen is prepared.
4. The opsonised pathogen can be centrifuged 5000× g for 5 min and washed twice with saline.
5. The yeast cells and purified haemocytes can be mixed in 2 ml in a 1:1 ratio in a 50 ml tube with the lid loosely closed and incubated at 200 rpm and 37 °C.
6. A 20 µl aliquot should be taken at 20-min intervals and diluted 1 in 100 and plated onto an agar that facilitates the chosen pathogens growth such as YEPD for *C. albicans*.
7. The plates should be incubated at 37 °C and colony forming units can be counted the following day.

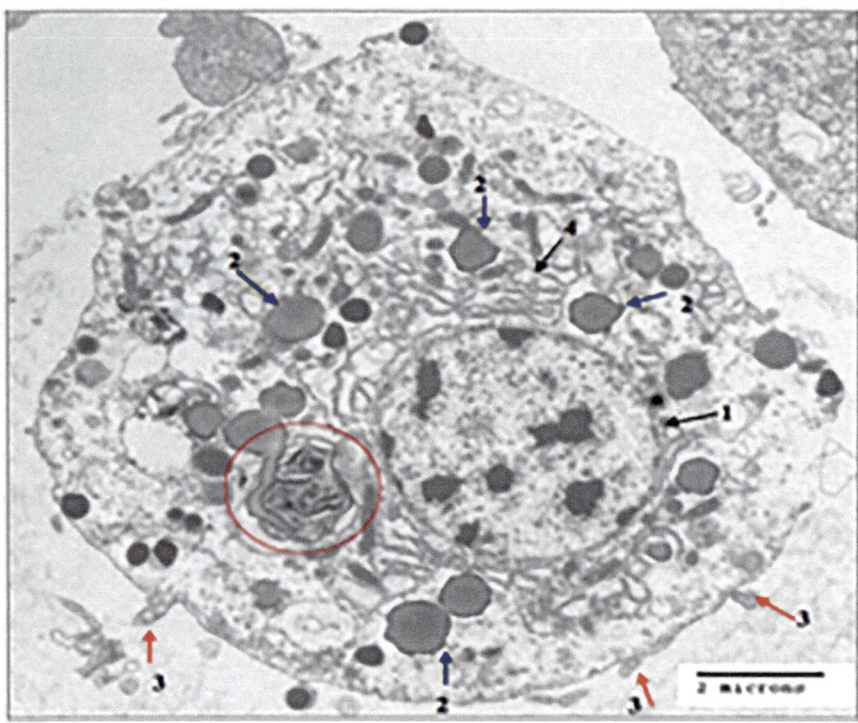

Fig. 12.7 Electron Micrograph of Galleria haemocyte. (1) Multi-lobbed nucleus, (2) Granular cytoplasm, (3) Pseudopodia, (4) Digested fungal cell

8.6   Proteomic Analysis

1. The diluted haemolymph can be acetone precipitated by mixing with 5 times the volume of ice-cold acetone and storing the samples at −20 °C overnight.
2. Remove samples from the freezer and centrifuge at 13,000× g for 10 min at 4 °C.
3. Pour acetone out gently and allow the remaining acetone to evaporate.
4. Resuspend samples in 25 μl sample resuspension Buffer (6M Urea, 2M Thiourea, 0.1M Tris-HCl (pH 8.0)). The pellet may take time to resuspend. Sonicate samples for 5 min in a water bath and vortex for 1 min to aid sample resuspension.
5. Quantify the protein content using a Qubit kit. Set amount of protein to 2 μl and then read samples.
6. To remainder of sample add 105 μl 50 mM Ammonium bicarbonate.
7. Add 1 μl 0.5M DTT to the samples and incubate at 56 °C for 20 min.
8. Remove samples from 56 °C water bath and cool to room temperature. Add 2.7 μl of 0.5 M IAA and incubate at room temperature for 15 min in the dark.

9. Following reduction and alkylation, add 1 μl of ProteaseMAX (stock (1% (w/v)): and 1 μl of trypsin (0.5 μg/μl): to each of the samples. Incubate at 37 °C for 18 h.
10. After digestion quench the reaction by adding 1 μl of Trifluoroacetic acid and incubate for 5 min.
11. Centrifuge the samples at 13,000× g for 10 min at room temperature. Remove the supernatant and place into a new tube.
12. Divide each sample into 2 aliquots (65 μl in each). Use one aliquot for C18 clean-up and store the remaining aliquot at −20 °C.
13. The eluted proteins can be dried down using a speedyvac for about 2 h at 40 °C and stored at −80 °C until loading on the mass spectrometer
14. The samples should be resuspended in a buffer compatible with the instrument used at a concentration of 750 ng of protein per microlitre.
15. For further details please consult (Sheehan et al. 2018a).

8.7 HPLC Quantification of Metabolites

1. At the timepoint of interest grind the inoculated larvae in mortar and pestle with liquid nitrogen
2. Wash the ground larvae with 5 ml 6 M HCL twice.
3. Split into 2 × 50 ml centrifuge tubes and add 25 ml of HPLC grade chloroform to each tube.
4. Mix the samples constantly for 30 min.
5. Centrifuge at 1260 g for 5 min.
6. Remove chloroform layer using a transfer pipette into a fresh centrifugation tube
7. Add an additional 25 ml of Chloroform to the organic extract and mix constantly for a further 30 min.
8. Centrifuge at 1260 g for 5 min
9. Remove chloroform layer add to first extract.
10. Rotovap samples at 56 °C until completely evaporated
11. Resuspend in 4 ml HPLC grade methanol
12. Freeze overnight
13. Rotovap at 64.7 °C
14. Resuspend in 200 ul HPLC grade Methanol
15. Before loading on HPLC centrifuge at 10,000× g for 10 min
16. Transfer top volume to HPLC vials
17. Load 20 ul onto C-18 Shimadzu HPLC column with diode array detection of 273 nm The mobile phase consisted of 34.9% (v/v) acetonitrile (Fisher Scientific), 0.1% (v/v) trifluoroacetic acid (Sigma Aldrich) and 65% (v/v) HPLC-grade water (ddH$_2$O).
18. For further details please consult Curtis et al. (2024).

References

Admella J, Torrents E (2022) A straightforward method for the isolation and cultivation of Galleria mellonella haemocytes. Int J Mol Sci 23(21):13483

Ahmad S, Hopkins T (1993) β-glucosylation of plant phenolics by phenol β-glucosyltransferase in larval tissues of the tobacco hornworm, Manduca sexta (L.). Insect Biochem Mol Biol 23:581–589

Asai M, Li Y, Spiropoulos J, Cooley W, Everest DJ, Kendall SL, Martín C, Robertson BD, Langford PR, Newton SM (2022) Galleria mellonella as an infection model for the virulent Mycobacterium tuberculosis H37Rv. Virulence 13:1543–1557

Bergin D, Reeves EP, Renwick J, Wientjes FB, Kavanagh K (2005) Superoxide production in Galleria mellonella haemocytes: identification of proteins homologous to the NADPH oxidase complex of human neutrophils. Infect Immun 73:4161–4170

Boraschi D, Alijagic A, Auguste M, Barbero F, Ferrari E, Hernadi S, Mayall C, Michelini S, Navarro Pacheco NI, Prinelli A, Swart E, Swartzwelter BJ, Bastús NG, Canesi L, Drobne D, Duschl A, Ewart M-A, Horejs-Hoeck J, Italiani P, Kemmerling B, Kille P, Prochazkova P, Puntes VF, Spurgeon DJ, Svendsen C, Wilde CJ, Pinsino A (2020) Addressing nanomaterial immunosafety by evaluating innate immunity across living species. Small 16:2000598

Borgognoni CF, Mormann M, Qu Y, Schäfer M, Langer K, Öztürk C, Wagner S, Chen C, Zhao Y, Fuchs H, Riehemann K (2015) Reaction of human macrophages on protein corona covered TiO_2 nanoparticles. Nanomedicine 11:275–282

Brennan M, Thomas DY, Whiteway M, Kavanagh K (2002) Correlation between virulence of Candida albicans mutants in mice and Galleria mellonella larvae. FEMS Immunol Med Microbiol 34:153–157

Brinkmann V, Reichard U, Goosmann C, Fauler B, Uhlemann Y, Weiss DS, Weinrauch Y, Zychlinsky A (2004) Neutrophil extracellular traps kill bacteria. Science 303:1532–1535

Browne N, Heelan M, Kavanagh K (2013) An analysis of the structural and functional similarities of insect haemocytes and mammalian phagocytes. Virulence 4:597–603

Burden N, Clift MJD, Jenkins GJS, Labram B, Sewell F (2021) Opportunities and challenges for integrating new in vitro methodologies in hazard testing and risk assessment. Small 17:2006298

Büyükgüzel E, Büyükgüzel K, Snela M, Erdem M, Radtke K, Ziemnicki K, Adamski Z (2013) Effect of boric acid on antioxidant enzyme activity, lipid peroxidation, and ultrastructure of midgut and fat body of Galleria mellonella. Cell Biol Toxicol 29:117–129

Campbell JS, Pearce JC, Bebes A, Pradhan A, Yuecel R, Brown AJP, Wakefield JG (2024) Characterising phagocytes and measuring phagocytosis from live Galleria mellonella larvae. Virulence 15:2313413

Carvalho GC, Marena GD, Leonardi GR, Sábio RM, Corrêa I, Chorilli M, Bauab TM (2022) Lycopene, mesoporous silica nanoparticles and their association: a possible alternative against vulvovaginal candidiasis? Molecules [Online] 27(23):8558

Cé R, Silva RC, Trentin DS, Marchi JGB, Paese K, Guterres SS, Macedo AJ, Pohlmann AR (2020) Galleria mellonella larvae as an in vivo model to evaluate the toxicity of polymeric nanocapsules. J Nanosci Nanotechnol 20:1486–1494

Cerenius L, Lee BL, Söderhäll K (2008) The proPO-system: pros and cons for its role in invertebrate immunity. Trends Immunol 29:263–271

Cerenius L, Kawabata S-I, Lee BL, Nonaka M, Söderhäll K (2010) Proteolytic cascades and their involvement in invertebrate immunity. Trends Biochem Sci 35:575–583

Champion OL, Titball RW, Bates S (2018) Standardization of G. mellonella larvae to provide reliable and reproducible results in the study of fungal pathogens. J Fungi (Basel) 4(3):108

Chen RY, Keddie BA (2021) Galleria mellonella (Lepidoptera: Pyralidae) haemocytes release extracellular traps that confer protection against bacterial infection in the hemocoel. J Insect Sci 21(6):17

Curtis A, Binder U, Kavanagh K (2022) Galleria mellonella larvae as a model for investigating fungal-host interactions. Front Fungal Biol 3:893494

Curtis A, Dobes P, Marciniak J, Hurychova J, Hyrsl P, Kavanagh K (2024) Characterization of Aspergillus fumigatus secretome during sublethal infection of Galleria mellonella larvae. J Med Microbiol 73(6):001844. https://doi.org/10.1099/jmm.0.001844

Danielsen PH, Bendtsen KM, Knudsen KB, Poulsen SS, Stoeger T, Vogel U (2021) Nanomaterial- and shape-dependency of TLR2 and TLR4 mediated signaling following pulmonary exposure to carbonaceous nanomaterials in mice. Part Fibre Toxicol 18:40

De La Harpe KM, Kondiah PPD, Choonara YE, Marimuthu T, Du Toit LC, Pillay V (2019) The hemocompatibility of nanoparticles: a review of cell-nanoparticle interactions and hemostasis. Cells 8(10):1209

Demirtürk Z, Uçkan F, Mert S (2024) Interactions of alumina and polystyrene nanoparticles with the innate immune system of Galleria mellonella. Drug Chem Toxicol 47:483–495

Donaldson K, Murphy F, Schinwald A, Duffin R, Poland CA (2011) Identifying the pulmonary hazard of high aspect ratio nanoparticles to enable their safety-by-design. Nanomedicine (London) 6:143–156

Driscoll KE, Borm PJA (2020) Expert workshop on the hazards and risks of poorly soluble low toxicity particles. Inhal Toxicol 32:53–62

Dushay MS (2009) Insect haemolymph clotting. Cell Mol Life Sci 66:2643–2650

Eskin A (2022) Effects of silicon dioxide nanoparticles (SiO2 NPs) on total haemocyte count and haemocyte viability of Galleria mellonella. Int J Trop Insect Sci 42:2617–2623

Eskin A, Nurullahoğlu ZU (2022) Effects of zinc oxide nanoparticles (ZnO NPs) on the biology of Galleria mellonella L. (Lepidoptera: Pyralidae). J Basic Appl Zool 83:54

Eskin A, Nurullahoğlu ZU (2023) Influence of zinc oxide nanoparticles (ZnO NPs) on the haemocyte count and haemocyte-mediated immune responses of the Greater Wax Moth Galleria mellonella (Lepidoptera: Pyralidae). Drug Chem Toxicol 46:1176–1186

Eskin A, Öztürk Ş, Körükçü M (2019) Determination of the acute toxic effects of zinc oxide nanoparticles (ZnO NPs) in total haemocytes counts of Galleria mellonella (Lepidoptera: Pyralidae) with two different methods. Ecotoxicology 28:801–808

Eskin A, Ekremoglu M, Altinkaynak C, Özdemir N (2022) Effects of organic-inorganic hybrid nanoflowers' framework on haemocytes and enzymatic responses of the model organism, Galleria mellonella (Lepidoptera: Pyralidae). Int J Trop Insect Sci 42:333–344

Fuchs BB, O'brien E, Khoury JB, Mylonakis E (2010) Methods for using Galleria mellonella as a model host to study fungal pathogenesis. Virulence 1:475–482

Genç TT, Kaya S, Günay M, Çakaloğlu Ç (2024) Humoral immune response of Galleria mellonella after mono- and co-injection with Hypericum perforatum extract and Candida albicans. APMIS 132:358–370

Gupta A (1985) The identity of the so-called crescent cell in the haemolymph of the cockroach, Gromphadorhina portentosa (Schaum)(Dictyoptera: Blaberidae). Cytologia 50:739–746

Hamamoto H, Tonoike A, Narushima K, Horie R, Sekimizu K (2009) Silkworm as a model animal to evaluate drug candidate toxicity and metabolism. Comp Biochem Physiol C Toxicol Pharmacol 149:334–339

Han JJ (2023) FDA Modernization Act 2.0 allows for alternatives to animal testing. Artif Organs 47:449–450

Hartung T (2009) A toxicology for the 21st century—mapping the road ahead. Toxicol Sci 109:18–23

Hofer S, Hofstätter N, Punz B, Hasenkopf I, Johnson L, Himly M (2022) Immunotoxicity of nanomaterials in health and disease: current challenges and emerging approaches for identifying immune modifiers in susceptible populations. WIREs Nanomed Nanobiotechnol 14:e1804

Hoffmann JA, Reichhart J-M, Hetru C (1996) Innate immunity in higher insects. Curr Opin Immunol 8:8–13

Jander G, Rahme LG, Ausubel FM (2000) Positive correlation between virulence of Pseudomonas aeruginosa mutants in mice and insects. J Bacteriol 182:3843–3845

Kavanagh K, Reeves EP (2004) Exploiting the potential of insects for in vivo pathogenicity testing of microbial pathogens. FEMS Microbiol Rev 28(1):101–112. https://doi.org/10.1016/j.femsre.2003.09.002

Kelly J, Kavanagh K (2011) Caspofungin primes the immune response of the larvae of Galleria mellonella and induces a non-specific antimicrobial response. J Med Microbiol 60:189–196

Kopácek P, Weise C, Götz P (1995) The prophenoloxidase from the wax moth Galleria mellonella: purification and characterization of the proenzyme. Insect Biochem Mol Biol 25:1081–1091

Kwadha CA, Ong'amo GO, Ndegwa PN, Raina SK, Fombong AT (2017) The biology and control of the greater wax moth galleria mellonella. Insects 8(2):61. https://doi.org/10.3390/insects8020061

Luque T, Okano K, O'reilly DR (2002) Characterization of a novel silkworm (Bombyx mori) phenol UDP-glucosyltransferase. Eur J Biochem 269:819–825

Maguire R, Duggan O, Kavanagh K (2016) Evaluation of Galleria mellonella larvae as an in vivo model for assessing the relative toxicity of food preservative agents. Cell Biol Toxicol 32:209–216

Maguire R, Kunc M, Hyrsl P, Kavanagh K (2017a) Analysis of the acute response of Galleria mellonella larvae to potassium nitrate. Comp Biochem Physiol C Toxicol Pharmacol 195:44–51

Maguire R, Kunc M, Hyrsl P, Kavanagh K (2017b) Caffeine administration alters the behaviour and development of Galleria mellonella larvae. Neurotoxicol Teratol 64:37–44

Mantica F, Iñiguez LP, Marquez Y, Permanyer J, Torres-Mendez A, Cruz J, Franch-Marro X, Tulenko F, Burguera D, Bertrand S, Doyle T, Nouzova M, Currie PD, Noriega FG, Escriva H, Arnone MI, Albertin CB, Wotton KR, Almudi I, Martin D, Irimia M (2024) Evolution of tissue-specific expression of ancestral genes across vertebrates and insects. Nat Ecol Evol 8:1140–1153

Matsumoto H, Tanaka K, Noguchi H, Hayakawa Y (2003) Cause of mortality in insects under severe stress. Eur J Biochem 270:3469–3476

Mowlds P, Barron A, Kavanagh K (2008) Physical stress primes the immune response of Galleria mellonella larvae to infection by Candida albicans. Microbes Infect 10:628–634

Moya-Andérico L, Vukomanovic M, Cendra MDM, Segura-Feliu M, Gil V, Del Río JA, Torrents E (2021) Utility of Galleria mellonella larvae for evaluating nanoparticle toxicology. Chemosphere 266:129235

Mukherjee K, Raju R, Fischer R, Vilcinskas A (2013) Galleria mellonella as a model host to study gut microbe homeostasis and brain infection by the human pathogen listeria monocytogenes. Adv Biochem Eng Biotechnol 135:27–39

Naz S, Gul A, Zia M (2020) Toxicity of copper oxide nanoparticles: a review study. IET Nanobiotechnol 14:1–13

Niwa R, Sakudoh T, Matsuya T, Namiki T, Kasai S, Tomita T, Kataoka H (2011) Expressions of the cytochrome P450 monooxygenase gene Cyp4g1 and its homolog in the prothoracic glands of the fruit fly Drosophila melanogaster (Diptera: Drosophilidae) and the silkworm Bombyx mori (Lepidoptera: Bombycidae). Appl Entomol Zool 46:533–543

Orihel TC (1975) The peritrophic membrane: its role as a barrier to infection of the arthropod host. In: Invertebrate immunity. Academic, pp 65–73

Pagán AJ, Ramakrishnan L (2018) The formation and function of granulomas. Annu Rev Immunol 36:639–665

Pavan C, Escolano-Casado G, Bellomo C, Cananà S, Tomatis M, Leinardi R, Mino L, Turci F (2022) Nearly free silanols drive the interaction of crystalline silica polymorphs with membranes: implications for mineral toxicity. Front Chem 10:1092221

Pearce JC, Campbell JS, Prior JL, Titball RW, Wakefield JG (2024) PiggyBac mediated transgenesis and CRISPR/Cas9 knockout in the greater waxmoth, Galleria mellonella. bioRxiv, 2024.09.17.613535

Pereira MF, Rossi CC, Vieira De Queiroz M, Martins GF, Isaac C, Bosse JT, Li Y, Wren BW, Terra VS, Cuccui J (2015) Galleria mellonella is an effective model to study Actinobacillus pleuropneumoniae infection. Microbiology 161:387–400

Pereira MF, Rossi CC, Da Silva GC, Rosa JN, Bazzolli DMS (2020) Galleria mellonella as an infection model: an in-depth look at why it works and practical considerations for successful application. Pathog Dis 78(8):ftaa056

Ramarao N, Nielsen-Leroux C, Lereclus D (2012) The insect Galleria mellonella as a powerful infection model to investigate bacterial pathogenesis. J Vis Exp 70:e4392

Renwick J, Reeves EP, Wientjes FB, Kavanagh K (2007) Translocation of proteins homologous to human neutrophil p47phox and p67phox to the cell membrane in activated haemocytes of Galleria mellonella. Dev Comp Immunol 31:347–359

Sabella S, Carney RP, Brunetti V, Malvindi MA, Al-Juffali N, Vecchio G, Janes SM, Bakr OM, Cingolani R, Stellacci F, Pompa PP (2014) A general mechanism for intracellular toxicity of metal-containing nanoparticles. Nanoscale 6:7052–7061

Sezer Tuncsoy B, Tuncsoy M, Gomes T, Sousa V, Teixeira MR, Bebianno MJ, Ozalp P (2019) Effects of copper oxide nanoparticles on tissue accumulation and antioxidant enzymes of Galleria mellonella L. Bull Environ Contam Toxicol 102:341–346

Sheehan G, Clarke G, Kavanagh K (2018a) Characterisation of the cellular and proteomic response of Galleria mellonella larvae to the development of invasive aspergillosis. BMC Microbiol 18:63

Sheehan G, Garvey A, Croke M, Kavanagh K (2018b) Innate humoral immune defences in mammals and insects: the same, with differences ? Virulence 9:1625–1639

Sheehan G, Farrell G, Kavanagh K (2020a) Immune priming: the secret weapon of the insect world. Virulence 11:238–246

Sheehan G, Konings M, Lim W, Fahal A, Kavanagh K, Van De Sande WWJ (2020b) Proteomic analysis of the processes leading to Madurella mycetomatis grain formation in Galleria mellonella larvae. PLoS Negl Trop Dis 14:e0008190

Shokal U, Eleftherianos I (2017) Evolution and function of thioester-containing proteins and the complement system in the innate immune response. Front Immunol 8:759

Smith DFQ, Casadevall A (2021) Fungal immunity and pathogenesis in mammals versus the invertebrate model organism Galleria mellonella. Pathog Dis 79(3):ftab013

Sugeçti S, Tunçsoy B, Büyükgüzel E, Özalp P, Büyükgüzel K (2021) Ecotoxicological effects of dietary titanium dioxide nanoparticles on metabolic and biochemical parameters of model organism Galleria mellonella (Lepidoptera: Pyralidae). J Environ Sci Health C 39:423–434

Szebeni J (2014) Complement activation-related pseudoallergy: a stress reaction in blood triggered by nanomedicines and biologicals. Mol Immunol 61:163–173

Thomaz L, Gustavo De Almeida L, Silva FRO, Cortez M, Taborda CP, Spira B (2020) In vivo activity of silver nanoparticles against Pseudomonas aeruginosa infection in Galleria mellonella. Front Microbiol 11:582107

Tojo S, Naganuma F, Arakawa K, Yokoo S (2000) Involvement of both granular cells and plasmatocytes in phagocytic reactions in the greater wax moth, Galleria mellonella. J Insect Physiol 46:1129–1135

Tuncsoy B, Mese Y (2021) Influence of titanium dioxide nanoparticles on bioaccumulation, antioxidant defense and immune system of Galleria mellonella L. Environ Sci Pollut Res 28:38007–38015

Tuncsoy, B, Tuncsoy M (2023) Particle Size Effect of Micro and Nano Aluminium Oxides on Antioxidant Defence System of Model Organism Galleria mellonella. Bull Environ Contam Toxicol 110(4):75. https://doi.org/10.1007/s00128-023-03715-7

Tunçsoy B, Sugeçti S, Büyükgüzel E, Özalp P, Büyükgüzel K (2021) Effects of copper oxide nanoparticles on immune and metabolic parameters of Galleria mellonella L. Bull Environ Contam Toxicol 107:412–420

Wadman M (2023) FDA no longer has to require animal testing for new drugs. Science 379:127–128

Wang X, Zhang Y, Zhang R, Zhang J (2019) The diversity of pattern recognition receptors (PRRs) involved with insect defense against pathogens. Curr Opin Insect Sci 33:105–110

Wittwer D, Franchini A, Ottaviani E, Wiesner A (1999) Presence of il-1- and Tnf-like molecules in galleria Mellonella (lepidoptera) haemocytes and in an insect cell line Fromestigmene Acraea (lepidoptera). Cytokine 11:637–642

Wrońska AK, Kaczmarek A, Sobich J, Grzelak S, Boguś MI (2022) Intracellular cytokine detection based on flow cytometry in haemocytes from Galleria mellonella larvae: a new protocol. PLoS One 17:e0274120

Wu G, Liu Y, Ding Y, Yi Y (2016) Ultrastructural and functional characterization of circulating haemocytes from Galleria mellonella larva: cell types and their role in the innate immunity. Tissue Cell 48:297–304

Xu M-N, Li L, Pan W, Zheng H-X, Wang M-L, Peng X-M, Dai S-Q, Tang Y-M, Zeng K, Huang X-W (2021) Zinc oxide nanoparticles prime a protective immune response in Galleria mellonella to defend against Candida albicans. Front Microbiol 12:766138

Yamamoto K, Ichinose H, Aso Y, Fujii H (2010) Expression analysis of cytochrome P450s in the silkmoth, Bombyx mori. Pestic Biochem Physiol 97:1–6

Zhang P, Cao M, Chetwynd AJ, Faserl K, Abdolahpur Monikh F, Zhang W, Ramautar R, Ellis L-JA, Davoudi HH, Reilly K, Cai R, Wheeler KE, Martinez DST, Guo Z, Chen C, Lynch I (2024) Analysis of nanomaterial biocoronas in biological and environmental surroundings. Nat Protoc 19(10):3000–3047

Zorlu T, Nurullahoğlu ZU, Altuntaş H (2018) Influence of dietary titanium dioxide nanoparticles on the biology and antioxidant system of model insect, Galleria mellonella (L.) (Lepidoptera: Pyralidae). J Entomol Res Soc 20:89–103

Open Access This chapter is licensed under the terms of the Creative Commons Attribution-NonCommercial-NoDerivatives 4.0 International License (http://creativecommons.org/licenses/by-nc-nd/4.0/), which permits any noncommercial use, sharing, distribution and reproduction in any medium or format, as long as you give appropriate credit to the original author(s) and the source, provide a link to the Creative Commons license and indicate if you modified the licensed material. You do not have permission under this license to share adapted material derived from this chapter or parts of it.

The images or other third party material in this chapter are included in the chapter's Creative Commons license, unless indicated otherwise in a credit line to the material. If material is not included in the chapter's Creative Commons license and your intended use is not permitted by statutory regulation or exceeds the permitted use, you will need to obtain permission directly from the copyright holder.

Drosophila Model Unveils Nanoparticle Interactions: Implications for Safety

13

Ghada Tagorti and Bülent Kaya

Abstract

Nanoparticles are increasingly integrated to advancements in technology, medicine, and environmental science, offering unique properties that enhance drug delivery systems, diagnostic methods, and material efficiency. However, the widespread use of nanoparticles also raises concerns regarding their potential toxicity and environmental impact. Addressing these concerns necessitates robust safety assessments and a comprehensive understanding of nanoparticle interactions within biological systems. *Drosophila melanogaster*, commonly known as the fruit fly, serves as an invaluable model for studying the biological implications of nanoparticles. Its well-documented genetic background, short life cycle, and biological similarities to mammals make it ideal for evaluating nanoparticle toxicity. This chapter explores the role of *Drosophila* in nanoparticle research, particularly in assessing safety and elucidating the mechanisms of nanoparticle interaction at the cellular and behavioral levels. Studies using *Drosophila* have demonstrated that nanoparticles can induce genetic, cellular, and behavioral changes, which vary depending on the composition, size, shape, and exposure level of the nanoparticle. These findings highlight the necessity of safety assessments and the development of standardized testing protocols. In conclusion, as nanoparticles use become more prevalent across multiple sectors, the importance of rigorous safety evaluations grows. Using *Drosophila* for these assessments not only deepens our understanding of nanoparticle toxicity but also supports the development of safer nanotechnologies.

Keywords

Drosophila · DNA damage · Nanoparticles · Nanosafety · Oxidative stress

---

G. Tagorti · B. Kaya (✉)
Department of Biology, Akdeniz University, Antalya, Türkiye
e-mail: bkaya@akdeniz.edu.tr

© The Author(s) 2025
E. Alfaro-Moreno, F. Murphy (eds.), *Nanosafety*,
https://doi.org/10.1007/978-3-031-93871-9_13

1 Introduction

Drosophila melanogaster was first time described by Johann Wilhelm Meigen in 1830, a German entomologist known for his pioneering work in the study of European Diptera. Following Meigen, the American entomologist Charles Woodworth was interested in *Drosophila* (Villegas 2019). Later, Woodworth suggest using *Drosophila* as a model organism in genetic research. Thomas Hunt Morgan brought these flies into his laboratory at a following introduction made by Frank Lutz, an American entomologist, who extensively studied the basic biology of this fly (Markow 2015). In 1910, Morgan identified his initial mutant of *Drosophila*, a male with white eyes and subsequently established its location on the X chromosome. This discovery was the first instance of a specific gene being localized to a specific chromosome (Green 2010). In 1933, Morgan awarded the Nobel Prize in Physiology or Medicine for discovering the role of chromosomes in inheritance. Several other Nobel Prizes have been awarded for research involving *Drosophila*. For instance, Hermann Joseph Muller, in 1946, awarded the prize for discovering how mutations could be induced using X-ray irradiation. Edward B. Lewis, Christiane Nüsslein-Volhard, and Eric F. Wieschaus received the 1995 prize for their work on the genetic regulation of early embryonic development. Most recently, in 2017, Jeffrey C. Hall, Michael Rosbash, and Michael W. Young were awarded the prize for uncovering the molecular mechanisms controlling circadian rhythms (Markow 2015; Huang 2018). Interest in *D. melanogaster* across various scientific fields, particularly for toxicity assessments, has led to the development of a new research area termed drosophotoxicology (Rand 2010).

In this chapter, we explored the role of *D. melanogaster* in the study of nanomaterials. As well as its application in toxicology to understand how it contributes to safety assessments. Each section aims to highlight the utility and significance of *Drosophila* in these rapidly advancing areas of research.

2 Genetics and Biology of *Drosophila melanogaster*

D. melanogaster classified within the Family Drosophilidae, is commonly known as the fruit fly, have originated in sub-Saharan Africa between 10,000 and 15,000 years ago, before migrating to Europe and Asia. Later, facilitated by human movement and trade, *Drosophila* has dispersed widely, establishing populations on every continent (Markow 2015). Females typically measures approximately 2.5 mm in length, generally larger than males. Males can be distinguished by several morphological features, including the sex-comb (i.e., a row of dark hairs on the tarsus of the first leg) and distinctive body coloration that differentiates them from females. This species exhibited a relatively short lifespan of approximately 80–100 days at an optimal room temperature of 25 °C. *D. melanogaster* exhibits holometabolous development, undergoing a complete metamorphosis from larva to adult through four distinct developmental stages: embryos, larvae, pupae, and adults. The embryos developed into first-stage larvae within approximately 24 h. These larvae subsequently grew through two additional stages, each lasting about 24 h. During its life cycle, *Drosophila* undergoes substantial morphological transformations that are marked by significant physiological changes and growth patterns. For instance, during the larval stages, there is heightened glycolytic activity, increased lactate production, and considerable synthesis of glycogen and triglycerides. These metabolic activities are essential for supporting the energy-intensive process of metamorphosis. In this way, larvae increase in size by more than 20-fold from the first to the third instar within approximately 4 days, indicating rapid growth and development. Additionally, during these larval stages, specific tissues known as imaginal discs undergo significant changes. These discs are mitotic tissues that grow and differentiate extensively during the larval period to develop into critical adult structures, such as wings and eyes (Malita and Rewitz 2021). The larva contains 19 discs, comprising nine bilateral pairs that form epidermal structures and one genital disc. The pupal stage, spanning 3–5 days, involves the transformation of tissues into adult structures, leading to eclosion from the pupae (Beira and Paro 2016).

The *Drosophila* genome was sequenced and published in 2000 (Adams et al. 2000). It is estimated to contain over 14,000 protein-coding genes (Brown et al. 2014). Remarkably, approximately 75% of those related to human diseases have homologs, underscoring its utility in both basic and applied genetic research (Yamamoto et al. 2014). Furthermore, *Drosophila* has a comprehensive complement of genes that encode enzymes involved in xenobiotic metabolism, including well-conserved enzymes across the various phases of detoxification: Phase I (e.g., cytochrome P450), Phase II (e.g., glutathione-S-transferases), and Phase III enzymes (e.g., ABC ATP-dependent transporters) (Dermauw and Van Leeuwen 2014).

Drosophila primarily defends against pathogens through its innate immune response, which is supported by physical barriers analogous to mammals (Lemaitre and Hoffmann 2007). The innate immune system in *Drosophila* is classified into humoral and cellular responses. Cellular immunity is mediated by hemocytes, (analogous to mammalian blood cells): (i) plasmatocytes represents 95% of hemocytes, are comparable to mammalian phagocytes and are responsible for producing antimicrobial peptides (AMPs); (ii) crystal cells, which constitute 5% of hemocytes, play a role in melanization and wound healing, and (iii) lamellocytes are engaging in encapsulation of pathogens (Vlisidou and Wood 2015). Moreover, plasmatocytes play a crucial role in activating the humoral immune response, with significant interplay between the two responses. The humoral response is characterized by the synthesis of AMPs and anti-pathogenic factors, orchestrated through the Toll, Imd, c-Jun N-terminal kinase (JNK), and Janus kinase/signal transducer and activator of transcription (JAK/STAT) signaling pathways (Lemaitre and Hoffmann 2007).

Drosophila also exhibits apoptosis, a programmed cell death mechanism critical for maintaining cellular integrity and tissue homeostasis. This conserved biological process plays a pivotal role in development and immune regulation, close to its function in mammalian systems. In *Drosophila*, the caspase family comprises seven members, classified into three initiator caspases: Death related ced-3/Nedd2-like caspase (Dredd), Death regulator Nedd2-like caspase (Dronc), and Ser/Thr-rich caspase (Strica); and four effector caspases, specifically *Drosophila* caspase interleukin 1β-converting enzyme (Drice), Death-associated molecule related to Mch2 (Damm), Death executioner caspase related to apopain/yama (Decay), and *Drosophila* caspase-1 (Dcp-1) (Kietz and Meinander 2023). All caspases in *Drosophila* are associated with apoptotic signalling; however, specific variants such as the caspase-8 homolog Dredd primarily function as regulators of the inflammatory response to Gram-negative bacteria (Leulier et al. 2000).

The brain of an adult *Drosophila* contains about 200,000 neurons that organize into specific circuits, which facilitate complex behaviors including courtship, sleep, circadian rhythms, learning, and memory (Zheng et al. 2018). In addition, glial cells make up 5 to 10% of the total cell population in the central nervous system. Morphologically, the structure of neurons in mammals are primarily multipolar, whereas in *Drosophila* neurons are predominantly unipolar. Despite these structural differences, neuronal circuits exhibit plasticity, adapting their structure and function in response to stimuli during development stage and adult life. In *Drosophila*, as in humans, the blood-brain barrier regulates paracellular diffusion and manages the influx and efflux of soluble molecules (Santarelli et al. 2023).

3   Advantages of Using *Drosophila* in Scientific Research

Drosophila is a well-known model organism owing to its short life span, easy maintenance in the laboratory, high fecundity (30–50 eggs/day), and the presence of various mutant (Pandey and Nichols 2011). It shares nearly 80% of its functional protein domain identity with humans, and 75% of human disease-related genes have homologs in *Drosophila*. Additionally, the fruit fly is an excellent example of the 3R principle (Replace, Refine, Reduce) (Pandey and Nichols 2011). *Drosophila* exhibits several physiological similarities to humans including the role of insulin signaling and energy homeostasis in metabolic disorders (Garofalo 2002; Chatterjee and Perrimon 2021). Moreover, *Drosophila* possesses the midgut, which plays a similar role in nutrient absorption, and the hindgut in water and electrolyte absorption. Considering that oral ingestion is a key route for nanoparticles (NPs) to enter the human body, it is significant that the *Drosophila* model can mimic this entry route of exposure through the intestinal barrier.

The larval midgut is organized into five distinct regions with varying pH levels: it begins with a neutral anterior midgut, progresses to an acidic region, returns to a neutral state, transitions through an intermediate area, and culminates in an alkaline region. These segments are broadly analogous to the main divisions found in the adult gut. Similar to humans, the low pH region in *Drosophila* serves a protective role, guarding against pathogenic bacteria and managing the levels of non-pathogenic bacterial populations. Disruption of this acidic region leads to an increase in bacterial populations, which can potentially impact the overall health and homeostasis of the gut environment (Overend et al. 2016).

Furthermore, *Drosophila* has a complex brain and neuronal network, making it an ideal model for studying neurodegenerative diseases, such as Alzheimer's disease, Parkinson's disease, Huntington's disease and neurological effects following NPs ingestion (Bolus et al. 2020; Nitta and Sugie 2022). It also features robust detoxification and DNA repair systems comparable to those in humans. Additionally, spermatogenesis in both *Drosophila* and mammals demonstrates a high degree of conservation. The anatomical configuration of *Drosophila* testes enables accurate delineation of specific stages altered by compounds during spermatogenesis. Notably, *D. melanogaster* displays less genomic redundancy, enhancing the clarity of toxicological assessments (Yuan et al. 2019).

Notably, the involvement of distinct life cycle stages in *Drosophila* offers unique opportunities for assessing the toxicity of nanomaterials providing distinct advantages over other model systems. The impact of NPs on development serves as an initial indicator of toxic endpoints, which can be quantitatively assessed by measuring changes in body length or mass, as well as delays in development. During the larval stages, growth-regulating pathways are particularly active; thus, toxic exposures can disrupt signalling mechanisms that influence cell fate and differentiation, particularly within the imaginal discs of the larva. Disruptions in these tissues during the larval stages can lead to abnormal morphogenesis in the corresponding adult structures.

In addition, multiple tools are available to provide extensive information and data regarding *Drosophila* such as FlyBase which offers high-quality datasets related to *Drosophila* genes and genomes. This platform compiles various published phenotypes and gene expression data facilitating detailed assessment into genetic networks and protein functions. Moreover, numerous global stock centers, such as the Bloomington *Drosophila* Stock Center (https://bdsc.indiana.edu), the Kyoto Stock Center (http://www.dgrc.kit.ac.jp), and the Vienna Stock Center (https://stockcenter.vdrc.at), provide access to a wide variety of necessary fly strains (Fischer et al. 2023).

4 Overview of Nanotoxicology and Safety Assessment Principles

Nanoparticles (NPs) are particles with a size ranging between 1 and 100 nm, encompassing various types such as polymeric NPs, liposomes, lipid micelles, quantum dots, dendrimers, and metallic NPs (Horikoshi and Serpone 2013). In 2022, the European Commission refined the definition of nanomaterials to align with legislation. Based on EC Recommendation 2022/C 229/01, a nanomaterial consists of '1) solid particles; 2) 50 % or more of its constituent particles fulfil at least one of the following conditions: (a) One or more external dimensions of the particle are in the size range 1 nm to 100 nm (b) The particle has an elongated shape, such as a rod, fibre or tube, where two external dimensions are smaller than 1 nm and the other dimension is larger than 100 nm (c) The particle has a plate-like shape, where one external dimension is smaller than 1 nm and the other dimensions are larger than 100 nm; (3) the volume specific surface area (VSSA), which can be used to demonstrate that a given particulate material is not a nanomaterial. The corresponding exclusion criterion is a VSSA of less than 6 m^2/cm^3.' (European Commission 2022).

Nanomaterials have the potential to provide novel strategies to improve the quality and design of products (Rambaran and Schirhagl 2022; Joseph et al. 2023). The advent of nanotechnology has a special impact on various fields such as tissue engineering, agriculture, energy, environmental, biomedical and pharmaceutical sectors (Kuppusamy et al. 2016). Based upon the unique physicochemical properties, thermal stability, optical characteristics and widespread biological activities, the nanotechnological intervention have significantly revolutionized the current knowledge of conventional pharmaceutical and biomedical settings with improved efficacy. This explains the widespread use of NPs in several applications leading to increased human exposure, and raising concerns regarding potential toxicity (Yang et al. 2021).

NPs interact with biological systems and elicit various responses *in vivo*. The cellular uptake of NPs is influenced by various physicochemical properties, resulting in complex cellular mechanisms. For instance, the biological medium (pH) and the concentration of the NPs could interfere with proteins absorption on NP surface known as the protein coronae. This process can alter the properties of NPs such as the dispersion and the functionalities yielding detrimental or beneficial properties which can also induce different biological responses (Sengottiyan et al. 2023).

Nanoparticles induce inflammation by activating pro-inflammatory signaling pathways and stimulating cytokine production. This inflammatory response leads to oxidative stress and diminishes the cellular antioxidant defenses (Khansari et al. 2009; Manke et al. 2013; Maku et al. 2024). Oxidative stress refers to the imbalance between the generation and elimination of reactive oxygen and nitrogen species in cells (Zarkovic 2020). The predominant reactive oxygen species (ROS) are superoxide, hydroxyl radicals, and hydrogen peroxide (H_2O_2), whereas the major reactive nitrogen species are peroxynitrite and nitrogen dioxide (Fröhlich 2013). Superoxide ($O_2^{\bullet-}$), a primary source of ROS, is formed after xenobiotic exposure and/or via multiple physiological enzymatic (cytochrome P450, NADPH oxidase (NOX), cytoplasmic xanthine oxidase (OX), and mitochondrial electron transport chain (ETC)) and non-enzymatic processes (glycation under hyperglycemic conditions) (Fujii et al. 2022). In contrast, the hydroxyl radical is the most aggressive ROS with a strong oxidization potential to damage biomolecules (Hou et al. 2020). It is generated via the Fenton or the Haber-Weiss reaction involving H_2O_2 and superoxide. Excessive oxidative stress causes DNA damage, lipid peroxidation, protein fragmentation, and aggregation. Antioxidants act as the first line of defense to neutralize ROS, such as superoxide dismutases (SODs), catalase (CAT), and the glutathione peroxidase (GPX) family (Fröhlich 2013). Three superoxide dismutase isoforms (SOD1–3) are produced in humans, are localized in distinct compartments, and are responsible for the elimination of superoxide anions. SOD1 (Cu/Zn SOD) is present in the cytoplasm and extracellular compartments and is considered the most abundant SOD. SOD2 (Mn-SOD) is ubiquitously expressed in the mitochondrial matrix, whereas SOD3 (Cu/Zn-SOD) is localized in the extracellular milieu, especially in vascular and pulmonary systems (Starkov 2008). The catalytic reaction of SODs produces H_2O_2, whereas CAT catalyzes the conversion of H_2O_2 to oxygen and water molecules in peroxisomes. In addition to CAT, the GPX family (GPX1–8) reduces H_2O_2 levels by oxidizing GSH to GSSG (Liang et al. 2011). Under a consistent increase in oxidative stress, disruption of mitochondrial function and apoptosis can occur (Fröhlich 2013). Oxidative stress causes tumorigenesis (Hayes et al. 2020), neurodegenerative diseases (Barnham et al. 2004), acute and chronic kidney diseases (Rapa et al. 2019; Tomsa et al. 2019), diabetes mellitus (Asmat et al. 2016), and autoimmune diseases (Ramani et al. 2020; Lin et al. 2021).

NPs can directly generate ROS by interfering with the electron transfer process, mitochondrial function, and NADP+/NADPH ratio (Asharani et al. 2009; Wang et al. 2017; Lee et al. 2018). Upon the initiation of the oxidative stress response, the activation of nuclear factor erythroid 2-related factor 2 (Nrf2) initiates the upregulation of antioxidant expression, followed by the induction of inflammation mediated by nuclear factor kappa-light-chain-enhancer of activated B cells (NF-κB) and activator protein 1 (Ap-1) (Li et al. 2008; Huang et al. 2010).

With the advent of carbon nanotubes, the early 2000s marked the beginning of intensive research into the health hazards induced by NPs. Since this discovery, researchers have extensively explored the potential negative health impacts of NP exposure by using diverse model organisms (Liu et al. 2013). Despite numerous studies on NPs toxicity, significant gaps remain, especially in understanding the health toxicology, mechanisms, and assessment models related to NPs adverse effects (Xuan et al. 2023). Nanomaterials vary significantly, indicating that subtle differences in material properties can lead to varied biological responses (Costa and Fadeel 2016). This variation, coupled with inconsistent results such as nanomaterial behavior in biological systems and a lack of verified data, reduces effective safety evaluations of nanomaterials (Zielińska et al. 2020).

Assessing the toxicity of the vast array of nanomaterials (NMs) would require an extensive number of safety evaluations. Consequently, there is an urgent need to adopt alternative methods that can produce mechanistic data derived from model organisms like *Drosophila*, to apply the principles of reducing, refining, or replacing (3Rs) in toxicity assessments. While a single alternative testing method can offer basic insights into mechanisms or toxicity, it might not be adequate for comprehensive hazard assessments. Thus, integrating multiple alternative testing approaches into a battery testing strategy could enhance our understanding of NMs behavior and toxicity in human health and the environment (Shatkin and Ong 2016).

Initially, the majority of NMs consisted of silicon dioxide (SiO_2), carbon black, silver (Ag), and titanium dioxide (TiO_2). Over time, these were progressively substituted with a range of advanced materials such as fullerenes, carbon nanotubes, graphene, nanocellulose, polymers, nanofibrils, and dendrimers, reflecting a significant evolution in the composition and application of nanomaterials. The complexity and unique characteristics of these materials necessitate thorough toxicological assessments to guarantee their safety across various applications, thus protecting public health and preserving environmental integrity (Tirumala et al. 2021).

Several factors influence nanotoxicity, particularly the physical and chemical properties, such as size, shape, surface charge, catalytic activity, and the presence or absence of active groups on the surface. Additionally, the mode of internalization and subsequent intracellular trafficking mechanisms are also critical. Although various factors contribute to NPs toxicity, oxidative stress and the production of reactive oxygen species (ROS) are considered fundamental mechanisms driving the toxic effects of NPs, irrespective of their size and shape (Sukhanova et al. 2018; Mishra and Panda 2021).

It has been noted that NPs, which closely resemble airborne pollutants, predominantly affect the lungs and heart. These NPs disperse extensively throughout the lung regions, inducing both pulmonary and systemic effects. The initial responses, characterized by inflammation and oxidative stress, may progress to fibrosis, granulomatous reactions, coagulopathies, and cardiac disturbances, ultimately culminating in organ damage. An additional critical determinant of respiratory nanotoxicity is the dosage regimen. The production of reactive oxygen species (ROS) and reactive nitrogen species (RNS), as well as the expression of pro-inflammatory cytokines in cells exposed to nanomaterials, is generally dependent on the dose (Zhang et al. 2022; Thu et al. 2023; Zhou et al. 2023). Metallic NPs with dimensions less than 50 nm can bypass the mucociliary clearance system and penetrate the alveoli, where they may either be localized within the alveolar epithelium or enter the systemic circulation. Conversely, larger particles, typically greater than 5 μm, are generally blocked by the mucociliary system and subsequently eliminated from the body via the lymphatic system. The probability for metallic NMs to deposit in the respiratory tract is predominantly controlled by their aerodynamic or thermodynamic diameter (Thu et al. 2023).

The subsequent phase in NPs toxicity research involves developing strategies to minimize the hazards associated with NPs exposure. This includes the development of safe and effective NP synthesis techniques, the identification of biomarkers indicative of NPs toxicity, and the implementation of protective protocols for personnel employed in industries using NPs (Xuan et al. 2023).

The European Chemicals Agency has implemented various guidelines and recommendations to facilitate compliance with EU regulations concerning NMS. Under the "REACH" framework (Registration, Evaluation, Authorization and Restriction of Chemical Substances), the risk assessment of NMs should align with the methodology used for traditional chemicals, beginning with effect evaluation. A risk quotient below 1 is deemed acceptable if the estimated exposure is lower than the concentration at which no adverse effects were observed in experimental studies. To comprehensively evaluate the effects, conducting *in vitro* and/or *in vivo* experiments might be necessary, along with other methods to characterize the NMs in question. This could involve gathering data on critical physicochemical properties that affect toxicity, including size distribution, aggregation status, shape, surface area, reactivity, water solubility, surface properties, and long-term stability. The next step involves identifying all potential exposure sources, necessitating a thorough understanding of the manufacturing process and likely exposure routes. This knowledge is crucial for selecting suitable testing methods and formulating risk prevention strategies (Zielińska et al. 2020).

Similarly, the U.S. Environmental Protection Agency (EPA) has formulated guidelines to assist companies and researchers in evaluating the risks and ensuring the safety of NPs. Notably, the EPA's risk assessment framework for NPs includes evaluations of their physical and chemical properties, exposure pathways, bioaccessibility, toxicity, and environmental impacts. Additionally, the International Organization for Standardization (ISO) has established multiple standards aimed at aiding the risk assessment and safety evaluation of NPs, such as ISO/TS 12901-2:2018, which provides specific guidance for assessing NPs toxicity (Xuan et al. 2023).

Furthermore, the National Institute for Occupational Safety and Health (NIOSH), the Industrial Technology Development Organization (NEDO), and the American Conference of Governmental Industrial Hygienists (ACGIH), have issued a limited number of guidelines for nanomaterials and established occupational exposure limits (OELs). These measures are designed to mitigate the risk of toxicity associated with exposure to NMs (Rodríguez-Ibarra et al. 2020; Tirumala et al. 2021).

5 Techniques for Evaluating Nanomaterial Toxicity in *Drosophila*

The predominant method for NPs administration in *Drosophila* involves ingestion, followed by inhalation and injection. Ingestion is particularly favored for larval and adult stages due to the ability to precisely control compound concentrations within the dietary medium. However, it is important to acknowledge that *Drosophila* exhibit variability in feeding behavior, as described by Garlapow et al. (2017). When exposure is limited to early developmental stages, such as the first and second instars, the growth dilution effect must be considered. For instance, a dosage deemed toxic during the first instar may be mitigated to non-toxic levels by the time the third instar is attained. Continuous exposure through feeding across all larval stages is commonly practiced, enabling the attainment of a steady internal state of the toxicant by the third instar, which persists through the pupal stages (Rand et al. 2019, 2023). In certain cases, alternative methods such as injection or direct microtransfer are preferred over ingestion due to the unpredictable amount and stability of NPs in food. Microinjection provides a more consistent delivery mechanism, facilitating precise dosage measurements in the nanogram range. This method enhances the reproducibility of experiments and allows for accurate identification of the specific developmental stage at which mortality can occur (Simpson et al. 2014). Furthermore, microinjection is particularly useful for introducing water-soluble chemicals into the pupa (Rand et al. 2023).

Of note, the daily food consumption of *Drosophila* larvae is notable, averaging around 3 μL, in contrast to adult *Drosophila*, which consume about 1.5 μL of food medium each day. This difference highlights the varying nutritional needs and feeding behaviors between the developmental stages of these organisms, reflecting their metabolic and physiological changes (Jovanović et al. 2018). To measure the quantity of food wetted with NPs, several approaches are applied. For instance, the measurement using dye, however the intensity of the dye signal reaches a plateau rapidly over time. Other assay consists of measuring the percentage of flies that extend their proboscis into the medium. However, the percentage of flies extending their proboscis is not equivalent to the volume of consumed medium (Shell et al. 2018). Medium intake can also be measured using capillary feeders suspended from a cotton plug. To take in account of evaporation, the decrease in liquid medium was quantified using feeders placed in vials without *Drosophila* (Songvorawit et al. 2022). The drawbacks of these methods include their inability to monitor individual medium intake over time, the requirement for *Drosophila* to eat from specialized devices in constrained positions, or the necessity to use dyes or radioactive labels. These could restrict the viability of conducting high throughput, unbiased studies on feeding behaviors, as well as hinder the identification of key behavioral parameters that influence food selection and consumption (Itskov et al. 2014). Another way to measure is by using FlyPAD, a method that operates on the principle of measuring the capacitance between two electrodes. A sensor with one electrode positioned under *Drosophila* and another under the medium. When *Drosophila* makes contact with the medium using its proboscis or leg, it changes the dielectric constant between the electrodes, leading to a detectable alteration in capacitance (Itskov et al. 2014). This advanced method quantifies several crucial parameters related to feeding in *Drosophila*, however, it is not intended for measuring the consumption of solid food media in standard laboratory settings (Shell et al. 2018).

The inhalation route offers a practical method to administer substances that possess unpleasant tastes, as *Drosophila* tend not to consume food mediums mixed with bad flavored substances (Pandey and Nichols 2011). Additionally, the inhalation pathway facilitates rapid delivery, enhances the availability of the drug for instance, and increases the rate of absorption through the cuticle, bypassing the digestive tract. This administration method also permits simultaneous treatment of numerous individuals, contrasting with injection techniques that necessitate individual application. Therefore, *Drosophila* used in the evaluation of air contaminants, NMs and drugs via this route of exposure (Posgai et al. 2009; Eom et al. 2017; Santalla et al. 2021; Cho et al. 2024).

A diverse array of techniques to examine the impacts of various nanoparticles using *Drosophila* are available. These assays encompass behavioral, biochemical and genetic tests, each assessing distinct toxicological endpoint. Typical developmental abnormalities caused by toxic exposure include developmental delays, stage-specific stalling, or mortality. Key endpoints to monitor these effects are the time to pupariation, duration of metamorphosis, and the success of eclosion. Comparing these endpoints provides insights into the potential toxic effects of NPs during various developmental stages. Additionally, toxic exposure can also affect morphological features such as wings, eyes, bristles, and overall body size.

The lifespan assay is a robust method for evaluating the impacts of toxicity (Linford et al. 2013). The typical lifespan for *Drosophila* ranges from a median of approximately 80 days to a maximum of around 100 days, but this can fluctuate significantly due to factors like genetic backgrounds and culture conditions (Piper and Partridge 2018). Therefore, maintaining consistent culture conditions including temperature, food, and the density of flies per vial is essential for generating reliable outcomes. The quality of the media, particularly the condition of the food vial surfaces, is the most critical aspect of the lifespan assay. The food surface should be sufficiently moist to ensure optimal feeding by *Drosophila*, yet not overly wet, which could cause the surface to become sticky, trapping *Drosophila* or stimulating the growth of bacterial species (Landis et al. 2020). To evaluate longevity accurately, replicate cultures need to be prepared, and the deceased flies are counted across a span of 100 days, with the medium being refreshed every 2 days to mitigate the risks of dietary deprivation and contamination that could cause the death (Linford et al. 2013).

Locomotion in *Drosophila* larvae, known as the crawling behavior, is orchestrated through a series of peristaltic contractions. These movements are regulated by a network of interneurons, comprising both excitatory cholinergic interneurons and inhibitory interneurons that are either GABAergic or glutamatergic (Clark et al. 2018; Hunter et al. 2021). The anatomical structure of *Drosophila* larvae includes three thoracic segments (T1-T3) and eight abdominal segments (A1–A8), with each segment hosting approximately 30 body wall muscles. These muscles are categorized into two primary groups: longitudinal muscles, which align along the body axis, and transverse muscles, which are oriented perpendicular to the body axis (Gallio et al. 2011; Frank et al. 2015). The propulsive mechanism involves a sequential wave of muscle contractions that travel from the posterior to the anterior end of the organism, facilitating both forward and backward movement. Each complete cycle of these contractions lasts about 1 s. The initiation and coordination of these muscle contractions are mediated by central pattern generator interneurons located within the ventral nerve cord, similar to the function of the mammalian spinal cord (Hunter et al. 2021). The crawling velocity of *Drosophila* larvae serves as a quantitative metric for evaluating neuronal damage. By analyzing alterations in locomotor speed, it is possible to determine the extent to which a compound may impair crawling ability, thus providing insights into its neurotoxic potential (Dan et al. 2019).

The negative geotaxis, ring assay, or climbing assay evaluates the neuromuscular performance of adult *Drosophila*. In this method, groups of flies are placed into empty vials, which are then tapped to make the flies fall to the bottom. Subsequently, when these vials are tapped, the flies instinctively climb upwards. The assessment is based on counting the number of flies that reach a predetermined height within a specified time limit, and the results are expressed as a percentage (Taylor and Tuxworth 2019).

During the process of mitochondrial aerobic respiration, reactive oxygen species (ROS) are generated, which initiate mitochondrial damage and play a crucial role in various toxicological mechanisms, ultimately contributing to the pathogenesis of numerous diseases. As described in the preceding sections, NMs induce the production of ROS. Consequently, the monitoring and regulation of ROS levels have become fundamental practices within research communities.

The Oxidative Stress Index (OSI), calculated as the ratio of the Total Oxidation Status (TOS) to the Total Antioxidant Status (TAS), offers a comprehensive measure of the balance between oxidative and antioxidative levels within biological systems. This index is particularly valuable as both oxidative and antioxidative parameters are represented into a single, interpretable parameter, where a higher

OSI indicates a predominance of oxidation relative to antioxidant capacity. This assay is simple and cost-effectiveness. The assessment of TOS and TAS through standardized assays allows for automation, which facilitates high-throughput analysis (Çetinkaya et al. 2023).

Nanomaterials penetrate the nucleus and interact with DNA, causing disruptions to its structure and function. This interaction results in DNA strand breaks, alterations in base pairs, and chromosomal damage. Furthermore, NMs may interfere with the assembly of microtubules during mitosis, resulting in clastogenic effects. These genetic alterations can be quantitatively analyzed using the comet assay (Azqueta and Dusinska 2015). The comet assay was initially described in 1984 as a technique for detecting DNA strand breaks induced by radiation in individual mammalian cells (Ostling and Johanson 1984). Later, slight modifications has been applied to this method, including the alkalinization of the electrophoresis buffer, facilitated the development of the alkaline comet assay which is now the most commonly employed comet (Singh et al. 1988). This sensitive method is capable of detecting DNA strand breaks and alkali-labile sites, such as apurinic/apyrimidinic sites, across various eukaryotic cell types (Collins et al. 2023). Its utility extends to model organisms such as *Drosophila*, initially involving cells from brain ganglia. Within the context of DNA repair analysis, the *Drosophila* comet assay has proven effective in distinguishing between the DNA damage responses (DDRs) of different mutant strains, revealing pathway-specific repair capabilities *in vivo* within somatic cells (Gaivão and Sierra 2014; Rodríguez et al. 2023). Furthermore the comet assay facilitates the concurrent evaluation of various types and concentrations of NMs, significantly diminishing the experimental workload, augmenting productivity, and minimizing variabilities.

To detect a spectrum of genetic alterations, including point mutations, deletions, non-disjunction, and mitotic recombination events, the Somatic Mutation and Recombination Test (SMART), commonly referred to as the wing spot test, is used on *Drosophila*. This assay detects mutant wing spots in adult flies, indicative of genetic damage in wing imaginal disc cells of transheterozygous larvae. This damage leads to a loss of heterozygosity (Graf et al. 1984). The rapid proliferation of cells during larval development increases the probability of interaction between genotoxins and the larval genome. The manifestation of genotoxicity in this assay is typically observed as mutant spots on the adult fly wings, appearing either as single or twin spots (Graf et al. 1984; Pitchakarn et al. 2021; Tagorti and Kaya 2022). Single spots, which are induced by either mutation or recombination events, manifest in two distinct forms according to Graf et al. 1984. The small spots, comprises just one or two cells, each exhibiting multiple trichomes on the same cell (known as *mwh* type); (2) large spots that include two or more cells from the *mwh* type or encompass four or more cells from the deformed trichome cells (known as flr^3 type). On the other hand, twin spots result solely from mitotic recombination events. These are characterized by the adjacent presence of both *mwh* and flr^3 cells (Graf et al. 1984; Vales et al. 2013; Demir et al. 2011). Furthermore, using specific strains and crosses enables the detection of potential promutagens, known as inactive precursors, that convert into active mutagens through metabolic processes in an organism, potentially leading to DNA damage and increased risk of mutations and cancer (Saner et al. 1996).

More complex behaviors like mating and fecundity, dietary preferences, and even associative learning and memory can be used to evaluate NPs toxicity in *Drosophila* (Rand et al. 2023).

6　Importance of Standardized Protocols and Validation in Nanosafety Studies

The outcomes of NMs safety studies have demonstrated considerable inconsistency, with results varying significantly across different studies and laboratories. It is broadly acknowledged that many techniques used for characterizing NMs might introduce artefacts that are dependent on the chosen method for measuring a specific endpoint, the sample preparation procedures, and the settings of the instrumentation, all of which can substantially affect the results. Consequently, there is an imperative need to develop innovative and unconventional methods and assays for the accurate and consistent assessment of NMs safety (Exner et al. 2023).

Validation of assay protocols involves verifying that the methods produce accurate and reliable results when applied repeatedly under different conditions. This step is essential to provides confidence in the results, which is paramount when the data may influence regulatory decisions and public health policies. Moreover, validated protocols can facilitate regulatory harmonization, making it easier for NMs to be approved for market entry across different regions without redundant testing.

7　Impact of Nanoparticles on *Drosophila*

7.1　Mechanisms of Nanoparticle Uptake, Distribution, and Clearance in *Drosophila*

Drosophila is a suitable model for exploring the uptake mechanism and internalization of NMs via the intestinal barrier. For instance, the chitinous peritrophic matrix within the gut of *Drosophila* serves a role analogous to that of the human gut mucus in defending enterocytes (Apidianakis and Rahme 2011). Consequently, the *Drosophila* intestine serves as a viable alternative model to the human intestine for investigating the interactions and internalization of nanoparticles. NPs can adhere to the microvilli and be internalized into the cytoplasm of midgut epithelial cells for translocation via the hemolymph (similar to mammalian blood). For instance, polylactic acid nanoplastics are encased by semi-vacuole membranes, displacing them close to the peritrophic membrane to ensure the internatlization process (Alaraby et al. 2024). This process has also been reported for TiO_2 nanoparticles (Alaraby et al. 2021). Once translocated into the circulatory system, NPs can be dispersed in various tissues and organs, including Malpighian tubules. This translocation phenomenon has been reported for ZnO-NPs (Alaraby et al. 2015) and polylactic acid nanoplastics (Alaraby et al. 2024). Interestingly, polylactic acid nanoplastics can accumulate within the gut bacteria of *Drosophila* compared with other NPs (Alaraby et al. 2024).

7.2 Cellular and Molecular Responses of *Drosophila* to Nanoparticle Exposure

Chronic exposure to TiO_2 NPs (20 nm; 10–20 mg/kg) damages the morphology of the neuromuscular junction (NMJ) by altering the expression of genes associated with NMJ development, leading to deficits in locomotor behavior in *Drosophila* (Zhang et al. 2023). Exposure to TiO_2 nanoparticles (100 nm) at 0.9 and 1.8 mg/mL over a period of 30 days disrupted the elongation of spermatids in *Drosophila*. The comprehensive RNA sequencing analysis indicated that exposure to TiO_2 nanoparticles influenced several metabolic pathways, such as carbohydrate metabolism and the activity of cytochrome P450 enzymes (Cheng et al. 2024).

Dietary exposure to citrate-coated silver NPs (AgNPs, 20 nm) at a concentration of 30 μg/mL in *Drosophila* led to a systemic increase in reactive oxygen species (ROS) and activated the Nrf2-dependent antioxidant pathway. Furthermore, exposure at a higher concentration of 50 μg/mL triggered caspase 3 activation yielding apoptosis, and double-stranded DNA breaks across various larval tissues. Additionally, this higher concentration promoted autophagy activation in both the larval brain and fat bodies (Mao et al. 2018). AgNPs (20 nm) induced DNA damage in *Drosophila* larval hemocytes at 4 mg/mL concentration evaluated by using the comet assay. Similar results were recorded following the exposure of *Drosophila* S2 Schneider cells at 100 μg/mL AgNPs concentration. AgNPs exposure increased the oxidative stress and ROS production. Interestingly, the RNA-seq analysis reveal the increased level of mRNA of metallothionein suggesting that *Drosophila* react to AgNPs exposure by the production of metallothioneins to ensure detoxification (Wang et al. 2023). Rod-shaped hydroxyapatite NPs (length of 70–80 nm, and diameter of 40–50 nm) at low concentrations (5 mg/L) induced more oxidative stress in the larval gut compared to highest concentration (80 mg/L). This could be explained by the agglomeration of NPs at high concentration yielding loss of the properties recorded with small size (Pappus et al. 2017). Two forms of hydroxyapatite NPs round (length: 35–60 nm) and rod (length: 45–90 nm) were administered to third instar larvae. Genotoxic effect has been reported for the round form in the wing spot assay whereas both forms of NPs caused DNA damage based on comet assay results. Interestingly both forms induced upregulation of genes related to mismatch repair genes. In addition, an increase in caspases 8 and 9 levels have been reported on both forms of NPs at 200 ppm concentrations suggesting apoptotic potential of hydroxyapatite NPs on *Drosophila* (Güneş et al. 2023). Platinum NPs (Pt, 100 nm) are proposed to be used for wound healing in *Drosophila*. Pt-NPs were found to scavenge the ROS with no internal gut and without causing any damage to the hemocytes (Bag et al. 2023). Copper NPs can accumulate in *Drosophila* midgut yielding to ROS production (Baeg et al. 2018).

Table 13.1 An overview of studies reporting detrimental cellular effects of NPs on *Drosophila*

Name	Size (nm)	Concentration	Observed effects	References
AgNPs	20	4 mg/mL	DNA damage	Wang et al. (2023)
Citrate-coated AgNPs	20	50 µg/mL	Autophagy Apoptosis DNA damage	Mao et al. (2018)
Copper NPs	~40	0.15 mg/mL	ROS production	Baeg et al. (2018)
Cobalt NPs	50	10 mM	DNA damage	Kurşun et al. (2022)
Hydroxyapatite NPs	Round (L: 35–60) rod (L: 45–90)	200 ppm	DNA damage Apoptosis	Güneş et al. (2023)
Hydroxyapatite NPs	Rod (L: 70–80, D: 40–50)	5 mg/L	Oxidative stress	Pappus et al. (2017)
Nickel ferrite magnetic NPs	25	10 mM	DNA damage	Burgazlı et al. (2024)
TiO_2 NPs	20	10–20 mg/kg	Damage in the neuromuscular junction	Zhang et al. (2023)
TiO_2 NPs	100	0.9 and 1.8 mg/mL	Alteration in the elongation of spermatids	Cheng et al. (2024)

Three nanocomposite materials: Graphene Oxide-Adamantylamine (GO/ADMA), Montmorillonite-Adamantylamine (Clay/ADMA), and Montmorillonite-Carbon Nanotubes (Clay/CNTs) were evaluated using *Drosophila* at concentrations between 100–4000 µg/mL. None of these composite NMs exhibited mutagenic or recombinogenic properties, as no induction of higher frequencies has been recorded in wing spots across a broad concentration spectrum (Efthimiou et al. 2022). A summary of research detailing the harmful cellular impacts of nanoparticles on *Drosophila* is presented in Table 13.1.

7.3 Effects of Nanoparticles on *Drosophila* Development, Behavior, and Physiology

The exposure of nanoribbons of imidacloprid (insecticide) induced a substantial decrease in both viability and olfactory potential in *Drosophila* (Vidal et al. 2024). Similarly, a study conducted using bifenthrin NPs (insecticide) reported developmental delays, significant decreases in viability, and diminished olfactory capacity (Cruces et al. 2023). The uptake of magnetite (Fe_3O_4) NPs interfered with the oogenesis period, causes ovarian abnormalities, and slows the development of egg chambers, potentially leading to reduced fertility in female *Drosophila* (Chen et al. 2015).

AgNPs (20 nm, 4 mg/mL) reduced *Drosophila* fertility of females. The pupation and eclosion ratios have been altered following AgNPs exposure that could be due to the upregulation of *Ilp8* mRNA levels, an insulin like peptide responsible for delay in metamorphosis by inhibiting the biosynthesis of ecdysone. Abnormal phenotype has been reported for newly emerged flies with a white body color because of interference of melanin synthesis. Furthermore, impairing in climbing ability has been reported (Wang et al. 2023). A developmental delay has been reported after administration of hydroxyapatite NPs to *Drosophila* mostly noticed with the pupa count reduction. In addition, body weight decreases with impaired walking behavior, and phenotypic defect in the wings and eyes has been observed for newly emerged flies. It is worth mentioning that crawling speed decrease with more turns suggesting poor coordination between mechanosensory neurons and the brain (Pappus et al. 2017). Interestingly, *Drosophila* exposed to Copper NPs (50–80 nm) at concentration higher than 0.3 mg/mL were overactive and jumped more frequently (Budiyanti et al. 2022).

Lethal concentrations of dietary of citrate-capped AgNPs significantly impeded development and induced extensive mortality in *Drosophila*. Conversely, sublethal concentrations, although not fatal, reduced the lifespan of adults. This study demonstrates that exposure to lethal doses (≥30 µg/mL of citrate-capped AgNPs) not only increases mortality but also disrupts normal larval development (Mao et al. 2018). *Drosophila* fertility has been decreased following AgNPs exposure for three consecutive generations (Panacek et al. 2011). In addition, AgNPs altered spermatogenesis in *Drosophila* (Ong et al. 2016). When *Drosophila* were co-exposed to AgNPs with an average diameter of approximately 10 nm and polystyrene NPs of about 50 nm, an antagonistic effect was observed. Specifically, co-exposure led to a reduction in the pigmentation of the emerging adult *Drosophila* following silver exposure. Additionally, co-exposure reduced the delay in emergence typically caused by exposure to silver alone (Alaraby et al. 2022).

TiO_2 NPs rutile type had no effect on pupation and eclosion rates at 200 mg/kg concentration (Posgai et al. 2011) whereas the anatase type at 20 mg/kg concentration decreased both rates (Jovanović et al. 2016). Similarly, TiO_2 NPs anatase (10–25 nm) at 20 mg/kg reduced the pupation and eclosion rates of *Drosophila* but only at the F4 generation (Zhang et al. 2023).

Nanocomposites have seen a significant rise in use across health, energy, and environmental sectors, attributed to their enhanced physicochemical properties and minimized aggregation. Metal/metal oxide NPs can be produced through both chemical and biological methods using various metal salts and exhibit surface plasmon resonance, which allows for versatile surface modifications. Their biocompatibility also broadens their application in fields such as anticancer, antibacterial, antiviral, antifungal, and anti-insecticidal treatments. However, their adoption has been limited by slow degradation and tendency to aggregate. To counteract this, NPs have been polymer-coated to create nanocomposites that improve these properties. For instance, the $ZnFe_2O_4$@poly(tBGE-alt-PA) metallopolyester nanocomposite was evaluated using *Drosophila*, showing that larvae exposed to this nanocomposite failed to mature into adults and exhibited abnormal development processes. Additionally, exposure led to an irregular crawling pattern in third instar larvae, deviating from the usual straight trajectory to a zigzag motion and reduced speed, especially as nanocomposite concentrations increased in their diet (Chauhan et al. 2024).

Nanoplastics are ubiquitous in the environment and originate from various sources, including industrial processes, packaging materials, and the degradation of larger plastic debris. The potential health effects of long-term exposure to low concentrations of nanoplastics are not well understood. Moreover, once organisms are exposed to nanoplastics, removing these particles from the environment or from within the organisms themselves is particularly challenging. This is due to their widespread distribution and persistence, as well as their ability to integrate into biological systems and accumulate over time. Consequently, even environmentally realistic concentrations of nanoplastics can pose significant risks, as they are not readily biodegradable and can persist in ecosystems for extended periods. Given these challenges, scientific studies have increasingly turned to model organisms to better understand the impacts of nanoplastics (Lai et al. 2022; Allen et al. 2022).

Exposure to clean spherical polystyrene nanoplastics (20 nm) in the diet resulted in minimal toxicity but pronounced sublethal impacts on *Drosophila*. While there were significant alterations in locomotion, other factors such as mortality, development, and fertility remained largely unaffected. Although circadian rhythms were not influenced by the nanoplastics, there was an observed increase in daily activity at a concentration of 50 ppm. This increase is likely a compensatory response to decreased nutrient absorption, potentially caused by dietary changes or intestinal damage (Matthews et al. 2021). To evaluate the long-term toxicity of polystyrene nanoplastics (PS-NPs) on multi-generations, 100 nm PS-NPs were administered to five successive generations of *Drosophila*. Observations revealed significant accumulation of PS-NPs in the gut and ovaries after 5 days. Additionally, there was a notable decline in egg production and eclosion rates, alongside developmental delays, across multiple generations exposed to PS-NPs (Tu et al. 2023). Of note the effect of nanoplastics is not solely based on the size but is also attributed to harmful microorganims and metals that can accumulate on their surfaces in the intestines. Furthermore, nanoplastics exhibit endocrine-disrupting properties thereby enhancing the potential of toxic effect of these nanomaterials (Santos et al. 2023; Ullah et al. 2023).

Parkinson's disease (PD) is a prevalent progressive neurodegenerative disorder characterized by bradykinesia and various metabolic complications. The lack of effective drug candidates represents a significant therapeutic challenge for this long-term disease. Recently, piperine-coated gold NPs have been evaluated using *Drosophila*. Paraquat, an herbicide known for its neurotoxic effects that closely simulate Parkinson's symptoms was used. *Drosophila* exposed to paraquat were treated with these NPs and have exhibited increased survival rates and enhanced climbing activity. This suggests that piperine-coated gold NPs could mitigate the neurotoxic effects of paraquat, offering a potential avenue for alleviating neurotoxicity in PD models (Srivastav et al. 2020).

Ropinirole (RP), a non-ergot D2/D3 dopamine agonist, has been reported to exhibit anti-parkinsonian effects in primate models of Parkinson's Disease and was the inaugural drug approved by the FDA for the management of Restless Legs

Syndrome. With advancements in nanotechnology, neurological disorders are now being approached through the development of drug nanocomposites. A study on the RP silver nanocomposite (RPAgNC) was conducted using a transgenic *Drosophila* model of PD, comparing its effects to those of the standard RP drug. Administration of RPAgNC to *Drosophila* substantially mitigated neuronal degeneration more effectively than RP alone. The results confirm that RPAgNC exhibits a heightened neuroprotective effect. Additionally, PD model flies treated with RPAgNC demonstrated marked improvements in motor functions, specifically in jumping and flying abilities, compared to those treated solely with RP. Parallel investigations involving other nanocomposites, such as the bromocriptine alginate nanocomposite, have also shown promising results in alleviating cognitive impairments in PD model transgenic *Drosophila* (Siddique et al. 2016; Naz et al. 2020).

Using *Drosophila* as a model, we discovered that NMs can exhibit beneficial effects by enhancing biological functions and drug discovery. However, NMs can also pose harmful effects, potentially causing toxicity and physiological disruptions at cellular levels (Table 13.2).

8   Conclusion

The increasing use of *Drosophila* in nanotoxicological studies is supported by a substantial and expanding collection of experimental research that assess several endpoints including fecundity, development and survival, behaviour change. These studies also explore the uptake pathway of NPs, the localization within subcellular structures, cytotoxic and genotoxic potential as well as gene expression analysis. There is broad consensus regarding the efficacy of the *Drosophila* model, with general agreement that oxidative stress induced by reactive oxygen species generation, leading to genotoxic damage, is a primary concern following NMs ingestion. However, the reported effects of NMs exposure are often minimal, and discrepancies in study findings are common, potentially due to the reliance on overall effects in the organism that may not directly reflect the cellular damage caused by NMs. Furthermore, challenges in determining effective doses, variations in the physicochemical properties of NMs in dietary formulations, and difficulties in characterizing NM uptake and distribution within the gastrointestinal tract has been observed. The genetic and physiological differences between *Drosophila* and humans complicate the direct extrapolation of findings from *Drosophila* to human contexts. Metabolic and pharmacokinetic disparities further influence the exposure levels and bioavailability of NMs, thereby impacting the resultant effects and constraining the applicability of dosage and treatment protocols derived from *Drosophila* studies. Moreover, the multifactorial nature of human diseases presents further challenges for *Drosophila* models to fully elucidate disease mechanisms and capture the complexity of human conditions. To address these limitations, an integrative approach that combines insights from *Drosophila* models with other alternative approaches is required. Such a comprehensive methodology can narrow the translational gap, enhancing the reliability of different findings.

Table 13.2 Overview of studies reporting adverse effects of NPs on development, behavior, and physiology of *Drosophila*

Name	Size (nm)	Concentration	Observed effects	References
AgNPs	20	4 mg/mL	Abnormal phenotypes Reduction in fertility in female *Drosophila* Reduction in climbing activity	Wang et al. (2023)
AgNPs	~30	20 mg/L	Reduction in fertility	Panacek et al. (2011)
AgNPs	~20	5 mg/L	Alteration in spermatogenesis	Ong et al. (2016)
Bifenthrin NPs	2–20	10 ppm	A developmental delay, reduction in viability, and olfactory capacity	Cruces et al. (2023)
Bromocriptine alginate nanocomposite	20	1.5 µM	Alleviation in cognitive impairments in PD model transgenic *Drosophila*	Siddique et al. (2016)
Citrate-capped AgNPs	20	≥30 µg/mL	Interference with larval development	Mao et al. (2018)
Copper NPs	50–80	>0.3 mg/mL	Alteration of climbing activity	Budiyanti et al. (2022)
Imidacloprid nanoribbons	W:160–470	0.03 ppm	Reduction in viability and olfactory ability	Vidal et al. (2024)
Magnetite NPs	15	300 µg/g	Reduction in fertility in female *Drosophila*	Chen et al. (2015)
Piperine-coated gold NPs	5–10	10 µM	Reduction in the neurotoxic effects of paraquat	Srivastav et al. (2020)
Polystyrene nanoplastics	20	50 ppm	Alteration in locomotion Increase daily activity	Matthews et al. (2020)
Polystyrene nanoplastics	100	100 mg/L	Reduction in egg production and eclosion rates A developmental delay across multiple generations	Tu et al. (2023)
Rod hydroxyapatite NPs	L:70–80; D:40–50	80 mg/L	A developmental delay Abnormal phenotypes Poor crawling activity	Pappus et al. (2017)
Ropinirole silver nanocomposite	10–70	1 mM	Reduction in neuronal degeneration	Naz et al. (2020)
TiO_2 NPs anatase	167 ± 50	20 mg/kg	A developmental delay	Jovanović et al. (2016)
TiO_2 NPs anatase	10–25	20 mg/kg	A developmental delay at F4 generation	Zhang et al. (2023)
$ZnFe_2O_4$@poly(tBGE-alt-PA) metallopolyester nanocomposite	17.55	200 µM	An abnormal and developmental delay Poor crawling activity	Chauhan et al. (2024)

References

Adams MD, Celniker SE, Holt RA, Evans CA, Gocayne JD, Amanatides PG, Scherer SE, Li PW, Hoskins RA, Galle RF, George RA, Lewis SE, Richards S et al (2000) The genome sequence of *Drosophila melanogaster*. Science 287:2185–2195. https://doi.org/10.1126/science.287.5461.2185

Alaraby M, Annangi B, Hernández A, Creus A, Marcos R (2015) A comprehensive study of the harmful effects of ZnO nanoparticles using *Drosophila melanogaster* as an in vivo model. J Hazard Mater 296:166–174. https://doi.org/10.1016/j.jhazmat.2015.04.053

Alaraby M, Hernández A, Marcos R (2021) Novel insights into biodegradation, interaction, internalization and impacts of high-aspect-ratio TiO2 nanomaterials: a systematic in vivo study using *Drosophila melanogaster*. J Hazard Mater 409:124474. https://doi.org/10.1016/j.jhazmat.2020.124474

Alaraby M, Abass D, Villacorta A, Hernández A, Marcos R (2022) Antagonistic in vivo interaction of polystyrene nanoplastics and silver compounds. A study using *Drosophila*. Sci Total Environ 842:156923. https://doi.org/10.1016/j.scitotenv.2022.156923

Alaraby M, Abass D, Farre M, Hernández A, Marcos R (2024) Are bioplastics safe? Hazardous effects of polylactic acid (PLA) nanoplastics in *Drosophila*. Sci Total Environ 919:170592. https://doi.org/10.1016/j.scitotenv.2024.170592

Allen S, Allen D, Karbalaei S, Maselli V, Walker TR (2022) Micro(nano)plastics sources, fate, and effects: what we know after ten years of research. J Hazard Mater Adv 6:100057. https://doi.org/10.1016/j.hazadv.2022.100057

Apidianakis Y, Rahme LG (2011) *Drosophila melanogaster* as a model for human intestinal infection and pathology. Dis Model Mech 4:21–30. https://doi.org/10.1242/dmm.003970

Asharani, P.V., Low Kah Mun, G., Hande, M.P., Valiyaveettil, S., 2009. Cytotoxicity and genotoxicity of silver nanoparticles in human cells. ACS Nano 3, 279–290. ://doi.org/https://doi.org/10.1021/nn800596w

Asmat U, Abad K, Ismail K (2016) Diabetes mellitus and oxidative stress—a concise review. Saudi Pharmaceut J 24:547–553. https://doi.org/10.1016/j.jsps.2015.03.013

Azqueta A, Dusinska M (2015) The use of the comet assay for the evaluation of the genotoxicity of nanomaterials. Front Genet 6. https://doi.org/10.3389/fgene.2015.00239

Baeg E, Sooklert K, Sereemaspun A (2018) Copper oxide nanoparticles cause a dose-dependent toxicity via inducing reactive oxygen species in Drosophila. Nano 8:824. https://doi.org/10.3390/nano8100824

Bag J, Mukherjee S, Tripathy M, Mohanty R, Shendha PK, Hota G, Mishra M (2023) Platinum as a novel nanoparticle for wound healing model in *Drosophila melanogaster*. J Clust Sci 34:1087–1098. https://doi.org/10.1007/s10876-022-02292-9

Barnham KJ, Masters CL, Bush AI (2004) Neurodegenerative diseases and oxidative stress. Nat Rev Drug Discov 3:205–214. https://doi.org/10.1038/nrd1330

Beira JV, Paro R (2016) The legacy of drosophila imaginal discs. Chromosoma 125:573–592. https://doi.org/10.1007/s00412-016-0595-4

Bolus H, Crocker K, Boekhoff-Falk G, Chtarbanova S (2020) Modeling Neurodegenerative Disorders in Drosophila melanogaster. IJMS 21:3055. https://doi.org/10.3390/ijms21093055

Brown JB, Boley N, Eisman R, May GE, Stoiber MH, Duff MO, Booth BW, Wen J, Park S, Suzuki AM, Wan KH, Yu C, Zhang D, Carlson JW, Cherbas L, Eads BD et al (2014) Diversity and dynamics of the *Drosophila* transcriptome. Nature 512:393–399. https://doi.org/10.1038/nature12962

Budiyanti DS, Moeller ME, Thit A (2022) Influence of copper treatment on bioaccumulation, survival, behavior, and fecundity in the fruit fly *Drosophila melanogaster*: toxicity of copper oxide nanoparticles differ from dissolved copper. Environ Toxicol Pharmacol 92:103852. https://doi.org/10.1016/j.etap.2022.103852

Burgazlı AY, Güneş M, Yalcin B, Tagorti G, Kaya B (2024) Demir Bazlı Manyetik Nanopartiküllerin Genotoksik Etkilerinin Drosophila melanogaster'de Araştırılması (investigation of the genotoxic effects of iron-based magnetic nanoparticles in Drosophila melanogaster). GÜFFD 5:39–51

Çetinkaya Z, Güneş E, Şavkliyildiz İ (2023) Investigation of biochemical properties of flash sintered ZrO_2–SnO_2 nanofibers. Mater Chem Phys 293:126900. https://doi.org/10.1016/j.matchemphys.2022.126900

Chatterjee N, Perrimon N (2021) What fuels the fly: energy metabolism in *Drosophila* and its application to the study of obesity and diabetes. Sci Adv 7:eabg4336. https://doi.org/10.1126/sciadv.abg4336

Chauhan S, Naik S, Kumar R, Ruokolainen J, Kesari KK, Mishra M, Gupta PK (2024) *In vivo* toxicological analysis of the $ZnFe_2O_4$@poly(*t*BGE-*alt*-PA) nanocomposite: a study on fruit Fly. ACS Omega 9:6549–6555. https://doi.org/10.1021/acsomega.3c07111

Chen H, Wang B, Feng W, Du W, Ouyang H, Chai Z, Bi X (2015) Oral magnetite nanoparticles disturb the development of *Drosophila melanogaster* from oogenesis to adult emergence. Nanotoxicology 9:302–312. https://doi.org/10.3109/17435390.2014.929189

Cheng X, Jiang T, Huang Q, Ji L, Li J, Kong X, Zhu X, He X, Deng X, Wu T, Yu H, Shi Y, Liu L, Zhao X, Wang X, Chen H, Yu J (2024) Exposure to titanium dioxide nanoparticles leads to specific disorders of spermatid elongation via multiple metabolic pathways in *Drosophila* testes. ACS Omega 9:23613–23623. https://doi.org/10.1021/acsomega.4c01140

Cho Y, Park CM, Heo Y-J, Park H-B, Kim M-S (2024) *Drosophila melanogaster* as potential alternative animal model for evaluating acute inhalation toxicity. J Toxicol Sci 49:49–53. https://doi.org/10.2131/jts.49.49

Clark MQ, Zarin AA, Carreira-Rosario A, Doe CQ (2018) Neural circuits driving larval locomotion in Drosophila. Neural Dev 13:6. https://doi.org/10.1186/s13064-018-0103-z

Collins, A., Møller, P., Gajski, G., Vodenková, S., Abdulwahed, A., Anderson, D., Bankoglu, E.E., Bonassi, S., Boutet-Robinet, E., Brunborg, G., Chao, C., Cooke, M.S et al., 2023. Measuring DNA modifications with the comet assay: a compendium of protocols. Nat Protoc 18, 929–989. ://doi.org/https://doi.org/10.1038/s41596-022-00754-y

Costa PM, Fadeel B (2016) Emerging systems biology approaches in nanotoxicology: towards a mechanism-based understanding of nanomaterial hazard and risk. Toxicol Appl Pharmacol 299:101–111. https://doi.org/10.1016/j.taap.2015.12.014

Cruces MP, Pimentel E, Vidal LM, Jiménez E, Suárez H, Camps E, Campos-González E (2023) Genotoxic action of bifenthrin nanoparticles and its effect on the development, productivity, and behavior of *Drosophila melanogaster*. J Toxic Environ Health A 86:661–677. https://doi.org/10.1080/15287394.2023.2234408

Dan P, Sundararajan V, Ganeshkumar H, Gnanabarathi B, Subramanian AK, Venkatasubu GD, Ichihara S, Ichihara G, Sheik Mohideen S (2019) Evaluation of hydroxyapatite nanoparticles – induced in vivo toxicity in *Drosophila melanogaster*. Appl Surf Sci 484:568–577. https://doi.org/10.1016/j.apsusc.2019.04.120

Demir E, Vales G, Kaya B, Creus A, Marcos R (2011) Genotoxic analysis of silver nanoparticles in *Drosophila*. Nanotoxicology 5:417–424. https://doi.org/10.3109/17435390.2010.529176

Dermauw W, Van Leeuwen T (2014) The ABC gene family in arthropods: comparative genomics and role in insecticide transport and resistance. Insect Biochem Mol Biol 45:89–110. https://doi.org/10.1016/j.ibmb.2013.11.001

Efthimiou I, Vlastos D, Ioannidou C, Tsilimigka F, Drosopoulou E, Mavragani-Tsipidou P, Potsi G, Gournis D, Antonopoulou M (2022) Assessment of the genotoxic potential of three novel composite nanomaterials using human lymphocytes and the fruit fly Drosophila melanogaster as model systems. Chem Eng J Adv 9:100230. https://doi.org/10.1016/j.ceja.2021.100230

Eom H-J, Liu Y, Kwak G-S, Heo M, Song KS, Chung YD, Chon T-S, Choi J (2017) Inhalation toxicity of indoor air pollutants in Drosophila melanogaster using integrated transcriptomics and computational behavior analyses. Sci Rep 7:46473. https://doi.org/10.1038/srep46473

European Commission (2022) Commission Recommendation of 10 June 2022 on the definition of nanomaterial. Available at https://eur-lex.europa.eu/legal-content/EN/TXT/PDF/?uri=CELEX:32022H0614(01)

Exner TE, Papadiamantis AG, Melagraki G, Amos JD, Bossa N, Gakis GP, Charitidis CA, Cornelis G, Costa AL, Doganis P, Farcal L, Friedrichs S, Furxhi I, Klaessig FC, Lobaskin V, Maier D, Rumble J, Sarimveis H, Suarez-Merino B, Vázquez S, Wiesner MR, Afantitis A, Lynch I (2023) Metadata stewardship in nanosafety research: learning from the past, preparing for an "on-the-fly" FAIR future. Front Phys 11:1233879. https://doi.org/10.3389/fphy.2023.1233879

Fischer FP, Karge RA, Weber YG, Koch H, Wolking S, Voigt A (2023) Drosophila melanogaster as a versatile model organism to study genetic epilepsies: an overview. Front Mol Neurosci 16:1116000. https://doi.org/10.3389/fnmol.2023.1116000

Frank DD, Jouandet GC, Kearney PJ, Macpherson LJ, Gallio M (2015) Temperature representation in the *Drosophila* brain. Nature 519:358–361. https://doi.org/10.1038/nature14284

Fröhlich E (2013) Cellular targets and mechanisms in the cytotoxic action of non-biodegradable engineered nanoparticles. CDM 14:976–988. https://doi.org/10.2174/1389200211314090004

Fujii J, Homma T, Osaki T (2022) Superoxide radicals in the execution of cell death. Antioxidants 11:501. https://doi.org/10.3390/antiox11030501

Gaivão I, Sierra LM (2014) *Drosophila* comet assay: insights, uses, and future perspectives. Front Genet 5. https://doi.org/10.3389/fgene.2014.00304

Gallio M, Ofstad TA, Macpherson LJ, Wang JW, Zuker CS (2011) The coding of temperature in the Drosophila brain. Cell 144:614–624. https://doi.org/10.1016/j.cell.2011.01.028

Garlapow ME, Everett LJ, Zhou S, Gearhart AW, Fay KA, Huang W, Morozova TV, Arya GH, Turlapati L, Armour G, Hussain YN, McAdams SE, Fochler S, Mackay TFC (2017) Genetic and genomic response to selection for food consumption in Drosophila melanogaster. Behav Genet 47:227–243. https://doi.org/10.1007/s10519-016-9819-x

Garofalo RS (2002) Genetic analysis of insulin signaling in drosophila. Trends Endocrinol Metab 13:156–162. https://doi.org/10.1016/S1043-2760(01)00548-3

Graf U, Würgler FE, Katz AJ, Frei H, Juon H, Hall CB, Kale PG (1984) Somatic mutation and recombination test in *Drosophila melanogaster*. Environ Mutagen 6:153–188. https://doi.org/10.1002/em.2860060206

Green MM (2010) A century of Drosophila genetics through the prism of the *white* Gene. Genetics 184:3–7. https://doi.org/10.1534/genetics.109.110015

Güneş M, Yalçın B, Burgazlı AY, Tagorti G, Yavuz E, Akarsu E, Kaya N, Marcos R, Kaya B (2023) Morphologically different hydroxyapatite nanoparticles exert differential genotoxic effects in *Drosophila*. Sci Total Environ 904:166556. https://doi.org/10.1016/j.scitotenv.2023.166556

Hayes JD, Dinkova-Kostova AT, Tew KD (2020) Oxidative stress in cancer. Cancer Cell 38:167–197. https://doi.org/10.1016/j.ccell.2020.06.001

Horikoshi S, Serpone N (2013) Introduction to nanoparticles. In: Serpone N (ed) Horikoshi, S. Wiley, Microwaves in Nanoparticle Synthesis, pp 1–24. https://doi.org/10.1002/9783527648122.ch1

Hou, W., Zhu, X., Liu, J., Map, J., 2020. Inhibition of miR-153 ameliorates ischemia/reperfusion-induced cardiomyocytes apoptosis by regulating Nrf2/HO-1 signaling in rats. Biomed Eng Online 19, 15. ://doi.org/https://doi.org/10.1186/s12938-020-0759-6

Huang R-C (2018) The discoveries of molecular mechanisms for the circadian rhythm: the 2017 nobel prize in physiology or medicine. Biom J 41:5–8. https://doi.org/10.1016/j.bj.2018.02.003

Huang, Y.-W., Wu, C., Aronstam, R.S., 2010. Toxicity of transition metal oxide nanoparticles: recent insights from in vitro studies. Materials 3, 4842–4859. ://doi.org/https://doi.org/10.3390/ma3104842

Hunter I, Coulson B, Zarin AA, Baines RA (2021) The *Drosophila* larval locomotor circuit provides a model to understand neural circuit development and function. Front Neural Circuits 15:684969. https://doi.org/10.3389/fncir.2021.684969

Itskov PM, Moreira J-M, Vinnik E, Lopes G, Safarik S, Dickinson MH, Ribeiro C (2014) Automated monitoring and quantitative analysis of feeding behaviour in *Drosophila*. Nat Commun 5:4560. https://doi.org/10.1038/ncomms5560

Joseph T, Kar Mahapatra D, Esmaeili A, Piszczyk Ł, Hasanin M, Kattali M, Haponiuk J, Thomas S (2023) Nanoparticles: taking a unique position in medicine. Nano 13:574. https://doi.org/10.3390/nano13030574

Jovanović, B., Cvetković, V.J., Mitrović, T.Lj., 2016. Effects of human food grade titanium dioxide nanoparticle dietary exposure on Drosophila melanogaster survival, fecundity, pupation and expression of antioxidant genes. Chemosphere 144, 43–49. ://doi.org/https://doi.org/10.1016/j.chemosphere.2015.08.054

Jovanović B, Jovanović N, Cvetković VJ, Matić S, Stanić S, Whitley EM, Mitrović TL (2018) The effects of a human food additive, titanium dioxide nanoparticles E171, on *Drosophila melanogaster* - a 20 generation dietary exposure experiment. Sci Rep 8:17922. https://doi.org/10.1038/s41598-018-36174-w

Khansari N, Shakiba Y, Mahmoudi M (2009) Chronic inflammation and oxidative stress as a major cause of age- related diseases and cancer. IAD 3:73–80. https://doi.org/10.2174/187221309787158371

Kietz C, Meinander A (2023) Drosophila caspases as guardians of host-microbe interactions. Cell Death Differ 30:227–236. https://doi.org/10.1038/s41418-022-01038-4

Kuppusamy P, Yusoff MM, Maniam GP, Govindan N (2016) Biosynthesis of metallic nanoparticles using plant derivatives and their new avenues in pharmacological applications—an updated report. Saudi Pharmaceut J 24:473–484. https://doi.org/10.1016/j.jsps.2014.11.013

Kurşun AY, Yalci NB, Güneş M, Tagortí G, Kaya B (2022) *Drosophila melanogaster*'in somatik hücrelerinde kobalt nanopartiküllerinin indüklediği genotoksisiteye karşı resveratrol'ün anti-genotoksik etkisi. (Antigenotoxic effect of resveratrol against genotoxicity induced by cobalt nanoparticles in somatic cells of Drosophila melanogaster). Eur J Biol Chem Sci 5:50–55. https://doi.org/10.46239/ejbcs.1069388

Lai H, Liu X, Qu M (2022) Nanoplastics and human health: hazard identification and biointerface. Nanomaterials 12:1298. https://doi.org/10.3390/nano12081298

Landis GN, Doherty D, Tower J (2020) Analysis of *Drosophila melanogaster* lifespan. In: Curran SP (ed) Aging, methods in molecular biology. Springer US, New York, pp 47–56. https://doi.org/10.1007/978-1-0716-0592-9_4

Lee A-R, Lee S-J, Lee M, Nam M, Lee S, Choi J, Lee H-J, Kim D-U, Hoe K-L (2018) Editor's highlight: a genome-wide screening of target genes against silver nanoparticles in fission yeast. Toxicol Sci 161:171–185. https://doi.org/10.1093/toxsci/kfx208

Lemaitre B, Hoffmann J (2007) The host defense of *Drosophila melanogaster*. Annu Rev Immunol 25:697–743. https://doi.org/10.1146/annurev.immunol.25.022106.141615

Leulier F, Rodriguez A, Khush RS, Abrams JM, Lemaitre B (2000) The *Drosophila* caspase Dredd is required to resist gram-negative bacterial infection. EMBO Rep 1:353–358. https://doi.org/10.1093/embo-reports/kvd073

Li N, Xia T, Nel AE (2008) The role of oxidative stress in ambient particulate matter-induced lung diseases and its implications in the toxicity of engineered nanoparticles. Free Radic Biol Med 44:1689–1699. https://doi.org/10.1016/j.freeradbiomed.2008.01.028

Liang Q, Sheng Y, Jiang P, Ji L, Xia Y, Min Y, Wang Z (2011) The gender-dependent difference of liver GSH antioxidant system in mice and its influence on isoline-induced liver injury. Toxicology 280:61–69. https://doi.org/10.1016/j.tox.2010.11.010

Lin W, Shen P, Song Y, Huang Y, Tu S (2021) Reactive oxygen species in autoimmune cells: function, differentiation, and metabolism. Front Immunol 12:635021. https://doi.org/10.3389/fimmu.2021.635021

Linford NJ, Bilgir C, Ro J, Pletcher SD (2013) Measurement of lifespan in *Drosophila melanogaster*. JoVE 50068. https://doi.org/10.3791/50068

Liu Y, Zhao Y, Sun B, Chen C (2013) Understanding the toxicity of carbon nanotubes. Acc Chem Res 46:702–713. https://doi.org/10.1021/ar300028m

Maku AM, Buba AB, Oyewole OA, Alhassan AM, Isibor PO (2024) Oxidative stress and inflammation induced by nanoparticles. In: Isibor PO, Devi G, Enuneku AA (eds) Environmental nanotoxicology. Springer Nature, Cham, pp 121–133. https://doi.org/10.1007/978-3-031-54154-4_7

Malita A, Rewitz K (2021) Interorgan communication in the control of metamorphosis. Curr Opin Insect Sci 43:54–62. https://doi.org/10.1016/j.cois.2020.10.005

Manke A, Wang L, Rojanasakul Y (2013) Mechanisms of nanoparticle-induced oxidative stress and toxicity. Biomed Res Int 2013:1–15. https://doi.org/10.1155/2013/942916

Mao B-H, Chen Z-Y, Wang Y-J, Yan S-J (2018) Silver nanoparticles have lethal and sublethal adverse effects on development and longevity by inducing ROS-mediated stress responses. Sci Rep 8:2445. https://doi.org/10.1038/s41598-018-20728-z

Markow, T.A., 2015. The secret lives of *Drosophila* flies. eLife 4, e06793. https://doi.org/10.7554/eLife.06793

Matthews S, Xu EG, Roubeau Dumont E, Meola V, Pikuda O, Cheong RS, Guo M, Tahara R, Larsson HCE, Tufenkji N (2021) Polystyrene micro- and nanoplastics affect locomotion and daily activity of *Drosophila melanogaster*. Environ Sci Nano 8:110–121. https://doi.org/10.1039/D0EN00942C

Mishra M, Panda M (2021) Reactive oxygen species: the root cause of nanoparticle-induced toxicity in *Drosophila melanogaster*. Free Radic Res 55:919–935. https://doi.org/10.1080/10715762.2021.1914335

Naz F, Rahul, Fatima M, Naseem S, Khan W, Mondal AC, Siddique YH (2020) Ropinirole silver nanocomposite attenuates neurodegeneration in the transgenic *Drosophila melanogaster* model of Parkinson's disease. Neuropharmacology 177:108216. https://doi.org/10.1016/j.neuropharm.2020.108216

Nitta Y, Sugie A (2022) Studies of neurodegenerative diseases using *Drosophila* and the development of novel approaches for their analysis. Fly 16:275–298. https://doi.org/10.1080/19336934.2022.2087484

Ong C, Lee QY, Cai Y, Liu X, Ding J, Yung L-YL, Bay B-H, Baeg G-H (2016) Silver nanoparticles disrupt germline stem cell maintenance in the *Drosophila* testis. Sci Rep 6:20632. https://doi.org/10.1038/srep20632

Ostling O, Johanson KJ (1984) Microelectrophoretic study of radiation-induced DNA damages in individual mammalian cells. Biochem Biophys Res Commun 123:291–298. https://doi.org/10.1016/0006-291X(84)90411-X

Overend G, Luo Y, Henderson L, Douglas AE, Davies SA, Dow JAT (2016) Molecular mechanism and functional significance of acid generation in the *Drosophila* midgut. Sci Rep 6:27242. https://doi.org/10.1038/srep27242

Panacek A, Prucek R, Safarova D, Dittrich M, Richtrova J, Benickova K, Zboril R, Kvitek L (2011) Acute and chronic toxicity effects of silver nanoparticles (NPs) on *Drosophila melanogaster*. Environ Sci Technol 45:4974–4979. https://doi.org/10.1021/es104216b

Pandey UB, Nichols CD (2011) Human disease models in *Drosophila melanogaster* and the role of the Fly in therapeutic drug discovery. Pharmacol Rev 63:411–436. https://doi.org/10.1124/pr.110.003293

Pappus SA, Ekka B, Sahu S, Sabat D, Dash P, Mishra M (2017) A toxicity assessment of hydroxyapatite nanoparticles on development and behaviour of *Drosophila melanogaster*. J Nanopart Res 19:136. https://doi.org/10.1007/s11051-017-3824-8

Piper MDW, Partridge L (2018) *Drosophila* as a model for ageing. Biochim Biophys Acta (BBA) Mol Basis Dis 1864:2707–2717. https://doi.org/10.1016/j.bbadis.2017.09.016

Pitchakarn P, Inthachat W, Karinchai J, Temviriyanukul P (2021) Human Hazard assessment using *Drosophila* wing spot test as an alternative in vivo model for genotoxicity testing—a review. IJMS 22:9932. https://doi.org/10.3390/ijms22189932

Posgai R, Ahamed M, Hussain SM, Rowe JJ, Nielsen MG (2009) Inhalation method for delivery of nanoparticles to the *Drosophila* respiratory system for toxicity testing. Sci Total Environ 408:439–443. https://doi.org/10.1016/j.scitotenv.2009.10.008

Posgai R, Cipolla-McCulloch CB, Murphy KR, Hussain SM, Rowe JJ, Nielsen MG (2011) Differential toxicity of silver and titanium dioxide nanoparticles on Drosophila melanogaster development, reproductive effort, and viability: size, coatings and antioxidants matter. Chemosphere 85:34–42. https://doi.org/10.1016/j.chemosphere.2011.06.040

Ramani S, Pathak A, Dalal V, Paul A, Biswas S (2020) Oxidative stress in autoimmune diseases: an under dealt malice. CPPS 21:611–621. https://doi.org/10.2174/1389203721666200214111816

Rambaran T, Schirhagl R (2022) Nanotechnology from lab to industry—a look at current trends. Nanoscale Adv 4:3664–3675. https://doi.org/10.1039/D2NA00439A

Rand MD (2010) Drosophotoxicology: the growing potential for *Drosophila* in neurotoxicology. Neurotoxicol Teratol 32:74–83. https://doi.org/10.1016/j.ntt.2009.06.004

Rand MD, Vorojeikina D, Peppriell A, Gunderson J, Prince LM (2019) Drosophotoxicology: elucidating kinetic and dynamic pathways of methylmercury toxicity in a *Drosophila* model. Front Genet 10:666. https://doi.org/10.3389/fgene.2019.00666

Rand MD, Tennessen JM, Mackay TFC, Anholt RRH (2023) Perspectives on the *Drosophila melanogaster* model for advances in toxicological science. Current Protocols 3:e870. https://doi.org/10.1002/cpz1.870

Rapa SF, Di Iorio BR, Campiglia P, Heidland A, Marzocco S (2019) Inflammation and oxidative stress in chronic kidney disease—potential therapeutic role of minerals, vitamins and plant-derived metabolites. IJMS 21:263. https://doi.org/10.3390/ijms21010263

Rodríguez R, Gaivão I, Aguado L, Espina M, García J, Martínez-Camblor P, Sierra LM (2023) The comet assay in *Drosophila*: a tool to study interactions between DNA repair systems in DNA damage responses in vivo and ex vivo. Cells 12:1979. https://doi.org/10.3390/cells12151979

Rodríguez-Ibarra C, Déciga-Alcaraz A, Ispanixtlahuatl-Meráz O, Medina-Reyes EI, Delgado-Buenrostro NL, Chirino YI (2020) International landscape of limits and recommendations for occupational exposure to engineered nanomaterials. Toxicol Lett 322:111–119. https://doi.org/10.1016/j.toxlet.2020.01.016

Saner C, Weibel B, Würgler FE, Sengstag C (1996) Metabolism of promutagens catalyzed by *Drosophila melanogaster* CYP6A2 enzyme in *Saccharomyces cerevisiae*. Environ Mol Mutagen 27:46–58. https://doi.org/10.1002/(SICI)1098-2280(1996)27:1<46::AID-EM7>3.0.CO;2-C

Santalla M, Gómez I, Valverde C, Ferrero P (2021) A low-cost portable device to deliver smoke, volatile or vaporized substances to *Drosophila melanogaster* , useful for research and/or educational assays. Bio-protocol 11. https://doi.org/10.21769/BioProtoc.4244

Santarelli S, Londero C, Soldano A, Candelaresi C, Todeschini L, Vernizzi L, Bellosta P (2023) *Drosophila melanogaster* as a model to study autophagy in neurodegenerative diseases induced by proteinopathies. Front Neurosci 17:1082047. https://doi.org/10.3389/fnins.2023.1082047

Santos A, Oliveira M, Venâncio C (2023) Concomitant presence of nanosized plastics and metal(loid)s: is there cause for alarm? State-of-the-art and recommendations for future studies. TrAC Trends Anal Chem 164:117110. https://doi.org/10.1016/j.trac.2023.117110

Sengottiyan S, Mikolajczyk A, Jagiełło K, Swirog M, Puzyn T (2023) Core, coating, or corona? The importance of considering protein coronas in nano-QSPR modeling of zeta potential. ACS Nano 17:1989–1997. https://doi.org/10.1021/acsnano.2c06977

Shatkin JA, Ong KJ (2016) Alternative testing strategies for nanomaterials: state of the science and considerations for risk analysis. Risk Anal 36:1564–1580. https://doi.org/10.1111/risa.12642

Shell BC, Schmitt RE, Lee KM, Johnson JC, Chung BY, Pletcher SD, Grotewiel M (2018) Measurement of solid food intake in *Drosophila* via consumption-excretion of a dye tracer. Sci Rep 8:11536. https://doi.org/10.1038/s41598-018-29813-9

Siddique YH, Khan W, Fatima A, Jyoti S, Khanam S, Naz F, Rahul A, Singh BR, Naqvi AH (2016) Effect of bromocriptine alginate nanocomposite (BANC) on a transgenic *Drosophila* model of Parkinson's disease. Dis Model Mech 9:63–68. https://doi.org/10.1242/dmm.022145

Simpson DA, Thompson AJ, Kowarsky M, Zeeshan NF, Barson MSJ, Hall LT, Yan Y, Kaufmann S, Johnson BC, Ohshima T, Caruso F, Scholten RE, Saint RB, Murray MJ, Hollenberg LCL (2014) In vivo imaging and tracking of individual nanodiamonds in *Drosophila melanogaster* embryos. Biomed Opt Express 5:1250. https://doi.org/10.1364/BOE.5.001250

Singh NP, McCoy MT, Tice RR, Schneider EL (1988) A simple technique for quantitation of low levels of DNA damage in individual cells. Exp Cell Res 175:184–191. https://doi.org/10.1016/0014-4827(88)90265-0

Songvorawit N, Phengphuang P, Khongkhieo T (2022) Fluorescent silica nanoparticles as an internal marker in fruit flies and their effects on survivorship and fertility. Sci Rep 12:19745. https://doi.org/10.1038/s41598-022-24301-7

Srivastav S, Anand BG, Fatima M, Prajapati KP, Yadav SS, Kar K, Mondal AC (2020) Piperine-coated gold nanoparticles alleviate paraquat-induced neurotoxicity in *Drosophila melanogaster*. ACS Chem Neurosci 11:3772–3785. https://doi.org/10.1021/acschemneuro.0c00366

Starkov AA (2008) The role of mitochondria in reactive oxygen species metabolism and signaling. Ann N Y Acad Sci 1147:37–52. https://doi.org/10.1196/annals.1427.015

Sukhanova A, Bozrova S, Sokolov P, Berestovoy M, Karaulov A, Nabiev I (2018) Dependence of nanoparticle toxicity on their physical and chemical properties. Nanoscale Res Lett 13:44. https://doi.org/10.1186/s11671-018-2457-x

Tagorti G, Kaya B (2022) Publication trends of somatic mutation and recombination tests research: a bibliometric analysis (1984–2020). Genom Inform 20:e10. https://doi.org/10.5808/gi.21083

Taylor MJ, Tuxworth RI (2019) Continuous tracking of startled *Drosophila* as an alternative to the negative geotaxis climbing assay. J Neurogenet 33:190–198. https://doi.org/10.1080/01677063.2019.1634065

Thu HE, Haider M, Khan S, Sohail M, Hussain Z (2023) Nanotoxicity induced by nanomaterials: a review of factors affecting nanotoxicity and possible adaptations. OpenNano 14:100190. https://doi.org/10.1016/j.onano.2023.100190

Tirumala MG, Anchi P, Raja S, Rachamalla M, Godugu C (2021) Novel methods and approaches for safety evaluation of nanoparticle formulations: a focus towards in vitro models and adverse outcome pathways. Front Pharmacol 12:612659. https://doi.org/10.3389/fphar.2021.612659

Tomsa AM, Alexa AL, Junie ML, Rachisan AL, Ciumarnean L (2019) Oxidative stress as a potential target in acute kidney injury. PeerJ 7:e8046. https://doi.org/10.7717/peerj.8046

Tu Q, Deng J, Di M, Lin X, Chen Z, Li B, Tian L, Zhang Y (2023) Reproductive toxicity of polystyrene nanoplastics in *Drosophila melanogaster* under multi-generational exposure. Chemosphere 330:138724. https://doi.org/10.1016/j.chemosphere.2023.138724

Ullah S, Ahmad S, Guo X, Ullah S, Ullah S, Nabi G, Wanghe K (2023) A review of the endocrine disrupting effects of micro and nano plastic and their associated chemicals in mammals. Front Endocrinol 13:1084236. https://doi.org/10.3389/fendo.2022.1084236

Vales G, Demir E, Kaya B, Creus A, Marcos R (2013) Genotoxicity of cobalt nanoparticles and ions in *Drosophila*. Nanotoxicology 7:462–468. https://doi.org/10.3109/17435390.2012.689882

Vidal LM, Pimentel E, Escobar-Alarcón L, Cruces MP, Jiménez E, Suárez H, Leyva Y (2024) Toxicity evaluation of novel imidacloprid nanoribbons, using somatic mutation and fitness indexes in *Drosophila melanogaster*. J Toxic Environ Health A 87:398–418. https://doi.org/10.1080/15287394.2024.2316649

Villegas SN (2019) One hundred years of *Drosophila* cancer research: no longer in solitude. Dis Model Mech 12:dmm039032. https://doi.org/10.1242/dmm.039032

Vlisidou I, Wood W (2015) *Drosophila* blood cells and their role in immune responses. FEBS J 282:1368–1382. https://doi.org/10.1111/febs.13235

Wang G, Jin W, Qasim AM, Gao A, Peng X, Li W, Feng H, Chu PK (2017) Antibacterial effects of titanium embedded with silver nanoparticles based on electron-transfer-induced reactive oxygen species. Biomaterials 124:25–34. https://doi.org/10.1016/j.biomaterials.2017.01.028

Wang Z, Zhang L, Wang X (2023) Molecular toxicity and defense mechanisms induced by silver nanoparticles in *Drosophila melanogaster*. J Environ Sci 125:616–629. https://doi.org/10.1016/j.jes.2021.12.027

Xuan L, Ju Z, Skonieczna M, Zhou P, Huang R (2023) Nanoparticles-induced potential toxicity on human health: applications, toxicity mechanisms, and evaluation models. MedComm 4:e327. https://doi.org/10.1002/mco2.327

Yamamoto S, Jaiswal M, Charng W-L, Gambin T, Karaca E, Mirzaa G, Wiszniewski W, Sandoval H, Haelterman NA, Xiong B, Zhang K, Bayat V, David G, Li T et al (2014) A *Drosophila* genetic resource of mutants to study mechanisms underlying human genetic diseases. Cell 159:200–214. https://doi.org/10.1016/j.cell.2014.09.002

Yang J, Liu J, Wang P, Sun J, Lv X, Diao Y (2021) Toxic effect of titanium dioxide nanoparticles on corneas in vitro and in vivo. Aging 13:5020–5033. https://doi.org/10.18632/aging.202412

Yuan X, Zheng H, Su Y, Guo P, Zhang X, Zhao Q, Ge W, Li C, Xi Y, Yang X (2019) Drosophila Pif1A is essential for spermatogenesis and is the homolog of human CCDC157, a gene associated with idiopathic NOA. Cell Death Dis 10:125. https://doi.org/10.1038/s41419-019-1398-3

Zarkovic N (2020) Roles and functions of ROS and RNS in cellular physiology and pathology. Cells 9:767. https://doi.org/10.3390/cells9030767

Zhang Y, Liang J, Cao N, Gao J, Song L, Tang X (2022) Coal dust nanoparticles induced pulmonary fibrosis by promoting inflammation and epithelial-mesenchymal transition via the NF-κB/NLRP3 pathway driven by IGF1/ROS-mediated AKT/GSK3β signals. Cell Death Discov 8:500. https://doi.org/10.1038/s41420-022-01291-z

Zhang X, Song Y, Wang J, Wu C, Xiang H, Hu J, Gong H, Sun M (2023) Chronic exposure to titanium dioxide nanoparticles induces deficits of locomotor behavior by disrupting the development of NMJ in *Drosophila*. Sci Total Environ 888:164076. https://doi.org/10.1016/j.scitotenv.2023.164076

Zheng Z, Lauritzen JS, Perlman E, Robinson CG, Nichols M, Milkie D, Torrens O, Price J, Fisher CB, Sharifi N, Calle-Schuler SA, Kmecova L, Ali IJ, Karsh B, Trautman ET, Bogovic JA, Hanslovsky P, Jefferis GSXE, Kazhdan M, Khairy K, Saalfeld S, Fetter RD, Bock DD (2018) A complete electron microscopy volume of the brain of adult *Drosophila melanogaster*. Cell 174:730–743.e22. https://doi.org/10.1016/j.cell.2018.06.019

Zhou X, Jin W, Ma J (2023) Lung inflammation perturbation by engineered nanoparticles. Front Bioeng Biotechnol 11:1199230. https://doi.org/10.3389/fbioe.2023.1199230

Zielińska A, Costa B, Ferreira MV, Miguéis D, Louros JMS, Durazzo A, Lucarini M, Eder P, Chaud V, Morsink M, Willemen N, Severino P, Santini A, Souto EB (2020) Nanotoxicology and Nanosafety: safety-by-design and testing at a glance. IJERPH 17:4657. https://doi.org/10.3390/ijerph17134657

Open Access This chapter is licensed under the terms of the Creative Commons Attribution-NonCommercial-NoDerivatives 4.0 International License (http://creativecommons.org/licenses/by-nc-nd/4.0/), which permits any noncommercial use, sharing, distribution and reproduction in any medium or format, as long as you give appropriate credit to the original author(s) and the source, provide a link to the Creative Commons license and indicate if you modified the licensed material. You do not have permission under this license to share adapted material derived from this chapter or parts of it.

The images or other third party material in this chapter are included in the chapter's Creative Commons license, unless indicated otherwise in a credit line to the material. If material is not included in the chapter's Creative Commons license and your intended use is not permitted by statutory regulation or exceeds the permitted use, you will need to obtain permission directly from the copyright holder.

Zebrafish as a Model to Investigate Nanoparticles

14

Lina Lundin and Steffen H. Keiter

Abstract

The rapid development of nanoparticles (NPs) has raised significant concerns regarding their potential toxic effects on both human health and the environment. Zebrafish (*Danio rerio*) have emerged as a valuable *in vivo* model for nanotoxicity studies due to their physiological and genetic similarities to humans, high fecundity, transparent embryos, and rapid development. This chapter provides a comprehensive review of zebrafish as a model organism for investigating NP toxicity. It highlights key advantages, including their suitability for high-throughput screening and real-time visualization of NP biodistribution. The chapter discusses various NP uptake pathways, such as the gills, gastrointestinal tract, and blood-brain barrier, and explores the biological barriers that influence NP accumulation. Furthermore, it summarizes toxicological findings on teratogenic, immunotoxic, neurotoxic, and hepatotoxic effects of NPs across different zebrafish life stages. The use of zebrafish allows for the investigation of both acute and chronic NP exposure, offering insights into developmental and reproductive toxicity, oxidative stress, and genotoxicity. By bridging the gap between simple in vitro tests and more complex mammalian models, zebrafish serve as an essential model for assessing the potential risks of nanomaterials for human and environmental health.

Keywords

Zebrafish · Nanoparticles · Exposure · Uptake · Accumulation · Toxicity

L. Lundin · S. H. Keiter (✉)
Man-Technology-Environment Research Centre (MTM), Biology, School of Science and Technology, Örebro University, Örebro, Sweden
e-mail: steffen.keiter@oru.se

© The Author(s) 2025
E. Alfaro-Moreno, F. Murphy (eds.), *Nanosafety*,
https://doi.org/10.1007/978-3-031-93871-9_14

1 Introduction

Over the past decades, the field of nanoparticle (NP) synthesis and application has rapidly advanced, leading to the creation of multifunctional NPs with various properties (Haque and Ward 2018). In addition, the potential benefits of incorporating engineered nanomaterials into commercial products have significantly driven research and production efforts to develop new nanoscale materials. Therefore, NPs are used in a wide range of applications, such as chemically inert additives, polymer fillers, pigments, dyes, UV protectors, and processing aids (Stark et al. 2015). They are used for creating functional surfaces and membranes with properties like antimicrobial and UV protection (Das et al. 2013b). In medicine, NPs are used for biosensing, bioimaging, drug delivery, tissue engineering, and therapeutic applications due to their small size and ability to be functionalized for specific targeting (Das et al. 2013a; Mcnamara and Tofail 2015). Consequently, due to the huge variety of properties and applications it is inevitable that this leads to human and environmental exposure that potentially cause adverse effects (Lin et al. 2013); therefore, measuring NP toxicity is crucial (Lin et al. 2013; Friedman et al. 2013).

In the last decade, a large variety of studies have been published that investigated the potential toxicological effects of NPs (Subramanian et al. 2024). These studies utilized a wide range of experiments including *in vitro* assays using biomolecules, organelles, and cells, to *in vivo* studies using multi-cellular organisms like daphnia and sea urchins, and more complex animal models such as rodents and primates covering multiple toxicological endpoints such as reproductive toxicity, neurotoxicity, genotoxicity, embryotoxicity, and endocrine disruption (Moutabian et al. 2022; Janzadeh et al. 2022; Di Ianni et al. 2022; Siivola et al. 2022; Caixeta et al. 2020; Kose et al. 2023; Klein et al. 2023; Leso et al. 2023). While higher animal models provide valuable insights due to their biological complexity, they are limited by ethical concerns, high costs, slow and inaccessible embryo development, and the large quantities of materials required. In contrast, simpler models such as cell lines or multi-cellular organisms, though useful for initial toxicity and genotoxicity screenings, may lack the depth required for understanding complex physiological interactions. However, given the rapidly increasing number of nanomaterials produced each year, there is a critical need for experimental models that balance biological relevance with high-throughput screening capabilities. Zebrafish have emerged as a valuable *in vivo* model for nanotoxicity studies due to their unique features. These include high fecundity, transparent embryos, fast and well-characterized development, low maintenance costs, and the ability to manipulate genes. These advantages allow zebrafish to serve as an intermediate model between *in vitro* assays and mammalian models, enabling the validation of *in vitro* results and the prioritization of more complex *in vivo* studies.

This chapter focuses on the current advances in using zebrafish for toxicity assessments of engineered NPs. It highlights the benefits that make zebrafish a popular and efficient model organism, describes various nanotoxicological assessment methods, and summarizes key findings from studies using zebrafish. Additionally, it discusses the challenges and prospects of using zebrafish in nanotoxicity research, emphasizing the potential of this model to bridge the gap between simple *in vitro* tests and more complex animal studies.

2 The Zebrafish Model

In the field of nano(eco)toxicology, zebrafish represents a cost-effective and advanced model for evaluating *in vivo* toxicity, offering advantages over more traditional models like rodents and primates (Haque and Ward 2018). For instance, cardiotoxicity of compounds to human can be predicted in zebrafish embryos with more than 80% success (Brown et al. 2016).

Zebrafish is a tropical freshwater fish native to Southeast Asia and can be found in both stagnant and flowing waters. Their life cycle is divided into four stages: embryo, larvae, juvenile, and adult, with full development occurring over 90 days (Spence et al. 2008). Throughout these stages, zebrafish exhibit varying sensitivities to stressors and toxins. For instance, they are more vulnerable to toxins before the liver and Blood-Brain Barrier are fully developed, with the latter leading to increased brain sensitivity (Xie et al. 2010). In this context, the chorion, a membrane with pores around the embryo, plays a crucial role in protecting the zebrafish embryos from pollutants in their early life-stages (Cheng et al. 2007). Zebrafish undergo rapid development, forming key organs such as the heart, blood circulation, and fins within the first 120 hours (Kimmel et al. 1995). In their natural environment, juvenile zebrafish face increased exposure to high concentrations of various pollutants due to their nursery habitats often being accumulation sites for such contaminants (Chae and An 2017). Moreover, early life stages are particularly susceptible to predator-induced mortality (Atherton and McCormick 2017). Early-stage behavior is critical for ecological fitness, influencing growth, reproduction, and survival (Smith and Blumstein 2008). Given these developmental differences, it is important to conduct studies across various life stages and account for them in data analysis.

For (eco-)toxicological research, zebrafish represents a very popular *in vivo* model for studying effects caused by pollutants for several reasons. They are easy to maintain and breed, and their transparent embryos and larvae stages facilitate the study of teratogenic and embryotoxic effects (Langheinrich 2003). Fertilization and embryonic development are external, and together with the transparency of the chorion and the developmental stages it enables high-throughput visual analyses, including imagining of fluorescent or colored contaminants, including NPs, within the body (Batel et al. 2018). Additionally, their ability to spawn every 2–3 days, producing several hundred eggs, combined with rapid organ development, makes them particularly favorable to research. Behavioral changes such as food seeking and predator avoidance can be investigated within only a few days (Spence et al. 2008). Zebrafish are particularly valuable for research due to their physiological similarities to humans, making them a relevant model for understanding human health impacts and serving as an indicator species for potential environmental and ecological consequences. As a vertebrate model, zebrafish share approximately 70% genetic similarities with humans, offering more reliable insights into potential health effects of pollutants when compared with invertebrate model organisms (Spence et al. 2008; Bhagat et al. 2020). The zebrafish genome is completely sequenced which makes it easy to study genetic changes induced by pollutants (Howe et al. 2013). Additionally, numerous transgenic zebrafish lines are available that express fluorescent reporter genes, such as green fluorescent protein (GFP) or red fluorescent protein, in specific cells, tissues, or organs (Bai and Tang 2023).

Importantly, the non-feeding embryonic stage of zebrafish is not considered as a protective test organism and, thus, falls outside the scope of EU-directive 2010/63 which makes it more accessible for a wide range of studies (Bhagat et al. 2020). Therefore, the zebrafish (*Danio rerio*) is one of the most utilized test organisms for assessing environmental pollution and is frequently employed to investigate the toxic effects of NPs (Haque and Ward 2018).

3 Exposure of Zebrafish to NPs

The chemistry of manufactured NPs in aquatic environments has been extensively studied (Elimelech et al. 2013; Petosa et al. 2010; Grasso et al. 2002; Lead and Wilkinson 2006; Stolpe and Hassellöv 2010; Handy et al. 2008a). Key aspects such as the physico-chemical properties, behavior, and methods for measuring manufactured NPs in aqueous environments, as well as key chemistry issues for ecotoxicologists, are discussed in detail elsewhere (Handy et al. 2008b; Christian et al. 2008). This includes topics such as NP dispersion in various natural waters, surface chemistry, reactivity, and aggregation behavior.

Unlike many common organic and inorganic pollutants, manufactured NPs generally do not dissolve in true aqueous solutions, making solubility a less relevant concept, except for the gradual dissolution of atoms or ions from the NP surface into the water (Handy et al. 2008a). Due to this instability of NPs in aqueous media, particularly of metal NPs, dispersants or stabilizing agents (e.g., citrate) or surface coatings like polyethylene glycol (PEG) or polyvinylpyrrolidone (PVP) are added to the exposure medium (Pereira et al. 2019). When commercial NPs are synthesized with these added features, they should be considered integral to the NP, as their behavior and stability can vary significantly depending on the testing medium. The diversity of manufactured materials also makes it difficult to generalize their chemistry in natural waters. In addition, many NPs exist as emulsions or suspensions, often without any consideration of water during the manufacturing process (Handy et al. 2008a). However, some NPs are surface-coated or functionally modified for specific biological applications, such as enhancing miscibility in physiological salines or solvents used in drug delivery for clinical purposes.

The biological consequences of exposure to NP aggregates, compared to colloidal suspensions, depend in part on the epithelial biology of the organisms. However, it is important not to assume that aggregated material is not bioavailable; instead, aggregation alters the bioavailability of NPs, changing their mode of delivery from respiratory uptake in the water column to dietary exposure through sediment (Ji et al. 2024). Additionally, aggregation may change the size of the NPs ingested by organisms, transforming nanoscale particles into larger, micron-scale aggregates, which could influence how organisms absorb or process the material (Baker et al. 2014).

Several of these aspects of NP physico-chemistry are particularly important for understanding the routes of uptake into fish. A key assumption in chemical ecotoxicity is that the physical or chemical form of a substance in the environment is closely related to its absorption by organisms and its potential toxic effects. This assumption also applies to NPs in fish toxicology, but it is essential to consider colloid and aggregation chemistry in addition to the usual abiotic factors, such as temperature, dissolved oxygen, salinity, and water hardness. For instance, according to OECD guideline 236, zebrafish embryo tests require a water temperature of 26 ± 1 °C, though studies have used a range of 25–30 °C, with 59% at 28 ± 1 °C (Pereira et al. 2019). It was shown that higher temperatures, above 28 °C, can accelerate embryo development and increase malformations (Beekhuijzen et al. 2015). The pH also plays a key role, as it affects NP behavior in the medium, especially when combined with ionic strength and NP properties like surface charge (Clemente et al. 2017). Adjusting pH or ionic strength can reduce NP aggregation but may cause stress to the fish (Petersen et al. 2015). These factors not only influence NP dispersion and aggregation but can also affect the interactions between NPs and biological systems, potentially altering their toxicity and bioavailability.

In conclusion, the stability and behavior of NPs in aquatic environments are influenced by various parameters, such as surface coatings, dispersants, and environmental conditions like pH, temperature, and ionic strength. Together, these parameters are making it difficult to generalize their behavior and toxicity. Therefore, the interaction of physico-chemical factors with biological systems emphasizes the need for more adapted methodologies for studying NP toxicity in zebrafish, particularly in accounting for the role of colloid and aggregation chemistry.

4 Uptake and Accumulation of NPs—Biological Barriers

In general, the uptake of NPs into zebrafish depends primarily on size. Smaller NPs can reach internal organs such as the eyes, liver, pancreas, heart, and brain, while larger particles (>200 nm) tend to accumulate in the gut, gills, and skin and their mucosa (Handy and Eddy 2004; Torres-Ruiz et al. 2021). Additionally, there is strong protection to prevent NPs from entering the bloodstream and reaching the brain. However, before NPs can enter the circulatory system or reach their target tissues, they must first overcome several biological barriers, such as the gastrointestinal barrier, blood-brain barrier, and blood-retinal barrier, among others (Li et al. 2017). The transport of NPs across these barriers is influenced by various factors, including particle size and surface properties (Kulkarni and Feng 2013). Zebrafish serve as an ideal model for studying NP uptake across biological barriers because, unlike *in vitro* models, they possess the complex metabolism and physiological characteristics of an *in vivo* system. While mice and rats represent some of the most used animal models to evaluate NPs for barrier permeability, these models are expensive, labor-intensive, and time-consuming, making them less suitable for high-throughput screening of NP permeability (Geldenhuys et al. 2012). When studying the uptake and accumulation of NPs in zebrafish, the differentiation between the various life-stages is crucial.

4.1 The Chorion as Barrier

In zebrafish embryos, but also for other fish species, the chorion acts as a natural barrier until hatching, effectively blocking various pollutants, including NPs, from entering the embryo. The outer chorion membrane is approximately 1.5 ± 2.5 mm thick and is composed of three distinct layers (Rawson et al. 2000). The chorion is perforated by cone-shaped pore canals with a diameter of up to 500–700 nm with a larger diameter at the inner surface (Cheng et al. 2007; Rawson et al. 2000). The pores allow essential substances like oxygen and salt ions to pass through the chorion, as well as some nano-sized pollutants. Several studies found that NPs were effectively blocked by the embryo's chorion and have highlighted the protective role of the embryonic chorion against pollutants, showing increased sensitivity to xenobiotics after the chorion is removed (Henn and Braunbeck 2011; Van Pomeren et al. 2017; Vranic et al. 2019). However, research also demonstrated the toxicological effects of small-sized NPs capable of penetrating the zebrafish chorion and reach the yolk sac (Pitt et al. 2018a; Qiang and Cheng 2019; Pitt et al. 2018b; Xu et al. 2021; Gu et al. 2021). This is likely due to the size of the chorion's pores allowing only particles smaller than this threshold to pass.

Apart from pore size, other physicochemical properties of pollutants also play a key role in the chorion's barrier function. For example, the lipophilicity of test compounds has been shown to significantly impact this function (Braunbeck et al. 2005). Nevertheless, even larger NPs that cannot penetrate the chorion may still adhere to its surface due to their high affinity for the embryonic chorion, potentially causing biological effects because they impair the exchange of oxygen and salts (Duan et al. 2020).

Research has shown that under environmentally relevant exposure conditions, NP aggregation in aqueous suspensions increases their hydrodynamic diameter, which may reduce their uptake by the zebrafish chorion (Cheng et al. 2007; Pereira et al. 2019; Chao et al. 2018). Additionally, interactions between NPs and macromolecules in the test medium alter their uptake and behavior in zebrafish embryos. For example, natural organic matter has been shown to reduce NP toxicity. Kteeba et al. (2017) found that natural organic matter, isolated from rivers in Milwaukee, Yukon, and Suwannee, mitigated the toxic effects of zinc oxide NPs (ZnO-NPs; 10–30 nm), leading to fewer delays in hatching, as well as reduced mortality and malformations.

4.2 Uptake via the Gills

As the essential respiratory organ of fish, gills not only facilitate gas exchange but also is essential for ion and water transfer, ammonia excretion, and osmoregulation (Evans and Nunez 2015; Koppang et al. 2015; Karlsson 1983). To support these diverse physiological functions, gills have evolved intricate morphological structures, including the gill arch, filaments, operculum, petals, flakes, and rakes. These structures adapt to changes in environmental factors such as water flow, temperature, ion concentration, and salinity (Chen et al. 2023b). Additionally, gills are highly sensitive to environmental changes and pollution (Evans and Nunez 2015). The relationship between the gills' structural characteristics, their multiple physiological roles, and their ability to react to environmental changes has gained significant attention (Chen et al. 2012).

Fish are typically exposed to NPs in suspension, which, due to their small size, can easily reach the gills and pose a respiratory risk. The structure of fish gill epithelium is like that of other vertebrates (e.g., rat gut or lung), meaning concerns for fish may also apply to other animals (Handy et al. 2008a). Fish gills are covered by a boundary layer of unstirred water layer and mucus, which can interact with NPs. NPs move through water via Brownian motion and diffusional processes, and their movement into the unstirred layer is influenced by forces like shear stress, depending on factors such as water flow, particle size, and the viscosity of the unstirred layer (Handy et al. 2008a). This behavior is like atoms or small molecules, although the properties of NPs, such as size and shape, may differ (Pedley and Fischbarg 1978). Fish mucus is mainly water, mucoproteins, and other materials, and like metal ions, NPs may interact with mucus due to their surface charge. Some NPs might bind strongly to mucus, getting trapped more easily than smaller molecules (Handy and Eddy 2004). This happens, for example, with carbon nanotubes (CNTs) in fish gills (Smith et al. 2007). While mucus helps protect fish by trapping harmful substances and shedding them, this protective mechanism may be overwhelming if too much mucus is used up, as could happen with the exposure to NPs. If NPs bypass the mucus, they might enter the body through either cell membranes or between tight junctions of the gill cells.

For respiratory exposure, fish have a much tighter gill epithelium compared to the lung epithelium of mammals and so the paracellular route is less vulnerable in fish (Karlsson 1983). Therefore, NPs are less likely to penetrate these tight junctions because of their size (Chen et al. 2023b). The common uptake, for instance, of metals across biological membranes typically occurs through carrier-mediated transport via metal ion transporters (Bury and Handy 2010; Handy and Eddy 2004). Unlike metal ions, NPs are too large to pass through ion transporters or paracellular diffusion pathways, making endocytosis the most likely route of NP uptake (Shaw and Handy 2011). The clathrin-mediated endocytosis represents a receptor-mediated process where clathrin forms coated pits on the membrane, leading to the formation of endocytic vesicles and the uptake of NPs (Gehr 2018). However, it was shown that cationic NPs can enter cells in an energy-independent fashion, bypassing the common endocytosis route, which is described as direct translocation process (Lin and Alexander-Katz 2013). For very small NPs (<20 nm), it is theoretically possible to diffuse through the cell membrane. This can occur if they acquire a hydrophilic surface coating, such as through steric effects where particles are coated with phospholipids, or if they become hydrophobic by achieving no net charge, for instance, when positively charged nanometals are incidentally coated with biological anions like those found in mucus (Lead et al. 2018). In another study by Zanella et al. (2017) it was shown that uncoated iron oxide NPs (FeO-NPs) was surrounded by a protein corona, and by this could cross plasma membranes. In some cases, NPs can cause inflammation and damage to tissues like the gills or gut. When tissues are damaged, substances, including NPs, could enter the bloodstream more easily, though NPs might aggregate and precipitate in these situations, reducing their uptake (Federici et al. 2007; Smith et al. 2007).

4.3 Uptake of NPs via the Gastrointestinal Tract

The adult zebrafish gastrointestinal tract (GIT) is a three-part folded tube extending from the esophagus to the anus, occupying much of the abdominal cavity. Lacking a stomach, the anterior intestine, or intestinal bulb, functions as a reservoir with a wider lumen than the posterior intestine. The transparent intestinal wall reveals large, randomly shaped epithelial folds, which are significantly larger than mammalian villi. These folds become progressively shorter from the anterior to the posterior intestine (Prochaska et al. 2020). The intestinal epithelium consists of three main cell types: absorptive enterocytes, goblet cells, and enteroendocrine cells. Enterocytes in the anterior intestine possess long microvilli (up to 7.5 μm), which decrease in length throughout the gut and are connected by tight junctions, measuring 200–400 nm (Cheng et al. 2016). The absence of Paneth cells, crypts, and organized lymphoid structures is notable, and contain specialized antigen-presenting enterocytes in the posterior midintestine, which are analogous to microfold cells in mammals (Wallace et al. 2005). The midintestine is the primary site of protein absorption and features an abundance of goblet cells and microvilli, while the posterior intestine, with fewer microvilli and goblet cells, plays a role in osmoregulation. Smooth muscle and connective tissue layers surrounding the gut are simpler than in mammals, and enteric ganglia are located between the circular and longitudinal smooth muscle layers (Wallace et al. 2005).

The structure of the GIT impacts the uptake of NPs, with distinct patterns of bioaccumulation and toxic effects depending on particle size and the developmental stage of the fish. In zebrafish embryos, NPs predominantly accumulate in organs such as the gut, yolk, and brain, with studies suggesting a correlation between NP size and their organ localization. Torres-Ruiz et al. (2021) describes that NPs in the size range of 20–200 nm, when waterborne, were internalized without a clear relationship between size or exposure duration and the specific organs affected. Larger NPs (>200 nm), however, tend to accumulate in the gut, gills, and skin, particularly in embryos (Parenti et al. 2019; Van Pomeren et al. 2017). When NPs accumulate in the gut, they frequently spread to adjacent organs, such as the liver, gall bladder, and pancreas. Zebrafish embryos and larvae exhibit significant sensitivity to NP exposure, which can lead to developmental and behavioral abnormalities. For example, aqueous exposure to polystyrene NPs (PS-NPs) caused rapid uptake and widespread distribution across the head, gill arches, and pectoral fins in early life stages (24–144 hours post-fertilization), whereas dietary exposure localized NPs primarily to the gut (Skjolding et al. 2017). Zhang et al. (2022) also observed that the predominant uptake route for PS-NPs in embryos was oral ingestion, with minimal dermal uptake.

As zebrafish mature into juveniles, NP uptake patterns shift, with dietary exposure playing a more significant role in NP accumulation in the GIT. For instance, studies using fluorescent polystyrene and gold NPs (Au-NPs) revealed that dietary exposure resulted in NP deposition in the stomach and intestines, whereas aqueous exposure targeted gills and intestines (Skjolding et al. 2017). Notably, the size of NPs is likely influencing their accumulation patterns. Larger particles (25–50 nm) were found to accumulate in the eye, likely because of intestinal uptake and subsequent biodistribution (Van Pomeren et al. 2017).

In adult zebrafish, dietary uptake continues to play a critical role in NP accumulation. Studies by Shi et al. (2020) and Huang et al. (2020) demonstrated that dietary NPs tend to accumulate in the intestine and liver. The intestine is often the initial site of NP accumulation, where particles may translocate across the intestinal membranes, entering circulation and contributing to systemic biodistribution. This gastrointestinal absorption is a major route for NP bioaccumulation, and the liver plays an important role in processing these particles during exposure. Interestingly, dietary exposure appears to facilitate greater NP accumulation in internal tissues compared to aqueous exposure, as demonstrated in studies of fullerenol and PS-NPs (Habumugisha et al. 2023; Shi et al. 2020). Similarly, an acute toxicology pathological study of NPs to adult tilapia also showed that the major translocation pathway to enter the circulation of NPs in fish may be via the gill-blood route and/or the intestine-blood route (Srinonate et al. 2015), because the gill and gastrointestinal organs are principal routes that are directly exposed to and take up the toxicants from ambient water into the fish body. Besides, Hem-NPs (hematite NPs) predominantly accumulated in the intestinal tract in both larval and adult zebrafish during dietary exposure, with lesser amounts in other tissues, such as the head and gills (Huang et al. 2020). Adult zebrafish exposed to PS-NPs in a study by Habumugisha et al. (2023) showed that the highest bioaccumulation occurred in the intestine, followed by the liver, gill, muscle, and brain. The bioaccumulation of PS-NPs was concentration-, tissue-, and time-dependent, with the intestine acting as a major uptake site. After 16 days of depuration, residual NPs were still present, particularly in the brain, indicating that clearance from tissues could take several weeks.

In addition to accumulation, NP exposure has been linked to pro-inflammatory responses, particularly in the GIT. Brun et al. (2018) showed that zebrafish exposed to metal and plastic NPs displayed transcriptional alterations in genes regulating immune responses, with copper NPs eliciting a particularly strong inflammatory response in the intestines. This suggests that NPs not only accumulate in the GI system but also trigger immune system activation. Additionally, developmental exposure to NPs can also result in long-term physiological and behavioral effects. For instance, Parenti et al. (2019) found that waterborne exposure to PS-NPs caused developmental abnormalities, reduced growth rates, and hypoactivity in zebrafish, with significant transfer to the brain. These toxic effects underline the significance of understanding how exposure routes and NP characteristics influence their biological effects. In conclusion, NP uptake via the gastrointestinal system in zebrafish varies depending on the life stage, with early life stages (embryos and larvae) being particularly vulnerable to waterborne NPs, while dietary exposure seems to play a larger role in juveniles and adults.

4.4 The Blood-Brain-Barrier

The blood-brain barrier (BBB) plays a critical role in protecting the brain from toxic substances and pathogens by acting as a selective diffusion barrier. Its primary function is to restrict the passage of molecules from the bloodstream into the brain parenchyma, maintaining a tightly regulated environment essential for neuronal function. Structurally, the BBB consists of endothelial cells interconnected by tight junctions, which are supported by astrocytes, pericytes, and neurons. These cellular components ensure that substances, especially larger and hydrophilic molecules, cannot passively diffuse into the brain (Abbott et al. 2006; Miner and Diamond 2016). The absence of fenestrations in the endothelial cells further restricts the diffusion of small molecules, including water-soluble substances, which are also impeded by inter-endothelial junctions such as tight junctions (Liu et al. 2012). These junctions regulate the paracellular permeability of the BBB and contribute to its highly selective nature. Molecules can cross the BBB via two primary pathways: the paracellular pathway, where ions and small molecules diffuse passively between cells, and the transcellular pathway, which involves mechanisms such as receptor-mediated transcytosis and active transport (O'Keeffe and Campbell 2016). Several properties, including molecular weight, surface charge, lipophilicity, and solubility, influence the ability of a substance or particle to permeate the BBB (Almutairi et al. 2016). Molecules smaller than 400 Da and those that are lipid-soluble are generally capable of crossing the BBB, while larger or hydrophilic compounds face significant barriers (Dong 2018).

In recent years, nanotechnology has emerged as a potential solution to overcome the limitations posed by the BBB in drug delivery. NPs can be engineered with precise physical and chemical properties that enhance their ability to cross the BBB, either by increasing circulation time, improving target specificity, or utilizing receptor-mediated pathways (Pardridge 2006). In this context, zebrafish have gained prominence as a model organism for studying NP interactions with the BBB due to their unique advantages, including transparent larvae that allow real-time visualization of BBB interactions (Saleem and Kannan 2021).

Today, zebrafish is an established model organism for investigating the permeability of the BBB to NPs. The transparency of zebrafish's early-life stages allows researchers to directly observe the biodistribution and localization of fluorescently labeled NPs in real-time, providing crucial insights into how these particles interact with brain vasculature and cross the BBB (Goldsmith and Jobin 2012). Moreover, zebrafish larvae possess a cerebral microvasculature that mimics the structural and functional properties of the mammalian BBB, making them a relevant model for studying BBB dynamics and NP-mediated delivery (Saleem and Kannan 2021). While zebrafish offer many advantages as a model system, they do have limitations. Their nervous system is less complex than that of humans, and while rodents and mice have more developed nervous systems, they still aren't perfect analogs for studying human brain diseases. Previously, adult zebrafish were thought to lack liver macrophages (Kupffer cells), but recent research has confirmed their presence, enhancing the model's utility in liver studies (Shwartz et al. 2019). Nonetheless, higher vertebrate models better mimic human pathologies, and there's ongoing debate about how well animal models translate to human clinical outcomes, as all models have inherent limitations.

A study by Gomez-Garcia et al. (2018) used zebrafish to evaluate the impact of NP size, charge, and surface chemistry on BBB permeability. They found that negatively charged PS-NPs were less likely to interact with the endothelial glycocalyx, allowing for better circulation and reduced clearance compared to positively charged particles. Moreover, liposomes and other soft NPs accumulated in specific regions of the brain vasculature, indicating the potential for targeted delivery using engineered NPs (Gomez-Garcia et al. 2018). Similarly, Rabanel et al. (2021) demonstrated that intravenously injected NPs in zebrafish larvae rapidly accumulated in the brain vasculature, suggesting their ability to cross the BBB. Over time, NPs were found in extravascular spaces within the brain, indicating successful delivery (Rabanel et al. 2021). This study also emphasized the role of surface modifications such as PEGylation (polyethylene glycol) in enhancing NP stability and reducing immunogenicity, crucial factors in ensuring successful BBB penetration (Veronese and Mero 2008). Kalaiarasi et al. (2016) developed functionalized poly N-isopropyl acrylamide nanogels with 20 nm size for the delivery of the Alzheimer's drug donepezil across the BBB in zebrafish. Their study revealed that these nanogels could overcome the BBB, demonstrating sustained drug release and reduced neurotoxicity, making them a promising vehicle for treating neurodegenerative diseases (Kalaiarasi et al. 2016). While NPs offer great promise for overcoming the BBB, their potential toxicity must be carefully evaluated.

In conclusion, the use of zebrafish as a model for studying NP interactions with the BBB has provided important insights into NP permeability and brain-targeted drug delivery. The transparency of zebrafish larvae allows real-time visualization of NP behavior, highlighting the influence of factors such as size, charge, and surface chemistry on BBB penetration. Additionally, various studies demonstrate that surface modifications can enhance NP stability and reduce clearance, promoting successful delivery across the BBB. However, despite these promising findings, challenges remain, including species-specific differences in BBB complexity between zebrafish and humans. Further, while engineered NPs show a high potential for treating neurodegenerative diseases, their long-term toxicity must be rigorously assessed to guarantee human and environmental health.

5   Toxicological Effects of NPs

The toxicological effects of NPs between early-life stages and adult zebrafish differ significantly, due to physiological differences, such as organ development and sensitivity. For example, copper oxide NPs (CuO-NPs) were found to cause neurotoxicity in zebrafish embryos, delaying retinal neuro-differentiation and reducing locomotor abilities (Sun et al. 2016). While NPs like PS-NPs can cross the BBB in both larvae and adults, their accumulation is higher in earlier stages (Bhagat et al. 2020). In terms of reproductive toxicity, embryonic exposure to titanium dioxide

NPs (TiO$_2$-NPs) led to disrupted thyroid endocrine function, while in adults, reproductive toxicity manifested through direct damage to germ cells and ovarian tissue (Chakraborty et al. 2016). Chronic exposure to TiO$_2$-NPs in adults has been particularly concerning, with studies showing reproductive and genotoxic impacts, while larvae showed a heightened sensitivity to these NPs (Wang et al. 2022a). Therefore, this section describes the effects of NPs on early-life stages and adult zebrafish separately.

5.1 Teratogenic and Developmental Effects in Early-Life Stages of Zebrafish

To the best of our knowledge, the first study investigating toxic effects of NPs on zebrafish embryos was conducted by Cheng et al. (2007), who tested the toxicity of carbon nanotubes (CNTs), observing delayed hatching at 120 mg/L due to CNT accumulation in the chorion. Since then, investigations using zebrafish embryos has become an increasingly popular tool in nanotoxicological research. This growth correlates with advances in zebrafish molecular biology, including the sequencing of its genome (Howe et al. 2013). These developments have allowed the identification of zebrafish genes homologous to those of humans, enabling the use of various molecular techniques, including RT-PCR, RNA-Seq, and ELISA to evaluate the biological responses of zebrafish embryos to NP exposure (Pereira et al. 2019).

Nanotoxicological research has increasingly focused on understanding the molecular mechanisms of NP toxicity. While earlier studies primarily assessed morphological changes, delayed hatching, and mortality (Asharani et al. 2008; Bar-Ilan et al. 2009), advances in molecular biology allowed for a deeper understanding of the underlying biological processes. Recent work has utilized techniques like qRT-PCR, RNAseq, RNAWestern blotting, and intracellular Reactive Oxygen Species (ROS) assays to assess oxidative stress, DNA damage, and the activation of apoptosis pathways following NP exposure (Zheng et al. 2018; Bar-Ilan et al. 2012; Zhao et al. 2016; Hu et al. 2021; Massarsky et al. 2014; Pereira et al. 2019). These molecular insights have revealed that both NPs and their dissolved ions contribute to ROS formation, leading to damage in lipids, proteins, and DNA, eventually resulting in cell apoptosis and developmental malformations in zebrafish embryos (Huang et al. 2024).

The interaction of NPs with the chorion has been widely studied as a key factor influencing NP toxicity, including its role as a barrier for the uptake of NPs (compare Sect. 4.1). For instance, Fent et al. (2010) demonstrated that fluorescent silica NPs (FS-NPs), ranging from 60 to 200 nm, did not cross the chorion and, consequently, did not induce malformations or interfere with gas exchange during zebrafish development. Similarly, Cheng et al. (2007) found that CNTs (11 nm; 360 mg/L) were also prevented from crossing the chorion, resulting in a lack of toxicity and developmental defects during 96 hours of exposure. In contrast, silver NPs (Ag-NPs; 5–55 nm) were able to penetrate the chorion and accumulate in the brain, heart, yolk, and blood, leading to severe morphological deformities, including pericardial edema, mouth and notochord malformations, and tail tissue degradation (Asharani et al. 2008, 2011; Chen et al. 2017b). The interaction of NPs with macromolecules in the test medium can also influence their toxicity. For example, the interaction of ZnO-NPs with natural organic matter has been shown to reduce toxicity; for instance, mitigating delayed hatching, mortality, and malformations (Kteeba et al. 2018).

Accumulation of NPs in zebrafish embryo tissues has been a common endpoint in several studies. Several studies observed accumulation in the yolk sac (Duan et al. 2020; Duan et al. 2017; Chen et al. 2023a; Gu et al. 2021; Sun et al. 2021), while accumulation in the gastrointestinal tract was documented by Pitt et al. (2018a), Brun et al. (2018), and Sarasamma et al. (2020). Similar accumulation was observed by Khan et al. (2024), who reported time-dependent accumulation of lanthanide-doped upconversion nanoparticles (Ln-UC-NPs) in the chorion, eyes, skin, pharynx, esophagus, and intestines of zebrafish larvae. Liu et al. (2019) also demonstrated that ingestion of 40 nm Ag-NPs beginning at 96 hpf (hours post fertilization) led to mortality, suggesting the internalization of NPs within the intestinal lumen as a key factor for lethality.

Brun et al. (2018) investigated the potential sites of NP accumulation, specifically the skin and intestine, using zebrafish embryos due to their transparency. They evaluated the inflammatory responses elicited by PS-NPs and copper NPs (Cu-NPs), comparing their effects in the skin and intestine. While copper is well-known for causing inflammation, the mechanisms behind PS-NP toxicity remain largely unexplored. Nevertheless, their study demonstrated that both NPs induced pro-inflammatory responses in the skin and intestine, with more pronounced effects observed in the skin. Transcriptional changes in immune system genes were noted for both PS-NPs and Cu-NPs, with Cu-NPs inducing the strongest response. The study also suggested that NP-induced damage to neuromasts in the lateral line could lead to behavioral changes, ultimately impacting population-level outcomes. Their work highlights the potential of skin-based immune responses as a screening tool for NP toxicity and as an important element in the development of Adverse Outcome Pathways (AOPs).

The toxicological impact of NPs on zebrafish extends to specific organs and tissues, including the skeleton. Although the effects of environmental NPs on bones and cartilages have been less studied, Huang et al. (2021) highlighted that SiO_2 and Ni-NPs are among the few NP types that have been shown to cause skeletal abnormalities. SiO_2-NPs (15 and 50 nm) were found to disrupt chondrogenesis in the heads and jaws of zebrafish embryos (Wang et al. 2014), while Ni-NPs induced cartilage defects in the same regions (Ispas et al. 2009). This suggests a potential for broader toxic effects of various NPs on craniofacial skeletons but is requiring further investigations.

Oxidative stress represents another prominent mechanism through which NPs induce toxicity. Increased ROS levels after exposure to various NPs were reported by Duan et al. (2016), Feng et al. (2022), Gu et al. (2021), and Sun et al. (2021). Jia et al. (2019) reviewed several studies that pointed to the role of oxidative stress in NP-induced teratogenicity. CuONPs were found to induce ROS-mediated teratogenicity, leading to cell apoptosis and embryo deformities, while ZnO-NPs caused delayed hatching, increased malformations, and disruptions in neuronal and vascular development. These toxic effects were driven by oxidative stress, DNA damage, and altered enzymatic activities, including those of superoxide dismutase (SOD) and catalase (CAT).

Behavioral changes in zebrafish exposed to NPs have been widely reported. Khan et al. (2024) assessed neurobehavioral changes, showing that Ln-UCNP exposure significantly reduced locomotor response at both low and high concentrations (1 and 100 mg/L). Decreased activity during dark cycles in a light-dark photomotor test was observed, indicating impaired neurobehavioral functioning. These findings align with results from Gu et al. (2021), where zebrafish exposed to TiO_2-NPs exhibited reduced swimming rates and other behavioral alterations. Jia et al. (2019) also discussed the neurotoxic effects of ZnO and Ag-NPs, with these materials causing disruptions in locomotion and normal behavioral development.

In terms of cardiovascular effects, bradycardia was a common outcome of NP exposure for early-life stages of zebrafish (Pitt et al. 2018a, b; Feng et al. 2022; Duan et al. 2016). Similarly, Khan et al. (2024) found a significant reduction in heart rate and blood flow in zebrafish embryos exposed to 100 mg/L Ln-UC-NPs. This effect, however, was not observed at lower concentrations. Jia et al. (2019) also noted that TiO_2-NPs and Ag-NPs could cause cardiac abnormalities, supporting the idea that cardiovascular and developmental toxicity is common across different NP types.

In conclusion, embryonic development in zebrafish is impacted in multiple ways by exposure to various NPs. Delayed and decreased hatching was observed by Feng et al. (2022), with Khan et al. (2024) similarly reporting significant delays in hatching at 1 and 10 mg/L UC-NPs, along with reduced survival at 100 mg/L. Jia et al. (2019) emphasized the teratogenicity of ZnO-NPs, which caused delayed hatching, reduced survival, and malformations. They also pointed to the long-term toxicity of TiO_2-NPs, which, while relatively inert in short-term exposures, showed significant effects over time, including reproductive impairment and vascular toxicity.

5.2 Toxic Effects of Nanoparticles in Adult Zebrafish

NP exposure has been shown to affect various systems in adult zebrafish, making them a valuable model for studying NP-induced toxicity. This section focuses on the major toxicological endpoints observed in adult zebrafish, including immunotoxicity, hepatotoxicity, reproductive toxicity, neurotoxicity, oxidative stress, and genotoxicity.

5.2.1 Immunotoxicity

Immunotoxicity has been observed in adult zebrafish following NP exposure. The immune system of zebrafish shares various similarities with higher vertebrates, such as the presence of a thymoid structure and head kidney functioning similarly to human bone marrow. However, zebrafish lack lymph nodes, with the spleen, liver, and intestine compensating as secondary immune organs (Bai and Tang 2023).

Ag-NPs have been shown to induce oxidative stress and immunotoxicity in adult zebrafish, with downregulation of immune-related genes such as TLR4, IL1B, CEBP, TRF and TLR22 in adult zebrafish (Krishnaraj et al. 2016). PS-NPs exposure upregulated immune system-related genes like IL1A and IL1B, contributing to inflammation and immune dysregulation (Bhagat et al. 2020). Research on TiO_2-NPs has revealed broader systemic effects, including impacts on the gut microbiota, which may weaken immune defenses and increase susceptibility to bacterial infections (D'amora et al. 2022). Besides, a chronic co-exposure to TiO_2-NPs and bisphenol A demonstrated increased bacterial load and inflammation in a dose-dependent manner, further compromising immune function (Chen et al. 2018). A study by Kuang et al. (2023) shows that co-exposure to ammonia and TiO_2-NPs in zebrafish results in significant immune system impacts. Both substances cause damage to intestinal cells, including cell vacuolation, lymphocyte infiltration, and goblet cell hyperplasia. Additionally, the combination of ammonia and TiO_2-NPs activates inflammatory pathways (TLR4/5-mediated MyD88-dependent), leading to intestinal inflammation. Exposure to copper II oxide NPs (CuO-NPs) affect the intestinal microbiota, particularly microbes involved in the metabolism of short-chain fatty acids (SCFAs) and lipopolysaccharides (LPS), which play critical roles in immune regulation (Xu et al. 2024). Peng et al. (2023) show that oral exposure to graphene oxide nanoparticles (GO-NP) in adult zebrafish modulated the composition of the gut microbiome in adult zebrafish, which triggered the induction of ILC2-like cells with regulatory attributes, showing how nanomaterials can influence the crosstalk between the microbiome and immune system.

5.2.2 Hepatotoxicity

As in other vertebrates, the liver in zebrafish is an essential organ involved in metabolism, detoxification, and maintaining homeostasis, making it a primary objective for toxicological studies. It was shown that nanoparticles can induce hepatotoxic effects through mechanisms such as oxidative stress, inflammation, apoptosis, and metabolic disturbances. For instance, Choi et al. (2010) observed significant oxidative stress and apoptosis in the liver of adult zebrafish after exposure to Ag-NPs. Specifically, malondialdehyde levels, a marker of lipid peroxidation, and elevated total glutathione content, suggesting elevated oxidative stress were increased. This was accompanied by reduced expression of antioxidant enzymes, such as catalase and glutathione peroxidase. The histological examination of liver tissues showed disrupted hepatic cell cords and signs of apoptosis. Similarly, Krishnaraj et al. (2016) investigated Ag-NPs, reporting oxidative stress and genotoxic effects in the liver. Ag-NPs were found to induce cytological changes, including irregularly shaped parenchymal cells and loss of cellular contact. Szudrowicz et al. (2022) found that Ag-NPs coated by PvP (polyvinylpyrrolidone) caused proliferation of hepatocytes in combination with a reduction of the expression levels of the *cat*, *gpx1a*, and *sod1* genes. Additionally, it was demonstrated that various Ag-NPs, particularly silver nanospheres, disrupted lipid levels and caused oxidative stress in the liver. Key lipids such as phosphatidylcholine, phosphatidylethanolamine, phosphatidylinositol, and phosphatidylserine were differentially affected depending on the morphology of the NPs (Xie et al. 2024). Zheng et al. (2024) showed Ag-NPs induce liver inflammation through ferrotopsis in zebrafish. Ferroptosis is an iron-dependent cell death characterized by increased ROS and associated with differentially regulated genes, such as *arrdc3*, *txnip* and *egfr* under Ag-NPs exposure. Furthermore, ferroptosis is linked to glucose metabolism diseases. Chronic exposure (28 days) to AgNPs in adult zebrafish, both alone and in combination with ZnO-NPs, caused significant liver damage, primarily in the form of necrosis (Mahjoubian et al. 2023). In the same study it was discovered that ZnO-NPs reduced acute toxicity of Ag-NPs; however, they enhanced liver toxicity during chronic exposure by increasing Ag-NP bioaccumulation and worsening tissue damage. Wang et al. (2022b) emphasizes that while both sexes of zebrafish experience comparable bioenergetic impairments, measured as decreased activities of mitochondrial Complex I and V, due to ZnO-NPs exposure, their metabolic responses in terms of fatty acid and amino acid metabolism differ significantly.

Chen et al. (2011) explored the chronic toxicity of TiO_2-NPs and their accumulation in zebrafish organs, including the liver. While the study primarily highlighted the accumulation of TiO_2-NPs in the liver without pronounced histopathological changes, there was a reduction in liver weight over time. This suggests potential sublethal effects, particularly considering the long-term ecological risk posed by TiO_2-NPs. Mahjoubian et al. (2024) compared the toxicity of TiO_2 and Sn-doped TiO_2-NPs in zebrafish after acute and chronic exposure; overall, Sn-doped TiO_2-NPs induced greater liver toxicity than pure TiO_2-NPs, characterized by oxidative stress, tissue damage, genotoxicity, and apoptotic changes. In a study on GO-NPs, Chen et al. (2016) demonstrated that while acute liver toxicity was not evident, prolonged exposure led to oxidative stress and cellular damage. Increased malondialdehyde levels, superoxide dismutase and catalase activities, as well as a reduced glutathione content indicated oxidative stress. Histologically, the liver displayed vacuolation, cell disintegration, and loose cellular arrangements. Correspondingly, Souza et al. (2017) demonstrated that that sub-lethal doses of GO-NPs can lead to significant liver damage in zebrafish, including non-uniform hepatocyte shape, pyknotic nuclei, vacuole formation, and cell rupture and necrosis. Lu et al. (2016) examined the effects of 70 nm PS-NPs on zebrafish liver, revealing inflammation and lipid accumulation. Both oxidative stress and metabolic disruptions were observed, as shown by increased activity of superoxide dismutase and catalase and alterations in lipid and energy metabolism. Findings by Rehman et al. (2024) that PS-NP exposure, both low (0.1 mg/L) and high (10 mg/L) concentrations, leads to severe liver damage, including oxidative stress, inflammation, and fibrosis, with impacts on lipid metabolism and liver tissue health.

5.2.3 Reproductive Toxicity

Reproductive toxicity from NP exposure has been extensively studied in zebrafish. The mechanisms include disrupted oogenesis, hormonal imbalances, oxidative stress, apoptosis, and inflammatory responses. These impacts are not only limited to the directly exposed generation but can also extend to their offspring, affecting overall fertility, fecundity, and reproductive health.

Exposure to Ag-NPs caused oxidative stress and apoptosis in germ cells through a mitochondrial-dependent pathway, ultimately impairing reproductive abilities (Ma et al. 2018). Orbea et al. (2017) investigated the effects of polyvinylpyrrolidone/polyethylenimine-coated Ag-NPs (~5 nm) on zebrafish reproduction and embryo development. Waterborne exposure of adult zebrafish to these Ag-NPs at environmentally relevant concentrations (100 ng Ag/L) led to a significant reduction in fecundity by the second week of treatment. Similarly, Au-NPs between 10 and 50 nm in size were shown to enter the ovaries of female zebrafish, causing DNA strand breaks in ovarian cells (Dayal et al. 2016). Another study demonstrated that TiO_2-NPs exposure induced autophagy and necrosis in testicular cells, particularly Sertoli cells, which negatively impacted spermatogenesis (Kotil et al. 2017). Additionally, chronic exposure to TiO_2-NPs has been linked to significant reproductive toxicity. A 91-day exposure led to reduced egg production and increased embryo mortality. Histological analysis showed disruptions in folliculogenesis, while gene expression changes affected over 3000 ovarian genes after chronic exposure to TiO_2-NP (Wang et al. 2011). Structural damage to ovarian follicles and impaired spermatogenesis further emphasizes the reproductive risks associated with TiO_2-NP exposure, leading to alterations in population dynamics and aquatic ecosystem balance.

In a study by Mawed et al. (2022), exposure to ZnO-NPs resulted in cytotoxicity in maturing oocytes of female zebrafish, triggered by oxidative stress, autophagy, and apoptosis. ROS generation elevated the mutated ovarian tP53 protein, causing follicular developmental retardation and necroptosis, which mimicked features of apoptosis and necrosis. This led to altered ovulation and a reduction in fecundity, underlining the reproductive risks of ZnO-NP exposure. Rashidian et al. (2023) investigated the chronic exposure to small-sized SiO_2-NPs in adult zebrafish, revealing decreases in gonad weight and absolute fecundity. High concentrations of SiO_2-NPs led to a decline in larval survival rates and triggered a variety of biochemical changes, including altered cholesterol, cortisol, and glucose levels. These findings suggest that SiO_2-NP exposure has significant toxic effects on reproductive performance, reducing fertility and larval viability. Wu et al. (2023) examined the effects of low-dose CuO-NPs exposure on both female and male zebrafish over 20 days. The study revealed that CuO-NPs caused structural damage to ovaries and testes, alongside hormonal imbalances. Both male and female zebrafish exhibited increased levels of 17β-estradiol and decreased testosterone levels, indicating significant disruptions in sex hormone regulation. The expression of various genes along the hypothalamic-pituitary-gonadal (HPG) axis was affected, including up-regulation of *erα/er2β* and *cyp19a* in females, and *erα/er2β*, *lhr*, *hmgra/hmgrb*, *3βhsd*, and *17βhsd* in males. Zhang et al. (2024b) investigated the impact of PS-NPs on female zebrafish and their offspring. PS-NP exposure disrupted the HPG axis, leading to a reduction in gonadal 17-estradiol and an increase in testosterone, which in turn reduced the number of mature oocytes. The effects extended to the next generation, as maternal PS-NP exposure caused reproductive dysfunctions in both directly exposed zebrafish and their offspring, impairing fertility and disrupting endocrine function.

Various studies have also investigated the effects of NPs when co-exposed to other chemicals that are known to cause adverse effects in zebrafish. Kandaswamy et al. (2024) showed that co-exposure to PS-NPs and the pharmaceutical diclofenac (DCF) resulted in reduced hatching rates and increased larval mortality. The findings indicated that PS-NPs, in combination with DCF, amplified oxidative stress and apoptosis, leading to significant reproductive toxicity in zebrafish. Liu et al. (2021) investigated the combined effects of Si-NPs and cadmium chloride ($CdCl_2$) on zebrafish ovaries. The study revealed that co-exposure to Si-NPs (25 µg/mL) and $CdCl_2$ (1 µmol/L) over 30 days led to significant structural and functional alterations in the ovaries, including a reduced number of mature oocytes and malformation of offspring. The combination of pollutants amplified the adverse effects on lipid metabolism, estrogen metabolism, and oocyte maturation processes, highlighting the synergistic reproductive toxicity of these two pollutants. These findings provide critical insights into the potential environmental hazards of co-exposure to Si-NPs and $CdCl_2$ in aquatic ecosystems. Xian et al. (2024) studied co-exposure to PS-NPs and 4-MBC (4-methylbenzylidene camphor), which intensified the accumulation of 4-MBC in zebrafish tissues. This led to sex-specific reproductive impairments, including disrupted vitellogenesis and oocyte maturation in females and spermatogenesis disruption in males.

The numerous studies demonstrate that NPs impair reproductive health through pathways such as oxidative stress, apoptosis, and hormonal disruption. The results indicate that these adverse effects not only compromise the reproductive capacity of directly exposed zebrafish but also extend to subsequent generations, leading to diminished fertility and developmental anomalies in offspring. Additional studies indicate that co-exposure scenarios further amplify the risks associated with NPs, revealing how interactions with environmental pollutants, such as pharmaceuticals and heavy metals, exacerbate reproductive dysfunction.

5.2.4 Neurotoxicity

Neurotoxicity is generally characterized by disruptions in neural tissues resulting in significant irregular activity of the nervous system. Zebrafish have been proven to be a sensitive model for the investigation of NP-induced neurotoxicity due to their well-differentiated brain structures and high genetic and anatomical homology to mammals. The zebrafish nervous system shares developmental processes, including neurogenesis and axon guidance, with mammals, making it a suitable model for investigating neurotoxic effects on behavior and brain function (Zhao et al. 2022). Moreover, zebrafish possess distinct brain subdivisions, telencephalon, diencephalon, mesencephalon, and rhombencephalon, analogous to those in mammals.

As described in Sect. 4.4, one critical concern with NPs is their ability to cross the BBB, enabling them to directly interact with neural tissues. PS-NPs, for instance, can penetrate the BBB, causing neurotoxic effects such as seizure-like behavior and reduced swimming ability in adult zebrafish (summarized in Bhagat et al. 2020). Similar neurotoxic effects have been noted with other NP types, including titanium dioxide TiO_2-NPs, which are linked to oxidative stress in the brain. While the exact mechanisms remain unclear, TiO_2 exposure can impair spatial recognition and induce dopaminergic neuronal loss, indicating a potential risk for neurodegenerative conditions like Parkinson's disease (Zhao et al. 2022). Further, TiO_2-NPs have been reported to activate expressions of key neurotoxic genes, such as *bdnf c-fos* and *c-jun*, while suppressing others like *p38*, *ngf*, and *cre*, leading to brain damage in zebrafish (Sheng et al. 2016).

Several classes of NPs, including metallic, carbon-based, and silica-based, have demonstrated neurotoxic effects in zebrafish. For example, Ag-NPs induce size-dependent neurotoxic effects: 10 nm particles cause reduced locomotion, while 50 nm particles result in hyperactivity (Powers et al. 2011). Similarly, Au-NPs have been found to elevate acetylcholinesterase activity, affecting neurotransmission and leading to behavioral disturbances (Dedeh et al. 2015). Acetylcholinesterase (AChE) activity, a key regulator of brain function, is an important biomarker for neurotoxicity. Studies have shown a decrease in AChE activity in zebrafish exposed to NPs (Sarasamma et al. 2020; Chen et al. 2017a). Inhibition of neurotransmitters such as dopamine, melatonin, GABA, serotonin, and oxytocin has also been reported in zebrafish exposed to PS-NPs, along with upregulation of α-1 tubulin, myelin basic protein (mbp) mRNA, mesencephalic astrocyte-derived neurotrophic factor (manf) mRNA, and dopamine content, all of which play crucial roles in neurological processes (Sarasamma et al. 2020).

Oxidative stress-induced apoptosis is a common response to NP exposure, observable through methods such as acridine orange staining (Parng et al. 2007). Moreover, transcriptomics, proteomics, and metabolomics technologies have advanced our understanding of how NPs affect molecular regulation in the brain (Bai and Tang 2023). In terms of behavioral alterations, NPs have been shown to affect zebrafish locomotor activity, sensory responses, and complex behaviors such as anxiety and learning. Locomotor changes occur at various levels of the nervous system, from neuromuscular junctions to higher brain centers, and are influenced by factors such as NP size and coating (Zhao et al. 2022). Other behavioral assays, such as the novel tank test, have been employed to assess anxiety-like behavior, revealing the nuanced effects of neurotoxicants on adult zebrafish behavior (Egan et al. 2009).

Inorganic NPs such as silica (SiO_2) and graphene oxide (GO) have also demonstrated neurotoxic effects in zebrafish. SiO_2-NPs can induce apoptosis in the central nervous system and disrupt locomotion, despite not causing overt developmental defects (Pham et al. 2016). Ag-NPs, due to their widespread use in medical and consumer products, have been extensively studied for their neurotoxic effects. Exposure to TiO_2-NPs similarly results in impaired motor neuron development, reduced swimming speed, and oxidative stress in the brain, contributing to neurodegeneration in different life stages of zebrafish (Gu et al. 2021; Hu et al. 2017). In adult zebrafish, neurotoxic effects of NPs manifest as altered social behaviors, anxiety, aggression, and learning deficits. For example, exposure to $AgNO_3$ impairs social recognition and memory without affecting anxiety or aggression, suggesting specific neural circuit dysfunction (Fu et al. 2021). In zebrafish, assays that measure learning and memory behaviors, including habituation and sensitization tests, provide further insight into the cognitive impairments that might be caused by NP exposure (Zhao et al. 2022). Jin et al. 2023 demonstrated that zeolitic imidazolate framework-8 NPs (ZIF-8 NPs) at a concentration of 90 mg/L could significantly decrease the locomotor activity (i.e., hypoactivity) of both sexes. After a ball falling stimulation, zebrafish exposed to ZIF-8 NPs (9.0 and 90 mg/L) exhibited more freezing states (i.e., temporary cessations of movement), and males were more sensitive than females.

Studies by Anila et al. (2021) also highlight the toxicity of less commonly studied NPs, such as palladium NPs (Pd-NPs). Pd-NPs, widely used as catalytic promoters, have raised concerns about their environmental impact due to their release into aquatic ecosystems. Recent research reports concentration-specific toxicity in zebrafish larvae and adults, with high doses of Pd-NPs influencing hatching rates, embryo survival, and motor functions. Long-term exposure to Pd-NPs results in erratic swimming patterns, inhibited AChE activity, and significant oxidative stress, as evidenced by changes in antioxidant enzyme activity and lipid peroxidation (Anila et al. 2021).

Zebrafish offer a powerful model for evaluating the neurotoxicity of nanoparticles. With their conserved neuroanatomy and behavior, they allow for the investigation of complex neural and behavioral endpoints. A combination of molecular, genetic, and behavioral assessments provides a comprehensive toolkit for understanding how different NPs affect neural function and behavior, offering valuable insights for NP safety assessments.

5.2.5 Oxidative Stress and Genotoxicity

Oxidative stress is a key mechanism through which NPs exert their toxic effects in zebrafish. This stress is primarily driven by the overproduction of ROS, which overwhelms the antioxidant defense system, and finally leading to cellular damage. Several types of NPs, including metal-based, metal-organic frameworks, and carbon-based particles, have been shown to induce oxidative stress in various tissues of adult zebrafish, with particularly severe effects on vital organs such as the liver, brain, gills, and muscles.

Exposure to zinc oxide NPs (ZnO-NPs) significantly elevates hydroxyl radical levels, particularly in the liver, leading to lipid peroxidation and oxidative damage as marked by increased malondialdehyde (MDA) content. This has been corroborated by Al-Zahaby et al. (2023), who found that long-term exposure to ZnO-NPs induces severe histological and genetic damage in the olfactory organs, demonstrating that these particles can impair sensory functions and induce apoptosis through ROS generation. Similarly, iron oxide NPs (IO-NPs) at higher concentrations caused oxidative stress in the liver of zebrafish, as evidenced by differential expression of nearly 1000 transcripts related to ion binding and oxidative stress pathways (Villacis et al. 2017). A significant reduction in the activities of key antioxidant enzymes such as catalase (CAT), glutathione S-transferase (GST), and superoxide dismutase (SOD) was observed, rendering the liver more vulnerable to oxidative damage. TiO_2-NPs further exacerbate this by decreasing antioxidant enzyme activities in both the liver and gills, underscoring the sensitivity of zebrafish to NP exposure. Tang et al. (2019) highlighted that the depletion of these critical enzymes diminishes the organism's ability to neutralize ROS, resulting in pronounced cellular and tissue damage.

Metal-organic framework NPs, such as Zeolitic imidazolate framework-8 (ZIF-8-NPs), have emerged as potential environmental and biomedical agents. However, their toxicological impact, especially in the brain, is concerning. Jin et al. (2023) demonstrated that exposure to ZIF-8-NPs significantly impaired antioxidant enzyme activity in the brain of adult zebrafish. A decrease in SOD, CAT, and GST activities suggests that the brain is highly susceptible to oxidative stress, which can lead to neurobehavioral toxicity. Ag-NPs, commonly used in various consumer products, were shown by to release free Ag + ions post-exposure, causing significant oxidative stress in the liver (Choi et al. 2010). The elevated malondialdehyde (MDA) levels, combined with suppressed antioxidant enzyme activities and increased DNA damage markers (e.g., p53 protein and Bax gene expression), point to a severe apoptotic response induced by ROS accumulation. The chronic oxidative stress also led to significant histological changes, including apoptotic alterations in liver tissues (Choi et al. 2010). Carbon-based NPs like C70 fullerenes (C70-NPs) have been shown to cause oxidative stress, particularly in the brain and muscle tissues of zebrafish. Sarasamma et al. (2019) demonstrated that exposure to C70-NPs significantly increased ROS levels in these tissues, leading to DNA damage and upregulation of inflammatory markers such as TNF-α and IL-1β. Additionally, catalase and MDA levels were significantly altered, indicating that chronic exposure to carbon-based NPs can compromise both antioxidant defense and tissue integrity. Other types of NPs, such as SiO_2-NPs, have also been shown to disrupt antioxidant defenses. Rashidian et al. (2023) reported that long-term exposure to SiO_2-NPs led to significant increases in oxidative stress biomarkers like catalase and SOD in multiple tissues, including the liver and gonads. This was accompanied by an adverse impact on reproductive performance, indicating that oxidative stress extends its harmful effects beyond somatic cells, affecting reproductive health and offspring survival. Finally, PS-NPs in combination with pharmaceuticals, such as diclofenac (DCF), demonstrated a synergistic toxic effect in zebrafish. Co-exposure to PS-NPs and DCF exacerbated oxidative stress and histopathological alterations in the intestinal region, as shown by Kandaswamy et al. (2024). The marked impairment of antioxidant enzymes like SOD and CAT, along with increased expression of pro-inflammatory markers (TNF-α and IL-1β), reveals that NPs can potentiate the toxic effects of other environmental contaminants (Kandaswamy et al. 2024).

In summary, oxidative stress plays a central role in NP-induced toxicity in adult zebrafish. The imbalance between ROS production and antioxidant defenses leads to extensive cellular damage across multiple organs, with potential long-term effects such as apoptosis, genotoxicity, and impaired organ function. Further research is needed to fully elucidate the mechanisms of NP-induced oxidative stress and its ecological and biological ramifications.

5.2.6 Toxicity on the Gills

As in other aquatic organisms, the gills of zebrafish are essential for gas exchange, ion and water transfer, ammonia excretion, and osmoregulation. In adult zebrafish, gills also represent a primary exposure pathway for NP contamination. A growing body of research has demonstrated that various NPs can cause significant toxicity to gill tissues, leading to structural damage and oxidative stress.

Graphene oxide (GO) nanoparticles have been shown to induce a range of histopathological changes in zebrafish gills. In a study by Souza et al. (2017), chronic exposure to GO resulted in dose-dependent morphological alterations of the gills. Concentrations between 10 and 20 mg/L caused severe injuries, including swollen mucocytes, epithelial lifting, and aneurysms, as well as extensive necrosis. Cu-NPs have also been identified as highly toxic to zebrafish gills. According to Griffitt et al. (2007), exposure to Cu-NPs caused significant damage to the gill lamellae, characterized by epithelial cell proliferation and edema in both primary and secondary gill filaments. The extent of damage was dose-dependent, with higher concentrations of Cu-NPs causing greater epithelial hypertrophy. Notably, Cu-NPs was found to be 40% more toxic than dissolved copper at equivalent concentrations. TiO_2-NPs are commonly used in consumer products, and their impact on zebrafish gills has been study in various research projects. Tang et al. (2019) demonstrated that long-term exposure to TiO_2-NPs caused oxidative damage in the gills, with significant reductions in the activity of key antioxidant enzymes, such as superoxide dismutase (SOD), catalase (CAT), and glutathione S-transferase (GST). In the high concentration exposure group (100 mg/L), enzyme activity levels dropped to less than half of normal control levels, indicating severe oxidative stress. These results suggest that prolonged exposure to TiO_2-NPs can compromise the gills' ability to counter oxidative damage, potentially leading to tissue degradation and impaired functionality. Further research by Mahjoubian et al. (2021) demonstrated that silver-doped TiO_2-NPs (Ag-TiO_2-NPs) were even more toxic than pure TiO_2-NPs. Ag-TiO_2-NPs induced severe histological lesions, higher oxidative stress markers, and increased apoptosis in zebrafish gills. Gene expression analysis revealed that Ag-doping significantly upregulated stress-related genes, such as Hsp70 and Bax, while downregulating protective genes like Bcl-2. This suggests that silver doping can amplify the toxic effects of TiO_2-NPs, leading to more pronounced damage to gill tissues.

The toxicity of Ag-NPs varies based on their size and morphology. Osborne et al. (2015) demonstrated that smaller 20 nm Ag-NPs caused significantly more damage to zebrafish gills compared to larger 110 nm particles. Gills exposed to the 20 nm Ag-NPs exhibited higher silver content and more severe ultrastructural damage, particularly in the basolateral membranes, where silver disrupted the Na^+/K^+ ion channel. This was evidenced by reduced ATPase activity, indicating a critical functional disruption. In contrast, the larger particles caused less damage and were less efficiently deposited in the gill tissue, highlighting the importance of nanoparticle size in determining toxicity. Similarly, Xie et al. (2024) found that the morphology of nanosilver also plays a key role in its toxicity. In their study, silver nanospheres (Ag-NSs) exhibited the highest toxicity toward adult zebrafish gills, leading to lipid profile disruptions, oxidative stress, and inflammation. Silver nanoflakes (Ag-NFs) were more toxic than silver nanowires (Ag-NWs), but less so than Ag-NSs.

Pd-NPs have also been studied for their toxic effects on zebrafish gills. Anila et al. (2021) found that exposure to Pd-NPs at concentrations as low as 22 ng/L induced oxidative stress, apoptosis, and structural disruption in gill tissues. ROS generation was a primary mechanism of damage, leading to oxidative stress and cellular apoptosis in gill tissue. C60 fullerenes (C60-NPs) have been shown to induce both behavioral and physiological changes in zebrafish. Sarasamma et al. (2018) reported that exposure to low concentrations of C60 NPs not only reduced locomotion and altered circadian rhythm behaviors but also led to significant oxidative stress and DNA damage in gill tissues. The biochemical data demonstrated reduced levels of antioxidant enzymes, along with increased markers of hypoxia, inflammation, and stress in the gills. These findings indicate that C60 NPs can disrupt the normal function of gill tissue, even at relatively low exposure levels.

In conclusion, the various studies summarized here demonstrate that different NPs can cause significant damage to the gills of adult zebrafish. The common mechanisms of toxicity include histological alterations, oxidative stress, apoptosis, necrosis, and impaired antioxidant defenses. Given the central role of gills in zebrafish physiology, the observed damage emphasizes the potential ecological risks posed by nanoparticle contamination in aquatic environments.

5.2.7 Effects on GIT and the Microbiome

As described in Sect. 4.3, the gastrointestinal tract (GIT) of zebrafish plays a crucial role in the uptake and transport of nutrients, but also serves as a major uptake route for environmental pollutants such as NPs. Therefore, the exposure of GIT tissues to NPs can result in significant physiological disruptions. In addition to direct tissue alterations, it is known that NPs can influence the gut microbiome, which is densely populated with a diverse community of microorganisms that plays a fundamental role for the health of an organism, including zebrafish. The gut microbiome is involved in numerous critical biological processes, including nutrient metabolism, immune function, and energy regulation (Zhang et al. 2024a). Disruptions of the microbial ecosystem, often referred to as dysbiosis, have been linked to various diseases, including metabolic disorders such as obesity and diabetes (Jia et al. 2021). Given the role of gut microbiota in maintaining homeostasis, the potential for NPs to alter microbial composition is a major concern. Therefore, research has increasingly highlighted the sensitivity of the gut microbiota to environmental pollutants, with nanoparticles being of particular concern (Evariste et al. 2019). The potential for NPs to induce oxidative stress and immune dysfunction in zebrafish intestines through mechanisms such as ROS production, disruption of ion channels, and alteration of enzyme activities has been documented in several studies (compare Sects. 5.2.1 and 5.2.5). Furthermore, exposure to NPs may result in long-term changes to gut microbial communities, potentially leading to chronic inflammation and other health impairments.

For instance, Tang et al. (2019) demonstrated that adult zebrafish exposed to TiO_2-NPs showed notable accumulation of NPs in the digestive tract. While the fish efficiently excreted most of the NPs, there was evidence of oxidative stress in intestinal tissues, with a particular increase in ROS under high exposure conditions. This suggests that NPs can trigger oxidative damage, affecting biological redox systems and disrupting cellular function, even in tissues where they do not bioaccumulate over time.

Similar findings have been reported for Ag-NPs. Osborne et al. (2015) explored the size-dependent toxicity of 20 nm and 110 nm citrate-coated Ag-NPs in adult zebrafish, focusing on their effects on the gills and intestines. The study revealed that smaller 20 nm particles were retained in the intestinal tissue even after a depuration period, leading to significant histopathological changes and impaired Na^+/K^+ ion channel function. Bao et al. (2020) explored sex-dependent effects of Ag-NP exposure on adult zebrafish. They exposed male and female zebrafish to two sizes of uncoated Ag-NPs (20 nm and 80 nm) and found significant sex differences in response to the NPs. Males exhibited greater sensitivity to Ag-NP toxicity than females, with 20 nm Ag-NPs showing higher toxicity than the larger 80 nm particles. Specifically, intestinal Na^+/K^+-ATPase activity and superoxide dismutase levels were significantly altered in male zebrafish, highlighting both size-related and sex-specific toxicity. Interestingly, their study further revealed that the intestines were more severely impacted by Ag-NP exposure than the liver. Lacave et al. (2018) showed after long-term exposure that Ag NPs and ionic silver can accumulate in the liver and intestines of zebrafish, with inflammatory and histopathological damage persisting for up to 6 months post-exposure. Transcriptomic analysis further revealed that Ag-NPs provoke more extensive immune and DNA damage-related responses compared to ionic silver, emphasizing the distinctive hazards posed by the nanoparticle form. Together, this demonstrates the critical role of NP size in determining their toxicological and physiological effects, particularly in tissues like the intestines, where they can interact with vital ion-regulating mechanisms. Research on other metal-based NP types, such as metal-organic frameworks (MOF-NPs) (Yang et al. 2024) and lead-halide perovskite NPs (Patsiou et al. 2020), has revealed that these materials can induce significant disruptions to the gut microbiome and intestinal health in zebrafish, often through oxidative stress and inflammation. Alterations in the gut microbial community can result in long-term health consequences, as seen in studies where persistent inflammatory markers and shifts in microbial composition were observed long after initial exposure.

Recent research has also addressed the interactions between NPs and other environmental contaminants, which may exacerbate their toxic effects. Liu et al. (2024) demonstrated that the co-exposure of adult zebrafish to PS-NPs and 2,4-di-tert-butylphenol (2,4-DTBP) altered intestinal histopathology and microbiota composition in a sex-dependent manner, with male zebrafish showing aggravated inflammation and disrupted microbial balance.

In conclusion, the GIT of zebrafish serves as a key entry point for NPs, making it particularly vulnerable to their toxic effects. Research highlights that NPs not only induce oxidative stress and disrupt ion-regulating mechanisms in intestinal tissues but also alter gut microbiota composition, leading to potential long-term health impairments. Studies on Ag-NPs indicate size- and sex-dependent differences in toxicity, with smaller particles and male zebrafish showing higher sensitivity. Moreover, co-exposure to NPs and other environmental pollutants impairs the toxic impact of NPs, further impairing intestinal health and microbial balance.

6 Conclusion

In the field of nanotoxicology, zebrafish have become a widely used model to assess the toxicity of various NPs. Studies on zebrafish employ a broad range of methods, reflecting the versatility of this model in addressing different aspects of NP toxicity. These methods include both acute and chronic exposure experiments across different life stages, from embryos to adult fish. Techniques such as morphological analysis, behavioral assays, histopathology, and molecular biology approaches allow to investigate how NPs affect zebrafish at both cellular and organismal levels. Real-time visualization of NP uptake and distribution within the transparent embryos of zebrafish further enhances the ability to monitor NP bioaccumulation and its effects over time. Besides, various types of NPs are studied in these contexts, including metallic NPs such as Ag-NPs, Au-NPs, and TiO_2-NPs, as well as carbon-based materials like carbon nanotubes and polymeric NPs such as PS-NPs. Each type of these NPs comes with its own set of characteristics, such as size, shape, surface charge, and coating, which directly influence its behavior, bioavailability, and toxicity in zebrafish. This diversity in NP types introduces considerable complexity into the investigations of their toxic effects.

Toxicological endpoints assessed in zebrafish vary widely depending on the type of study. Commonly investigated effects include developmental and teratogenic impacts, such as delayed hatching and malformations, alongside other systemic toxicities like immunotoxicity, hepatotoxicity, neurotoxicity, and reproductive toxicity. Additional endpoints include the study of oxidative stress and genotoxicity, where NPs induce DNA damage or interfere with normal cellular processes, often resulting in long-term effects. The various studies also investigate how NPs disrupt critical physiological systems, including the gut microbiome, where they may alter microbial communities and influence immune system responses. However, comparing the outcomes of different studies is not without significant challenges. One of the primary issues lies in the variability of experimental designs. Differences in the NPs used, including their size, coating, surface charge, and shape, play a crucial role in how they interact with biological systems. For instance, smaller NPs tend to penetrate biological barriers more easily than larger particles, leading to greater bioaccumulation and, potentially, more pronounced toxic effects. Another challenge stems from the variety of exposure routes used in these studies. Zebrafish are often exposed to NPs via waterborne exposure or dietary intake, each affecting the absorption and distribution of NPs within the organism. Waterborne exposure may primarily affect external tissues such as the gills, whereas dietary exposure could lead to accumulation in internal organs like the gastrointestinal tract. This variation in exposure methods can result in significant differences in observed toxic effects, even when similar NPs are used. Moreover, environmental conditions during NP exposure, such as water temperature, pH, and ionic strength, can influence the behavior of NPs, affecting their aggregation and bioavailability. Even slight changes in these parameters can result in NPs behaving differently in aquatic environments, complicating attempts to compare results across studies that do not standardize such conditions. The life stage of zebrafish used in the studies also introduces complexity. Embryos and larvae are far more sensitive to NP exposure than adults due to their undeveloped protective barriers, such as the blood-brain barrier. Consequently, studies focusing on early life stages may report more severe effects, while adult zebrafish studies might show lower toxicity, making direct comparisons across developmental stages difficult. Finally, the selection of toxicological endpoints varies widely between studies. Some investigations focus on molecular and genetic changes, while others emphasize physiological or behavioral effects. This variability in endpoint selection, coupled with differences in NP doses and exposure durations, makes it challenging to draw clear, consistent conclusions across different studies. Higher doses of NPs, often used to observe more immediate toxic effects, may not reflect environmentally relevant exposures, further limiting the ecological applicability of some findings.

In conclusion, zebrafish provide an invaluable model for studying NP toxicity; however, the diversity of experimental designs, NP properties, and toxicological endpoints across studies poses significant challenges for comparison. Standardizing testing protocols and methods, including NP characterization, exposure conditions, and endpoint selection, will be essential to improving the comparability of results. Despite these challenges, zebrafish remain an effective tool for advancing our understanding of NP safety, particularly in the context of environmental and human health risk assessments.

References

Abbott NJ, Rönnbäck L, Hansson E (2006) Astrocyte–endothelial interactions at the blood–brain barrier. Nat Rev Neurosci 7:41–53

Al-Zahaby SA, Farag MR, Alagawany M, Taha HSA, Varoni MV, Crescenzo G, Mawed SA (2023) Zinc oxide nanoparticles (ZnO-NPs) induce cytotoxicity in the zebrafish olfactory organs via activating oxidative stress and apoptosis at the ultrastructure and genetic levels. Animals (Basel) 13

Almutairi MMA, Gong C, Xu YG, Chang Y, Shi H (2016) Factors controlling permeability of the blood–brain barrier. Cell Mol Life Sci 73:57–77

Anila PA, Sutha J, Nataraj D, Ramesh M (2021) In vivo evaluation of Nano-palladium toxicity on larval stages and adult of zebrafish (Danio rerio). Sci Total Environ 765:144268

Asharani P, Wu YL, Gong Z, Valiyaveettil S (2008) Toxicity of silver nanoparticles in zebrafish models. Nanotechnology 19:255102

Asharani PV, Lianwu Y, Gong Z, Valiyaveettil S (2011) Comparison of the toxicity of silver, gold and platinum nanoparticles in developing zebrafish embryos. Nanotoxicology 5:43–54

Atherton JA, McCormick MI (2017) Kin recognition in embryonic damselfishes. Oikos 126:1062–1069

Bai C, Tang M (2023) Progress on the toxicity of quantum dots to model organism-zebrafish. J Appl Toxicol 43:89–106

Baker TJ, Tyler CR, Galloway TS (2014) Impacts of metal and metal oxide nanoparticles on marine organisms. Environ Pollut 186:257–271

Bao S, Tang W, Fang T (2020) Sex-dependent and organ-specific toxicity of silver nanoparticles in livers and intestines of adult zebrafish. Chemosphere 249:126172

Bar-Ilan O, Albrecht RM, Fako VE, Furgeson DY (2009) Toxicity assessments of multisized gold and silver nanoparticles in zebrafish embryos. Small 5:1897–1910

Bar-Ilan O, Louis KM, Yang SP, Pedersen JA, Hamers RJ, Peterson RE, Heideman W (2012) Titanium dioxide nanoparticles produce phototoxicity in the developing zebrafish. Nanotoxicology 6:670–679

Batel A, Borchert F, Reinwald H, Erdinger L, Braunbeck T (2018) Microplastic accumulation patterns and transfer of benzo[a]pyrene to adult zebrafish (Danio rerio) gills and zebrafish embryos. Environ Pollut 235:918–930

Beekhuijzen M, De Koning C, Flores-Guillén M-E, De Vries-Buitenweg S, Tobor-Kaplon M, Van De Waart B, Emmen H (2015) From cutting edge to guideline: A first step in harmonization of the zebrafish embryotoxicity test (ZET) by describing the most optimal test conditions and morphology scoring system. Reprod Toxicol 56:64–76

Bhagat J, Zang L, Nishimura N, Shimada Y (2020) Zebrafish: An emerging model to study microplastic and nanoplastic toxicity. Sci Total Environ 728:138707

Braunbeck T, Boettcher M, Hollert H, Kosmehl T, Lammer E, Leist E, Rudolf M, Seitz N (2005) Towards an alternative for the acute fish LC(50) test in chemical assessment: the fish embryo toxicity test goes multi-species – an update. ALTEX 22:87–102

Brown DR, Samsa LA, Qian L, Liu J (2016) Advances in the study of heart development and disease using zebrafish. J Cardiovasc Develop Dis 3:13

Brun NR, Koch BEV, Varela M, Peijnenburg WJGM, Spaink HP, Vijver MG (2018) Nanoparticles induce dermal and intestinal innate immune system responses in zebrafish embryos. Environ Sci Nano 5:904–916

Bury N, Handy R (2010) Copper and iron uptake in teleost fish. In: Handy NBAR (ed) Surface chemistry, Bioavailabil Metal Homeost Aquat Organ: An integrated approach. Essential Reviews in Experimental Biology

Caixeta MB, Araújo PS, Gonçalves BB, Silva LD, Grano-Maldonado MI, Rocha TL (2020) Toxicity of engineered nanomaterials to aquatic and land snails: A scientometric and systematic review. Chemosphere 260:127654

Chae Y, An YJ (2017) Effects of micro- and nanoplastics on aquatic ecosystems: current research trends and perspectives. Mar Pollut Bull 124:624–632

Chakraborty C, Sharma AR, Sharma G, Lee S-S (2016) Zebrafish: A complete animal model to enumerate the nanoparticle toxicity. J Nanobiotechnol 14:65

Chao S-J, Huang CP, Chen P-C, Chang S-H, Huang C (2018) Uptake of BDE-209 on zebrafish embryos as affected by SiO2 nanoparticles. Chemosphere 205:570–578

Chen J, Dong X, Xin Y, Zhao M (2011) Effects of titanium dioxide nano-particles on growth and some histological parameters of zebrafish (Danio rerio) after a long-term exposure. Aquat Toxicol 101:493–499

Chen L, Guo Y, Hu C, Lam PKS, Lam JCW, Zhou B (2018) Dysbiosis of gut microbiota by chronic coexposure to titanium dioxide nanoparticles and bisphenol A: implications for host health in zebrafish. Environ Pollut 234:307–317

Chen M, Yin J, Liang Y, Yuan S, Wang F, Song M, Wang H (2016) Oxidative stress and immunotoxicity induced by graphene oxide in zebrafish. Aquat Toxicol 174:54–60

Chen PJ, Tan SW, Wu WL (2012) Stabilization or oxidation of nanoscale zerovalent iron at environmentally relevant exposure changes bioavailability and toxicity in medaka fish. Environ Sci Technol 46:8431–8439

Chen Q, Cao X, Yan B, Guo Z, Xi Z, Li J, Ci N, Yan M, Ci L (2023a) Ecotoxicological evaluation of functional carbon nanodots using zebrafish (Danio rerio) model at different developmental stages. Chemosphere 333:138970

Chen Q, Yin D, Jia Y, Schiwy S, Legradi J, Yang S, Hollert H (2017a) Enhanced uptake of BPA in the presence of nanoplastics can lead to neurotoxic effects in adult zebrafish. Sci Total Environ 609:1312–1321

Chen SX, Yang XZ, Deng Y, Huang J, Li Y, Sun Q, Yu C-P, Zhu Y, Hong WS (2017b) Silver nanoparticles induce oocyte maturation in zebrafish (Danio rerio). Chemosphere 170:51–60

Chen X, Liu S, Ding Q, Teame T, Yang Y, Ran C, Zhang Z, Zhou Z (2023b) Research advances in the structure, function, and regulation of the gill barrier in teleost fish. Water Biol Secur 2:100139

Cheng D, Shami GJ, Morsch M, Chung RS, Braet F (2016) Ultrastructural mapping of the zebrafish gastrointestinal system as a basis for experimental drug studies. Biomed Res Int 2016:8758460–8758413

Cheng J, Flahaut E, Cheng SH (2007) Effect of carbon nanotubes on developing zebrafish (danio Rerio) embryos. Environ Toxicol Chem 26:708–716

Choi JE, Kim S, Ahn JH, Youn P, Kang JS, Park K, Yi J, Ryu D-Y (2010) Induction of oxidative stress and apoptosis by silver nanoparticles in the liver of adult zebrafish. Aquat Toxicol 100:151–159

Christian P, Von Der Kammer F, Baalousha M, Hofmann T (2008) Nanoparticles: structure, properties, preparation and behaviour in environmental media. Ecotoxicology 17:326–343

Clemente Z, Castro VLSS, Franqui LS, Silva CA, Martinez DST (2017) Nanotoxicity of graphene oxide: assessing the influence of oxidation debris in the presence of humic acid. Environ Pollut 225:118–128

D'amora M, Schmidt TJN, Konstantinidou S, Raffa V, De Angelis F, Tantussi F (2022) Effects of metal oxide nanoparticles in zebrafish. Oxidative Med Cell Longev 2022:3313016–3313037

Das S, Dowding JM, Klump KE, McGinnis JF, Self W, Seal S (2013a) Cerium oxide nanoparticles: applications and prospects in nanomedicine. Nanomedicine (Lond) 8:1483–1508

Das S, Mitra S, Khurana SMP, Debnath N (2013b) Nanomaterials for biomedical applications. Front Life Sci 7:90–98

Dayal N, Thakur M, Patil P, Singh D, Vanage G, Joshi DS (2016) Histological and genotoxic evaluation of gold nanoparticles in ovarian cells of zebrafish (Danio rerio). J Nanopart Res 18:291

Dedeh A, Ciutat A, Treguer-Delapierre M, Bourdineaud J-P (2015) Impact of gold nanoparticles on zebrafish exposed to a spiked sediment. Nanotoxicology 9:71–80

Di Ianni E, Jacobsen NR, Vogel UB, Møller P (2022) Systematic review on primary and secondary genotoxicity of carbon black nanoparticles in mammalian cells and animals. Mutat Res Rev Mutat Res 790:108441

Dong X (2018) Current strategies for brain drug delivery. Theranostics 8:1481–1493

Duan J, Hu H, Feng L, Yang X, Sun Z (2017) Silica nanoparticles inhibit macrophage activity and angiogenesis via VEGFR2-mediated MAPK signaling pathway in zebrafish embryos. Chemosphere 183:483–490

Duan J, Hu H, Li Q, Jiang L, Zou Y, Wang Y, Sun Z (2016) Combined toxicity of silica nanoparticles and methylmercury on cardiovascular system in zebrafish (Danio rerio) embryos. Environ Toxicol Pharmacol 44:120–127

Duan Z, Duan X, Zhao S, Wang X, Wang J, Liu Y, Peng Y, Gong Z, Wang L (2020) Barrier function of zebrafish embryonic chorions against microplastics and nanoplastics and its impact on embryo development. J Hazard Mater 395:122621

Egan RJ, Bergner CL, Hart PC, Cachat JM, Canavello PR, Elegante MF, Elkhayat SI, Bartels BK, Tien AK, Tien DH, Mohnot S, Beeson E, Glasgow E, Amri H, Zukowska Z, Kalueff AV (2009) Understanding behavioral and physiological phenotypes of stress and anxiety in zebrafish. Behav Brain Res 205:38–44

Elimelech M, Gregory J, Jia X (2013) Particle deposition and aggregation: measurement, modelling and simulation. Butterworth-Heinemann

Evans AN, Nunez BS (2015) Fresh water acclimation elicits a decrease in plasma corticosteroids in the euryhaline Atlantic stingray, Dasyatis sabina. Gen Comp Endocrinol 222:167–172

Evariste L, Barret M, Mottier A, Mouchet F, Gauthier L, Pinelli E (2019) Gut microbiota of aquatic organisms: A key endpoint for ecotoxicological studies. Environ Pollut 248:989–999

Federici G, Shaw BJ, Handy RD (2007) Toxicity of titanium dioxide nanoparticles to rainbow trout (Oncorhynchus mykiss): gill injury, oxidative stress, and other physiological effects. Aquat Toxicol 84:415–430

Feng M, Luo J, Wan Y, Zhang J, Lu C, Wang M, Dai L, Cao X, Yang X, Wang Y (2022) Polystyrene Nanoplastic exposure induces developmental toxicity by activating the oxidative stress response and base excision repair pathway in zebrafish (Danio rerio). ACS Omega 7:32153–32163

Fent K, Weisbrod CJ, Wirth-Heller A, Pieles U (2010) Assessment of uptake and toxicity of fluorescent silica nanoparticles in zebrafish (Danio rerio) early life stages. Aquat Toxicol 100:218–228

Friedman AD, Claypool SE, Liu R (2013) The smart targeting of nanoparticles. Curr Pharm Des 19:6315–6329

Fu CW, Horng JL, Tong SK, Cherng BW, Liao BK, Lin LY, Chou MY (2021) Exposure to silver impairs learning and social behaviors in adult zebrafish. J Hazard Mater 403:124031

Gehr P (2018) Interaction of nanoparticles with biological systems. Colloids Surf B: Biointerfaces 172:395–399

Geldenhuys WJ, Allen DD, Bloomquist JR (2012) Novel models for assessing blood–brain barrier drug permeation. Expert Opin Drug Metab Toxicol 8:647–653

Goldsmith JR, Jobin C (2012) Think small: zebrafish as a model system of human pathology. Biomed Res Int 817341:1–12

Gomez-Garcia MJ, Doiron AL, Steele RRM, Labouta HI, Vafadar B, Shepherd RD, Gates ID, Cramb DT, Childs SJ, Rinker KD (2018) Nanoparticle localization in blood vessels: dependence on fluid shear stress, flow disturbances, and flow-induced changes in endothelial physiology. Nanoscale 10:15249–15261

Grasso D, Subramaniam K, Butkus M, Strevett K, Bergendahl J (2002) A review of non-DLVO interactions in environmental colloidal systems. Rev Environ Sci Biotechnol 1:17–38

Griffitt RJ, Weil R, Hyndman KA, Denslow ND, Powers K, Taylor D, Barber DS (2007) Exposure to copper nanoparticles causes gill injury and acute lethality in zebrafish (Danio rerio). Environ Sci Technol 41:8178–8186

Gu J, Guo M, Huang C, Wang X, Zhu Y, Wang L, Wang Z, Zhou L, Fan D, Shi L, Ji G (2021) Titanium dioxide nanoparticle affects motor behavior, neurodevelopment and axonal growth in zebrafish (Danio rerio) larvae. Sci Total Environ 754:142315

Habumugisha T, Zhang Z, Fang C, Yan C, Zhang X (2023) Uptake, bioaccumulation, biodistribution and depuration of polystyrene nanoplastics in zebrafish (Danio rerio). Sci Total Environ 893:164840

Handy, R. D. & Eddy, F. B. 2004. Transport of solutes across biological membranes in eukaryotes: an environmental perspective. Physicochem Kinet Transp Biointerf, 9, 337–356, Wiley

Handy RD, Henry TB, Scown TM, Johnston BD, Tyler CR (2008a) Manufactured nanoparticles: their uptake and effects on fish—a mechanistic analysis. Ecotoxicology 17:396–409

Handy RD, Owen R, Valsami-Jones E (2008b) The ecotoxicology of nanoparticles and nanomaterials: current status, knowledge gaps, challenges, and future needs. Ecotoxicology 17:315–325

Haque E, Ward AC (2018) Zebrafish as a model to evaluate nanoparticle toxicity. Nanomaterials (Basel) 8

Henn K, Braunbeck T (2011) Dechorionation as a tool to improve the fish embryo toxicity test (FET) with the zebrafish (Danio rerio). Compar Biochem Physiol Part C Toxicol Pharmacol 153:91–98

Howe K, Clark MD, Torroja CF, Torrance J, Berthelot C, Muffato M, Collins JE, Humphray S, Mclaren K, Matthews L, Mclaren S, Sealy I, Caccamo M, Churcher C, Scott C, Barrett JC, Koch R, Rauch G-J, White S, Chow W, Kilian B, Quintais LT, Guerra-Assunção JA, Zhou Y, Gu Y, Yen J, Vogel J-H, Eyre T, Redmond S, Banerjee R, Chi J, Fu B, Langley E, Maguire SF, Laird GK, Lloyd D, Kenyon E, Donaldson S, Sehra H, Almeida-King J, Loveland J, Trevanion S, Jones M, Quail M, Willey D, Hunt A, Burton J, Sims S, Mclay K, Plumb B, Davis J, Clee C, Oliver K, Clark R, Riddle C, Elliott D, Threadgold G, Harden G, Ware D, Begum S, Mortimore B, Kerry G, Heath P, Phillimore B, Tracey A, Corby N, Dunn M, Johnson C, Wood J, Clark S, Pelan S, Griffiths G, Smith M, Glithero R, Howden P, Barker N, Lloyd C, Stevens C, Harley J, Holt K, Panagiotidis G, Lovell J, Beasley H, Henderson C, Gordon D, Auger K, Wright D, Collins J, Raisen C, Dyer L, Leung K, Robertson L, Ambridge K, Leongamornlert D, Mcguire S, Gilderthorp R, Griffiths C, Manthravadi D, Nichol S, Barker G et al (2013) The zebrafish reference genome sequence and its relationship to the human genome. Nature 496:498–503

Hu Q, Guo F, Zhao F, Fu Z (2017) Effects of titanium dioxide nanoparticles exposure on parkinsonism in zebrafish larvae and PC12. Chemosphere 173:373–379

Hu Q, Wang H, He C, Jin Y, Fu Z (2021) Polystyrene nanoparticles trigger the activation of p38 MAPK and apoptosis via inducing oxidative stress in zebrafish and macrophage cells. Environ Pollut 269:116075

Huang B, Cui Y-Q, Guo W-B, Yang L, Miao A-J (2020) Waterborne and dietary accumulation of well-dispersible hematite nanoparticles by zebrafish at different life stages. Environ Pollut 259:113852

Huang W, Mo J, Li J, Wu K (2024) Exploring developmental toxicity of microplastics and nanoplastics (MNPS): insights from investigations using zebrafish embryos. Sci Total Environ 933:173012

Huang W, Wu T, Au WW, Wu K (2021) Impact of environmental chemicals on craniofacial skeletal development: insights from investigations using zebrafish embryos. Environ Pollut 286:117541

Ispas C, Andreescu D, Patel A, Goia DV, Andreescu S, Wallace KN (2009) Toxicity and developmental defects of different sizes and shape nickel nanoparticles in zebrafish. Environ Sci Technol 43:6349–6356

Janzadeh A, Behroozi Z, Saliminia F, Janzadeh N, Arzani H, Tanha K, Hamblin MR, Ramezani F (2022) Neurotoxicity of silver nanoparticles in the animal brain: a systematic review and meta-analysis. Forensic Toxicol 40:49–63

Ji Y, Wang Y, Wang X, Lv C, Zhou Q, Jiang G, Yan B, Chen L (2024) Beyond the promise: exploring the complex interactions of nanoparticles within biological systems. J Hazard Mater 468:133800

Jia H-R, Zhu Y-X, Duan Q-Y, Chen Z, Wu F-G (2019) Nanomaterials meet zebrafish: toxicity evaluation and drug delivery applications. J Control Release 311–312:301–318

Jia P-P, Junaid M, Wen P-P, Yang Y-F, Li W-G, Yang X-G, Pei D-S (2021) Role of germ-free animal models in understanding interactions of gut microbiota to host and environmental health: A special reference to zebrafish. Environ Pollut 279:116925

Jin L, Wang S, Chen C, Qiu X, Wang CC (2023) ZIF-8 nanoparticles induce behavior abnormality and brain oxidative stress in adult zebrafish (Danio rerio). Antioxidants (Basel) 12

Kalaiarasi S, Arjun P, Nandhagopal S, Brijitta J, Iniyan AM, Vincent SGP, Kannan RR (2016) Development of biocompatible nanogel for sustained drug release by overcoming the blood brain barrier in zebrafish model. J Appl Biomed 14:157–169

Kandaswamy K, Guru A, Panda SP, Antonyraj APM, Kari ZA, Giri J, Almutairi BO, Arokiyaraj S, Malafaia G, Arockiaraj J (2024) Polystyrene nanoplastics synergistically exacerbate diclofenac toxicity in embryonic development and the health of adult zebrafish. Compar Biochem Physiol Part C Toxicol Pharmacol 281:109926

Karlsson L (1983) Gill morphology in the zebrafish, Brachydanio rerio (Hamilton- Buchanan). J Fish Biol 23:511–524

Khan IA, Yu T, Li Y, Hu C, Zhao X, Wei Q, Zhong Y, Yang M, Liu J, Chen Z (2024) In vivo toxicity of upconversion nanoparticles (NaYF4:Yb, Er) in zebrafish during early life stages: developmental toxicity, gut-microbiome disruption, and proinflammatory effects. Ecotoxicol Environ Saf 284:116905

Kimmel CB, Ballard WW, Kimmel SR, Ullmann B, Schilling TF (1995) Stages of embryonic development of the zebrafish. Dev Dyn 203:253–310

Klein J-P, Mery L, Boudard D, Ravel C, Cottier M, Bitounis D (2023) Impact of nanoparticles on male fertility: what do we really know? A systematic review. Int J Mol Sci 24:576

Koppang EO, Kvellestad A, Fischer U (2015) Fish mucosal immunity: gill. In: Mucosal health in aquaculture. Elsevier

Kose O, Mantecca P, Costa A, Carrière M (2023) Putative adverse outcome pathways for silver nanoparticle toxicity on mammalian male reproductive system: a literature review. Part Fibre Toxicol 20:1

Kotil T, Akbulut C, Yön ND (2017) The effects of titanium dioxide nanoparticles on ultrastructure of zebrafish testis (Danio rerio). Micron 100:38–44

Krishnaraj C, Harper SL, Yun S-I (2016) In vivo toxicological assessment of biologically synthesized silver nanoparticles in adult zebrafish (Danio rerio). J Hazard Mater 301:480–491

Kteeba SM, El-Adawi HI, El-Rayis OA, El-Ghobashy AE, Schuld JL, Svoboda KR, Guo L (2017) Zinc oxide nanoparticle toxicity in embryonic zebrafish: mitigation with different natural organic matter. Environ Pollut 230:1125–1140

Kteeba SM, El-Ghobashy AE, El-Adawi HI, El-Rayis OA, Sreevidya VS, Guo L, Svoboda KR (2018) Exposure to ZnO nanoparticles alters neuronal and vascular development in zebrafish: acute and transgenerational effects mitigated with dissolved organic matter. Environ Pollut 242:433–448

Kuang Y, Guo H, Ouyang K, Wang X, Li D, Li L (2023) Nano-TiO2 aggravates immunotoxic effects of chronic ammonia stress in zebrafish (Danio rerio) intestine. Compar Biochem Physiol Part C Toxicol Pharmacol 266:109548

Kulkarni SA, Feng S-S (2013) Effects of particle size and surface modification on cellular uptake and biodistribution of polymeric nanoparticles for drug delivery. Pharm Res 30:2512–2522

Lacave JM, Vicario-Parés U, Bilbao E, Gilliland D, Mura F, Dini L, Cajaraville MP, Orbea A (2018) Waterborne exposure of adult zebrafish to silver nanoparticles and to ionic silver results in differential silver accumulation and effects at cellular and molecular levels. Sci Total Environ 642:1209–1220

Langheinrich U (2003) Zebrafish: a new model on the pharmaceutical catwalk. BioEssays 25:904–912

Lead JR, Batley GE, Alvarez PJJ, Croteau M-N, Handy RD, Mclaughlin MJ, Judy JD, Schirmer K (2018) Nanomaterials in the environment: behavior, fate, bioavailability, and effects—An updated review. Environ Toxicol Chem 37:2029–2063

Lead JR, Wilkinson KJ (2006) Aquatic colloids and nanoparticles: current knowledge and future trends. Environ Chem 3:159–171

Leso V, Battistini B, Vetrani I, Reppuccia L, Fedele M, Ruggieri F, Bocca B, Iavicoli I (2023) The endocrine disrupting effects of nanoplastic exposure: A systematic review. Toxicol Ind Health 39:613–629

Li Y, Miao X, Chen T, Yi X, Wang R, Zhao H, Lee SM-Y, Wang X, Zheng Y (2017) Zebrafish as a visual and dynamic model to study the transport of nanosized drug delivery systems across the biological barriers. Colloids Surf B: Biointerfaces 156:227–235

Lin J, Alexander-Katz A (2013) Cell membranes open "doors" for cationic nanoparticles/biomolecules: insights into uptake kinetics. ACS Nano 7:10799–10808

Lin S, Zhao Y, Nel AE, Lin S (2013) Zebrafish: an in vivo model for nano EHS studies. Small 9:1608–1618

Liu P, Zhao Y, Wang S, Xing H, Dong W-F (2021) Effect of combined exposure to silica nanoparticles and cadmium chloride on female zebrafish ovaries. Environ Toxicol Pharmacol 87:103720

Liu R, Gao H, Liang X, Zhang J, Meng Q, Wang Y, Guo W, Martyniuk CJ, Zha J (2024) Polystyrene nanoplastics alter intestinal toxicity of 2,4-DTBP in a sex-dependent manner in zebrafish (Danio rerio). J Hazard Mater 478:135585

Liu W-Y, Wang Z-B, Zhang L-C, Wei X, Li L (2012) Tight junction in blood-brain barrier: An overview of structure, regulation, and regulator substances. CNS Neurosci Ther 18:609–615

Liu X, Dumitrescu E, Kumar A, Austin D, Goia D, Wallace KN, Andreescu S (2019) Differential lethal and sublethal effects in embryonic zebrafish exposed to different sizes of silver nanoparticles. Environ Pollut 248:627–634

Lu Y, Zhang Y, Deng Y, Jiang W, Zhao Y, Geng J, Ding L, Ren H (2016) Uptake and accumulation of polystyrene microplastics in zebrafish (Danio rerio) and toxic effects in liver. Environ Sci Technol 50:4054–4060

Ma Y-B, Lu C-J, Junaid M, Jia P-P, Yang L, Zhang J-H, Pei D-S (2018) Potential adverse outcome pathway (AOP) of silver nanoparticles mediated reproductive toxicity in zebrafish. Chemosphere 207:320–328

Mahjoubian M, Naeemi AS, Moradi-Shoeili Z, Tyler CR, Mansouri B (2023) Toxicity of silver nanoparticles in the presence of zinc oxide nanoparticles differs for acute and chronic exposures in zebrafish. Arch Environ Contam Toxicol 84:1–17

Mahjoubian M, Naeemi AS, Sheykhan M (2021) Toxicological effects of ag(2)O and ag(2)CO(3) doped TiO(2) nanoparticles and pure TiO(2) particles on zebrafish (Danio rerio). Chemosphere 263:128182

Mahjoubian M, Naeemi SADAT, A. & Sheykhan, M. (2024) Comparative toxicity of TiO2 and Sn-doped TiO2 nanoparticles in zebrafish after acute and chronic exposure, vol 202. Biol Trace Elem Res, pp 1–19

Massarsky A, Strek L, Craig PM, Eisa-Beygi S, Trudeau VL, Moon TW (2014) Acute embryonic exposure to nanosilver or silver ion does not disrupt the stress response in zebrafish (Danio rerio) larvae and adults. Sci Total Environ 478:133–140

Mawed SA, Marini C, Alagawany M, Farag MR, Reda RM, El-Saadony MT, Elhady WM, Magi GE, Cerbo DI, A. & El-Nagar, W. G. (2022) Zinc oxide nanoparticles (Zno-NPs) suppress fertility by activating autophagy, apoptosis, and oxidative stress in the developing oocytes of female zebrafish. Antioxidants (Basel) 11

Mcnamara K, Tofail SA (2015) Nanosystems: the use of nanoalloys, metallic, bimetallic, and magnetic nanoparticles in biomedical applications. Phys Chem Chem Phys 17:27981–27995

Miner JJ, Diamond MS (2016) Mechanisms of restriction of viral neuroinvasion at the blood–brain barrier. Curr Opin Immunol 38:18–23

Moutabian H, Ghahramani-Asl R, Mortezazadeh T, Laripour R, Narmani A, Zamani H, Ataei G, Bagheri H, Farhood B, Sathyapalan T, Sahebkar A (2022) The cardioprotective effects of nano-curcumin against doxorubicin-induced cardiotoxicity: A systematic review. Biofactors 48:597–610

O'Keeffe E, Campbell M (2016) Modulating the paracellular pathway at the blood–brain barrier: current and future approaches for drug delivery to the CNS. Drug Discov Today Technol 20:35–39

Orbea A, González-Soto N, Lacave JM, Barrio I, Cajaraville MP (2017) Developmental and reproductive toxicity of PVP/PEI-coated silver nanoparticles to zebrafish. Compar Biochem Physiol Part C Toxicol Pharmacol 199:59–68

Osborne OJ, Lin S, Chang CH, Ji Z, Yu X, Wang X, Lin S, Xia T, Nel AE (2015) Organ-specific and size-dependent ag nanoparticle toxicity in gills and intestines of adult zebrafish. ACS Nano 9:9573–9584

Pardridge WM (2006) Molecular Trojan horses for blood–brain barrier drug delivery. Curr Opin Pharmacol 6:494–500

Parenti CC, Ghilardi A, Della Torre C, Magni S, del Giacco L, Binelli A (2019) Evaluation of the infiltration of polystyrene nanobeads in zebrafish embryo tissues after short-term exposure and the related biochemical and behavioural effects. Environ Pollut 254:112947

Parng C, Roy NM, Ton C, Lin Y, Mcgrath P (2007) Neurotoxicity assessment using zebrafish. J Pharmacol Toxicol Methods 55:103–112

Patsiou D, Del Rio-Cubilledo C, Catarino AI, Summers S, Fahmi MOHD, A., Boyle, D., Fernandes, T. F. & Henry, T. B. (2020) Exposure to Pb-halide perovskite nanoparticles can deliver bioavailable Pb but does not alter endogenous gut microbiota in zebrafish. Sci Total Environ 715:136941

Pedley TJ, Fischbarg J (1978) The development of osmotic flow through an unstirred layer. J Theor Biol 70:427–447

Peng G, Sinkko HM, Alenius H, Lozano N, Kostarelos K, Bräutigam L, Fadeel B (2023) Graphene oxide elicits microbiome-dependent type 2 immune responses via the aryl hydrocarbon receptor. Nat Nanotechnol 18:42–48

Pereira AC, Gomes T, Ferreira Machado MR, Rocha TL (2019) The zebrafish embryotoxicity test (ZET) for nanotoxicity assessment: from morphological to molecular approach. Environ Pollut 252:1841–1853

Petersen EJ, Diamond SA, Kennedy AJ, Goss GG, Ho K, Lead J, Hanna SK, Hartmann NB, Hund-Rinke K, Mader B, Manier N, Pandard P, Salinas ER, Sayre P (2015) Adapting OECD aquatic toxicity tests for use with manufactured nanomaterials: key issues and consensus recommendations. Environ Sci Technol 49:9532–9547

Petosa AR, Jaisi DP, Quevedo IR, Elimelech M, Tufenkji N (2010) Aggregation and deposition of engineered nanomaterials in aquatic environments: role of physicochemical interactions. Environ Sci Technol 44:6532–6549

Pham DH, de Roo B, Nguyen XB, Vervaele M, Kecskés A, Ny A, Copmans D, Vriens H, Locquet JP, Hoet P, de Witte PAM (2016) Use of zebrafish larvae as a multi-endpoint platform to characterize the toxicity profile of silica nanoparticles. Sci Rep 6

Pitt JA, Kozal JS, Jayasundara N, Massarsky A, Trevisan R, Geitner N, Wiesner M, Levin ED, Giulio DI, R. T. (2018a) Uptake, tissue distribution, and toxicity of polystyrene nanoparticles in developing zebrafish (Danio rerio). Aquat Toxicol 194:185–194

Pitt JA, Trevisan R, Massarsky A, Kozal JS, Levin ED, Giulio DI, R. T. (2018b) Maternal transfer of nanoplastics to offspring in zebrafish (Danio rerio): a case study with nanopolystyrene. Sci Total Environ 643:324–334

Powers CM, Slotkin TA, Seidler FJ, Badireddy AR, Padilla S (2011) Silver nanoparticles alter zebrafish development and larval behavior: distinct roles for particle size, coating and composition. Neurotoxicol Teratol 33:708–714

Prochaska M, Li J, Wallace KN (2020) Chapter 13 – it takes guts: development of the embryonic and juvenile zebrafish digestive system. In: Cartner SC, Eisen JS, Farmer SC, Guillemin KJ, Kent ML, Sanders GE (eds) The zebrafish in biomedical research. Academic

Qiang L, Cheng J (2019) Exposure to microplastics decreases swimming competence in larval zebrafish (Danio rerio). Ecotoxicol Environ Saf 176:226–233

Rabanel J-M, Faivre J, Zaouter C, Patten SA, Banquy X, Ramassamy C (2021) Nanoparticle shell structural cues drive in vitro transport properties, tissue distribution and brain accessibility in zebrafish. Biomaterials 277:121085

Rashidian G, Mohammadi-Aloucheh R, Hosseinzadeh-Otaghvari F, Chupani L, Stejskal V, Samadikhah H, Zamanlui S, Multisanti CR, Faggio C (2023) Long-term exposure to small-sized silica nanoparticles (SiO_2-NPs) induces oxidative stress and impairs reproductive performance in adult zebrafish (Danio rerio). Comparat Biochem Physiol Part C Toxicol Pharmacol 273:109715

Rawson DM, Zhang T, Kalicharan D, Jongebloed WL (2000) Field emission scanning electron microscopy and transmission electron microscopy studies of the chorion, plasma membrane and syncytial layers of the gastrula-stage embryo of the zebrafish Brachydanio rerio: a consideration of the structural and functional relationships with respect to cryoprotectant penetration. Aquac Res 31:325–336

Rehman A, Huang F, Zhang Z, Habumugisha T, Yan C, Shaheen U, Zhang X (2024) Nanoplastic contamination: impact on zebrafish liver metabolism and implications for aquatic environmental health. Environ Int 187:108713

Saleem S, Kannan RR (2021) Zebrafish: A promising real-time model system for nanotechnology-mediated Neurospecific drug delivery. Nanoscale Res Lett 16:135

Sarasamma S, Audira G, Juniardi S, Sampurna BP, Lai YH, Hao E, Chen JR, Hsiao CD (2018) Evaluation of the effects of carbon 60 nanoparticle exposure to adult zebrafish: A behavioral and biochemical approach to elucidate the mechanism of toxicity. Int J Mol Sci 19

Sarasamma S, Audira G, Samikannu P, Juniardi S, Siregar P, Hao E, Chen JR, Hsiao CD (2019) Behavioral impairments and oxidative stress in the brain, muscle, and gill caused by chronic exposure of C(70) nanoparticles on adult zebrafish. Int J Mol Sci 20

Sarasamma S, Audira G, Siregar P, Malhotra N, Lai Y-H, Liang S-T, Chen J-R, Chen KH-C, Hsiao C-D (2020) Nanoplastics cause neurobehavioral impairments, reproductive and oxidative damages, and biomarker responses in zebrafish: throwing up alarms of wide spread health risk of exposure. Int J Mol Sci 21:1410

Shaw BJ, Handy RD (2011) Physiological effects of nanoparticles on fish: A comparison of nanometals versus metal ions. Environ Int 37:1083–1097

Sheng L, Wang L, Su M, Zhao X, Hu R, Yu X, Hong J, Liu D, Xu B, Zhu Y, Wang H, Hong F (2016) Mechanism of TiO_2 nanoparticle-induced neurotoxicity in zebrafish (anio rerio). Environ Toxicol 31:163–175

Shi Q, Fang C, Zhang Z, Yan C, Zhang X (2020) Visualization of the tissue distribution of fullerenols in zebrafish (Danio rerio) using imaging mass spectrometry. Anal Bioanal Chem 412:7649–7658

Shwartz A, Goessling W, Yin C (2019) Macrophages in zebrafish models of liver diseases. Front Immunol 10:2840

Siivola KM, Burgum MJ, Suárez-Merino B, Clift MJD, Doak SH, Catalán J (2022) A systematic quality evaluation and review of nanomaterial genotoxicity studies: a regulatory perspective. Part Fibre Toxicol 19:59

Skjolding LM, Ašmonaitė G, Jølck RI, Andresen TL, Selck H, Baun A, Sturve J (2017) An assessment of the importance of exposure routes to the uptake and internal localisation of fluorescent nanoparticles in zebrafish (Danio rerio), using light sheet microscopy. Nanotoxicology 11:351–359

Smith BR, Blumstein DT (2008) Fitness consequences of personality: a meta-analysis. Behav Ecol 19:448–455

Smith CJ, Shaw BJ, Handy RD (2007) Toxicity of single walled carbon nanotubes to rainbow trout, (Oncorhynchus mykiss): respiratory toxicity, organ pathologies, and other physiological effects. Aquat Toxicol 82:94–109

Souza JP, Baretta JF, Santos F, Paino IMM, Zucolotto V (2017) Toxicological effects of graphene oxide on adult zebrafish (Danio rerio). Aquat Toxicol 186:11–18

Spence R, Gerlach G, Lawrence C, Smith C (2008) The behaviour and ecology of the zebrafish, Danio rerio. Biol Rev 83:13–34

Srinonate A, Banlunara W, Maneewattanapinyo P, Thammacharoen C, Ekgasit S, Kaewamatawong T (2015) Acute toxicity study of nanosilver particles in tilapia (Oreochromis niloticus): pathological changes, particle bioaccumulation and metallothionien protein expression. Thai J Veter Med 45:81–89

Stark WJ, Stoessel PR, Wohlleben W, Hafner A (2015) Industrial applications of nanoparticles. Chem Soc Rev 44:5793–5805

Stolpe B, Hassellöv M (2010) Nanofibrils and other colloidal biopolymers binding trace elements in coastal seawater: significance for variations in element size distributions. Limnol Oceanogr 55:187–202

Subramanian D, Manogaran PONNUSAMY, G. & Dharmadurai, D. (2024) A systematic review on the impact of micro-nanoplastics on human health: potential modulation of epigenetic mechanisms and identification of biomarkers. Chemosphere 363:142986

Sun M, Ding R, Ma Y, Sun Q, Ren X, Sun Z, Duan J (2021) Cardiovascular toxicity assessment of polyethylene nanoplastics on developing zebrafish embryos. Chemosphere 282:131124

Sun Y, Zhang G, He Z, Wang Y, Cui J, Li Y (2016) Effects of copper oxide nanoparticles on developing zebrafish embryos and larvae. Int J Nanomedicine 11:905–918

Szudrowicz H, Kamaszewski M, Adamski A, Skrobisz M, Frankowska-Łukawska J, Wójcik M, Bochenek J, Kawalski K, Martynow J, Bujarski P, Pruchniak P, Latoszek E, Bury-Burzymski P, Szczepański A, Jaworski S, Matuszewski A, Herman AP (2022) The effects of seven-day exposure to silver nanoparticles on fertility and homeostasis of zebrafish (Danio rerio). Int J Mol Sci 23:11239

Tang T, Zhang Z, Zhu X (2019) Toxic effects of TiO_2 NPs on zebrafish. Int J Environ Res Public Health 16

Torres-Ruiz M, de la Vieja A, de Alba Gonzalez M, Esteban Lopez M, Castaño Calvo A, Cañas Portilla AI (2021) Toxicity of nanoplastics for zebrafish embryos, what we know and where to go next. Sci Total Environ 797:149125

Van Pomeren M, Brun NR, Peijnenburg WJGM, Vijver MG (2017) Exploring uptake and biodistribution of polystyrene (nano)particles in zebrafish embryos at different developmental stages. Aquat Toxicol 190:40–45

Veronese FM, Mero A (2008) The impact of PEGylation on biological therapies. BioDrugs 22:315–329

Villacis RAR, Filho JS, Piña B, Azevedo RB, Pic-Taylor A, Mazzeu JF, Grisolia CK (2017) Integrated assessment of toxic effects of maghemite (γ-Fe2O3) nanoparticles in zebrafish. Aquat Toxicol 191:219–225

Vranic S, Shimada Y, Ichihara S, Kimata M, Wu W, Tanaka T, Boland S, Tran L, Ichihara G (2019) Toxicological evaluation of SiO2 nanoparticles by zebrafish embryo toxicity test. Int J Mol Sci 20:882

Wallace KN, Akhter S, Smith EM, Lorent K, Pack M (2005) Intestinal growth and differentiation in zebrafish. Mech Dev 122:157–173

Wang J, Zhu X, Zhang X, Zhao Z, Liu H, George R, Wilson-Rawls J, Chang Y, Chen Y (2011) Disruption of zebrafish (Danio rerio) reproduction upon chronic exposure to TiO2 nanoparticles. Chemosphere 83:461–467

Wang S, Alenius H, El-Nezami H, Karisola P (2022a) A new look at the effects of engineered ZnO and TiO2 nanoparticles: evidence from Transcriptomics studies. Nano 12:1247

Wang X, Chen S, Qin Y, Wang H, Liang Z, Zhao Y, Zhou L, Martyniuk CJ (2022b) Metabolomic responses in livers of female and male zebrafish (Danio rerio) following prolonged exposure to environmental levels of zinc oxide nanoparticles. Aquat Toxicol 253:106333

Wang Y-P, Li X, Xue J-Y, Zhang Y-S, Feng X-Z (2014) Developmental and cartilaginous effects of protein-coated SiO 2 nanoparticle corona complexes on zebrafish larvae. RSC Adv 4:18541–18548

Wu G, Gao L, Zhang S, Du D, Xue Y (2023) Effects of copper oxide nanoparticles on reproductive system of zebrafish. Ecotoxicol Environ Saf 263:115252

Xian H, Li Z, Bai R, Ye R, Feng Y, Zhong Y, Liang B, Huang Y, Guo J, Wang B, Dai M, Tang S, Ren X, Chen X, Chen D, Yang X, Huang Z (2024) From cradle to grave: deciphering sex-specific disruptions of the nervous and reproductive systems through interactions of 4-methylbenzylidene camphor and nanoplastics in adult zebrafish. J Hazard Mater 470:134298

Xie J, Farage E, Sugimoto M, Anand-Apte B (2010) A novel transgenic zebrafish model for blood-brain and blood-retinal barrier development. BMC Dev Biol 10:76

Xie Q, Li Z, Chen Y, Zhao Y, Xu Y, Hong Z, Chen Z, Zhang Z, Xu H, Yin Z, Wu X (2024) Mass spectrometry imaging reveals the morphology-dependent toxicological effects of Nanosilvers on multiple organs of adult zebrafish (Danio rerio). Environ Sci Technol 58:10015–10027

Xu B, Zhang L, Wu D, Qi Z, Cao J, Li W, Fan L, Shi Y, Wu Y, Li G (2024) CuO nanoparticles elicit intestinal immunotoxicity in zebrafish based on intestinal microbiota dysbiosis. Food Funct 15:7619–7630

Xu K, Zhang Y, Huang Y, Wang J (2021) Toxicological effects of microplastics and phenanthrene to zebrafish (Danio rerio). Sci Total Environ 757:143730

Yang L, Chen H, Kaziem AE, Miao X, Huang S, Cheng D, Xu H, Zhang Z (2024) Effects of exposure to different types of metal-organic framework nanoparticles on the gut microbiota and liver metabolism of adult zebrafish. ACS Nano 18:25425–25445

Zanella D, Bossi E, Gornati R, Bastos C, Faria N, Bernardini G (2017) Iron oxide nanoparticles can cross plasma membranes. Sci Rep 7:11413

Zhang B, Yang H, Cai G, Nie Q, Sun Y (2024a) The interactions between the host immunity and intestinal microorganisms in fish. Appl Microbiol Biotechnol 108:30

Zhang C, Li L, Alava JJ, Yan Z, Chen P, Gul Y, Wang L, Xiong D (2024b) Female zebrafish (Danio rerio) exposure to polystyrene nanoplastics induces reproductive toxicity in mother and their offspring. Aquat Toxicol 273:107023

Zhang P, Wang Y, Zhao X, Ji Y, Mei R, Fu L, Man M, Ma J, Wang X, Chen L (2022) Surface-enhanced Raman scattering labeled nanoplastic models for reliable bio-nano interaction investigations. J Hazard Mater 425:127959

Zhao X, Ren X, Zhu R, Luo Z, Ren B (2016) Zinc oxide nanoparticles induce oxidative DNA damage and ROS-triggered mitochondria-mediated apoptosis in zebrafish embryos. Aquat Toxicol 180:56–70

Zhao Y, Yang Q, Liu D, Liu T, Xing L (2022) Neurotoxicity of nanoparticles: insight from studies in zebrafish. Ecotoxicol Environ Saf 242:113896

Zheng M, Lu J, Zhao D (2018) Toxicity and transcriptome sequencing (RNA-seq) analyses of adult zebrafish in response to exposure carboxymethyl cellulose stabilized iron sulfide nanoparticles. Sci Rep 8:8083

Zheng Y, Song J, Qian Q, Wang H (2024) Silver nanoparticles induce liver inflammation through ferroptosis in zebrafish. Chemosphere 362:142673

Open Access This chapter is licensed under the terms of the Creative Commons Attribution-NonCommercial-NoDerivatives 4.0 International License (http://creativecommons.org/licenses/by-nc-nd/4.0/), which permits any noncommercial use, sharing, distribution and reproduction in any medium or format, as long as you give appropriate credit to the original author(s) and the source, provide a link to the Creative Commons license and indicate if you modified the licensed material. You do not have permission under this license to share adapted material derived from this chapter or parts of it.

The images or other third party material in this chapter are included in the chapter's Creative Commons license, unless indicated otherwise in a credit line to the material. If material is not included in the chapter's Creative Commons license and your intended use is not permitted by statutory regulation or exceeds the permitted use, you will need to obtain permission directly from the copyright holder.

Part IV

Nanosafety Related to Nanomaterials in the Environment

Environmental Nanosafety

15

Begoña Espiña and Laura Rodriguez-Lorenzo

Abstract

During the last decade, the environmental risk assessment (ERA) of nanomaterials (NMs) has advanced towards a better understanding the differential methodologies and tools needed to evaluate their behaviour and potential impact. Despite being regulated by agencies devoted to chemicals, such as the European Chemicals Agency (ECHA), NMs present inherent characteristics that make implementing the gold standard methods for quantification and ecotoxicology tests initially designed for chemicals challenging. In this chapter, we revise the advances of the last 10 years in knowledge, methods and instruments to assess the potential risks of NMs, including nanoplastics, in the environment: the development of specific standards and guidelines, analytical methods and instruments, high-throughput tests, frameworks for grouping and read-across or the exponential increase on the development of NM-specific computational tools and *in silico* models. We also identify the new trends in NM development, as the materials developed today can be the new contaminants tomorrow and stress the importance of the recently released safe and sustainable by-design framework for NMs. Finally, we will revise the current gaps that the researchers, regulatory bodies, and industry must fill to attain a good framework for NMs ERA.

Keywords

Environmental risk assessment · Econanotoxicity · Eco-corona · Life cycle analysis · Safe and sustainable-by-design

B. Espiña (✉) · L. Rodriguez-Lorenzo
Water Quality Group, International Iberian Nanotechnology Laboratory, Braga, Portugal
e-mail: begona.espina@inl.int

© The Author(s) 2025
E. Alfaro-Moreno, F. Murphy (eds.), *Nanosafety*,
https://doi.org/10.1007/978-3-031-93871-9_15

1 Introduction

Owing to their unique qualities and the almost infinite tailoring possibilities, engineered nanomaterials (ENMs) have been exponentially explored for many applications, and medium-term forecasts predict that ENM manufacturing will grow constantly over the next decades. Industrial production of ENMs worldwide is estimated to be more than 1,000,000 tons/year, and according to the European Union, the expected median annual growth in production is 5%. However, nanomaterials are also produced in nature, for example, in dust or volcanic ash. They can also result unintentionally from human activity (e.g. car exhaust, burning candles) or as products of degradation of human-engineered materials such as secondary nanoplastics (NPLs), produced mainly by the fragmentation of secondary microplastics. NPLs are nowadays identified as highly relevant in concentration and, more importantly, toxicological impact.

Consequently, the volume and variety of NMs released into the environment will increase accordingly. However, so far, experimental environmental risk assessment studies have lagged behind studies on applications or human toxicology risks. This necessitates a comprehensive approach to nanosafety that includes eco-design strategies, advanced testing methods, and life cycle assessments.

Material Flow Analyses (MFAs) have now incorporated factors like nanoparticle size and dynamic release, which are essential in estimating environmental concentrations (Keller et al. 2024). On the other hand, Environmental fate models (EFMs) now consider processes such as aggregation and transformation, providing more accurate predictions of nanomaterial behaviour in ecosystems (Keller et al. 2024). Furthermore, the rise of computational nanotoxicology has shifted risk assessment from traditional experimental methods to *in silico* models, which are faster and more cost-effective (Tang et al. 2024).

These models include multimedia environmental models and quantitative structure–activity relationships, addressing the complexities of ENM interactions with the environment (Tang et al. 2024).

In this chapter, we intend to give a general overview of the current state of knowledge on the environmental safety of nanomaterials, data gaps, regulatory framework, and future perspectives.

2 Environmental Impact of Nanomaterials

2.1 Nanomaterials Occurrence in the Environment

2.1.1 Air

Nanoparticles, which are ultrafine particles (<100 nm), are found in the air from both natural and human-made sources. They can deposit in the respiratory tract and enter the body, raising concerns about their health impacts (Rabajczyk et al. 2020).

About 10.8% of consumer products, including ENMs, have the potential to release nanomaterials into the air, leading to inhalation exposure. Silver-containing products are particularly prevalent; exposure can occur through wet and dry aerosols. Indoor environments are significant exposure sites, especially with products like cleaning agents that can release nanoparticles during use (Vance and Marr 2015).

On the other hand, micro/nano plastics (MNPs) are detected globally, even in remote areas, indicating atmospheric transport as a key dissemination pathway. This widespread presence poses a global health risk. The health risks associated with inhaled MNPs include immune response issues, oxidative stress, and potential links to cardiovascular and developmental diseases (Luo et al. 2024).

2.1.2 Water

The occurrence of nanomaterials in aquatic environments is a complex issue influenced by various factors, including the types of nanomaterials, their transformations, and their interactions with aquatic ecosystems.

TiO_2, Ag, CuO, and CNT NMs present predicted environmental concentrations (PECs) in waters in the ranges at 0–16,000, 0–619, 0.02–6, and 2×10^{-5}–1.82 ng/L, respectively. PECs in sediments are higher than those in surface waters; TiO_2, Ag, CuO, and CNT NMs are ranged at 0–186, 0–0.47, 3.5×10^{-3}–2.1, and 1×10^{-4}–2.66×10^{-2} mg/kg, respectively. The reason is that NMs released into waters could quickly adsorb to the suspended minerals, finally depositing into the sediment environments. Actually, sediment is the compartment in which most NMs entering surface water end up (Zhao et al. 2021).

ENMs suffer physical, chemical, and biological transformations in the environment, impacting their behaviour, exposure concentration, and effects on biota. The transformations depend on their initial properties and environmental conditions (Fig. 15.1).

However, not all environmental NMs are engineered. Some naturally present nanoparticles often result from mineral weathering and exhibit diverse shapes and compositions. For instance, it is very common to find Ti-based nanoparticles in marine environments and their distribution is influenced by natural inputs, such as river estuaries (Liu et al. 2023).

Nanomaterials are also utilized in water for; i) treatment processes, such as particularly for microplastic remediation (Saleem et al. 2022); ii) to enhance fish growth and health in aquaculture (Nguyen et al. 2024); iii) as antimicrobials (Vijayaram et al. 2023). This utilisation increases the likelihood of a significant release to the water.

2.1.3 Soil

ENMs accumulate in soil from various sources, including wastewater and consumer products, leading to potential soil contamination. The toxicity of these materials, particularly those containing heavy metals, raises concerns about bioaccumulation and long-term environmental effects (Rath et al. 2024).

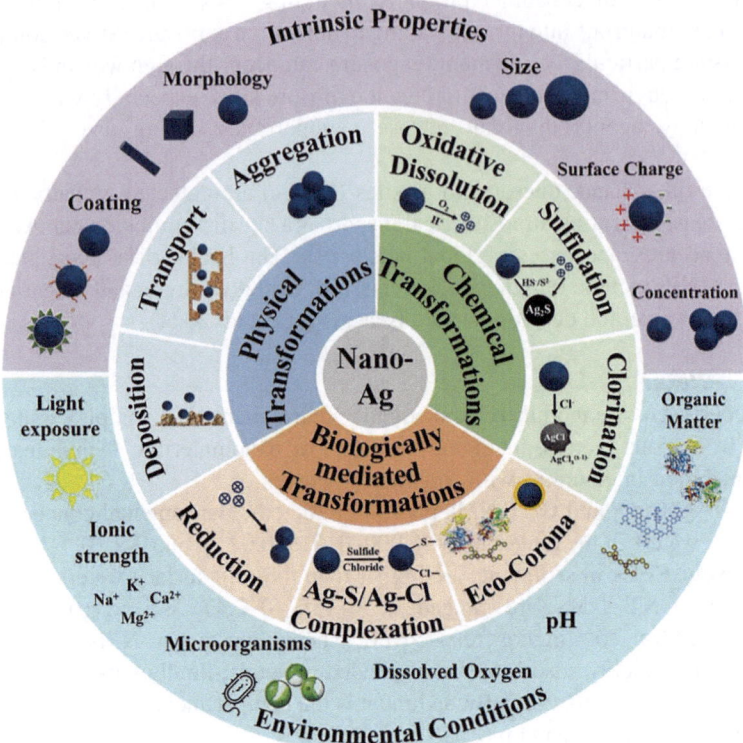

Fig. 15.1 Environmental behaviour of silver nanomaterials in aquatic environments. (From Yang et al. 2024. Reprinted with permission from Elsevier)

PECs for all the NMs are 2–4 orders of magnitude lower in soils than that in sediments. However, a high PEC for CuO NMs was detected in soil in Denmark, (0.039–0.13 mg/kg CuO and $CuCO_3$ NMs) explained by their use as wood preservatives. This case exemplifies why the risk assessment based on exposure level should have a very strong base on local analysis of human activities. PECs of graphene NMs are speculated to be at a similar concentration level with (or even higher than) carbon nanotubes (CNTs) based on its functional and surface properties and production volumes (Zhao et al. 2021).

However, ENMs are also widely used in agriculture and aquaculture, where they can enhance soil health and agricultural productivity, but also pose environmental concerns, particularly regarding microbial communities and soil contamination.

Nanoparticles can alter soil microbial communities, affecting beneficial bacteria essential for nutrient cycling and plant health (Tamuk et al. 2024). However, some studies show that certain metal-based nanomaterials can suppress harmful soil-borne pathogens while promoting beneficial microorganisms (Li et al. 2024).

2.2 Environmental Fate and Transport

2.2.1 Persistence and Degradation

Many nanomaterials are non-degradable and can accumulate in environmental compartments (Tran et al. 2023). For instance, carbon nanotubes and nanoscale zerovalent iron are known for their stability and resistance to degradation, enabling a prolonged environmental presence (Ahmed et al. 2022). Other materials such as some TiO_2 remain in biological systems, raising concerns about chronic exposure (Sohal et al. 2018).

Some nanomaterials can be remediated through photocatalysis and adsorption, however, a more rational approach for ecological sustainability is synthesising biodegradable nanomaterials (Ahmed et al. 2022; Rossi 2022).

2.2.2 Bioaccumulation and biomagnification

NMs can bioaccumulate and potentially biomagnify in food chains, although current studies show biomagnification factors (BMF) of less than 1, indicating limited impact. Nanomaterial properties, ion variations, and environmental conditions like pH and dissolved oxygen influence NMs toxicity (Zhang et al. 2024). The physicochemical properties of NMs, such as size and coating, significantly influence their bioaccumulation potential (Handy et al. 2021).

Engineered carbon nanomaterials (CNMs) and nanoplastics exhibit variable bioaccumulation patterns, with limited absorption across epithelial surfaces in multicellular organisms (Petersen et al. 2022). A meta-analysis revealed that zooplankton accumulate ENMs more than phytoplankton and fish, with a mean biomagnification factor (BMF) of 17.4 for zooplankton, although trophic transfer to fish was minimal (Zheng and Nowack 2022).

Recent works have also studied the accumulation of ENMs in aquaculture species extensively used for human consumption such as clams, mussels, seabream, seabass, turbot or seaweed. These studies have particular relevance as well in food safety. Results indicate a limited accumulation of PVP-Ag and citrate-TiO_2 NPs in edible parts of fish (muscle and skin); however, differential accumulation and biodistribution were reported on other organs such as the liver or kidney and, interestingly, not only depending on the NP's type, but also on their size. Values up to 6 μg/g and 120 μg/g PVP-Ag NPs were found in the liver and kidney, respectively, while up to 1 μg/g and 6 μg/g citrate-TiO_2. Clams accumulated more PVP-Ag and citrate-TiO_2 NPs among the filter-feeding molluscs than mussels. This difference could be explained by their distinct clearance rates and/or the size range of the agglomerates of NPs being more similar to clams' feed (Nakamura 2001; Jacobs et al. 2015). Seaweeds were the organisms where more NPs were found: 10^9 and 10^{10} NP/g of PVP-Ag and citrate-TiO_2 NPs, respectively.

In plants, NMs can translocate from roots to aboveground tissues, affecting their growth and potentially leading to bioaccumulation in herbivores (Elbehiry et al. 2022). Interspecies differences should also be taken into account not only regarding the bioaccumulation potential but also regarding the transformations that the NMs suffer once accumulated. In this regard, a recent study reported similar accumulation of PVP-Ag NPs in different species of seaweeds used for human consumption but a differential biodistribution and transformation within the tissues (Quarato et al. 2024) (Fig. 15.2).

Hence, while some studies suggest limited biomagnification of certain nanomaterials, the complexity of their interactions within food webs necessitates further research to fully understand their ecological impacts.

2.3 Ecotoxicology

The most common toxic effects induced by NMs include oxidative stress, immune system disruption, and cellular damage. NMs can penetrate cell membranes, reaching mitochondria and nuclei, where they can cause gene mutations and inhibit mitochondrial processes involved in cell metabolism (Jain et al. 2023).

2.3.1 Toxic Effects on Aquatic Life

It has been reported that the introduction of nanomaterials into aquatic ecosystems can lead to severe toxicity, causing mortality in various aquatic species (Zafar et al. 2023). However, there is a general scarcity of ecotoxicity tests conducted with environmentally relevant concentrations of NMs with the exception of TiO_2 NPs. *Daphnia magna* was shown to accumulate Ag NPs, which caused oxidative stress, altered its growth and reproduction, and even resulted in lethality at ng/L concentrations (LD50 = 0.18 µg/L). Ag- and Ti-based NM were shown to induce genotoxicity and increase oxidative stress-related markers in mussels. Oxidative stress seems to be the most sensitive toxicity mechanism probably due to the extensive profusion of studies that evaluate this effect. Although, most of the species employed for aquatic ecotoxicity of NMs are freshwater with a few exceptions for fish, mussels, and shrimp.

Very recent studies reported the exposure of marine fish species extensively used in aquaculture for human consumption, such as turbot and seabream, to Ag and TiO_2 NMs. These studies showed alterations in hepatic tissue structure and alterations in the expression of proteins related to lipid and fatty acid metabolism, lipid breakdown for energy, lipid transport, and homeostasis (Araújo et al. 2022, 2024; Fonseca et al. 2023). Interestingly, these authors found size related differential

Fig. 15.2 (continued) fraction was observed in *P. palmata* after 28 days of exposure. To understand this, (**b**) shows a low-magnification STEM image, which shows the localisation of AgNPs within the tissue. This image was processed by merging 3 individual STEM images at the same magnification using Affinity Photo (version 1.10.5.1342). STEM images at higher magnification indicate the position of the EDX maps were acquired. The cyan circular frames indicate the area analyzed, where the AgNPs were localised, whose spectra are reported. Both in (**a**) and in (**b**), analyzed AgNPs are labeled by (1), (2), and (3), respectively. (From Quarato et al. 2024. Reprinted with permission from Elsevier)

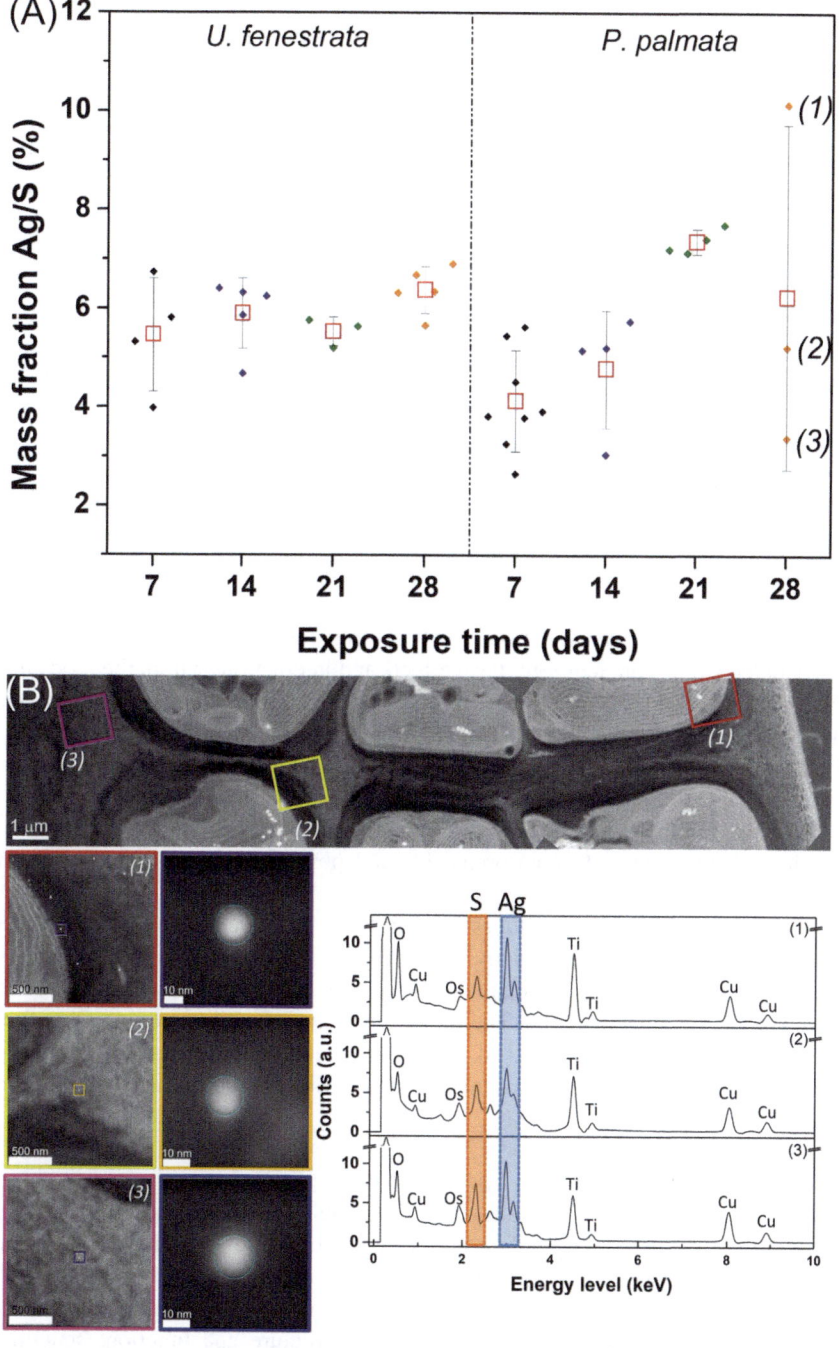

Fig. 15.2 (a) Mass fraction percentage of Ag/S ratio estimated at different times of exposure in *Ulva fenestrata* and *Palmaria palmata*, respectively. The highest variability in the Ag/S mass

effects; they found a variable abundance of lipid droplets (LDs) in hepatocytes dependent on TiO_2 NPs size, an increase in turbot exposed to smaller TiO_2 NPs and a depletion with larger TiO_2 NPs. The expression patterns of genes related to oxidative and immune responses and lipid metabolism (nrf2, nfκb1, and cpt1a) were dependent on the presence of TiO_2 NPs and time of exposure supporting the variance in hepatic LDs distribution over time with the different NPs. In this case, citrate was the coating, and it was proposed as the main responsible, reinforcing the idea that the full composition of the NM should be taken into account for the hazard characterization(Fonseca et al. 2023).

Nanoparticles can also exhibit bactericidal and mutagenic effects, impacting microbial communities and disrupting ecological balance (Shinde et al. 2023). ZnO nanoparticles can decrease fertility rates in aquatic animals and inhibit photosynthesis in plants due to heavy metal release upon degradation (Umair Hassan et al. 2024).

2.3.2 Impact on Terrestrial Organisms

ENMs accumulate in soil, potentially affecting soil ecology, crop production, and human health. Their toxic mechanisms include oxidative stress, usually through ion release (Wang et al. 2023). Different terrestrial organisms respond variably to nanomaterials, with some exhibiting resilience and others showing significant toxicity (Morgado et al. 2022).

Soil properties influence the transport and transformation of ENMs, necessitating detailed studies on their long-term effects and interactions within the food chain (Wang et al. 2023). Some reports have shown that MNPs disrupt soil microbial structure and diversity, which are essential for biogeochemical cycles [37], and that their presence can hinder beneficial microorganisms, reducing their populations and bioactive molecule synthesis (Yamini et al. 2023).

ENMs exhibit varying toxicity based on their properties and environmental conditions, affecting lower trophic organisms and potentially disrupting food chains (Kumar et al. 2022).

NMs can induce oxidative stress by generating reactive oxygen species (ROS) such as O_2^- and H_2O_2. This oxidative stress can decrease antioxidant enzyme activity, resulting in growth reduction and pathological abnormalities in organisms like earthworms (Zhou et al. 2023). The immune system's inability to promptly eliminate synthetic particles can lead to chronic inflammation and increase the risk of inflammatory disorders in terrestrial organisms' digestive, respiratory, and reproductive systems.

2.3.3 Effects on Microbial Communities

NMs may impact microbial communities, influencing their structure, function, and ecosystem interactions. Depending on the type, concentration, and environmental conditions, these effects can be either beneficial or detrimental.

Metal nanoparticles can enhance microbial growth and nutrient cycling by increasing nutrient availability improving soil structure and function, benefiting plant growth and agricultural productivity (Mahapatra et al. 2024; Upadhayay et al. 2024). On the other hand, some nanomaterials can promote biofilm formation, which is beneficial in biotechnological applications. This is due to their ability to interact with microbial cells and influence the production of the exopolymer matrix and quorum sensing systems (Maksimova and Zorina 2024).

However, high concentrations of certain nanoparticles, such as silver and graphene-based nanomaterials, can be toxic to microorganisms. This toxicity can reduce microbial diversity and disrupt soil functions, potentially damaging the microbial communities (Cao et al. 2024; Mahapatra et al. 2024). A global meta-analysis revealed that metal NMs, particularly silver nanoparticles (Ag NPs), have significant negative effects on soil microbial diversity, biomass, and function. These effects are more pronounced than those of carbon-based NMs, with Ag NMs showing the greatest negative impact (Chen et al. 2023). Nanoparticles can induce oxidative stress in microbial cells by generating ROS, leading to cell damage and reduced microbial biomass. This can negatively impact soil health and its ability to support plant growth (Maksimova and Zorina 2024; Tamuk et al. 2024). Recent studies on nano-polystyrene (nPS) and its modified forms show that surface modifications can significantly alter their effects on sediment microbial communities. While bacterial diversity remains stable, fungal communities are notably inhibited, and nitrogen metabolism processes such as nitrification and denitrification are affected (Zhao et al. 2023).

ZnO NPs are commonly used in agriculture and have been shown to affect soil microbiota. At low concentrations, they can stimulate plant growth and microbial activity, but higher disrupt key microbial processes such as mineralisation and nitrogen fixation (Иванова et al. 2024). On the other hand, CuO NPs have been found to cause persistent changes in microbial community structure and function, particularly affecting Gram-positive bacterial and fungal communities. These changes can last extended periods, indicating long-term ecological risks (Borymski et al. 2023).

While nanomaterials offer significant potential for enhancing agricultural productivity and environmental management, comprehensive regulations and sustainable practices for their use in agriculture must be developed to mitigate potential adverse effects (Tamuk et al. 2024).

3   Assessment of EcoNanosafety

3.1   Environmental Risk Assessment Frameworks

The environmental risk assessment (ERA) of NMs integrates various methodologies to evaluate their potential impacts on ecosystems. This assessment involves understanding NMs' environmental concentrations, exposure scenarios, and toxicological effects.

Despite advances in understanding nanomaterial fate and effects, uncertainties persist in hazard and exposure assessments. Specific proposals include accounting for exposure concentrations in aquatic toxicity tests, addressing the availability of dissolving nanomaterials in test systems and deriving effect data for soil organisms. Reliable environmental risk assessments for nanomaterials remain uncertain. These challenges are significant for regulators and industries, necessitating ongoing research and adaptation of regulatory frameworks (Schwirn et al. 2020).

Computational nanotoxicology models include material flow analysis, multimedia environmental models, and physiologically based toxicokinetics models, which help predict nanomaterials' environmental fate and transport. Quantitative nanostructure-activity relationships and meta-analysis are also employed to predict the toxicological effects of nanomaterials, offering a more efficient alternative to traditional experimental methods (Tang et al. 2024). Probabilistic risk assessment methods, particularly those that separately quantify variability and uncertainty, are effective for assessing the NMs' environmental hazards. The SimpleBox4Nano model for exposure assessment, combined with Monte Carlo simulations to estimate chronic critical effect concentrations (CEC) and predicted-no-effect concentrations (PNEC) has been reported as good alternative for risk assessment considering long-term ecological impacts (Jacobs et al. 2016).

3.1.1 Hazard Identification

Current ecotoxicity test methods, such as those involving algae (OECD 201) and Daphnia (OECD 202), are generally appropriate for assessing the environmental hazards of nanomaterials but significant uncertainties related to nanoparticle exposure characterisation in test systems remain. To improve assessments, it is crucial to establish clear dosing methods, accurately measure nanoparticles in the test medium, and report abiotic factors influencing their behaviour. Additionally, generating chronic ecotoxicity data and understanding bioaccumulation potential is essential for evaluating the long-term ecological impacts of engineered nanoparticles (Crane et al. 2008).

High-throughput laboratory methods are valuable for assessing the ecotoxicological impact of nanomaterials. These methods allow for rapid screening of large libraries of engineered nanomaterials (ENMs) to identify potential hazards at cellular and organismal levels (Godwin et al. 2009).

The University of California Center for the Environmental Implications of Nanotechnology (UC CEIN) has developed predictive toxicological approaches to assess ENM libraries, focusing on material properties that could lead to environmental hazards (Thomas et al. 2011).

However, ecotoxicity testing cannot pace the rate of new NMs production, and strategies, such as grouping and read-across frameworks, have been developing during the last years to minimize the need for experimental tests. One of the most advanced is the GRACIOUS Framework approach, which is built upon developing a hypothesis, collecting only the data needed to support the hypothesis that links NMs properties to fate and hazard characteristics relevant to risk assessment. For a set of pre-defined hypotheses, integrated approaches for testing and assessment (IATAs) are needed to support the efficient and effective collection of data that enables supporting or rejecting each hypothesis. IATAs were initially developed for human toxicology. For ERA, one of the most recently developed is for surface water (Cross et al. 2024).

3.1.2 Exposure Assessment

The environmental exposure assessment of nanomaterials is a complex and evolving field that involves extensive data acquisition of the environmental occurrence of NMs or, in its majority, predicting environmental concentrations (PEC) and characterizing nanomaterials in various environmental matrices.

Life cycle analysis is essential to understanding the environmental implications of nanomanufacturing, including the fate, transport, and toxicity of nanomaterials. This approach helps identify potential environmental release points and potential impact (Elzey et al. 2009; Martínez et al. 2020).

Material Flow Analyses (MFAs) and Environmental Fate Models (EFMs) are used for predicting the environmental concentrations of nanomaterials and take into account factors such as nanoparticle size distribution, release dynamics, and environmental transformations. EFMs, including multimedia compartment models and watershed models, provide estimates of mass flows and concentrations in various environmental compartments, aiding in exposure assessment (Keller et al. 2024).

Multimedia compartment models and watershed models are used to predict concentrations in different environmental compartments, such as air, water, and soil. These models can be adapted to consider dynamic conditions, such as new nanomaterial introductions or seasonal variations (Keller et al. 2024).

3.2 Analytical Methods for Detection

Due to the great variety of NMs that can be delivered in the environment, the development of analytical techniques for their identification and quantification has increased in the last decades. However, there is not a universal analytical tool that can be used for all kinds of NMs and for that, the selection of a set of analytical techniques based on different NMs physicochemical properties is of vital importance to be able to identify with high precision and accuracy what NM was found in the specific environmental matrix and reduce the "false positives" in this field. Thus, the detection of nanomaterials as environmental pollutants has advanced significantly through various innovative methods, taking into account the unique properties of NMs to enhance sensitivity and selectivity. Table 15.1 aims to provide an overview of the most used analytical techniques to identify and quantify NMs, including plastics, in the environment, discussing each method's specific advantages and disadvantages.

Elemental analysis techniques are mainly based on mass spectrometry (MS), and a variety of microscopic techniques are commonly used for NM detection and/or quantification as they are relatively easy to apply and provide a rapid understanding of the behaviour of NPs in the environment matrix. Inductively coupled plasma (ICP) (López-Mayán et al. 2023) and chromatography (Okoffo and Thomas 2024) separation techniques coupled with MS are powerful tools for detecting and quantifying trace and ultra-trace elements. They yield elemental information of not only the dissolved fraction but also particulate elements, allowing the quantification of NMs in environmental liquid matrices without additional labelling. However, these

Table 15.1 Summary of the analytical techniques used for the identification and quantification of NMs, discussing the specific advantages and disadvantages of each method as well as the information extracted from each technique

Technique	Information given	Advantages	Limitations
Spectroscopy techniques			
Raman confocal microscopy	Chemical composition and crystallinity for several inorganic nanomaterials, carbon-based nanomaterials and nanoplastics Typical spatial resolution: 0.5–1 μm. Using optical tweezers down to 20 nm (Khosravi and Gordon 2024). In combination with AFM: down to 9 nm. (Bereczki et al. 2023) Quantification of particles per area Localization in complex environments	Non-destructive and non-contact Possibility for (semi)automation of analysis Potential to identify and size simultaneously (optical tweezer-Raman) Compatibility with flow-field-fractionation makes the setup well-suited for nanoparticles	Time-consuming data collection Particle signals are highly altered by the presence of other molecules (e.g., dye) and matrix interference (i.e., fluorescence) Optical trapping can be destabilized by Brownian motion of particles Limitation of the colocalisation to find big particle close to small particles
SERS	Chemical composition and crystallinity Potential for single particle analysis with a size down to 2 nm Localization in complex environments Quantification down to ng/L (Chaisrikhwun et al. 2023)	Non-destructive Highly sensitive technique with potential for single particle detection	Requires controlled creation of well-defined substrates; otherwise, substrate variability can lead to poor reproducibility of measurements Time-consuming data collection
XPS	Chemical composition Detection limited to samples containing >0.1 wt% of the element of interest	Non-destructive Analysis of the oxidation state of the particles Potential speciation	Time-consuming sample preparation and data analysis Surface composition only

NMR	Elemental composition Size and shape as well as the surface area Mass and surface composition Potential quantification	Non-destructive Relatively inexpensive, fast, and easy sample measurements Detection limits are size and shape independent	Well purified NMs are required. Substantial matrix interference (i.e., organic matter) In some cases, dissolving the particles is required prior to analysis Careful selection of the solvent to avoid interference (e.g. in the case of nanoplastic, solvent signal may overlap with the polymer signal) Requires a calibration curve before sample analysis Higher concentrations are needed for the detection (range 19–21 µg/mL) The limit of quantification in the range of mg/mL
Microscopy techniques			
SEM; SEM–EDX, FIB-SEM, Raman-SEM	Particle size Size distributions Particle shape Particle surface characteristics Combine with EDS/EDX, typically, and Raman (Schmidt et al. 2021), recently, for chemical composition Combination with FIB: Potential for 3D images of whole particle structure Localization in complex environments (e.g., particles adsorbed on the tissue' surface)	Non-destructive Simple sample preparation High resolution images Different options for obtaining chemical fingerprints with SEM images Combination with FIB: Possibility for 3D imaging	Charging effects Spatial resolution limit of >1 nm Relatively limited chemical information can be obtained (only using complimentary techniques) Must sacrifice some spatial resolution to obtain a chemical fingerprint (EDX special resolution >1 µm; and Raman: lateral resolution of 200–300 nm) Limit of detection with chemical composition >1 wt% Use FIB: sample destruction to obtain 3D images

(continued)

Table 15.1 (continued)

Technique	Information given	Advantages	Limitations
TEM, HRTEM, STEM-EDX	Particle size Size distributions Particle shape Combine with EDS/EDX for chemical composition Spatial resolution limit of >0.1 nm Localization in complex environments (e.g., association and or cellular uptake)	Non-destructive High resolution images	Chemical information can only be obtained via the use of complimentary techniques Sample preparation can be extensive
Fluorescence Microscopy; Confocal laser scanning microscopy (CLSM)	Presence of (fluorescent) additives Localization in complex environments Potential for 3D images Size ≥120 nm	Non-destructive Possibility for 3D imaging Allows for particle tracking in-situ	Limited to intrinsically fluorescent and fluorescently labelled NMs Risk of photobleaching No chemical confirmation Highly dependent upon microscopes used for measurements
Hyperspectral-enhanced Dark Field Microscopy	Localization and identification in complex environments Particle shape Particle size Imaging of NMs with size down to 5 nm (Fakhrullin et al. 2021)	Non-destructive Relatively simple and rapid technique Potential to detect particles in-situ with minimal sample preparation Rapid, label-free imaging High reproducibility and reliable particle identification Possibility for (semi)automation of analysis	Diffraction limited spatial resolution Difficult to discern particle scattering from other highly scattering material in complex samples Complete chemical composition only in combination with other techniques Requires calibration or creation of a reference library prior to sample analysis

AFM	Particle size Size distributions Particle shape Particle surface characteristics Viable for particles with sizes down to 1 nm Combination with vibrational spectroscopies, i.e. Raman and IR, for chemical fingerprint	Non-destructive High resolution images Can measure particles in dried or suspended states Possibility for 3D imaging	Do not give chemical information Time consuming data collection Artefacts due to particle movement during measurement Highly dependent upon microscopes used for measurements
Scattering techniques			
DLS, DDLS	Particle size Colloidal stability Viable for particles with sizes between 10 μm – 1 nm at concentrations of ~10^8–10^{12} particles mL^{-1} Using DDLS: Detection in complex environments	Non-destructive Simple Fast analysis Applicable for a relatively broad range of particle sizes and types DDLS reduces the matrix interference (minimum sample preparation for liquid samples)	Must limit the concentration of particle suspensions to prevent multi-scattering events No chemical confirmation Scattering strongly biased towards large particles or aggregates Errors when estimating the size of non-spherical particles (e.g. fibers or lines, platelets or films) For DDLS: Relatively complex data processing
NTA	Particle size Size distributions Colloidal stability Concentrations Viable for particles with sizes between 1 μm – 10 nm at concentrations of ~10^7–10^9 particles/mL	Non-destructive Single particle tracking limits scattering biases Good accuracy with polydispesed samples	No chemical information Applicable for a relatively narrow range of particle sizes

(continued)

Table 15.1 (continued)

Technique	Information given	Advantages	Limitations
AF4-MALS	Particle size Size distributions Potential for detection in complex environments Sample separation Viable for particles with sizes between 1 μm – 1 nm	Non-destructive Simultaneous sample separation and particle detection	No chemical information Risk of particle interaction with column membrane
TDA	Particle size Give concentrations Viable for particles with sizes <127 nm [136]. Particle size is limited by the diameter of the capillary used for measurements	Non-destructive Highly sensitive technique including nanoplastics Requires small sample volumes	Particle size is limited by the capillary used for measurements No chemical composition information Risk of particle interaction with the capillary
Mie Scattering Technique (Niskanen et al. 2019; Mou et al. 2024)	Size down to 10 nm Limit of detection limit of 4.2 μg/L for particles as small as 25 nm Potential quantification	Non-destructive Simple technique	Errors when estimating the size of non-spherical particles (e.g. fibers or lines, platelets or films) Relatively complex data processing
Thermal Techniques			
DSC	Transition temperatures (i.e. T_g, T_c, T_m) Particle mass Particle crystallinity Limited to a minimum sample mass of >0.2 mg	Allows for mass quantification of particles present in a sample Relatively simple sample preparation	Limited to organic NMs (e.g. nanoplastics) Sample destruction High signal variability for particles that are different sizes or have different degrees of crystallinity Difficult to distinguish between polymers whose melting points are similar Not well suited for analysis of particles composed of polymers with broad thermal transition temperature ranges Highly dependent upon the machine used for the measurements No information of size/shape

TGA	Particle mass Thermal stability Detection of particle degradation Possibility to combine with IR or Py-GC–MS to obtain chemical composition information Limited to use for samples with masses between 100 mg – 1 mg	Viable for higher sample masses, thus allowing particle analysis in-situ Relatively simple sample preparation	Sample destruction Limited to organic NMs Most polymer decomposition products are formed in the same temperature range Need well-purified NMs (overlap in the thermal degradation temperatures with other organic matter) Requires calibration prior to sample analysis Highly dependent upon the machine used for the measurements No information of size and shape
Py-GC–MS (Okoffo and Thomas 2024) or Py-GC–MS-ToF	Chemical Composition Characterization of additives and coatings Viable for use with samples containing >1 μg of the material of interest. However, there is high variation for the LOD of different materials	Chemical analysis of multiple particles at a time is possible Highly sensitive detection of additional chemicals or additives	Sample destruction Limited to organic NMs Relatively complex data processing Requires dry, preconcentrated samples Highly variable LODs Limited to small quantities (i.e. <0.5 mg of material No information about size and shape
TED-GC–MS	Chemical Composition Viable for samples with masses of <20 mg that contain >0.5 wt% of the material of interest	Highly sensitive chemical analysis Possibility to analyze particles in-situ without removal of surrounding matrix	Sample destruction Limited to organic NMs No information about size and shape Limited to organic NMs Requires calibration prior to sample analysis
MALDI-ToF	Chemical fingerprint Particle size Viable for use with samples containing >25 ng [143] of the material of interest.	Highly sensitive technique which can provide information on molecular structures and molecular weights Potential for direct identification of additives or adsorbed chemicals	Sample pretreatment (i.e. sample destruction through thermal means or dissolution in organic solvents) required to obtain reliable data

(continued)

Table 15.1 (continued)

Technique	Information given	Advantages	Limitations
SP-ICP-MS	Elemental composition Size Quantitative technique Limit of sizes >15 nm Combination with centrifugal ultrafiltration can be distinguished between nanoparticles and ionic metals (<10% error in detecting nanoparticles (Henke et al. 2024))	Potential for simultaneous collection of data on particle size, concentration, and chemical composition Can be coupled with AF4 to obtain size distribution Simultaneous information about dissolved and particulate element Combination with laser ablation can be analyzed also solid samples (Laborda et al. 2023)	Sample destruction Information about the element mass, but not about size/shape or composition Difficult analysis of organic particles caused by low ionization efficiency of carbon Incomplete volatilization can cause an underestimation of the size Requires calibration prior to sample analysis
ToF-SIMS Imaging	Chemical information Size Lateral resolution limit of >70 nm and depth resolution limit of >9 nm	Potential for simultaneous collection of data on particle size and chemical composition Potential to detect differences in particles because of degradation	Sample destruction Difficult to differentiate plastics from other organics present in samples. More complex data analysis required Dependent upon the machine used for measurement
Others			
QCM	Particle mass and concentrations Viable for samples containing >1 ng/mL of the material of interest	Non-destructive Highly sensitive and accurate detection of sample mass	No chemical information Cannot differentiate between mass of NMs and mass of other organic matter Requires relatively pure samples
Particle-Impact Electrochemistry	Particle size Particle concentration Viable for samples containing >1 pM of interest material	Non-destructive Highly sensitive and accurate technique with potential for single particle detection Capable of measuring polydisperse samples	Focused more on studying the interactions between particles and the chosen electrodes Highly specialized equipment and data interpretation required

The information included in this table was extracted mainly from the reviews published by Caldwell et al. (2022), Labuda et al (2023), Tawfik A. Saleh (2020) and Abbas et al. (2024)

methods do not provide any information about NM distribution in the matrix, which is important when studying bioaccumulation. Pyrolysis gas chromatography-MS (PyGC-MS) and thermal desorption-proton transfer reaction-MS (TD-PTR-MS) have been used incrementally to identify nanoplastics in environmental matrix. Despite their extraordinary sensitivity (< 1 μg) and quantification capability, they are destructive techniques that prevent measurement repetition and cannot be used to gain insight into exact particle properties such as size or shape.

Raman spectroscopy has been a tool in characterising the natural and man-made NMs based on their vibrational modes (Nivaz et al. 2022). This technique offers the great advantage of not being invasive and can detect the presence of NMs in tissues of several millimetres in thickness. Additionally, as Raman spectroscopy is unaffected by water, it is a highly suitable technology in the environment field. However, Raman spectroscopy has a reported resolution limit of ~0.5 μm when Raman confocal microscopy is used, which hinders its use for detecting single particles at nano range (1–100 nm) even in simple samples. Moreover, this technique in pone to fluorescence interference and has an inherently low signal to noise ratio. Unlike Raman spectroscopy, surface-enhanced Raman scattering (SERS) overcomes limitation due to weak signals via electromagnetic enhancement mechanism to Raman signal enhancement (Terry et al. 2022). Additionally, the ability of metal surfaces to quench fluorescence background, makes SERS becoming sensitive enough to detect traces analysis even at the single level particle. However, only few applications on NMs detection have been reported with exception of nanoplastics (Quarato et al. 2021; Caldwell et al. 2024). As many of the reports have focused on the detection of a limited number of nanoplastics, often in very clean samples matrices such as ultrapure water. This leaves many uncertainties regarding the influence of parameters such as plastic type and sample matrix content on particle detection with SERS.

For the localization of NMs within a cell and tissue as well as the morphological analysis of these NMs (single particle or aggregates, shape and size), microscopic techniques such as light and transmission electron microscopy must be applied. Light microscopy provides visualization of NMs ether by their intrinsic physicochemical properties (e.g. scattering (Fakhrullin et al. 2021)and chemical composition (Vieira et al. 2023)) or by labelling them (e.g. fluorescently tagged (Nguyen and Tufenkji 2022)) but with its resolution limited to approximately 200–500 nm (Verdaasdonk et al. 2014) cannot resolve the NMs (size 100–1 nm) and distinguish between a single particle or agglomerates. Transmission electron microscopy (TEM) with a resolution range from Ångstrom to nanometer is the method of choice for resolving electron dense particles but requires in the case of localize NMs into tissue a complex and long sample section preparation, including fixation of the tissue. Scanning electron microscopy (SEM) offers a fast visualization with a simpler sample preparation; however, its size resolution depends to the nature of the NM and the presence of non-conductive materials provokes drifting in the image acquisition. Beside imaging NMs, elemental composition can be determined using energy dispersive X-ray spectroscopy (EDS/EDX)(Kim et al. 2014) in both TEM and SEM and recently Raman spectroscopy in SEM (Schmidt et al. 2021). For these microscopic techniques, stereological approaches are required to understand the 3D distribution of NMs within a defined reference volume (Gimenez et al. 2016).

To understand the colloidal stability of NMs, light scattering techniques are commonly used. Dynamic light scattering (DLS) is particularly popular and widely adopted for this purpose because it's easy to use and its highly quantitative nature. However, the analysis of non-spherical NMs and polydisperse samples are not straightforward by DLS. Additionally, low-scatted NMs (e.g. plastics) are difficult to distinguish from the matrix and the hydrodynamic size can be overestimated for polydisperse samples since the presence of small population of big particles have a strong impact due to Rayleigh approximation used in DLS (Caldwell et al. 2022). Dynamic depolarized light scattering (DDLS) overcomes part of the limitation of DLS because this method offers scattering information originating exclusively from the NMs on an essentially zero-background as reported by Balog et al (Balog et al. 2015). Another alternative to DLS is nanoparticle tracking analysis (NTA). The size values in NTA are number-weighted and consequently theresults are less biased by the presence of a population of larger particles. In addition, NTA provides information about the concentration of particles/mL and has been reported as a promising tool to characterize polydispersed NMs in complex matrices (Mehrabi et al. 2017). This is highly relevant in the study of nanoparticles released from or in natural samples, as non-desirable particles can be discriminated, and nanoparticles with different sizes in a polydisperse sample can be visualised by simply adjusting the camera level and the screen gain (Sorasan et al. 2024).

4 Regulatory and Policy Aspects

4.1 International Guidelines and Standards

The regulation of nanomaterials in the environment involves multiple key regulatory bodies and frameworks at both regional and global levels. These bodies ensure the safe use, handling, and disposal of nanomaterials to mitigate potential environmental and health risks. Some of the most relevant regulatory bodies and frameworks involved in regulating nanomaterials are briefly described next.

4.1.1 United States
The United States Environmental Protection Agency (EPA) is responsible for ensuring that nanomaterials in products are safe for both human health and the environment. Under Toxic Substances Control Act (TSCA), nanomaterials are subject to pre-manufacture notifications, which require information on their production and intended use. The EPA can impose restrictions or bans on nanomaterials that pose significant risks.

4.1.2 Canada
Health Canada, under Canadian Environmental Protection Act (CEPA), is responsible for assessing and managing the risks posed by nanomaterials in Canada. The

government has specific provisions that require risk assessments for new nanomaterials before they can be introduced into the market. It collaborates closely with the Canadian Food Inspection Agency (CFIA) for products that may impact the environment.

4.1.3 European Union

The European Union's Registration, Evaluation, Authorisation, and Restriction of Chemicals (REACH) regulation includes specific provisions for nanomaterials. These provisions, effective since January 2020, aim to ensure comprehensive risk assessments of nanomaterials, addressing challenges in hazard and exposure assessments (Schwirn et al. 2020; Tschiche et al. 2022).

EFSA provides scientific advice on the safety of nanomaterials used in food and feed within the EU. They evaluate potential environmental impacts, especially regarding nanomaterials' use in food packaging and processing. This includes setting specifications and restrictions to minimize consumer exposure (Gazsó et al. 2019). A proposal by organizations like CIEL and ClientEarth suggests a horizontal regulation to address gaps in existing EU legislation, ensuring comprehensive hazard, risk, and exposure assessments for nanomaterials (Azoulay and Buonsante 2014).

4.1.4 Global Regulatory Framework

At the global level, UN Strategic Approach to International Chemicals Management (SAICM) has developed nanospecific resolutions and activities to manage nanomaterials, integrating them into its Global Plan of Action. These efforts align with World Trade Organization (WTO) agreements, ensuring compatibility with international trade laws (Karlaganis and Liechti 2013).

At the national level, the National Institute for Public Health and the Environment (RIVM) is noteworthy. It is actively involved in assessing the risks associated with nanomaterials, especially concerning environmental health. They provide recommendations for regulating nanomaterials and have worked with the Dutch government and the European Commission to develop safety guidelines.

4.2 Existing Guidelines and Standards for Nanomaterials

The guidelines and standards for nanomaterials in the environment focus on their characterization, risk assessment, and regulatory compliance, considering their distinct properties and potential environmental impacts. Developing standardised methods and protocols is essential to ensuring consistent and reliable assessments of nanomaterials in various environmental contexts.

The European Union's REACH legislation has been updated to include specific provisions for nanomaterials, requiring detailed physicochemical characterization and environmental safety assessments. This includes methods for determining adsorption/desorption, degradation, and exposure scenarios, although some areas still lack standardized methods (Nielsen et al. 2021).

The dependence of NMs' properties and hence (eco)toxicology dependence on size, morphology, composition, surface coatings, and environmental conditions have stimulated several major initiatives internationally to develop standard reference materials and protocols. Within the U.S., the National Institute of Standards and Technology (NIST) has validated standard reference materials and standard protocols for characterising the physicochemical properties of these materials. In addition, the International Alliance for NanoEHS Harmonization, established in 2007, is committed to developing and distributing standard reference materials, methods, and procedures for studying the environmental health and safety of NMs.

Organisation for Economic Co-operation and Development (OECD) plays a critical role in providing guidelines and testing protocols for nanomaterials, especially regarding environmental safety. They focus on the potential risks of nanomaterials and collaborate with member countries to develop policies for the safe use and disposal of nanomaterials. The OECD has developed Test Guideline No. 318, which focuses on the dispersion stability of nanomaterials in simulated environmental media. This guideline is designed to be executed in standard laboratories and provides a systematic approach to compare different engineered nanomaterials (ENMs) under environmentally relevant conditions (Monikh et al. 2018). There is a lack of standardized dispersion protocols for ecotoxicity testing of nanomaterials in the marine environment. Although some studies have characterised nanomaterials in ecotoxicological media, many do not follow standardised procedures, highlighting the need for more consistent methodologies (Brunelli et al. 2024). ISO develops international standards for nanotechnology, which are used by regulatory bodies worldwide. ISO standards help define safe practices for handling nanomaterials, their characterisation, and their environmental and health risks. While ISO is not a regulatory agency, it provides essential national and international regulation frameworks.

The existing guidelines and standards for nanomaterials in the environment are based on traditional chemical testing protocols, with adaptations needed due to the unique properties of nanomaterials. The OECD and ISO provide guidelines for assessing the toxicity of nanomaterials, emphasizing the importance of specific preparation, delivery, and metrology for accurate testing. Current test guidelines focus on terrestrial toxicity testing using various organisms like *Enchytraeus crypticus*, *E. albidus*, *Caenorhabditis elegans*, and rodents to evaluate the environmental impacts of nanomaterials. Adaptations to existing guidelines and the development of novel tools are ongoing to enhance the accuracy of toxicity assessments for nanomaterials in the environment (Mendonça et al. 2017).

Harmonizing testing media for nanomaterial research to increase data comparability is very important. It has been suggested that a minimum set of medium characteristics should be reported in five categories: aquatic, soil/sediment, biological, engineered systems, and product matrix testing media. This harmonization will support predictive utility assessments, mechanistic understanding, and categorization strategies in nanomaterial environmental, health, and safety research (Geitner et al. 2020).

The European Chemical Agency (ECHA) is revising its guidance documents on how the industry is to complete chemical safety assessments to address the challenges that nanoparticles pose for ecotoxicological testing. Progress has been made in updating guidance documents on sample preparation and characterization, but more specific guidance is needed on nanospecific sample preparation, characterization techniques, and interpreting results of ecotoxicological testing of nanoparticles.

Recommendations from the EnvNano project include the need for multiple characterisation methods to describe dispersion and dissolution rates over time and for various test concentrations, supplementing existing algal and daphnia tests with additional tests for nanomaterials that dissolve in testing media, and the importance of considering shading in algal tests and determining uptake, depuration, and trophic transfer of nanomaterials for each commercialised functionalisation (Hansen et al. 2017).

The existing guidelines and standards for nanomaterials in the environment do not currently incorporate the impact of an "eco-corona" formed by the adsorption of natural biomolecules onto nanomaterials. This eco-corona alters nanomaterials' stability, identity, and toxicity, particularly to organisms like *Daphnia magna*. The paper suggests updating toxicity testing policies to include the formation of an eco-corona before testing to ensure a more realistic representation of environmental conditions and to accurately assess the ecological relevance of nanomaterial toxicity (Nasser et al. 2020).

On the other hand, advanced analytical techniques are essential for the detection and characterisation of nanomaterials in complex environmental matrices. Current methods include microscopic, spectroscopic, and mass spectrometric techniques, but challenges remain in improving sensitivity and resolution for in situ analysis (Jiang et al. 2022).

Regulatory frameworks, such as those developed by the OECD, have been established to address nanomaterials' potential risks to human health and the environment. These frameworks include guidelines for assessing risks from inhalation and other exposure routes, reflecting nanomaterials' unique properties (Rehman and Moore 2021).

While significant progress has been made in developing guidelines and standards for nanomaterials, challenges remain in achieving comprehensive standardization across all environmental contexts. Continued efforts are needed to refine existing methods and develop new protocols to address the dynamic and complex nature of nanomaterials in the environment.

5   Strategies for Ensuring Environmental Nanosafety

5.1   Safe Design of Nanomaterials

5.1.1   Principles of Green Nanotechnology

Safe-and-Sustainable-by-Design (SSbD) concept integrates safety and sustainability into the design of nanomaterials, ensuring that they are both effective and environmentally benign (Wyrzykowska et al. 2022). SSbD of NMs is a critical area of research, focusing on integrating safety and sustainability principles from the early stages of material innovation. This approach, known as Safe and Sustainable by Design (SSbD), aims to address potential hazards and environmental impacts throughout the lifecycle of nanomaterials. The following sections outline key aspects of SSbD, drawing insights from recent research.

The European Commission emphasises the need to integrate of SSbD principles early in the innovation process to facilitate market introduction and reduce costs. This involves considering ecological, social, and economic factors, and aligning regulatory aspects to foster informed decision-making and innovation (Cassee et al. 2024).

The development of categorization tools, such as the AdMaCat, supports the integration of SSbD principles by providing a framework for evaluating the environmental impact of nano-enabled materials during the design phase (Medina et al. 2024).

The Safe-by-Design (SbD) approach offers valuable insights for developing SSbD frameworks. Key recommendations include preserving knowledge from existing approaches, incorporating lifecycle thinking, and developing high-throughput screening models to operationalize SSbD (Sudheshwar et al. 2023).

While the SSbD framework provides a comprehensive approach to designing safe and sustainable nanomaterials, challenges remain in harmonizing standards and ensuring regulatory preparedness. Continuous collaboration among stakeholders and regular updates to the framework are essential to address these challenges and achieve sustainable innovation in nanomaterials (Cassee et al. 2024).

5.1.2 Design Strategies to Minimize Environmental Impact

A comprehensive design strategy is essential to minimize nanomaterials' environmental impact. This strategy focuses on eco-design principles, understanding toxicity mechanisms, and leveraging computational models. These strategies aim to balance nanomaterials' benefits in environmental applications with their potential risks to ecosystems and human health.

Incorporating ecological risk assessment at the design stage can help identify and mitigate hazardous features of nanomaterials. This approach involves evaluating the ecotoxicity of nanomaterials alongside their performance, allowing for the design of safer materials from the outset (Ilaria et al. 2022).

In marine environments, eco-design strategies should consider the interactions of NMs with natural and anthropogenic pollutants. This includes understanding the formation of eco- and bio-coronas and their impact on marine organisms (Corsi et al. 2021).

Understanding the molecular mechanisms of nanoparticle toxicity, such as cell surface interactions, ion dissolution, and reactive oxygen species generation, is crucial. Redesign strategies can then be tailored to mitigate these specific toxic effects. Strategies to reduce toxicity include altering surface charge, using protective coatings, and incorporating less toxic elements. These modifications aim to maintain the functional properties of nanomaterials while reducing their environmental impact (Buchman et al. 2019).

Utilizing nanoinformatics and predictive models can optimize the design of nanomaterials by identifying hazardous features early in the development process. This involves using nanodescriptors to predict the environmental and health impacts of nanomaterials (Wyrzykowska et al. 2022).

While these strategies offer pathways to reduce the environmental impact of nanomaterials, it is important to consider the broader context of their application. The effectiveness of these strategies depends on continuous research and adaptation to new findings in nanotoxicology and environmental science. Additionally, balancing the trade-offs between functionality and safety remains a critical challenge in the sustainable development of nanotechnologies.

6 Future Perspectives and Research Needs

6.1 Emerging Technologies and Applications

Emerging technologies and applications of nanomaterials are revolutionising various fields, particularly in healthcare, environmental monitoring, and analytical chemistry. The following sections explore key areas where nanomaterials are making significant impacts and from where we can expect emerging risks in the next years if significant release into the environment is verified.

- **Biomedical Applications:** Nanomaterials are being used to enhance drug delivery systems, allowing for targeted and controlled release of therapeutics. This includes programmable drug delivery systems that can interface with human tissue to deliver drugs precisely where needed (Malik et al. 2023; Hughes et al. 2024). Nanotechnology is advancing diagnostics and treatment in fields such as oncology, gene therapy, and regenerative medicine. For instance, metal–organic frameworks (MOFs) have been developed for effective osteoarthritis treatment, demonstrating low cytotoxicity and high efficiency in delivering therapeutic microRNAs (Jia et al. 2023). Nanomaterials are also being integrated into new medical tools and processes, enhancing the sensitivity and specificity of diagnostic procedures (Malik et al. 2023).
- **Environmental Monitoring and Pollution Control:** Nanomaterials like metal oxides and graphene are being utilized for the detection and removal of air pollutants. These materials offer high sensitivity and efficiency, capable of detecting pollutants at parts-per-billion levels and effectively removing them from the environment (Samriti et al. 2023). Fe-based MOFs have shown promise in degrading hazardous medical waste, providing a sustainable solution to environmental pollution from medical sources (Jia et al. 2023).
- **Analytical Chemistry and Detection Technologies:** Nanomaterial-Based Films are being developed for sensitive detection of analytes in agriculture, food safety, and clinical diagnostics. They take advantage of the porosity and flexibility of

nanomaterials to facilitate rapid and efficient analyte detection. Emerging applications include the development of wearable devices and detection chips that utilize nanomaterials for real-time monitoring and diagnostics (Ma et al. 2022).
- **Sustainable Agriculture and Crop Productivity:** Emerging technologies and applications of nanomaterials in agriculture and food are revolutionizing the industry by enhancing crop productivity, improving food quality, and promoting sustainable practices. These advancements address critical challenges such as food security, environmental sustainability, and efficient resource utilization. The integration of nanotechnology in agriculture offers innovative solutions across various stages of crop production and food processing. Nano-enabled delivery systems optimize the application of fertilizers and pesticides, minimizing waste and maximizing efficacy (Shah et al. 2024). Some developed nanocarriers enhance crop productivity and environmental management by reducing ecotoxicity and enabling controlled release of agrochemicals. On the other hand, nanoparticles such as Ag, Fe, Zn, and TiO_2 are employed in modern farming for seed treatment and disease diagnostics, transforming agricultural ecosystems to ensure future food security (Mohanty et al. 2024). Nanotechnology facilitates precision agriculture through nanosensors and nanobiosensors that monitor crop health, water usage, and soil quality. These technologies also improve food processing by enhancing preservation, smart packaging, and quality control, reducing post-harvest losses and extending shelf life (Shah et al. 2024). Nanomaterials are being utilised as advanced diagnostic tools and in crop disease-resistance strategies. They enable rapid pathogen detection and targeted disease management, reducing reliance on chemical inputs and promoting sustainable agricultural practices. Nano-fungicides, nano-bactericides, and nano-pesticides are nanomaterials that help manage emerging pathogens and enhance plant health by delivering plant hormones and signaling molecules (Jabran et al. 2024). Nanomaterials support climate-smart agriculture by equipping plants to combat abiotic and biotic stresses, optimizing nutrient supply, and improving soil and water quality. These advancements aim to achieve sustainable food production amidst climate challenges (Otari et al. 2024). While nanomaterials improve crop productivity, their impact on plant-associated microorganisms, crucial for nutrient uptake and disease prevention, requires further exploration. Sustainable practices must consider the effects of nanoparticles on these microorganisms to ensure long-term agricultural health.
- **Nanomaterials in Bioenergy Production:** Nanoparticles are pivotal in optimising biomass conversion for biofuel production. They enhance enzyme immobilisation and improve pre-treatment and catalytic processes, making biofuel production more efficient and cost-effective (Ghosh et al. 2024). Using metallic nanomaterials, nanocomposites, and carbon-based nanomaterials significantly boosts biodiesel, biogas, biohydrogen, and bioethanol production and purification. These materials improve feedstock conversion rates and product quality (Awogbemi and Von Kallon 2024).
- **Hydrogen Production and Storage:** Nanomaterials are central to hydrogen production, mainly through photocatalytic reactions and microbial fuel cells.

Their unique properties enhance the efficiency of hydrogen synthesis and storage, addressing challenges in renewable energy systems. Developing nanomaterials from waste for hydrogen production is an emerging area, focusing on sustainable and eco-friendly energy solutions (Anish et al. 2023).
- **Renewable Energy Systems:** In solar energy, nanotechnology enhances the efficiency and lifespan of photovoltaic cells, leading to increased electricity production. This is crucial for the advancement of solar energy technologies. Nanomaterials are also being used in wind energy applications to improve fault detection in wind turbines, thereby increasing the reliability and efficiency of wind energy systems (Tirth et al. 2023).
- **Green Nanoparticles in Energy Applications:** Green nanoparticles, synthesized through eco-friendly methods, are being explored for their potential in energy storage devices like batteries and supercapacitors. These materials offer a sustainable approach to energy production and storage (El-Esawy et al. 2023).

6.2 Research Gaps

The environmental risk assessment of nanomaterials (NMs) is a blooming field with significant research gaps which span various domains, including the environmental impact of nanoparticles, the challenges in assessing their effects in agricultural settings, the complexities of nanoplastics, and the limitations of current assessment models.

More studies using diverse biological models and biomarkers are needed to assess the environmental impact, as current research predominantly uses immortalised cell lines and primary cells (Kumah et al. 2023).

The environmental risk assessment of nanomaterials in agriculture is still in its infancy. Critical knowledge gaps include understanding the environmental concentrations of nanoparticles and their effects on soil life. Some processes such as adsorption/desorption, dissolution, and biological degradation of nanoparticles in soil are poorly understood, highlighting the need for field-based studies to assess these factors (Singh and Gurjar 2022).

Research on nanoplastics is limited, with few studies focusing on their presence in biotic matrices. Most research has concentrated on the effects of nanoplastics on aquatic organisms, leaving a gap in understanding their impact on terrestrial organisms. To address these gaps, a tailored environmental risk assessment framework for nanoplastics is needed (Masseroni et al. 2022).

Traditional experimental methods for assessing the risks of engineered nanomaterials are time-consuming and costly. Although promising, computational models face challenges such as the need for more accurate data and improved model validation (Tang et al. 2024). The disconnection between simplified experimental systems and real-world conditions poses a challenge in translating findings into practical applications. A framework for evaluating the environmental relevance of experimental designs is necessary to bridge this gap (Surette et al. 2021).

Despite advancements in understanding the environmental fate and effects of ENMs, there are still significant knowledge gaps. These gaps arise due to the challenge of translating findings from simplified experimental systems to complex real-world scenarios. To address this, the Framework for Relevance And Methods Evaluation (FRAME) was proposed, focusing on three pillars: ENM properties, experimental conditions, and exposure scenarios. By utilising FRAME, researchers can assess the environmental relevance of their experimental designs, identify gaps, and prevent overgeneralisation of results, ultimately advancing the field of environmental risk assessment of nanomaterials (Surette et al. 2021).

While these research gaps highlight the challenges in ERA of nanomaterials, they also present opportunities for advancing the field. Addressing these gaps through interdisciplinary research and improved methodologies can lead to more effective risk management strategies for nanomaterials.

There is a gap regarding the need to consider chemical transformations and surface functionalization when grouping nanoparticles for assessment. It is very important assessing how transformations affect toxicity levels, especially in different environmental conditions. While pristine nanoparticles may show similar toxicity, transformations can significantly alter their hazard potential, making it challenging to group them based solely on manufacturing properties. Therefore, future research should focus on understanding how these transformations impact the ERA of nanomaterials (Schultz et al. 2020).

Eco-design strategies are crucial for integrating nanosafety into the development of MNMs. By considering ecological risk assessments at the design stage, hazardous features of nanomaterials can be identified and mitigated early on. This approach not only enhances the safety of nanotechnologies but also supports their sustainable application in environmental remediation (Corsi et al. 2021; Ilaria et al. 2022).

Traditional toxicological methods are insufficient for assessing the risks associated with nanomaterials. New approaches, such as nanotoxicology, are being developed to evaluate potential differential toxicity, immunotoxicity, and genotoxicity. Technologies like organs-on-chips and sophisticated sensors are also being adapted to understand better the interactions of nanomaterials with biological systems (Lebre et al. 2022).

For instance, the testing of biofunctionalized nanosilver has shown that chronic exposure scenarios provide a more accurate assessment of ecotoxicity compared to acute exposure. This highlights the need for ecologically relevant testing strategies to ensure the environmental safety of nanomaterials (Bellingeri et al. 2022).

6.2.1 Analytical Challenges and Prospects

Due to their low concentrations and structural heterogeneity, analysing and characterising nanomaterials in the environment is challenging. Although the advanced methods described in Sect. 3.2 are employed to overcome these challenges, improved analytical sensitivity and resolution are needed to facilitate the standardisation of nanomaterial analysis methods, which is crucial for accurate environmental risk assessment (Jiang et al. 2022).

Acknowledgments The authors acknowledge funding from NANOCULTURE Interreg Atlantic Area project (EAPA_590/2018) and SbDToolBox- Nanotechnology-based tools and tests for Safe-by-Design nanomaterials (NORTE-01-0145-FEDER-000047) supported by North Portugal Regional Operational Programme (NORTE2020) under the PORTUGAL 2020 Partnership Agreement through the European Regional Development Fund (ERDF). L.R.-L. acknowledges funding to FCT (Fundação para a Ciência e Tecnologia) for the Scientific Employment Stimulus Program (2020.04021.CEECIND).

References

Abbas Q et al (2024) Recent advances in the detection and quantification of manufactured nanoparticles (MNPs) in complex environmental and biological matrices. J Clean Prod 471:143454. https://doi.org/10.1016/J.JCLEPRO.2024.143454

Ahmed SN et al. (2022) Application of nanomaterials in environmental pollution abatement and their impact on ecological sustainability: recent status and future perspective. 629–654. https://doi.org/10.1201/9781003161158-26

Anish M et al (2023) Utilization of nano materials in hydrogen production - emerging technologies and its advancements: an overview. Int J Hydrog Energy. https://doi.org/10.1016/j.ijhydene.2023.05.223

Araújo MJ et al (2022) Proteomics reveals multiple effects of titanium dioxide and silver nanoparticles in the metabolism of turbot, Scophthalmus maximus. Chemosphere 308:136110. https://doi.org/10.1016/j.chemosphere.2022.136110

Araújo MJ et al (2024) Diving into the metabolic interactions of titanium dioxide nanoparticles in "Sparus aurata" and "Ruditapes philippinarum". Environ Pollut 360:124665. https://doi.org/10.1016/J.ENVPOL.2024.124665

Awogbemi O, Von Kallon DV (2024) Recent advances in the application of nanomaterials for improved biodiesel, biogas, biohydrogen, and bioethanol production. Fuel. https://doi.org/10.1016/j.fuel.2023.130261

Azoulay D, Buonsante VA (2014) Regulation of nanomaterials in the EU: proposed measures to fill in the gap. Eur J Risk Regul 5(2):228–235. https://doi.org/10.1017/S1867299X00003652

Balog S et al (2015) Characterizing nanoparticles in complex biological media and physiological fluids with depolarized dynamic light scattering. Nanoscale 7(14):5991–5997. https://doi.org/10.1039/C4NR06538G

Bellingeri A et al (2022) Ecologically based methods for promoting safer nanosilver for environmental applications. J Hazard Mater 438:129523. https://doi.org/10.1016/j.jhazmat.2022.129523

Bereczki A et al (2023) Sub-10 nm nanoparticle detection using multi-technique-based micro-raman spectroscopy. Polymers 15(24):4644. https://doi.org/10.3390/POLYM15244644/S1

Borymski S et al (2023) Copper-oxide nanoparticles exert persistent changes in the structural and functional microbial diversity: a 60-day mesocosm study of zinc-oxide and copper-oxide nanoparticles in the soil-microorganism-nanoparticle system. Microbiol Res 274:127395. https://doi.org/10.1016/j.micres.2023.127395

Brunelli A et al (2024) An overview on dispersion procedures and testing methods for the ecotoxicity testing of nanomaterials in the marine environment. Sci Total Environ:171132. https://doi.org/10.1016/j.scitotenv.2024.171132

Buchman JT et al (2019) Understanding nanoparticle toxicity mechanisms to inform redesign strategies to reduce environmental impact. Acc Chem Res 52(6):1632–1642. https://doi.org/10.1021/ACS.ACCOUNTS.9B00053

Caldwell J et al (2022) The micro-, submicron-, and nanoplastic hunt: a review of detection methods for plastic particles. Chemosphere 293:133514. https://doi.org/10.1016/J.CHEMOSPHERE.2022.133514

Caldwell J et al (2024) Detection of submicron- and nanoplastics spiked in environmental fresh- and saltwater with Raman spectroscopy. Mar Pollut Bull 203:116468. https://doi.org/10.1016/J.MARPOLBUL.2024.116468

Cao H et al (2024) Effects of graphene-based nanomaterials on microorganisms and soil microbial communities. Microorganisms. https://doi.org/10.3390/microorganisms12040814

Cassee FR et al (2024) Roadmap towards safe and sustainable advanced and innovative materials. Comput Struct Biotechnol J 25:105–126. https://doi.org/10.1016/j.csbj.2024.05.018

Chaisrikhwun B, Ekgasit S, Pienpinijtham P (2023) Size-independent quantification of nanoplastics in various aqueous media using surfaced-enhanced Raman scattering. J Hazard Mater 442:130046. https://doi.org/10.1016/J.JHAZMAT.2022.130046

Chen S et al (2023) Threats to the soil microbiome from nanomaterials: a global meta and machine-learning analysis. Soil Biol Biochem. https://doi.org/10.1016/j.soilbio.2023.109248

Corsi I et al (2021) Eco-interactions of engineered nanomaterials in the marine environment: towards an eco-design framework. Nanomaterials 11(8). https://doi.org/10.3390/NANO11081903

Crane M et al (2008) Ecotoxicity test methods and environmental hazard assessment for engineered nanoparticles. Ecotoxicology 17(5):421–437. https://doi.org/10.1007/S10646-008-0215-Z

Cross RK et al (2024) An integrated approach to testing and assessment (IATA) to support grouping and read-across of nanomaterials in aquatic systems. Nano Today 54:102065. https://doi.org/10.1016/j.nantod.2023.102065

Elbehiry F, Elbasiouny H, El-Ramady H (2022) Bioaccumulation and biotransformation of metal-based nanoparticles in plants. In: Toxicity of nanoparticles in plants. Elsevier, pp 299–315. https://doi.org/10.1016/B978-0-323-90774-3.00001-5

El-Esawy MA et al (2023) Recent advances of green nanoparticles in energy and biological applications. Mater Today. https://doi.org/10.1016/j.mattod.2023.12.001

Elzey, S. et al. (2009) Nanoscience and nanotechnology: environmental and health impacts:681–727. https://doi.org/10.1002/9780470523674.CH21

Fakhrullin R, Nigamatzyanova L, Fakhrullina G (2021) Dark-field/hyperspectral microscopy for detecting nanoscale particles in environmental nanotoxicology research. Sci Total Environ 772:145478. https://doi.org/10.1016/J.SCITOTENV.2021.145478

Fonseca E et al (2023) Getting fat and stressed: effects of dietary intake of titanium dioxide nanoparticles in the liver of turbot Scophthalmus maximus. J Hazard Mater 458:131915. https://doi.org/10.1016/j.jhazmat.2023.131915

Gazsó A et al (2019) Regulating Nanotechnological applications for food contact materials. Eur J Risk Regul 10(1):219–226. https://doi.org/10.1017/ERR.2019.9

Geitner NK et al (2020) Harmonizing across environmental nanomaterial testing media for increased comparability of nanomaterial datasets. Environ Sci Nano 7(1):13–36. https://doi.org/10.1039/C9EN00448C

Ghosh P et al (2024) Recent advances of nanotechnology in ameliorating bioenergy production: a comprehensive review. Sustain Chem Pharm. https://doi.org/10.1016/j.scp.2023.101392

Gimenez Y et al (2016) 3D imaging of nanoparticle distribution in biological tissue by laser-induced breakdown spectroscopy. Scient Reports 6(1):1–9. https://doi.org/10.1038/srep29936

Godwin HA et al (2009) The University of California Center for the environmental implications of nanotechnology. Environ Sci Technol 43(17):6453–6457. https://doi.org/10.1021/ES8034544

Handy RD et al (2021) The bioaccumulation testing strategy for manufactured nanomaterials: physico-chemical triggers and read across from earthworms in a meta-analysis. Environ Sci Nano 8(11):3167–3185. https://doi.org/10.1039/D1EN00444A

Hansen SF et al (2017) Revising REACH guidance on information requirements and chemical safety assessment for engineered nanomaterials for aquatic ecotoxicity endpoints: recommendations from the EnvNano project. Environ Sci Europe 29(1):14. https://doi.org/10.1186/S12302-017-0111-3

Henke AH et al (2024) Interlaboratory comparison of centrifugal ultrafiltration with ICP-MS detection in a first-step towards methods to screen for nanomaterial release during certification of drinking water contact materials. Sci Total Environ 912:168686. https://doi.org/10.1016/J.SCITOTENV.2023.168686

Hughes KJ et al (2024) Unveiling trends: nanoscale materials shaping emerging biomedical applications. ACS Nano. https://doi.org/10.1021/acsnano.4c04514

Ilaria C et al (2022) Environmental safety of nanotechnologies: the eco-design of manufactured nanomaterials for environmental remediation. Sci Total Environ 864:161181. https://doi.org/10.1016/j.scitotenv.2022.161181

Jabran M et al (2024) Exploring the potential of nanomaterials (NMs) as diagnostic tools and disease resistance for crop pathogens. Chem Biol Technol Agric. https://doi.org/10.1186/s40538-024-00592-y

Jacobs P et al (2015) Length- and weight-dependent clearance rates of juvenile mussels (Mytilus edulis) on various planktonic prey items. Helgol Mar Res 69(1):101–112. https://doi.org/10.1007/S10152-014-0419-Y/FIGURES/6

Jacobs R et al (2016) Combining exposure and effect modeling into an integrated probabilistic environmental risk assessment for nanoparticles. Environ Toxicol Chem 35(12):2958–2967. https://doi.org/10.1002/ETC.3476

Jain AK et al (2023) Nanomaterials for toxicity constraints and risk assessment. In: Nanomaterials for sustainable development. Springer Nature Singapore, Singapore, pp 65–99. https://doi.org/10.1007/978-981-99-1635-1_3

Jia J, Mu Q, Zhou H (2023) Editorial: biomedical applications and health impacts of emerging nanostructured materials. Front Bioeng Biotechnol. https://doi.org/10.3389/fbioe.2023.1282946

Jiang C et al (2022) Current methods and prospects for analysis and characterization of nanomaterials in the environment. Environ Sci Technol 56(12):7426–7447. https://doi.org/10.1021/acs.est.1c08011

Karlaganis G, Liechti R (2013) The regulatory framework for nanomaterials at a global level: SAICM and WTO insights. Rev Eur Comparat Intl Environ Law 22(2):163–173. https://doi.org/10.1111/REEL.12031

Keller AA et al (2024) Predicting environmental concentrations of nanomaterials for exposure assessment – a review. NanoImpact:100496. https://doi.org/10.1016/j.impact.2024.100496

Khosravi B, Gordon R (2024) Accessible double Nanohole Raman tweezer analysis of single nanoparticles. J Phys Chem C 128(36):15048–15053. https://doi.org/10.1021/ACS.JPCC.4C03536/ASSET/IMAGES/MEDIUM/JP4C03536_0006.GIF

Kim H-A et al (2014) Nanometrology and its perspectives in environmental research. Environ Health Toxicol 29:e2014016. https://doi.org/10.5620/EHT.E2014016

Kumah EA et al (2023) Human and environmental impacts of nanoparticles: a scoping review of the current literature. BMC Public Health 23(1). https://doi.org/10.1186/s12889-023-15958-4

Kumar CMV et al (2022) The impact of engineered nanomaterials on the environment: release mechanism, toxicity, transformation, and remediation. Environ Res 212:113202. https://doi.org/10.1016/j.envres.2022.113202

Laborda F et al (2023) Catching particles by atomic spectrometry: benefits and limitations of single particle - inductively coupled plasma mass spectrometry. Spectrochim Acta B At Spectrosc 199:106570. https://doi.org/10.1016/J.SAB.2022.106570

Labuda J et al (2023) Analytical chemistry of engineered nanomaterials: part 1. Scope, regulation, legislation, and metrology (IUPAC technical report). Pure Appl Chem 95(2):133–163. https://doi.org/10.1515/PAC-2021-1001/ASSET/GRAPHIC/J_PAC-2021-1001_FIG_010.JPG

Lebre F et al (2022) Nanosafety: an evolving concept to bring the safest possible nanomaterials to society and environment. Nanomaterials 12(11):1810. https://doi.org/10.3390/nano12111810

Li H et al (2024) Commonly used engineered nanomaterials improve soil health via suppressing soil-borne fusarium and positively altering soil microbiome. https://doi.org/10.1021/acsestengg.3c00501

Liu R et al (2023) Discovery of environmental nanoparticles in a mineral water spring from Yiyuan County, Shandong Province, Eastern China: a new form of elements in mineral water. Water. https://doi.org/10.3390/w15193497

López-Mayán JJ et al (2023) Bioaccumulation of titanium dioxide nanoparticles in green (Ulva sp.) and red (Palmaria palmata) seaweed. Microchim Acta 190(8):287. https://doi.org/10.1007/s00604-023-05849-1

Luo D et al (2024) Micro- and nano-plastics in the atmosphere: a review of occurrence, properties and human health risks. J Hazard Mater:133412. https://doi.org/10.1016/j.jhazmat.2023.133412

Ma T et al (2022) Recent advances in determination applications of emerging films based on nanomaterials. Adv Colloid Interf Sci 311:102828. https://doi.org/10.1016/j.cis.2022.102828

Mahapatra A et al (2024) Role of nanoparticles on soil microbial community and functionality. Advan Environ Eng Green Technol Book Ser:43–56. https://doi.org/10.4018/979-8-3693-1471-5.ch003

Maksimova YG, Zorina AS (2024) Antibiofilm and Probiofilm effects of nanomaterials on microorganisms (review). Appl Biochem Microbiol 60:1–16. https://doi.org/10.1134/s0003683824010125

Malik S, Muhammad K, Waheed Y (2023) Emerging applications of nanotechnology in healthcare and medicine. Molecules. https://doi.org/10.3390/molecules28186624

Martínez G et al (2020) 'Environmental impact of nanoparticles' application as an emerging technology: a review. Materials 14(1). https://doi.org/10.3390/MA14010166

Masseroni A et al (2022) Nanoplastics: status and knowledge gaps in the finalization of environmental risk assessments. Toxics 10(5):270. https://doi.org/10.3390/toxics10050270

Medina AR et al (2024) A design-phase environmental safe-and-sustainable-by-design categorization tool for the development and innovation of Nano-enabled advanced materials (AdMaCat). Environ Sci Nano. https://doi.org/10.1039/d4en00068d

Mehrabi K et al (2017) Improvements in nanoparticle tracking analysis to measure particle aggregation and mass distribution: a case study on engineered nanomaterial stability in incineration landfill leachates. Environ Sci Technol 51(10):5611–5621. https://doi.org/10.1021/ACS.EST.7B00597/ASSET/IMAGES/LARGE/ES-2017-00597Y_0006.JPEG

Mendonça MCP et al (2017) Nanomaterials in the environment: perspectives on in vivo terrestrial toxicity testing. Front Environ Sci 5. https://doi.org/10.3389/FENVS.2017.00071

Mohanty P et al (2024) Biofabricated nanomaterials in sustainable agriculture: insights, challenges and prospects. Biofabrication. https://doi.org/10.1088/1758-5090/ad60f7

Monikh FA et al (2018) Scientific rationale for the development of an OECD test guideline on engineered nanomaterial stability. NanoImpact 11:42–50. https://doi.org/10.1016/J.IMPACT.2018.01.003

Morgado RG et al (2022) Terrestrial organisms react differently to nano and non-nano cu(OH)2 forms. Sci Total Environ 807. https://doi.org/10.1016/J.SCITOTENV.2021.150699

Mou L et al (2024) A powerful method for in situ and rapid detection of trace nanoplastics in water—Mie scattering. J Hazard Mater 470:134186. https://doi.org/10.1016/J.JHAZMAT.2024.134186

Nakamura Y (2001) Filtration rates of the Manila clam, Ruditapes philippinarum: dependence on prey items including bacteria and picocyanobacteria. J Exp Mar Biol Ecol 266(2):181–192. https://doi.org/10.1016/S0022-0981(01)00354-9

Nasser F, Constantinou JK, Lynch I (2020) Nanomaterials in the environment acquire an "eco-Corona" impacting their toxicity to Daphnia magna-a call for updating toxicity testing policies. Proteomics 20(9). https://doi.org/10.1002/PMIC.201800412

Nguyen B, Tufenkji N (2022) Single-particle resolution fluorescence microscopy of Nanoplastics. Environ Sci Technol 56(10):6426–6435. https://doi.org/10.1021/ACS.EST.1C08480/ASSET/IMAGES/LARGE/ES1C08480_0007.JPEG

Nguyen QM et al (2024) Review—nanotechnology in aquaculture: applications and challenges. J Electrochem Soc. https://doi.org/10.1149/1945-7111/ad48c2

Nielsen MB et al (2021) Nanomaterials in the European chemicals legislation – methodological challenges for registration and environmental safety assessment. Environ Sci Nano 8(3):731–747. https://doi.org/10.1039/D0EN01123A

Niskanen I et al (2019) Determination of nanoparticle size using Rayleigh approximation and Mie theory. Chem Eng Sci 201:222–229. https://doi.org/10.1016/J.CES.2019.02.020

Nivaz SR et al (2022) Utilization of Raman spectroscopy in nanomaterial/bionanomaterial detection. In: Handbook of microbial nanotechnology, pp 145–156. https://doi.org/10.1016/B978-0-12-823426-6.00015-2

Okoffo ED, Thomas KV (2024) Quantitative analysis of nanoplastics in environmental and potable waters by pyrolysis-gas chromatography–mass spectrometry. J Hazard Mater 464:133013. https://doi.org/10.1016/J.JHAZMAT.2023.133013

Otari S et al (2024) Advancements in bionanotechnological applications for climate-smart agriculture and food production. Biocatal Agric Biotechnol. https://doi.org/10.1016/j.bcab.2024.103117

Petersen EJ et al (2022) Evaluation of bioaccumulation of nanoplastics, carbon nanotubes, fullerenes, and graphene family materials. Environ Int 173:107650. https://doi.org/10.1016/j.envint.2022.107650

Quarato M et al (2021) Detection of silver nanoparticles in seawater using surface-enhanced raman scattering. Nano 11(7):1711. https://doi.org/10.3390/NANO11071711/S1

Quarato M et al (2024) Bioaccumulation, biodistribution, and transformation of polyvinylpyrrolidone-coated silver nanoparticles in edible seaweeds. Sci Total Environ 949:174914. https://doi.org/10.1016/J.SCITOTENV.2024.174914

Rabajczyk A et al (2020) Metal nanoparticles in the air: state of the art and future perspectives. Environ Sci Nano 7(11):3233–3254. https://doi.org/10.1039/D0EN00536C

Rath SK, Harshavardhan M, Srivastava P (2024) A concise overview of effect of nanomaterials in soil and associated microbiota. J Environ Nanotechnol 13(2):183–193. https://doi.org/10.13074/jent.2024.06.241544

Rehman N, Moore S (2021) An overview of the state of the regulatory and preclinical requirements for nanomaterials including medical devices. Integr Environ Assess Manag 17(6):1098–1104. https://doi.org/10.1002/IEAM.4426

Rossi G (2022) Nanoparticles and their role in environmental decontamination technologies:127–140. https://doi.org/10.1007/978-981-19-1384-6_7

Saleem H et al (2022) Advances of nanomaterials for air pollution remediation and their impacts on the environment. Chemosphere 287. https://doi.org/10.1016/J.CHEMOSPHERE.2021.132083

Saleh TA (2020) Trends in the sample preparation and analysis of nanomaterials as environmental contaminants. Trends Environ Analyt Chem 28:e00101. https://doi.org/10.1016/J.TEAC.2020.E00101

Samriti et al (2023) Emerging nanomaterials in the detection and degradation of air pollutants. Curr Opin Environ Sci Health 35:100497. https://doi.org/10.1016/j.coesh.2023.100497

Schmidt R et al (2021) Correlative SEM-Raman microscopy to reveal nanoplastics in complex environments. Micron 144:103034. https://doi.org/10.1016/J.MICRON.2021.103034

Schultz CL et al (2020) Chemical transformation and surface functionalisation affect the potential to group nanoparticles for risk assessment. Environ Sci Nano 7(10):3100–3107. https://doi.org/10.1039/D0EN00578A

Schwirn K et al (2020) Environmental risk assessment of nanomaterials in the light of new obligations under the REACH regulation – which challenges remain and how to approach them? Integr Environ Assess Manag 16(5):706–717. https://doi.org/10.1002/IEAM.4267

Shah MA et al (2024) Application of nanotechnology in the agricultural and food processing industries: a review. Sustain Mater Technol:e00809–e00809. https://doi.org/10.1016/j.susmat.2023.e00809

Shinde SA, More PR, Ingle AP (2023) Hazardous effects of nanomaterials on aquatic life. Nanotechnol Agricult Agroecosyst:423–450. https://doi.org/10.1016/B978-0-323-99446-0.00012-X

Singh D, Gurjar BR (2022) Nanotechnology for agricultural applications: facts, issues, knowledge gaps, and challenges in environmental risk assessment. J Environ Manag 322:116033. https://doi.org/10.1016/j.jenvman.2022.116033

Sohal IS et al (2018) Dissolution behavior and biodurability of ingested engineered nanomaterials in the gastrointestinal environment. ACS Nano 12(8):8115–8128. https://doi.org/10.1021/ACSNANO.8B02978

Sorasan C et al (2024) New versus naturally aged greenhouse cover films: degradation and micronanoplastics characterization under sunlight exposure. Sci Total Environ 918:170662. https://doi.org/10.1016/J.SCITOTENV.2024.170662

Sudheshwar A et al (2023) Learning from safe-by-design for safe-and-sustainable-by-design: mapping the current landscape of safe-by-design reviews, case studies, and frameworks. Environ Int 183:108305. https://doi.org/10.1016/j.envint.2023.108305

Surette MC et al (2021) What is "environmentally relevant"? A framework to advance research on the environmental fate and effects of engineered nanomaterials. Environ Sci Nano 8(9):2414–2429. https://doi.org/10.1039/D1EN00162K

Tamuk P, Kharshiing B, Yomso J (2024) Role of nanoparticles in perturbation of soil microbial communities:16–22. https://doi.org/10.58532/v3becs24p1ch2

Tang W et al (2024) Computational nanotoxicology models for environmental risk assessment of engineered nanomaterials. Nanomaterials. https://doi.org/10.3390/nano14020155

Terry LR et al (2022) Applications of surface-enhanced Raman spectroscopy in environmental detection. Analyt Sci Adv 3(3–4):113–145. https://doi.org/10.1002/ANSA.202200003

Thomas CR et al (2011) Nanomaterials in the environment: from materials to high-throughput screening to organisms. ACS Nano 5(1):13–20. https://doi.org/10.1021/NN1034857

Tirth V et al (2023) Emerging nano-engineered materials for protection of wind energy applications photovoltaic based nanomaterials. Sustain Energy Technol Assess 56:103101. https://doi.org/10.1016/j.seta.2023.103101

Tran T-K et al (2023) Review on fate, transport, toxicity and health risk of nanoparticles in natural ecosystems: emerging challenges in the modern age and solutions toward a sustainable environment. Sci Total Environ:169331. https://doi.org/10.1016/j.scitotenv.2023.169331

Tschiche HR et al (2022) Environmental considerations and current status of grouping and regulation of engineered nanomaterials. Environ Nanotechnol Monitor Manage 18:100707. https://doi.org/10.1016/j.enmm.2022.100707

Umair Hassan M et al (2024) Application of zinc oxide nanoparticles to mitigate cadmium toxicity: mechanisms and future prospects. Plants 13(12):1706. https://doi.org/10.3390/PLANTS13121706

Upadhayay VK et al (2024) Role of nanomaterials on soil microbial community and functionality. In: Advances in environmental engineering and green technologies book series, pp 333–353. https://doi.org/10.4018/979-8-3693-1471-5.ch017

Vance ME, Marr LC (2015) Exposure to airborne engineered nanoparticles in the indoor environment. Atmos Environ 106(106):503–509. https://doi.org/10.1016/J.ATMOSENV.2014.12.056

Verdaasdonk JS et al (2014) Bending the rules: Widefield microscopy and the Abbe limit of resolution. J Cell Physiol 229(2):132–138. https://doi.org/10.1002/JCP.24439

Vieira A et al (2023) Innovative antibacterial, photocatalytic, titanium dioxide microstructured surfaces based on bacterial adhesion enhancement. ACS Appl Bio Mater 6(2):754–764. https://doi.org/10.1021/ACSABM.2C00956/ASSET/IMAGES/LARGE/MT2C00956_0006.JPEG

Vijayaram S et al (2023) Inorganic nanoparticles for use in aquaculture. Rev Aquac. https://doi.org/10.1111/raq.12803

Wang C et al (2023) Environmental behaviors and toxic mechanisms of engineered nanomaterials in soil. Environ Res:117820. https://doi.org/10.1016/j.envres.2023.117820

Wyrzykowska E et al (2022) Representing and describing nanomaterials in predictive nanoinformatics. Nat Nanotechnol 17(9):924–932. https://doi.org/10.1038/s41565-022-01173-6

Yamini V et al (2023) Environmental effects and interaction of nanoparticles on beneficial soil and aquatic microorganisms. Environ Res:116776. https://doi.org/10.1016/j.envres.2023.116776

Yang Y et al (2024) Environmental behavior of silver nanomaterials in aquatic environments: an updated review. Sci Total Environ. https://doi.org/10.1016/j.scitotenv.2023.167861

Zafar I et al (2023) Toxic effects of nanomaterials on aquatic animals and their future prospective Xenobiot Aquat Anim Reproduct Develop Impacts:325–351. https://doi.org/10.1007/978-981-99-1214-8_16

Zhang P et al (2024) Analysis of nanomaterial biocoronas in biological and environmental surroundings. Nat Protoc:1–48. https://doi.org/10.1038/S41596-024-01009-8/METRICS

Zhao J et al (2021) Engineered nanomaterials in the environment: are they safe? Crit Rev Environ Sci Technol 51(14):1443–1478. https://doi.org/10.1080/10643389.2020.1764279

Zhao J et al (2023) Surface modification significantly changed the effects of nano-polystyrene on sediment microbial communities and nitrogen metabolism. J Hazard Mater 460:132409. https://doi.org/10.1016/j.jhazmat.2023.132409

Zheng Y-R, Nowack B (2022) Meta-analysis of bioaccumulation data for nondissolvable engineered nanomaterials in freshwater aquatic organisms. Environ Toxicol Chem 41(5):1202–1214. https://doi.org/10.1002/etc.5312

Zhou Y et al (2023) Nanoplastics alter ecosystem multifunctionality and may increase global warming potential. Glob Chang Biol 29(14):3895–3909. https://doi.org/10.1111/gcb.16734

Иванова ЕИ, Perfileva AI, Krutovsky KV (2024) Zinc oxide nanoparticles in the "soil–bacterial community–plant" system: impact on the stability of soil ecosystems. Agronomy 14(7):1588. https://doi.org/10.3390/agronomy14071588

Open Access This chapter is licensed under the terms of the Creative Commons Attribution-NonCommercial-NoDerivatives 4.0 International License (http://creativecommons.org/licenses/by-nc-nd/4.0/), which permits any noncommercial use, sharing, distribution and reproduction in any medium or format, as long as you give appropriate credit to the original author(s) and the source, provide a link to the Creative Commons license and indicate if you modified the licensed material. You do not have permission under this license to share adapted material derived from this chapter or parts of it.

The images or other third party material in this chapter are included in the chapter's Creative Commons license, unless indicated otherwise in a credit line to the material. If material is not included in the chapter's Creative Commons license and your intended use is not permitted by statutory regulation or exceeds the permitted use, you will need to obtain permission directly from the copyright holder.

The Environmental Impacts of Nanoplastics in Marine Ecosystems

16

Giulia Galani, Aline Nunes, Isadora Piccinin, Lucas Fazardo, Suelen Goulart, Thais Alberti, Vânia Vilas-Boas, and Marcelo Maraschin

Abstract

Nanoplastics (NPs) are generated through the degradation of plastic and microplastics, becoming smaller due to mechanical actions and environmental weathering. Their diminutive size endows NPs with unique physical properties, allowing them to penetrate biological barriers, accumulate in tissues, and contribute to biomagnification, thereby affecting food chains and biodiversity. NPs exhibit distinct ecotoxicological mechanisms, leading to increased uptake and bioaccumulation. Their properties arise from complex interactions between the plastic source material and transformations caused by biotic and abiotic factors. While NPs are pervasive, their detection is challenging. In marine ecosystems, organisms like phytoplankton and fish facilitate NP transport through ingestion. This trophic transfer impacts the growth, reproduction, and survival of aquatic

---

G. Galani (✉)
UFSC, Federal University of Santa Catarina, Florianópolis, Brazil

INL, International Iberian Nanotechnology Laboratory, Braga, Portugal
e-mail: giulia.galani@posgrad.ufsc.br

A. Nunes
UNESP, São Paulo State University, Institute of Biosciences, São Paulo, Brazil

I. Piccinin · L. Fazardo · S. Goulart
UFSC, Federal University of Santa Catarina, Florianópolis, Brazil

T. Alberti
UCS, University of Caxias do Sul, Caxias do Sul, Brazil

V. Vilas-Boas
INL, International Iberian Nanotechnology Laboratory, Braga, Portugal

M. Maraschin
UFSC, Federal University of Santa Catarina, Florianópolis, Brazil

UCS, University of Caxias do Sul, Caxias do Sul, Brazil

© The Author(s) 2025
E. Alfaro-Moreno, F. Murphy (eds.), *Nanosafety*,
https://doi.org/10.1007/978-3-031-93871-9_16

species and raises concerns about human health, as NPs can accumulate in vital organs. Monitoring NPs is essential due to their significant environmental impacts and the health risks they pose. Effective detection methodologies are crucial for informing legislation, such as the European Commission's EU Plastics Strategy. Current solutions focus on removing plastic waste, yet this is often the least effective strategy. A comprehensive understanding of the plastic ecosystem is needed to develop preventative policies and collaborative international projects for sustainable plastic management.

Keywords

Nanoplastics · Bioaccumulation · Environment · Plastics strategy

1 Nanoplastics—Introduction

Plastic overproduction and pollution are the defining challenges of our times (UNEP 2018). It is occurring at a higher rate than it is possible to conduct safety assessments of their use and disposal. There is already extensive literature on the impressive amount of plastic waste in the world and its impact (Park et al. 2024). For instance, in 2010, plastic particles smaller than 200 mm represented at least 1% of plastic waste deposited in the oceans, corresponding to 15 to 51 trillion particles, or 93 to 236 thousand tons (van Sebille et al. 2015). Currently, due to the exponential increase in the production, use, and disposal of plastic products, especially single-use ones, such numbers have grown. Most of the estimations are based on macro and microplastics (MPs) because they are straightforward to detect and quantify, not taking into account the nanoplastic (NP) fraction. Thompson et al. proposed the term microplastic for the first time in 2004, referring to polymeric residue with less then 5 mm found in aquatic bodies.

It is conventionally accepted that a NP measures up to 1 μm across any one dimension (SAPEA 2019). Some authors consider larger sizes to still be MPs (OECD 2021) since the definition of nanomaterials comprehends particles ranging in size up to 100 nm (EC 2023a, b, c). In the environment, the formation of nanoplastics (NPs) occurs from the physical-chemical degradation of plastics and MPs, which have their particle size reduced due to mechanical actions and weathering. Due to the efficiency of NP formation mechanisms and the difficulty in quantifying small-scale particles, it is likely that the amount of nanosized plastics present on a global scale has been underestimated throughout the years (Peller et al. 2022).

The nanometric scale has a significant impact on the characteristics and functionality of nanomaterials, particularly the dramatic increase in surface area. There is a substantial increase in the number of surface atoms compared to the total volume of the particle, modifying its chemical reactivity and functionality. At this scale, physicochemical properties tend to behave differently due to quantum effects and surface phenomena, presenting unique optical, thermal, and mechanical properties (Auffan et al. 2009). Even though NPs are polymeric nanomaterials, and many nanomaterials are made of polymers, not all polymers are plastic. This distinction is important due to the interchangeability of the terminology between plastic and polymers (Brewer et al. 2020). Therefore, since NPs behave differently than their micro size contaminants due to the nanometric scale, they should be analyzed as nanomaterials.

Although the polymers represent most of the plastic composition, it is proven that their toxic and sublethal effects are directly related to the release of additives after chemical and physical degradation processes that reduce the particle size, with a size-dependency relationship being perceived between the degree of tissue contamination and size of NPs (Lu et al. 2022). These particles can penetrate through biological barriers, depositing in tissues and participating in biomagnification processes, resulting in the alarming spread of their effects throughout food chains, escalating their impacts on biodiversity (Joppien et al. 2022; Bilal et al. 2023).

NPs may originate from primary sources, in which they are intentionally fabricated into nanometric scales, such as being used mostly in cosmetics, pharmaceuticals, paints and coatings, construction, water treatment, and farming. They are widely employed due to the range of functions they may serve, e.g., abrasive, emulsifier, binder, filler, fluid absorbent, thickening agent, flocculant, dispersing agent, anti-static agent, and mechanical resistance (EC 2023a). On the other hand, secondary NPs are created when macroplastics are broken down into smaller debris in both aquatic and terrestrial environments, when they undergo a range of biological (i.e., enzymatic) and abiotic degradation processes, such as thermal, physical, chemical, and photodegradation. Eventually, NPs will contaminate water supplies, food chains, and the air, translocating into living organisms and affecting all trophic levels, especially marine fauna (EC 2023a).

Even though their impact is clear and growing, to date there is no specific or unified global legislation for nanomaterials, as primary NPs have been registered through specific regulatory agencies of each country. It is noteworthy that in the EU, public and private institutions are responsible for studying the repercussions that such a nanotechnology can have on the environment and human health (EC 2019a).

Several strategies are required for monitoring and managing secondary NP production and pollution, which do not encompass the unique characteristics of NPs. Therefore, it is crucial to invest in biomonitoring studies to further determine NPs' toxicity and their environmental impacts. This is vital for setting international guidelines for uncovering and regulating contaminants, as well as determining mitigation strategies. Furthermore, it is essential to prevent contamination in the first place. To this end, it is important to know more about the structure, properties and consequences of NPs accumulation, to understand and create knowledge that can stimulate studies to prevent and resolve the effects of their bioaccumulation.

2 Physicochemical Properties and Behavior of Nanoplastics

2.1 Physical and Chemical Properties

The properties of NPs are predominantly inherited from their plastic source material, which can be further altered by chemical, biological, and environmental factors (Kiran et al. 2022; Atugoda et al. 2022). Due to their diminutive dimensions, NPs exhibit unique physical properties compared to larger plastic particles. One of the key physical characteristics of NPs is their high surface area-to-mass ratio. This large specific surface area confers NPs with increased reactivity and ability to interact with other substances present in the environment, which can significantly affect their mobility, bioavailability, and potential for contaminant adsorption (Ramirez Arenas et al. 2021; Sharma et al. 2022).

Another relevant physical property is the morphology of NPs, which can present various shapes, e.g., spheres, fibers, and irregular fragments. This morphological diversity directly influences the behavior and transport of these materials in the environment (Prüst et al. 2020; Liu et al. 2021a). The zeta potential, which reflects the electrical charge on the particle surface, is also an important physical property of nanosized materials, as it affects the stability of NP suspensions and their interactions with other charged species in the medium (Liu et al. 2021a; Pradel et al. 2023).

In addition to their unique physical characteristics, NPs also exhibit distinct chemical properties compared to larger-scale plastic particles. These chemical properties greatly influence the behavior and environmental impacts of these nanomaterials (Shen et al. 2019). One of the primary chemical traits of NPs is the adsorption of contaminants on their surface. Due to their high specific surface area, NPs have a considerable capacity to adsorb various persistent organic pollutants, heavy metals, and even microorganisms present in the environment. This adsorption process can affect the bioavailability and transport of these contaminants within the ecosystem (Shen et al. 2019).

Another relevant chemical property is the photodegradation capacity of NPs. When exposed to solar radiation, these materials can undergo chain scission and oxidation processes, directly influencing their residence time and ultimate fate in the environment (Jansen et al. 2024). Furthermore, monomers and additives used in the manufacture of plastics, such as bisphenols and phthalates, can also affect their reactivity and toxicity. Certain chemical substances present in the structure can be released and interact with organisms, potentially causing toxic effects (Hahladakis et al. 2018; Liu et al. 2021b).

Thus, the properties of NPs result from complex interactions between the traits inherited from the plastic source material and the transformations they undergo due to biotic and abiotic factors. Understanding these properties is crucial to assessing the fate, behavior, and potential impacts of NPs on the environment.

2.2 Transport and Distribution of NPs in the Environment

The transport and distribution of NP within environmental systems is a crucial aspect to understand their impacts, as it greatly influence their fate, behavior, and toxicity. NPs can be transported through a variety of physical, chemical, and biological processes, which ultimately determine how they move, accumulate, and interact with organisms and habitats (Huang et al. 2020; Yu et al. 2024). The traits of the NPs themselves, such as size, shape, and surface properties, can significantly affect their transport and distribution. Environmental conditions like temperature, salinity, and pH can also influence the behavior and transport of NPs in the marine environment (Huang et al. 2022; Le et al. 2023).

In the terrestrial environment, plastics primarily come from agricultural plastic film coverage, application of a wide range of types of compounds, wastewater irrigation, and automobile tire residues. In the atmosphere, they are derived mainly from synthetic fibers (e.g., clothing) or coating materials. In the aquatic environment, plastics generally originate from terrestrial sources, municipal drainage networks, fishing, shipbuilding, and tourism (Huang et al. 2020). Plastics in the terrestrial environment are washed away by surface runoff, followed by flow into the ocean along streams and rivers. Physical factors such as wind and water currents play a significant role in the movement and distribution of NPs in the environment. These processes can lead to their horizontal and vertical transport, with sedimentation and resuspension being particularly important for their vertical distribution. As result, NPs may accumulate in certain habitats and hotspots, depending on local environmental conditions (Huang et al. 2020, 2022).

NPs are present everywhere, but detection is difficult due to their small size and chemical composition (Trevisan et al. 2022). They are found in soil, especially on farmland, due to plastic breakdown (Ng et al. 2018). Lightweight NPs can travel long distances in the air (Atugoda et al. 2022), as they occur in urban and remote areas, like the Swiss Alps (Materić et al. 2020), North Atlantic, and Siberian waters (Allen et al. 2022). This widespread presence shows the scale of the problem, requiring urgent mitigation efforts to address NP impacts on organisms and food webs.

In marine environments, organisms such as phytoplankton, fish, and marine mammals can also influence the transport of NPs through ingestion and incorporation into the food web. This trophic transfer from primary to secondary consumers and ultimately to top predators is associated with compromised growth, reproduction, behavior, and mortality of aquatic organisms. Additionally, the process of biofouling, where NPs become coated with organic matter, can alter their buoyancy and transport patterns along the water column (Zaki and Aris 2022). Thus, NPs can be associated with other pollutants, which could cause even more damages to the environment.

2.3 Interaction of NPs with Other Pollutants

In addition to the intrinsic properties of NPs and the additives present in their polymeric structures, their interactions with other types of pollutants can significantly influence their environmental fate and ecological impacts. These form homo- or hetero-aggregates that affect the dynamics, translocation, deposition, bioavailability, and interaction with other chemicals in the environment. Thus, NPs can act as vectors, transporting and concentrating other contaminants, such as heavy metals, persistent organic pollutants (POPs), and even pathogens, through adsorption or incorporation processes (Town and van Leeuwen 2020; Bhagat et al. 2021; Atugoda et al. 2022).

The large surface area-to-volume ratio of NPs makes them effective sorbents for other contaminants present in the environment (Bhagat et al. 2021; Trevisan et al. 2022). Studies have shown that plastic pellets can accumulate and concentrate pollutants such as polychlorinated biphenyls (PCBs), polycyclic aromatic hydrocarbons (PAHs), and pesticides (Taniguchi et al. 2016), as well as heavy metals (Vedolin et al. 2018). The interaction of NPs with other contaminants can increase the adverse effects on organisms, for example, resulting in growth inhibition of species (Zhang and Goss 2020; Zheng et al. 2024), affecting neurogenic pathways (Varshney et al. 2024), triggering overproduction of ROS and cell membrane damage (Narayanan et al. 2024), causing metabolic abnormalities, and histopathological changes with multigenerational effects (Junaid et al. 2023).

In addition to chemical pollutants, it has been discovered that NPs also interact with and transport microorganisms, including bacteria, viruses, and fungi. The attachment of these pathogens to NPs can facilitate their dispersal in the environment and increase the risk of disease transmission (Tang et al. 2022; Nath et al. 2023).

Therefore, these complex interactions between NPs and other environmental pollutants demonstrate how crucial it is to accurately assess the overall ecological and human health risks associated with NP pollution. Understanding the mechanisms and implications of these interactions will enable the development of effective mitigation strategies.

3 Toxicity and Environmental Impacts of Nanoplastics

(Micro) nanoplastics (MNPs) interact with human organisms and other animals through the processes of absorption, distribution, and excretion. Studies including metabolism are restricted due to the low biodegradability of plastics (Prata 2023). These materials are mainly absorbed via inhalation and ingestion, with skin absorption being reduced due to restrictions on the use of MNPs in personal care products, the epidermal barrier, particle size, and physical-chemical properties (Cheung and Fok 2016; Jeong et al. 2024).

Ingestion is a significant route of exposure, especially for aquatic organisms and humans who consume contaminated water or food (Jeong et al. 2024). Smaller particles are more prone to absorption in the gastrointestinal tract (Hoang and

Felix-Kim 2020). In studies with rodents and fish, polyethylene and polystyrene particles have been observed in the digestive tract, and even in vital organs after ingestion (Nobuyuki et al. 2020; Amanda Pereira and Malafaia 2021). In addition, MNPs can be transferred between different trophic levels, increasing the risk of exposure along the food chain. Importantly, depending on the size and surface charge, nanoparticles can enter the circulatory system after translocation through the intestinal barrier (Messika et al. 2018; Kangas et al. 2023). Although inhalation of MNPs has been less investigated, particles smaller than 10 μm can penetrate deep into the lungs, where cleaning mechanisms are less efficient, increasing the risk of adverse impacts (Wright and Kelly 2017).

The ability of NPs to penetrate biological barriers increases the toxicity potential of the adsorbed toxins. In the body, particles >200 nm are mostly removed by the reticuloendothelial system (liver, spleen, macrophages, monocytes), while those <70 nm can be physically captured by tissues. The spatial scale of surface properties like electrostatic and steric interactions determines initial particle adsorption, involving both attractive (e.g., hydrogen bonds, ions, Van Der Waals forces) and repulsive (e.g., configurational entropy, osmotic, electrostatic) forces, resulting in complex and dynamic interaction profiles (Lewis 2004). Due to this behavior, NPs possess distinct ecotoxicological modes of action, due to increased uptake and bioaccumulation. Additionally, the bioaccumulation of NP's chemical additives and plasticizers, such as bisphenol A or phthalates, even at low nanogram concentrations, can cause endocrine disruption, requiring refined techniques for their detection and quantification (Mitrano et al. 2021).

After absorption, MNPs can move to various organs in the body (Jeong et al. 2024). Studies in rodents show that particles of 0.5 and 5 μm can accumulate in the liver, spleen, kidneys, heart, and lungs in a size-dependent manner (Zhang et al. 2024). In fish, 70 nm particles were found in the liver and brain after aquatic exposure (Ding et al. 2020). The intracellular absorption of MNPs varies with the size and shape of the particles. Particles smaller than 700 nm are absorbed by receptor-mediated endocytosis, while larger particles are absorbed by phagocytosis (Aderem and Underhill 1999). Recently, it was observed that 50 nm NPs particles are internalized by clathrin-, caveolin- and macropinocytosis-mediated pathways, while 500 nm particles are internalized only by macropinocytosis, as 5 μm particles could not be internalized into cells (Liu et al. 2021c).

Regarding the excretion process, the majority of MNPs (>90%) are eliminated via feces (Jeong et al. 2024). Studies in humans and different terrestrial and aquatic animals show that the elimination time of MNPs through feces can vary considerably. For NPs, this time varies from 25 to 36.9 h, depending on the type and size of the particles (Peng et al. 2022). However, these excreted MNPs can undergo aggregation and sedimentation in the environment due to the coating with intestinal fluid, increasing the potential for re-ingestion and impacting ecosystems (Hoang and Felix-Kim 2020).

Studies on the impact of MNPs in the human body have shown that those polymers can accumulate in vital organs such as liver, lungs, and reproductive organs. Similarly, the accumulation of MNPs has been detected in the brain, intestine, liver, and gonads of fish species (Mattsson et al. 2017; Amanda Pereira et al. 2022). Although most MNPs are eventually excreted, those that remain in the body can cause prolonged exposure and increase health risks (Jon et al. 2019).

3.1 Bioaccumulation and Biomagnification in the Marine Food Chain

The formation of nanoparticles alters the chemical and physical characteristics of the particles and, consequently, their availability and biological impact on aquatic organisms (Mattsson et al. 2015). Biological reactivity often increases with decreasing size, considering that the smaller the particle, the greater the ability to pass through cell boundaries and accumulate in organisms, as well as the reactivity of the particles (Mattsson et al. 2015; Worm et al. 2017). Importantly, the knowledge currently available about the effects of NPs in the bioaccumulation and biomagnification processes is still scarce, especially in relation to marine biota, due to their size and the lack of unified protocols for investigation (Ferreira et al. 2019). Because of this, many of the studies relating to the transfer and accumulation of NPs are based on the transfer of MPs (Huang et al. 2022).

Ecotoxicology studies have revealed that MPs and NPs have been accumulating in a variety of organisms (Huang et al. 2022). Evidences indicate that MPs and NPs can contribute to the bioaccumulation and consequent toxicity of other contaminants, such as persistent organic pollutants, pharmaceuticals, and heavy metals (Sun et al. 2022). Several factors can affect bioaccumulation, such as the size, shape, and surface properties of the (macro)nanoparticles, but, in general, smaller particles tend to cross biological barriers more easily and accumulate in tissues (Huang et al. 2022). Additionally, the particle's shape also plays a relevant role, with non-spherical ones showing a longer residence time in some organisms (Qiao et al. 2019). Some *in situ* studies have found no significant evidence of bioaccumulation, suggesting that controlled laboratory conditions may not fully reflect what happens in natural ecosystems (Garcia et al. 2021). In addition, the methods used to quantify environmental contamination by bioplastics, especially for NPs, still have important methodological gaps (Huang et al. 2022).

Considering biomagnification, which requires NPs to be concentrated along trophic levels, this potential is not well-established (Bour et al. 2015). Laboratory and some *in situ* studies have demonstrated that NPs can undergo trophic transfer through the food chain (Fig. 16.1), where shrimps, mussels, and crustaceans transferred plastic particles to their predators (Farrell and Nelson 2013; Outi et al. 2014). Current studies on the kinetics of particle clearance in organisms are more focused on lower trophic levels, such as bivalves and crustaceans, with a scarcity of data on animals at higher levels (Huang et al. 2022).

The residence time of plastic particles in the digestive organs varies according to the species, and they are removed through feces or specific systems, depending in the food chain level. Secondary consumers retain the particles longer, while higher

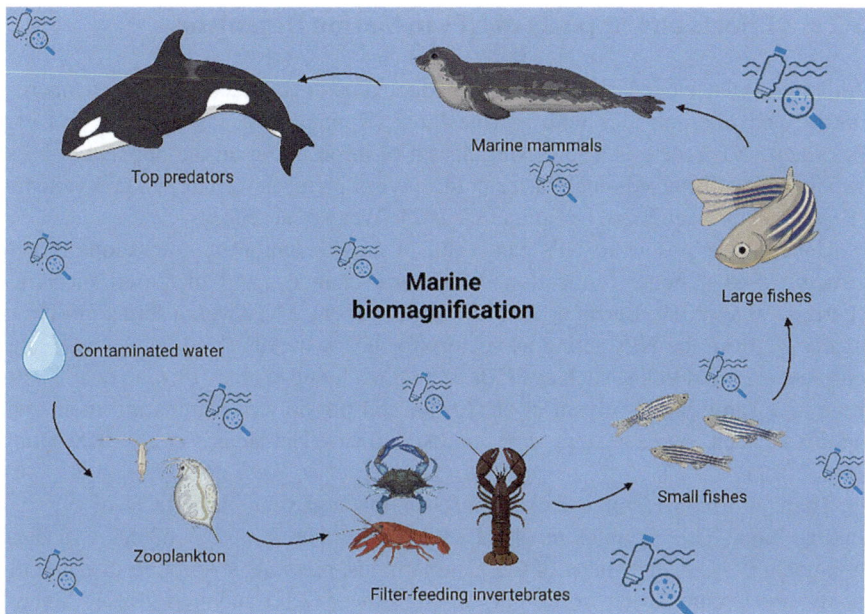

Fig. 16.1 Marine biomagnification. (Created with BioRender.com)

trophic-level predators can excrete them quickly (Huang et al. 2022; Zantis et al. 2022). For example, some experiments found no biomagnification, with lower concentrations in predators such as crabs compared to their prey, i.e., mussels, a fact attributed to the excretion ability of the plastic particles by the predators (Wang et al. 2021). Smaller particles like the size of natural proteins, can go through gut cells using active and passive diffusion, moving to other tissues using the bloodstream. However, there is not enough research on how plastics are moved and removed in these systems.

There may be discrepancies in the bioaccumulation and biomagnification of NPs between aquatic and terrestrial ecosystems. The partitioning of NPs plays an important role in aquatic organisms, while the predatory relationship in terrestrial systems seems to lead to more significant biomagnification effects (Huang et al. 2020).

Due to the size difference, NPSs tend to accumulate more than MP, dispersing and accumulating in other tissues (Qiao et al. 2019). Despite MNPs being able to accumulate in lower trophic levels, the phenomenon of nutrient dilution can prevent their effective biomagnification along the food chain. As MNPs move through the different trophic levels, their concentration tends to become diluted due to the increase in biomass of the predators. This dilution effect can significantly limit the ability of MP/NP to accumulate in high concentrations in organisms at the top of the food chain (He et al. 2022; Huang et al. 2022).

3.2 Effects and Impacts of NPs in Marine Organisms

Although extensive research has been conducted on the interaction between macro-, meso-, and microplastics with marine biota (Roman et al. 2019; Kühn and van Franeker 2020), there is a surprising dearth of information on the impacts of NPs, which can be dispersed much faster in the environment, as well as inside organisms (Boyle and Örmeci 2020; DeLoid et al. 2021; Wang et al. 2023).

Generally, exposure to NPs can result in growth inhibition, alterations in cell structure and genome replication, behavioral changes, and increased mortality (Ferreira et al. 2019; Barría et al. 2020) (Table 16.1). Moreover, a series of effects can come from the NPs acting as vectors for heavy metals (lead, manganese, and chrome, e.g.) and POPs, such as PCBs and DDT (Jeong et al. 2018; Gao et al. 2019; Liao and Yang 2020; Selvam et al. 2021); and possibly disseminating their own additives, such as plasticizers, dyes, and stabilizers (Hahladakis et al. 2018; Dong et al. 2022).

There is a considerable lack of information regarding the effects of NPs on marine biota. The variables involved are innumerable: a huge number of species, composition, size, and shape of the nanoparticles, previous absorption of contaminants, and accumulation in the gut, among others (Zaki and Aris 2022). A short summary of biological effects of NPs in marine organisms published in recent years can be found in Table 16.1.

3.2.1 Bacteria

Among the various ways NPs can affect bacteria in the marine ecosystem, the most harmful is their interaction with the cell membrane (Table 16.1). An important property of NPs, as demonstrated by Frère et al. (2018), is their ability to absorb nutrients from the environment, making them attractive to bacteria and other microorganisms. Combined with the pollutant's low density, these microorganisms can be dispersed in the environment at much higher speeds than they otherwise would (Khatmullina and Isachenko 2017). The reduced density of NPs is a crucial factor that facilitates their mobility and dispersal throughout the marine ecosystem. This low density allows the NPs to remain suspended in the water column rather than quickly sinking, making them easily transported by currents and waves. As a result, microorganisms attached to or incorporated into these low-density NPs are also disseminated much more efficiently, amplifying the harmful impacts of pathogenic species across wider areas (Kirstein et al. 2016). Moreover, the presence of NPs in biofilms, and microbial communities, can stimulate the already accelerated horizontal gene transfer between different bacterial species, accelerating the formation of new genotypes and, consequently, the selection of bacterium cells resistant to antibiotics, for instance (Guo et al. 2020; Uruén et al. 2020). Microorganisms, including bacteria and fungi, play a crucial role in nutrient cycles and are subject to the effects of plastic nanoparticles, which can have antimicrobial activity, affecting their metabolic pathways and vital cellular functions (Fringer et al. 2020).

Table 16.1 Impacts of nanoplastics on marine organisms

Group	Species	Effects	References
Bacteria	*Escherichia coli*	Decreased growth	Kim et al. (2022)
	Bacillus sp.	Decreased cell viability	
	Halomonas alkaliphila	Decreased growth, increased ROS formation	Sun et al. (2021)
Phytoplankton	*Platymonas helgolandica*	Decreased growth and photosynthesis efficiency, increased mortality and membrane permeability	Wang et al. (2020)
	Planktothrix agardhii	Decreased growth, increased hetero-aggregates formation	Schampera et al. (2021)
	Thalassiosira pseudonana, Skeletonema grethae, Phaeodactylum tricornutum, Dunaliella tertiolecta	Decreased growth	Shiu et al. (2020)
	Cladocopium sp., Symbiodinium tridacnidorum	Decreased growth	Ripken et al. (2020)
	Chaetoceros neogracile	Decreased chlorophyll content, esterase activity, cellular growth, and photosynthetic efficiency	González-Fernández et al. (2019)
	Rhodomonas báltica	Increased growth rate and oxidative stress responses, reduced photosynthetic efficiency and cell viability	Gomes et al. (2020)
	Symbiodiniaceae	Impairment of photosynthesis, induced oxidative stress	Marangoni et al. (2022)
Plants/Algae	*Elodea sp.*	Increased growth, size of side shoots and total mass, decreased root-shoot ratio	van Weert et al. (2019)
	Myriophyllum spicatum	Increased root mass, decreased root-shoot ratio, and total size	
	Cymodocea nodosa	Increased leaf loss rate, affected seagrass photosynthetic machinery, and oxidative stress	Menicagli et al. (2022)

(continued)

Table 16.1 (continued)

Group	Species	Effects	References
Invertebrates	*Brachionus plicatilis*	Significantly suppressed population growth and negative impact on the life span	Li et al. (2023)
		Inhibited ingestion, decreased body volume, delayed spawning, and reduced eggs and offspring in the F0 and F1	Dong et al. (2022)
		Induce mortality at concentrations higher than 4.69 mg/L	Venâncio et al. (2019)
	Brachionus koreanus	Adverse effects just in combined exposure: Increased gene expression and oxidative stress	Jeong et al. (2021)
	Stylophora pistillata	Significant bleaching and oxidative stress	Marangoni et al. (2022)
	Diaphanosoma celebensis	Increased mortality rate by 25% and lipid peroxidation (induce oxidative damage)	Yoo et al. (2021)
	Artemia franciscana	Increased mortality, local accumulation	Bergami et al. (2017)
		Reduced mobility, absorption and accumulation	Machado et al. (2021)
	Euphausia superba	Slow embryo development, increased moulting, change in swimming behavior	Bergami et al. (2020)
	Hediste diversicolor	Increased burrowing time, decreased cholinesterase activity	Silva et al. (2020a, b)
	Sterechinus neumayeri	Oxidative stress, inflammatory response, and gene modulation (sod, cat, Mt., NF-kB, Centr, LBP/BPI, GNBP-1, Grp78, Grp170, Hsp70 and Bcl-2)	Bergami et al. (2019)
	Mytilus galloprovincialis	Decrease motility, ROS levels, and phagocytic activity. Increased apoptosis	Sendra et al. (2020)
	Crassostrea gigas	Lower fertilization and embryo-larval development	Tallec et al. (2020)

Fish	Sparus aurata	Increase in plasma cholesterol and triglyceraldehyde	Brandts et al. 2021
		Transcriptional level of genes and antioxidant response inhibited, increased anti-inflammatory response	Balasch et al. 2021
	Larimichthys crocea	Increased oxidative stress and mortality, augmented gut microbiota	Gu et al. (2020)
		Decreased survival, growth rate and cell viability, increased oxidative stress, disrupted lipid metabolism of hepatocytes	Lai et al. (2021)
	Dicentrarchus labrax	Increase in genes linked to lipid metabolism, no effect to genes linked to immune response	Brandts et al. (2018)
	Oryzias melastigma	Increase in apoptotic cells, oxidative stress, and activation levels of antioxidants	Kang et al. (2021)

3.2.2 Phytoplankton

The impacts on NPs in marine primary producers are diverse (Table 16.1). Plankton, a key part of the aquatic food chain, are also impacted by plastic nanoparticles (Atugoda et al. 2022). The variables at play regarding phytoplankton are numerous (Gao et al. 2023), making the effects difficult to predict. It has been documented that the presence of NPs interferes with the interaction between the cyanobacterium *Planktothrix agardhii* and its parasite *Rhizophydium megarrhizum* (Schampera et al. 2021). In this study, the particles affected the growth rate and heteroaggregation of the host, factors that have a strong correlation with the parasite success. Studies have shown that exposure to plastic nanoparticles can cause damage to the cell membrane and morphology and organelles of microalgae, while obstructing photosynthesis (Wang et al. 2020). However, the cell walls of algae can act as barriers to the penetration of nanoparticles, reducing their absorption and bioaccumulation (Sjollema et al. 2016).

3.2.3 Invertebrates

The rotifers, small primary consumer organisms inhabiting the plankton, are capable of fragmenting MPs into NPs (Zhao et al. 2023) and the toxicity of these nanoparticles has been evidenced in *Brachionus plicatilis* and *B. koreanus* (Table 16.1). Venâncio et al. (2019) have shown higher sensitivity of *B. plicatilis* to polymethylmethacrylate compared to other planktonic species, i.e., microalgae. In turn, the population growth and gene expression related to metabolism and transcriptionof *B. koreanus* was only impacted when NP exposure was combined withpolycyclic aromatic hydrocarbons (PAH) (Jeong et al. (2021). Similarly, increased mortality or slow embryo development where found in *Artemia fransiscana* and *Euphasia superba* following NP exposure (Table 16.1).

Krill (*Euphasia superba*), as the rotifers, can ingest MPs and fragment them into NPs within their digestive tract (Bergami et al. 2020). Once expelled, NPs lowers the density of the fecal matter, decreasing its sinking speed and potentially interfering with the trophic chain (Bergami et al. 2020). In sessile species, such as corals, exposure to NPs can stimulate bleaching and increased mucus production, posing an additional threat to these organisms already threatened by climate change (Marangoni et al. 2022; Valerio 2022). The benthic crustaceans (*Artemia fransiscana*), polychaetes (*Hediste diversicolor*), echinoderms (*Sterechinus neumayeri*), and mollusks (*Mytilus galloprovincialis*) have also shown sensitivity to NP exposure (Table 16.1). Bivalves are particularly at risk, as their filter-feeding mechanism makes them susceptible to ingesting and retaining small particles, leading to serious toxic effects (Ward et al. 2019; Huffman Ringwood 2021; Sendra et al. 2021). Additionally, NPs can cause restricted regeneration of body segments, increased production of reactive oxygen species, and lysosomal damage, suggesting changes in the immune system in bivalves and polychaetes (Canesi et al. 2015; Silva et al. 2020a, b).

3.2.4 Vertebrates

The zebrafish (*Danio rerio*), a freshwater fish, has been subject of studies on the impacts of NPs contamination. The results have demonstrated so far damages on larval development, reproduction, locomotion, and immunity (Bhagat et al. 2020). Among the few studies on marine fish, the impacts on the Gilt-head bream (*Sparus aurata*) and the large yellow croaker (*Larimichthys crocea*) can be highlighted. Those species displayed oxidative stress, suppression of immunological response, and a significant impact on gut biota (Gu et al. 2020; Balasch et al. 2021; Brandts et al. 2021) (Table 16.1). Fishes can accumulate NPs in various organs, causing toxicity related to oxidative stress, reproduction, and DNA damage (Bhagat et al. 2020; Atugoda et al. 2022).

In marine tetrapods, such as seabirds, sea turtles, and marine mammals, the effects of NPs exposure are poorly known (Fossi et al. 2020; da Silva Videla and Vieira de Araujo 2021). MPs have been commonly found in the gastrointestinal tract and feces of marine mammals, including cetaceans and pinnipeds (Meaza et al. 2021; Zantis et al. 2021). While MPs appear to be transitory in the animal's gut (Nelms et al. 2019), NPs can have a deep effects (Banerjee and Shelver 2021), suggesting that nanoparticles can be internalized by mammal species.

Despite seabirds being commonly used as sentinel species in plastic contamination studies, particles <1 mm are difficult to detect (O'Hanlon et al. 2017). It is known that seabirds can act as vectors, transporting marine-derived plastics to terrestrial environments. Studies with *Ardenna carneipes* (flesh-footed shearwaters) in Australia and with Arctic seabird species revealed substantial amounts of plastic debris deposited in their breeding colonies, rendering these locations sinks for marine plastic pollution (Bourdages et al. 2020; Grant et al. 2021). Moreover, in *A. carneipes*, the ingestion of plastic residue has been associated with fibrosis development, being this pathophysiology named plasticosis (Charlton-Howard et al. 2023).

A study in Australia detected the presence of NPs in *A. carneipes*, but no correlation between the volume of MPs in the animal's gut and that of the nanoparticles in their feces was found, suggesting that bioaccumulation was the source of the particles (Keys et al. 2022). Turtles are highly susceptible to plastic ingestion (Rizzi et al. 2019; Camedda et al. 2022). In one of the few studies found, the intestinal microbiota of the Chinese soft-shelled turtle (*Pelodiscus sinensis*) was impacted by NPs, were able to penetrate the eggshell and enter the developing embryo, causing a decrease in the embryo's heart rate, however no significant changes were observed in the morphology or behavior of the hatchlings (Gao et al. 2024).

As previously mentioned, NPs pose a great threat as vectors of their chemical additives, heavy metals, and POPs (Clukey et al. 2018; López-Berenguer et al. 2020; Muñoz et al. 2021; Yamashita et al. 2021). In particular, the impact of phthalates—a type plasticizer—on marine fauna is a growing concern due to their toxic effects, including endocrine disruption, oxidative stress, and metabolic disorders (Zhang et al. 2021). Those esters of the phthalic acid have been detected in sea turtles, marine mammals, and seabirds in adipose tissue, plasma, uropygial gland, muscle, liver and gonads (Routti et al. 2021; Vanstreels et al. 2023; Savoca et al.

2023; Lucas Fazardo et al. 2024). Recently, four phthalate esters (DMP, DEP, DBP, and DEHP) along with cholesterol and fatty acid were investigated in *P. blainvillei* (fransiscana dolphins) blubber samples in Southern Brazil. A pervasive contamination by those chemicals was found, particularly DEHP, which was detected in all samples. Additionally, positive correlations between DEP contents and the size and weight of the dolphins were found, as well as between DEHP concentrations and the fatty acid C17: 1 (Lucas Fazardo et al. 2024). Plastic pollution causes lethal and sub-lethal effects on marine megafauna, but the extent and magnitude of population-level impacts remain largely unknown (Senko et al. 2020), addressing that a lot more needs to be done if humanity's goal to preserve marine biodiversity is to be attained.

4 Techniques for Identifying and Monitoring Nanoplastics

4.1 Fundamental Concepts of Identification and Monitoring Techniques

The identification and monitoring of NPs are fundamental due to the great environmental impact of this type of material on biodiversity the health of living beings, including humans and the stability of the ecosystems. This is due to the size of these particles which, because they are very small (<100 nm), are able to cross biological barriers, accumulating in the tissues of living organisms and releasing xenobiotics with potential toxicity (Bhagat et al. 2020; Pelegrini et al. 2023). Because of these factors, the standardization of techniques for identifying and monitoring NPs is essential for understanding the state of contamination of bodies of water, soil, and other natural resources essential to life and biodiversity, thus enabling the development of strategies and public policies to mitigate the impacts of this pollutants in the environment (Schwaferts et al. 2019).

Therefore, it is essential that the techniques developed undergo validation processes to assess the precision, standardization, reproducibility, and quantification limits of the methods used to identify NPs. Validation critically and scientifically ensures the accuracy and reliability of the method, guaranteeing reproducible and accurate results, regardless of the matrix used in the monitoring process (Correia and Loeschner 2018; Enyoh et al. 2021).

Many guidelines worldwide establish the criteria and parameters to be considered when validating laboratory analytical methods, such as selectivity, linearity, matrix effect, working range, precision, accuracy, detection limit, quantification limit, and robustness. The selection of a guideline depends on directly of the method to be used, but it is also important to be attentive to the different terminologies and parameters present in the literature, since it can be relatable (Raposo and Ibelli-Bianco 2020).

4.2 Sampling and Collection of NPs

The techniques used to sample and collect NPs from biological and environmental sources are critical to the successful application of further identification techniques and to correctly understand the distribution and impact of these contaminants. Handling and analyzing nanometric particles requires additional care due to their propensity to agglomerate and the difficulty of separating them from other substances present in the samples. One of the main challenges is sample contamination during collection and analysis, which can lead to false positives (Fu et al. 2020; Li et al. 2022).

All collection protocols must guarantee the absence of external contaminants in order to give reliability to the results obtained from the detection analyses. For this reason, the instruments used must undergo standardized cleaning protocols, preferably with solvents of different polarities. In addition to the instruments, the storage containers must also be free of contaminants that could interfere with the quantification of a given analyze (Pelegrini et al. 2023; Li et al. 2022; Shorny et al. 2023).

For biological samples, it is also important that the instruments used are rinsed with organosolvent(s) throughout the process of removing the tissues of interest, minimizing possible external contamination, including paying attention to the transport of the material between the collection point and the analysis center. Environmental conditions, such as temperature and salinity of the water, for example, should also be recorded in order to put the results obtained into context (Pelegrini et al. 2023).

The separation, isolation, and purification of NPs from the samples collected are processes that can vary depending on the type of sample and the scale of the study. When collecting water, for example, the filtration process with a membrane of suitable porosity is essential in order to retain the material of interest. Sediments may require chemical digestion processes to remove organic matter before filtration. Biological samples are often subjected to enzymatic or chemical digestion to allow the separation of NPs from organic tissues (Fu et al. 2020; Enyoh et al. 2021).

In all cases, it is crucial to avoid cross-contamination and ensure that samples are not exposed to plastic sources during processing. In large-scale studies, automated methods can be used to process large volumes of samples efficiently. Techniques such as ultracentrifugation and flotation are also employed to separate NPs from particles of similar density (Mariano et al. 2021).

The final purification of the samples involves removing interferents that could affect subsequent analysis and may include additional filtration, washing, or centrifugation steps. Sample purity is also a crucial step for the accuracy of advanced analytical techniques used later, such as electron microscopy and spectroscopy (Fu et al. 2020; Mariano et al. 2021). Each stage of the process, from collection to purification, must be carefully documented and controlled to ensure the quality and integrity of the data obtained, as well as allow more precise discussions on the profile of the contamination identified.

Table 16.2 Analytical techniques used for nanoplastics identification, classification, and functioning principle

Techniques	Classification	Functioning principle	Size range
Hydrodynamic chromatography (HDC)	Separation	Utilize hydrodynamic and surface forces to separate particles in liquid	10 nm ~ 1000 nm
Field flow fraction (FFF)	Separation	Apply external fields into microfluidic environments for separation of dispersed particles	10 nm ~ 400 nm
Fluorescence microscopy (FM)	Characterization	Collect fluorescent emission from samples that are excited by the excitation wavelengths with proper selection of the filter cubes or lasers	> 0.05 μm
Scanning electron microscope (SEM)	Characterization	Scans surfaces to create high resolution images and can be coupled with Raman spectroscopy	< 400 nm
Dynamic light scattering (DLS)	Characterization	Uses a laser beam to pass through a liquid suspension containing the analyte particles that scatter the incident laser at different scattering angles	1 nm ~ 10 μm
Nanoparticle tracking analysis (NTA)	Characterization	Applies intense laser light to illuminate free diffusing particles to track their Brownian motion with monochrome imaging	10 nm ~ 1000 nm
Hyperspectral imaging (HSI)	Characterization	Based on the unique interaction between different chemical species in the sample and the incident light with different wavelengths	< 300 μm
Atomic force microscopy (AFM)	Characterization	Operates by scanning across the sample surface with a tiny tip that approaches the sample surface	5 nm ~ 10 nm
Flow cytometry (FCM)	Characterization	Analyzes particles within a moving flow based on the detection of light scattering and fluorescence by laser excitation and enables rapid high-resolution imaging	< 1 μm
X-ray photoelectron spectroscopy (XPS)	Identification	Identify the energy distribution of photoelectrons emitted from elements at the surface of polymeric materials	100 nm ~ 5 mm
μ-Fourier-transform infrared spectroscopy (μ-FTIR)	Identification	Acquires chemical information through detecting the vibration modes of sample molecules at different infrared frequencies over a wide spectral range	<10 μm

(continued)

Table 16.2 (continued)

Techniques	Classification	Functioning principle	Size range
Raman spectroscopy	Identification	Utilizes the frequency shift of inelastic scattered light from the sample based on Raman effect to obtain the vibrational modes and identify chemical species of samples	>1 μm
Thermogravimetric analysis (TGA)	Identification	Thermal technique that identifies polymers physical and chemical properties and microplastic concentration.	–
Differential scanning calorimetric (DSC)	Identification	Separation between endothermic and exothermic events that do not have wight loss, is complementary to TGA	–
Gas chromatography— Mass spectrometry (GC-MS)	Identification	Utilizes a capillary column, in which the stationary column medium materials have different relative affinity toward the analyte molecules and result in differentiated travel times	–
Inductively coupled plasma mass spectrometry (ICP-MS)	Identification	Allows the determination of particle concentrations, size and agglomeration state on very dilute suspensions by its low sensitivity of carbon	>45 nm
High performance liquid chromatography (HPLC)	Identification	The different solubility and polarity of the sample components in the mobile and stationary phases cause different flow rates for each component as they are flushed out of the column, and thus separating these components	–

Source: Prepared by the author based on information contained in Abbasi et al. (2023), Easton et al. (2020), Fu et al. (2020), and Maragliano et al. (2021)

4.3 Analytical Techniques for NP Identification and Characterization

The identification and characterization of NPs present a number of methodological and technical challenges related to the handling of very small particles (<100 nm) which require highly sensitive analytical techniques. In addition to this, the heterogeneity of environmental samples also represents a point of attention to the need for precision in analysis, making it necessary to employ techniques that make it possible to differentiate NPs from other compounds in order to reduce the chances of misinterpretations about the degree of contamination of the sites sampled (Battistini et al. 2021; Mariano et al. 2021; Zhou et al. 2021a, b).

In addition, the variability in the shape, size, and chemical composition of NPs is a limiting factor in the development of techniques to identify and characterize such nanomaterials requiring a combination of advanced analytical techniques. These techniques make it possible to detect particles as small as 1 nm (Table 16.2), which in some cases also helps to differentiate the particles, allowing us to get around the main challenges of characterizing and identifying NPs in environmental samples (Caputo et al. 2021; Enyoh et al. 2021; Zhou et al. 2021a, b).

Among the most widely used analytical techniques for identifying and characterizing NPs are spectroscopy and microscopy, each of which has its particularities and applications that allow identifying different traits of these materials to be explored in environmental and biological samples.

4.3.1 Spectroscopy

Spectroscopic techniques analyse the interaction between matter and electromagnetic radiation, allowing the study of spectral characteristics of a given sample like light scattering, absorption, and reflection to obtain information about its chemical composition. Analytical techniques like Fourier transform infrared spectroscopy (FTIR) and Raman spectroscopy have been widely used to identify polymer types in whole samples, providing unique spectral signatures for different plastic materials. In its turn, mass spectroscopy methods, such as pyrolysis coupled with gas chromatography-mass spectrometry (Py-GC-MS), are used to analyze the detailed chemical composition of NPs after thermal decomposition, providing insights into the origin and transformation of those nanomaterials (Enyoh et al. 2021; Zhou et al. 2021a, b; Li et al. 2022; Wang et al. 2024).

4.3.2 Microscopy Techniques

Microscopy techniques are powerful tools for visualizing very small particles, such as scanning electron microscopy (SEM) and transmission electron microscopy (TEM), providing high-resolution images of particles and, revealing morphological details thereof; important traits for further analysis of the potential impact of NPs in the marine ecosystems, for instance. In addition, atomic force microscopy (AFM) makes it possible to obtain detailed three-dimensional images of the surfaces of NPs, which is interesting from the point of view of studies into the degradation of the source material. These microscopy techniques are essential for differentiating NPs from other particles with similar sizes, making their use even more relevant for samples of environmental origin (Murray and Örmeci 2020; Yang et al. 2021; Akhatova et al. 2022; Shorny et al. 2023).

The identification and characterization of NPs involves advanced analytical techniques that provide detailed information on the morphology, composition, and origin of these particles. Despite the challenges associated with sample contamination and heterogeneity, these methods are essential for understanding the extent of NP pollution and its environmental and health impacts. The relevance of this topic is highlighted by the need to protect ecosystems and human health, underlining the importance of ongoing researches aiming at the development of monitoring and mitigation technologies that allow better understanding the impacts of NP

Table 16.3 Principal projects of EU and their aims

Project	Objective
EU plastic strategy (PS)	To decouple economic growth from resource use, while protecting natural resources and promoting sustainable development
Plastics	To transform how plastic products are designed, produced, used and recycled
Biobased, biodegradable and compostable plastic	To promote biobased, biodegradable and compostable plastic
Global action on plastic	To take part in global action on plastics through an intergovernmental negotiating committee
Microplastics	To reduce MPs by 30% until 2030
Plastic bags	To prevent the unsustainable consumption and use of lightweight plastic carrier
Packaging waste	To deal quantities of packaging waste
Plastic waste shipments	To end the export of plastic waste to third countries
CE4Plastics	To develop advanced methods for identifying and qualifying NPs to mitigate its presence in drinking water from single-use and reusable plastic bottles and disseminate knowledge to the public about the risks of NPs
VORTEX	To investigate the microbial transformation of plastics in the ocean
In-no-plastic	To develop technology and social initiatives to remove plastic from aquatic ecosystems
MIGMIPS	To use gut barrier models to study the translocation and cell absorption of MPs and NPs
MS6Plastics	To develop pretreatment and pre-concentration protocols to detect and characterize low ranges of MPs and NPs in fish and shellfish samples
EUROqCHARM	To evaluate existing methodologies for assessing plastic pollution and provide standardized European-level techniques
MonPlas	To develop technology for early detection of plastic particles in water
POLYRISK	To examine the toxic effects of MNPs particles on the immune system, to understand the mechanisms of toxicity using *in vitro* study models and identify biomarkers of toxicity in blood and saliva
AURORA	To develop a framework for evaluating the impact of MNPs pollution during pregnancy and early life
IMPTOX	To develop an analytical platform for investigating the combined effects and toxicity of MNPs, along with other environmental contaminants and how these impact allergic diseases in both preclinical and clinical studies
PLASTICHEAL	To create a new and reliable methodology and scientific evidence base to provide regulator with the necessary knowledge to address the short and long-term potential health impacts
PlasticsFatE	To implement a comprehensive measurement and testing program to improve and validate methods for identifying MNPs

Source: Prepared by the author based on information contained in EC (2019a)

pollution in a given marine ecosystem. Moreover, precise and standardized methodologies for detection and analysis of NPs across diverse sample types is crucial to support the development of effective legislation and suitable public policies.

5   Legislation and Regulation of Nanoplastics

To understand the behavior, hazards, and potential remediation of NPs, besides creating new policies and legislation to control the environmental impact of those nanomaterials (Table 16.3), the European Union (EU) has supported research projects in the last few years. In January 2018, the European Commission adopted the EU Plastics Strategy as part of the Circular Economy Action Plan, aiming to decouple economic growth from resource use, while protecting Europe's natural resources and promoting sustainable development. This strategy will help EU striving to reduce its consumption footprint and double its circular material use rate within the next decade. This includes revised legislative proposals on plastic waste. Also addressing concerns about the production and recycling of those polymers (EC 2024a, b). The EU policy aims to transform how plastic products are designed, produced, used, and recycled within the EU (EC 2024c). The strategy includes specific policies, such as promoting biobased, biodegradable, and compostable plastics, contributing to a more sustainable plastic economy and bringing environmental benefits with these materials (EC 2024d).

The EU also takes part in global actions on plastics through an intergovernmental negotiating committee (EC 2024e). The strategy aims to reduce MPs release by 30% by 2030 (E, 2024f) and prevent the unsustainable consumption and use of lightweight plastic carrier bags (EC 2024g). Additionally, the strategy deals with the quantities of packaging waste (EC 2024h), ends the export of plastic waste to third countries (EC 2024i), and seeks to prevent and reduce the impact of single-use plastic products on the environment (EC 2024j).

The EU has launched several research projects to address the MNPs problem. For instance, the CE4Plastics project (2022–2024) aims to develop advanced methods for identifying and qualifying NPs to mitigate its presence in drinking water from single-use and reusable plastic bottles. Furthermore, the project plans to disseminate knowledge to the public about the risks of NPs. In its turn, the VORTEX project (2018–2023) was designed to investigate the microbial transformation of plastics in the ocean. The In-No-Plastic project (2020–2023) is focused on developing technology and social initiatives to remove plastic from aquatic ecosystems (EC 2023a). Additionally, the MIGMIPS project (2021–2023) propose the use of gut barrier models to study the translocation and cell absorption of MPs and NPs. MS6Plastics (2022–2024) has been developing pre-treatment and pre-concentration protocols to detect and characterize low ranges of MPs and NPs in fish and shellfish samples. Moreover, the EUROqCHARM project aims to evaluate existing methodologies for assessing plastic pollution and provide standardized European-level techniques. Finally, the MonPlas project is working on developing technology for early detection of plastic particles in water (EC 2023a).

These projects are all part of the EU's RESPONSE project (2020–2023), which brings together expertise in oceanography, environmental chemistry, ecotoxicology, experimental ecology, and modeling. The overarching goal of the RESPONSE project is to answer key questions about the impact of MPs and NPs, and to develop strategies for understanding the distribution pathway and biological effects of these particles within marine ecosystems (EC 2023a). Since some research gaps were identified in the European Strategy for Plastics in a Circular Economy, the European Cluster on Health Impacts of Micro and Nanoplastics (CUSP) was implemented. The CUSP projects aim to contribute to the health-relevant objectives of the Bioeconomy Strategy by providing relevant evidences for new preventive policies. The CUSP initiative includes five projects taking place between 2021 and 2025, as follows POLYRISK, AURORA, IMPTOX, PLASTICHEAL, and PlasticsFatE (CUSP 2021).

The POLYRISK project examines the toxic effects of micro and nanoplastics particles on the immune system, to understand the mechanisms of toxicity using *in vitro* study models and identify biomarkers of toxicity in blood and saliva. The AURORA (Actionable European Roadmap for Early-life Health Risk Assessment of Micro and Nanoplastics) project intends to develop a framework for evaluating the impact of MPs and NPs pollution during pregnancy and early life, utilizing in-depth tests and epidemiological data to understand the implications for child development and health (CUSP 2021).

The IMPTOX project aims to develop an analytical platform for investigating the combined effects and toxicity of MPs and NPs, along with other environmental contaminants, and how these affects allergic diseases in both preclinical and clinical studies. The PLASTICHEAL, on the other hand, aims to create a new and reliable methodology and scientific evidence base to provide regulators with the necessary knowledge to address the short and long-term potential health impacts of MPs and NPs. Finally, the PlasticsFatE (Plastics Fate and Effects in the Human Body) project aims to implement a comprehensive measurement and testing program to improve and validate methods for identifying those micro and nanomaterials. This work is expected to lead to a new risk assessment strategy that will support the health-related goals of the European strategies for plastics (CUSP 2021).

Besides, the German Federal Institute for Materials Research and Testing intends to produce small quantities of NPs particles as reference materials for use in the POLYRISK and PlasticsFatE projects. Some projects have created NP particles with detectable characteristics, such as chemically entrapped metals that act as tracers (Mitrano et al. 2021) or fluorescent dyes (Catarino et al. 2019), to facilitate their identification and tracking in research.

In 2022, the European Commission created a draft proposal to restrict the use of synthetic polymer microparticles below 5 mm and fiber-like particles below 15 mm in various products. However, the proposal did not include restrictions on smaller-sized plastic particles, such as NPs. The draft legislation aims to ban the use of MPs in cosmetics, cleaning products, pesticides, and other applications within 3–12 years (ECHA 2019).

The European Chemicals Agency (ECHA) has proposed additional detailed measures for the European Commission to evaluate, including restricting the placement of plastic particles (on their own or in mixtures) on the market, with a transitional period to allow time for compliance by the industrial sector. ECHA also recommended a labeling requirement for the use of MPs and NPs, as well as a reporting obligation to ECHA to facilitate monitoring and control of residual releases. Furthermore, ECHA proposed a size limit of 100 nm, while the Committee for Socioeconomic Analysis (SEAC) suggested a lower size limit, i.e. 1 nm, for the scope of the restrictions of use of nanomaterials (ECHA 2019).

The ECHA's dossier identified over 400 plastic additives, such as fillers, UV stabilizers, and plasticizers that contribute to the toxicity of plastic particles. These additives have a production volume exceeding 100 t/year (ECHA 2019). However, a recent review found that there are over 6000 chemicals used in plastic (Aurisano et al. 2021). Since the implementation of the REACH regulation in 2007, more than 50 plasticizers, including phthalates like DEHP, BBP, DBP, and DIBP, have been registered and prohibited. Additional restrictions have also been put in place, such as the ban on the use of phthalates in children's toys since 1999. Several EU member states have implemented their own bans on specific products or types of products (EC 2023b).

In November 2023, the European Commission restricted MPs intentionally added to products under the EU's REACH chemical legislation. This regulation prohibits the sale of MPs and the placement of products intentionally containing MPs on the market, where MPs are defined as polymer particles below 5 mm that are organic, insoluble, and resistant to degradation (CR 2023; EC 2023b).

Beyond the EU, countries have acted to address the issue of MPs. Until 2019, Australia had ongoing voluntary industry initiatives. Brazil expressed intentions to ban the manufacturing and sale of personal care products containing plastic microbeads. Canada implemented a ban on the manufacturing, import, and sale of any toiletries for cleaning or hygiene that contain plastic microbeads. The USA also banned the manufacturing and sale of rinse-off products with exfoliating or cleansing functions on the human body (ECHA 2019).

Regulatory actions worldwide struggle to keep pace with the rapid release and proliferation of NPs in the environment. Recent and ongoing initiatives to regulate intentionally added MPs have sparked debate over the most appropriate lower size limit for classification. This debate is primarily driven by the enforcement challenges posed by the detection of smaller particles. However, it is worth mentioning that scientific evidences do not support setting a specific size limit, as the environmental and human impacts of plastics occur across a wide range of particle sizes. Therefore, it is important to establish new policies worldwide to evaluate, prevent, and mitigate the effects of MPs and NPs in the environment.

Preventing MPs from entering the environment should take priority over cleanup efforts, which are often too expensive, technically challenging, or not environmentally beneficial. While upstream solutions are crucial, downstream efforts should not be overlooked, as they are also necessary components of an effective strategy. Robust and practical legislation and regulations are essential to successfully prevent MPs and NPs from contaminating the environment. The path forward for regulating plastic pollution in the coming decades involves strengthening policies to prevent MNPs from entering the environment and enforcing existing top-down regulations. Transitioning to a circular economy emerges as the primary solution to reduce contamination by these pollutants (Munhoz et al. 2023).

6   Gaps and Perspectives

Although the impact of plastic particles is clear and growing worldwide, there is a significant lack of data on the impacts of NPs on marine life. The variables involved are numerous: the diversity of species affected, the composition of the plastic polymers, the characteristics of the particles, the previous absorption of pollutants, and their accumulation in the digestive system, among other factors (EC 2023a). However, the available knowledge is sufficient to understand the severity of the impact. The complex interaction of NPs with natural ecosystems reflects an emerging challenge in contemporary environmental science.

While studies have traditionally focused on the effects of macro- and microplastics, research on NPs reveals a distinct and multifaceted dynamic (Boyle and Örmeci 2020; DeLoid et al. 2021; Wang et al. 2023). These tiny particles, often measuring less than 100 nanometers (EC 2023b), possess a vast specific surface area and an unique ability to transport chemical contaminants, thereby amplifying potential risks to terrestrial and aquatic ecosystems (Shen et al. 2019; Town and van Leeuwen 2020; Bhagat et al. 2021; Atugoda et al. 2022). In addition to directly affecting biodiversity and habitat structure, NPs can alter fundamental processes such as nutrient cycling in soil and oxygen availability in aquatic environments, impacting microscopic organisms and more complex species along trophic chains (Fringer et al. 2020; Bhagat et al. 2020).

The global dissemination of NPs through physical, chemical, and biological processes, moving horizontally and vertically via factors such as wind, aquatic currents, and turbulence (Huang et al. 2020; Huang et al. 2022; Yu et al. 2024), also underscores their ubiquity and the complexity of associated environmental challenges. While some particles are directly absorbed by filter-feeding organisms or ingested by marine species, others accumulate in sediments and soils, potentially affecting decomposition processes and the quality of water resources used by human communities (Sendra et al. 2021; Ng et al. 2018). The interaction between NPs and natural microbiomes also warrants attention, as these particles can alter the composition and function of essential microbial communities crucial for ecosystem health, influencing the stability of fragile ecosystems in the face of additional environmental pressures such as climate change and industrial pollution (Fringer et al. 2020; Marangoni et al. 2022).

On the other hand, the potential risks of NPs to human and animal health have factors not well-understood, such as concentration, size, shape, polymer type, weathering state, and exposure route, which can significantly impact their effects. Additionally, plastics contain not just polymers, but also small organic chemicals (e.g., additives, residual monomers, polymerization aids) that can leach out when MNPs are released into the environment. As a result, laboratory-based studies focusing solely on the plastics themselves may be theoretical and not entirely reliable. To address this gap, there is a need for improved analytical methods and sampling procedures that can enable the generation of reliable data on the concentrations and properties of MNPs in the environment. This would help providing a more comprehensive understanding of the potential impacts of these contaminants on ecological systems and human health (EC 2019b).

The challenge with environmental analyzes, regarding the analytical techniques currently adopted for MPs and NPs investigations mainly SEM, TEM, and AFM microscopy, as well as FTIR/Raman spectroscopies and Pyr-GC-MS (Yang et al. 2021; Wang et al. 2024), is the need for effective and reliable extraction and detection methods that avoid interferences from non-plastic nanomaterials. These two aspects are interdependent, creating a dilemma that must be addressed for accurate and reliable environmental analysis of nanomaterials (EC 2019b). Importantly, research on plastic pollution has primarily focused on characterizing the problem rather than contributing to solutions. Addressing the knowledge gaps in plastic pollution is vital, but accumulating more descriptive evidence alone often does not influence public opinion or policymaking, leading to a gap in new legislation regarding this contaminant (EC 2019a).

The EU has, since the Circular Economy in 2018, launched and supported projects to produce reusable and recyclable plastic products and set goals to eliminate single-use plastics. Some other countries, such as Canada, New Zealand, South Korea, and the USA, are implementing restrictions on the intentional use of MPs in cosmetics and hygiene products. Additionally, international organizations have policies to reduce and prevent plastic pollution, such as the G7 Ocean Plastics Charter, the Arctic Council Desktop Study, the G20 Framework, the UN Global Partnership on Marine Litter, and the UN Sustainable Development Goals (Boyle and Örmeci 2020).

Regarding technological solutions for this concern, in most cases the focus has been on removing plastic debris from the environment, such as the Ocean Cleanup Project and the International Coastal Cleanup. Although this approach may offer some benefits in terms of raising awareness and population education, clean-up measures are the least effective solution when considering the waste management hierarchy, since it addresses the symptoms instead of the causes of the problem, i.e., the sources of plastic pollution. The FP7 CLEAN SEA project has policy options and highlights that "priority should be given to those stages that lead to waste prevention (in terms of reduction and preparation for re-use)" (EC 2019a). In this scenario, to solve this huge problem, it is necessary to understand the entire plastics ecosystem and how its various stages (design, production, usage, and post-use handling) are interconnected and affect one another. Therefore, for real prevention and remediation of this problem, all knowledge should be applied to develop new policies and projects in association with international governments, making plastic debris management a precautionary approach in the future (EC 2019b).

7 Conclusions

The information herein presented reveals the scarcity of research that can accurately depict the current situation on NPs in the environment and their impacts on animal health, for instance. In fact, among other reasons for this picture, the need for developing appropriate and standardized analytical protocols to effectively detect and characterize NPs has been emphasized.

With a substantial investment of resources and concerted efforts of researchers in this field, international governments can gain a more realistic understanding to the creation of new and more suitable policies and initiatives targeting the NPs issue. Furthermore, it is important to implement a comprehensive and sustainable approach to plastic management, to determine when the use of plastics is truly necessary, and how to properly dispose them at the end of their life cycle.

Beyond that, the search for alternatives to single-use plastics should be a priority, to completely replace them within society. Lastly, educating the general population is the key to increasing awareness and enabling international organizations and governments to minimize the overall use of plastics moving forward. With a multifaceted approach addressing research, policy, product design, and public awareness, meaningful progress can be envisaged in tackling the growing challenge of NPs and plastic pollution as a whole in both terrestrial and marine ecosystems.

Funding Declaration Research was supported by grant 88887.714746/2022-00 to G.G. from Coordenação de Aperfeiçoamento de Pessoal de Nível Superior (CAPES); 2023/03886-1 to A.N. from São Paulo Research Foundation (FAPESP); 471/SED/2021 to I.P from Programa de Bolsas Universitárias de Santa Catarina/Fundo Estadual de Apoio à Manutenção e ao Desenvolvimento da Educação Superior (UNIEDU/FUMDES); 88887.001911/2024-00 to L.L. from CAPES; 88887.832606/2023-00 to T.A from CAPES; N.857253 by the European Union's H2020 project Sinfonia and N.101057510 by Horizon Europe Project LEARN to V.V.

Competing Interests We declare that there are no competing interests.

References

Abbasi S, Razeghi N, Yousefi MR, Podkościelna B, Oleszczuk P (2023) Microplastics identification in water by TGA–DSC method: Maharloo Lake, Iran. Environ Sci Pollut Res 30(25):67008–67018. https://doi.org/10.1007/s11356-023-27214-8

Aderem A, Underhill DM (1999) Mechanism of phagocytosis in macrophages. Annu Rev Immunol 17(1):593–623. https://doi.org/10.1146/annurev.immunol.17.1.593

Akhatova F, Ishmukhametov I, Fakhrullina G, Fakhrullin R (2022) Nanomechanical atomic force microscopy to probe cellular microplastics uptake and distribution. Intl J Mol Sci [online] 23(2):806. https://doi.org/10.3390/ijms23020806

Allen D, Allen S, Abbasi S, Baker A, Bergmann M, Brahney J, Butler T, Duce RA, Eckhardt S, Evangeliou N, Jickells T, Kanakidou M, Kershaw P, Laj P, Levermore J, Li D, Liss P, Liu K, Mahowald N, Masque P (2022) Microplastics and nanoplastics in the marine-atmosphere environment. Nature Rev Earth Environ 3(6):393–405. https://doi.org/10.1038/s43017-022-00292-x

Amanda Pereira, da Costa Araújo Guilherme, Malafaia (2021) Microplastic ingestion induces behavioral disorders in mice: A preliminary study on the trophic transfer effects via tadpoles and fish Journal of Hazardous Materials 401123263. https://doi.org/10.1016/j.jhazmat.2020.123263

Amanda Pereira da Costa, Araújo Thiarlen Marinho da, Luz Thiago Lopes, Rocha Mohamed Ahmed Ibrahim, Ahmed Daniela de Melo e, Silva Md Mostafizur, Rahman Guilherme, Malafaia (2022) Toxicity evaluation of the combination of emerging pollutants with polyethylene microplastics in zebrafish: Perspective study of genotoxicity mutagenicity and redox unbalance Journal of Hazardous Materials 432128691. https://doi.org/10.1016/j.jhazmat.2022.128691

Atugoda T, Piyumali H, Wijesekara H, Sonne C, Lam SS, Mahatantila K, Vithanage M (2022) Nanoplastic occurrence, transformation and toxicity: a review. Environ Chem Lett 21(1):363–381. https://doi.org/10.1007/s10311-022-01479-w

Auffan M, Rose J, Bottero J-Y, Lowry GV, Jolivet J-P, Wiesner MR (2009) Towards a definition of inorganic nanoparticles from an environmental, health and safety perspective. Nat Nanotechnol [online] 4(10):634–641. https://doi.org/10.1038/nnano.2009.242

Aurisano N, Weber R, Fantke P (2021) Enabling a circular economy for Chemicals in Plastics. Curr Opin Green Sustain Chem 31(1):100513. https://doi.org/10.1016/j.cogsc.2021.100513

Balasch JC, Brandts I, Barría C, Martins MA, Tvarijonaviciute A, Tort L, Oliveira M, Teles M (2021) Short-term exposure to Polymethylmethacrylate Nanoplastics alters muscle antioxidant response, development and growth in Sparus Aurata. Mar Pollut Bull 172(1):112918. https://doi.org/10.1016/j.marpolbul.2021.112918

Banerjee A, Shelver WL (2021) Micro- and Nanoplastic induced cellular toxicity in mammals: a review. Sci Total Environ 755(2):142518. https://doi.org/10.1016/j.scitotenv.2020.142518

Barría C, Brandts I, Tort L, Oliveira M, Teles M (2020) Effect of Nanoplastics on fish health and performance: a review. Mar Pollut Bull 151(1):110791. https://doi.org/10.1016/j.marpolbul.2019.110791

Battistini B, Petrucci F, Bocca B (2021) In-house validation of AF4-MALS-UV for polystyrene Nanoplastic analysis. Anal Bioanal Chem 413(11):3027–3039. https://doi.org/10.1007/s00216-021-03238-2

Bergami E, Pugnalini S, Vannuccini ML, Manfra L, Faleri C, Savorelli F, Dawson KA, Corsi I (2017) Long-term toxicity of surface-charged polystyrene Nanoplastics to marine planktonic species Dunaliella Tertiolecta and Artemia Franciscana. Aquat Toxicol 189(1):159–169. https://doi.org/10.1016/j.aquatox.2017.06.008

Bergami E, Krupinski Emerenciano A, González-Aravena M, Cárdenas CA, Hernández P, Silva JRMC, Corsi I (2019) Polystyrene nanoparticles affect the innate immune system of the Antarctic Sea urchin Sterechinus Neumayeri. Polar Biol 42(4):743–757. https://doi.org/10.1007/s00300-019-02468-6

Bergami E, Manno C, Cappello S, Vannuccini ML, Corsi I (2020) Nanoplastics affect Moulting and Faecal pellet sinking in Antarctic krill (Euphausia superba) juveniles. Environ Int 143(1):105999. https://doi.org/10.1016/j.envint.2020.105999

Bhagat J, Zang L, Nishimura N, Shimada Y (2020) Zebrafish: an emerging model to study microplastic and Nanoplastic toxicity. Sci Total Environ 728(1):138707. https://doi.org/10.1016/j.scitotenv.2020.138707

Bhagat J, Nishimura N, Shimada Y (2021) Toxicological interactions of microplastics/nanoplastics and environmental contaminants: current knowledge and future perspectives. J Hazard Mater 405(1):123913. https://doi.org/10.1016/j.jhazmat.2020.123913

Bilal M, Ul Hassan H, Taj M, Rafiq N, Nabi G, Ali A, Gabol K, Shah MIA, Ghaffar RA, Sohail M, Arai T (2023) Biological magnification of microplastics: a look at the induced reproductive toxicity from simple invertebrates to complex vertebrates. Water [online] 15(15):2831. https://doi.org/10.3390/w15152831

Bour A, Mouchet F, Silvestre J, Gauthier L, Pinelli E (2015) Environmentally relevant approaches to assess nanoparticles ecotoxicity: a review. J Hazard Mater 283(1):764–777. https://doi.org/10.1016/j.jhazmat.2014.10.021

Bourdages MPT, Provencher JF, Baak JE, Mallory ML, Vermaire JC (2020) Breeding seabirds as vectors of microplastics from sea to land: evidence from colonies in Arctic Canada. Sci Total Environ 764(1):142808. https://doi.org/10.1016/j.scitotenv.2020.142808

Boyle K, Örmeci B (2020) Microplastics and Nanoplastics in the freshwater and terrestrial environment: a review. Water 12(9):2633. https://doi.org/10.3390/w12092633

Brandts I, Teles M, Tvarijonaviciute A, Pereira ML, Martins MA, Tort L, Oliveira M (2018) Effects of Polymethylmethacrylate Nanoplastics on Dicentrarchus Labrax. Genomics [online] 110(6):435–441. https://doi.org/10.1016/j.ygeno.2018.10.006

Brandts I, Barría C, Martins MA, Franco-Martínez L, Barreto A, Tvarijonaviciute A, Tort L, Oliveira M, Teles M (2021) Waterborne exposure of gilthead seabream (Sparus aurata) to Polymethylmethacrylate Nanoplastics causes effects at cellular and molecular levels. J Hazard Mater 403(1):123590. https://doi.org/10.1016/j.jhazmat.2020.123590

Brewer A, Dror I, Berkowitz B (2020) The mobility of plastic nanoparticles in aqueous and soil environments: a critical review. ACS ES&T Water 1(1):48–57. https://doi.org/10.1021/acsestwater.0c00130

Camedda A, Matiddi M, Vianello A, Coppa S, Bianchi J, Silvestri C, Palazzo L, Massaro G, Atzori F, Ruiu A, Piermarini R, Cocumelli C, Briguglio P, Hochscheid S, Brundu R, de Lucia GA (2022) Polymer composition assessment suggests prevalence of single-use plastics among items ingested by loggerhead sea turtles in the western mediterranean sub-region. Environ Pollut [online] 292(A):118274. https://doi.org/10.1016/j.envpol.2021.118274

Canesi L, Ciacci C, Bergami E, Monopoli MP, Dawson KA, Papa S, Canonico B, Corsi I (2015) Evidence for immunomodulation and apoptotic processes induced by cationic polystyrene nanoparticles in the hemocytes of the marine bivalve Mytilus. Mar Environ Res 111(1):34–40. https://doi.org/10.1016/j.marenvres.2015.06.008

Caputo F, Vogel R, Savage J, Vella G, Law A, Della Camera G, Hannon G, Peacock B, Mehn D, Ponti J, Geiss O, Aubert D, Prina-Mello A, Calzolai L (2021) Measuring particle size distribution and mass concentration of Nanoplastics and microplastics: addressing some analytical challenges in the sub-micron size range. J Coll Interf Sci [online] 588(1):401–417. https://doi.org/10.1016/j.jcis.2020.12.039

Catarino AI, Frutos A, Henry TB (2019) Use of fluorescent-labelled Nanoplastics (NPs) to demonstrate NP absorption is inconclusive without adequate controls. Sci Total Environ 670(1):915–920. https://doi.org/10.1016/j.scitotenv.2019.03.194

Charlton-Howard HS, Bond AL, Rivers-Auty J, Lavers JL (2023) 'Plasticosis': characterising macro- and microplastic-associated fibrosis in seabird tissues. J Hazard Mater [online] 450(1):131090. https://doi.org/10.1016/j.jhazmat.2023.131090

Cheung PK, Fok L (2016) Evidence of microbeads from personal care product contaminating the sea. Mar Pollut Bull [online] 109(1):582–585. https://doi.org/10.1016/j.marpolbul.2016.05.046

Clukey KE, Lepczyk CA, Balazs GH, Work TM, Li QX, Bachman MJ, Lynch JM (2018) Persistent organic pollutants in fat of three species of Pacific pelagic longline Caught Sea turtles: accumulation in relation to ingested plastic marine debris. Sci Total Environ 610-611:402–411. https://doi.org/10.1016/j.scitotenv.2017.07.242

Correia M, Loeschner K (2018) Detection of nanoplastics in food by asymmetric flow field-flow fractionation coupled to multi-angle light scattering: possibilities, challenges and analytical limitations. Analyt Bioanalyt Chem [online] 410(22):5603–5615. https://doi.org/10.1007/s00216-018-0919-8

CR (2023) https://single-market-economy.ec.europa.eu/document/download/9edd7d2c-2ca4-44dd-b50f-0ffc679becdd_en?filename=C_2023_6419_F1_COMMISSION_REGULATION_UNDER_ECT_EN_V5_P1_2620969.PDF

CUSP (2021) CUSP cluster – the European research cluster to understand the health impacts of micro- and nanoplastics. [online]. CUSP Cluster. Available at: https://cusp-research.eu/. Accessed 20 June 2024

da Silva Videla E, Vieira de Araujo F (2021) Marine debris on the Brazilian coast: which advances in the last decade? A literature review. Ocean Coastal Manage 199(1):105400. https://doi.org/10.1016/j.ocecoaman.2020.105400

DeLoid GM, Cao X, Bitounis D, Singh D, Llopis PM, Buckley B, Demokritou P (2021) Toxicity, uptake, and nuclear translocation of ingested micro-nanoplastics in an in vitro model of the small intestinal epithelium. Food Chem Toxicol 158(1):112609. https://doi.org/10.1016/j.fct.2021.112609

Ding J, Huang Y, Liu S, Zhang S, Zou H, Wang Z, Zhu W, Geng J (2020) Toxicological effects of nano- and micro-polystyrene plastics on red tilapia: are larger plastic particles more harmless? J Hazard Mater 396(1):122693. https://doi.org/10.1016/j.jhazmat.2020.122693

Dong, Liu Zhao-Feng, Guo Yao-Yang, Xu Faith, Ka Shun Chan Yu-Yao, Xu Matthew, Johnson Yong-Guan, Zhu (2022) Widespread occurrence of microplastics in marine bays with diverse drivers and environmental risk Environment International 168107483. https://doi.org/10.1016/j.envint.2022.107483

Dong, Wang Shaoguo, Ru Wei, Zhang Zhenzhong, Zhang Yuejiao, Li Lingchao, Zhao Lianxu, Li Jun, Wang (2022) Impacts of nanoplastics on life-history traits of marine rotifer (Brachionus plicatilis) are recovered after being transferred to clean seawater Environmental Science and Pollution Research 29(28):42780-42791 https://doi.org/10.1007/s11356-021-18121-x

Easton CD, Kinnear C, McArthur SL, Gengenbach TR (2020) Practical guides for x-ray photoelectron spectroscopy: analysis of polymers. J Vac Sci Technol A 38(2):023207. https://doi.org/10.1116/1.5140587

ECHA (2019). Annex XV restriction report proposal for a restriction substance name(s): intentionally added microplastics IUPAC name(s): n/a ec number(s): n/a cas number(s): n/a. [online] Available at: https://echa.europa.eu/documents/10162/05bd96e3-b969-0a7c-c6d0-441182893720. Accessed 20 June 2024

Enyoh CE, Wang Q, Chowdhury T, Wang W, Lu S, Xiao K, Chowdhury MAH (2021) New analytical approaches for effective quantification and identification of Nanoplastics in environmental samples. PRO 9(11):2086. https://doi.org/10.3390/pr9112086

European Commission (2019a) A CIRCULAR ECONOMY FOR PLASTICS: insights from research and innovation to inform policy and funding decisions. Publications Office of the European Union. [online] https://doi.org/10.2777/269031

European Commission (2019b) Microplastics [online]. Publications Office of the European Union. Available at: https://environment.ec.europa.eu/topics/plastics/microplastics_en. Accessed 4 June 2024

European Commission (2023a) Amending Annex XVII to Regulation (EC) No 1907/2006 of the European Parliament and of the Council concerning the Registration, Evaluation, Authorisation and Restriction of Chemicals (REACH) as Regards Synthetic Polymer Microparticles (Text with EEA relevance). [online] Brussels: Publications Office of the European Union. Available at: https://single-market-economy.ec.europa.eu/document/download/9edd7d2c-2ca4-44dd-b50f-0ffc679becdd_en?filename=C_2023_6419_F1_COMMISSION_REGULATION_UNDER_ECT_EN_V5_P1_2620969.PDF. Accessed 20 June 2024

European Commission (2023b) Nanoplastics—state of knowledge and environmental and human health impacts. [online] Publications Office of the European Union. Available at: https://op.europa.eu/en/publication-detail/-/publication/a9088790-ace5-11ed-8912-01aa75ed71a1/language-en

European Commission (2023c) Guidance on the implementation of the commission recommendation 2022/C 229/01 on the definition of nanomaterial. [online] Publications Office of the European Union. Available at: https://data.europa.eu/doi/10.2760/143118

European Commission (2024a) Circular economy. [online]. Publications Office of the European Union. Available at: https://environment.ec.europa.eu/topics/circulareconomy_en. Accessed 4 June 2024

European Commission (2024b) Global action on plastics. [online]. Publications Office of the European Union. Available at: https://environment.ec.europa.eu/topics/plastics/global-action-plastics_en. Accessed 4 June 2024

European Commission (2024c) Plastic waste shipments [online]. Publications Office of the European Union. Available at: https://environment.ec.europa.eu/strategy/biodiversity-strategy-2030_en.%20Accessed%204%20jun%202024. Accessed 4 June 2024

European Commission (2024d) Biobased, biodegradable and compostable plastics [online]. Publications Office of the European Union. Available at: https://environment.ec.europa.eu/topics/plastics/biobased-biodegradable-and-compostable-plastics_en. Accessed 4 June 2024

European Commission (2024e) Plastics. [online] Publications Office of the European Union. Available at: https://environment.ec.europa.eu/topics/plastics_en. Accessed 4 June 2024

Farrell P, Nelson K (2013) Trophic level transfer of microplastic: Mytilus Edulis (L.) to Carcinus Maenas (L.). Environ Pollut 177(1):1–3. https://doi.org/10.1016/j.envpol.2013.01.046

Ferreira I, Venâncio C, Lopes I, Oliveira M (2019) Nanoplastics and marine organisms: what has been studied? Environ Toxicol Pharmacol [online] 67(1):1–7. https://doi.org/10.1016/j.etap.2019.01.006

Fossi MC, Baini M, Simmonds MP (2020) Cetaceans as ocean health indicators of marine litter impact at global scale. Front Environ Sci 8(1). https://doi.org/10.3389/fenvs.2020.586627

Frère L, Maignien L, Chalopin M, Huvet A, Rinnert E, Morrison H, Kerninon S, Cassone A-L, Lambert C, Reveillaud J, Paul-Pont I (2018) Microplastic bacterial communities in the bay of Brest: influence of polymer type and size. Environ Pollut 242(A):614–625. https://doi.org/10.1016/j.envpol.2018.07.023

Fringer VS, Fawcett LP, Mitrano DM, Maurer-Jones MA (2020) Impacts of Nanoplastics on the viability and riboflavin secretion in the model bacteria Shewanella Oneidensis. Biogeochem Dynam 8(1). https://doi.org/10.3389/fenvs.2020.00097

Fu W, Min J, Jiang W, Li Y, Zhang W (2020) Separation, characterization and identification of microplastics and Nanoplastics in the environment. Sci Total Environ 721(1):137561. https://doi.org/10.1016/j.scitotenv.2020.137561

Gao F, Li J, Sun C, Zhang L, Jiang F, Cao W, Zheng L (2019) Study on the capability and characteristics of heavy metals enriched on microplastics in marine environment. Mar Pollut Bull 144(1):61–67. https://doi.org/10.1016/j.marpolbul.2019.04.039

Gao S, Huang G, Zhang P, Xin X, Yin J, Han D, Rosendahl S, Read S (2023) Toxicity and mechanism of Nanoplastics to phytoplankton in high-latitude aquatic ecosystems of Canadian prairie: effects of multiple environmental factors. Sci Total Environ 893(1):164676–164676. https://doi.org/10.1016/j.scitotenv.2023.164676

Gao S, Zhang S, Sun J, He X, Xue S, Zhang W, Li P, Lin L, Qu Y, Ward-Fear G, Chen L, Li H (2024) Nanoplastic pollution changes the intestinal microbiome but not the morphology or behavior of a freshwater turtle. Sci Total Environ 934(1):173178–173178. https://doi.org/10.1016/j.scitotenv.2024.173178

Garcia F, de Carvalho AR, Riem-Galliano L, Tudesque L, Albignac M, ter Halle A, Cucherousset J (2021) Stable isotope insights into microplastic contamination within freshwater food webs. Environ Sci Technol 55(2):1024–1035. https://doi.org/10.1021/acs.est.0c06221

Gomes T, Almeida AC, Georgantzopoulou A (2020) Characterization of cell responses in Rhodomonas Baltica exposed to PMMA Nanoplastics. Sci Total Environ 726(1):138547. https://doi.org/10.1016/j.scitotenv.2020.138547

González-Fernández C, Toullec J, Lambert C, Le Goïc N, Seoane M, Moriceau B, Huvet A, Berchel M, Vincent D, Courcot L, Soudant P, Paul-Pont I (2019) Do transparent exopolymeric particles (TEP) affect the toxicity of Nanoplastics on Chaetoceros neogracile? Environ Pollut 250(1):873–882. https://doi.org/10.1016/j.envpol.2019.04.093

Grant ML, Lavers JL, Hutton I, Bond AL (2021) Seabird breeding Islands as sinks for marine plastic debris. Environ Pollut 276(1):116734. https://doi.org/10.1016/j.envpol.2021.116734

Gu H, Wang S, Wang X, Yu X, Hu M, Huang W, Wang Y (2020) Nanoplastics impair the intestinal health of the juvenile large yellow croaker Larimichthys Crocea. J Hazard Mater 397(1):122773. https://doi.org/10.1016/j.jhazmat.2020.122773

Guo X, Sun X, Chen Y, Hou L, Liu M, Yang Y (2020) Antibiotic resistance genes in biofilms on plastic wastes in an estuarine environment. Sci Total Environ 745(1):140916. https://doi.org/10.1016/j.scitotenv.2020.140916

Hahladakis JN, Velis CA, Weber R, Iacovidou E, Purnell P (2018) An overview of chemical additives present in plastics: migration, release, fate and environmental impact during their use, disposal and recycling. J Hazard Mater [online] 344(344):179–199. https://doi.org/10.1016/j.jhazmat.2017.10.014

He S, Chi H-Y, Li C, Gao Y, Li Z-C, Zhou X-X, Yan B (2022) Distribution, bioaccumulation, and trophic transfer of palladium-doped Nanoplastics in a constructed freshwater ecosystem. Environ Sci Nano 9(4):1353–1363. https://doi.org/10.1039/d1en00940k

Hoang TC, Felix-Kim M (2020) Microplastic consumption and excretion by fathead minnows (Pimephales promelas): influence of particles size and body shape of fish. Sci Total Environ 704(1):135433. https://doi.org/10.1016/j.scitotenv.2019.135433

Huang D, Tao J, Cheng M, Deng R, Chen S, Yin L, Li R (2020) Microplastics and Nanoplastics in the environment: macroscopic transport and effects on creatures. J Hazard Mater 407(1):124399. https://doi.org/10.1016/j.jhazmat.2020.124399

Huang D, Chen H, Shen M, Tao J, Chen S, Yin L, Zhou W, Wang X, Xiao R, Li R (2022) Recent advances on the transport of microplastics/nanoplastics in abiotic and biotic compartments. J Hazard Mater [online] 438(1):129515. https://doi.org/10.1016/j.jhazmat.2022.129515

Huffman Ringwood A (2021) Bivalves as biological sieves: bioreactivity pathways of microplastics and Nanoplastics. Biol Bull 241(2):185–195. https://doi.org/10.1086/716259

Jansen MAK, Andrady AL, Bornman JF, Aucamp PJ, Bais AF, Banaszak AT, Barnes PW, Bernhard GH, Bruckman LS, Busquets R, Häder, Donat P, Hanson ML, Heikkilä AM, Hylander S, Lucas RM, Mackenzie R, Madronich S, Neale PJ, Neale RE, Olsen CM (2024) Plastics in the environment in the context of UV radiation, climate change and the Montreal protocol: UNEP environmental effects assessment panel, update 2023. Photochem Photobiol Sci [online] 23(4):629–650. https://doi.org/10.1007/s43630-024-00552-3

Jeong C-B, Kang H-M, Lee YH, Kim M-S, Lee J-S, Seo JS, Wang M, Lee J-S (2018) Nanoplastic ingestion enhances toxicity of persistent organic pollutants (POPs) in the Monogonont rotifer Brachionus Koreanus via multixenobiotic resistance (MXR) disruption. Environ Sci Technol 52(19):11411–11418. https://doi.org/10.1021/acs.est.8b03211

Jeong C-B, Kang H-M, Byeon E, Kim M-S, Ha SY, Kim M, Jung J-H, Lee J-S (2021) Phenotypic and transcriptomic responses of the rotifer Brachionus Koreanus by single and combined exposures to nano-sized microplastics and water-accommodated fractions of crude oil. J Hazard Mater 416(1):125703. https://doi.org/10.1016/j.jhazmat.2021.125703

Jeong J, Im J, Choi J (2024) Integrating aggregate exposure pathway and adverse outcome pathway for micro/nanoplastics: a review on exposure, toxicokinetics, and toxicity studies. Ecotoxicol Environ Saf 272(1):116022–116022. https://doi.org/10.1016/j.ecoenv.2024.116022

Jon S. B., de Vlieger Daan J. A., Crommelin Katherine, Tyner Daryl C., Drummond Wenlei, Jiang Scott E., McNeil Sesha, Neervannan Rachael M., Crist Vinod P., Shah (2019) Report of the AAPS Guidance Forum on the FDA Draft Guidance for Industry: "Drug Products Including Biological Products that Contain Nanomaterials" The AAPS Journal 21(4). https://doi.org/10.1208/s12248-019-0329-7

Joppien M, Westphal H, Chandra V, Stuhr M, Doo SS (2022) Nanoplastic incorporation into an organismal skeleton. Sci Rep 12(1):14771. https://doi.org/10.1038/s41598-022-18547-4

Junaid M, Abbas Z, Siddiqui JA, Liu S, Tabraiz S, Yue Q, Wang J (2023) Ecotoxicological impacts associated with the interplay between micro(nano)plastics and pesticides in aquatic and terrestrial environments. Trends Anal Chem 165(1):117133–117133. https://doi.org/10.1016/j.trac.2023.117133

Kang H-M, Byeon E, Jeong H, Kim M-S, Chen Q, Lee J-S (2021) Different effects of nano- and microplastics on oxidative status and gut microbiota in the marine Medaka Oryzias Melastigma. J Hazard Mater 405(1):124207. https://doi.org/10.1016/j.jhazmat.2020.124207

Kangas A, Setälä O, Kauppi L, Lehtiniemi M (2023) Trophic transfer increases the exposure to microplastics in Littoral predators. Mar Pollt Bull [online] 196(1):115553. https://doi.org/10.1016/j.marpolbul.2023.115553

Keys BC, Grant ML, Rodemann T, Mylius KA, Pinfold TL, Rivers-Auty J, Lavers JL (2022) New methods for the quantification of ingested Nano- and Ultrafine plastics in seabirds. Environ Sci Technol 57(1):310–320. https://doi.org/10.1021/acs.est.2c06973

Khatmullina L, Isachenko I (2017) Settling velocity of microplastic particles of regular shapes. Mar Pollut Bull 114(2):871–880. https://doi.org/10.1016/j.marpolbul.2016.11.024

Kim SY, Kim YJ, Lee S-W, Lee E-H (2022) Interactions between bacteria and Nano (micro)-sized polystyrene particles by bacterial responses and microscopy. Chemosphere 306(1):135584. https://doi.org/10.1016/j.chemosphere.2022.135584

Kiran BR, Kopperi H, Venkata Mohan S (2022) Micro/nano-plastics occurrence, identification, risk analysis and mitigation: challenges and perspectives. Rev Environ Sci Biotechnol 21(1):169–203. https://doi.org/10.1007/s11157-021-09609-6

Kirstein IV, Kirmizi S, Wichels A, Garin-Fernandez A, Erler R, Löder M, Gerdts G (2016) Dangerous hitchhikers? Evidence for potentially pathogenic vibrio spp. on microplastic particles. Mar Environ Res 120(1):1–8. https://doi.org/10.1016/j.marenvres.2016.07.004

Kühn S, van Franeker JA (2020) Quantitative overview of marine debris ingested by marine megafauna. Marine Pollut Bull [online] 151(1):110858. https://doi.org/10.1016/j.marpolbul.2019.110858

Lai W, Xu D, Li J, Wang Z, Ding Y, Wang X, Li X, Xu N, Mai K, Ai Q (2021) Dietary polystyrene Nanoplastics exposure alters liver lipid metabolism and muscle nutritional quality in carnivorous marine fish large yellow croaker (Larimichthys crocea). J Hazard Mater 419(1):126454. https://doi.org/10.1016/j.jhazmat.2021.126454

Le V-G, Nguyen M-K, Nguyen H-L, Lin C, Hadi M, Hung NTQ, Hoang H-G, Nguyen KN, Tran H-T, Hou D, Zhang T, Bolan NS (2023) A comprehensive review of micro- and nano-plastics in the atmosphere: occurrence, fate, toxicity, and strategies for risk reduction. Sci Total Environ [online] 904(1):166649. https://doi.org/10.1016/j.scitotenv.2023.166649

Lewis JA (2004) Colloidal processing of ceramics. J Am Ceram Soc [online] 83(10):2341–2359. https://doi.org/10.1111/j.1151-2916.2000.tb01560.x

Li Q, Lai Y, Li P, Liu X, Yao Z, Liu J, Yu S (2022) Evaluating the occurrence of polystyrene nanoparticles in environmental waters by agglomeration with alkylated Ferroferric oxide followed by micropore membrane filtration collection and Py-GC/MS analysis. Environ Sci Technol 56(12):8255–8265. https://doi.org/10.1021/acs.est.2c02033

Li X, Lu L, Ru S, Eom J, Wang D, Samreem and Wang, J. (2023) Nanoplastics induce more severe multigenerational life-history trait changes and metabolic responses in marine rotifer Brachionus plicatilis: comparison with microplastics. J Hazard Mater [online] 449(1):131070. https://doi.org/10.1016/j.jhazmat.2023.131070

Liao Y, Yang J (2020) Microplastic serves as a potential vector for Cr in an in-vitro human digestive model. Sci Total Environ 703(1):134805. https://doi.org/10.1016/j.scitotenv.2019.134805

Liu G, Cai W, Liu H, Jiang H, Bi Y, Wang H (2021a) The Association of Bisphenol a and phthalates with risk of breast cancer: a meta-analysis. Intl J Environ Res Public Health [online] 18(5):2375. https://doi.org/10.3390/ijerph18052375

Liu L, Xu K, Zhang B, Ye Y, Zhang Q, Jiang W (2021b) Cellular internalization and release of polystyrene microplastics and Nanoplastics. Sci Total Environ 779(1):146523. https://doi.org/10.1016/j.scitotenv.2021.146523

Liu Y, Shao H, Liu J, Cao R, Shang E, Liu S, Li Y (2021c) Transport and transformation of microplastics and Nanoplastics in the soil environment: a critical review. Soil Use Manag 37(2):224–242. https://doi.org/10.1111/sum.12709

López-Berenguer G, Peñalver P, Martínez-López E (2020) A critical review about neurotoxic effects in marine mammals of mercury and other trace elements. Chemosphere [online] 246(1):125688. https://doi.org/10.1016/j.chemosphere.2019.125688

Lu Y-Y, Li H, Ren H, Zhang X, Huang F, Zhang D, Huang Q, Zhang X (2022) Size-dependent effects of polystyrene nanoplastics on autophagy response in human umbilical vein endothelial cells. J Hazard Mater 421(1):126770. https://doi.org/10.1016/j.jhazmat.2021.126770

Lucas F, de Lima Suelen, Goulart Giulia Galani, Martha Susane, Lopes Marzia, Antonelli Daphne Wrobel, Goldberg Sandro, Sandri Isadora Nicole Lara, Piccinin Cristiane Kiyomi Miyaji, Kolesnikovas Marcelo, Maraschin (2024) Detection of phthalate esters and targeted metabolome analysis in Franciscana dolphin (Pontoporia blainvillei) blubber in the coast of Santa Catarina southern Brazil Marine Pollution Bulletin 205116598. https://doi.org/10.1016/j.marpolbul.2024.116598

Machado AJT, Mataribu B, Serrão C, da Silva Silvestre L, Farias DF, Bergami E, Corsi I, Marques-Santos LF (2021) Single and combined toxicity of amino-functionalized polystyrene nanoparticles with potassium dichromate and copper sulfate on brine shrimp Artemia Franciscana larvae. Environ Sci Pollut Res 28(33):45317–45334. https://doi.org/10.1007/s11356-021-13907-5

Marangoni LFB, Beraud E, Ferrier-Pagès C (2022) Polystyrene Nanoplastics impair the photosynthetic capacities of Symbiodiniaceae and promote coral bleaching. Sci Total Environ 815(1):152136. https://doi.org/10.1016/j.scitotenv.2021.152136

Mariano S, Tacconi S, Fidaleo M, Rossi M, Dini L (2021) Micro and Nanoplastics identification: classic methods and innovative detection techniques. *Frontiers*. Toxicology 3(1). https://doi.org/10.3389/ftox.2021.636640

Materić D, Kasper-Giebl A, Kau D, Anten M, Greilinger M, Ludewig E, van Sebille E, Röckmann T, Holzinger R (2020) Micro- and Nanoplastics in alpine snow: a new method for chemical identification and (semi)quantification in the Nanogram range. Environ Sci Technol 54(4):2353–2359. https://doi.org/10.1021/acs.est.9b07540

Mattsson K, Hansson L-A, Cedervall T (2015) Nano-plastics in the aquatic environment. Environ Sci: Processes Impacts 17(10):1712–1721. https://doi.org/10.1039/c5em00227c

Mattsson K, Johnson EV, Malmendal A, Linse S, Hansson L-A, Cedervall T (2017) Brain damage and Behavioural disorders in fish induced by plastic nanoparticles delivered through the food chain. Sci Rep [online] 7(1). https://doi.org/10.1038/s41598-017-10813-0

Meaza I, Toyoda JH, Wise JP Sr (2021) Microplastics in sea turtles, marine mammals and humans: a one environmental health perspective. *Frontiers in environmental*. Science 8(1). https://doi.org/10.3389/fenvs.2020.575614

Menicagli V, Castiglione MR, Balestri E, Giorgetti L, Bottega S, Sorce C, Spanò C, Lardicci C (2022) Early evidence of the impacts of microplastic and Nanoplastic pollution on the growth and physiology of the seagrass Cymodocea Nodosa. Sci Total Environ 838(3):156514. https://doi.org/10.1016/j.scitotenv.2022.156514

Messika, Revel Amélie, Châtel Catherine, Mouneyrac (2018) Micro(nano)plastics: A threat to human health? Current Opinion in Environmental Science & Health 117-23. https://doi.org/10.1016/j.coesh.2017.10.003

Mitrano DM, Wohlleben W (2020) Microplastic regulation should be more precise to incentivize both innovation and environmental safety. Nat Commun [online] 11(1):5324. https://doi.org/10.1038/s41467-020-19069-1

Mitrano DM, Wick P, Nowack B (2021) Placing Nanoplastics in the context of global plastic pollution. Nature Nanotechnol [online] 16(5):491–500. https://doi.org/10.1038/s41565-021-00888-2

Munhoz DR, Harkes P, Beriot N, Larreta J, Basurko OC (2023) Microplastics: a review of policies and responses. Microplastics [online] 2(1):1–26. https://doi.org/10.3390/microplastics2010001

Muñoz CC, Hendriks AJ, Ragas AMJ, Vermeiren P (2021) Internal and maternal distribution of persistent organic pollutants in sea turtle tissues: a meta-analysis. Environ Sci Technol 55(14):10012–10024. https://doi.org/10.1021/acs.est.1c02845

Murray A, Örmeci B (2020) Removal effectiveness of Nanoplastics (<400 nm) with separation processes used for water and wastewater treatment. Water 12(3):635. https://doi.org/10.3390/w12030635

Narayanan G, Talib M, Singh N, Darbha GK (2024) Toxic effects of polystyrene Nanoplastics and polycyclic aromatic hydrocarbons (chrysene and fluoranthene) on the growth and physiological characteristics of Chlamydomonas Reinhardtii. Aquat Toxicol 268(1):106838–106838. https://doi.org/10.1016/j.aquatox.2024.106838

Nath J, De J, Sur S, Banerjee P (2023) Interaction of microbes with microplastics and nanoplastics in the agroecosystems—impact on antimicrobial resistance. Pathogens [online] 12(7):888. https://doi.org/10.3390/pathogens12070888

Nelms SE, Barnett J, Brownlow A, Davison NJ, Deaville R, Galloway TS, Lindeque PK, Santillo D, Godley BJ (2019) Microplastics in marine mammals stranded around the British coast: ubiquitous but transitory? Sci Rep 9(1):1075. https://doi.org/10.1038/s41598-018-37428-3

Ng E-L, Huerta Lwanga E, Eldridge SM, Johnston P, Hu H-W, Geissen V, Chen D (2018) An overview of microplastic and Nanoplastic pollution in agroecosystems. Sci Total Environ 627(1):1377–1388. https://doi.org/10.1016/j.scitotenv.2018.01.341

Nobuyuki, OM, Ito T, Hano K, Kono Kazuhiko, Mochida (2020) Estimation of the uptake and gut retention of microplastics in juvenile marine fish: Mummichogs (Fundulus heteroclitus) and red seabreams (Pagrus major) Marine Pollution Bulletin 160111630. https://doi.org/10.1016/j.marpolbul.2020.111630

O'Hanlon NJ, James NA, Masden EA, Bond AL (2017) Seabirds and marine plastic debris in the northeastern Atlantic: a synthesis and recommendations for monitoring and research. Environ Pollut 231(2):1291–1301. https://doi.org/10.1016/j.envpol.2017.08.101

OECD (2021) Policies to reduce microplastics pollution in water. OECD Publishing, Paris. https://doi.org/10.1787/7ec7e5ef-en

Outi, Setälä Vivi, Fleming-Lehtinen Maiju, Lehtiniemi (2014) Ingestion and transfer of microplastics in the planktonic food web Environmental Pollution 18577–83. https://doi.org/10.1016/j.envpol.2013.10.013

Park BC, Brown A, Laubinger F, Börkey P (2024) Monitoring trade in plastic waste and scrap. OECD Environ Working Pap 233(1). https://doi.org/10.1787/013bcfdd-en

Pelegrini K, Pereira TCB, Maraschin TG, Teodoro LDS, Basso NRDS, De Galland GLB, Ligabue RA, Bogo MR (2023) Micro- and Nanoplastic toxicity: a review on size, type, source, and test-organism implications. Sci Total Environ [online] 878(1):162954. https://doi.org/10.1016/j.scitotenv.2023.162954

Peller JR, Mezyk SP, Shidler S, Castleman J, Kaiser S, Faulkner RF, Pilgrim CD, Wilson A, Martens S, Horne GP (2022) Facile Nanoplastics formation from macro and microplastics in aqueous media. Environ Pollut 313(1):120171. https://doi.org/10.1016/j.envpol.2022.120171

Peng C, He N, Wu Y, Lü Y, Sun H, Wang L (2022) Excretion characteristics of nylon microplastics and absorption risk of Nanoplastics in rats. Ecotoxicol Environ Saf 238(1):113586–113586. https://doi.org/10.1016/j.ecoenv.2022.113586

Pradel A, Catrouillet C, Gigault J (2023) The environmental fate of nanoplastics: what we know and what we need to know about aggregation. NanoImpact 29(1):100453. https://doi.org/10.1016/j.impact.2023.100453

Prata JC (2023) Microplastics and human health: integrating pharmacokinetics. Crit Rev Environ Sci Technol 53(16):1–23. https://doi.org/10.1080/10643389.2023.2195798

Prüst M, Meijer J, Westerink RHS (2020) The plastic brain: neurotoxicity of micro- and nanoplastics. Part Fibre Toxicol 17(1):24. https://doi.org/10.1186/s12989-020-00358-y

Qiao R, Deng Y, Zhang S, Wolosker MB, Zhu Q, Ren H, Zhang Y (2019) Accumulation of different shapes of microplastics initiates intestinal injury and gut microbiota Dysbiosis in the gut of zebrafish. Chemosphere 236(1):124334. https://doi.org/10.1016/j.chemosphere.2019.07.065

Ramirez Arenas L, Ramseier Gentile S, Zimmermann S, Stoll S (2021) Nanoplastics adsorption and removal efficiency by granular activated carbon used in drinking water treatment process. Sci Total Environ 791(1):148175. https://doi.org/10.1016/j.scitotenv.2021.148175

Raposo F, Ibelli-Bianco C (2020) Performance parameters for analytical method validation: controversies and discrepancies among numerous guidelines. Trends Analyt Chem [online] 129(1):115913. https://doi.org/10.1016/j.trac.2020.115913

Ripken C, Khalturin K, Shoguchi E (2020) Response of coral reef dinoflagellates to nanoplastics under experimental conditions suggests downregulation of cellular metabolism. Microorganisms 8(11):1759. https://doi.org/10.3390/microorganisms8111759

Rizzi M, Rodrigues FL, Medeiros L, Ortega I, Rodrigues L, Monteiro DS, Kessler F, Proietti MC (2019) Ingestion of plastic marine litter by sea turtles in southern Brazil: abundance, characteristics and potential selectivity. Mar Pollut Bull 140(1):536–548. https://doi.org/10.1016/j.marpolbul.2019.01.054

Roman L, Hardesty BD, Hindell MA, Wilcox C (2019) A quantitative analysis linking seabird mortality and marine debris ingestion. Sci Rep 9(1):3202. https://doi.org/10.1038/s41598-018-36585-9

Routti H, Harju M, Lühmann K, Aars J, Ask A, Goksøyr A, Kovacs KM, Lydersen C (2021) Concentrations and endocrine disruptive potential of phthalates in marine mammals from the Norwegian Arctic. Environ Int 152(1):106458. https://doi.org/10.1016/j.envint.2021.106458

SAPEA (2019) https://doi.org/10.26356/microplastics

Savoca D, Barreca S, Lo Coco R, Punginelli D, Orecchio S, Maccotta A (2023) Environmental aspect concerning phthalates contamination: analytical approaches and assessment of biomonitoring in the aquatic environment. Environments [online] 10(6):99. https://doi.org/10.3390/environments10060099

Schampera C, Wolinska J, Bachelier JB, de Souza Machado AA, Rosal R, González-Pleiter M, Agha R (2021) Exposure to Nanoplastics affects the outcome of infectious disease in phytoplankton. Environ Pollut 277(1):116781. https://doi.org/10.1016/j.envpol.2021.116781

Schwaferts C, Niessner R, Elsner M, Ivleva NP (2019) Methods for the analysis of submicrometer- and nanoplastic particles in the environment. TrAC Trends Anal Chem 112(1):52–65. https://doi.org/10.1016/j.trac.2018.12.014

Selvam S, Jesuraja K, Venkatramanan S, Roy PD, Jeyanthi Kumari V (2021) Hazardous microplastic characteristics and its role as a vector of heavy metal in groundwater and surface water of coastal South India. J Hazard Mater 402(1):123786. https://doi.org/10.1016/j.jhazmat.2020.123786

Sendra M, Saco A, Yeste MP, Romero A, Novoa B, Figueras A (2020) Nanoplastics: from tissue accumulation to cell translocation into Mytilus Galloprovincialis hemocytes. Resilience of immune cells exposed to Nanoplastics and Nanoplastics plus vibrio Splendidus combination. J Hazard Mater 388(1):121788. https://doi.org/10.1016/j.jhazmat.2019.121788

Sendra M, Sparaventi E, Novoa B, Figueras A (2021) An overview of the internalization and effects of microplastics and Nanoplastics as pollutants of emerging concern in bivalves. Sci Total Environ 753(1):142024. https://doi.org/10.1016/j.scitotenv.2020.142024

Senko J, Nelms S, Reavis J, Witherington B, Godley B, Wallace B (2020) Understanding individual and population-level effects of plastic pollution on marine megafauna. Endanger Species Res 43(1):234–252. https://doi.org/10.3354/esr01064

Sharma VK, Ma X, Lichtfouse E, Robert D (2022) Nanoplastics are potentially more dangerous than microplastics. Environ Chem Lett 21(4):1933–1936. https://doi.org/10.1007/s10311-022-01539-1

Shen M, Zhang Y, Zhu Y, Song B, Zeng G, Hu D, Wen X, Ren X (2019) Recent advances in toxicological research of Nanoplastics in the environment: a review. Environ Pollut 252(1):511–521. https://doi.org/10.1016/j.envpol.2019.05.102

Shiu R-F, Vazquez CI, Chiang C-Y, Chiu M-H, Chen C-S, Ni C-W, Gong G-C, Quigg A, Santschi PH, Chin W-C (2020) Nano- and Microplastics trigger secretion of protein-rich extracellular polymeric substances from phytoplankton. Sci Total Environ 748(1):141469. https://doi.org/10.1016/j.scitotenv.2020.141469

Shorny A, Steiner F, Hörner H, Skoff SM (2023) Imaging and identification of single Nanoplastic particles and agglomerates. Scientif Rep [online] 13(1):10275. https://doi.org/10.1038/s41598-023-37290-y

Silva MSS, Oliveira M, Lopéz D, Martins M, Figueira E, Pires A (2020a) Do nanoplastics impact the ability of the polychaeta hediste diversicolor to regenerate? Ecolog Indicat., [online] 110(1):105921. https://doi.org/10.1016/j.ecolind.2019.105921

Silva MSS, Oliveira M, Valente P, Figueira E, Martins M, Pires A (2020b) Behavior and biochemical responses of the polychaeta hediste diversicolor to polystyrene nanoplastics. Sci Total Environ 707(1):134434. https://doi.org/10.1016/j.scitotenv.2019.134434

Sjollema SB, Redondo-Hasselerharm P, Leslie HA, Kraak MHS, Vethaak AD (2016) Do plastic particles affect microalgal photosynthesis and growth? Aquat Toxicol 170(1):259–261. https://doi.org/10.1016/j.aquatox.2015.12.002

Sun H, Jiao R, Wang D (2021) The difference of aggregation mechanism between microplastics and nanoplastics: role of Brownian motion and structural layer force. Environ Pollut 268(B):115942. https://doi.org/10.1016/j.envpol.2020.115942

Sun T, Wang S, Ji C, Li F, Wu H (2022) Microplastics aggravate the bioaccumulation and toxicity of coexisting contaminants in aquatic organisms: a synergistic health Hazard. J Hazard Mater 424(B):127533. https://doi.org/10.1016/j.jhazmat.2021.127533

Tallec K, Paul-Pont I, Boulais M, Le Goïc N, González-Fernández C, Le Grand F, Bideau A, Quéré C, Cassone A-L, Lambert C, Soudant P, Huvet A (2020) Nanopolystyrene beads affect motility and reproductive success of oyster spermatozoa (Crassostrea gigas). Nanotoxicology [online] 14(8):1039–1057. https://doi.org/10.1080/17435390.2020.1808104

Tang M, Ding G, Lu X, Huang Q, Du H, Xiao G, Wang D (2022) Exposure to Nanoplastic particles enhances Acinetobacter survival, biofilm formation, and serum resistance. Nano 12(23):4222–4222. https://doi.org/10.3390/nano12234222

Taniguchi S, Colabuono FI, Dias P, Renato, Fisner M, Turra A, Izar GM, Moledo D, Saha M, Hosoda J, Yamashita R, Takada H, Lourenço RA, Magalhães CA, Bícego MC, Montone RC (2016) Spatial variability in persistent organic pollutants and polycyclic aromatic hydrocarbons found in beach-stranded pellets along the coast of the state of São Paulo, Southeastern Brazil. Mar Pollut Bullet 106(1–2):87–94. https://doi.org/10.1016/j.marpolbul.2016.03.024

Town RM, van Leeuwen HP (2020) Uptake and release kinetics of organic contaminants associated with micro- and Nanoplastic particles. Environ Sci Technol 54(16):10057–10067. https://doi.org/10.1021/acs.est.0c02297

Trevisan R, Ranasinghe P, Jayasundara N, Di Giulio RT (2022) Nanoplastics in aquatic environments: impacts on aquatic species and interactions with environmental factors and pollutants. Toxics [online] 10(6):326. https://doi.org/10.3390/toxics10060326

UNEP (2018) Single-use plastics: a roadmap for sustainability. [online] UNEP – UN Environment Programme. Available at: https://www.unep.org/resources/report/single-use-plastics-roadmap-sustainability

Uruén C, Chopo-Escuin G, Tommassen J, Mainar-Jaime RC, Arenas J (2020) Biofilms as promoters of bacterial antibiotic resistance and tolerance. Antibiotics 10(1):3. https://doi.org/10.3390/antibiotics10010003

Valerio I (2022) Plastic pollution in coral reefs: interaction patterns between primary and secondary micro and Nano plastic particles and tropical corals in controlled environments. [online]. Università Degli Studi Di Milano-Bicocca. Milano. Available at: https://boa.unimib.it/retrieve/9162b130-36c1-42ef-8221-0df3c6d8d047/phd_unimib_745913.pdf. Accessed 24 Oct 2024

van Sebille E, Wilcox C, Lebreton L, Maximenko N, Hardesty BD, van Franeker JA, Eriksen M, Siegel D, Galgani F, Law KL (2015) A global inventory of small floating plastic debris. Environ Res Lett [online] 10(12):124006. https://doi.org/10.1088/1748-9326/10/12/124006

van Weert S, Redondo-Hasselerharm PE, Diepens NJ, Koelmans AA (2019) Effects of Nanoplastics and microplastics on the growth of sediment-rooted Macrophytes. Sci Total Environ 654(1):1040–1047. https://doi.org/10.1016/j.scitotenv.2018.11.183

Vanstreels RET, Piccinin IN, Maraschin M, Gallo L, Serafini PP, Pereira AS, Santos AP, Egert L, Uhart M (2023) Phthalate esters (plasticizers) in the Uropygial gland and their relationship to plastics ingestion in seabirds along the coast of Espírito santo, eastern Brazil. J Zoo Wildl Med 53(4). https://doi.org/10.1638/2022-0053

Varshney S, Hegstad-Pettersen MM, Siriyappagouder P, Olsvik PA (2024) Enhanced neurotoxic effect of PCB-153 when co-exposed with polystyrene nanoplastics in zebrafish larvae. Chemosphere 355(1):141783. https://doi.org/10.1016/j.chemosphere.2024.141783

Vedolin MC, Teophilo CYS, Turra A, Figueira RCL (2018) Spatial variability in the concentrations of metals in beached microplastics. Mar Pollut Bull 129(2):487–493. https://doi.org/10.1016/j.marpolbul.2017.10.019

Venâncio C, Ferreira I, Martins MA, Soares AMVM, Lopes I, Oliveira M (2019) The effects of nanoplastics on marine plankton: a case study with Polymethylmethacrylate. Ecotoxicol Environ Safety [online] 184(1):109632. https://doi.org/10.1016/j.ecoenv.2019.109632

Wang S, Liu M, Wang J, Huang J, Wang J (2020) Polystyrene Nanoplastics cause growth inhibition, morphological damage and physiological disturbance in the marine microalga Platymonas Helgolandica. Mar Pollut Bull 158(1):111403. https://doi.org/10.1016/j.marpolbul.2020.111403

Wang T, Hu M, Xu G, Shi H, Leung JYS, Wang Y (2021) Microplastic accumulation via trophic transfer: can a predatory crab counter the adverse effects of microplastics by body defence? Sci Total Environ 754(1):142099. https://doi.org/10.1016/j.scitotenv.2020.142099

Wang R, Li X, Li J, Dai W, Luan Y (2023) Bacterial interactions with Nanoplastics and the environmental effects they cause. Fermentation [online] 9(11):939. https://doi.org/10.3390/fermentation9110939

Wang JJ, Hill C, Li D, Shi Y, Yang L, Draper S, Xiao L, Boland J (2024) A simple spectral method for Nanoplastic identification and characterisation. Res Square (Research Square) 1(1). https://doi.org/10.21203/rs.3.rs-3988674/v1

Ward JE, Rosa M, Shumway SE (2019) Capture, ingestion, and egestion of microplastics by suspension-feeding bivalves: a 40-year history. Anthr Coasts 2(1):39–49. https://doi.org/10.1139/anc-2018-0027

Worm B, Lotze HK, Jubinville I, Wilcox C, Jambeck J (2017) Plastic as a persistent marine pollutant. Annu Rev Environ Resour 42(1):1–26. https://doi.org/10.1146/annurev-environ-102016-060700

Wright SL, Kelly FJ (2017) Plastic and human health: a micro issue? Environ Sci Technol, [online] 51(12):6634–6647. https://doi.org/10.1021/acs.est.7b00423

Yamashita R, Hiki N, Kashiwada F, Takada H, Mizukawa K, Hardesty BD, Roman L, Hyrenbach D, Ryan PG, Dilley BJ, Muñoz-Pérez JP, Valle CA, Pham CK, Frias J, Nishizawa B, Takahashi A, Thiebot J-B, Will A, Kokubun N, Watanabe YY (2021) Plastic additives and legacy persistent organic pollutants in the preen gland oil of seabirds sampled across the globe. Environ Monitor Contam Res [online] 1(1):97–112. https://doi.org/10.5985/emcr.20210009

Yang T, Luo J, Nowack B (2021) Characterization of Nanoplastics, fibrils, and microplastics released during washing and abrasion of polyester textiles. Environ Sci Technol 55(23):15873–15881. https://doi.org/10.1021/acs.est.1c04826

Yoo J, Cho H, Jeon M, Jeong C-B, Jung J-H, Lee Y-M (2021) Effects of polystyrene in the brackish water flea Diaphanosoma celebensis: size-dependent acute toxicity, ingestion, egestion, and antioxidant response. Aquat Toxicol 235(1):105821–105821. https://doi.org/10.1016/j.aquatox.2021.105821

Yu Z, Xu X, Guo L, Jin R, Lu Y (2024) Uptake and transport of micro/nanoplastics in terrestrial plants: detection, mechanisms, and influencing factors. Sci Total Environ [online] 907(1):168155. https://doi.org/10.1016/j.scitotenv.2023.168155

Zaki MRM, Aris AZ (2022) An overview of the effects of Nanoplastics on marine organisms. Sci Total Environ 831(1):154757. https://doi.org/10.1016/j.scitotenv.2022.154757

Zantis LJ, Carroll EL, Nelms SE, Bosker T (2021) Marine mammals and microplastics: a systematic review and call for standardisation. Environ Pollut 269(1):116142. https://doi.org/10.1016/j.envpol.2020.116142

Zantis LJ, Bosker T, Lawler F, Nelms SE, O'Rorke R, Constantine R, Sewell M, Carroll EL (2022) Assessing microplastic exposure of large marine filter-feeders. Sci Total Environ [online] 818(1):151815. https://doi.org/10.1016/j.scitotenv.2021.151815

Zhang Y, Goss GG (2020) Potentiation of polycyclic aromatic hydrocarbon uptake in zebrafish embryos by Nanoplastics. Environ Sci Nano 7(6):1730–1741. https://doi.org/10.1039/d0en00163e

Zhang Y, Jiao Y, Li Z, Tao Y, Yang Y (2021) Hazards of phthalates (PAEs) exposure: a review of aquatic animal toxicology studies. Sci Total Environ 771(1):145418. https://doi.org/10.1016/j.scitotenv.2021.145418

Zhang Z, Chen W, Chan H, Peng J, Zhu P, Li J, Jiang X, Zhang Z, Wang Y, Tan Z, Peng Y, Zhang S, Lin K, Yung KK-L (2024) Polystyrene microplastics induce size-dependent multi-organ damage in mice: insights into gut microbiota and fecal metabolites. J Hazard Mater 461(1):132503–132503. https://doi.org/10.1016/j.jhazmat.2023.132503

Zhao J, Lan R, Wang Z, Su W, Song D, Xue R, Liu Z, Liu X, Dai Y, Yue T, Xing B (2023) Microplastic fragmentation by rotifers in aquatic ecosystems contributes to global nanoplastic pollution. Nat Nanotechnol 19(3):406–414. https://doi.org/10.1038/s41565-023-01534-9

Zheng Q, Wu H, Yan L, Zhang Y, Wang J (2024) Effects of polystyrene Nanoplastics and PCB-44 exposure on growth and physiological biochemistry of chlorella vulgaris. Sci Total Environ 918(1):170366–170366. https://doi.org/10.1016/j.scitotenv.2024.170366

Zhou C-Q, Lu C-H, Mai L, Bao L-J, Liu L-Y, Zeng EY (2021a) Response of Rice (Oryza Sativa L.) roots to Nanoplastic treatment at seedling stage. J Hazard Mater [online] 401(1):123412. https://doi.org/10.1016/j.jhazmat.2020.123412

Zhou X-X, Liu R, Hao L-T, Liu J-F (2021b) Identification of polystyrene Nanoplastics using surface enhanced Raman spectroscopy. Talanta 221(1):121552. https://doi.org/10.1016/j.talanta.2020.121552

Open Access This chapter is licensed under the terms of the Creative Commons Attribution-NonCommercial-NoDerivatives 4.0 International License (http://creativecommons.org/licenses/by-nc-nd/4.0/), which permits any noncommercial use, sharing, distribution and reproduction in any medium or format, as long as you give appropriate credit to the original author(s) and the source, provide a link to the Creative Commons license and indicate if you modified the licensed material. You do not have permission under this license to share adapted material derived from this chapter or parts of it.

The images or other third party material in this chapter are included in the chapter's Creative Commons license, unless indicated otherwise in a credit line to the material. If material is not included in the chapter's Creative Commons license and your intended use is not permitted by statutory regulation or exceeds the permitted use, you will need to obtain permission directly from the copyright holder.

Part V

In Silico Approaches to Assess Nanosafety

Molecular Docking in Nanotoxicology

17

Michael González-Durruthy, Ana S. Moura, and M. Natália D. S. Cordeiro

Abstract

Molecular docking has become a significant approach in drug discovery and nanotoxicology, as it enables large-scale genomic, proteomic, and nanomaterial data studies, through virtual screening, lead optimization, and side-effect prediction. This chapter outlines the fundamental aspects and significant applications of molecular docking simulation, emphasizing nano-biointeractions and potential side-effect/nanotoxicity relationship prediction. It discusses the expert curation and state-of-the-art structure-based methods to systemically address these challenges, focusing on key nanoparticle interactions at the molecular level, particularly regarding DNA interaction, towards a rational nanomaterial design with the desired lowest nanotoxicity possible for safer biomedical applications. Finally, the gaps and challenges of the field are identified and integrated solutions are proposed.

Keywords

Computational nanotoxicity · Molecular docking · Scoring functions · Nanoparticles

M. González-Durruthy (✉)
INL-International Iberian Nanotechnology Laboratory, Nanosafety Department, Braga, Portugal

A. S. Moura · M. Natália D. S. Cordeiro
LAQV@REQUIMTE/Department of Chemistry and Biochemistry, Faculty of Sciences, University of Porto, Porto, Portugal

© The Author(s) 2025
E. Alfaro-Moreno, F. Murphy (eds.), *Nanosafety*,
https://doi.org/10.1007/978-3-031-93871-9_17

1 Introduction

Reliable mechanistic insight in Nanotoxicology is the game-changer in nanosafety and related sub-disciplines. Pinpointing not only the key nanodescriptors, e.g., physico-chemical properties linked to a given nanotoxicological profile, either it may be for environmental, cytotoxicity, or immunotoxicity profiling, but also being able to sort their intra-significance and weight for the nano-biointeractions under probing sustains the production and risk assessment of new nanomaterials (Meldrum et al. 2017; Ventura et al. 2022; John et al. 2022; Gupta et al. 2021). The complexity of the interactions at the nanoscale level can significantly limit the knowledge being directly acquired via a sheer experimental approach. This, on the other hand, causes not only a gap between the production of novel materials and their eventual release in and subsequent continuous assessment for its interaction with the environment and human systems, and longitudinal living chains, but also curtails the chance for preventive toxicological correction in industrial and health context (Gupta et al. 2021; Kumah et al. 2023; Tang et al. 2024; Forest 2022).

In fact, the detailed assessment of nanoparticle (NP) interactions with biological targets (BT) represents a major challenge in Health and Environmental safety, where there is the urgent and continuous need to devise systematic strategies for integrating increasingly diverse bio-datasets as to provide insights into disease pathogenesis, nanodrug discovery, and the safe application of nanomaterials (Arathi et al. 2023; Auría-Soro et al. 2019; Runa et al. 2018). Such an assessment usually not only involves extensive experimental research for a period between a decade and a decade-and-a half per drug candidate, including nanobiotechnological products. Moreover, it often presents a cost higher than 1 billion US dollars/ ≈700 million €, adding the economic factor to the time urgency that is not being met (Patra et al. 2018; Thapa and Kim 2022; Schlander et al. 2021; Mullard 2020).

Within the NP/BT interface mechanistic interpretation (nano-toxicodynamics), the understanding of protein-ligand interaction is a field that has witnessed significant progress through in silico approaches, such as Molecular Docking Simulations, i.e., the technique to probe the spatial orientation for the interaction between two or more molecular fragments (e.g., protein-ligand interaction). This is particularly important as experimental techniques face reported difficulties in understanding ligand-target interactions at the molecular level, which sustains an increase of the widespread use of high-performing computational platforms versus the available (and abundant in many cases) structural data, for effective, reliable, cost and time-saving mechanistic insight approaches. Furthermore, structure-based molecular docking has a predictive nature regarding the structure of complexes formed by interacting molecules, facilitating the screening of potential drug-targets complex, nanomaterials, and elucidating biomolecular interactions (Ferreira et al. 2015; Pinzi and Rastelli 2019; Jaiswal et al. 2023; Trisciuzzi et al. 2018). As such, it is with no surprise one finds how crucial this in silico technique is in drug discovery, virtual screening, lead optimization, side effect predictions, and nanotoxicology, while also playing a role in drug repositioning, i.e., repurposing existing nanoparticles for new therapeutic targets, thus accelerating the identification of novel applications for approved drugs.

In fact, public projects, such as OpenZika (Ekins et al. 2016), have evidenced the large-scale potential of docking for screening compounds against protein models, while sustained publication track-record indicates the capacity for the docking to predict biochemical coupling underlying adverse reactions in nanomaterials, initiated by on-target or off-target interactions before in vitro and in vivo experimental evaluation (Chen et al. 2024; Fan et al. 2023). In addition, regulatory organisms encourage resourcing to in silico techniques as a pre-assessment stage (Pappalardo et al. 2022; Fischer et al. 2020), prior to the accepted experimental good practices, as the industry, namely the pharmaceutical industry (Pappalardo et al. 2022; Fischer et al. 2020), stresses the inadequacy of the current toxicity assessment experimental methods to be cost and time-effective for continuous toxicological evaluation, and updated mechanistic insight. Indeed, there is a consensus emphasis for the need of implementing in silico docking techniques in drug and nanomaterials development (Pinzi and Rastelli 2019; Arjmand et al. 2024; Murugan et al. 2022; Gangrade et al. 2016).

This chapter proposes to provide insight into molecular docking simulations, emphasizing its applications in nanotoxicology, such as predicting mechanistic toxicodynamics effects, and applications in repurposing nanodrugs to mitigate nanotoxicity. As such, and with molecular docking in nanotoxicology acting as the backbone of this chapter, it presents a transversal analysis for the current docking software tools and online services, comparing their performance on benchmark datasets while addressing the present challenges of in silico docking approaches. We also focus, through curated and extensive literature review, the docking based-python tools and chemoinformatic machine learning systems, compiling with the remaining analysis for building a collaborative effort in the research quality of Computational Nanotoxicology, and fostering nanoligand-protein interaction mechanistic insight breakthroughs.

2 Molecular Docking and Nanotoxicology: Concepts, Principles, and Background

Molecular docking techniques and algorithms have developed in conjunction with, and in synergy with, the advancements of experimental techniques such as X-ray crystallography, nuclear magnetic resonance spectroscopy, and high-throughput protein purification in the early years of the twenty-first century (Pinzi and Rastelli 2019; Sethi et al. 2020). The synergy comes from the data provided by these techniques being essential for the predictive goal of docking simulation processes, i.e., the *in silico* prediction of interaction between ligand and protein receptors is sustained by the experimental probing of the target receptor three-dimensional structure, or the atoms attached to carbon and hydrogen within the receptor structure binding site.

This data pool has increased throughout the years, as novel ligand materials are developed (e.g., carbon nanotubes, fullerenes, graphene) for the relevant target receptors (e.g., proteins, enzymes, RNA, DNA) (Pinzón et al. 2011; Lu 2022). The molecular docking approach gathered attention in these first two decades, as its computational results provided solid ground for designing strong binding affinity chemicals with the prospective bio-targets, a potential that attracted the attention of pharmatoxicology. In fact, these computational techniques can be resourced for the prediction and mitigation of potential nanotoxic effects, through the mechanistic interpretation of the involved interaction, something achieved by virtually testing several poses, i.e., ligand binding conformations, between nanoparticles and proteins, followed by establishing a ranking through scoring functions quantifying the free energy of binding. Furthermore, the prediction in made for interactive behaviour in the protein biophysical environments, i.e., specific binding sites, thus conveying information for designing safer, less toxic, and more effective nanomaterials.

2.1 Molecular Docking Categorization

A broadly categorization of molecular docking can be made as per the degree of conformational flexibility of its intervenient ligand/receptor, i.e., as per the ability of ligand (or nanoparticles) and receptor to adopt different conformations while remaining structural faithful (Ewing et al. 2001). The rigid body docking allows no conformational flexibility for both, the semi-flexible ligand docking allows internal bond rotation in the ligand while the receptor is fixed or partially flexible, and in the third case, flexible docking, both fragments have conformational flexibility.

Regarding simulation time, rigid docking has preference due to its lower computational demands, but the higher time-consuming calculations of flexible docking are compensated by more accurate predictions of binding geometries, something crucial for understanding the physicochemical idiosyncrasies of the nanomaterial/bio-target interaction under investigation (Ewing et al. 2001).

In addition, molecular docking research can be optimized by conjugation with a wide variety of computational techniques, such as genetic algorithms (GA), i.e., adaptive search algorithms for randomly exploiting which are the most fitted for the solution space (Gu et al. 2015), evolutionary programming (EP), i.e., optimizing the solution space by stochastically removing less desired solutions, and introducing small random changes (Li et al. 2010), or molecular dynamics, i.e., simulating how the nanoligand-protein system dynamically evolves in its interaction during a fixed period of time (González-Durruthy et al. 2016).

2.2 Knowledge-Based Docking Methods

The stability of the protein structural conformation, as per the theory of the protein energy landscape, i.e., free-energy surface, is related with its n-state folding conformation and associated energy, as its folding properties and corresponding conformations relate with free energy (Zheng et al. 2007). The energy landscape approach has been popular in protein structural problem-solving but meets hardship, due to increased complexity calculation, when considering interactions between two molecules and the quest to determine global energy minima (Verkhivker et al. 2000). As such, present protocols are relying on both physics' principles (e.g., steric complementarity), and computer science and engineering-borrowed techniques, such as pattern recognition, optimization, and machine learning algorithms (Crampon et al. 2022).

To predict the interaction between nanomaterials, namely nanoparticles, with bio-macromolecules, the field of Nanotoxicology has adapted some of these approaches, with *ab initio* docking methods resourcing to sheer physical principles for predicting the binding modes, without using prior knowledge of existing complexes (González-Durruthy et al. 2016). *Ab initio* approaches can be particularly useful where experimental data might be limited, hard to access from a mechanistic point of view, or inexistent. Notwithstanding, knowledge-based docking approaches, i.e., docking approaches based on scoring functions developed through statistical analysis of known protein-ligand structures, deriving as result the potentials of mean force (PMF), which are the potentials of atom pair interaction of the complex (Zheng et al. 2007). Such approach includes sequence alignment, threading, and structure-based methods, and can explore the available nanoparticle interaction data for predicting novel nanoparticle toxicological profiles as per the already studied corpus of nanoparticles.

Albeit the limited number of protein-protein and protein-nanoparticle complexes currently stored in both databases and nanorepositories, docking templates can often be found for most interactions, provided that the components have known or homology-built structures (Zhang and Skolnick 2005). For example, the template modelling alignment (TM-align) methodology identifies the best structural alignment between protein pairs, combining template modelling score (TM-score), i.e., the degree of similitude between two protein structures, rotation matrix and dynamic programming, which serves as a foundation for template-based docking.

The TM-align approach is an algorithm for sequence-independent protein structure comparison, first generating an optimized residue-to-residue alignment based on structural similarity for two protein structures of unknown equivalence, through near-optimal, i.e., approximate, iterations. It then provides an optimal superposition of the two structures based on the detected alignment, as well as the TM-score, which gauges structural similarity. The TM-score value ranges from 0, i.e., absolute unmatched, to 1, indicating perfect match, with protein database orthodoxy considering scores below 0.2 corresponding to randomly chosen unrelated proteins, while those above 0.5 generally assume the same fold in Structural Classification of Proteins (SCOP) and Class Architecture Topology Homology (CATH), the two most prominent protein structure classification schemes. Therefore, the TM-align can predict how nanoparticle-antibody or antibody-antibody interaction systems might bind, something essential for nanotoxicology mechanistic insight.

Furthermore, as the translational, rotational, and conformational degrees of freedom allow for numerous binding modes between ligands and target molecules, databases such as Protein Data Bank (PDB) (Burley et al. 2022), ZINC (Irwin et al. 2012), or PubChem (Kim et al. 2022), provide the structural and affinity data necessary to develop and validate feasible conformation-generating docking algorithms for the several fields of nanotoxicology (e.g., inmunonanotoxicity research) (Abdelsattar et al. 2021).

3 Molecular Docking Approach and Scoring Functions

Within the framework of molecular docking, one defines a scoring function as a mathematical function used to determine the binding affinity of a protein-ligand interaction through physics-based molecular mechanics force fields (Goodsell et al. 1996), something critical as molecular docking aims at determine the binding affinity and spatial conformation interaction between fragments, the identification of thermodynamically favoured binding poses and orientations of the ligand versus the protein's binding site. Resourcing to a combination of physicochemical parameters of the nearest protein-ligand interacting atoms (e.g., van der Waals forces or electrostatic interactions), the scoring function quantitatively predicts the binding free energy, ΔG, allowing for the ranking of protein-ligand docking complex poses (Trott and Olson 2010). As such, the ranking is a selective tool for promising binding modes. In Fig. 17.1 we can see displayed an example of the graphical representation for a thermodynamic scoring function.

A broad classification of scoring functions can be made as per the used physicochemical parameters and intended probing objectives regarding the ligand-protein interaction (Murugan et al. 2022). Force-field-based scoring functions resource to

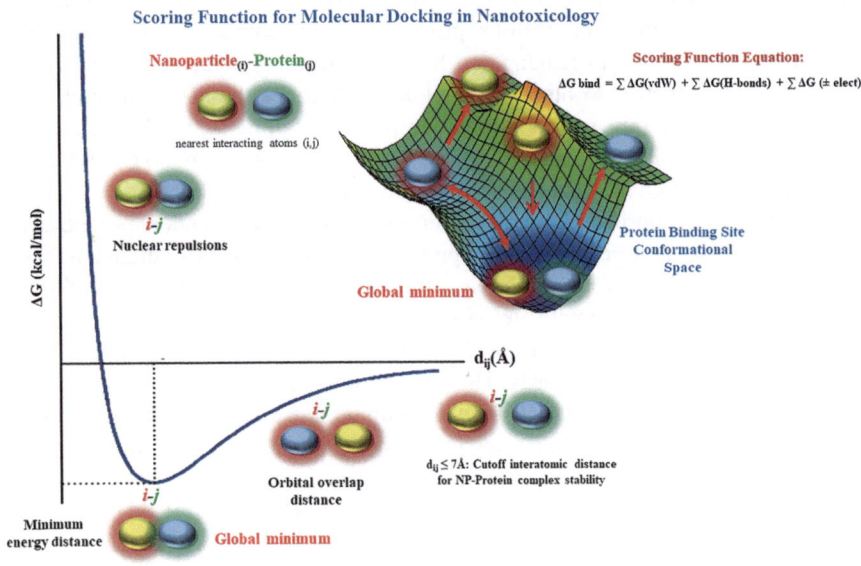

Fig. 17.1 Graphical representation of a thermodynamic scoring function for nanobiointeractions between a nanoparticle (NP), in yellow, and a protein (P), in blue (the spheres represent the interacting atoms from NP and P). (*Left side*) The x-axis represents the interatomic distances between the NP and P, respectively i and j, both measured in angstrom (Å), and the y-axis represents the Gibbs free energy of binding affinity, ΔG, for the docking complexes, in kcal/mol. (*Right side*) The 3D grid-surface represents the conformational potential energy surface where the scoring function evaluate the best-ranked docking poses fitting the global minimum and lowest ΔG values as function of van der Waals forces, hydrogen bonding, and electrostatic interactions

van der Waals and electrostatic terms, providing an estimation of the binding free energy, and are crucial for understanding the complex electrostatic environment of the docking complexes (Goodsell et al. 1996). Empirical scoring functions rely on assessing the contribution from hydrogen bonds, binding entropy, ionic interactions, and hydrophobic effects, to predict how nanoparticles will interact with different biological targets based on their surface properties and specific functional groups (Yu et al. 2021). Finally, knowledge-based scoring functions access the data pool of co-crystallized complexes, either ligand-protein or nanoparticle-protein, and determine through statistical analyses the preferred contact frequencies and distances, which provide insights into the likely binding modes of nanoparticles (González-Durruthy et al. 2017). Notwithstanding, whichever scoring function approach is chosen the selection of the appropriate software for the docking simulation is also critical. The specific needs of the research must be cross-referenced with available in silico tools, and in particular the handling of the unique properties of nanoparticles and their interactions with biological targets must be met in nanotoxicology investigations.

There are at present over 107 developed scoring functions (Feher 2006), each with unique advantages for toxicodynamics studies, including those investigating nano-bio-targets, providing they are efficiently integrated in a given simulation context. In fact, the choice of scoring function can significantly impact the accuracy of predictions, especially in nanotoxicology, challenged by the diversity of size, shapes, charge, or chemical functions of the nanomaterial structures. Indeed, consensus scoring (Nhat Phuong et al. 2023), which combines multiple scoring functions, has been proposed to improve prediction accuracy in this in silico landscape, alongside the encouragement to use a weighted scoring system in nanotoxicology context, as it accounts for the complex NP/BT interactions, thus enhancing predictive reliability. Nevertheless, it should be emphasized that regardless of the chosen approach and software, the resulting predictions should be validate with experimental data.

3.1 Nanotoxicology Versus Scoring Function I: Autodock Vina Scoring Function

One of the most used molecular docking software is AutoDock Vina (Tao et al. 2018). Its popularity is due to its acknowledged accuracy and speed, with scoring function based on a hybrid *empirical + knowledge-based* approach, which evaluates the binding affinity of ligands, including nanoparticles, to biological targets by considering factors such as hydrogen bonding, hydrophobic interactions, and the spatial arrangement of atoms, and the weighted sum of energy terms, such as van der Waals steric, electrostatic, or hydrogen interactions. In nanotoxicology, the AutoDock Vina is resourced as identification tool for potential toxicological impacts, generating multiple docking poses and ranking them based on their predicted binding affinities, allowing for the efficient screening of large libraries of nanoparticles, and facilitating the identification of those with the highest binding affinity and lowest potential toxicity. Moreover, it can incorporate flexibility in both the ligand and the target protein enhances the reliability of the predictions (Jaghoori et al. 2016).

3.2 Nanotoxicology Versus Scoring Function II: GOLD Scoring Function

Another prominent scoring algorithm is the Genetic Optimization for Ligand Docking (GOLD) (Verdonk et al. 2003), recognized for its robust and flexible scoring function, and based on a genetic algorithm that optimizes the fit of nanoparticles, either inorganic or organic, into the binding sites of target proteins (Verdonk et al. 2003). GOLD resources to various interaction parameters, such as hydrogen bonds, van der Waals forces, and hydrophobic contacts, assessing multiple docking poses, and optimizing them through iterative processes which identify the most energetically favorable interactions, i.e., $\Delta G < 0$ (kcal/mol). The GOLD scoring function also incorporates terms for solvation and entropy, which are essential for accurately predicting the behavior of nanoparticles in the biophysical environment of the binding sites of enzymes, and nucleic acids, i.e., DNA and RNA (González-Durruthy et al. 2017), thus aiding in the rational design of nanomaterials through its detailed scoring and ranking. This approach provides a more realistic simulation of nanoparticle interactions, allowing researchers to use GOLD to perform *in silico* high-throughput screening of nanoparticle libraries from a big-data perspective, identifying nanotherapeutic candidates with the most favorable profile for efficiency and safety biomedical applications and environmental interface.

3.3 Nanotoxicology Versus Scoring Function III: Glide Scoring Function

The Glide software for molecular docking presents the Glide scoring function which uses a series of hierarchical filters to search for possible nanoligand locations in the receptor binding-site (González-Durruthy et al. 2017; Friesner et al. 2004; Wang et al. 2024). In Nanotoxicology, these hierarchical filters are composed by a series of tests the ligand, e.g., a NP, must pass to be considered a potential match for the target receptor, and are applied in a specific order, with each subsequent filter being more stringent than the previous one. The first process is a relatively fast high throughput virtual screening (HTVS) mode for an initial and broad search of possible protein binding site locations, screening large numbers of ligands. The fragments passing this stage undergo the standard precision (SP) docking mode, which involves a more detailed and accurate scoring of the ligand binding pose, including energy minimization and refinement in the field of the receptor. The final stage witnesses the implementation of an extra precision (XP) algorithm for the final selection of poses, involving an even more rigorous scoring function and a different functional form for GlideScore allowing for efficient and accurate identification of potential ligand-receptor interactions by progressively refining the search and scoring process. The Glide scoring function is particularly useful when intending to gain mechanistic insight of the potential toxic effects of the ligands (Friesner et al. 2004).

4 Addressing Nano-genotoxicity: Modeling Nanoparticle-DNA Interactions

Nanotoxicology is an aggregating field, composed by many specific toxicological areas. A genotoxic ligand-fragment potentially leads to structural DNA damage, either through direct or indirect interaction, with great potential for causing cancer or birth defects [53]. Within nanotoxicological context, is focused in how nanomaterials interact with DNA, identifying the interaction mechanisms, and with that information designing less genotoxic nanomaterials (e.g., through implementing protective coatings or developing targeted delivery systems that minimize exposure to genetic material) (Kohl et al. 2020; Feng et al. 2022).

The interactions between conventional ligands and nucleic acids are recognized as critical cellular processes and one would not expect the little attention the *in silico* research community has historically given to docking studies focusing nanomaterials and nucleic acids interactions (Feng et al. 2022). Though there has been the adaptation of originally protein-protein developed docking tools (e.g., HADDOCK) to the include nucleic acids as simulation input, the specific nature, i.e., their unique physicochemical properties, of nanomaterial-nucleic acid interaction requires a specific approach (Honorato et al. 2024).

The interaction between nanoparticles and DNA often occurs at the minor and major grooves of the DNA helix, i.e., the unequally sized spaces between the helical strands of DNA where ligand-DNA binding for activities such as replication, transcription, and repair happen (Honorato et al. 2024), with NP groove-binding might interfere with the normal function of DNA, potentially leading to mutations or disruptions in cellular processes (Honorato et al. 2024). In fact, NP-DNA binding can obstruct the access of essential regulatory proteins, such as transcription factors and polymerases, disrupting normal gene expression and DNA replication, which leads to DNA synthesis error-based mutations, capable of time-dependent accumulation and therefore contributing to carcinogenesis (Kakisaka et al. 2015). In addition, this binding can induce structural changes endangering the integrity of genetic material, such as the bending or unwinding of the DNA helix (Kakisaka et al. 2015).

Notwithstanding, nanomaterials interaction with cellular components can also generate reactive oxygen species (ROS), becoming a source for DNA oxidative damage, a known factor in genotoxicity, which results severe genetic damage by causing strand breaks, base modifications, and cross-linking (Dey et al. 2012). Cellular damage persistence of this type can lead to cell death or malignant transformation if not addressed.

As per the abovementioned, the adaptation of molecular docking protocols for nanomaterials-nucleic acids interactions mandatorily implies the assessing of NP-DNA binding, and its subsequent (or not) nucleic acid structural disruption, namely at the groove sites, evaluating the mutagenic or carcinogenic outcome potential (Feng et al. 2022). It should also be noted that NP-DNA grooves interaction involves several mechanisms, which need to be taken in account by the in silico approaches (e.g., the electrostatic interactions, i.e., how charge surfaces of NP are attracted by the oppositely DNA charged regions, as in the case of positively charged NPs interacting strongly with the DNA backbone negatively charged phosphate groups, which allow them to bind both the major and minor grooves).

Recent advances to improve docking methodologies included developing new scoring functions and enhancing the accuracy of existing tools to better predict the binding affinities and alpha helix conformational changes induced by NP-DNA interaction (Feng et al. 2022).

4.1 Nanoparticle-DNA Major and Minor Groove Interactions

As the DNA major groove is wider and deeper than the minor groove, it provides a larger surface area for binding, and NPs can interact with this prospective binding area through various mechanisms, such as electrostatic interactions, hydrogen bonding, and van der Waals forces (Feng et al. 2022; Rahman et al. 2019). Alongside the electrostatic examples already mentioned, the presence of nanoparticles in the major groove can block the access of essential proteins, disrupting normal cellular processes and contributing to genotoxic effects (Feng et al. 2022; DNA Grooves Encyclopedia of Genetics 2008; Rahman et al. 2019).

Another important NP-DNA interaction regards the binding of ultrafine (≈3–5 nm) carbon nanoparticles to the major groove, which can induce conformational changes in the DNA structure, such as bending or unwinding of the helix (Privalov et al. 2011), something that can prevent the proper function of both DNA polymerases and other enzymes involved in DNA replication and repair, potentially leading to mutations or chromosomal aberrations. Examples of carbon-based NP and mitochondrial DNA (mtDNA) major groove are displayed in Fig. 17.2.

The narrower and shallower minor groove, with width typically around 1.5–2.5 nanometers (nm), is also a critical site for molecular interactions, namely those involving ultrafine carbon nanoparticles, as the size of this groove allows to

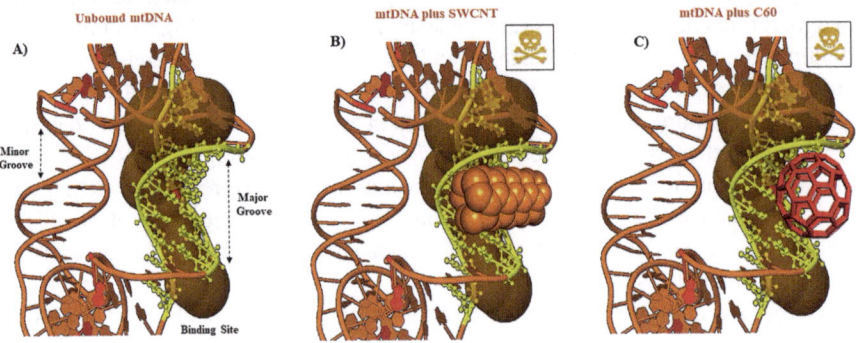

Fig. 17.2 Representative examples of ultrafine carbon-based NP- mtDNA major groove binding interactions (Feng et al. 2022; Rahman et al. 2019). (**a**) Scheme of unbound mtDNA with uninteracting grooves and the van der Waals surface region predicted binding site (yellow); (**b**) Scheme of mtDNA - single-walled carbon nanotube (SWCNT) docking complex (genotoxic potential); (**c**) Scheme of mtDNA – fullerene (C60) docking complex (genotoxic potential). Note: (**b**) and (**c**) illustrate how relevant molecular docking can be in nanogenotoxicity prediction as per structural binding complex analysis

accommodate smaller NP molecules or partially bind fragments of larger compounds. The central region of the minor groove is often compressed to around 1.3–1.5 nm, while the flanking regions are wider, typically around 2–3 nm, and ultrafine carbon nanoparticles, particularly those with specific surface modifications (oxidized-Gr or defect-Gr), can fit into the minor groove and establish stable thermodynamic docking interactions (DNA Grooves Encyclopedia of Genetics 2008).

This type of NPs interacts with the minor groove through hydrogen bonds and van der Waals forces, stabilizing or destabilizing the DNA structure and function as per the nature of the NP and its surface chemical reactivity (Holmannova et al. 2022). Hydrophobic NPs can physically insert into the minor groove, displacing water molecules and altering the local hydration shell of the DNA helix under exposure, affecting the overall stability of the DNA helix and influence the binding affinity of minor groove-binding proteins. Additionally, pathological changes in the minor groove's structure can also affect the recognition and binding of DNA by various regulatory proteins, potentially leading to aberrant gene expression or impaired DNA repair mechanisms (Privalov et al. 2011). Examples of these interactions are displayed in Fig. 17.3.

4.2 Nanoparticle-DNA Interactions Mechanisms

Four main NP-DNA interaction mechanisms must be considered when approaching genotoxicity prediction: hydrogen bonding, hydrophobic bonding, van der Waals forces, and π-π interactions.

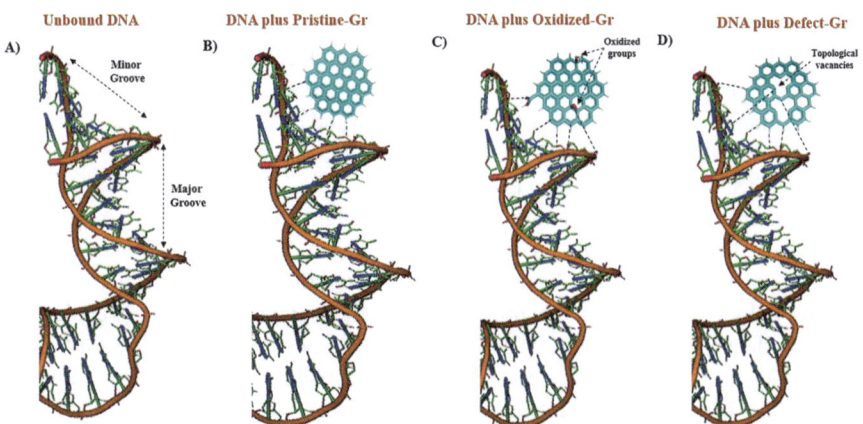

Fig. 17.3 Molecular docking examples of ultrafine graphene NPs interacting with the DNA minor groove, with black-doted lines representing relevant regions involved on stabilizing the docking complexes (Privalov et al. 2011). A) Unbound DNA, with identification of major and minor grooves; B) DNA - pristine-Gr complex binding in minor groove site; C) DNA - oxidized-Gr complex, where oxidized groups provide additional potential for interacting; D) DNA plus defect-Gr complex, where vacancies provide additional potential for interacting

4.2.1 Hydrogen Bonding Nanoparticle-DNA Interactions Mechanism

The specificity, i.e., the specificity of base pairing, and stabilization, through their strength, of hydrogen bonds provide stability to the nanoparticle-DNA docking complex, assuring interaction remains intact even under physiological conditions through their strength (Privalov et al. 2011). The specificity, in particular, allows for NP selectively target specific DNA sequences, a rational-design cornerstone in gene therapy or drug delivery systems used to minimize associated toxicological risk (Privalov et al. 2011; DNA Grooves Encyclopedia of Genetics 2008).

The interaction through hydrogen bonds between NP and the nucleotide bases exposed in the DNA grooves depends on the presence of hydrogen bond donors and acceptors on the NP surface, something that can be engineered through surface modifications, to include specific chemical groups (Holmannova et al. 2022). Such groups can act as hydrogen bond donors (e.g., hydroxyl (-OH) groups) or acceptors (e.g., carbonyl (C=O) groups), while there are groups that act as providers for hydrogen bonding sites (e.g., amino (NH_2) groups).

Furthermore, when a NP encounters a DNA strand, it could selectively recognize and bind to specific base pairs based on its surface chemistry, since the nucleotide bases in DNA, i.e., adenine (A), thymine (T), cytosine (C), and guanine (G), have complementary pairing rules, with A pairing with T via two hydrogen bonds, and C pairing with G via three hydrogen bonds. As such, if the NP has hydroxyl groups, it can form hydrogen bonds with A or thymine T bases, while having amino groups on its NP surface will allow interaction with C or G bases (Holmannova et al. 2022).

4.2.2 Hydrophobic Nanoparticle-DNA Interactions Mechanism

Hydrophobic, i.e., water-repelling, nanomaterials, namely carbon-based nanomaterials, exhibit a strong affinity for hydrophobic regions in DNA, being one critical site the minor groove of DNA (Holmannova et al. 2022). Unlike the major groove, the minor groove provides a more confined environment where water molecules are less abundant due to the proximity of the DNA strands (DNA Grooves Encyclopedia of Genetics 2008). As such, the minor groove is considered less hydrated compared to other DNA regions (Rahman et al. 2019).

The ultrafine hydrophobic graphene NPs (GrNPs) exploit this feature by snugly fitting into the minor groove, with their hydrophobic surfaces align with the hydrophobic patches of the DNA bases, creating a stable thermodynamic interaction. As hydrophobic GrNPs enter the minor groove, they displace water molecules that would otherwise occupy this space. The water displacement induced by GrNPs not only leads to a dehydration effect, but also affects the local environment around the DNA (Rahman et al. 2019; Holmannova et al. 2022).

Moreover, the reduction of hydration affects the intrinsic conformational dynamics of DNA structures and impacts its biochemical functions, with possible gene expression, DNA replication, and repairs processes being influenced. As it may affects accessibility of transcription factors and other regulatory proteins to specific DNA sequences as well, the application potential for hydrophobic NPs interacting with DNA (e.g., biosensing) also raises concerns about potential nanotoxicity, which should be critically evaluated (Rahman et al. 2019; Holmannova et al. 2022).

4.2.3 Van der Waals Nanoparticle-DNA Interactions Mechanism

The weak NPs-DNA binding affinity interactions due to the van der Waals forces arise from fluctuations in electron density within molecules, leading to temporary dipoles, and involve several mechanisms (Holmannova et al. 2022).

When the NPs approach DNA, their electronic clouds will induce temporary dipoles in the DNA bases, which in turn interact with the NP electron cloud. This interaction is an attractive dispersive wan der Waals force, whose magnitude depends on the size and shape of the NPs (Holmannova et al. 2022). Larger NP generally exhibit stronger dispersion forces, while shape complementarity enhances the NP—DNA grooves binding stability, which in turn strengthens the van der Waals interactions (DNA Grooves Encyclopedia of Genetics 2008).

Another mechanism relates to the van der Waals forces associated with the changes in DNA groove's hydration environment, due to the chemical properties and hydrophobicity of the NP-DNA interactions (Wu et al. 2018). Graphene-based NPs graphene often have hydrophobic surfaces due to their carbon-based composition, which interact with hydrophobic patches on DNA (Holmannova et al. 2022; Wu et al. 2018), something that can lead to the displacement of water molecules from the minor groove and subsequent change in the local hydration environment, with this van der Waals force significantly contributing to the binding affinity (Wu et al. 2018).

Notwithstanding, the interactions between nanoparticles and DNA raise concerns about potential nanogenotoxicity, as the NP-DNA close contact can cause structural alterations, base pair disruptions, epigenetic modifications, and oxidative damage. In fact, as cells recognize the NPs are foreign, stress responses are activating, which impact the cell viability, and ultimately the NPs can trigger inflammatory responses, further affecting DNA stability (Wu et al. 2018).

4.2.4 π-π Nanoparticle-DNA Interactions Mechanism

Aromatic interactions, i.e., the noncovalent force that attracts two aromatic rings through the alignment of opposite electrostatic potentials of the rings, also known as π-π interactions, are also an important interaction in the context of NP-DNA binding affinity. In the NP-DNA π-π interaction, the parallel alignment occurs between aromatic rings present in the NP surface and the nucleotide bases of DNA with aromatic rings (e.g., adenine, thymine, guanine, and cytosine). This alignment allows for the overlap of π-electron clouds, leading to thermodynamically stable non-covalent interactions, with the respective strength influenced by distance, orientation, and environment around the interacting aromatic systems. In fact, some NPs, such as graphene or carbon-based NPs, are constructed with aromatic molecules or containing intrinsic aromatic structures, to interact with the similar conjugated DNA π systems.

The π-π interactions between carbon-based NPs and DNA impact through the following induced effects on the DNA structure and function: stabilization of

NP-DNA complexes, i.e., the π-π interactions can enhance the binding affinity of NPs to DNA by providing additional stabilization to the complex, influencing the overall conformation of the DNA molecule, and potentially leading to structural changes (Schneider et al. 2013; Xu et al. 2009; Zhao et al. 2015; Sivasakthi et al. 2013; Bellier et al. 2022); induced conformational change, as following the NP-DNA binding through π-π interactions can cause the DNA helix to bend, twist, or undergo other non-physiological conformational changes, which affects the accessibility of DNA to regulatory proteins and enzymes, potentially interfering with processes such as transcription, replication, and repair (Schneider et al. 2013; Xu et al. 2009; Zhao et al. 2015; Sivasakthi et al. 2013; Bellier et al. 2022); intercalation and groove binding, that is, since NPs can interact with DNA either by intercalating between the base pairs or by binding within the major or minor grooves, with the π-π interactions playing a crucial role in both scenarios, namely in the intercalation, where the aromatic systems of the nanoparticle slip between the stacked base pairs of DNA, stabilizing the intercalated structure (Schneider et al. 2013; Xu et al. 2009; Zhao et al. 2015; Sivasakthi et al. 2013; Bellier et al. 2022); and disruption of protein-DNA interactions, as the NP-DNA binding via π-π interactions can disrupt the binding of native proteins that recognize specific DNA sequences, leading to altered gene expression and potentially genotoxic effects if critical regulatory proteins are inhibited from accessing their target sites on DNA (Zhao et al. 2015).

The understanding of NP-DNA complexes through π-π interactions allow for the rational design of NPs to specifically target DNA sequences, enabling applications in gene therapy, drug delivery, and diagnostics (Schneider et al. 2013; Xu et al. 2009; Zhao et al. 2015; Sivasakthi et al. 2013; Bellier et al. 2022). Controlling the nature and extent of these interactions, engineered NP will selectively bind to desired DNA regions, modulating or enhancing therapeutic efficacy while minimizing off-target effects (Schneider et al. 2013; Xu et al. 2009; Zhao et al. 2015; Sivasakthi et al. 2013; Bellier et al. 2022). Nevertheless, the potential genotoxicity associated with strong π-π interactions must be carefully evaluated since prolonged or irreversible NP-DNA binding could lead to mutations or disruptions in normal cellular functions. As such, the NP design must find a balance between the strength of π-π interactions needed to achieve effective binding without compromising DNA integrity.

5 Implementing Binding Site Prediction in Computational Nanotoxicology

The *in silico* approach to nanotoxicology implies knowing the availability and characteristics of established nanotoxicology -associated software, and the understanding of the most recent advancements in the area. As perceived by the overview displayed in Table 17.1, there are several reliable online software that are a valuable resource to researchers probing nanotoxicity *in silico*. The choice depends of the specificity of application and the datasets (size, type).

In addition to these already established approaches, there have been recent methodological advancements in the precision of active site molecular docking

Table 17.1 Overview of molecular docking online servers in nanotoxicological context

Online server	Scoring Function	Applications	Advantages	Disadvantages
ezCADD[a]	Vina and Smina scoring function (empirical, 2D/3D visualization)	2D/3D molecular modeling Drug discovery Nanotoxicology	Rapid visualization Integrates 2D/3D modelling User-friendly interface	Limited flexibility for large datasets Computationally intensive for complex models
ZDock[b]	Efficient global search coupled with shape complementarity, electrostatics, desolvation	Rigid-body docking Protein-protein interactions (e.g., antibody-protein interactions)	Protein-protein interactions prediction with high accuracy	Limited flexibility May not account for dynamic interactions
PatchDock[c]	Geometric hashing, shape complementarity	Generates candidate transformations for protein-nanoparticle docking through complementary patches	Handles large molecules Efficient in identifying binding sites	Geometric approach may miss flexible binding sites
ClusPro[d]	Electrostatic, desolvation, clustering function with fast Fourier transform (FFT) correlation approach with knowledge-based or statistical potentials	Protein-ligand interactions Clustering solutions	Automated Handles large nanoparticle-protein complexes datasets High success rate	May not accurately predict flexible nanoparticle-protein interactions
RosettaDock[e]	Monte Carlo energy minimization	Refinement of docking solutions	Refines docking solutions Improves prediction accuracy	High computational demand - limited to refinement
GRAMM-X[f]	Scoring function based on smoothed Lennard-Jones knowledge-based potential followed by optimization and rescoring	Potentials applications on rigid-body docking Protein-DNA interactions	Efficient rigid-body docking Fast processing	Limited flexibility May not account for protein dynamics
HexServer[g]	FFT-based correlation	Protein-protein docking Interaction studies	Fast processing Handles large protein and nanoparticles datasets	Limited to rigid-body (may miss flexible interactions)

(continued)

Table 17.1 (continued)

Online server	Scoring Function	Applications	Advantages	Disadvantages
SwissDock[b]	Empirical, EADock DSS engine	Small molecule docking Nanodrug discovery	Accessible Integrates various docking steps High success for small ligands	Less effective for large, flexible ligands
PharmMapper[i]	Reverse pharmacophore mapping	Nanoparticle target identification Small molecule interactions	Identifies potential drug targets High-throughput	Limited to known pharmacophores (may miss novel targets)

[a] https://www.dxulab.org/software; [b]http://zdock.umassmed.edu; [c]https://www.cs.tau.ac.il/~ppdock/PatchDock/; [d] https://cluspro.org/help.php; https://www.rosettacommons.org; [f] https://gramm.compbio.ku.edu; [g]https://hex.loria.fr; [h]https://www.swissdock.ch; [i]http://lilab-ecust.cn/pharmmapper/index.html

prediction. The resource to algorithms based on Delaunay triangulation with weighted points or fast Voronoi tessellation, which analyse the concave surfaces of enzymes to identify relevant binding pockets, is one key factor (Le Guilloux et al. 2009).

Employing a grid-based approach, these algorithms calculate various properties of the enzyme surface (e.g., as hydrophobicity, polarity, and shape) and result in the detection of junction cavities, the cartesian coordinates of the associated cavity centre, and the cavity volumes, which are ranked as per its ligand-target binding probability (Feinstein and Brylinski 2015). A 3D visualization of the detected cavities allows the observation of the binding pockets as 3D-volumetric maps for specific receptors evaluated, a crucial detail for setting up future docking box simulations with a higher degree of precision (Feinstein and Brylinski 2015).

Furthermore, by identifying concave surfaces like protuberances, subcavities, and tunnels, while excluding convex surfaces during the prediction of active sites on proteins, it is assured that only the most relevant binding sites are considered, improving the accuracy of docking predictions and sustaining mechanistic interpretation (Le Guilloux et al. 2009).

These prediction algorithms have gained additional sophistication through the integration of artificial intelligence, particularly deep learning techniques, such as Artificial neural networks (ANNs) that have increasingly been applied in the last 5 years to binding site prediction, offering a high degree of accuracy and efficiency (Jiménez et al. 2017). The advantage that ANNs offer in contrast with traditional algorithms relays especially in its processing of large input datasets, either there are structural and functional data or learning complex patterns and relationships (Jiménez et al. 2017), thus enhancing binding site prediction, identifying toxicological hotspots on proteins with unprecedented precision (Jiménez et al. 2017). From a technical point of view, the ANNs predictive enhancement is the outcome

Table 17.2 Examples of ANN activation functions for protein binding site prediction

Activation Function	Application/Use	Range	Objective	Research potential
Sigmoid $f(x) = \dfrac{1}{1+e^{-x}}$	Output layer of binary classification problems	0 to 1	Prediction of protein surface binding site absence/presence	Determines the likelihood of specific protein regions - NP binding
ReLU (rectified linear unit) $f(x) = \max(0, x)$	Hidden layers of ANNs	0 to $+\infty$	Enabling the network to learn complex patterns more effectively	Allows the capture of intricate NP- protein surface interactions
$\tanh = \dfrac{e^x - e^{-x}}{e^x - e^x}$	Need for outputs (binding site coordinates) to be centered around zero	−1 to 1	Provides a smoother gradient than the sigmoid function, often leading to faster convergence	Modelling the relative affinity of NPs per different binding sites
Softmax $\sigma(z)_i = \dfrac{e^{z_i}}{\sum_k e^{z_j}}$	Output layer for multi-class classification problems	0 to 1 (sum to 1)	Outputs a probability distribution over multiple classes	Facilitates the identification of the primary site of interaction for nanoparticles
Leaky ReLU $f(x)\{\begin{array}{l}x\ if\ x > 0 \\ ax\ if\ x \leq 0\end{array}$ $a = 0.01$	Similar to ReLU, but allows a small, non-zero gradient when the input is negative	−∞ to +∞	Mitigates the dying ReLU problem, improving model robustness	Enhances prediction accuracy, particularly in cases where traditional ReLU might fail
ELU (exponential linear unit) $f(x)\{\begin{array}{l}x\ if\ x > 0 \\ a(e^{x-1})\ if\ x \leq 0\end{array}$ $a: 0.01-1$	Variant of ReLU, allows for negative values to avoid dead neurons	−∞ to ∞	Combines the benefits of ReLU and negative values handling, allowing more complex model behaviors	Enables more accurate and flexible modeling of detailed nanotoxicological assessments
Swish $f(x) = x \cdot \sigma(x)$	Self-gated activation function often used in hidden layers	−∞ to ∞	Improves prediction accuracy by combining linear and non-linear properties	Enhances the network's ability to model complex, non-linear interactions

from combining several activation functions (*vide* Table 17.2 for function examples) to model non-linear relationships between input features and output predictions, which may include the physicochemical properties of the query protein surface, existing interaction data, and also environmental conditions.

To implement ANNs for binding site prediction in molecular docking, the following five steps could be considered as a standard (Jiménez et al. 2017):

1. *Data Pre-processing*: Collecting structural data on protein, physicochemical properties, and known interaction binding sites from relevant repositories to build a curated dataset.
2. *Feature Extraction*: Identifying relevant features such as hydrophobicity, electrostatic potential, and geometric properties of protein surfaces.
3. *Model Training*: Using large datasets to train the ANN, adjusting weights through backpropagation to minimize prediction errors.
4. *Model Validation*: Evaluating the model's performance on unseen data to ensure accuracy and generalizability.
5. *Prediction and Analysis*: Applying the trained ANN to predict potential binding sites for nanoparticles, and analyzing the results to identify toxicological hotspots

6 State-of-the-Art Molecular Docking Approaches in Nanotoxicology

A field as Molecular Docking applied to nanotoxicology is continuously in evolution, scaling up or proposing new approaches to apply this in silico methodology to nano science and nano technology, namely in the context of nanotoxicology. The four main approaches in the state of the art are presented in this section: blind docking, fragment-based, reverse structure-based, and ensemble-based.

6.1 Blind Docking Approach

Blind docking differs from traditional docking methods as its NP—PT predictions are not sustained by prior knowledge of the binding site location (Vorobjev 2010), scanning the entire protein surface to identify potential binding pockets instead of focusing a known active site. The global search blind docking algorithm explores all possible binding regions, ensuring that even unexpected or novel binding sites are considered, an extremely useful tool for when the binding site is unknown or when studying proteins with multiple potential interaction sites. Since NPs often interact with proteins in ways that are not entirely predictable, due to their diverse surface chemistries and the dynamic nature of biological environments, blind docking methods provide a comprehensive map of nanointeractions by identifying all possible binding sites on a target protein (Hassan et al. 2017). As such, by identifying potential toxicological hotspots, blind docking sustains the design of safer nanomaterials, guiding mitigation modifications for hypothetical adverse interactions (Hassan et al. 2017).

6.2 Fragment-Based Approach

One of the challenges of molecular docking techniques is addressing the complexity of large interactive systems, such as complex nanoparticles with different structural constitutional attributes, with associated computational time-consuming and high economic-cost processes (Hall et al. 2011; Stefano et al. 2024). The fragment-based docking is an innovative approach that decomposes ligands (e.g., advanced NP composite) into smaller and manageable fragments to be individually docked into the binding site of the target protein. Since this approach involves identifying key NP molecular fragments, and subsequentially docking them into the protein binding site, with each fragment's binding mode is analyzed and optimized, and re-assembled afterwards to form a complete ligand (Hall et al. 2011; Stefano et al. 2024), it probes specific subfragment-target receptor interactions that might be missed when considering the whole nanosystem.

The breaking down of the supramolecular nanosystems into its constituents parts for a binding affinity investigation allows to pinpoint which fragments are responsible for strong or weak interactions, understanding how specific NP contribute to its overall binding affinity and toxicity, and proposing a template for rational design of nanomaterials indicating which toxicophore moieties (i.e., potential structural alerts) should be modified or even eliminated (Yanagisawa et al. 2022). Moreover, the fragment-based docking can reveal how different NP fragments might interact with multiple binding sites, for a more comprehensive insight of the nanomaterial toxicodynamics behaviour (Yanagisawa et al. 2022), and facilitate identification of novel binding sites that could be pharmacologically targeted or avoided to mitigate toxicological effects (Stefano et al. 2024; Yanagisawa et al. 2022).

6.3 Reverse Structure-Based Approach

Another powerful technique that has emerged in molecular docking is revers, or inverse, docking methodologies. Instead of predicting how a single ligand binds to a specific protein, reverse docking screens a single ligand against a large database of protein structures to identify potential binding targets in a wide array of potential biological targets (McCarthy et al. 2022; da Santos-Junior et al. 2024). Regarding the construction of a nanotoxicological profile, reverse docking can identify unintended NP interactions with various proteins and uncover unexpected binding sites and interactions that might not be evident through conventional docking methods (McCarthy et al. 2022), which allows the prediction of possible off-target effects, i.e., adverse biological responses. This technique relies on both advanced algorithms and nanorepositories for the rapid identification of potential toxicological hotspots, and the integration of reverse docking with other computational techniques, such as machine learning, can further refine the predictive outcomes for implementing rational nanomaterial design.

6.4 Ensemble-Based Docking

In nanotoxicology, where the interactions between target receptors and NPs, presenting a wide range of properties, can be highly complex and variable. The ensemble-based docking is an advanced approach that accounts for the dynamic nature of crystallographic protein structures by considering multiple conformations of the target during the docking process (Korb et al. 2012; Mohammadi et al. 2022). In contrast with traditional docking methods, which typically resource to a single and static conformation of the protein, something that might overlook important nanointeractions, the ensemble-based docking could provide a landscape view by simulating a range of protein conformational states, by mimicking a more realistic scenario.

The first phase generating an ensemble of protein structures (e.g., derived from molecular dynamics simulations or NMR spectroscopy) to represent the flexibility and movement of the protein based on its conformational normal modes. Each conformation in the ensemble is then used as a target for docking studies, thus including the NPs interaction with different conformational states of the protein in the study (Mohammadi et al. 2022). As such, it becomes possible to identify hidden or transient binding sites due to inaccessibility under certain physiological or pathological conditions (e.g., flexible or intrinsically disordered loops in the protein secondary structure, which are common in biological systems) (Mohammadi et al. 2022; Acharya et al. 2020; Falcon et al. 2019). In addition, this method also helps in understanding the conformational changes induced by NP-protein binding, providing toxicodynamic mechanism insight. The integration of ensemble-based docking with other techniques, such as high-throughput screening, can improve its predictive power and reliability (Acharya et al. 2020; Falcon et al. 2019; Huang and Zou 2007).

7 Predicting Off-Target Effects by Using Molecular Docking

One of the major nanomaterials challenges in predicting off-target effects, i.e. NPs interacting with unintended targets, is ligand promiscuity, which refers to the ability of a single ligand to bind to multiple targets (Bender et al. 2007). This is especially relevant when dealing with complex mixtures of NPs that might interact with an unsuspected proteins, nucleic acids, and other biomolecules range, as ligand promiscuity usually leads to potentially causing undesirable side effects or nanotoxicity.

The reverse docking technique is one of the advanced screening docking algorithms that help mapping the potential binding sites, and predict adverse interactions, as it identifies all possible preferences on binding interactions, either on-target and off-target (McCarthy et al. 2022). Moreover, the inherent NP flexibility and heterogeneity complicates the prediction of their behaviour in biological environments, and resourcing to ensemble-based docking, which considers multiple conformations of target proteins, and fragment-based docking, provides more comprehensive predictions. Integrating this technique with ANNs or deep learning models will further enhance the side effects prediction due to algorithm combination, as ANNs can process extensive datasets and identify complex patterns in NP interactions that traditional algorithms might miss (Mohammadi et al. 2022).

Furthermore, the dynamic nature of complex biological systems and the presence of various cellular and subcellular components demands that both the direct NP-target protein interaction and their indirect effects within a given biochemical network are considered (Tang et al. 2024; Forest 2022). An example is the conformational change NP-induced in proteins, leading to altered biological functions or interactions with other cellular components. These secondary effects can be significant in understanding the full spectrum of nanotoxicity induced by a given nanoparticle. The holistic approach of employing a combination of advanced docking techniques, machine learning models, and extensive biological databases, allows for safer design and application of nano biomaterials.

8 Importance of Experimental Validation for *In Silico* Nanotoxicology

Experimental validation is the ultimate validation and an unescapable feature in in silico nanotoxicology techniques and approaches. Notwithstanding, this does not invalidate per se in silico models when experimental techniques or data is scarce or unavailable, as these can resource to validated techniques or hybrid approaches in specific frameworks for a pre-experimental validation stage (Forest 2022).

Albeit the convergence of *in silico* predictions with experimental data strengthens the reliability of the docking studies and enhances their applicability in real-world scenarios, there are key aspects of the experimental validation versus the in silico results: (i) accuracy and reproducibility, i.e., the experimental validation confirms the accuracy of docking simulations, ensuring that the predicted NP binding modes and affinities are consistent with actual biological interactions, and reproducible results across different experimental setups further substantiate the robustness of the in silico models (Li et al. 2022); (ii) mechanistic insights as per docking predictions should be further elucidated through experimental techniques (e.g., X-ray crystallography, NMR spectroscopy, and surface plasmon resonance (SPR), or Small-Angle X-ray Scattering (SAXS), to validate the precise NP binding sites and NP-induced conformational changes (González-Durruthy et al. 2020); (iii) integration of in vitro and in vivo testing, as experimental validation bridges the gap between in silico and these studies, confirming docking predictions in cellular or animal models, and enables to assess the physiological and pathological impacts of nanobiointeractions within complex biochemical systems (González-Durruthy et al. 2017); and iv) sustaining the rational design of nanomaterials, i.e., empirical validation should be applied to the optimization of NP properties, such as size, shape, surface chemistry, and functionalization, for nanomaterial rational design aiming to enhance therapeutic efficacy versus minima adverse effects (González-Durruthy et al. 2017; González-Durruthy et al. 2020).

Therefore, the experimental validation of molecular docking simulations, by bridging between computational predictions and real-world applications, validates in vivo the docking predictions, assuring the results have physiological significance, which is essential for assessing the NP biocompatibility, and shed light on the cascading effects of NP binding (e.g., alterations in signal transduction, gene expression, and metabolism), which are vital for grasping the full scope of unexplored nanotoxicity (Forest 2022). As such, empirical evidence not only corroborates the precision of *in silico* forecasts but also provides crucial knowledge for both mechanistic inquiries and the strategic crafting of novel nanomaterials.

9 Current Gaps, Challenges, and Opportunities for *In Silico* Nanotoxicology

Science is a continuum of overcoming gaps and challenges in the several fields of knowledge, and molecular docking is no exception. As such, there are still limitations in this in silico approaches to nanotoxicology, namely those that affect the accuracy and reliability of docking predictions, particularly in the context of nanoparticle interactions with biological systems.

The first limitation is pervasive to all scientific models, i.e., the inherent simplification of physicochemical reality, which in this case regards the molecular models of the fragment's interactions. Docking tools often use rigid or semi-rigid representations of both NP and target proteins, failing to capture their intrinsic dynamic and that can lead to inaccurate predictions of binding modes and affinities, thus limiting its potential extrapolations and generalizations in real scenario (Furxhi et al. 2020). This explains the importance of novel docking techniques, such as ensemble-based, that might account for molecular dynamics and conformational profile of the molecules isolated and under interaction. Moreover, the complex nature of biological environments, including the presence of multiple interacting molecules and varying conditions, must be addressed, namely by multiscale modelling approaches that integrate molecular docking with other computational techniques (e.g., coarse-grained simulations) Yan et al. 2023).

The scoring functions in docking simulations are another challenge, as these functions, designed to quantitatively predict binding affinities (ΔG), often fail to accurately represent the complex interaction energies involved in NP-protein binding, especially evaluating complex nano-mixture interactions (Denton 2017; Casalini et al. 2019). In fact, current scoring functions may not adequately consider factors such as solvation effects, entropic contributions, and the unique NP surface properties new emerging modular nanomaterials (e.g., metal-organic framework nanoparticles (MOF-NPs), which include diverse chemical functionalities, and complex porosity (Yang et al. 2017). Developing more accurate and comprehensive docking scoring functions that can incorporate these factors is a research path for improving the reliability of in silico predictions.

The selection of appropriate docking tools and parameters is also critical. Different docking programs may yield varying results for the same system, and the choice of algorithm, scoring function, and search parameters can significantly influence the outcomes. In nanotoxicology, where the interactions are complex and multifaceted, the selection of appropriate docking tools and parameters is also critical, as different docking programs may yield dissimilar results for the same nanosystem.

As such, the tool selection needs to be specific for handling the unique properties of nanoparticles (e.g., the NP size, shape, and surface chemistry), which can all affect their binding stability with biological targets.

As previously discussed, the in silico results validation is one of the most significant challenges. The necessary experimental validation to confirm the accuracy of docking predictions is often limited by the availability of crystallographic high-resolution structural data, which provides detailed atomic coordinates of biomolecules, and the complexity of biological target under investigation (Tang et al. 2024; Forest 2022). This level of detail is essential for accurate docking simulations, allowing prediction for binding sites, binding modes, and interaction affinities with a high degree of confidence, and crystallographic data serve as the gold standard for validating computational models (Tang et al. 2024; Forest 2022). Enhanced collaboration interfacing computational and experimental approaches is needed to diminish the gap between computational predictions and real-world nanotoxicity interactions.

Docking simulations typically provide a static snapshot of the interactions between NPs and targets, unable to account for dose-dependent and time-dependent processes, which presents a major gap in the techniques, as nanotoxicological are often influenced by the concentration of NPs and their exposure duration (Pinzi and Rastelli 2019). Traditional docking studies do not incorporate these variables, thus limiting their predictive power regarding the cumulative and temporal effects of nanoparticle exposure, which can impact the binding affinity and stability of nanoparticle-protein complexes (Pinzi and Rastelli 2019; González-Durruthy et al.,2016). A possible avenue to overcome this aspect could be integrating molecular dynamics simulations and kinetic modelling in docking studies.

Finally, there are fringes in the current understanding of NP interaction that need to be probed. An example is the possible influence of the protein corona formed when the NPs interact with biological fluids, which is not fully understood and is often not considered in docking simulations (Nishihira et al. 2019; Park 2020). As the dynamic formation of the protein corona can significantly alter the NP surface properties and their subsequent interactions with biological targets, the incorporation of the formation and impact of the protein corona in docking models is a critical area for future research.

10 Conclusions

Currently, over 160 docking tools and 109 scoring functions are available, sustaining the results of 43,000 published research works resourcing to molecular docking as reliable investigation approach (including Nanotoxicology). The significance of the *in silico* field is undisputed; however, like all scientific endeavors, it still presents challenges that offer opportunities for further improvement and advancement on exploring nano-biointeractions. Advancement of new structure-based algorithms and its selection, the inclusion of time and concentration-dependent relationships, or even the degree of target conformational flexibility allowed in the simulations are among these challenges. The integrated approach of experimental validation interfacing the computational mechanistic insight is often lacking.

Despite the challenges, *in silico* nanotoxicology has achieved significant successes in predicting the toxicodynamic mechanisms of nanoparticles, evaluating potential toxicological effects, and steering the development of safer, novel nanomaterials with potential biomedical applications and the lowest environmental impact.

Acknowledgments M.G.D thanks European Union's H2020 project Sinfonia (N.857253). Also, EU Project iCare (Grant agreement ID: 101092971) and EU Project LEARN (Grant agreement ID: 101057510). This work received support and help from FCT/MCTES (LA/P/0008/2020 DOI 10.54499/LA/P/0008/2020, UIDP/50006/2020 DOI 10.54499/UIDP/50006/2020 and UIDB/50006/2020 DOI 10.54499/UIDB/50006/2020), through national funds. Ana S. Moura further acknowledges FCT/MECS for the contract IF CEECIND/03631/2017.

References

Abdelsattar AS, Dawoud A, Helal MA (2021) Interaction of nanoparticles with biological macromolecules: a review of molecular docking studies. Nanotoxicology 15(1):66–95. https://doi.org/10.1080/17435390.2020.1842537

Acharya A, Agarwal R, Baker MB, Baudry J, Bhowmik D, Boehm S, Byler KG, Chen SY, Coates L, Cooper CJ, Demerdash O, Daidone I, Eblen JD, Ellingson S, Forli S, Glaser J, Gumbart JC, Gunnels J, Hernandez O, Irle S (2020) Supercomputer-based ensemble docking drug discovery pipeline with application to COVID-19. J Chem Inform Model [online] 60(12):5832–5852. https://doi.org/10.1021/acs.jcim.0c01010

Arathi KBM, Joseph X, Mohanan PV (2023) Biological safety and cellular interactions of nanoparticles. In: Mohanan PV, Kappalli S (eds) Biomedical applications and toxicity of nanomaterials. Springer, Singapore, pp 559–587. https://doi.org/10.1007/978-981-19-7834-0_21

Arjmand F, Tabassum S, Khan HY (2024) Molecular docking and computational. In: *In Silico* investigations of metal-based drug agents, in: advances and prospects of 3-d metal-based anticancer drug candidates. Springer, Singapore, pp 149–168. https://doi.org/10.1007/978-981-97-0146-9_6

Auría-Soro C, Nesma T, Juanes-Velasco P, Landeira-Viñuela A, Fidalgo-Gomez H, Acebes-Fernandez V, Gongora R, Parra MJA, Manzano-Roman R, Fuentes M (2019) Interactions of nanoparticles and biosystems: microenvironment of nanoparticles and biomolecules in nanomedicine. Nano 9(10):1365. https://doi.org/10.3390/nano9101365

Bellier N, Baipaywad P, Ryu N, Lee JY, Park H (2022) Recent biomedical advancements in graphene oxide- and reduced graphene oxide-based nanocomposite nanocarriers. Biomater Res 26(1):65. https://doi.org/10.1186/s40824-022-00313-2

Bender A, Scheiber J, Glick M, Davies JW, Azzaoui K, Hamon J, Urban L, Whitebread S, Jenkins JL (2007) Analysis of pharmacology data and the prediction of adverse drug reactions and off-target effects from chemical structure. ChemMedChem 2(6):861–873. https://doi.org/10.1002/cmdc.200700026

Burley SK, Bhikadiya C, Bi C, Bittrich S, Chao H, Chen L, Craig PA, Crichlow GV, Dalenberg K, Duarte JM, Dutta S, Fayazi M, Feng Z, Flatt JW, Ganesan S, Ghosh S, Goodsell DS, Green RK, Guranovic V, Henry J (2022) RCSB Protein Data Bank (RCSB.org): delivery of experimentally-determined PDB structures alongside one million computed structure models of proteins from artificial intelligence/machine learning. Nucl Acids Res [online] 51(D1):D488–D508. https://doi.org/10.1093/nar/gkac1077

Casalini T, Limongelli V, Schmutz M, Som C, Jordan O, Wick P, Borchard G, Perale G (2019) Molecular modeling for nanomaterial–biology interactions: opportunities, challenges, and perspectives. Front Bioeng Biotechnol Sect Nanobiotechnol 7. https://doi.org/10.3389/fbioe.2019.00268

Chen M, Hei J, Huang Y, Liu X, Huang Y (2024) In vivo safety evaluation method for nanomaterials for cancer therapy. Clin Transl Oncol 26(9):2126–2141. https://doi.org/10.1007/s12094-024-03466-9

Crampon K, Giorkallos A, Deldossi M, Baud S, Steffenel LA (2022) Machine-learning methods for ligand–protein molecular docking. Drug Discov Today 27(1):151–164. https://doi.org/10.1016/j.drudis.2021.09.007

da Santos-Junior PFS, de Batista VM, Nascimento S, Nunes IC, Silva LR, Costa CACB, De Freitas JD, Quintans-Júnior LJ, de Araújo-Junior JX, de Freitas MEG, Zhan P, Green KD, Garneau-Tsodikova S, Mendonça-Júnior FJB, Rodrigues-Junior VS, da Silva-Júnior EF (2024) A consensus reverse docking approach for identification of a competitive inhibitor of acetyltransferase enhanced intracellular survival protein from mycobacterium tuberculosis. Bioorg Med Chem 15(108):117774. https://doi.org/10.1016/j.bmc.2024.117774

Denton AD (2017) Effective electrostatic interactions in colloid-nanoparticle mixtures. Phys Rev E 96(6). https://doi.org/10.1103/physreve.96.062610

Dey B, Thukral S, Krishnan S, Chakrobarty M, Gupta S, Manghani C, Rani V (2012) DNA-protein interactions: methods for detection and analysis. Mol Cell Biochem 365(1–2):279–299. https://doi.org/10.1007/s11010-012-1269-z

DNA Grooves (2008) Encyclopedia of genetics, genomics, proteomics and informatics. In: Springer eBooks, vol 527. Springer, Dordrecht. https://doi.org/10.1007/978-1-4020-6754-9_4651

Ekins S, Perryman AL, Horta Andrade C (2016) OpenZika: an IBM world community grid project to accelerate Zika virus drug discovery. PLoS Negl Trop Dis 10(10):e0005023. https://doi.org/10.1371/journal.pntd.0005023

Ewing TJA, Makino S, Skillman AG, Kuntz ID (2001) DOCK 4.0: search strategies for automated molecular docking of flexible molecule databases. J Comput Aided Mol Des 15(5):411–428. https://doi.org/10.1023/a:1011115820450

Falcon WE, Ellingson SR, Smith JC, Baudry J (2019) Ensemble docking in drug discovery: how many protein configurations from molecular dynamics simulations are needed to reproduce known ligand binding? J Phys Chem B 123(25):5189–5195. https://doi.org/10.1021/acs.jpcb.8b11491

Fan D, Cao Y, Cao M, Wang Y, Cao Y, Gong T (2023) Nanomedicine in cancer therapy. Signal Transd Targeted Ther [online] 8(1):293. https://doi.org/10.1038/s41392-023-01536-y

Feher M (2006) Consensus scoring for protein–ligand interactions. Drug Discov Today 11(9–10):421–428. https://doi.org/10.1016/j.drudis.2006.03.009

Feinstein WP, Brylinski M (2015) Calculating an optimal box size for ligand docking and virtual screening against experimental and predicted binding pockets. J Chem 7(1):18. https://doi.org/10.1186/s13321-015-0067-5

Feng Y, Yan Y, He J, Tao H, Wu Q, Huang S-Y (2022) Docking and scoring for nucleic acid–ligand interactions: principles and current status. Drug Discov Today 27(3):838–847. https://doi.org/10.1016/j.drudis.2021.10.013

Ferreira LG, Dos Santos RN, Oliva G, Andricopulo A (2015) Molecular docking and structure-based drug design strategies. Molecules 20(7):13384–13421. https://doi.org/10.3390/molecules200713384

Fischer I, Milton C, Wallace H (2020) Toxicity testing is evolving! Toxicol Res [online] 9(2):67–80. https://doi.org/10.1093/toxres/tfaa011

Forest V (2022) Experimental and computational nanotoxicology—complementary approaches for nanomaterial hazard assessment. Nano 12(8):1346. https://doi.org/10.3390/nano12081346

Friesner RA, Banks JL, Murphy RB, Halgren TA, Klicic JJ, Mainz DT, Repasky MP, Knoll EH, Shelley M, Perry JK, Shaw DE, Francis P, Shenkin PS (2004) Glide: a new approach for rapid, accurate docking and scoring. 1. Method and assessment of docking accuracy. J Med Chem 47(7):1739–1749. https://doi.org/10.1021/jm0306430

Furxhi I, Murphy F, Mullins M, Arvanitis A, Poland CA (2020) Nanotoxicology data for in silico tools: a literature review. Nanotoxicology 14(5):612–637. https://doi.org/10.1080/17435390.2020.1729439

Gangrade D, Sawant G, Mehta A (2016) Re-thinking drug discovery: *in silico* method. J Chem Pharm Res 8(8):1092–1099

González-Durruthy M, Werhli AV, Seus V, Machado KS, Pazos A, Munteanu CR, González-Díaz H, Monserrat JM (2017) Decrypting strong and weak single-walled carbon nanotubes interactions with mitochondrial voltage-dependent anion channels using molecular docking and perturbation theory. Sci Rep 7(1):13271. https://doi.org/10.1038/s41598-017-13691-8

González-Durruthy M, Giri AK, Moreira I, Concu R, Melo A, Ruso JM, Cordeiro MNDS (2020) Computational modeling on mitochondrial channel nanotoxicity2. Nano Today 34:100913–100913. https://doi.org/10.1016/j.nantod.2020.100913

Goodsell DS, Morris GM, Olson AJ (1996) Automated docking of flexible ligands: applications of AutoDock. J Mol Recog 9(1):1–5. https://doi.org/10.1002/(sici)1099-1352(199601)9:1<1::aid-jmr241>3.0.co;2-6

Gu J, Yang X, Kang L, Wu J, Wang X (2015) MoDock: a multi-objective strategy improves the accuracy for 1782 molecular docking. Algorith Mol Biol 10(1):8. https://doi.org/10.1186/s13015-015-0034-8

Gupta D, Yadav P, Garg D, Gupta TK (2021) Pathways of nanotoxicity: modes of detection, impact, and challenges. Front Mater Sci 15(4):512–542. https://doi.org/10.1007/s11706-021-0570-8

Hall DR, Ngan CH, Zerbe BS, Kozakov D, Vajda S (2011) Hot spot analysis for driving the development of hits into leads in fragment-based drug discovery. J Chem Inf Model 52(1):199–209. https://doi.org/10.1021/ci200468p

Hassan NM, Alhossary AA, Mu Y, Kwoh C-K (2017) Protein-ligand blind docking using QuickVina-W with inter-process Spatio-temporal integration. Sci Rep 7:15451. https://doi.org/10.1038/s41598-017-15571-7

Holmannova D, Borsky P, Svadlakova T, Borska L, Fiala Z (2022) Carbon nanoparticles and their biomedical applications. Appl Sci [online] 12(15):7865. https://doi.org/10.3390/app12157865

Honorato RV, Trellet ME, Jiménez-García B, Schaarschmidt JJ, Giulini M, Reys V, Koukos PI, Rodrigues JP, Karaca E, Zundert GC, Roel-Touris J, Noort CW, Jandová Z, Melquiond AS, Bonvin A (2024) The HADDOCK2.4 web server: a leap forward in integrative modelling of biomolecular complexes. Nat Protoc 19:3219–3241. https://doi.org/10.1038/s41596-024-01011-0

Huang SY, Zou X (2007) Ensemble docking of multiple protein structures: considering protein structural variations in molecular docking. Prot Struct Funct Bioinform 66(2):399–421. https://doi.org/10.1002/prot.21214

Irwin JJ, Sterling T, Mysinger MM, Bolstad ES, Coleman RG (2012) ZINC: A free tool to discover chemistry for biology. J Chem Inf Model 52(7):1757–1768. https://doi.org/10.1021/ci3001277

Jaghoori MM, Bleijlevens B, Olabarriaga SD (2016) 1001 ways to run AutoDock Vina for virtual screening. J Comput Aided Mol Des 30(3):237–249. https://doi.org/10.1007/s10822-016-9900-9

Jaiswal C, Pant KK, Behera RKS, Chandra V (2023) Development of new molecules through molecular docking. In: Verma P (ed) Industrial microbiology and biotechnology. Springer, Singapore, pp 643–640. https://doi.org/10.1007/978-981-99-2816-3_22

Jiménez J, Doerr S, Martínez-Rosell G, Rose AS, De Fabritiis G (2017) DeepSite: protein-binding site predictor using 3D-convolutional neural networks. Bioinformatics [online] 33(19):3036–3042. https://doi.org/10.1093/bioinformatics/btx350

John AT, Wadhwa S, Mathur A (2022) Nanotoxicology: exposure, mechanism, and effects on human health. In: Jindal T (ed) New Frontiers in environmental toxicology. Springer, Cham, pp 35–77. https://doi.org/10.1007/978-3-030-72173-2_5

Kakisaka M, Sasaki Y, Yamada K, Kondoh Y, Hikono H, Osada H, Tomii K, Saito T, Aida Y (2015) A novel antiviral target structure involved in the RNA binding, dimerization, and nuclear export functions of the influenza A virus nucleoprotein. PLoS Pathog 11(7):e1005062. https://doi.org/10.1371/journal.ppat.1005062

Kim S, Chen J, Cheng T, Gindulyte A, He J, He S, Li Q, Shoemaker BA, Thiessen PA, Yu B, Zaslavsky L, Zhang J, Bolton EE (2022) PubChem 2023 update. Nucleic Acids Res 51(D1):D1373–D1380. https://doi.org/10.1093/nar/gkac956

Kohl Y, Rundén-Pran E, Mariussen E, Hesler M, El Yamani N, Longhin EM, Dusinska M (2020) Genotoxicity of nanomaterials: advanced in vitro models and high throughput methods for human hazard assessment – a review. Nano 10(10):1911. https://doi.org/10.3390/nano10101911

Korb O, Olsson TSG, Bowden SJ, Hall RJ, Verdonk ML, Liebeschuetz JW, Cole JC (2012) Potential and limitations of ensemble docking. J Chem Inf Model 52(5):1262–1274. https://doi.org/10.1021/ci2005934

Kumah EA, Fopa RD, Harati S, Boadu P, Zohoori FV, Pak T (2023) Human and environmental impacts of nanoparticles: a scoping review of the current literature. BMC Public Health 23:1059. https://doi.org/10.1186/s12889-023-15958-4

Le Guilloux V, Schmidtke P, Tuffery P (2009) Fpocket: an open source platform for ligand pocket detection. BMC Bioinform 10(1):168. https://doi.org/10.1186/1471-2105-10-168

Li X, Li Y, Cheng T, Liu Z, Wang R (2010) Evaluation of the performance of four molecular docking programs on a 1138 diverse set of protein-ligand complexes. J Comput Chem 31(11):2109–2125. https://doi.org/10.1002/jcc.21498

Li X, Huang Y, Chen J (2022) Advances in in Silico toxicity assessment of nanomaterials and emerging contaminants. In: Guo LH, Mortimer M (eds) Advances in toxicology and risk assessment of nanomaterials and emerging contaminants. Springer, Singapore, pp 325–347. https://doi.org/10.1007/978-981-16-9116-4_14

Lu X (2022) Connecting fullerenes with carbon nanotubes and graphene. In: Lu X, Akasaka T, Slanina Z (eds) Handbook of fullerene science and technology. Springer, Singapore, pp 265–270. https://doi.org/10.1007/978-981-16-8994-9_8

McCarthy MJ, Chushak Y, Gearhart JM (2022) Reverse molecular docking and deep-learning to make predictions of receptor activity for neurotoxicology. Computat Toxicol 24:100238. https://doi.org/10.1016/j.comtox.2022.100238

Meldrum K, Guo C, Marczylo EL, Gant TW, Smith R, Leonard MO (2017) Mechanistic insight into the impact of nanomaterials on asthma and allergic airway disease. Part Fibre Toxicol 14(1):45. https://doi.org/10.1186/s12989-017-0228-y

Mohammadi S, Narimani Z, Ashouri M, Firouzi R, Karimi-Jafari MH (2022) Ensemble learning from ensemble docking: revisiting the optimum ensemble size problem. Sci Rep 12(1):410. https://doi.org/10.1038/s41598-021-04448-5

Mullard A (2020) $1.3 billion per drug? Nat Rev Drug Discov 19(4):226. https://doi.org/10.1038/d41573-020-00043-x

Murugan NA, Podobas A, Gadioli D, Vitali E, Palermo G, Markidis S (2022) A review on parallel virtual screening softwares for high-performance computers. Pharmaceuticals 15(1):63. https://doi.org/10.3390/ph15010063

Nhat Phuong D, Flower DR, Chattopadhyay S et al (2023) Towards effective consensus scoring in structure-based virtual screening. Interdiscip Sci Comput Life Sci 15:131–145. https://doi.org/10.1007/s12539-022-00546-8

Nishihira, V.S.K., Rubim, A.M., Brondani, M., Tadiello, J., Pohl, A.R., Friedrich, J.F., De Lara, J.D., Nunes, C.M., Feksa, L.R., Simão, E., Vaucher, R. de A., Durruthy, M.G., Laporta, L.V. and Rech, V.C. (2019). "In vitro and in silico protein corona formation evaluation of curcumin and capsaicin loaded-solid lipid nanoparticles". Toxicol In Vitro, 61, 104598. doi:https://doi.org/10.1016/j.tiv.2019.104598

Pappalardo F, Wilkinson J, Francois Busquet F, Bril A, Palmer M, Walker B, Currelli C, Russo G, Marchal T, Toschi E, Alessandrello R, Costignola V, Klingmann I, Contin M, Staumont B, Woiczinski M, Kaddick C, Di Salvatore V, Aldieri A, Geris L (2022) Toward a regulatory pathway for the use of in Silico trials in the CE marking of medical devices. IEEE J Biomed Health Inform 26(11):5282–5286. https://doi.org/10.1109/jbhi.2022.3198145

Park SJ (2020) Protein-nanoparticle interaction: corona formation and conformational changes in proteins on nanoparticles. Int J Nanomedicine 15:5783–5802. https://doi.org/10.2147/IJN.S254808

Patra JK, Das G, Fraceto LF, Campos EVR, del Rodriguez-Torres MP, Acosta-Torres LS, Diaz-Torres LA, Grillo R, Swamy MK, Sharma S, Habtemariam S, Shin H-S (2018) Nano based drug delivery systems: recent developments and future prospects. J Nanobiotechnol [online] 16(1):71. https://doi.org/10.1186/s12951-018-0392-8

Pinzi L, Rastelli G (2019) Molecular docking: shifting paradigms in drug discovery. Int J Mol Sci 20(18):4331. https://doi.org/10.3390/ijms20184331

Pinzón JR, Villalta-Cerdas A, Echegoyen L (2011) Fullerenes, carbon nanotubes, and graphene for molecular electronics. In: Metzger R (ed) Unimolecular and supramolecular electronics I. Topics in current chemistry, vol 312. Springer, Berlin/Heidelberg, pp 127–174. https://doi.org/10.1007/128_2011_176

Privalov PL, Dragan AI, Crane-Robinson C (2011) Interpreting protein/DNA interactions: distinguishing specific from non-specific and electrostatic from non-electrostatic components. Nucleic Acids Res 39(7):2483–2491. https://doi.org/10.1093/nar/gkq984

Rahman A, O'Sullivan P, Rozas I (2019) Recent developments in compounds acting in the DNA minor groove. Med Chem Comm 10(1):26–40. https://doi.org/10.1039/c8md00425k

Runa S, Hussey MJ, Payne CK (2018) Nanoparticle-cell interactions: relevance for public health. J Phys Chem B, [online] 122(3):1009–1016. https://doi.org/10.1021/acs.jpcb.7b08650

Schlander M, Hernandez-Villafuerte K, Cheng CY, Mestre-Ferrandiz J, Baumann M (2021) How much does it cost to research and develop a new drug? A systematic review and assessment. Pharmaco Econom [online] 39(11):1243–1269. https://doi.org/10.1007/s40273-021-01065-y

Schneider GF, Xu Q, Hage S, Luik S, Spoor JNH, Malladi S, Zandbergen H, Dekker C (2013) Tailoring the hydrophobicity of graphene for its use as nanopores for DNA translocation. Nat Commun 4(1):2619. https://doi.org/10.1038/ncomms3619

Sethi A, Joshi K, Sasikala K, Alvala M (2020) Molecular docking in modern drug discovery: principles and recent applications. In: Gaitonde V, Karmakar P, Trivedi A (eds) Drug discovery and development – new advances. IntechOpen. https://doi.org/10.5772/intechopen.85991

Sivasakthi V, Anbarasu A, Ramaiah S (2013) Π-π interactions in structural stability: role in RNA binding proteins. Cell Biochem Biophys 67(3):853–863. https://doi.org/10.1007/s12013-013-9573-0

Stefano MD, Galati S, Piazza L, Gado F, Grancho C, Macchia M, Giordano A, Tuccinardi T, Poli G (2024) Watermelon: setup and validation of an in silico fragment-based approach. J Enzyme Inhib Med Chem 39(1). https://doi.org/10.1080/14756366.2024.2356179

Tang W, Zhang X, Hong H, Chen J, Zhao Q, Wu F (2024) Computational nanotoxicology models for environmental risk assessment of engineered nanomaterials. Nanomaterials [online] 14(2):155. https://doi.org/10.3390/nano14020155

Tao A, Huang Y, Shinohara Y, Caylor ML, Pashikanti S, Xu D (2018) ezCADD: A rapid 2D/3D visualization-enabled web modeling environment for democratizing computer-aided drug design. J Chem Inf Model 59(1):18–24. https://doi.org/10.1021/acs.jcim.8b00633

Thapa RK, Kim JO (2022) Nanomedicine-based commercial formulations: current developments and future prospects. J Pharm Investig 53(1):19–33. https://doi.org/10.1007/s40005-022-00607-6

Trisciuzzi D, Alberga D, Leonetti F, Novellino E, Nicolotti O, Mangiatordi GF (2018) Molecular docking for predictive toxicology. In: Nicolotti O (ed) Computational toxicology. Methods in molecular biology, vol 1800. Humana Press, New York, NY, pp 181–197. https://doi.org/10.1007/978-1-4939-7899-1_8

Trott O, Olson AJ (2010) AutoDock Vina: improving the speed and accuracy of docking with a new scoring function, efficient optimization, and multithreading. J Comput Chem 31(2):455–461. https://doi.org/10.1002/jcc.21334

Ventura C, Torres V, Vieira L, Gomes B, Rodrigues AS, Rueff J, Penque D, Silva MJ (2022) New "omics" approaches as tools to explore mechanistic Nanotoxicology. In: Louro H, Silva MJ (eds) Nanotoxicology in safety assessment of nanomaterials. Advances in experimental medicine and biology, vol 1357. Springer, Cham, pp 179–194. https://doi.org/10.1007/978-3-030-88071-2_8

Verdonk, M.L., Cole, J.C., Hartshorn, M.J., Murray, C.W. and Taylor, R.D. (2003). "Improved protein-ligand docking using GOLD". Proteins Struct Funct Bioinform, 52(4), 609–623. doi:://doi.org/https://doi.org/10.1002/prot.10465

Verkhivker GM, Bouzida D, Gehlhaar DK, Rejto PA, Arthurs S, Colson AB, Freer ST, Larson V, Luty BA, Marrone T, Rose PW (2000) Deciphering common failures in molecular docking of ligand-protein complexes. J Comput Aided Mol Design 14(8):731–751. https://doi.org/10.1023/A:1008158231558

Vorobjev YN (2010) Blind docking method combining search of low resolution binding sites with ligand pose refinement by molecular dynamics-based global optimization. J Comput Chem 31(5):1080–1092. https://doi.org/10.1002/jcc.21394

Wang L, He X, Ji B, Han F, Niu T, Cai L, Zhai J, Hao D, Wang J (2024) Geometry optimization algorithms in conjunction with the machine learning potential ANI-2x facilitate the structure-based virtual screening and binding mode prediction. Biomol Ther 14(6):648. https://doi.org/10.3390/biom14060648

Wu X, Mu F, Wang Y, Zhao H (2018) Graphene and graphene-based nanomaterials for DNA detection: a review. Molecules 23(8):2050. https://doi.org/10.3390/molecules23082050

Xu K, Huang J, Ye Z, Ying Y, Li Y (2009) Recent development of nano-materials used in DNA biosensors. Sensors 9(7):5534–5557. https://doi.org/10.3390/s90705534

Yan J, Zhang Z, Zhang K, Liu Q (2023) Multi-scale iterative refinement towards robust and versatile molecular docking. arXiv (Cornell University). https://doi.org/10.48550/arXiv.2311.18574

Yanagisawa K, Kubota R, Yoshikawa Y, Ohue M, Akiyama Y (2022) Effective Protein-Ligand Docking Strategy via Fragment Reuse and a Proof-of-Concept Implementation. ACS Omega 7(34):30265–30274. https://doi.org/10.1021/acsomega.2c03470

Yang Q, Xu Q, Jiang H-L (2017) Metal–organic frameworks meet metal nanoparticles: synergistic effect for enhanced catalysis. Chem Soc Rev 46(15):4774–4808. https://doi.org/10.1039/C6CS00724D

Yu AC, Lian H, Kong X, Lopez Hernandez H, Qin J, Appel EA (2021) Physical networks from entropy-driven non-covalent interactions. Nat Commun 12(1):746. https://doi.org/10.1038/s41467-021-21024-7

Zhang Y, Skolnick J (2005) TM-align: a protein structure alignment algorithm based 1440 on the TM-score. Nucleic Acids Res 33(7):2302–2309. https://doi.org/10.1093/nar/gki524

Zhao Y, Li J, Gu H, Wei D, Xu Y, Fu W, Yu Z (2015) Conformational preferences of π–π stacking between ligand and protein, analysis derived from crystal structure data geometric preference of π–π interaction. Interdiscip Sci: Comput Life Sci 7(3):211–220. https://doi.org/10.1007/s12539-015-0263-z

Zheng S, Robertson TA, Varani G (2007) A knowledge-based potential function predicts the specificity and relative binding energy of RNA-binding proteins. FEBS J 274(24):6378–6391. https://doi.org/10.1111/j.1742-4658.2007.06155.x

Open Access This chapter is licensed under the terms of the Creative Commons Attribution-NonCommercial-NoDerivatives 4.0 International License (http://creativecommons.org/licenses/by-nc-nd/4.0/), which permits any noncommercial use, sharing, distribution and reproduction in any medium or format, as long as you give appropriate credit to the original author(s) and the source, provide a link to the Creative Commons license and indicate if you modified the licensed material. You do not have permission under this license to share adapted material derived from this chapter or parts of it.

The images or other third party material in this chapter are included in the chapter's Creative Commons license, unless indicated otherwise in a credit line to the material. If material is not included in the chapter's Creative Commons license and your intended use is not permitted by statutory regulation or exceeds the permitted use, you will need to obtain permission directly from the copyright holder.

Carbon and GaN Nanomaterials for Environmental Contaminant Removal Through Ab Initio Simulations

18

Laura F. O. Vendrame, Mariana Z. Tonel, Paulo B. O. Lira Junior, Silvete Guerini, Mirkos O. Martins, Ivana Zanella, and Solange B. Fagan

Abstract

Nanoscience and nanotechnology can produce materials with extraordinary properties and functionalities, revolutionizing various industry segments, particularly in nanotoxicology and nanosafety for environmental applications. Different nanostructures and classes of nanomaterials are notable for their wide range of practical applications, particularly in adsorption processes. Carbon nanomaterials (CNMs) exhibit intriguing structures with great potential for developing highly selective and efficient adsorbent systems. These nanomaterials can effectively remove inorganic, organic, and biological contaminants from the environment. Similarly, GaN nanotubes (GaNNTs) have been used as sensors and filters due to their high sensitivity to specific pollutants, such as gases, heavy metals, and organic compounds. This chapter presents studies involving nanostructures as adsorbents for removing the herbicide glyphosate (GLYP), such as GaNNT and graphene oxide (GO), through *ab initio* studies using Density Functional Theory (DFT). Furthermore, literature studies from recent years are reported, addressing the fundamental aspects of the use of *ab initio* methodologies in the perception of nanotoxicity in the environment and for removing contaminants from water, such as the distinction between physical and chemical adsorption.

Keywords

Carbon nanostructures · DFT · Glyphosate · Nanotoxicology · Nanosafety

L. F. O. Vendrame · M. Z. Tonel · M. O. Martins · I. Zanella · S. B. Fagan (✉)
Área de Ciências Tecnológicas, Universidade Franciscana, Santa Maria, RS, Brazil

P. B. O. Lira Junior · S. Guerini
Departamento de Física, Universidade Federal do Maranhão, São Luís, MA, Brazil

© The Author(s) 2025
E. Alfaro-Moreno, F. Murphy (eds.), *Nanosafety*,
https://doi.org/10.1007/978-3-031-93871-9_18

Abbreviations

0D	Zero-dimensional
1D	One-dimensional
2D	Two-dimensional
3D	Bulk
CDs	Cyclodextrins
CNMs	Carbon Nanomaterials
CNTs	Carbon Nanotubes
DFT	Density Functional Theory
GaNNTs	GaN nanotubes
GLYP	Glyphosate
GO	Graphene Oxide
MWCNT	Multi-Walled Carbon Nanotube
QSARs	Quantitative Structure-Activity Relationships
QSPR	Quantitative Structure-Property Relationship
QSTR	Quantitative Structure Toxicity Relationship
SP	Spin Polarization
SWCNT	Single-Walled Carbon Nanotube
ΔH/L	ΔHOMO/LUMO difference

1 Introduction

Nanotechnology is one of the most significant and notable areas of technological advancement in the twenty-first century. It offers opportunities for study and innovation due to the unusual properties of organized matter and a range of techniques and methods for manipulating materials at the nanoscale (1–100 nm), to enhance their properties and introduce new functionalities (Wang et al. 2024).

In this context, nanomaterials have been widely used due to their potential role in nanosafety and nanotoxicity, mainly in environmental preservation, such as adsorbents to remove water contaminants (Wang et al. 2024). This application is highlighted for this type of material due to its unique properties. Some examples have high specific surface area, enhanced reactivity, selective adsorption, photocatalytic and magnetic properties, and amphiphilic structure in pristine or modified materials with polar groups (Kumar et al. 2020).

The characteristics of nanomaterials, especially about adsorption processes, are fundamental for recognizing dangers, especially in the face of nanotoxicity. This is because adsorption processes help to understand how nanomaterials interact with contaminants (physical or chemical adsorption), influence their distribution and fate in the environment, affect biocompatibility, and modify toxicity mechanisms (Kumar et al. 2020).

Some of the risks related to nanotoxicity include environmental pollution, especially water contamination from industry and agriculture. Industries dump millions of gallons of effluents full of toxic substances, including heavy metal ions, dyes, herbicides, highly stable aromatic compounds, and medicines (Kumar et al. 2020).

Some pollutants are invisible in the water, resistant to biodegradation, and remain in water bodies for long periods, representing several risks to the ecosystem. One example is glyphosate (GLYP), the active ingredient in many commercial herbicides and one of the most widely used in agriculture worldwide (Georgin et al. 2024).

Glyphosate contaminates water and enters the food chain, and can act as an endocrine disruptor, be carcinogenic, and cause birth defects when consumed by pregnant women (Georgin et al. 2024). *In vitro* studies have already shown that GLYP infiltrates the brain and increases the chances of neurodegenerative diseases (Winstone et al. 2022).

Several research groups are focusing on adsorbent CNMs for wastewater treatment and purification. Recent studies demonstrate that CNMs could solve the vast majority of environmental contamination problems, showing their efficiency compared with traditional adsorbents (Gusain et al. 2020; Mishra et al. 2023; Wang et al. 2024). At the same time, due to the properties of CNTs, a search began for new materials with promising properties. Recent studies show the potential of using GaNNTs as sensors or filters for atmospheric pollutants, such as gases, heavy metals, and organic compounds (Coutinho et al. 2009; Khan and Srivastava 2016).

An effective way to understand the adsorption of molecules on nanomaterials is through *ab initio* computational simulations using Density Functional Theory (DFT), such as the distinction between physical and chemical adsorption. Therefore, in this chapter, we present a review of articles published on these topics in recent years. In addition, we show unpublished studies from our research group on nanomaterials such as GO and GaNNTs interacting with the GLYP herbicide, using *ab initio* methods based on DFT to demonstrate the relation between adsorption properties and the environmental impact.

2 Carbon Nanomaterials and Their Applications as Nanoadsorbents

In recent years, the interest and significance of carbon-based nanomaterials have emerged exponentially. This fact is attributed to the positive impact of nanomaterials on improving our well-being in a wide variety of sectors due to their remarkable physicochemical properties and multiple applications (Choudhary et al. 2024; Gusain et al. 2020).

Among the various areas in which CNMs are being applied, environmental remediation, particularly with regard to nanotoxicity and water treatment, has been gaining significant attention (Malik 2024; Mishra et al. 2023). This is because many of the problems faced in these areas can be solved or mitigated using CNMs as nanoadsorbents (Aslam et al. 2024).

Due to their dimensionality and properties, among the different classes of nanomaterials, some carbon allotropes, such as graphene, CNTs (single-walled carbon nanotubes—SWCNTs and multi-walled carbon nanotubes—MWCNTs),

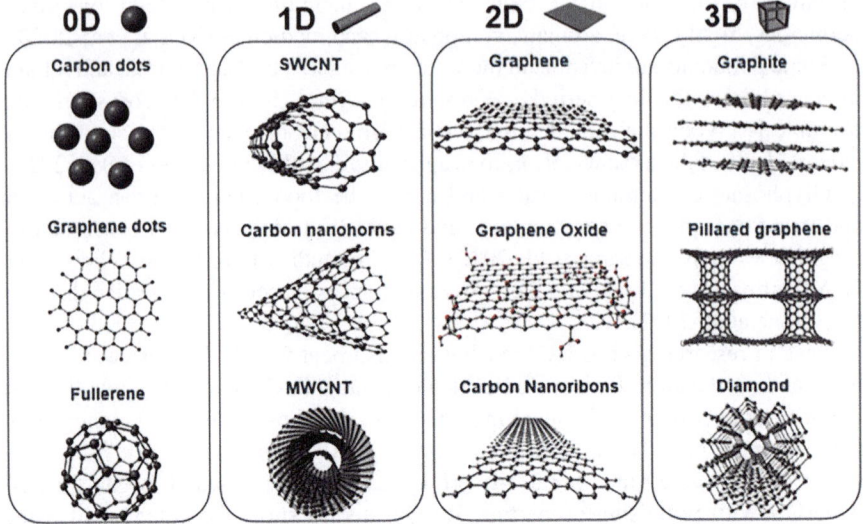

Fig. 18.1 Some examples for 0D, 1D, 2D, and 3D carbon nanomaterials

fullerenes, and derivatives, have gained prominence in scientific research on adsorption processes, mainly with contaminating molecules (Manimegalai et al. 2023).

These carbon allotropes differ in the molecular composition, as well as the packing arrangements of atoms in solid states or bonding orbitals, which influence their varied properties, and, consequently, provide different interactions between carbon-based nanoadsorbents and the adsorbate molecule (Aslam et al. 2024). Figure 18.1 shows examples of 0D, 1D, 2D, and 3D carbon nanomaterials.

The properties of CNMs include their exceptional porosities (pore volume and average pore diameter), high surface area, and internal and external morphology, such as hollow and layered structures. All of this is associated with different internal and external adsorption sites, and the ease of chemical activation, doping, amphoteric properties, and functionalization. The main focus of the research groups is to show the effectiveness of CNMs compared to conventional adsorbents, especially concerning nanotoxicity (Artiga et al. 2024; Vijeata et al. 2024).

One of these CNMs is the fullerene, a zero-dimensional (0D) material that is a highly symmetric cage-shaped molecule with sp^2-like hybridization due to the curvature effect and its finite number of atoms. The C_{60} molecule has interconnected hexagonal and pentagonal rings of carbon (12 pentagons and 20 hexagons), with approximately 7 Å of diameter (Kroto et al. 1985).

As the number of carbon atoms grows, fullerene can have multiple isomers with different shapes like hollow spheres, tubes, or ellipses (Kulkarni et al. 2024). Furthermore, the ability to chemically modify it inside and outside the cage has led to the generation of fullerene derivatives that exhibit improved physicochemical merits (Kulkarni et al. 2024).

The 0D nature of fullerenes, with their inner and outer surfaces covered by homogeneous π-electrons, facilitates their support or adsorption on a substrate through non-covalent bonds, such as π-π interactions (Rajkamal and Kim 2023). In fullerene clusters, there are distinct adsorption sites that can significantly contribute to the overall adsorption.

As an effective electron acceptor, fullerene can extract electrons from the nearby positions of the substrate to which it is attached. As a result, the electronic configuration surrounding the fullerene is modified. This essential feature of C_{60} molecules provides them with a potent capacity to adapt electronic configurations and engage positively with other compounds, offering distinct benefits as electrocatalysts that do not apply to graphene and CNTs.

In fullerene aggregates, the observed adsorption sites include the upper surface of the fullerene, which typically results in weaker physisorption, and the stronger adsorption sites located in the grooves, interstitial spaces, and between aggregates (Naderi et al. 2024). These characteristics of fullerenes make them highly attractive for environmental applications in scientific research, especially about to adsorption and nanotoxicity, being increasingly studied for their ability to adsorb pollutants in water, although to a lesser extent compared to CNTs and graphene derivatives (Naderi et al. 2024).

For example, DFT studies have explored the adsorption of pharmaceuticals like temozolomide, procarbazine, carmustine, and lomustine on the surface of pristine and modified fullerenes for water purification, observing physical adsorption between the systems (Samanta and Das 2017). Additionally, studies using DFT of magnetic fullerenes interacting with ciprofloxacin are reported showing that the adsorption process occurs through electrostatic and π-π stacking interactions (Elessawy et al. 2020). Other experimental studies focus on fullerenes as adsorbents for removal of heavy metals, dyes, and metalloid ions from water (Azzouz et al. 2023).

CNTs present a tubular architecture, sp^2-like hybridization, consisting of a cylindrical graphene layer arranged in a hexagonal beehive pattern. CNTs are categorized according to the number of layers: SWCNTs and MWCNTs, and exist in three forms depending on their geometry: chiral, zig-zag, and armchair (Ogunsola et al. 2024). Their electronic and mechanical properties vary according to chirality and diameter (0.4 to <3 nm for SWCNTs and 1.4 to 100 nm in the case of MWCNTs) (Ogunsola et al. 2024). Due to their distinct morphology and structure, combined with exceptional properties such as superior optical activity, electrical conductivity, thermal characteristics, mechanical strength, and chemical reactivity, CNTs have been extremely attractive for numerous applications (Vijeata et al. 2024).

Several studies have indicated that CNTs exhibit significant adsorption capabilities for a wide variety of organic and inorganic pollutants present in water, including herbicides (Vijeata et al. 2024; Yao et al. 2021), synthetic dyes (Shahidi and Moazzenchi 2018), heavy metals (Shahidi and Moazzenchi 2018), and pharmaceutical compounds (Duarte et al. 2022; Nayak and Pathan 2023).

2D nanosheets are regarded as the thinnest materials, featuring nanoscale thickness and macroscale lateral dimensions. Graphene, a 2D nanomaterial, consists of six-membered rings with sp^2 hybridization structured as a single carbon layer in a honeycomb graphite structure and has been applied in various studies due to its thermal, mechanical, and electrical properties (Kumar et al. 2024).

Since graphene is considered nonpolar, hydrophobic, and tends to aggregate rapidly, chemical modifications to graphene oxide (GO), reduced graphene oxide (rGO), and other derivatives with smaller hydrophilic functional groups have made its derivatives precursors for a range of applications, including its significant use in removing pollutants (Anegbe et al. 2024; Kumari et al. 2023). GO and rGO, for example, exhibit different morphological and chemical characteristics compared to graphene, primarily dependent on the various synthesis methods used, and possess remarkable adsorption capacity for aromatic molecules such as dyes and metals through hydrogen and π-π bonding and electrostatic interactions (Casabella-Font et al. 2024).

Graphene can also be functionalized with other materials, such as cellulose, carrageenan, and alginate, enhancing their properties, as well as magnetic metals and inorganic acids, among others (El-Shafai et al. 2021; Liu et al. 2023). All these functionalizations make graphene more selective towards different adsorbents, including contaminants. They can be applied in various ways, including wastewater treatment and pollutant adsorption.

Their extensive surface area (ranging from approximately 295.6 to 1205.8 m^2/g) and high chemical stability make them effective in removing pollutants from water and air (Deng et al. 2024). Researchers have documented the effectiveness of graphene derivatives (such as GO and reduced GO) in removing various contaminants from aqueous solutions, including antibiotics (Gahrouei et al. 2024), heavy metals (Yang et al. 2022), synthetic dyes (Gomes et al. 2022), and herbicides (Kumari et al. 2023).

The number of articles published on the topic in the last 10 years fully demonstrates the importance given to CNMs as nanoadsorbents. The results reveal annual results with significant increases in the number of scientific articles, as can be observed in Fig. 18.2.

For instance, in 2023 alone, there are over 2.189 articles in the literature on studies involving CNMs as adsorbents, that is approximately eight times more than the number of articles published in 2014. Studies of CNMs related to adsorption and toxicity have also shown a significant increase in the last 10 years, as shown in Fig. 18.2. In 2023 there were 1.347 articles, which is nine times more than in 2014.

Another relevant information (Fig. 18.2) is the percentage comparison between publications that focus exclusively on CNMs and adsorption and those that also address toxicity (see red line). It has been observed that, over the years, the inclusion of the term "toxicity" in studies involving CNMs and adsorption has become increasingly frequent.

18 Carbon and GaN Nanomaterials for Environmental Contaminant Removal... 517

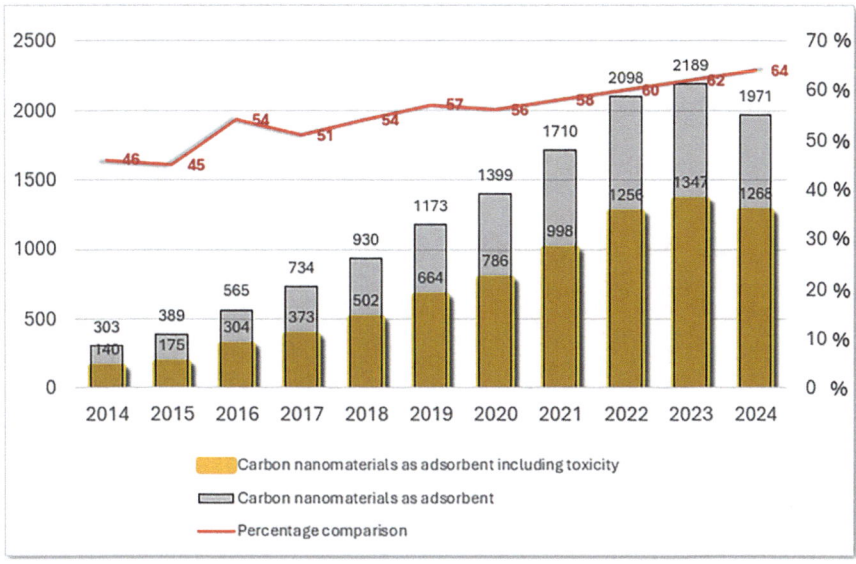

Fig. 18.2 Articles published since 2014 that address the use of CNMs as adsorbents, both alone (in gray) and including toxicity (in yellow), showing the percentage variation (in red) in the number of publications on both topics over the years. (www.sciencedirect.com, last accessed 23th July 2024)

3   *Ab Initio* Methods by Density Functional Theory as a Tool for Nanotoxicity Research

The importance of the DFT is shown mainly in the number of applications of formalism, which in just over two decades has grown surprisingly (Becke 2014). The volume of publications shows that DFT calculations represent a milestone of success, with significant progress in methodological issues and approximations, starting with the renowned articles by "Hohenberg and Kohn" (Kohn and Sham 1965) and "Kohn and Sham" (Hohenberg and Kohn 1964), with 42.382/50.549 citations now, dating from 1964 and 1965, respectively.

Subsequently, several approaches have been established and applied to current and past challenges in chemistry and physics. Although the origins are commonly associated with the articles by Hohenberg, Kohn, and Sham (Hohenberg and Kohn 1964; Kohn and Sham 1965), the widespread use of DFT methods began only after 1990, when new approaches were evaluated, and final efforts showed the success of DFT, especially in the molecule studies (Jones 2015).

Simultaneously, advances in DFT and their integration in calculations with molecular dynamics, molecular docking, Quantitative Structure-Activity Relationships (QSARs), and other methodologies have expanded, providing an effective way to analyze different biological systems such as toxicity (Štekláč and Breza 2021). In this way, DFT has emerged as a powerful computational tool in

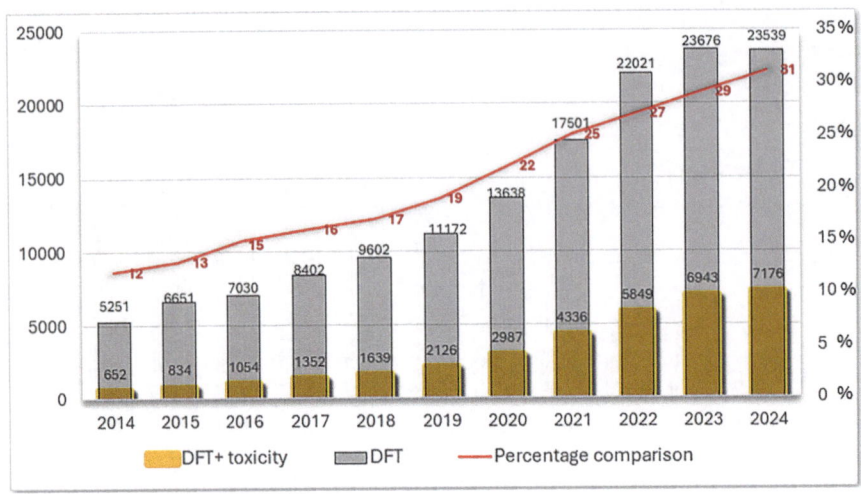

Fig. 18.3 Articles published since 2014 that address DFT studies, both in isolation (in gray) and including toxicity (in yellow), show the percentage variation (red line) in the number of publications on both topics over the years. (www.sciencedirect.com, last accessed August 4, 2024)

various fields of chemistry, including toxicology (Cronin and Schultz 2001; Geerlings et al. 2003; Koch and Holthausen 2008; Raies and Bajic 2016).

As of 2023, the literature includes over 23.676 articles on studies involving CNMs and DFT. This represents a roughly threefold increase from 2014, when there were 5.251 publications on this topic.

Furthermore, research on CNMs related to adsorption and toxicity has experienced significant growth over the past decade. As shown in Fig. 18.3, the number of articles on this topic has surged to 6.943, representing a tenfold increase from the 652 articles published in 2014. In 2014, 12% of the articles that mentioned "DFT" also included "toxicity". So far in 2024, this proportion has grown to 31%, while in 2023 it was 29%, evidencing a significant increase in the simultaneous focus on both topics.

DFT has been a valuable tool in toxicity studies. The application of DFT in toxicity studies has become prevalent in predicting and elucidating the sources of toxicity using quantum molecular descriptors. This approach is commonly employed in studies related to QSAR, Quantitative Structure-Property Relationship (QSPR) and Quantitative Structure Toxicity Relationship (QSTR) (Aslam et al. 2024; Malik 2024). It highlights one of DFT's key roles in understanding the fundamental chemistry underlying toxicological reactions in predictive toxicology (Artiga et al. 2024; Manimegalai et al. 2023).

Research on phenols has shown that DFT-derived descriptors like molecular complexity, valence path cluster, and topological diameter significantly influence toxicity (Vijeata et al. 2024), leading to the development of predictive QSTR models with high statistical significance and predictive power (Kroto et al. 1985). Overall, DFT plays a vital role in elucidating the relationship between chemical structures and cytotoxicity, offering a deeper understanding of toxicological reactions and aiding in environmental risk assessments.

As the application of DFT in nanotoxicity evaluation, it can be used to study the adsorption of biomolecules on the nanoparticle surface. DFT can help to elucidate the mechanisms of reactive oxygen species generation and the interaction of nanoparticles with biological antioxidants, and provide insights into the binding mechanisms and conformational changes induced by nanoparticle exposure. DFT-derived descriptors can be used to develop QSTR models for predicting nanoparticle toxicity (Kulkarni et al. 2024) based on their physicochemical properties.

DFT has also proven to be a highly useful tool for investigating toxicity of nanomaterials in adsorption processes, providing detailed insights into molecular interactions and properties that affect toxicity, such as the stability of complexes formed (Talha et al. 2024; Yang et al. 2024). DFT addresses detection and removal functions by analyzing the structural and electronic characteristics of the adsorbent-adsorbate systems, as well as the targeted adsorption processes (Liang et al. 2022; Talha et al. 2024).

Recent studies demonstrate how DFT is widely used to understand these mechanisms, providing clues about nanotoxicity, showing whether the nanomaterial becomes a good adsorbent, being able to eliminate unwanted materials from the environment, as well as the possible potentiation of the nanomaterial by chemical adsorption processes, increasing toxicity when the structure of the system is altered (Hamed Mashhadzadeh et al. 2018; Hsu et al. 2024; Li et al. 2022; Liu et al. 2020).

In physical adsorption processes, which involve Van der Waals forces and electrostatic interactions without the formation of strong chemical bonds, high adsorption capacity can lead to the release of pollutants, being effective in removing contaminants that are difficult to evaluate by other methods (Liang et al. 2022).

On the other hand, in chemical adsorption, where chemical bonds are formed between the adsorbent and the adsorbate, stable complexes can be formed that affect the release of pollutants and, consequently, the toxicity of the system and may pose additional risks (Krishna et al. 2023). Therefore, DFT is essential to evaluate and optimize adsorption conditions through structural and electronic studies, to contribute to minimizing toxicity risks and ensuring effective removal of pollutants.

4   Nanomaterials as Adsorbent for the Herbicide Glyphosate: An *Ab Initio* Study

Glyphosate [N-(phosphonomethyl)glycine, GLYP] is one of the most widely used broad-spectrum organophosphorus herbicides (Gill et al. 2017). This compound is extensively employed in agriculture to control perennial and annual weeds and in forestry, home gardens, and urban areas. GLYP effectively controls harmful or invasive weeds (Che and Liu 2014; Chen and Liu 2007). It is absorbed by the leaves and young stems of plants, being transported throughout the enzymatic system, where it inhibits the synthesis of amino acids such as phenylalanine, tryptophan, and tyrosine, as well as preventing the formation of lignin, alkaloids, flavonoids, and benzoic acids (Aitbali et al. 2018; Pereira et al. 2021). After application, the concentration of GLYP in the soil can persist for one to 26 weeks, becoming a potential contaminant for soil and groundwater (Gomes et al. 2016; Mesnage and Antoniou 2017; Yang et al. 2018). However, the degradation of GLYP in the environment is paramount to its fate and toxicity (Yang et al. 2018). Furthermore, GLYP has been classified as potentially carcinogenic since 2015 (Portier et al. 2016; Tarone 2018).

Graphene-based materials and their derivatives are highly effective in removing and remediating various pesticides from water. Theoretical and experimental studies show that the potential adsorption mechanisms and applications of these nanomaterials can be obtained facilitating the development of alternative methods for removing toxic contaminants (De Souza Antônio et al. 2021; Khoshnam et al. 2019; Yadav et al. 2019).

In the same way, GaNNTs are known for their applications in electronic, photonic and optoelectronic devices (Khan and Srivastava 2016). Recent studies show that GaNNTs are also excellent material to detect and remove environmental pollutants (Silva et al. 2021). GaNNTs have high sensitivity and selectivity and can detect environmental contaminants at low concentrations as well as distinguish different types of contaminants (Ogungbemiro et al. 2023).

In different configurations, the structural, electronic, and energetic properties of GO and GaNNT interacting with GLYP molecules were investigated using first-principles calculations based on density functional theory (DFT) and implemented in the SIESTA computational code (Soler et al. 2002). The energy shift was fixed at 0.04 eV, and the generalized gradient approximation (GGA) with the Perdew, Burke, and Ernzerhof (PBE) parameterization (Perdew et al. 1996) was used for the exchange and correlation term. Troullier Martins pseudopotentials described the interaction between core and valence electrons with a double-zeta plus polarization basis (DZP) to expand the pseudowave functions of the valence electrons (Ogungbemiro et al. 2023). To represent the charge density, a mesh-cutoff of 400 Ry was adopted for network integrations. All atomic structures were relaxed until the residual forces were less than 0.05 eV/Å for all atoms. To calculate the binding energy (E_b) is calculated according to Eq. (18.1):

$$E_b = -\left(E_{A+B} - E_A - E_B\right) \tag{18.1}$$

where E_{A+B} is the energy of the interacting systems and E_A and E_B are the isolated energy values. A similar methodology has already been published by our group (da Bruckmann et al. da Bruckmann et al. 2022a, b).

The electronic properties of the structures were analyzed individually to find their most reactive sites and make interactive systems from them. The isolated structures of GO, GaNNT, and GLYP and the energy levels/bands, are shown in Fig. 18.4. The GO isolated structure presents spin polarization of 1.00 µB, while GaNNT and GLYP pristine do not present spin polarization. The ΔHOMO/LUMO (ΔH/L) difference of GO and GLYP is respectively 0.49 and 4.15 eV, values of GO similar to those of Vargas et al. (2023) and de Oliveira et al. (2021) using a similar methodology. The type and number of functional groups directly influence the polarization of

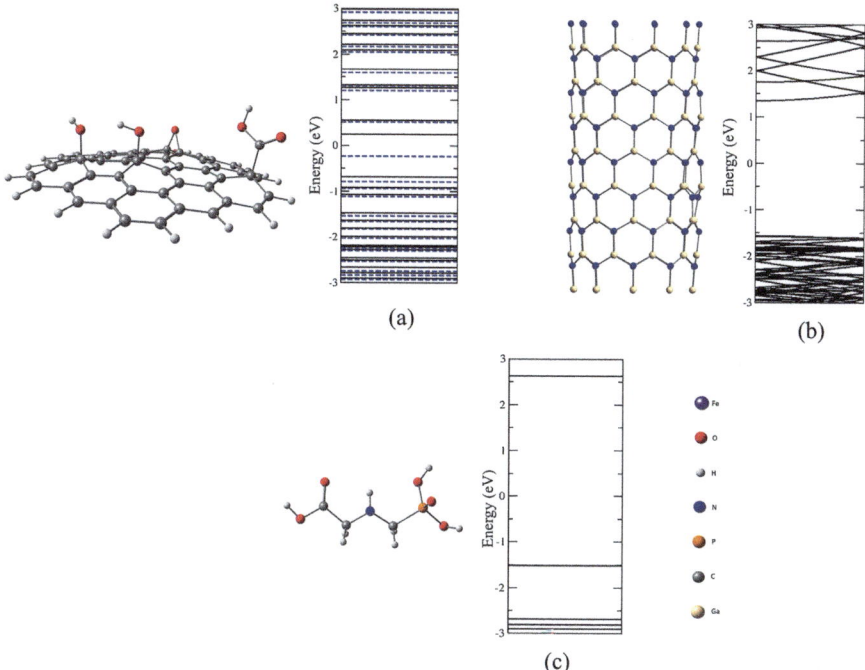

Fig. 18.4 Electronic energy levels for 0D GO (**a**), GaNNT (**b**), and GLYP (**c**). The up (blue dashed lines) and down (black filled lines) electronic bands are plotted for spin-polarized systems. Fermi level is set to zero

graphene (Tonel et al. 2017). On the other hand, the electronic band gap of GaNNT presents a value of 2.96 eV.

The most stable configurations of GO and GaNNT interacting with the herbicide GLYP are shown in Fig. 18.5. The values of binding energy (E_b), charge transfer (Δq), $\Delta H/L$, and spin polarization (SP) of the most stable configurations studied are presented in Table 18.1. For all configurations considered, the charge transfer (see Table 18.1) was obtained through the Mülliken population analysis (Mulliken 1955).

Considering the interaction between GO and GLY, the GO+GLYP-III configuration presented the most significant stability with a binding energy of 1.11 eV. In this configuration, the GLYP molecule is perpendicular to the GO, interacting directly with the carboxyl groups. Tonel et al. (2021) showed that the interaction of the -COOH type with -COOH is favored when the molecule interacts perpendicularly. In this stable configuration, we observed a charge transfer from GLYP to GO of 0.62 e$^-$, therefore, GLYP is an electronic charge donor molecule. This system has a spin polarization of 1.00 µB, similar to the isolated GO.

The study of the interaction of the GLYP molecule with the GaNNT considers different arrangements to determine which would be most energetically favorable. Figure 18.1 shows the three most stable configurations after structural relaxation.

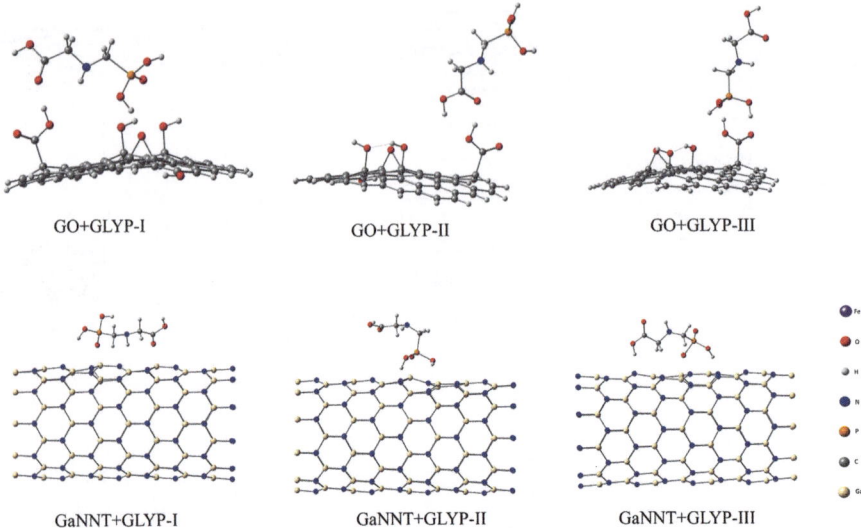

Fig. 18.5 Most stable configurations studied of nanostructures interacting with the GLYP molecule

Table 18.1 Most stable configurations studied, binding energy (E_b) (eV), spin polarization (SP) (μ_B), HOMO/LUMO difference ($\Delta H/L$) (eV), and charge transfer (CT) (e^-). The '+/−' sign indicates that GLYP has received/donor electronic charge, making it a charge acceptor/donor

Configuration	E_b (eV)	SP (μ_B)	$\Delta H/L$ (eV)	CT (e^-)
GO+GLYP-I	0.66	1.00	0.48	−0.59
GO+GLYP-II	0.47	1.00	0.48	−0.17
GO+GLYP-III	1.11	1.00	0.47	−0.62
GaNNT+GLYP-I	1.70	0	2.29	+0.45
GaNNT+GLYP-II	1.59	0	2.68	+0.09
GaNNT+GLYP-III	1.45	0	2.53	−0.03

Analyzing the binding energy values, we observe that configuration I is predicted as the most energetically favorable, with an absolute value of 1.70 eV, followed by configurations II and III, with 1.59 and 1.45 eV, respectively, as can be seen in Table 18.1. Observing the binding energy results we can infer that the interaction between the GLYP molecule and GaNNT occurs via a chemical process. Machado et al. (2016) considered that binding energy values less than or equal to 0.83 eV can be considered to be a physical process, while energy values above this value are considered associated with a chemical process (Machado et al. 2016). The shortest distance between the GLYP molecule and the GaNNT in configuration I is 1.93 Å, considering the distance between the gallium atom of the GaNNT and the oxygen atom of the GLYP molecule. For configurations II and III, the distances are 1.44 and 1.70 Å, respectively, considered between the nitrogen atom of the GaNNT and the hydrogen atom of the molecule, in both configurations. Analyzing Table 18.1, we notice that the GLYP molecule behaves as an electron acceptor in configurations I and II, while in configuration III, it behaves as an electron acceptor. These electronic charge transfer behaviors were directly associated with the molecule arrangements through a more/less electronegativity of the atom interacting with the GaN nanostructure.

Through calculations based on DFT and the interaction of GO and GaNNT with the herbicide GLYP, we can observe characteristics in the electronic, energetic, and structural properties that lead us to identify whether the adsorption can be physical or chemical. In the adsorption of the herbicide GLYP on GaNNT, we observed a considerable decrease in the energy gap to the isolated GaNNT. At the same time, we observed a charge transfer, with GLYP receiving a charge from GaNNT. In the interaction of GO with GLYP, we observed little change in ΔH/L, with the value being equal and/or very close to that of isolated GO. At the same time, we observed that the binding energy values of GO interacting with GLYP are lower than those of GaNNT.

The discussed results from the GLYP interacting with GO and GaNNT, associated with recent studies from the literature (Hamed Mashhadzadeh et al. 2018; Li et al. 2022; Liu et al. 2020) demonstrate the ability of DFT methods to provide insights about nanotoxicity, demonstrating that the nanomaterial is an adsorbent, being able to eliminate contaminants from the environment, or increasing the toxicity when the system is altered by chemical adsorption processes [50–53]. These indications are observed in theoretical-experimental studies (Bruckmann et al. b; De Matos et al. 2024; Machado et al. 2016; Rocha et al. 2024; Vargas et al. 2023) and confirm the importance of this technique to associate the interaction with the resulting modifications on the systems and reduce or improve the toxic impact on the environment.

5 Final Remarks

Ab initio simulations, specially on DFT, stands out in several areas of chemistry and physics, where its use has grown significantly due to its ability to provide insights into the atomic and molecular mechanisms of nanotoxicity.

In this chapter, we highlight how *ab initio* methods can contribute to evidence the nanotoxicity in studies with nanomaterials. We demonstrate, through a literature survey in the last 10 years, that DFT has accumulated several applications in this area, with a significant increase in recent years.

Finally, we present a study where we specifically evaluate how GO and GaNNT interact with the herbicide GLYP. The results of these calculations allow us to analyze the energetic, electronic, and structural characteristics, helping to clarify whether the adsorption process is of a physical or chemical nature and elucidate possible toxicity mechanisms.

Acknowledgments The authors are grateful to CENAPAD-SP (National Center for High-Performance Processing in São Paulo) and UFN (Franciscan University) for the computational space, and the Brazilian agencies CNPq (Grants: 443154/2023-6; 421701/2017-0; 309162/2021-1), CAPES (Grant: 88881.506898/2020-01), and FINEP (Grant: 01.22.0536.00) for financial support.

References

Aitbali Y, Ba-Mhamed S, Elhidar N, Nafis A, Soraa N, Bennis M (2018) Glyphosate based-herbicide exposure affects gut microbiota, anxiety and depression-like behaviors in mice. Neurotoxicol Teratol 67:44–49. https://doi.org/10.1016/j.ntt.2018.04.002

Anegbe B, Ifijen IH, Maliki M, Uwidia IE, Aigbodion AI (2024) Graphene oxide synthesis and applications in emerging contaminant removal: a comprehensive review. Environ Sci Eur 36:15. https://doi.org/10.1186/s12302-023-00814-4

Artiga Á, Lin H, Bianco A (2024) Interaction of industrial graphene and carbon nanotubes with human primary macrophages: assessment of nanotoxicity and immune responses. Carbon 223:119024. https://doi.org/10.1016/j.carbon.2024.119024

Aslam J, Verma C, Hussain CM, Aslam R (eds) (2024) Carbon allotropes: advanced materials for anticorrosive coatings. CRC Press, Boca Raton

Azzouz A, Hejji L, Kumar V, Kim K-H (2023) Nanomaterials-based aptasensors: an efficient detection tool for heavy-metal and metalloid ions in environmental and biological samples. Environ Res 238:117170. https://doi.org/10.1016/j.envres.2023.117170

Becke AD (2014) Perspective: fifty years of density-functional theory in chemical physics. J Chem Phys 140:18A301. https://doi.org/10.1063/1.4869598

Casabella-Font O, Riva M, Balcázar JL, Radjenovic J, Pijuan M (2024) Distinctive effects of graphene oxide and reduced graphene oxide on methane production kinetics and pharmaceuticals removal in anaerobic reactors. Bioresour Technol 403:130849. https://doi.org/10.1016/j.biortech.2024.130849

Che H, Liu S (2014) Contaminant detection using multiple conventional water quality sensors in an early warning system. Procedia Eng 89:479–487. https://doi.org/10.1016/j.proeng.2014.11.239

Chen S, Liu Y (2007) Study on the photocatalytic degradation of glyphosate by TiO2 photocatalyst. Chemosphere 67:1010–1017. https://doi.org/10.1016/j.chemosphere.2006.10.054

Choudhary F, Mudgal P, Parvez A, Sharma P, Farooqi H (2024) A review on synthesis, properties and prospective applications of carbon nanomaterials. Nano-Struct Nano-Objects 38:101186. https://doi.org/10.1016/j.nanoso.2024.101186

Coutinho SS, Lemos V, Guerini S (2009) Band-gap tunability of a (6,0) BN nanotube bundle under pressure: *Ab initio* calculations. Phys Rev B 80:193408. https://doi.org/10.1103/PhysRevB.80.193408

Cronin MTD, Schultz TW (2001) Development of quantitative structure–activity relationships for the toxicity of aromatic compounds to *Tetrahymena pyriformis* : comparative assessment of the methodologies. Chem Res Toxicol 14:1284–1295. https://doi.org/10.1021/tx0155202

da Bruckmann FS, Rossato Viana A, Tonel MZ, Fagan SB, da Garcia WJS, de Oliveira AH, Dorneles LS, Roberto Mortari S, da Silva WL, da Silva IZ, Rhoden CRB (2022a) Influence of magnetite incorporation into chitosan on the adsorption of the methotrexate and in vitro cytotoxicity. Environ Sci Pollut Res 29:70413–70434. https://doi.org/10.1007/s11356-022-20786-x

da Bruckmann FS, Zuchetto T, Ledur CM, dos Santos CL, da Silva WL, Binotto Fagan S, Zanella da Silva I, Bohn Rhoden CR (2022b) Methylphenidate adsorption onto graphene derivatives: theory and experiment. New J Chem 46:4283–4291. https://doi.org/10.1039/D1NJ03916D

De Matos CF, Leão MB, Vendrame LFO, Jauris IM, Zanella I, Fagan SB (2024) Unlocking the paracetamol adsorption mechanism in graphene tridimensional-based materials: an experimental-theoretical approach. Front Carbon 3:1305183. https://doi.org/10.3389/frcrb.2024.1305183

de Oliveira PV, Zanella I, Bulhões LOS, Fagan SB (2021) Adsorption of 17 β- estradiol in graphene oxide through the competing methanol co-solvent: experimental and computational analysis. J Mol Liq 321:114738. https://doi.org/10.1016/j.molliq.2020.114738

De Souza Antônio R, Guerra ACS, De Andrade MB, Nishi L, Baptista ATA, Bergamasco R, Vieira AMS (2021) Application of graphene nanosheet oxide for atrazine adsorption in aqueous solution: synthesis, material characterization, and comprehension of the adsorption mechanism. Environ Sci Pollut Res 28:5731–5741. https://doi.org/10.1007/s11356-020-10693-4

Deng Z, Ou H, He T, Zhang H, Jiang Y, Liang Q, Xiang X (2024) Soft-hard segments regulated framework of nitrogen-doped reduced graphene oxide aerogel for efficient adsorption of pollutants. J Mol Liq 403:124929. https://doi.org/10.1016/j.molliq.2024.124929

Duarte EDV, Oliveira MG, Spaolonzi MP, Costa HPS, da Silva TL, Silva MGCD, Vieira MGA (2022) Adsorption of pharmaceutical products from aqueous solutions on functionalized carbon nanotubes by conventional and green methods: a critical review. J Clean Prod 372:133743. https://doi.org/10.1016/j.jclepro.2022.133743

Elessawy NA, Elnouby M, Gouda MH, Hamad HA, Taha NA, Gouda M, Mohy Eldin MS (2020) Ciprofloxacin removal using magnetic fullerene nanocomposite obtained from sustainable PET bottle wastes: adsorption process optimization, kinetics, isotherm, regeneration and recycling studies. Chemosphere 239:124728. https://doi.org/10.1016/j.chemosphere.2019.124728

El-Shafai NM, Ramadan MS, Amin MA, El-Mehasseb IM (2021) Graphene oxide/cellulose derivative nanohybrid membrane with yttrium oxide: upgrading the optical and electrochemical properties for removing organic pollutants and supercapacitors implementations. J Energy Storage 44:103344. https://doi.org/10.1016/j.est.2021.103344

Gahrouei AE, Vakili S, Zandifar A, Pourebrahimi S (2024) From wastewater to clean water: recent advances on the removal of metronidazole, ciprofloxacin, and sulfamethoxazole antibiotics from water through adsorption and advanced oxidation processes (AOPs). Environ Res 252:119029. https://doi.org/10.1016/j.envres.2024.119029

Geerlings P, De Proft F, Langenaeker W (2003) Conceptual density functional theory. Chem Rev 103:1793–1874. https://doi.org/10.1021/cr990029p

Georgin J, Stracke Pfingsten Franco D, Gindri Ramos C, Nguyen Tran H, Benettayeb A, Imanova G, Ali I (2024) Recent advances in removing glyphosate herbicide and its aminomethylphosphonic acid metabolite in water. J Mol Liq 402:124786. https://doi.org/10.1016/j.molliq.2024.124786

Gill JPK, Sethi N, Mohan A (2017) Analysis of the glyphosate herbicide in water, soil and food using derivatising agents. Environ Chem Lett 15:85–100. https://doi.org/10.1007/s10311-016-0585-z

Gomes MP, Le Manach SG, Moingt M, Smedbol E, Paquet S, Labrecque M, Lucotte M, Juneau P (2016) Impact of phosphate on glyphosate uptake and toxicity in willow. J Hazard Mater 304:269–279. https://doi.org/10.1016/j.jhazmat.2015.10.043

Gomes BFML, De Araújo CMB, Do Nascimento BF, Freire EMPDL, Da Motta Sobrinho MA, Carvalho MN (2022) Synthesis and application of graphene oxide as a nanoadsorbent to remove cd (II) and Pb (II) from water: adsorption equilibrium, kinetics, and regeneration. Environ Sci Pollut Res 29:17358–17372. https://doi.org/10.1007/s11356-021-16943-3

Gusain R, Kumar N, Ray SS (2020) Recent advances in carbon nanomaterial-based adsorbents for water purification. Coord Chem Rev 405:213111. https://doi.org/10.1016/j.ccr.2019.213111

Hamed Mashhadzadeh A, Ghorbanzadeh Ahangari M, Salmankhani A, Fataliyan M (2018) Density functional theory study of adsorption properties of non-carbon, carbon and functionalized graphene surfaces towards the zinc and lead atoms. Physica E 104:275–285. https://doi.org/10.1016/j.physe.2018.08.010

Hohenberg P, Kohn W (1964) Inhomogeneous electron gas. Phys Rev 136:B864–B871. https://doi.org/10.1103/PhysRev.136.B864

Hsu CY, Saadh MJ, Taki AG, Mohammed SK, Bahair H, Adthab AH, Abduvalieva D, Mumtaz H, Salem-Bekhit MM, Mirzaei M, Dai M, Maaliw Iii RR, Mosaddad SA (2024) Density functional theorybased analyses of beryllium oxide fullerene assisted adsorptions of ammonia, phosphine, and arsine toxic gases towards sensing and removal prospective applications. Colloids Surf A Physicochem Eng Asp 692:133939. https://doi.org/10.1016/j.colsurfa.2024.133939

Jones RO (2015) Density functional theory: its origins, rise to prominence, and future. Rev Mod Phys 87:897–923. https://doi.org/10.1103/RevModPhys.87.897

Khan MS, Srivastava A (2016) NH3 and NO2 adsorption analysis of GaN nanotube: a first principle investigation. J Electroanal Chem 775:243–250. https://doi.org/10.1016/j.jelechem.2016.05.048

Khoshnam F, Zargar B, Moghadam MR (2019) Adsorption and removal of ametryn using graphene oxide nano-sheets from farm waste water and optimization using response surface methodology. J Iran Chem Soc 16:1383–1390. https://doi.org/10.1007/s13738-019-01621-6

Koch W, HolthausenMC (2008) A chemist's guide to density functional theory, 2nd edn, 5 reprint ed. Wiley-VCH, Weinheim

Kohn W, Sham LJ (1965) Self-consistent equations including exchange and correlation effects. Phys Rev 140:A1133–A1138. https://doi.org/10.1103/PhysRev.140.A1133

Krishna, R.H., Chandraprabha, M.N., Samrat, K., Krishna Murthy, T.P., Manjunatha, C., Kumar, S.G., 2023. Carbon nanotubes and graphene-based materials for adsorptive removal of metal ions—a review on surface functionalization and related adsorption mechanism. Appl Surf Sci Adv 16, 100431. ://doi.org/https://doi.org/10.1016/j.apsadv.2023.100431

Kroto HW, Heath JR, O'Brien SC, Curl RF, Smalley RE (1985) C60: Buckminsterfullerene. Nature 318:162–163. https://doi.org/10.1038/318162a0

Kulkarni K, Moghe Y, Tangadpalliwar A, Kaur J, Vohra R (2024) A review on the smallest carbon fullerene C20: applications and device formation. Materials Today Proc S2214785324004450. https://doi.org/10.1016/j.matpr.2024.05.147

Kumar V, Lee Y-S, Shin J-W, Kim K-H, Kukkar D, Fai Tsang Y (2020) Potential applications of graphene-based nanomaterials as adsorbent for removal of volatile organic compounds. Environ Int 135:105356. https://doi.org/10.1016/j.envint.2019.105356

Kumar R, Juneja P, Rathi R, Kumar A (2024) A structured review on properties of graphene based composites. Materials Today Proc S2214785324003274. https://doi.org/10.1016/j.matpr.2024.05.030

Kumari K, Singh MB, Tomar N, Kumar A, Kumar V, Dabodhia KL, Singh P (2023) Adsorption of pesticides using graphene oxide through computational and experimental approach. J Mol Struct 1291:136043. https://doi.org/10.1016/j.molstruc.2023.136043

Li H, Wu X, Cheng K, Zhu M, Wang L, Yu H, Yang J (2022) One-pot modified "grafting-welding" preparation of graphene/polyimide carbon films for superior thermal management. Carbon 186:738. https://doi.org/10.1016/j.carbon.2021.10.036

Liang W, Wang G, Peng C, Tan J, Wan J, Sun P, Li Q, Ji X, Zhang Q, Wu Y, Zhang W (2022) Recent advances of carbon-based nano zero valent iron for heavy metals remediation in soil and water: a critical review. J Hazard Mater 426:127993. https://doi.org/10.1016/j.jhazmat.2021.127993

Liu Y, Cai Z, Sheng L, Ma M, Wang X (2020) A magnetic relaxation switching and visual dual-mode sensor for selective detection of Hg2+ based on aptamers modified Au@Fe3O4 nanoparticles. J Hazard Mater 388:121728. https://doi.org/10.1016/j.jhazmat.2019.121728

Liu T, Aniagor CO, Ejimofor MI, Menkiti MC, Wakawa YM, Li J, Akbour RA, Yap P-S, Lau SY, Jeevanandam J (2023) Recent developments in the utilization of modified graphene oxide to adsorb dyes from water: a review. J Ind Eng Chem 117:21–37. https://doi.org/10.1016/j.jiec.2022.10.008

Machado FM, Carmalin SA, Lima EC, Dias SLP, Prola LDT, Saucier C, Jauris IM, Zanella I, Fagan SB (2016) Adsorption of alizarin red S dye by carbon nanotubes: an experimental and theoretical investigation. J Phys Chem C 120:18296–18306. https://doi.org/10.1021/acs.jpcc.6b03884

Malik MI (2024) Handbook of nanomaterials, volume 2, biomedicine, environment, food, and agriculture. Elsevier, Amsterdam

Manimegalai S, Vickram S, Deena SR, Rohini K, Thanigaivel S, Manikandan S, Subbaiya R, Karmegam N, Kim W, Govarthanan M (2023) Carbon-based nanomaterial intervention and efficient removal of various contaminants from effluents—a review. Chemosphere 312:137319. https://doi.org/10.1016/j.chemosphere.2022.137319

Mesnage R, Antoniou MN (2017) Facts and fallacies in the debate on glyphosate toxicity. Front Public Health 5:316. https://doi.org/10.3389/fpubh.2017.00316

Mishra RK, Mentha SS, Misra Y, Dwivedi N (2023) Emerging pollutants of severe environmental concern in water and wastewater: a comprehensive review on current developments and future research. Water-Energy Nexus 6:74–95. https://doi.org/10.1016/j.wen.2023.08.002

Mulliken RS (1955) Electronic population analysis on LCAO–MO molecular wave functions. II. Overlap populations, bond orders, and covalent bond energies. J Chem Phys 23:1841–1846. https://doi.org/10.1063/1.1740589

Naderi N, Ganjali F, Eivazzadeh-Keihan R, Maleki A, Sillanpää M (2024) Applications of hollow nanostructures in water treatment considering organic, inorganic, and bacterial pollutants. J Environ Manag 356:120670. https://doi.org/10.1016/j.jenvman.2024.120670

Nayak T, Pathan A (2023) Environmental remediation and application of carbon-based nanomaterials in the treatment of heavy metal-contaminated water: a review. Mater Today Proc 92:1659–1670. https://doi.org/10.1016/j.matpr.2023.06.227

Ogungbemiro FO, Louis H, Benjamin I, Okon GA, Okon IE, Agwupuye JA, Adeyinka AS (2023) Metals (cu, ag, au) encapsulated gallium nitride nanotubes (GaNNTs) as sensors for hexabromodiphenyl ether (HBDE) emerging organic pollutant: a computational study. J Saudi Chem Soc 27:101667. https://doi.org/10.1016/j.jscs.2023.101667

Ogunsola SS, Oladipo ME, Oladoye PO, Kadhom M (2024) Carbon nanotubes for sustainable environmental remediation: a critical and comprehensive review. Nano-Struct Nano-Objects 37:101099. https://doi.org/10.1016/j.nanoso.2024.101099

Perdew JP, Burke K, Ernzerhof M (1996) Generalized gradient approximation made simple. Phys Rev Lett 77:3865–3868. https://doi.org/10.1103/PhysRevLett.77.3865

Pereira HA, Hernandes PRT, Netto MS, Reske GD, Vieceli V, Oliveira LFS, Dotto GL (2021) Adsorbents for glyphosate removal in contaminated waters: a review. Environ Chem Lett 19:1525–1543. https://doi.org/10.1007/s10311-020-01108-4

Portier CJ, Armstrong BK, Baguley BC, Baur X, Belyaev I, Bellé R, Belpoggi F, Biggeri A, Bosland MC, Bruzzi P, Budnik LT, Bugge MD, Burns K, Calaf GM, Carpenter DO, Carpenter HM, López-Carrillo L, Clapp R, Cocco P, Consonni D, Comba P, Craft E, Dalvie MA, Davis D, Demers PA, De Roos AJ, DeWitt J, Forastiere F, Freedman JH, Fritschi L, Gaus C, Gohlke JM, Goldberg M, Greiser E, Hansen J, Hardell L, Hauptmann M, Huang W, Huff J, James MO, Jameson CW, Kortenkamp A, Kopp-Schneider A, Kromhout H, Larramendy ML, Landrigan PJ, Lash LH, Leszczynski D, Lynch CF, Magnani C, Mandrioli D, Martin FL, Merler E, Michelozzi P, Miligi L, Miller AB, Mirabelli D, Mirer FE, Naidoo S, Perry MJ, Petronio MG, Pirastu R, Portier RJ, Ramos KS, Robertson LW, Rodriguez T, Röösli M, Ross MK, Roy D, Rusyn I, Saldiva P, Sass J, Savolainen K, Scheepers PTJ, Sergi C, Silbergeld EK, Smith MT, Stewart BW, Sutton P, Tateo F, Terracini B, Thielmann HW, Thomas DB, Vainio H, Vena JE, Vineis P, Weiderpass E, Weisenburger DD, Woodruff TJ, Yorifuji T, Yu IJ, Zambon P, Zeeb H, Zhou S-F (2016) Differences in the carcinogenic evaluation of glyphosate between the International Agency for Research on Cancer (IARC) and the European Food Safety Authority (EFSA). J Epidemiol Community Health 70:741–745. https://doi.org/10.1136/jech-2015-207005

Raies AB, Bajic VB (2016) *In silico* toxicology: computational methods for the prediction of chemical toxicity. WIREs Comput Mol Sci 6:147–172. https://doi.org/10.1002/wcms.1240

Rajkamal A, Kim H (2023) Theoretical verification on adsorptive removal of caffeine by carbon and nitrogen-based surfaces: role of charge transfer, π electron occupancy, and temperature. Chemosphere 339:139667. https://doi.org/10.1016/j.chemosphere.2023.139667

Rocha ECMD, Rocha JAPD, Da Costa RA, Costa ADSSD, Barbosa EDS, Josino LPC, Brasil LDSNDS, Vendrame LFO, Machado AK, Fagan SB, Brasil DDSB (2024) High-throughput molecular modeling and evaluation of the anti-inflammatory potential of Açaí constituents against NLRP3 Inflammasome. IJMS 25:8112. https://doi.org/10.3390/ijms25158112

Samanta PN, Das KK (2017) Noncovalent interaction assisted fullerene for the transportation of some brain anticancer drugs: a theoretical study. J Mol Graph Model 72:187–200. https://doi.org/10.1016/j.jmgm.2017.01.009

Shahidi S, Moazzenchi B (2018) Carbon nanotube and its applications in textile industry—a review. J Textile Inst 109:1653–1666. https://doi.org/10.1080/00405000.2018.1437114

Silva AAJAD, Caetano CV, Guerini S (2021) Energetic and electronic properties of NH3, NO2 and SO2 interacting with GaN nanotube: a DFT study. J Mol Model 27:234. https://doi.org/10.1007/s00894-021-04826-w

Soler JM, Artacho E, Gale JD, García A, Junquera J, Ordejón P, Sánchez-Portal D (2002) The SIESTA method for *ab initio* order- *N* materials simulation. J Phys Condens Matter 14:2745–2779. https://doi.org/10.1088/0953-8984/14/11/302

Štekláč M, Breza M (2021) DFT studies of the toxicity of 4-substituted 1,2-benzoquinones. Polyhedron 210:115532. https://doi.org/10.1016/j.poly.2021.115532

Talha A, Shihab FH, Ahmed MT, Al Roman A, Kowser Z, Roy D (2024) Density functional theory study of the adsorption and dissociation of OF2 and O3 gases on the surface of pristine and Al, Ti and Cr doped graphene. AIP Adv 14:075008. https://doi.org/10.1063/5.0214735

Tarone RE (2018) On the International Agency for Research on Cancer classification of glyphosate as a probable human carcinogen. Eur J Cancer Prev 27:82–87. https://doi.org/10.1097/CEJ.0000000000000289

Tonel MZ, Lara IV, Zanella I, Fagan SB (2017) The influence of the concentration and adsorption sites of different chemical groups on graphene through first principles simulations. Phys Chem Chem Phys 19:27374–27383. https://doi.org/10.1039/C7CP05549H

Tonel MZ, Zanella I, Fagan SB (2021) Theoretical study of small aromatic molecules adsorbed in pristine and functionalised graphene. J Mol Model 27:193. https://doi.org/10.1007/s00894-021-04806-0

Vargas GO, Schnorr C, Nunes FB, Da Rosa Salles T, Tonel MZ, Fagan SB, Zanella Da Silva I, Silva LFO, Mortari SR, Dotto GL, Rhoden CRB (2023) Highly furosemide uptake employing magnetic graphene oxide: DFT modeling combined to experimental approach. J Mol Liq 379:121652. https://doi.org/10.1016/j.molliq.2023.121652

Vijeata A, Chaudhary GR, Chaudhary S, Ibrahim AA, Umar A (2024) Recent advancements and prospects in carbon-based nanomaterials derived from biomass for environmental remediation applications. Chemosphere 357:141935. https://doi.org/10.1016/j.chemosphere.2024.141935

Wang H, Jafir M, Irfan M, Ahmad T, Zia-ur-Rehman M, Usman M, Rizwan M, Hamoud YA, Shaghaleh H (2024) Emerging trends to replace pesticides with nanomaterials: recent experiences and future perspectives for ecofriendly environment. J Environ Manag 360:121178. https://doi.org/10.1016/j.jenvman.2024.121178

Winstone JK, Pathak KV, Winslow W, Piras IS, White J, Sharma R, Huentelman MJ, Pirrotte P, Velazquez R (2022) Glyphosate infiltrates the brain and increases pro-inflammatory cytokine TNFα: implications for neurodegenerative disorders. J Neuroinflam 19:193. https://doi.org/10.1186/s12974-022-02544-5

Yadav S, Goel N, Kumar V, Singhal S (2019) Graphene oxide as proficient adsorbent for the removal of harmful pesticides: comprehensive experimental cum DFT investigations. Anal Chem Lett 9:291–310. https://doi.org/10.1080/22297928.2019.1629999

Yang Y, Deng Q, Yan W, Jing C, Zhang Y (2018) Comparative study of glyphosate removal on goethite and magnetite: adsorption and photo-degradation. Chem Eng J 352:581–589. https://doi.org/10.1016/j.cej.2018.07.058

Yang J, Shojaei S, Shojaei S (2022) Removal of drug and dye from aqueous solutions by graphene oxide: adsorption studies and chemometrics methods. NPJ Clean Water 5:5. https://doi.org/10.1038/s41545-022-00148-3

Yang J, Zhang M, Zhang Y, Gao M, Gao M, Wang K, Song Z, Liu Z, Wang Z, Shen B (2024) Density functional theory study of adsorption and dissociation of CH2Cl2 on the surfaces of transition metal (Fe, Co, Ni, and Cu)-doped carbon nanotubes. Chem Phys Impact 8:100437. https://doi.org/10.1016/j.chphi.2023.100437

Yao T, Liu L, Tan S, Li H, Liu X, Zeng A, Pan L, Li X, Bai L, Liu K, Xing B (2021) Can the multi-walled carbon nanotubes be used to alleviate the phytotoxicity of herbicides in soils? Chemosphere 283:131304. https://doi.org/10.1016/j.chemosphere.2021.131304

Open Access This chapter is licensed under the terms of the Creative Commons Attribution-NonCommercial-NoDerivatives 4.0 International License (http://creativecommons.org/licenses/by-nc-nd/4.0/), which permits any noncommercial use, sharing, distribution and reproduction in any medium or format, as long as you give appropriate credit to the original author(s) and the source, provide a link to the Creative Commons license and indicate if you modified the licensed material. You do not have permission under this license to share adapted material derived from this chapter or parts of it.

The images or other third party material in this chapter are included in the chapter's Creative Commons license, unless indicated otherwise in a credit line to the material. If material is not included in the chapter's Creative Commons license and your intended use is not permitted by statutory regulation or exceeds the permitted use, you will need to obtain permission directly from the copyright holder.

Quasi-SMILES as a Tool for Simulation of Endpoints Related to Nanomaterials

Andrey A. Toropov, Alla P. Toropova, Alessandra Roncaglioni, and Emilio Benfenati

Abstract

A simplified molecular input-line entry system (SMILES) is a well-known approach to representing the molecular structure of different substances. Quasi-SMILES is an extension of traditional SMILES by adding codes reflecting experimental conditions. The Monte Carlo technique of constructing and using quasi-SMILES for modeling nanomaterial toxicity is discussed. Examples of using quasi-SMILES for modeling nanooxide toxicity are presented. The influence of the so-called index of ideality of correlation (*IIC*) and correlation intensity index (*CII*) on stochastic Monte Carlo optimization processes is considered. The ability of the mentioned new statistical criteria of predictive potential to improve nanooxide toxicity models is studied. The advantages and disadvantages of the models calculated with quasi-SMILES are discussed. The possible perspectives of the approach in the aspect of the *in silico* simulation are listed.

Keywords

QSAR · Monte Carlo method · Metal nanooxide · Index of ideality of correlation · Correlation intensity index · CORAL software

1 Introduction

The development of new materials and technologies for their production and processing is currently recognized as a very important factor for the scientific development of improved products in several sectors, with high economic value. Indeed,

---

A. A. Toropov (✉) · A. P. Toropova · A. Roncaglioni · E. Benfenati
Department of Environmental Health Science, Laboratory of Environmental Chemistry and Toxicology, Istituto di Ricerche Farmacologiche Mario Negri IRCCS, Milan, Italy
e-mail: andrey.toropov@marionegri.it

© The Author(s) 2025
E. Alfaro-Moreno, F. Murphy (eds.), *Nanosafety*,
https://doi.org/10.1007/978-3-031-93871-9_19

one of the priority areas for the development of modern materials science is nanomaterials and nanotechnologies (Costa et al. 2023; Åkerlund et al. 2019; Ortiz-Galvez et al. 2024; Chatterjee and Alfaro-Moreno 2023; Meneses et al. 2023; Lebre et al. 2022). The development of fundamental and applied ideas about nanomaterials and nanotechnologies in the coming years can lead to fundamental changes in many areas of human activity: materials science, energy, electronics, computer science, mechanical engineering, medicine, agriculture, and ecology. Along with computer information technologies and biotechnologies, nanotechnologies are the foundation of the scientific and technological revolutions.

It is difficult to imagine modern science without the widespread use of mathematical modeling, which consists of replacing the original object with its "image"—a mathematical model—and further studying this model using computer systems. This method combines many advantages related to both the improvement of the theory exploration and exmploitation of the experimental results. Working not with the object itself, but with its model makes it possible to painlessly, relatively quickly and without significant costs explore its properties and behavior in any conceivable situation. At the same time, computational experiments with object models make it possible, relying on the power of modern computational tools of computer science, to study objects in detail and in-depth to a sufficient extent that is inaccessible to purely theoretical approaches.

Both theoretical and experimental methods are used to study the electronic, optical and emission properties of solids. The use of experimental methods leads to large time, financial and labor costs. Therefore, the use of methods for simulation of the properties of nanostructures will significantly reduce these costs. The use of computer simulation for nanosystems has fundamental difficulties. Firstly, there is no long-range order, which is characteristic of crystals and makes it possible to reduce the number of independent degrees of freedom of the system; secondly, the short-range order characteristic of liquids does not allow one to determine all the functional properties of nanomaterials. Third, there are technical difficulties associated with simulation macro-objects at the atomic level. Direct simulation of such systems in the approximation of molecular dynamics and, especially, quantum mechanics is difficult even with the use of modern supercomputer technology. A solution may be to use a hierarchical multiscale approach in simulation, when at each lower level the parameters and variables necessary to build upper-level models are calculated. Currently, when developing nanomaterials with given properties, experimental methods are mainly used, which does not always allow for finding the optimal solution and increases the cost of development. Therefore, it is advisable to more actively involve mathematical simulation methods that make it possible to predict the composition, characteristics and properties of future nanomaterials. To implement mathematical simulation of physical and chemical processes, it is necessary to have mathematical models based on those or other theoretical approaches. Computational nanotechnology is critical to prototyping nanomaterials, devices, systems, and a variety of applications. At the same time, it can be used not only to understand and characterize systems obtained as a result of experiments but also to predict the properties of new materials, since there is a strong relationship between structural, mechanical, chemical and electrical properties in the nano-scale region.

The nanoindustry is qualitatively different from traditional industries, since the traditional, macroscopic technologies for handling matter are often inapplicable, and microscopic phenomena, negligibly weak on conventional scales, become much more significant due to the properties and interaction of individual atoms and molecules or aggregates of molecules. In practical terms, these are technologies for the production of devices and their components necessary for the creation, processing and manipulation of atoms, molecules and particles, the sizes of which range from 1 to 100 nanometers.

The following main directions in nanotechnology can be distinguished. Molecular design, i.e. dissection of existing molecules and synthesis of new molecules. The field of materials science is characterized by the development of "defect-free" high-strength and lightweight materials and materials with high electrical and thermal conductivity. Equipment design involves the creation of atomic force microscopes, magnetic force microscopes, multi-jet systems for molecular design, miniature ultrasensitive sensors, nanomotors, and nanorobots. Electronics involves the design of nanometre element bases for next-generation computers: nanowires, transistors, ultra-dense memory elements, optical and quantum processors, displays, and acoustic systems. In the sphere of optics, an example is the development of nanolasers.

Diversity of medical problems desire design and creation of nanorobots for recognizing and destroying viruses. The performing local surgical operations that preserve the integrity of the skin, controlled micro containers for high-precision delivery of doses of drugs to specific places in a living organism. Online micro laboratories for analysing the state of a living organism, as well as nanostructures for the diagnosis and gentle treatment of malignant diseases tumours, and production of biological tissues for transplantation.

In traditional *in silico* analysis, the prediction of endpoints is usually based on the molecular structure of various substances. However, there are also approaches based on the analysis of the numerical values of various physicochemical parameters and their combinations. In any case, models are built to evaluate the physicochemical or biochemical properties of interest.

At first glance, the question is natural: why using such indirect models? Is not it easier to measure the property of interest in a direct experiment? However, substances often have high costs. Obtaining them during synthesis or by some other means besides cost may require considerable time, as well as high qualifications of the corresponding scientists.

Chemical quantitative structure-activity relationship (QSAR ≈ *in silico*) models, designed to predict the biological activity of chemical compounds based on their molecular descriptors, may not be suitable for nano-QSAR due to the unique physicochemical properties and behavior of nanoparticles, which require specialized models that can take into account their size, shape, surface chemistry and interaction with biological systems. Integrating artificial intelligence and computational chemistry for nano-QSAR can provide valuable predictive results in the analysis of chemical systems. *In silico* algorithms help develop models on traditional substances that can predict the activity or properties of chemical compounds based on their molecular structure. These models could help identify and understand the mechanisms of potential toxicity and optimize the properties of nanomaterials. However, for nanomaterials such models are practically unattainable owing to their complexity (Fig. 19.1).

Fullerenes

Attribute	Codes of attributes
Dark or Irradiation	0 = Dark 1 = Irradiation
Mix S9	+ = with Mix S9 − = without Mix S9
Dose (g/plate)	A = 50 B = 100 C = 200 D = 400 E = 1000

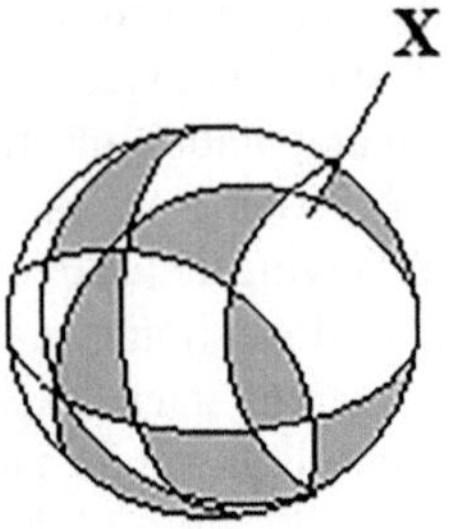

MWCNTs

Attribute	Codes of attributes
Preincubation	'y' = with preincubation 'n' = without preincubation
Mix S9	'+' = with Mix S9 '−' = without Mix S9
Dose (µg/plate)	'A' = 0 'B' = 50 'C' = 158 'D' = 500 'E' = 1581 'F' = 5000

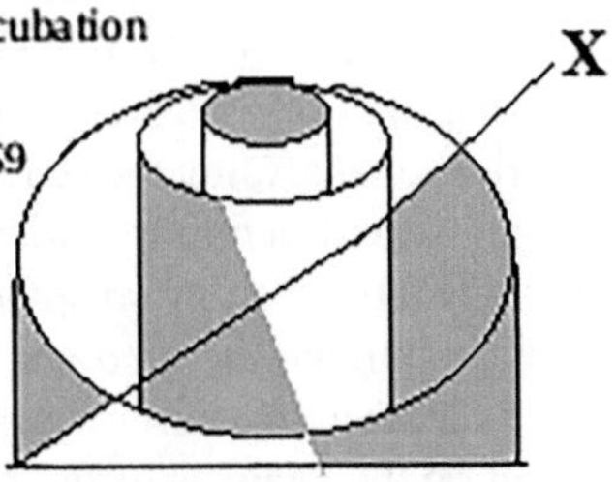

Fig. 19.1 Historically first version of the quasi-SMILES codes

With the development of nanotechnology, the problem of their conceptual understanding arises. It is important to note that without fulfilling the fundamental conceptual tasks, it is impossible to resolve the social, ethical, and psychological problems that arose with the advent of this nanoscientific perspective. To do this, it is necessary to pay attention on the way that nanotechnologies influence the human world order. The future of humanity appears as an existence that is created by man using increasingly powerful science-intensive technologies. This general, philosophical evaluation allows a person to create nanotechnology as a continuation of his being, as something that is not alienated from him. This is one of the aspects of the humanization of nanotechnology.

The only worldview today that takes into account the existence of nanotechnology is transhumanism. It is defined as "a rational and cultural movement asserting the possibility and desirability of fundamental changes in the human condition through the achievements of research, especially the use of technology, to eliminate ageing and significantly enhance the mental, physical and psychological capabilities of man".

The philosophy of technology finds its continuation in the twenty-first century due to the advent of nanotechnologies, which have sufficient capabilities for this, and a number of specific features that distinguish them from past technologies and require philosophical understanding. The changes associated with nanotechnology are greater than the changes from past technologies. Nanotechnologies penetrate into the principles of life of nature, which gives them the opportunity to change them, that is, the essence of life is transformed, and not just its form.

Nanotechnologies act as "root technology" and in this, they are likened to philosophy as the beginning of all sciences.

There are four dangers here: (1) Technocracy is often seen as the dominance of technical experts (expertocracy); (2) Technocracy is an orientation towards technology as the so-called "technological imperative": everything that can be produced is produced, to satisfy certain needs; (3) Technocracy is understood as the dominance of objective necessity until the emergence of a total "technical state", in which people still govern, but political decisions are no longer made; (4) Technocracy acts as a tendency towards an information- and system-controlled society in a more general form: towards an information system technocracy." Any technocracy arises because more power is given to technology, not to people. Thus, technocracy is gaining importance because there is ignorance.

The increasing role of nanotechnology should not be the beginning of technocracy; philosophy as a comprehensive integrative science prevents it. It is necessary to strictly distinguish between the technical and the non-technical so that a technocracy does not arise that enslaves people. Perhaps technology will be the basis of social and individual existence, but this does not mean that it will have a decisive influence on it. Nanotechnologies create the conditions for human life, and the philosophy of technology creates the conditions for the existence of nanotechnologies themselves. In the first case, these are technical conditions, in the second, non-technical conditions. Nanotechnology and philosophy condition each other only so that man becomes freer.

The Russian theoretical physicist L.D. Landau wrote that the greatest achievement of human genius is that "a person can understand things that he can no longer imagine."

The process of cognition is endless due to the infinity and diversity of matter, but in recent decades it has been noticed that the quality of our ignorance has changed. If in the recent past biologists, studying life processes, asked: "How can this be?" Today, from the heights of knowledge, they ask: "Of the many ways how this can be, which way has nature itself chosen?" Only universal laws are retained in science for a long time, and even then, over time, amendments or additions are made to them, and they appear to us in a completely different form, updated and transformed.

The history of scientific knowledge is a wave-like process, a process of replacing seemingly established, "normal" science with revolutionary breakthroughs. In recent decades, in the context of the extraordinary growth of world science, technology, technology and education, along with the concepts of "noosphere", "technosphere", and "ethosphere", the concept of "nanosphere" has been formed and filled with content.

Humanity is rapidly creating and entering an artificial technological world. In a broad sense, we have a certain opposition between the natural and the artificial. In the current century, engineering and biotechnological tasks are moving to the forefront.

Nanotechnology is the path to the formation of a new civilization with new values and ideals. It leads to the same revolution in the manipulation of matter that computers produced in the manipulation of information. The world of nanotechnology is a world of new possibilities, and artificial objects that allow us to equip our natural abilities in a new way.

In its essence, the nano approach represents the self-organization of systems, which has been the focus of so-called synergy for 40 years.

Nanotechnology is becoming a special subculture, an element of cyberculture and a sign of a change in technological paradigms, movement from the micro to the nanolevel with local and global civilizational consequences. Problems of ethics and bioethics become inevitable here. Nanotechnology has just begun to develop, but there are already serious debates between nanophiles and nanophobes. In general, nanotechnologists are faced with the most important humanistic task: to obtain maximum benefit from nanotechnology and at the same time show how to minimize and take control of the possible negative consequences of this new reality. The rule should be an axiom: the successes of science, divorced from the personal morality of the scientist and his social responsibility, are pushing humanity into the abyss.

Meanwhile, the gap between traditional moral values and the technical capabilities of modern nanotechnology is complicated by the growing moral pluralism and dynamism of society. It is not a coincidence that at the end of the last century, the "Ethics, Science and Technology" program was launched, within the framework of which the International Commission (COMEST) was formed. Its task is not only to resolve ethical issues but also to anticipate them.

For this purpose, special centres are also being created. For example, they are available at Arizona State University (ASU), at the University of California, Santa Barbara and other research universities. Their goal is to study the interaction of nanotechnologies with the social environment, and their impact on society, as well as to understand the directions of development of this area that are set by society itself. ASU's Real-Time Technology Assessment (RTTA) program will monitor the dynamics of nanotechnology research and discuss its possible future.

Since 2005, an international working group has been functioning, which focuses on studying the social consequences of the development of nanotechnology. In October 2006, the International Nanotechnology Council published a review article advocating mandatory restrictions on the dissemination of information regarding nanotechnology research to ensure safety. It should be noted that nanotechnology is widely used in economics by many countries. Nanotechnology has undeniable advantages over modern industry. Take, for example, water scarcity, which is a growing threat in several countries. However, the bulk of water resources is used not by the population at all, but by industry and agriculture.

With the advent of nanotechnology, "agro-industrial" water consumption will be significantly reduced because significantly less water is required for the production of nanoproducts. But this is a huge saving of irreplaceable natural resources and a direct reduction in environmental costs!

Nanotechnology also promises amazingly low-cost production of computers and display devices. And the production of nanostructures for electrical equipment will make it possible to use solar heat as a practically inexhaustible source of energy.

The main thing is that in many regions of the world, a production infrastructure that is destructive to nature has been formed and nanotechnology can, with minimal physical parameters, produce a large-scale industrial revolution.

Nanotechnological wonders are truly inexhaustible; a cubic millimetre can fit a supercomputer, and the cost of its production will be only a fraction of a per cent of today's costs!

1.1 What Is Quasi-SMILES?

Quasi-SMILES is an extension of ordinary SMILES (simplified molecular input-line entry system) by adding codes reflecting the conditions under which experiments were carried out, the results of which one would like to simulate (Toropova and Toropov 2023). To date, a number of works have been published on the use of quasi-SMILES for the simulation of the physicochemical and biochemical behaviour of various nanomaterials (Toropov and Toropova 2015a; Toropov et al. 2016; Toropova and Toropov 2022; Trinh et al. 2018; Ahmadi et al. 2021; Ahmadi et al. 2022; Kumar et al. 2023).

The first attempts to build and use quasi-SMILES were made in the works (Toropov and Toropova 2015b; Toropova and Toropov 2015). The quasi-SMILES developed in these works were based on attempts to add information about certain experimental conditions through one or two symbols that satisfy the following conditions. Firstly, they should not appear in traditional SMILES (Weininger 1988); and secondly, their total number should be minimal. Otherwise, these symbols had the same functions as the symbols from traditional SMILES.

These quasi-SMILES were used in optimization procedures for the so-called correlation weights. This method was developed for traditional SMILES proposed by Weininger (Weininger, 1998). Thus, adding the mentioned codes reflecting the experimental conditions almost did not require changes to the algorithm embedded in the CORAL (http:/www.insilico.eu/coral) software.

Models based on quasi-SMILES have significant heuristic differences from approaches built on various descriptors used for *in silico* models. In fact, any model can be considered as a kind of organization designed to ensure consistency between the database and the constructed forecast. However, in most of the traditional models, the values of the descriptors used to calculate the forecast of interest do not depend on the division of the available data into a training set and a validation set. In the case of quasi-SMILES-based models, the values of the descriptors change depending on the chosen distribution into the training and validation sets. Similarly, for other organizations designed to provide some random statistical balance, optimal descriptors are more like questioning participants in some kind of production or social process, while most *in silico* models are more like banking operations in ranking clients according to their financial status/capabilities. The success of the survey depends on the extent to which the selected 'questions' are interrelated with the required efficiency of the organization. For example, if it is necessary to achieve efficiency in the actions of employees who determine the state of the manufacture of certain products. In such cases, their trust in each other, their faith in the success of the enterprise, and their willingness to compromise are more useful than their physical endurance, intellectual level, or courage. However, of course, it is important to take into account the goals of the organization itself; if it is a political party, then trust in each other will turn out to be more of a hindrance than a help.

A system is self-organizing if it acquires some kind of spatial, temporal or functional structure without specific external influence. It is known that the processes of self-organization and self-assembly are important criteria in determining nanoreality, that is the existence of isolated or interacting nanoparticles. In the discussion technique of using quasi-SMILES, it can be noted that the process of forming models by determining correlation weights for various experimental conditions is also a process of self-organization. However, the analogy is related to functionality (self-organization), but surely not to physical conditions. It is clear, that the optimization has no external influences, only the current database. The influences may happen if the database is changed.

To increase the number of possible codes for quasi-SMILES, an option was proposed to use the special sign %, which in classic SMILES is a marker to designate rings when there are more than 9 of them. In quasi-SMILES, this sign is used to create a greater variety of codes. Figure 19.2 contains an example of scheme code of quasi-SMILES vs. correlation weight of the code.

In addition, squared brackets in traditional SMILES are used to identify special fragments of the molecular structure. In the construction of quasi-SMILES, one can insert in squared brackets some information on experimental conditions. For

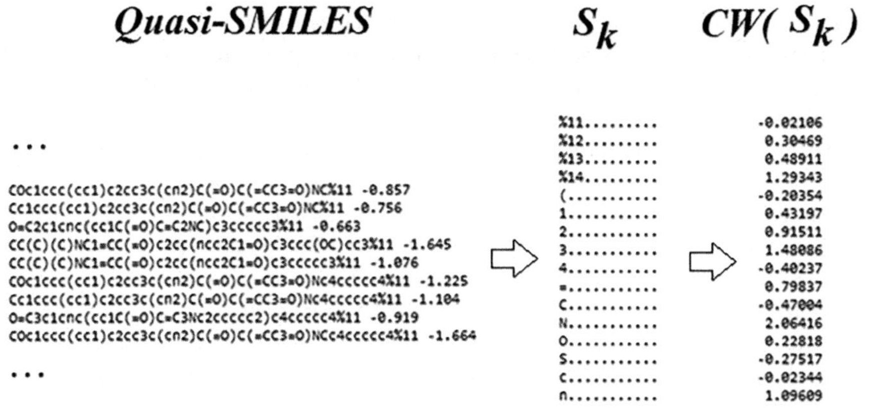

Quasi-SMILES S_k $CW(\ S_k\ )$

```
                                          %11..........   -0.02106
       ...                                %12..........    0.30469
                                          %13..........    0.48911
                                          %14..........    1.29343
COc1ccc(cc1)c2cc3c(cn2)C(=O)C(=CC3=O)NC%11  -0.857         (............   -0.20354
Cc1ccc(cc1)c2cc3c(cn2)C(=O)C(=CC3=O)NC%11   -0.756         1............    0.43197
O=C2c1cnc(cc1C(=O)C=C2NC)c3ccccc3%11        -0.663         2............    0.91511
CC(C)(C)NC1=CC(=O)c2cc(ncc2c1=O)c3ccc(OC)cc3%11  -1.645    3............    1.48086
CC(C)(C)NC1=CC(=O)c2cc(ncc2c1=O)c3ccccc3%11      -1.076    4............   -0.40237
COc1ccc(cc1)c2cc3c(cn2)C(=O)C(=CC3=O)Nc4ccccc4%11 -1.225   =............    0.79837
Cc1ccc(cc1)c2cc3c(cn2)C(=O)C(=CC3=O)Nc4ccccc4%11  -1.104   C............   -0.47004
O=C3c1cnc(cc1C(=O)C=C3NC2ccccc2)c4ccccc4%11       -0.919   N............    2.06416
COc1ccc(cc1)c2cc3c(cn2)C(=O)C(=CC3=O)NCc4ccccc4%11 -1.664  O............    0.22818
                                                           S............   -0.27517
       ...                                                 c............   -0.02344
                                                           n............    1.09609
```

Fig. 19.2 Practical use of symbol '%' to construct codes for quasi-SMILES

19 Quasi-SMILES as a Tool for Simulation of Endpoints Related to Nanomaterials

Table 19.1 Example of a comparison of two specifications of quasi-SMILES codes

	Examples of conditions	Codes using symbol %	Codes using squared brackets
Mutagenicity	TA100	%10	[TA100]
	TA98	%11	[TA98]
Coating	20-nm citrate	%12	[20cit]
	20-nm PVP	%13	[20PVP]
	50-nm citrate	%14	[50cit]
	50-nm PVP	%15	[50PVP]
	100-nm citrate	%16	[100cit]
	100-nm PVP	%17	[100PVP]
Doses (μg/plate)	0.0	%18	[d0,0]
	6.3	%19	[d6,3]
	12.5	%20	[d12,5]
	25	%21	[d25]
	50	%22	[d50]
	100	%23	[d100]

instance, it can be …[25'C][3 m/L][light]… to denote (i) temperature twenty-five Celsius; (ii) concentration three moles per litre; and (iii) fall of light on the experimental or studied object, respectively. Table 19.1 contains an example of a comparison of the two mentioned manners to define quasi-SMILES codes.

1.2 Build up a Model Using Quasi-SMILES

In the case of models of different endpoints for traditional substances represented via SMILES, the simulation is represented by equation

$$EndPoint = F(SMILES) = C_0 + C_1 \times DCW(T,N) \qquad (19.1)$$

In the case of models of different endpoints for untypical substances represented via quasi-SMILES, the simulation is represented by a quite similar equation

$$EndPoint = F(quasiSMILES) = C_0 + C_1 \times DCW(T,N) \qquad (19.2)$$

Equation 19.1 (as Eq. 19.2) is the regression model for the endpoint considered, where C_0 and C_1 are regression coefficients, and $DCW(T,N)$ is the descriptor which is equal to the sum of the so-called correlation weights of SMILES attributes. The numerical data on the correlation weights are calculated with the Monte Carlo method.

The essence of the algorithm for calculating correlation weights is to vary their values in order to obtain the maximum value of a certain target function. This may simply be a coefficient of determination or some quantity that allows an acceptable level of correlation to be achieved between the calculated and experimental values of the endpoint in consideration. The T and N are parameters of the Monte Carlo optimization. T is the integer in order to define rare SMILES attributes, which

$$\begin{pmatrix} Q_1 \\ Q_2 \\ Q_3 \\ \vdots \\ Q_m \end{pmatrix} \rightarrow \begin{pmatrix} S_{11} & S_{12} & \ldots & S_{N_1} \\ S_{21} & S_{22} & \ldots & S_{N_2} \\ S_{31} & S_{32} & \ldots & S_{N_3} \\ \vdots & \vdots & & \vdots \\ S_{m1} & S_{m2} & \ldots & S_{N_m} \end{pmatrix} \rightarrow \begin{pmatrix} CW(S_{11}) & CW(S_{12}) & \ldots & CW(S_{N_1}) \\ CW(S_{21}) & CW(S_{22}) & \ldots & CW(S_{N_2}) \\ CW(S_{31}) & CW(S_{32}) & \ldots & CW(S_{N_3}) \\ \vdots & \vdots & & \vdots \\ CW(S_{m1}) & CW(S_{m2}) & \ldots & CW(S_{N_m}) \end{pmatrix}$$

$$\begin{pmatrix} E_1 \\ E_2 \\ E_3 \\ \vdots \\ E_m \end{pmatrix} \sim \begin{pmatrix} DCW_1 = \sum_{j=1}^{N_1} CW(S_{1j}) \\ DCW_2 = \sum_{j=1}^{N_2} CW(S_{2j}) \\ DCW_3 = \sum_{j=1}^{N_3} CW(S_{3j}) \\ \vdots \\ DCW_m = \sum_{j=1}^{N_m} CW(S_{mj}) \end{pmatrix}$$

$$E_k = C_0 + C_1 \times DCW_k$$

Fig. 19.3 General scheme of building up a model using the CORAL software

should be ignored (i.e. which is not involved in the simulation process). N is the number of epochs of the optimization, i.e. the number of cycles to modify values of the correlation weights aimed to improve the target function value. The principal scheme of simulation of endpoint (E_k) for traditional SMILES and quasi-SMILES is the same. Figure 19.3 shows the scheme.

1.3 An Example of Using Quasi-SMILES Methodology

The systematic numerical data on metal oxide nanoparticle toxicity available in the literature (Gakis et al. 2023) is considered. Toxicity to human cells data (cell lines THP-1, A549, and Hatac) that are available in the specified source is considered here. Table 19.2 contains quasi-SMILES codes used to build up a model of the generalized toxicity of metal oxide nanoparticles. The format of the codes for size (nm) started with two symbols 'sz'; similarly for exposure times used the symbol 't'; for the endpoint used 'ep'. Nanoparticles and cell lines indicated by a group of symbols in squared brackets, e.g. [Al$_2$O$_3$], [Bi$_2$O$_3$], [CuO] (nanoparticles); and [A549], [THP-1], and [Hatac] (cell lines) (Table 19.3).

Table 19.2 Quasi-SMILES codes are used to simulate metal oxide nanoparticle toxicity

SAk	CW(SAk)	SAk	CW(SAk)	SAk	CW(SAk)
[A549]	−0.1065	[A549]	−0.1498	[A549]	0.1425
[Ag]	0.0	[Ag]	0.3824	[Ag]	0.0
[Al2O3]	0.0	[Al2O3]	−0.4330	[Al2O3]	−0.4770
[CeO2]	0.0	[Bi2O3]	0.0	[Bi2O3]	0.0
[CoO]	0.0	[CeO2]	0.0	[CeO2]	0.0
[Co]	0.0	[CoO]	0.0	[Cr2O3]	0.0
[Cu2O]	−0.1157	[Co]	0.2156	[CoO]	0.0
[CuO]	0.4372	[Cu2O]	0.0	[Co]	0.0
[Cu]	−0.4563	[CuO]	0.2617	[Cu2O]	−0.2276
[Fe2O3]	0.0	[Cu]	0.4066	[CuO]	0.2713
[Hacat]	−0.1295	[Fe2O3]	0.0	[Cu]	0.0421
[In2O3]	0.0	[Hacat]	0.1254	[Fe2O3]	0.0
[La2O3]	0.0	[In2O3]	0.0	[Hacat]	−0.1637
[MgO]	0.0	[La2O3]	0.0	[In2O3]	0.0
[Mn2O3]	0.0	[MgO]	0.0	[La2O3]	0.0
[MoO3]	0.0	[MoO3]	0.0	[Mn2O3]	0.0
[NiO]	−0.3673	[NiO]	−0.2797	[MoO3]	0.0
[Ni]	0.0	[Ni]	0.0	[NiO]	0.0
[V2O3]	0.0	[V2O3]	0.0	[Ni]	0.0
[THP-1]	0.1365	[THP-1]	−0.4856	[THP-1]	0.1870
[SiO2]	0.0	[Sb2O3]	0.0	[SiO2]	0.0
[SnO2]	0.4428	[SiO2]	0.0	[SnO2]	0.0
[Y2O3]	0.0	[SnO2]	0.3604	[Y2O3]	0.0
[TiO2]	0.0	[Y2O3]	0.0	[TiO2]	−0.1157
[WO3]	0.0	[TiO2]	−0.1960	[WO3]	0.0
[ZnO]	−0.4306	[WO3]	0.0	[ZnO]	−0.2626
[ZrO2]	0.0	[ZnO]	0.4058	[ZrO2]	0.0
[epEC50]	0.0	[ZrO2]	−0.3232	[epEC50]	0.0
[epIC50]	0.0291	[epEC50]	−0.3960	[epIC50]	0.3380
[epLC50]	0.0	[epIC50]	−0.1128	[epLC50]	0.3380
[epLD50]	0.0	[epLC50]	0.2189	[epLD50]	0.0
[epeC50]	0.0	[epLD50]	0.0	[epec50]	0.2042
[epec50]	0.1645	[epec50]	−0.3519	[t24h]	0.1774
[t24h]	0.0236	[t24h]	−0.3617	[t48h]	0.0
[t48h]	0.0	[t48h]	0.0	[sz103,6]	0.0
[sz103,6]	0.0	[sz103,6]	0.0	[sz100]	0.0
[sz100]	0.0	[sz100]	0.0	[sz12,2]	−0.1506
[sz12,2]	0.0	[sz12,2]	0.1837	[sz16,5]	0.0
[sz-]	0.0	[sz-]	0.0	[sz14]	0.0
[sz14]	0.0	[sz16,5]	0.0	[sz15]	0.0
[sz15]	0.0	[sz14]	0.0	[sz22,9]	−0.2820
[sz22,9]	−0.2142	[sz15]	0.0	[sz20]	0.0
[sz20]	0.0	[sz22,9]	−0.1126	[sz28,5]	0.0
[sz21]	0.0	[sz20]	0.0487	[sz29,8]	0.0
[sz29,8]	0.0	[sz21]	0.0	[sz30–50]	0.0
[sz30–70]	0.0	[sz28,5]	0.0	[sz312]	0.0
[sz312]	0.0	[sz29,8]	0.0	[sz33,4]	0.0
[sz33,4]	0.0	[sz30–50]	0.0	[sz30]	−0.2873
[sz30]	−0.1717	[sz312]	0.3168	[sz31]	0.0

(continued)

Table 19.2 (continued)

SAk	CW(SAk)	SAk	CW(SAk)	SAk	CW(SAk)
[sz31]	0.0	[sz33,4]	0.0	[sz32]	0.0
[sz32]	0.0293	[sz30]	0.0	[sz33]	0.0
[sz33]	0.3028	[sz32]	0.0	[sz39,9]	0.0
[sz39,9]	0.0	[sz33]	−0.4603	[sz40–68]	−0.2574
[sz40–68]	0.0	[sz39,9]	0.0	[sz39]	0.0
[sz42,3]	0.0	[sz40–68]	−0.2405	[sz45,4]	0.3178
[sz38]	0.0	[sz42,3]	0.0	[sz46,1]	0.0
[sz39]	0.0	[sz39]	0.0	[sz46,7]	0.0
[sz45,4]	−0.2492	[sz45,4]	−0.0996	[sz5—15]	0.4304
[sz46,1]	0.0	[sz46,7]	0.0	[sz48,9]	0.0
[sz5—15]	0.0	[sz5—15]	0.0	[sz44]	0.0
[sz48,9]	−0.0742	[sz48,9]	0.1938	[sz53,6]	0.3077
[sz44]	0.0	[sz53,6]	0.0	[sz64–69]	0.0
[sz48]	0.0	[sz64–69]	0.0	[sz60]	0.0
[sz53,6]	0.0547	[sz75,00]	0.0	[sz75,00]	0.0
[sz64–69]	0.0	[sz83–94]	0.0	[sz71]	0.0
[sz75,00]	0.0	[sz90–210]	0.0	[sz83–94]	−0.0115
[sz71]	0.0	[sz90]	0.0380	[sz90]	−0.2316
[sz83–94]	0.1883	[sz < 100]	0.0	[sz < 100]	0.0
[sz90]	−0.4181				
[sz < 100]	0.0				

Models are the following:

split-1: Endpoint = −3.79103 (± 0.02465) + 0.79135 (± 0.01809) * DCW(1,15)
split-2: Endpoint = −3.93366 (± 0.04635) + 0.73909 (± 0.02169) * DCW(1,15)
split-3: Endpoint = −3.72302 (± 0.03697) + 0.72687 (± 0.02099) * DCW(1,15)

Models are the following:

split-1: Endpoint = −5.79732 (± 0.06110) + 1.01126 (± 0.02361) * DCW(1,15).
split-2: Endpoint = −6.34159 (± 0.11559) + 0.85791 (± 0.02578) * DCW(1,15).
split-3: Endpoint = −4.17176 (± 0.04871) + 0.61211 (± 0.01794) * DCW(1,15).

The statistical quality of any model depends on the distribution of available data into training and validation. Therefore, it is quite advisable to study not one but several such distributions. Table 19.4 presents the percentages of identity across the five random splits considered here.

If i < J then the matrix element [i,j] means the percentage of identity for the validation sets (external sets). The i and j mean the numbering of 5 splits, examined.

The domain of applicability for the described model is defined via the so-called statistical defects of codes used in quasi-SMILES. These defects are calculated as:

$$d_k = \frac{|P(S_k) - P'(S_k)|}{N(S_k) + N'(S_k)} + \frac{|P(S_k) - P''(S_k)|}{N(S_k) + N''(S_k)} + \frac{|P'(S_k) - P''(S_k)|}{N'(S_k) + N''(S_k)} \quad (19.3)$$

Table 19.3 Comparison of lists of the codes for quasi-SMILES observed for different splits

Split 1		Split 2		Split 3	
SAk	CW(SAk)	SAk	CW(SAk)	SAk	CW(SAk)
[A549]	0.2237	[A549]	0.4338	[A549]	−0.1248
[Ag]	0.0	[Ag]	−0.2006	[Ag]	0.0
[Al2O3]	0.0	[Al2O3]	0.4132	[Al2O3]	0.3763
[CeO2]	0.0	[Bi2O3]	0.0	[Bi2O3]	0.0
[CoO]	0.0	[CeO2]	0.0	[CeO2]	0.0
[Co]	0.0	[CoO]	0.0	[Cr2O3]	0.0
[Cu2O]	−0.1372	[Co]	0.2041	[CoO]	0.0
[CuO]	0.3720	[Cu2O]	0.0	[Co]	0.0
[Cu]	0.4991	[CuO]	0.2509	[Cu2O]	−0.1465
[Fe2O3]	0.0	[Cu]	−0.4673	[CuO]	0.0744
[Hacat]	0.1787	[Fe2O3]	0.0	[Cu]	−0.0578
[In2O3]	0.0	[Hacat]	−0.2835	[Fe2O3]	0.0
[La2O3]	0.0	[In2O3]	0.0	[Hacat]	−0.1829
[MgO]	0.0	[La2O3]	0.0	[In2O3]	0.0
[Mn2O3]	0.0	[MgO]	0.0	[La2O3]	0.0
[MoO3]	0.0	[MoO3]	0.0	[Mn2O3]	0.0
[NiO]	−0.4635	[NiO]	−0.4025	[MoO3]	0.0
[Ni]	0.0	[Ni]	0.0	[NiO]	0.0
[V2O3]	0.0	[V2O3]	0.0	[Ni]	0.0
[THP-1]	−0.4820	[THP-1]	0.3528	[THP-1]	0.1187
[SiO2]	0.0	[Sb2O3]	0.0	[SiO2]	0.0
[SnO2]	−0.3874	[SiO2]	0.0	[SnO2]	0.0
[Y2O3]	0.0	[SnO2]	0.3082	[Y2O3]	0.0
[TiO2]	0.0	[Y2O3]	0.0	[TiO2]	0.2054
[WO3]	0.0	[TiO2]	−0.2873	[WO3]	0.0
[ZnO]	0.2205	[WO3]	0.0	[ZnO]	−0.1039
[ZrO2]	0.0	[ZnO]	−0.1006	[ZrO2]	0.0
[epEC50]	0.0	[ZrO2]	−0.2316	[epEC50]	0.0
[epIC50]	−0.0488	[epEC50]	0.1407	[epIC50]	−0.0757
[epLC50]	0.0	[epIC50]	−0.2352	[epLC50]	−0.1030
[epLD50]	0.0	[epLC50]	−0.4246	[epLD50]	0.0
[epeC50]	0.0	[epLD50]	0.0	[epec50]	0.4537
[epec50]	0.3011	[epec50]	−0.0922	[t24h]	0.4032
[t24h]	0.4862	[t24h]	−0.3786	[t48h]	0.0
[t48h]	0.0	[t48h]	0.0	[sz103,6]	0.0
[sz103,6]	0.0	[sz103,6]	0.0	[sz100]	0.0
[sz100]	0.0	[sz100]	0.0	[sz12,2]	−0.1940
[sz12,2]	0.0	[sz12,2]	0.1239	[sz16,5]	0.0
[sz-]	0.0	[sz-]	0.0	[sz14]	0.0
[sz14]	0.0	[sz16,5]	0.0	[sz15]	0.0
[sz15]	0.0	[sz14]	0.0	[sz22,9]	0.1421
[sz22,9]	0.2526	[sz15]	0.0	[sz20]	0.0
[sz20]	0.0	[sz22,9]	−0.2865	[sz28,5]	0.0
[sz21]	0.0	[sz20]	0.1824	[sz29,8]	0.0
[sz29,8]	0.0	[sz21]	0.0	[sz30–50]	0.0
[sz30–70]	0.0	[sz28,5]	0.0	[sz312]	0.0
[sz312]	0.0	[sz29,8]	0.0	[sz33,4]	0.0
[sz33,4]	0.0	[sz30–50]	0.0	[sz30]	0.0235

(continued)

Table 19.3 (continued)

Split 1		Split 2		Split 3	
SAk	CW(SAk)	SAk	CW(SAk)	SAk	CW(SAk)
[sz30]	−0.4287	[sz312]	−0.2999	[sz31]	0.0
[sz31]	0.0	[sz33,4]	0.0	[sz32]	0.0
[sz32]	−0.1572	[sz30]	0.0	[sz33]	0.0
[sz33]	−0.1188	[sz32]	0.0	[sz39,9]	0.0
[sz39,9]	0.0	[sz33]	−0.4918	[sz40–68]	0.2785
[sz40–68]	0.0	[sz39,9]	0.0	[sz39]	0.0
[sz42,3]	0.0	[sz40–68]	−0.1524	[sz45,4]	0.2225
[sz38]	0.0	[sz42,3]	0.0	[sz46,1]	0.0
[sz39]	0.0	[sz39]	0.0	[sz46,7]	0.0
[sz45,4]	0.1992	[sz45,4]	−0.2482	[sz5—15]	−0.1443
[sz46,1]	0.0	[sz46,7]	0.0	[sz48,9]	0.0
[sz5—15]	0.0	[sz5—15]	0.0	[sz44]	0.0
[sz48,9]	−0.2047	[sz48,9]	0.3389	[sz53,6]	0.4672
[sz44]	0.0	[sz53,6]	0.0	[sz64–69]	0.0
[sz48]	0.0	[sz64–69]	0.0	[sz60]	0.0
[sz53,6]	−0.0881	[sz75,00]	0.0	[sz75,00]	0.0
[sz64–69]	0.0	[sz83–94]	0.0	[sz71]	0.0
[sz75,00]	0.0	[sz90—210]	0.0	[sz83–94]	−0.3825
[sz71]	0.0	[sz90]	−0.0184	[sz90]	0.3365
[sz83–94]	−0.1772	[sz < 100]	0.0	[sz < 100]	0.0
[sz90]	0.3151				
[sz < 100]	0.0				

Table 19.4 The matrix of the percentages of identity across the five random splits is considered here

	Split 1	Split 2	Split 3	Split 4	Split 5
Split 1[a]	100	25.0	25.0	18.2	12.9
Split 2	6.3	100	25.0	30.3	19.4
Split 3	18.8	31.3	100	36.4	19.4
Split 4	48.5	36.4	36.4	100	31.3
Split 5	30.3	30.3	30.3	41.2	100

[a]If i > j then the matrix element [i,j] means the percentage of identity for the active training sets

where $P(S_k)$, $P'(S_k)$ $P''(S_k)$ are the probability of S_k in the active training, passive training, and calibration sets, respectively; $N(S_k)$, $N'(S_k)$, and $N''(S_k)$ are frequencies of S_k in the active training, passive training, and calibration sets, respectively. The statistical defects of quasi-SMILES (D_j) are calculated as:

$$D_j = \sum_{NA}^{k=1} d_k \tag{19.4}$$

where NA is the number of non-blocked codes in a quasi-SMILES.

A quasi-SMILES falls in the domain of applicability if

$$Dj < 2 * \bar{D} \tag{19.5}$$

19 Quasi-SMILES as a Tool for Simulation of Endpoints Related to Nanomaterials

Table 19.5 The statistical characteristics of the metal oxide nanoparticles toxicity model (Split 1)

CAS	1	2	3	Quasi-SMILES	Expr	Calc	Calc	Calc	AD	AD	AD
1	V	V	V	[ag][sz50][t24h][epLD50][Hacat]	−2.7320	−3.1401	−3.6400	−2.5115	Y	Y	Y
2	P	P	P	[ag][sz312][t24h][epIC50][THP-1]	−3.7470	−3.1646	−3.5279	−3.3996	Y	Y	Y
3	P	A	P	[ag][sz312][t24h][epIC50][A549]	−3.4030	−2.1617	−2.6542	−2.4395	Y	Y	Y
4	V	V	P	[ag][sz312][t24h][epIC50][THP-1]	−3.2880	−3.1646	−3.5279	−3.3996	Y	Y	Y
5	P	P	P	[ag][sz312][t24h][epIC50][THP-1]	−1.8850	−3.1646	−3.5279	−3.3996	Y	Y	Y
6	V	C	C	[ag][sz312][t24h][epIC50][A549]	−2.4530	−2.1617	−2.6542	−2.4395	Y	Y	Y
7	A	V	P	[ag][sz312][t24h][epIC50][THP-1]	−3.7700	−3.1646	−3.5279	−3.3996	Y	Y	Y
8	P	A	V	[ag][sz312][t24h][epIC50][THP-1]	−3.5930	−3.1646	−3.5279	−3.3996	Y	Y	Y
9	P	V	P	[Al2O3][sz44][t24h][epLC50][Hacat]	−1.8510	−1.6688	−2.1057	−1.8131	N	Y	Y
10	P	V	C	[Al2O3][sz31][t24h][epec50][Hacat]	−2.0090	−1.2324	−1.9083	−1.6965	Y	Y	Y
11	C	A	A	[Al2O3][sz40–68][t24h][epIC50][THP-1]	−3.0940	−2.5366	−2.5070	−2.3722	Y	N	Y
12	A	P	P	[Al2O3][sz40–68][t24h][epIC50][THP-1]	−2.0710	−2.5366	−2.5070	−2.3722	Y	N	Y
13	P	A	A	[Al2O3][sz40–68][t24h][epIC50][THP-1]	−2.1810	−2.5366	−2.5070	−2.3722	Y	N	Y
14	V	A	C	[Bi2O3][sz90][t24h][epLC50][Hacat]	−2.5040	−2.7320	−2.5078	−2.7478	Y	Y	Y
15	A	P	P	[CeO2][sz14][t24h][epec50][Hacat]	−2.2360	−2.2498	−2.4949	−2.0896	Y	N	Y
16	P	A	P	[CeO2][sz33,4][t24h][epIC50][THP-1]	−2.2110	−1.8602	−2.5852	−2.2847	Y	N	Y
17	V	P	A	[CeO2][sz33,4][t24h][epIC50][THP-1]	−2.5570	−1.8602	−2.5852	−2.2847	Y	N	Y
18	P	P	P	[co][sz20][t24h][epIC50][THP-1]	−4.6190	−2.5976	−2.8215	−3.2253	Y	Y	Y
19	V	V	C	[co][sz20][t24h][epIC50][THP-1]	−2.9280	−2.5976	−2.8215	−3.2253	Y	Y	Y
20	P	C	C	[co][sz20][t24h][epIC50][A549]	−1.7130	−1.5948	−1.9477	−2.2652	Y	Y	Y
21	P	A	A	[co][sz20][t24h][epIC50][A549]	−1.2880	−1.5948	−1.9477	−2.2652	Y	Y	Y
22	A	A	A	[co][sz20][t24h][epIC50][A549]	−1.6760	−1.5948	−1.9477	−2.2652	Y	Y	Y
23	V	C	C	[co][sz20][t24h][epIC50][A549]	−2.3880	−1.5948	−1.9477	−2.2652	Y	Y	Y

(continued)

Table 19.5 (continued)

CAS	1	2	3	Quasi-SMILES	Expr	Calc	Calc	Calc	AD	AD	AD
24	C	P	P	[co][sz20][t24h][epIC50][THP-1]	−2.5540	−2.5976	−2.8215	−3.2253	Y	Y	Y
25	C	C	P	[CoO][sz < 100][t24h][epLC50][Hacat]	−2.8340	−2.3113	−2.8333	−2.7244	N	N	Y
26	C	P	A	[CoO][sz75,00][t24h][epEC50][THP-1]	−3.0970	−2.7970	−3.5329	−3.1013	N	N	N
27	V	V	A	[Cr2O3][sz60][t24h][epLC50][Hacat]	−2.5240	−2.6714	−3.3718	−2.5128	Y	Y	Y
28	A	A	P	[cu][sz22,9][t24h][epIC50][THP-1]	−4.5890	−3.7558	−4.3295	−3.8749	Y	Y	Y
29	V	V	C	[cu][sz90][t24h][epIC50][THP-1]	−4.0830	−4.1982	−4.3986	−4.3680	Y	Y	Y
30	A	C	V	[cu][sz22,9][t24h][epIC50][THP-1]	−3.9960	−3.7558	−4.3295	−3.8749	Y	Y	Y
31	C	C	V	[cu][sz90][t24h][epIC50][THP-1]	−3.9870	−4.1982	−4.3986	−4.3680	Y	Y	Y
32	A	A	P	[cu][sz22,9][t24h][epIC50][A549]	−3.6760	−2.7529	−3.4557	−2.9148	Y	Y	Y
33	P	C	A	[cu][sz22,9][t24h][epIC50][A549]	−3.5750	−2.7529	−3.4557	−2.9148	Y	Y	Y
34	V	V	A	[cu][sz90][t24h][epIC50][A549]	−3.2780	−3.1954	−3.5249	−3.4079	Y	Y	Y
35	A	C	V	[cu][sz90][t24h][epIC50][A549]	−3.1240	−3.1954	−3.5249	−3.4079	Y	Y	Y
36	A	P	V	[cu][sz90][t24h][epIC50][A549]	−2.9310	−3.1954	−3.5249	−3.4079	Y	Y	Y
37	V	C	P	[cu][sz90][t24h][epIC50][A549]	−3.3110	−3.1954	−3.5249	−3.4079	Y	Y	Y
38	A	P	A	[cu][sz22,9][t24h][epIC50][A549]	−1.5760	−2.7529	−3.4557	−2.9148	Y	Y	Y
39	C	V	A	[cu][sz22,9][t24h][epIC50][A549]	−2.9560	−2.7529	−3.4557	−2.9148	Y	Y	Y
40	P	A	A	[cu][sz90][t24h][epIC50][THP-1]	−5.2150	−4.1982	−4.3986	−4.3680	Y	Y	Y
41	V	C	A	[cu][sz90][t24h][epIC50][THP-1]	−4.0500	−4.1982	−4.3986	−4.3680	Y	Y	Y
42	C	V	C	[cu][sz22,9][t24h][epIC50][THP-1]	−3.6870	−3.7558	−4.3295	−3.8749	Y	Y	Y
43	A	V	A	[cu][sz22,9][t24h][epIC50][THP-1]	−2.7470	−3.7558	−4.3295	−3.8749	Y	Y	Y
44	A	P	A	[Cu2O][sz83–94][t24h][epIC50][THP-1]	−4.6210	−4.1349	−5.0442	−3.7882	Y	N	Y
45	V	C	C	[Cu2O][sz83–94][t24h][epIC50][THP-1]	−4.0940	−4.1349	−5.0442	−3.7882	Y	N	Y
46	V	P	A	[Cu2O][sz83–94][t24h][epIC50][A549]	−3.5020	−3.1320	−4.1704	−2.8281	Y	N	Y

(continued)

Table 19.5 (continued)

CAS	1	2	3	Quasi-SMILES	Expr	Calc	Calc	Calc	AD	AD	AD
47	V	P	V	[Cu2O][sz83–94][t24h][epIC50][A549]	−3.4350	−3.1320	−4.1704	−2.8281	Y	N	Y
48	V	C	V	[Cu2O][sz83–94][t24h][epIC50][A549]	−3.5540	−3.1320	−4.1704	−2.8281	Y	N	Y
49	C	V	A	[Cu2O][sz83–94][t24h][epIC50][A549]	−2.9390	−3.1320	−4.1704	−2.8281	Y	N	Y
50	P	P	P	[Cu2O][sz83–94][t24h][epIC50][THP-1]	−4.6880	−4.1349	−5.0442	−3.7882	Y	N	Y
51	A	A	C	[Cu2O][sz83–94][t24h][epIC50][THP-1]	−4.2440	−4.1349	−5.0442	−3.7882	Y	N	Y
52	C	V	V	[CuO][sz48][t24h][epec50][Hacat]	−2.9010	−2.5993	−3.5489	−2.9542	Y	Y	Y
53	V	C	P	[CuO][sz28,5][t24h][epEC50][THP-1]	−3.7450	−3.8188	−4.2906	−3.1831	Y	Y	Y
54	A	V	C	[CuO][sz30][t24h][epIC50][THP-1]	−4.4230	−4.4016	−4.6541	−4.6642	Y	Y	Y
55	A	A	V	[CuO][sz45,4][t24h][epIC50][THP-1]	−4.3110	−3.9384	−3.9581	−3.6242	Y	Y	Y
56	V	C	C	[CuO][sz30][t24h][epIC50][THP-1]	−4.0370	−4.4016	−4.6541	−4.6642	Y	Y	Y
57	A	P	V	[CuO][sz30][t24h][epIC50][A549]	−3.8980	−3.3987	−3.7804	−3.7041	Y	Y	Y
58	P	A	V	[CuO][sz30][t24h][epIC50][A549]	−3.8660	−3.3987	−3.7804	−3.7041	Y	Y	Y
59	C	C	P	[CuO][sz45,4][t24h][epIC50][THP-1]	−3.4080	−3.9384	−3.9581	−3.6242	Y	Y	Y
60	P	A	A	[CuO][sz45,4][t24h][epIC50][A549]	−3.3490	−2.9355	−3.0844	−2.6641	Y	Y	Y
61	C	C	A	[CuO][sz45,4][t24h][epIC50][A549]	−3.2250	−2.9355	−3.0844	−2.6641	Y	Y	Y
62	A	P	C	[CuO][sz45,4][t24h][epIC50][A549]	−1.9800	−2.9355	−3.0844	−2.6641	Y	Y	Y
63	C	C	V	[CuO][sz45,4][t24h][epIC50][A549]	−2.8900	−2.9355	−3.0844	−2.6641	Y	Y	Y
64	P	V	V	[CuO][sz30][t24h][epIC50][A549]	−3.8420	−3.3987	−3.7804	−3.7041	Y	Y	Y
65	C	C	A	[CuO][sz30][t24h][epIC50][A549]	−3.5820	−3.3987	−3.7804	−3.7041	Y	Y	Y
66	A	C	C	[CuO][sz30][t24h][epIC50][THP-1]	−4.2030	−4.4016	−4.6541	−4.6642	Y	Y	Y
67	A	V	V	[CuO][sz45,4][t24h][epIC50][THP-1]	−3.8080	−3.9384	−3.9581	−3.6242	Y	Y	Y
68	C	C	C	[CuO][sz30][t24h][epIC50][THP-1]	−3.6480	−4.4016	−4.6541	−4.6642	Y	Y	Y
69	A	P	C	[Fe2O3][sz32][t24h][epLC50][Hacat]	−2.0330	−2.0188	−2.5666	−2.4223	N	N	Y

(continued)

Table 19.5 (continued)

CAS	1	2	3	Quasi-SMILES	Expr	Calc	Calc	Calc	AD	AD	AD
70	P	P	P	[Fe2O3][sz39][t24h][epec50][Hacat]	−2.2040	−1.8137	−2.5910	−2.1703	Y	N	Y
71	C	C	A	[In2O3][sz29,8][t24h][epLC50][Hacat]	−2.9230	−2.4780	−2.9720	−2.9352	N	N	Y
72	V	V	V	[La2O3][sz45,6][t24h][epLC50][Hacat]	−2.8670	−2.2393	−3.1610	−2.6850	Y	Y	Y
73	P	C	A	[La2O3][sz103,6][t48h][epLD50][A549]	−3.8090	−3.2976	−4.6582	−3.7964	N	N	N
74	P	A	V	[MgO][sz20][t24h][epec50][Hacat]	−1.6020	−1.3281	−1.5792	−2.4704	Y	Y	Y
75	C	V	C	[Mn2O3][sz29,8][t24h][epLC50][Hacat]	−2.6400	−2.2970	−3.2266	−2.7804	N	Y	Y
76	A	A	P	[MoO3][sz100][t24h][epec50][Hacat]	−2.1580	−2.1856	−2.1650	−2.1415	Y	N	Y
77	C	A	C	[Ni][sz64–69][t24h][epIC50][THP-1]	−2.8710	−3.1386	−2.9719	−3.6146	Y	N	Y
78	P	P	C	[Ni][sz64–69][t24h][epIC50][A549]	−1.7570	−2.1357	−2.0982	−2.6544	Y	N	Y
79	V	P	V	[Ni][sz64–69][t24h][epIC50][THP-1]	−2.6220	−3.1386	−2.9719	−3.6146	Y	N	Y
80	V	P	C	[Ni][sz64–69][t24h][epIC50][A549]	−2.0630	−2.1357	−2.0982	−2.6544	Y	N	Y
81	V	C	C	[NiO][sz20][t24h][epLC50][Hacat]	−2.4910	−0.5228	−2.0121	−2.2954	N	Y	Y
82	A	P	C	[NiO][sz48,9][t24h][epIC50][THP-1]	−3.5080	−2.9554	−3.2677	−3.4192	Y	Y	Y
83	A	A	C	[NiO][sz48,9][t24h][epIC50][A549]	−1.8770	−1.9525	−2.3940	−2.4591	Y	Y	Y
84	A	P	A	[NiO][sz48,9][t24h][epIC50][THP-1]	−1.6670	−2.9554	−3.2677	−3.4192	Y	Y	Y
85	P	A	P	[NiO][sz48,9][t24h][epIC50][THP-1]	−4.5020	−2.9554	−3.2677	−3.4192	Y	Y	Y
86	V	P	V	[Sb2O3][sz90—210][t24h][epLC50][Hacat]	−2.3100	−2.6714	−2.7530	−2.7170	Y	N	Y
87	P	A	P	[SiO2][sz15][t24h][epLC50][Hacat]	−2.1220	−1.9667	−2.1291	−2.4176	N	N	Y
88	C	V	C	[SnO2][sz46,1][t24h][epLC50][Hacat]	−2.6710	−1.9747	−3.0537	−2.4543	Y	Y	Y
89	C	P	V	[SnO2][sz21][t24h][epec50][Hacat]	−2.1790	−2.0345	−2.7140	−2.4488	Y	Y	Y
90	A	A	P	[SnO2][sz33][t24h][epIC50][THP-1]	−4.6570	−3.2457	−3.1638	−3.7071	Y	Y	Y
91	C	C	V	[SnO2][sz33][t24h][epIC50][THP-1]	−2.9380	−3.2457	−3.1638	−3.7071	Y	Y	Y

(continued)

Table 19.5 (continued)

CAS	1	2	3	Quasi-SMILES	Expr	Calc	Calc	Calc	AD	AD	AD
92	A	A	P	[SnO2][sz33][t24h][epIC50][THP-1]	−2.4290	−3.2457	−3.1638	−3.7071	Y	Y	Y
93	C	P	V	[TiO2][sz42,3][t24h][epLC50][Hacat]	−1.7600	−1.5900	−2.3093	−1.9027	N	Y	Y
94	V	A	A	[TiO2][sz30–50][t24h][epec50][Hacat]	−1.9030	−1.7357	−1.9102	−1.9000	Y	Y	Y
95	C	V	A	[TiO2][sz5—15][t24h][epec50][Hacat]	−1.9030	−1.2571	−1.9562	−1.9057	Y	Y	Y
96	C	P	A	[TiO2][sz5—15][t24h][epec50][Hacat]	−1.9030	−1.2571	−1.9562	−1.9057	Y	Y	Y
97	C	A	P	[TiO2][sz39,9][t24h][epEC50][THP-1]	−2.3980	−2.1632	−2.4287	−2.2881	N	Y	Y
98	V	C	A	[TiO2][sz12,2][t24h][epIC50][THP-1]	−2.3360	−1.8168	−2.4786	−2.4521	Y	Y	Y
99	C	C	P	[TiO2][sz12,2][t24h][epIC50][THP-1]	−2.2680	−1.8168	−2.4786	−2.4521	Y	Y	Y
100	P	A	C	[TiO2][sz12,2][t24h][epIC50][THP-1]	−1.9760	−1.8168	−2.4786	−2.4521	Y	Y	Y
101	A	A	A	[TiO2][sz12,2][t24h][epIC50][THP-1]	−2.9210	−1.8168	−2.4786	−2.4521	Y	Y	Y
102	C	V	V	[TiO2][sz12,2][t24h][epIC50][THP-1]	−2.5600	−1.8168	−2.4786	−2.4521	Y	Y	Y
103	P	V	C	[TiO2][sz12,2][t24h][epIC50][THP-1]	−1.8770	−1.8168	−2.4786	−2.4521	Y	Y	Y
104	P	P	V	[V2O3][sz-][t24h][epLC50][Hacat]	−2.2440	−1.8775	−3.0142	−2.7170	N	N	Y
105	A	V	V	[V2O3][sz-][t48h][epeC50][A549]	−3.8750	−3.8702	−5.0838	−3.6239	N	N	Y
106	C	V	V	[WO3][sz30–70][t24h][epLC50][Hacat]	−2.5630	−2.1959	−2.9465	−2.1088	N	Y	Y
107	P	C	A	[WO3][sz30][t24h][epec50][Hacat]	−2.3650	−2.0680	−2.5887	−2.3771	Y	N	Y
108	P	V	V	[Y2O3][sz38][t24h][epLC50][Hacat]	−2.2110	−2.1062	−2.9896	−2.4411	N	Y	Y
109	P	C	C	[Y2O3][sz33][t24h][epec50][Hacat]	−2.3540	−1.6711	−1.8340	−2.2576	Y	Y	Y
110	P	V	P	[ZnO][sz71][t24h][epLC50][Hacat]	−3.3170	−3.1332	−3.2855	−2.7475	Y	Y	Y
111	V	V	V	[ZnO][sz21][t24h][epec50][Hacat]	−2.9080	−2.7721	−2.9457	−2.8213	Y	Y	Y
112	C	V	V	[ZnO][sz21][t24h][epec50][Hacat]	−2.9080	−2.7721	−2.9457	−2.8213	Y	Y	Y
113	V	A	P	[ZnO][sz16,5][t24h][epEC50][THP-1]	−2.9210	−3.6559	−2.9493	−3.2664	Y	Y	Y
114	A	P	A	[ZnO][sz53,6][t24h][epIC50][THP-1]	−4.6880	−4.1407	−4.3242	−4.0685	Y	Y	Y

(continued)

Table 19.5 (continued)

CAS	1	2	3	Quasi-SMILES	Expr	Calc	Calc	Calc	AD	AD	AD
115	P	V	P	[ZnO][sz53,6][t24h][epIC50][THP-1]	−4.3010	−4.1407	−4.3242	−4.0685	Y	Y	Y
116	C	C	V	[ZnO][sz53,6][t24h][epIC50][A549]	−3.3730	−3.1378	−3.4505	−3.1083	Y	Y	Y
117	C	C	C	[ZnO][sz53,6][t24h][epIC50][A549]	−3.0880	−3.1378	−3.4505	−3.1083	Y	Y	Y
118	A	V	A	[ZnO][sz53,6][t24h][epIC50][THP-1]	−4.5900	−4.1407	−4.3242	−4.0685	Y	Y	Y
119	A	A	P	[ZnO][sz53,6][t24h][epIC50][THP-1]	−2.9990	−4.1407	−4.3242	−4.0685	Y	Y	Y
120	V	A	P	[ZrO2][sz46,7][t24h][epLC50][Hacat]	−2.0150	−2.2161	−2.0105	−1.9007	Y	Y	Y
121	V	V	V	[ZrO2][sz20–30][t24h][epec50][Hacat]	−2.0900	−1.9002	−2.6003	−2.0045	Y	Y	Y
122	P	P	C	[ZrO2][sz32][t24h][epIC50][A549]	−3.1710	−1.9508	−2.3614	−2.3090	Y	Y	Y
123	A	A	A	[ZrO2][sz32][t24h][epIC50][THP-1]	−2.3340	−2.9537	−3.2352	−3.2692	Y	Y	Y
124	V	C	C	[ZrO2][sz32][t24h][epIC50][A549]	−2.1300	−1.9508	−2.3614	−2.3090	Y	Y	Y

Table 19.5 contains the technical details on the model of toxicity for metal oxide nanoparticles observed in the case of the first random split of available data into the active and passive training sets, calibration and validation sets. In addition, Table 19.5 contains experimental and calculated metal oxide nanoparticle activity expressed in logarithmical units together with the statistical defects used to detect outliers (for split 1 outliers absent).

Figure 19.4 contains the graphical representation of the model (demonstrated in Table 19.5). As it has been shown (Toropova et al. 2024), there is a double cauterization (red and green) of correlations for both training sets. However, as shown in the figure, similar stratifications occur for the calibration and validation sets. However, for the latter, the differences in the values of the correlation coefficients in the "red" and "green" clusters are small.

A checking of the described approach for five random splits is demonstrated in Table 19.6. One can see, that all random splits confirm the good predictive potential.

1.4 Advantages of Quasi-SMILES Methodology

The construction of quasi-SMILES for groups of studied phenomena, where not only the structure but the experimental conditions are taken into account, is a fairly convenient process from the point of view of a potential user since in fact, it does not require additional information. All that is needed are descriptions of the experimental conditions and numerically expressed results of the experiment (values of the endpoint in question).

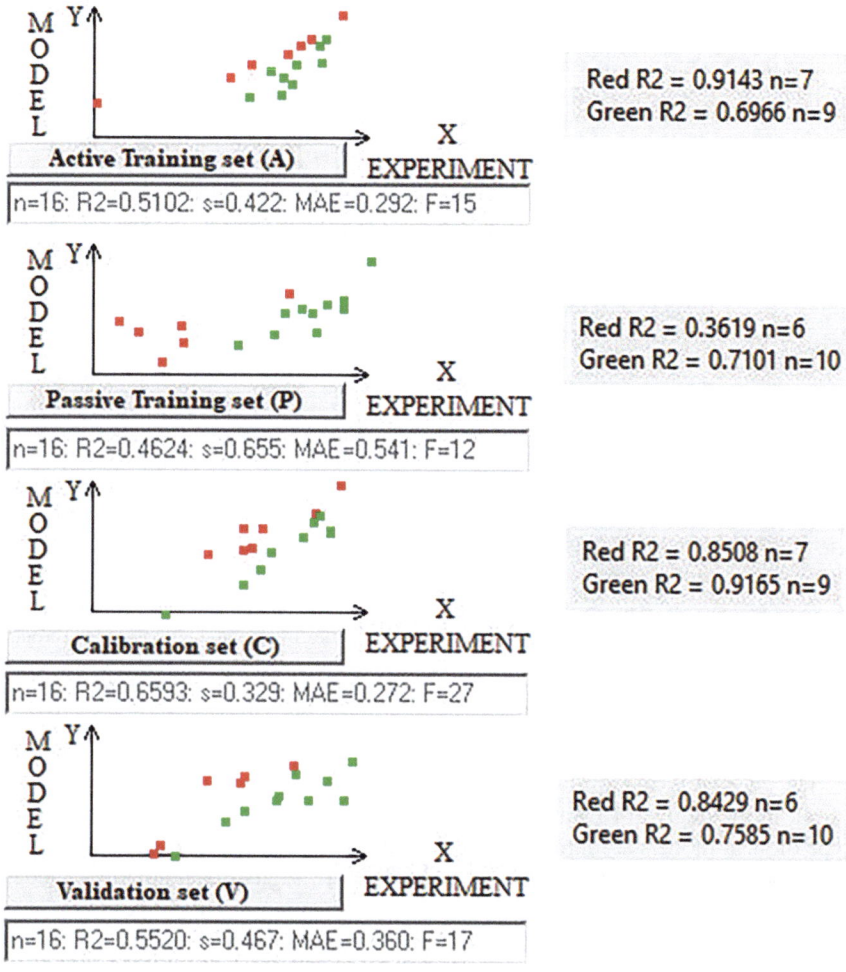

Fig. 19.4 The graphical representation of the model metal oxide nanoparticles toxicity (Split 1)

Modeling refers to the process of constructing, studying and applying models. The modeling process necessarily includes the construction of abstractions, inferences by analogy, and the construction of scientific hypotheses. Therefore, it is natural to ask the question: is modeling a special method of scientific knowledge, or is it not synonymous with the process of theoretical research or the process of cognitive activity in general.

To understand the essence of modeling, it is important not to lose sight of the fact that modeling is not the only source of knowledge about an object. The modeling process exists in the general process of cognition. This circumstance is taken into account not only at the stage of constructing the model but also at the final stage when the combination and generalization of research results obtained based on diverse means of cognition occurs.

Table 19.6 The statistical characteristics of models for five random splits

	n	R^2	IIC	CII	Q^2	CCCP	RMSE	F
A[a]	16	0.5102	0.5555	0.8044	0.2490	0.6644	0.422	15
P	16	0.4624	0.6739	0.7628	0.3572	0.2988	0.655	12
C	16	0.6593	0.8092	0.7813	0.5863	0.4633	0.329	27
V	16	0.5520	–	–	–	–	0.47	–
A	16	0.8819	0.9391	0.9313	0.8433	0.8673	0.192	104
P	14	0.7072	0.6011	0.7503	0.6206	−0.4895	0.373	29
C	18	0.5652	0.6211	0.7855	0.4613	0.3467	0.558	21
V	16	0.5215	–	–	–	–	0.45	–
A	16	0.7060	0.6535	0.8167	0.5601	0.6870	0.460	34
P	16	0.7882	0.8342	0.8453	0.7270	0.7300	0.438	52
C	16	0.5930	0.7692	0.8597	0.4999	0.5705	0.457	20
V	16	0.6396	–	–	–	–	0.36	–
A	17	0.6536	0.7186	0.7952	0.5336	0.4633	0.452	28
P	15	0.5760	0.6243	0.8186	0.3777	0.7662	0.470	18
C	15	0.7090	0.8409	0.7940	0.5840	0.2488	0.260	32
V	17	0.5547	–	–	–	–	0.45	–
A	15	0.5685	0.6598	0.6964	0.4277	−0.0049	0.515	17
P	17	0.3637	0.4574	0.6128	0.0576	−0.7391	0.623	9
C	15	0.7341	0.8528	0.8019	0.6289	0.2434	0.357	36
V	17	0.6468	–	–	–	–	0.43	–

[a] *A* active training set, *P* passive training set, *C* calibration set, *V* validation set, *n* the number of quasi-SMILES in a set, R^2 determination coefficient, *IIC* index of ideality of correlation (Toropova and Toropov 2017), *CII* correlation intensity index (Toropov and Toropova 2020), *CCCP* coefficient of conformism of a correlative prediction (Toropova and Toropov 2024), *RMSE* root mean squared error, *F* Fischer F-ratio

The class of ideal simplified models unites quite diverse models, differing, first of all, in the degree of formalization of reality. In scientific knowledge, the main type of ideal models are intuitive models that use a certain formalized language. In turn, the most important type of sign models are logical-mathematical models, which are expressed in the language of mathematics and logic. A logical-mathematical model is a certain system of mathematical relations and logical expressions (functions, equations, inequalities, algorithms, etc.) that reflect the essential properties of the object under study.

From this point of view, quasi-SMILES can be considered as a language for constructing a model. Firstly, the formalization of the model, that is the selection of an object (endpoint) and the basis for developing a forecast of how the endpoint will change if the quasi-SMILES is changed. Secondly, it is necessary to formulate a hypothesis about what experimental conditions and, in general, any eclectic data influence the value of the endpoint under consideration. In this sense, quasi-SMILES is almost a unique tool for the formation of the holistic reality of the phenomena being studied. The user can insert any observable data into quasi-SMILES, not just physical and chemical obviousness. The user can try to take into account the time of day (morning, afternoon, evening), the weather (warm, cold, rain, fog), and the mood of the experimenter. Before experience, it seems clear that eclectic data is unlikely to be useful for improving the model. However, after experience, it may turn out that completely illogical conditions lead to an improvement in the predictive potential of the model.

There are many examples of constructing visual and effectively operating subject-mathematical models for solving various physicochemical, biochemical, medical and economic problems.

To create strange, unexpected, paradoxical models, quasi-SMILES contains, firstly, the ability to include arbitrary codes for arbitrary conditions and circumstances; secondly, an option has been added to the SMILES (quasi-SMILES) survey system of the CORAL program to assess the chaos in the submitted data. However, this freedom in setting up the organization of the model, after conducting a sufficient number of computational experiments, is accompanied by the removal of non-informative components from consideration and the selection of a statistically justified set of conditions introduced into quasi-SMILES.

1.5 Disadvantages of Quasi-SMILES Methodology

However, experience shows that representative data are quite rare in practice. This leads to the fact that quasi-SMILES under construction often turn out to be duplicates since they include the same codes. This leads to the fact that statistical defects begin to prevail and the modeling system becomes unrepresentative. The choice of code lists for quasi-SMILES is subjective. Two users of the method may choose different lists of experimental conditions. This prevents standardization. In addition, it is possible to have two (or more) databases supplied with different lists of experimental conditions. This also prevents standardization. The stochastic processes underlying the construction of models for the same split of an available database on quasi-SMILES into training and validation sets yield close but not identical models characterized by different statistical quality for several runs of such processes.

1.6 Perspectives of Quasi-SMILES Methodology

A comparison of the number of publications corresponding to the keywords nano-QSAR and quasi-SMILES shows that there is some correlation between them (Fig. 19.5). There are many more of the former, but the declines and increases in these research activities almost coincide. In both cases, a basis for developing models is needed. A significant advantage of the quasi-SMILES technique is that quasi-SMILES themselves are a convenient language for both experimenters and model developers.

Unfortunately, quite often, models as well as computer programs (and many others) that look unnatural, disorganized, inadequately complex, and ugly, are most likely not useful (Borshchev et al. 1997). A necessary condition for creating internally harmonious and, ultimately, useful models is the "ability to communicate" with the model, that is, "the presence of languages for communication" with the model and the methods behind it, and their correct application. The quasi-SMILES technique may be a convenient language for the above dialogue user vs. model.

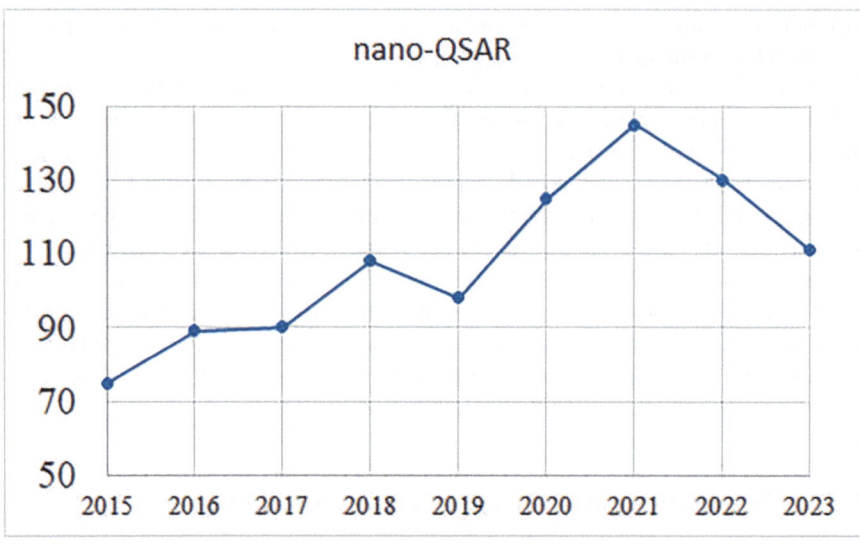

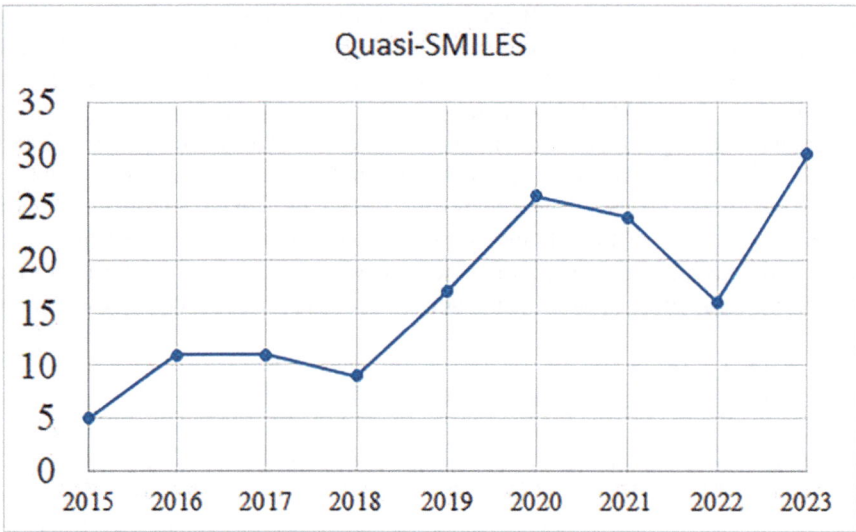

Fig. 19.5 The comparison of the numbers of publications on keywords 'Nano-QSAR' and 'quasi-SMILES'

To date, there is a collection of applications of quasi-SMILES for solving practical problems in modeling physicochemical and biochemical endpoints associated with nanomaterials (Ahmadi 2020; Ahmadi et al. 2021; Choi et al. 2019; Cheng et al. 2024; Pan et al. 2023; Ahmadi and Azimi 2023; Moinul et al. 2023; Azimi et al. 2024; Chatterjee et al. 2023; Hamidi et al. 2024; Jafari and Fatemi 2020, 2023; Kumar et al. 2023; Pang et al. 2023).

2 Conclusions

The value of a thought is its ability to provide convenience in solving a problem with minimal effort to understand how to do it. It is possible, the quasi-SMILES is a thought with value in the above sense. One could hope, the same applies to the index of ideality of correlation (*IIC*) and the correlation intensity index (*CII*), as the tool for improvement of the stochastic process aimed to simulate various endpoints related to nanomaterials.

References

Ahmadi S (2020) Mathematical modeling of cytotoxicity of metal oxide nanoparticles using the index of ideality correlation criteria. Chemosphere 242:125192. https://doi.org/10.1016/j.chemosphere.2019.125192

Ahmadi S, Azimi N (2023) Quasi-SMILES-based QSPR/QSAR modeling. In: Toropova AP, Toropov AA (eds) QSPR/QSAR analysis using SMILES and quasi-SMILES. Challenges and advances in computational chemistry and physics, vol 33. Springer, Cham, pp 191–210. https://doi.org/10.1007/978-3-031-28401-4_8

Ahmadi S, Aghabeygi S, Farahmandjou M, Azimi N (2021) The predictive model for band gap prediction of metal oxide nanoparticles based on quasi-SMILES. Struct Chem 32(5):1893–1905. https://doi.org/10.1007/s11224-021-01748-4

Ahmadi S, Ketabi S, Qomi M (2022) CO2 uptake prediction of metal-organic frameworks using quasi-SMILES and Monte Carlo optimization. New J Chem 46:8827–8837. https://doi.org/10.1039/d2nj00596d

Åkerlund E, Islam MS, McCarrick S, Alfaro-Moreno E, Karlsson HL (2019) Inflammation and (secondary) genotoxicity of Ni and NiO nanoparticles. Nanotoxicology 13(8):1060–1072. https://doi.org/10.1080/17435390.2019.1640908

Azimi A, Ahmadi S, Javan MJ, Rouhani M, Mirjafary Z (2024) QSAR models for the ozonation of diverse volatile organic compounds at different temperatures. RSC Adv 14(12):8041–8052. https://doi.org/10.1039/d3ra08805g

Borshchev, A.V., Karpov, Y.G., Roudakov, V.V., Filippov, A., Sintotskij, A., & Fedorenko, S. (1997). Analysis of a distributed election algorithm using COVERS 3.0-a case study. Lect Not Comput Sci, 1277, 175–188. DOI: https://doi.org/10.1007/3-540-63371-5_19, Springer, Berlin/Heidelberg

Chatterjee N, Alfaro-Moreno E (2023) In vitro cell transformation assays: a valuable approach for carcinogenic potentiality assessment of nanomaterials. Int J Mol Sci 24(9):8219. https://doi.org/10.3390/ijms24098219

Chatterjee M, Banerjee A, Tosi S, Carnesecchi E, Benfenati E, Roy K (2023) Machine learning – based q-RASAR modeling to predict acute contact toxicity of binary organic pesticide mixtures in honey bees. J Hazard Mater 460:132358. https://doi.org/10.1016/j.jhazmat.2023.132358

Cheng K, Pan Y, Yuan B (2024) Cytotoxicity prediction of nano metal oxides on different lung cells via Nano-QSAR. Environ Pollut 344:123405. https://doi.org/10.1016/j.envpol.2024.123405

Choi J-S, Trinh TX, Yoon T-H, Kim J, Byun H-G (2019) Quasi-QSAR for predicting the cell viability of human lung and skin cells exposed to different metal oxide nanomaterials. Chemosphere 217:243–249. https://doi.org/10.1016/j.chemosphere.2018.11.014

Costa S, Vilas-Boas V, Lebre F, Granjeiro JM, Catarino CM, Moreira Teixeira L, Loskill P, Alfaro-Moreno E, Ribeiro AR (2023) Microfluidic-based skin-on-chip systems for safety assessment of nanomaterials. Trends Biotechnol 41(10):1282–1298. https://doi.org/10.1016/j.tibtech.2023.05.009

Gakis GP, Aviziotis IG, Charitidis CA (2023) Metal and metal oxide nanoparticle toxicity: moving towards a more holistic structure-activity approach. Environ Sci Nano 10(3):761–780. https://doi.org/10.1039/d2en00897a

Hamidi E, Fatemi MH, Jafari K (2024) Thermal conductivity of carbon-based nanofluids; a theoretical modeling using nano-quantitative structure–property relationships. Chem Phys Lett 846:141344. https://doi.org/10.1016/j.cplett.2024.141344

Jafari K, Fatemi MH (2020) Application of nano-quantitative structure–property relationship paradigm to develop predictive models for thermal conductivity of metal oxide-based ethylene glycol nanofluids. J Therm Anal Calorim 142(3):1335–1344. https://doi.org/10.1007/s10973-019-09215-3

Jafari K, Fatemi MH (2023) Employing quasi-SMILES notation in development of nano-QSPR models for nanofluids. In: Toropova AP, Toropov AA (eds) QSPR/QSAR analysis using SMILES and quasi-SMILES. Challenges and advances in computational chemistry and physics, vol 33. Springer, Cham, pp 373–393. https://doi.org/10.1007/978-3-031-28401-4_15

Kumar P, Kumar A, Sindhu J, Lal S (2023) Quasi-SMILES as a basis for the development of QSPR models to predict the CO2 capture capacity of deep eutectic solvents using correlation intensity index and consensus modelling. Fuel 345:128237. https://doi.org/10.1016/j.fuel.2023.128237

Lebre F, Chatterjee N, Costa S, Fernández-De-Gortari E, Lopes C, Meneses J, Ortiz L, Ribeiro AR, Vilas-Boas V, Alfaro-Moreno E (2022) Nanosafety: an evolving concept to bring the safest possible nanomaterials to society and environment. Nano 12(11):1810. https://doi.org/10.3390/nano12111810

Meneses J, González-Durruthy M, Fernandez-de-Gortari E, Toropova AP, Toropov AA, Alfaro-Moreno E (2023) A nano-QSTR model to predict nano-cytotoxicity: an approach using human lung cells data. Part Fibre Toxicol 20(1):21. https://doi.org/10.1186/s12989-023-00530-0

Moinul M, Khatun S, Abdul Amin S, Jha T, Gayen S (2023) Quasi-SMILES as a tool for peptide QSAR modelling. In: Toropova AP, Toropov AA (eds) QSPR/QSAR analysis using SMILES and quasi-SMILES. Challenges and advances in computational chemistry and physics, vol 33. Springer, Cham, pp 269–294. https://doi.org/10.1007/978-3-031-28401-4_11

Ortiz-Galvez LM, Caballero-Guzman A, Lopes C, Alfaro-Moreno E (2024) Probabilistic material flow analysis of released nano titanium dioxide in Mexico. NanoImpact 35:100516. https://doi.org/10.1016/j.impact.2024.100516

Pan Y, Zhang X, Jiang J (2023) SMILES and quasi-SMILES descriptors in QSAR/QSPR modeling of diverse materials properties in safety and environment application. In: Toropova AP, Toropov AA (eds) QSPR/QSAR analysis using SMILES and quasi-SMILES. Challenges and advances in computational chemistry and physics, vol 33. Springer, Cham, pp 297–325. https://doi.org/10.1007/978-3-031-28401-4_12

Pang Y, Li R, Zhang Z, Ying J, Li M, Li F, Zhang T (2023) Based on the Nano-QSAR model: prediction of factors influencing damage to C. elegans caused by metal oxide nanomaterials and validation of toxic effects. Nano Today 52:101967. https://doi.org/10.1016/j.nantod.2023.101967

Toropov AA, Toropova AP (2015a) Quasi-SMILES and nano-QFAR: united model for mutagenicity of fullerene and MWCNT under different conditions. Chemosphere 139:18–22. https://doi.org/10.1016/j.chemosphere.2015.05.042

Toropov AA, Toropova AP (2015b) Quasi-QSAR for mutagenic potential of multi-walled carbon-nanotubes. Chemosphere 124(1):40–46. https://doi.org/10.1016/j.chemosphere.2014.10.067

Toropov AA, Toropova AP (2020) Correlation intensity index: building up models for mutagenicity of silver nanoparticles. Sci Total Environ 737:139720. https://doi.org/10.1016/j.scitotenv.2020.139720

Toropov AA, Achary PGR, Toropova AP (2016) Quasi-SMILES and nano-QFPR: the predictive model for zeta potentials of metal oxide nanoparticles. Chem Phys Lett 660:107–110. https://doi.org/10.1016/j.cplett.2016.08.018

Toropova AP, Toropov AA (2015) Mutagenicity: QSAR -quasi-QSAR -nano-QSAR. Mini-Rev Med Chem 15(8):608–621. https://doi.org/10.2174/1389557515666150219121652

Toropova AP, Toropov AA (2017) The index of ideality of correlation: a criterion of predictability of QSAR models for skin permeability? Sci Total Environ 586:466–472. https://doi.org/10.1016/j.scitotenv.2017.01.198

Toropova AP, Toropov AA (2022) Nanomaterials: quasi-SMILES as a flexible basis for regulation and environmental risk assessment. Sci Total Environ 823:153747. https://doi.org/10.1016/j.scitotenv.2022.153747

Toropova AP, Toropov AA (eds) (2023) QSPR/QSAR analysis using SMILES and quasi-SMILES. Challenges and advances in computational chemistry and physics, 33, 1–467. Springer, Cham. https://doi.org/10.1007/978-3-031-28401-4

Toropova AP, Toropov AA (2024) The coefficient of conformism of a correlative prediction (CCCP): building up reliable nano-QSPRs/QSARs for endpoints of nanoparticles in different experimental conditions encoded via quasi-SMILES. Sci Total Environ 927:172119. https://doi.org/10.1016/j.scitotenv.2024.172119

Toropova AP, Meneses J, Alfaro-Moreno E, Toropov AA (2024) The system of self-consistent models based on quasi-SMILES as a tool to predict the potential of nano-inhibitors of human lung carcinoma cell line A549 for different experimental conditions. Drug Chem Toxicol 47(3):306–313. https://doi.org/10.1080/01480545.2023.2174986

Trinh TX, Choi J-S, Jeon H, Byun H-G, Yoon T-H, Kim J (2018) Quasi-SMILES-based nano-quantitative structure-activity relationship model to predict the cytotoxicity of multiwalled carbon nanotubes to human lung cells. Chem Res Toxicol 31(3):183–190. https://doi.org/10.1021/acs.chemrestox.7b00303

Weininger D (1988) SMILES, a chemical language and information system: 1: introduction to methodology and encoding rules. J Chem Inf Comput Sci 28(1):31–36. https://doi.org/10.1021/ci00057a005

Open Access This chapter is licensed under the terms of the Creative Commons Attribution-NonCommercial-NoDerivatives 4.0 International License (http://creativecommons.org/licenses/by-nc-nd/4.0/), which permits any noncommercial use, sharing, distribution and reproduction in any medium or format, as long as you give appropriate credit to the original author(s) and the source, provide a link to the Creative Commons license and indicate if you modified the licensed material. You do not have permission under this license to share adapted material derived from this chapter or parts of it.

The images or other third party material in this chapter are included in the chapter's Creative Commons license, unless indicated otherwise in a credit line to the material. If material is not included in the chapter's Creative Commons license and your intended use is not permitted by statutory regulation or exceeds the permitted use, you will need to obtain permission directly from the copyright holder.

Life Cycle Assessment Towards Safe and Sustainable Biorefinery Systems of Marine Biomass—Focus on Engineered Nanoparticles

Carla Lopes, Véronique Adam, Luis Mauricio Ortiz-Galvez, Beatrice Salieri, Blanca Suarez Merino, Cyrille Durand, and Luis Taboada Antelo

Abstract

The biorefinery systems of marine biomass to recover valuable biocompounds is an emerging field in the context of efficient and sustainable management of resources. Among the distinct strategies with high potential at an industrial scale for valorisation, fish protein hydrolysates (FPH) stand out. These can be obtained from previously discarded biomass, fish processing by-products and by-catches. Such FPH contain unique bioactive peptides with a wide range of biological activities of interest in multiple areas. Therefore, evaluating the environmental impact of these valorisation processes to ensure their sustainability before industrial scale implementation is crucial. Life cycle assessment (LCA) is an appropriate tool. However, one of the most critical issues is properly accounting for the impact of certain pollutants released into the environment, such as engineered nanoparticles (ENPs). It has been stated that increased products embedded with ENPs could pose risks to biodiversity, particularly in the marine environment. The main objective of this chapter is to provide a perspective on the application of LCA in the context of emerging technologies, highlighting current limitations and the need to develop strategies to quantify the impacts linked to the release of ENPs into the marine environment, ensuring safety across the value chain of FPH.

C. Lopes (✉) · L. T. Antelo
Biosystems and Bioprocess Engineering Group (Bio2Eng), Institute of Marine Research – Spanish National Research Council (IIM-CSIC), Vigo, Spain
e-mail: clopes@iim.csic.es

V. Adam · B. Salieri · B. S. Merino · C. Durand
TEMAS Solutions (TEMASOL), Hausen, Switzerland

L. M. Ortiz-Galvez
Empa, Technology & Society Laboratory, St. Gallen, Switzerland

Keywords

Marine biomass · Biorefinery · Engineered nanoparticles · Life cycle assessment

1 Introduction

1.1 Biorefinery System and Its Importance Within SDGs

The concept of biorefinery systems has emerged during the last decade (Kokossis et al. 2014; Mountraki et al. 2016) as a sustainable processing of biomass conversion into various products that promotes a bio-based economy approach (Vlysidis et al. 2011). According to this broad definition, biorefinery can denote a concept, process, plant or even a set of facilities, where the key is the conversion of biomass into different product streams by integrating diverse processes and technologies (Moncada et al. 2015). Such integration is usually designed to maximise product outputs and income from handling a particular raw material source. Besides, other concerns should be accounted for, such as the competition with feedstock and environmental impacts, ensuring an effective sustainable biorefinery system (Venkata Mohan et al. 2016). In that context, improving the performance of the proposed bioprocesses by recommending more efficient alternative practices leading to a reduction of the environmental impacts (reuse of effluents or by-products, lower consumption of energy, water and reagents, enzymatic processes instead of chemical ones, and so on) is one of the principles of this type of systems (Vázquez et al. 2020).

On the way to a circular economy where it is intended to make the best use of the available resources and minimise waste production, the adoption of innovative sustainable production models, specifically biorefineries, has been proposed to enhance the value of marine biomass. These models aim to recover valuable biocompounds by converting low-value biomass into more useful products (Ferraro et al. 2010; Vázquez et al. 2022; Veríssimo et al. 2021). This is especially timely given that a major unsustainable fishing practice results in significant amounts of biomass being discarded in the ocean each year (Pérez Roda et al. 2019). Properly managing these resources could reduce ocean pollution, boost the competitiveness of seafood industry, and support a circular bio-economy. Biorefinery systems that process marine biomass to extract valuable components represent an emerging field focused on the efficient and sustainable management of natural resources towards the achievement of the objectives of the European Green Deal and the 2030 Agenda for the Sustainable Development Goals (SDGs) of the United Nations. However, like all emerging technologies, challenges remain in developing strategies that can be effectively scaled up for industrial use while also ensuring safety and sustainability.

1.2 Drivers for Environmental Impact Assessment of Biorefinery Systems

Marine ecosystem protection is crucial for sustaining fish stocks for present and future generations. However, unsustainable fishing practices have caused significant damage to oceans globally, threatening the long-term survival of fish species. One of the most harmful practices is the production of discards, as previously mentioned. It consists of returning unwanted catches to the sea, either dead or alive (most catches end up dying, and just a few particular species might survive), due to economic, regulatory or technical reasons (Ferraro et al. 2010; Lively and McKenzie 2023). Discards have declined in recent years due to the use of more selective gear and stricter regulation measures. However, they remain an unavoidable issue in almost all fisheries, with approximately 9.1 million tonnes of fish being discarded annually, corresponding to an average of 10.8% of the total catch (Pérez Roda et al. 2019).

The reduction of discards and unwanted by-catches has been established as one of the priority objectives within the Common Fisheries Policy (CFP) of the European Union (European Commission 2013) that aims to exploit fish stocks at sustainable levels to achieve the maximum sustainable yield. The reform also seeks to eliminate the practice of discarding fish, including the requirement to land all catches of species regulated by total allowable catches (TAC) and quotas (the so-called Landing Obligation). As a result of these measures, it is likely that the amount of biomass on land that cannot be used for direct human consumption will increase. This includes species regulated by quotas that do not meet the legal minimum size, as well as low-quality or deteriorated specimens that cannot be sold through fish auction channels. Besides discards now retained on board, the generation of secondary fish biomass (fish by-products and/or sidestreams) during the elaboration and transformation processes inland is also an issue that needs a proper management plan. It has been estimated that such by-products constitute up to 35% of the total captured fish (FAO 2022). These new sidestreams result from different types of fish pre-processing, such as eviscerating, cutting and filleting. Depending on the fish species and transformation process, the estimated percentage of by-products accounts for 15–20% muscle, 12–18% viscera, 9–15% spines, 9–12% heads and 1–3% skin (FAO 2022). Nowadays, the current fish by-products generated on land are mainly used for producing fishmeal and oil with medium-low quality. However, this process is not the most efficient or sustainable alternative, as it often results in low-value products that do not fully utilize the potential of the biomass. Additionally, the environmental impact of producing fishmeal and oil, including energy consumption and waste generation, further underscores the need for better alternatives. The proportion of by-products that cannot be used for these processes is treated as urban solid waste (Lopes et al. 2015).

So, new strategies are essential for the efficient management of marine biomass to fully utilize the potential of these by-products, ensuring a minimum environmental impact while stimulating both economic and social well-being embracing, this way, the three pillars of sustainability principles. This may include a basis for the

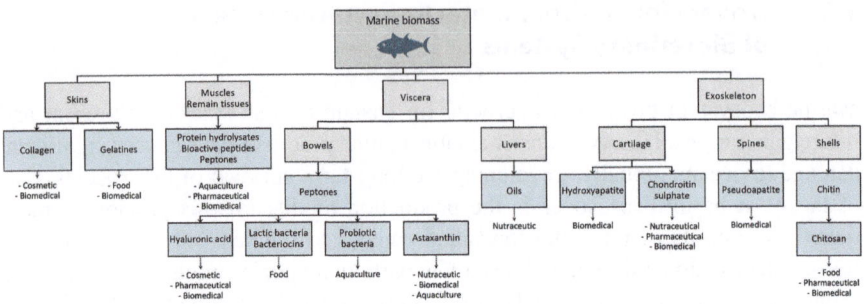

Fig. 20.1 Illustration of a potential biorefinery system of marine biomass

launch of value chains to obtain biocompounds with high added value that can generate other sources of wealth, always considering the full use of this new biomass towards zero-waste production systems/biorefineries. In this sense, devising new marine products of interest in the food, cosmetic, nutraceutical and pharmaceutical sectors has been the aim of intense research during the last years (Azelee et al. 2023; Ferraro et al. 2010; Giménez et al. 2009; Kim and Mendis 2006; Vázquez et al. 2019). There is a wide variety of biocompounds that can be obtained from the different marine biomass, such as protein hydrolysates, bioactive peptides, chitin/chitosan, collagen, and chondroitin sulphate, among others (Bae et al. 2008; Kurita 2006; Mizani et al. 2005; Vázquez et al. 2013). A prospective biorefinery strategy is represented in Fig. 20.1.

The feasibility of biorefinery systems still depends on the type, quantity, and quality of the available raw biomass, as well as the target industry they aim to serve. Therefore, during the conceptual design stage, an exhaustive knowledge of all the operational aspects (from the feedstock to the applied technology) including the limitations, is crucial for these emerging systems (Thirukumaran et al. 2022; Veríssimo et al. 2021). It is fundamental to explore and develop technologies that maximize the use of marine resources and support the sustainable growth of biorefinery systems, ensuring they can be effectively scaled to an industrial level.

Among the distinct groups of compounds with high potential at the industrial scale, fish protein hydrolysates (FPH) are particularly interesting in a wide range of applications, the health sector being the main beneficiary of such compounds. Their bioactive properties are well identified to exhibit numerous biological activities such as antioxidant, antimicrobial, anti-inflammatory, and anti-diabetic, among other benefits (Henriques et al. 2021; Ishak and Sarbon 2018; Nirmal et al. 2022; Phadke et al. 2021; Gao et al. 2021; Ortizo et al. 2023). FPH are obtained from hydrolysation of biomass, consisting of the breakage of proteins into peptides and finally into amino acids (Petrova et al. 2018). Despite the relevance of those molecules in fish, their functionality also depends on the sort of amino acids, hydrophobicity, and the location of peptides in the protein (Thirukumaran et al. 2022). The challenge to extract such bioactive compounds is directly linked to using appropriate downstream processes targeting specific components that enhance above-mentioned bioactive properties. Among the different extraction methods that can be used, enzymatic hydrolysis presents advantages over traditional procedures, such as chemical hydrolysis, due to the lower probability of undesired reactions and the lack of residual toxic chemicals, resulting in compounds with enhanced biological value (Bashiri et al. 2024). The principal disadvantages of enzymatic-based processes are the enzyme cost and enzyme stability and activity, since enzymes can be sensitive to environmental conditions such as pH, temperature and the presence of inhibitors. Therefore, maintaining optimal conditions to ensure enzyme stability and activity can be challenging and costly (Climent Barba et al. 2022). FPH can be obtained from different fish species through distinct hydrolysis conditions, for example, temperature and type and fraction of enzyme (Chalamaiah et al. 2012; Halim et al. 2016; Vázquez et al. 2017). Although enzymatic hydrolysis is a feasible alternative to recover FPH, it is also crucial to evaluate the environmental impacts of the processes, especially before an industrial scaling up.

To assess the environmental impacts of the bioprocesses, Life Cycle Assessment (LCA) proves to be a valuable tool (Bashiri et al. 2024; García-Santiago et al. 2021). It provides detailed insights into the environmental performance of biorefinery systems and helps optimize their processes (Righi 2019). LCA supports the evaluation of the sustainability of products and services, including marine biomass valorisation, from fishing activities to the final biorefinery product. Despite the positive impact of recycling fish waste in biorefinery systems on the circular economy (Kiehbadroudinezhad et al. 2023; Veríssimo et al. 2021), conducting an LCA is crucial to quantify the potential environmental impacts. In LCA, the entire life cycle of a service or product is examined. This includes identifying all the resources used and emissions generated at each stage.

Furthermore, the safe use of marine biomass is also a concern that must be addressed alongside the environmental impact evaluation. Within this aspect, tracking the potential presence of pollutants from the raw material sources/sidestreams to the final product is also fundamental. A wide sort of pollutants is released into air, water and soil during fish processing and valorisation processes. Besides direct sources of emission, pollutants might experience long-distance travel from the primary source of release (Bennett et al. 1999; Li et al. 2021). Therefore, it is foreseeable that some of those pollutants could end up in distant marine waters. Moreover, the rapid advancement of technology for developing novel products leads to the emergence of new pollutants. These emerging pollutants raise significant concerns about their environmental fate, potential toxicity, and long-term impacts on ecosystems and human health. As these substances become more prevalent, it is crucial to establish effective monitoring, regulatory measures, and mitigation strategies to manage their risks and ensure sustainable environmental practices. This is the case of engineered nanoparticles (ENPs), defined as a broad group of materials with dimensions ranging from 1 nm and 100 nm (European Commission 2022), used in a wide variety of daily-used products, such as cosmetics, sunscreens, sports clothes, and construction materials. As a result, ENPs can be found, for instance, in aquatic systems, sedimented or suspended, where aquatic organisms, such as fish, may be exposed through diet and drinking water, subsequently transferring these along the food chain (Dube and Okuthe 2023). Thus, it is imperative to assess if there is any potential risk due to the probable ENPs exposure pathway towards ensuring a safe use of the marine biomass. This concern is especially relevant for ENPs that are lipophilic, as some valorisation processes involve organic-rich phases, which could lead to the accumulation of these pollutants.

In this chapter, we focus on the particular case of ENPs in marine water that could be transferred to fish biomass and from there reach the final product, addressing the approaches to account for those releases during an LCA of biorefinery systems, where the potential impacts of these pollutants on human health and the environment are assessed using characterisation factors (CFs). CFs being specific to each substance, they are necessary for all chemicals and materials included in the LCA to enable their inclusion in the impact analysis.

Our goal is to address the potential environmental impacts and ensure that biorefinery processes are assessed comprehensively for their environmental safety, focusing on the fate and potential toxicity on marine organisms and human health.

2 Life Cycle Assessment

2.1 LCA Methodology

LCA is a standardized methodology (ISO 2006a, b) used to estimate the potential environmental impacts of a product or a service, from the raw materials extraction (cradle) to the stage at which the product is ready for use/consumption (gate) and to its end of life (grave) (Fig. 20.2). An LCA study quantifies all resources consumed and emissions generated at each stage of the life cycle of the product or service.

According to ISO 14040/44 (ISO 2006a, b) an LCA study shall include four phases (Fig. 20.3), as follows:

- Goal and scope definition: defining the purposes of the study, determining the boundaries of the system life cycle under study, and identifying important assumptions that will be made;
- Life cycle inventory (LCI) analysis: compiling a complete record of the relevant material and energy flows throughout the life cycle, in addition to any release of pollutants and other environmental aspects being studied;
- Life cycle impact assessment (LCIA): using the inventory compiled in the previous stage to calculate the environmental impacts across a limited set of understandable impact categories; and
- Interpretation: identifying the meaning of the results of the inventory and impact assessment relative to the study goal.

In a broader essence, according to the ISO 14040 (2006a) the goal and scope definition is the first phase in which the practitioner describes the intended application and reason to conduct the study, and defines the system boundaries of the study, such as temporal and geographical coverage. Basically, it is the description of the

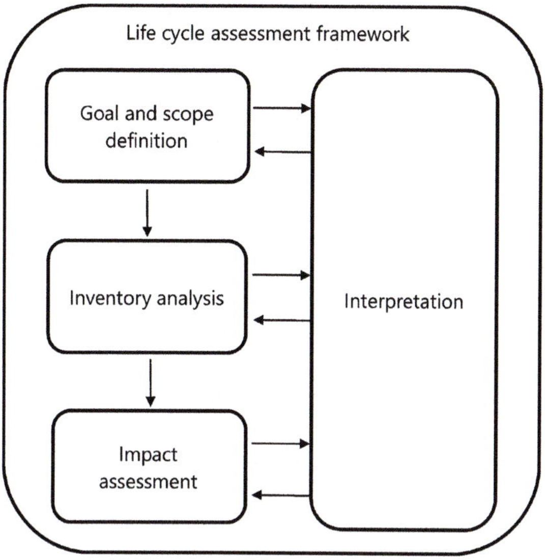

Fig. 20.2 General four iterative phase framework to perform an LCA (ISO 14040 2006a)

Fig. 20.3 Life cycle representation of a biorefinery system, from biowaste to end user, in this case from the production of FPH to a commercialized product. (**a**) Marine biomass; (**b**) Bioprocesses; (**c**) Bioactive compounds obtained from the breakage of proteins; (**d**) Pharmaceutical product. Designed in Canva, protein picture taken from the RCSB PDB (RCSB.org), part of the worldwide PDB (wwPDB.org), of PDB ID 1VZM. (Berman et al. 2003; Burley et al. 2023; Frazao et al. 2005a, b)

research question, objectives, the method/approach used, and the assumptions made. An important element in this step is the definition of the functional unit (FU), which is the reference unit related to the function(s) of the product and which should be the same for all product systems in a comparative LCA study. The base scenario(s) and alternatives (if used) are also described in detail.

Secondly, the inventory analysis is the phase in which quantitative and/or qualitative data of the corresponding product or service systems are collected, i.e., data of all the processes involved along the life cycle of the corresponding product or service according to the goal and scope. This links the system boundaries and flow charts to unit processes. It is important to mention that the emissions of substances and the extractions of natural resources over the entire product system are quantified in accordance with the system reference flows (amounts of goods or services purchased to fulfil the function and generate the FU).

Next, in the impact assessment phase, elementary flows are translated into contributions to environmental impact categories (i.e., global warming, acidification, and ecotoxicity) by means of so-called CF. In summary, it is the evaluation of the magnitude and significance of the impacts of a product system.

The final phase is the interpretation in which the results and the quality of the study (completeness, reliability, and consistency) are discussed.

LCA has been recognized as a valid methodology for emerging technologies (Bergerson et al. 2020); however, limitations and challenges have been identified in literature (Guinée 2016; Bergerson et al. 2020; van der Giesen et al. 2020) regardless of the sector of the product or service under investigation.

Among them the collection of reliable data, both in terms of their quality and quantity to generate the inventory of the study, might be challenging. Therefore, depending on the level of detail required, the process of data collection could be time-consuming (Jolliet et al. 2016). An LCA practitioner might need to make assumptions due to data unavailability, which can be caused by the novelty of the product or process, confidentiality issues or lack of control on upstream or downstream lifecycle stages such as end of life.

Even though ISO standards provide a general guidance, specific approaches can be adopted by the LCA practitioners, such as using different allocation approaches, selecting a specific database instead of another, or only considering some stages of the life cycle. Therefore, transparent communication of the results and their interpretation including disclosing the assumptions, data sources and methodological choices shall be reported, particularly for comparison purposes (Jolliet et al. 2016).

2.2 Review of LCA for Biorefinery Systems

Biomass, such as waste, can be used to obtain multiple products and for different purposes, e.g., production of biomaterials and bioenergy. Thus, the biomass or waste biorefinery principle for single and multiple waste streams/bioresources might be based on multiproduct and multi-process systems (Hosseinzadeh-Bandbafha et al. 2020). Therefore, performing an LCA would be beneficial to evaluate the feasibility according to the potential production, use, and waste management scenarios in a sustainable way.

Some researchers have already compiled review articles related to the use of LCA for waste biorefineries (Liu et al. 2021; Righi 2019; Ubando et al. 2020), e.g., using common biomass feedstock such as lignocellulosic food waste and aquatic organisms such as algae (Liu et al. 2021; Ubando et al. 2020), or biomass from marine processing by-products and by-catches (Sahu et al. 2016; Veríssimo et al. 2021; Zaky 2021).

For example, Iribarren et al. (2010) studied two alternative valorisation processes for mussel shells and organic remains in Galicia, Spain. They showed how waste management options could be assessed in the context of valorisation processes, while giving insights on the potential impacts of those alternatives. According to their LCA, the potential impacts linked to mussel meat by-product valorisation were lower than those regarding the rest of the systems. Other LCAs of marine biomass processing are listed in Table 20.1 and described below.

Amponsah et al. (2024) recently evaluated the environmental performance of a hypothetical seaweed biorefinery to produce packaging material. They used attributional LCA within six different scenarios, using different allocation strategies (i.e., system expansion, mass, and economic allocation) and comparing standard and green energy mixes. They showed that LCA can highlight potential trade-offs between environmental impact categories using multiple scenarios.

Golberg et al. (2021) performed an ex-ante LCA with three scenarios to evaluate the environmental impact of a hypothetical solar-seaweed biorefinery. They showed that although seaweed could be a potential competitor for conventional crops used, it is important to make some changes to reduce environmental impacts, such as new solutions for brine treatment and energy recovery, and transport of seaweed from the offshore cultivation location to the biorefinery.

López-Herrada et al. (2023) performed LCA using two scenarios of a biorefinery process for microalgal-based fungicide (with co-products) and another of the production of the fungicide without the biorefinery. In summary, they commented, based on their results and comparisons, that the microalgal bioprocess under study has potential benefits over conventional commercial alternatives, such as a lower carbon footprint.

Sadhukhan et al. (2019) presented an LCA of cultured macroalgae used to produce multiple products and compare which one could be the most beneficial (based on the technical-economic assessment and social LCA). They showed that, among the targeted products, producing 2,5-furandicarboxylic acid, followed by lactic acid, is a good option; however, it will depend on the energy mix used.

Barr and Landis (2018) used LCA and developed four scenarios to compare the environmental impacts of a biorefinery to produce multiple products from algae and from fish. According to their study, the environmental impact of the algae biorefinery system is comparable to the fish as an alternative to produce ω3-fatty acids. One challenge of this biorefinery is the energy-intensive filtration process (dewatering technology suggested).

Recently, Bashiri et al. (2024) performed an LCA of FPH and fish oil (as by-product) production via enzymatic hydrolysis, using economic allocation. They mentioned that electricity and fish are the main contributors to the environmental impact categories. While the hydrolysis stage contributes to water depletion and freshwater eutrophication, the preparation stage affects photochemical oxidant formation and terrestrial acidification.

Table 20.1 Examples of marine biomass LCA studies and methodological decisions

Reference	Marine waste	Use/Product	Functional unit	System boundary	Software used	Impact assessment method
López-Herrada et al. (2023)	Microalgae	Microalgal fungicide (and co-products)	1 L of fungicide	Biorefinery process; in the Mediterranean coast (in Spain)	Air.e LCA v3. 12.0.10	ReCiPe midpoint
Liu et al. (2021)	Microalgae	Biodiesel production	1 MJ of biodiesel	Valorisation process, cradle-to-tank (excluding biodiesel use and EoL); in China	GaBi 9.0	CML 2001
Amponsah et al. (2024)	Macroalgae	Co-production of fucoidan, laminarin, protein and alginate/cellulose packaging material	1 kg of packaging material	Valorisation stream; in the UK	OpenLCA (version 1.10.3)	ReCiPe (H) 2016
Golberg et al. (2021)	Macroalgae	Three scenarios to co-produce electricity, biofuels, and proteins	Basket of products derived from 1 ton of dry weight seaweed	Biorefinery process; in Israel	n.s.	ReCiPe midpoint (V1.13)
Sadhukhan et al. (2019)	Macroalgae	Protein, sugar and inorganic platforms and levulinic acid, FDCA, succinic acid and lactic acid production (from sugar)	n.s. (cultivated 1 kg dry brown macroalgae)	Valorisation process; n.s.	SimaPro 8.2.3.0	CML
Barr and Landis (2018)	Algae and fish	Production of one metric ton of ω3-fatty acids from algae (fish) with co-products of biofuel and protein	1 metric ton of ω3-fatty acids	Valorisation process; in USA	n.s.	TRACI 2.1
Bashiri et al. (2024)	Fish (by-products and side streams)	FPH, fish oil is the by-product	1 g of FPH	Valorisation process; in Norway	OpenLCA (version 1.11.0)	Recipe midpoint (H) (and comparison with other methods)
Monteiro et al. (2018)*	Fish canning wastewater	Three production processes of ω3-rich lipids	1 kg of ω3-rich lipids	Valorisation process; general assessment	n.s.	IMPACT 2002+

Note: Information from studies with an asterisk (*) was mainly collected from Caldeira et al. (2020)

n.s. not specified, FPH fish protein hydrolysate, EoL End-of-Life

2.3 Challenges for LCA of Emerging Technologies

Although there are a few studies already available, key challenges and limitations occur when performing LCAs of biorefineries (Ahlgren et al. 2015; Caldeira et al. 2020; Hosseinzadeh-Bandbafha et al. 2020; Righi 2019; Saavedra del Oso et al. 2023), as detailed below.

2.3.1 Goal and Scope Definition

The system boundaries depend on the goal and scope of the study, so they vary from study to study. Varying factors include spatial and temporal coverage (e.g., assessing one specific year rather than a long period of time) and decisions of including or excluding some processes or lifecycle stages. At early innovation stages, the use and end-of-life of products may lack specifications and be excluded from the system boundaries. To increase comparability between LCA studies, it is recommended to include all lifecycle stages and to set assumptions for those that are most uncertain, unless they are identical in the compared systems, in which case and only for comparative purposes, they can be excluded from the assessment (Thonemann et al. 2020).

Amposah et al. (2024) highlight the fact that it is difficult to define an optimal functional unit at early stages of a product development, since there lacks full characterisation of the product or material properties and of its intended functionality. The LCA practitioner then needs to define the FU based on the mass of product output, e.g., "1 kg of packaging tray material". The challenge of defining a FU is further complicated by the difficulty of identifying the alternative technologies that provide the same function(s), especially when several functions are provided by the same product (van der Giesen et al. 2020), or a new function is created (Moni et al. 2019).

The allocation of impacts among the co-products can be problematic if the biorefinery produces two or more products. In such cases, the process is usually associated with several functions (e.g., coproducing fucoidan, laminarin, protein and a composite packaging material, Amponsah et al. 2024) and it can be difficult to choose one. Then, looking at a specific process, from a certain feedstock to only one product, might be easier than having an overall assessment of multiple products of an entire biorefinery. Different allocation methodologies could be used, and the ISO guidelines (ISO, 2006b) recommend prioritizing system expansion before considering physical (e.g., mass or energy) and economic allocations.

2.3.2 Life Cycle Inventory

Inventory data might be incomplete due to the low technology readiness level and a lack of transparency and accountability within the supply chain of the specific industry. Several inventory items change as product development progresses, such as yield, energy efficiency and supply, amount of waste and emissions generated (Villares et al. 2017). A prospective assessment is then needed, based on assumptions of similar processes. It should be associated with scenario analysis, to enable the comparison of different potential processes. Various methods for scaling up

LCA from low to high technology readiness levels (so-called ex-ante LCA) exist. They are summarised by Tsoy et al. (2020), who also prioritise them according to problem formulation and data availability. These methods include process simulation, manual calculations, use of proxies, molecular structure models and stoichiometric calculations. Regarding LCA for biorefineries, de Souza et al. (2023) suggested the integration of scaling-up projections of biorefineries with the background projections of multiple socio-economic scenarios to produce bio-based polymers from forest residues. Saavedra del Oso et al. (2023) used upscaling methods similarly performed for other emerging technologies (i.e., chemical processes) to estimate utilities consumption.

Dynamic behaviour of feedstock availability and logistical issues, such as transportation routes, also impede data accuracy for prospective assessments and scenario analyses. The background system could also change, e.g., with the potential transition to renewable energy use. Uncertainty assessment could help decision making in such cases.

2.3.3 Impact Assessment

Biorefineries, as other emerging technologies, pose general challenges during the LCIA phase of an LCA. The long-term effect linked to emerging technologies could cause unknown impacts in the future, which might not be covered now due to missing suitable LCIA categories (Thonemann et al. 2020; van der Giesen et al. 2020).

The lack of CFs is also a challenge, particularly for those representative of prospective scenarios, where technical systems and variations over time could occur (van der Giesen et al. 2020). Furthermore, the impacts of new substances are neglected when associated CFs are not yet available. This is the case of ENPs, for example, where only a few CFs are available (see Sect. 3.3).

To overcome some of those challenges, the use of other methods alongside LCA could be an interesting strategy enabling an improved environmental analysis. For instance, the use of multicriteria decision analysis integrating LCA, risk assessment and economic analysis (Moni et al. 2019).

3 Calculating Characterisation Factors for ENPs

New ENPs are continuously being developed with increasing complexity and are being included in a broader range of products. Concerns have therefore been raised on their release from nano-enabled products and on their potential impact on marine ecosystems. ENPs may enter marine environments both via indirect routes, i.e., after their use and disposal (such as in discharges from wastewater treatment plants) and direct routes, e.g., emissions from ship hull paint.

In the phase of LCIA, the so-called CFs are used to quantify the environmental impacts of different activities or processes throughout their life cycle. A CF represents the relative contribution of a specific input or output parameter, such as greenhouse gas emissions, or water use, to a particular environmental impact category. The USEtox model is the consensus model to calculate CFs for human toxicity and ecotoxicity. It was developed for organic and inorganic substances (Rosenbaum et al. 2008) and in the last decades was applied by several authors to develop CFs for human toxicity and ecotoxicity (Buist et al. 2017; Deng et al. 2017; Ettrup et al. 2017; Miseljic and Olsen 2014; Salieri et al. 2015, 2019) but several adaptations are required to account for the unique characteristics of ENPs. There is an urgent need for a wider range of representative CFs that can account for different types and shapes of ENPs to support LCA practitioners in their assessments. However, in practice, studies tend to focus on only a few types of ENPs, and many groups of manufactured nanomaterials still lack proper definition and characterisation (Beloin-Saint-Pierre et al. 2018).

Following the USEtox principles, the characterisation of the ecotoxicity of a substance requires modelling its environmental fate, exposure and toxicity, which results in a fate factor (FF), an exposure factor (XF) and an effect factor (EF). Combined together, they give the final CF (Eq. 20.1, Fantke et al. 2015). A CF is specific to an impact category (e.g., human toxicity or ecotoxicity), an environmental compartment (e.g., aquatic ecotoxicity or terrestrial ecotoxicity) and a substance.

$$CF = FF \times XF \times EF \tag{20.1}$$

The specificities of fate, exposure and effect factors for ENPs and USEtox are detailed in the sections below.

3.1 Fate and Exposure Factors

3.1.1 Fate of ENPs in the Environment

ENPs undergo various types of transformation in aquatic systems that affect their location in environmental compartments (e.g., sediment vs. water column) and the way organisms are exposed to these materials. The main ENP fate processes include (i) physical processes such as agglomeration, aggregation, sedimentation and deposition (ii) chemical processes such as dissolution, speciation changes, redox reactions (oxidation and sulfidation), photochemical reactions and corona formation and (iii) biologically-mediated processes such as biodegradation and biomodification (e.g., by enzymes and microorganisms) (Lead et al. 2018; Peijnenburg et al. 2015). Surface modifications may also influence their behaviour. ENPs can be coated with biomolecules such as proteins, forming a so-called protein corona, which determines their recognition by specific cellular receptors and hence their transport in the body (Lundqvist et al. 2008; Lynch and Dawson 2008). Corona can form in environmental media, e.g., with dissolved organic matter (Peijnenburg et al. 2015).

Aggregation
Understanding ENP aggregation is important, as it affects their transport in environmental media: large and dense aggregates tend to settle more easily down the water column (Lead et al. 2018), so benthic organisms will be preferentially exposed to ENP aggregates (Lead et al. 2018; Selck et al. 2016). Aggregation also often reduces ENP bioavailability (Khan et al. 2012), as they are part of larger particles than freely

dispersed ENPs. It can however increase bioaccumulation, when it increases their accessibility (Ward and Kach 2009) or their ingestion rates (Croteau et al. 2014).

ENP aggregation can be described by the Derjaguin and Landau (1941), Verwey and Overbeek (1948) (DLVO) theory. The DLVO theory models the attractive van der Waals force and the repulsive double layer forces, counteracting each other. According to this theory, aggregation occurs due to deficiencies in stabilizing factors such as the energy barrier, which counteracts van der Waals forces (Peijnenburg et al. 2015). However, the DLVO theory fails to describe other forces at stake at very short distances (Cerbelaud et al. 2008), such as hydration forces (at distances shorter than 2 nm) (Ducker et al. 1991, 1992; Butt et al. 1995), steric repulsion forces resulting from adsorbed polymers, polyelectrolyte coating or natural organic matter (NOM) (Sander et al. 2004), hydrophobic forces (Christenson and Claesson 2001; Hoek and Agarwal 2006), bridging (Chen and Elimelech 2007), osmotic forces (Fritz et al. 2002; Phenrat et al. 2008) and magnetic forces (Phenrat et al. 2007), which are collectively known as non-DLVO forces (Peijnenburg et al. 2015).

Key factors influencing particle aggregation are pH, ionic strength, the presence of divalent cations, the type of organic matter and the ENP concentration (Handy et al. 2008; Bian et al. 2011; Baalousha et al. 2016). NOM can increase particle stability when replacing original coatings, thereby inducing electrosteric repulsion (Diegoli et al. 2008), and providing additional charge repulsion (Lead et al. 2018). It can also bridge particles together to form clusters (Corsi et al. 2021). In general, ENPs are more stable with higher dissolved organic carbon (DOC) concentrations and lower ionic strengths (Peijnenburg et al. 2015). Therefore, in seawater where ionic strength is high, ENPs are likely to undergo charge screening, aggregation and sedimentation (Klaine et al. 2008; Torre et al. 2015; Tallec et al. 2018; Bergami et al. 2016; Varó et al. 2019). However, ENPs can be transformed and remobilised to the water column by microbial and physicochemical activity (Galloway et al. 2010), and the sea-surface microlayer may also be an important sink for ENPs, due to their viscous properties (Lead et al. 2018). Gondikas et al. (2020) highlight the fact that ENP aggregation is strongly regulated by dissolved organic matter (Aiken et al. 2011), which is present in significant amounts in seawater. Indeed, ENPs may be stabilized for several days when algal blooms occur in seawater (Gondikas et al. 2020). Another limiting factor of aggregation in seawater is ENP concentration, which is expected to be low in oceans, therefore hampering high aggregation. It was indeed shown that dissolution (and not aggregation) drives the fate of silver (Ag) ENPs in seawater at environmentally relevant concentrations (Toncelli et al. 2017). Corsi et al. (2021) also showed that temperature significantly affects agglomeration rates in seawater, which are much lower in polar conditions than in temperate conditions (Bergami et al. 2016, 2017; Torre et al. 2014).

Chemical Transformation

If ENP dissolve, they are no longer nanoparticles, and their fate needs to be characterised in a conventional way. If they transform, their new chemical composition needs to be considered in the assessment. The potential transformation of their coating, if any, also needs to be considered as this drives their interactions with the organisms. It is therefore important to account for potential transformations of ENPs in aquatic ecosystems.

Transformation and dissolution depend on ENP physicochemical properties such as size, composition, surface energy and surface area, as well as on the chemistry of the surrounding media (Lead et al. 2018). The properties of the transformation products are key to understand their interactions with the environment and organisms (Lead et al. 2018) as they can result in changes in particle size, surface charge and solubility (Lead et al. 2018). Ag uptake rates increased by a factor of 2 or 10 for dissolved Ag compared to nanoparticulate Ag for several freshwater and marine species (Khan et al. 2012, 2015; Croteau et al. 2011).

Biodegradation

Biodegradation of ENPs can occur via various biological means such as enzymes, aquatic microorganisms, animals and plants (Peijnenburg et al. 2015; Lead et al. 2018). It affects both the fate and the potential effects of ENPs on organisms. For instance, it was shown that the biodegradation of poly-ethylene oxide as a capping agent induced the ENP aggregation (Kirschling et al., 2011 in Lead et al. 2018). Single-walled carbon nanotubes (SWCNTs) can be biodegraded by enzymatic catalysis, which can be interesting for regulating the fate of carbon nanotubes (CNTs) in biological systems (Allen et al., 2008; Allen et al., 2009 in Lead et al. 2018). Biodegradation also affects the toxicity of ENPs, as was shown for SWCNTs degraded by human neutrophil myeloperoxidase, which reduced their pulmonary inflammation (Kagan et al. 2010 in Lead et al. 2018).

3.1.2 Modelling ENPs Fate to Calculate Fate and Exposure Factors in LCIA

Parameters used to model the fate of traditional substances, such as partitioning coefficients, cannot describe the colloidal behaviour of ENPs in the environment. Several fate models were developed that describe the transformation and transport of ENPs in and between several environmental compartments, as required for the calculation of LCA characterisation factors. These include MendNano (Liu and Cohen 2014), SimpleBox4Nano (SB4N, Meesters et al. 2014, 2016) and nanoFATE (Garner et al. 2017). SB4N was used to adapt USEtox to ENPs, resulting in USEtox4Nano (Salieri et al. 2019), the most up-to-date strategy to calculate fate factors of ENPs for LCA. SB4N calculates ENP transport and concentrations in and between five environmental compartments (air, rain, surface water, soil and sediment) by considering nanospecific aggregation, attachment and dissolution using first-order rate constants for all processes. Model inputs are the ENP size, density, Hamaker constant, attachment efficiencies, dissolution rate constants, degradation and transformation rate constants and emission rates. It results in mass concentrations of ENPs in each environmental compartment, in time and at steady-state, in three different forms: freely dispersed particles, heteroaggreated with natural colloids (<450 nm) and attached to larger natural particles (>450 nm).

By combining SB4N with USEtox 2.0, Salieri et al. (2019) provided an advanced environmental fate model for ENPs, which they tested on titanium dioxide (TiO_2)

ENPs. The main adaptations of SB4N that enabled its combination with USEtox were 1) merging the air and rain compartments and 2) considering the sum of the three different ENP forms in concentration calculations. A dynamic analysis was also performed to estimate the time at which steady state is reached. This approach reduced the complexity of SB4N and made the USEtox model fit for ENP fate modelling.

3.2 Effect Factor

3.2.1 Ecotoxicity of ENPs

It is important to evaluate the potential toxicity of ENPs in fish since they are very sensitive to the presence of contaminants, which is also why they are used as biomarkers in toxicity assays (Gallego-Rios et al. 2021). ENPs can interact with marine organisms through various exposure routes: (i) ingestion of food or water (Kach and Ward 2008; Bergami et al. 2017), (ii) contact with gills (Abtahi et al. 2019; Bouallegui et al. 2017) and iii) body adsorption (Bergami et al. 2016; Eliso et al. 2020; Bellingeri et al. 2020). They can enter organisms either through diffusion or endocytic pathways, into the cells or across the epithelium (Geppert et al. 2021; Opršal et al. 2021; Rocha et al. 2015; Gaspar et al. 2018; Sendra et al. 2020; Noventa et al. 2018; Zande et al. 2020), and were identified in several organs such as kidney, gills, muscles, brain, gonad, hepatopancreas and liver (Dube and Okuthe 2023).

In the cells of marine organisms, ENPs were shown to induce direct catalytic mechanisms and indirect mechanisms to increase reactive oxygen species (ROS) levels (Zhang et al. 2020). ROS overproduction can result in oxidative damage and subsequent cytotoxicity and genotoxicity, up to apoptosis and cell death (Chowdhury and Saikia 2020; Lee et al. 2021; Fu et al. 2014). For example, Ag ENPs, when in direct contact with cell membranes, induce the production of ROS. This effect is reduced when the particles are coated, as the coating decreases the dissolution degree (Corsi et al. 2021). Mitigation of toxicity due to eco-coronas was demonstrated for polystyrene (Fadare et al. 2019, 2020) and metal oxide (Noventa et al. 2018) in seawater. The eco-corona had a buffering effect, limiting ENP surface reactivity and dissolution and hampering direct contact between ENP and biological membranes (Corsi et al. 2021).

High levels of cytokines, showing inflammation, were also detected in fish. They can be induced, e.g., by copper oxide (CuO) ENPs (Abdel-latif et al. 2021) or Ag ENPs (Speshock et al. 2016).

Direct interaction of ENPs with the nuclear membrane and DNA can occur, resulting in DNA damage (Gomes et al. 2013). Inflammatory responses and oxidative stress can also indirectly induce DNA damage (Dube and Okuthe 2023). Genotoxicity was reported in *Oreochromis mossambicus* upon exposure to silicon dioxide (SiO_2), aluminium oxide (Al_2O_3), TiO_2 and iron oxide (Fe_3O_4) ENPs (Vidya and Chitra 2018), in *Cyprino carpio* upon exposure to CuO ENPs (Nikdehghan et al. 2018) and *in vitro* (Vevers and Jha 2008; Reeves et al. 2008) and *in vivo* (Bobori et al. 2020; Rocco et al. 2015; Vicari et al. 2018) in fish exposed to TiO_2 ENPs.

Lipid peroxidation was induced by Ag ENPs (Xiang et al. 2020), causing damage to cell membranes. This can trigger immune responses and apoptosis (Rodriguez-Hernandez et al. 2020).

Autophagy is the process of maintaining cellular homeostasis by degrading cytoplasmic proteins and organelles and forming double-membrane vesicles (Aman et al. 2021). Its dysregulation can induce neurodegenerative diseases (Mao et al. 2016) such as suppressing fertility of zebrafish (zinc oxide (ZnO) ENPs, Mawed et al. 2022).

Behavioural and developmental effects of ENPs were reported in marine invertebrates. The feeding behaviour of the marine copepod *Paracyclopina nana*, its fecundity and growth rates were affected by polystyrene (PS) ENPs (10 µg.L^{-1}), which could have effects at population level (Jeong et al. 2017). PS ENPs were also shown to affect growth rate, fecundity, lifespan and reproduction of the rotifer *Brachionus koreanus* (Jeong et al. 2016). Shell formation was impaired and development delayed in *M. galloprovincialis* embryos by amino-modified nanopolystyrene (PS-NH$_2$) (0.001–20 µg.L^{-1}) and Ag ENPs (100 µg.L^{-1}) (Auguste et al. 2018). Major developmental effects, such as mantle, shell and hinge malformation to full developmental stop were shown in the oyster *Crassostrea gigas* exposed to PS and carboxyl-modified nanopolystyrene (PS-COOH) (0.1–25 µg.L^{-1}) (Tallec et al. 2018).

Different ENP properties are at stake when inducing toxicity. For example, surface charge of polymeric ENPs (Bergami et al. 2017; Torre et al. 2014; Manfra et al. 2017), crystal morphology of TiO$_2$ ENPs and dissolution of Ag ENPs (Kittler et al. 2010; Clément et al. 2013; Johnson et al. 2017; Dong et al. 2017) were shown to be the driving parameters for their toxicity (Corsi et al. 2021). Yang et al. (2012) highlight the fact that, although most studies reporting Ag ENP toxicity related to dissolved species focus on multicellular organisms, studies on single cell species often claim additional effects linked to the nano size.

ENPs accumulated in organisms can transfer up the trophic chain through diet. In the ocean, phytoplankton and algae are food for primary consumers such as zooplankton, crustaceans and small fish, which in turn feed secondary consumers such as larger fish. The trophic transfer of ENPs results in bioaccumulation and biomagnification (Lammel et al. 2019; Babaei et al. 2022). Ag ENPs were shown to transfer to *Dunaliella salina* (producer), *Artemia salina* (primary consumer) and *Poecilia reticulate* (secondary consumer) with bioconcentration factors (BCFs) of respectively 826, 131 and 1000 (Babaei et al. 2022). The same authors calculated biomagnification factors (BMFs) above 1 for fish (*Poecilia reticulate*) fed with shrimp (*Artemia salina*). Trophic transfers were also reported for ZnO ENPs (Skjolding et al. 2014), cobalt (Co) ENPs (Mei et al. 2021) and TiO$_2$ ENPs (Li et al. 2022). Bioaccumulation of ENPs can give rise to chronic effects even when aqueous concentrations are below toxic levels (Connolly et al. 2023). Following such pathways, ENPs can also be transferred at high concentrations in top predator species and humans (Connolly et al. 2023). In general, higher accumulation levels are observed with dietary exposure than with aqueous exposure (Zhu et al. 2010; Johnston et al. 2010).

In their review, Connolly et al. (2023) identified liver as a target organ following exposure to TiO_2, ZnO, cerium dioxide (CeO_2), Ag, selenium (Se), and cadmium sulphide (CdS) ENPs. The intestine also accumulates titanium (Ti) and zinc (Zn). Other target organs include the gills following exposure to iron dioxide (FeO_2) ENPs, the brain following TiO_2 ENPs exposure and muscle of fish exposed to Se ENPs. Size-dependent effects were also shown for multi-walled carbon nanotubes (MWCNTs) and ZnO ENPs, with smaller particles accumulating more than larger particles (Kaya et al. 2017; Cano et al. 2018; Sung et al. 2018). The presence of NOM also resulted in higher accumulation levels of graphene in *D. rerio* (Lu et al. 2017).

It is noteworthy that, although bioaccumulation and biomagnification are important parameters to consider in the impact assessment of ENPs in the value chain of FPH, they are currently not included in the USEtox model.

3.2.2 Human Toxicity Effects of ENPs

As previously mentioned, ENPs represent a wide group of materials with the only characteristic in common being that they are below 100 nm (European Commission 2022). ENPs have different sizes, coatings, charge and chemistries, leading to different toxicological profiles based on their physicochemical properties. During their life cycle, ENPs may also transform, adding more complexity to their toxicological assessment. As an example, in the field of nanomedicine, an intravenously injected ENP will transform immediately due to the attachment of serum proteins on its surface, the type of the proteins will depend on the ENP charge and will form what is called a protein corona, which may guide the biokinetics of the ENP (Monopoli et al. 2013) in the body. There are four main entry routes of ENPs, namely inhalation, oral, dermal and intravenous. Non-intravenously injected ENPs tend to distribute across the body, with those ≤ 2 nm being secreted through urine. Clearance time will vary based on the physicochemical characteristics of the ENPs, with some ENPs being able to cross biological barriers such as placenta and brain (Jia et al. 2020). In contrast, intravenously-injected ENPs are eliminated from the bloodstream more rapidly and accumulate mostly in the mononuclear-phagocytic system (MPS)-rich organs such as the liver and spleen (Kreyling et al. 2017). ENPs, due to their size, interact at cellular, subcellular and DNA levels, resulting in damaged cells or cell components, oxidative stress and physical interference with subcellular components such as lysosomes, mitochondria, proteins or DNA. Interactions with ENPs with cells can therefore, result in a wide range of downstream deleterious effects, such as cytomembrane leakage, mitochondria dysfunction, lysosome membrane permeabilization (LMP), endoplasmic reticulum (ER) stress, stimulated or blocked signalling pathways involving cell proliferation and death, cytoskeleton disruption, genotoxicity, etc., which may finally result in inflammatory response, cell cycle arrest, and cell death via different ways including apoptosis, necrosis or autophagy (Kumah et al. 2023). Since the deleterious outcomes of ENPs with cells very much depend on their physicochemical characteristics, evaluating the toxicity of ENPs becomes a challenge, and very often is based on case-by-case evaluation through *in vivo* studies. However, the large variety of ENPs being produced, together with their toxicological profiles being physicochemical dependent, translate into an unmanageable number of *in vivo* experiments (both from an ethical and economical perspective). Additionally, the current trend in the toxicology field is to move away from *in vivo* animal studies, which are based on phenomenological outcomes, to more cost-effective *in vitro* and *in silico* approaches, which are based on the mode of action of the substance under investigation.

3.3 Characterisation Factors of ENPs

Regarding human toxicity impacts, a CF is obtained by a combination of a FF (fate factor, occurrence in environmental compartments), an XF (exposure factor, representing intake) and an EF (effect factor, associated with disease incidence). Within the USETox 2.0 methodology, the EF are derived for inhalation and oral routes from the lifetime dose inducing non-cancer diseases in 50% of the population (ED50), considering 70% years of lifetime, a 70 kg body weight for ingestion and a 13 $m^3 \cdot d^{-1}$ inhalation rate. To obtain the EF for novel materials becomes challenging, because a particular population needs to be exposed so values can be obtained. In the absence of human population data, animal data may be used. However, in the case of ENPs, animal data are also scarce, and alternatives such as *in vitro* data are currently becoming an option following the international trend of reduce, replace and refine animal experimentation (3Rs) (Hubrecht and Carter 2019). *In vitro* data could be derived from human cells based on simple (monolayer) or complex (combination of different cell types mimicking organs) models, however current models mainly represent acute exposure so data from chronic exposure are still lacking. *In vitro* models are generally cheaper and allow for several types of ENPs and doses to be tested at the same time, speeding the EF calculation, as well as the identification of a benchmark dose (BMD) or ED50 values (rather than NOAEL and LOAEL values). If human cells are used, calculations will not need to include interspecies extrapolation factors, however it still becomes challenging to extrapolate from a cellular effect to a human effect. *In vitro* to *in vivo* extrapolation approaches have been developed to estimate human BMDs and subsequently *in vitro*-based EFs for inhaled, non-soluble nanoparticles (Romeo et al. 2022a, b), with various degrees of predictivity based on availability of *in vivo* data and selection of animal species. At present, however, the challenges regarding the extrapolation from an *in vitro* endpoint to an *in vivo* outcome (Salieri et al. 2020) represent a limiting factor to develop EFs from *in vitro* experiments. Nonetheless, in one such study, oxidative stress produced *in vitro* was correlated with inflammation effects in humans, with the ED50 measured in mg/million neutrophils directly corresponding to *in vivo* ED50 (Pu et al. 2017). Currently, most of the toxicity EF calculated for ENP have been produced following the USEtox model with adaptations to the nanosize (Pini et al. 2016; Tsang et al. 2017) (Table 20.2), which includes the use of surface area rather than mass, and the extrapolation from deposited dose to internal dose following Multiple Part Particle Dosimetry (MPPD) models (Miller et al. 2016). MPPD models calculate lung deposition of inhaled particles and clearance kinetics.

Table 20.2 Toxicity EF for non-cancer effects of nanomaterials (adapted from Romeo et al. 2022a, b) based on *in vitro* and *in vivo* data

References	ENM	Effect factor	Units	Exposure route	Differences from the USEtox methodology	Data source
Rodriguez-Garcia et al. (2014)	SWCNTs	5.3×10^{-2}	Cases/kg_{intake}	Inhalation	–	*In vivo*
	SWCNTs	1.1×10^{-3}	Cases/kg_{intake}	Ingestion	–	*In vivo*
	MWCNTs	1.4×10^{-2}	Cases/kg_{intake}	Inhalation	–	*In vivo*
	MWCNTs	13	Cases/kg_{intake}	Ingestion	–	*In vivo*
Buist et al. (2017)	MWCNTs	530	Cases/kg_{intake}	Inhalation	Dose descriptor calculated in mass deposited in the lungs	*In vivo*
	MWCNTs	2.5×10^{3}	Cases/kg_{intake}	Inhalation	Dose descriptor calculated in mass deposited in the lungs	*In vivo*
	Carbon black	2.9×10^{-2}	Cases/(m² g⁻¹ kg_{intake})	Inhalation	Surface area as dose metric	*In vivo*
Pini et al. (2016)	nano-TiO₂	1.72×10^{-2}	Cases/kg_{intake}	Inhalation	Indoor workplace exposure (45 years, 240 days per year)	*In vivo*
	nano-TiO₂	7.26×10^{-3}	Cases/kg_{intake}	Inhalation	–	*In vivo*
Ettrup et al. (2017)	nano-TiO₂	1.15	Cases/kg_{intake}	Inhalation	–	*In vivo*
	nano-TiO₂	2.94×10^{-2}	Cases/kg_{intake}	Ingestion	–	*In vivo*
Tsang et al. (2017)	nano-TiO₂	1.21×10^{6}	Cases/($kg_{deposited}$ g_{lung}^{-1})	Inhalation	Dose descriptor calculated in mass deposited per lung unit mass	*In vivo*
Buist et al. (2017)	nano-TiO₂	5.6×10^{-2}	Cases/(m² g⁻¹ kg_{intake})	Inhalation	Surface area as dose metric	*In vivo*
Fransman et al. (2017)	nano-TiO₂	5.6×10^{-2}	Cases/(m² g⁻¹ kg_{intake})	Inhalation	Surface area as dose metric	*In vivo*
Pu et al. (2017)	nano-Cu	5.96×10^{-1}	Cases/kg_{intake}	Ingestion	Calculated from *in vitro* experiments	*In vitro*
	nano-CuO	4.5×10^{-2}	Cases/kg_{intake}	Inhalation	Dose descriptor calculated *via* the relative potency approach	*In vitro*
	nano-CuO	7.5×10^{-3}	Cases/kg_{intake}	Ingestion	Dose descriptor calculated *via* the relative potency approach	*In vitro*

(continued)

Table 20.2 (continued)

References	ENM	Effect factor	Units	Exposure route	Differences from the USEtox methodology	Data source
Buist et al. (2017)	nano-Ag	6.5×10^{-1}	Cases/ $(m^2\,g^{-1}\,kg_{intake})$	Inhalation	Surface area as dose metric	*In vivo*
Salieri et al. (2020)	nano-Ag	1.2	Cases/ kg_{intake}	Inhalation	Dose descriptor calculated *via* the relative potency approach	*In vitro*
	nano-Ag	5.9×10^{-1}	Cases/ kg_{intake}	Ingestion	Dose descriptor calculated *via* the relative potency approach	*In vitro*
	nano-ZnO	2.9×10^{-2}	Cases/ kg_{intake}	Inhalation	Dose descriptor calculated *via* the relative potency approach	*In vitro*
	nano-ZnO	2.5×10^{-2}	Cases/ kg_{intake}	Ingestion	Dose descriptor calculated *via* the relative potency approach	*In vitro*

ENM engineered nanomaterial, *SWCNTs* single-walled carbon nanotubes, *MWCNTs* multi-walled carbon nanotubes, TiO_2 titanium dioxide, *Cu* copper, *CuO* copper oxide, *Ag* silver, *ZnO* zinc oxide

Since the amount of ENP types may be almost infinite, it becomes even a challenge to assess each individual form *in vitro*, hence strategies are being developed to group them in a way that one EF may be extrapolated to the whole group (European Commission 2022; Giusti et al. 2019) however this exercise is still challenging and under review by different regulatory agencies and international organisations such as the OECD (OECD 2018).

USEtox includes a multimedia fate model, where freshwater and seawater are present at global scale. The environmental media of sediment is included in the fate model with the transport of substances between water and sediment via adsorption/desorption and by sedimentation and resuspension of suspended particle matter. Overall, USEtox models CFs of substances for the impact of ecotoxicity in freshwater; no CF for sediment or marine environmental ecotoxicity is available.

Table 20.3 present the ecotoxicity and human toxicity CFs found in the literature for ENPs. On our knowledge there is no CF available for seawater. Notably, ENP CFs for toxicity-related impact categories are still scarce in the literature.

4 Conclusions

Numerous studies have highlighted marine biomass as a potential feedstock to recover valuable compounds, generating both ecological and economic benefits within the biorefinery concept model. This presents a prospective solution to

Table 20.3 Characterisation Factors developed for nanomaterials ecotoxicity and human toxicity using USEtox (adapted from Salieri et al. 2018)

References	Model used	ENM	Impact category	Characterisation factor
Eckelman et al. (2012)	USEtox®	CNT	Ecotoxicity	2.9×10^4 PAF m^3 day/kg$_{emitted}$ (worst scenario)
Rodriguez-Garcia et al. (2014)	USEtox®	SWCNT	Human toxicity	Non-cancer, urban air: 7.5×10^{-5} cases/kg$_{emitted}$ Non-cancer, rural air: 6.7×10^{-5} cases/kg$_{emitted}$
			Ecotoxicity	1.25×10^{-1} PAF m^3 day/kg$_{emitted}$
		MWCNT	Human toxicity	Non-cancer, urban air: 2.5×10^{-3} cases/kg$_{emitted}$. Rural air: 2.6×10^{-3} cases/kg$_{emitted}$
			Ecotoxicity	7.40×10^2 PAF m^3 day/kg$_{emitted}$
Mjesljic et al. (2014)	USEtox®	Nano-Ag	Ecotoxicity	8.6×10^3 PAF m^3 day/kg$_{emitted}$
		Nano-TiO$_2$	Ecotoxicity	2.6×10^1 PAF m^3 day/kg$_{emitted}$
Salieri et al. (2015)	USEtox®	Nano-TiO$_2$	Ecotoxicity	Free species (no aggregation): 0.28 PAF m^3 day/kg$_{emitted}$ Range of size distribution: 0.28 to 32.1 PAF m^3 day/kg$_{emitted}$
Pu et al. (2016)	USEtox®	Nano-CuO	Ecotoxicity	Seventeen freshwater CFs for nano-Cu are proposed as recommended values for subcontinental regions
Pini et al. (2016)	USEtox® and SB4N	Nano-TiO$_2$	Human toxicity	Cancer, inhalation, indoor: 1.43×10^{-2} cases/kg$_{emitted}$ Cancer, inhalation, outdoor: 1.34×10^{-4} cases/kg$_{emitted}$ Non-cancer, inhalation, indoor: 5.85×10^{-7} cases/kg$_{emitted}$ Non-cancer, inhalation, outdoor: 5.5×10^{-9} cases/kg$_{emitted}$
Deng et al. (2017)	USEtox®	Graphene oxide	Ecotoxicity	7.7×10^3 PAF day m^3/kg$_{emitted}$
Ettrup et al. (2017)	USEtox®	Nano-TiO$_2$	Human toxicity	Cancer, inhalation, emission to air: 1.90×10^{-6} cases/kg$_{emitted}$ Cancer, inhalation, emission to soil: 0 cases/kg$_{emitted}$ Cancer, inhalation, emission to water: 0 cases/kg$_{emitted}$ Non-cancer, inhalation, emission to air: 1.70×10^{-5} cases/kg$_{emitted}$ Non-cancer, inhalation, emission to soil: 1.42×10^{-8} cases/kg$_{emitted}$ Non-cancer, inhalation, emission to water: 1.25×10^{-6} cases/kg$_{emitted}$
			Ecotoxicity	Emission to air: 6.05×10^2 PAF m^3/kg$_{emitted}$ Emission to soil: 1.19 PAF m^3 d/kg$_{emitted}$ Emission to water: 1.55×10^3 PAF m^3 d/kg$_{emitted}$

(continued)

Table 20.3 (continued)

References	Model used	ENM	Impact category	Characterisation factor
Buist et al. (2017)	USEtox®	Nano-Ag; nano-TiO$_2$; carbon black, MWCNT	Human toxicity	Human effect factor values were calculated for the four ENPs under investigation
Salieri et al. (2019)	USEtox® and SB4N	Nano-TiO$_2$	Ecotoxicity	3.374 PAF m^{-3} kg^{-1}

PAF potentially affected species, *ENM* engineered nanomaterial, *SB4N* SimpleBox4Nano, *CNT* carbon nanotube, *SWCNT* single-walled carbon nanotube, *MWCNT* multi-walled carbon nanotube, *Ag* silver, *TiO$_2$* titanium dioxide, *CuO* copper oxide

mitigate one of the biggest ecological problems worldwide associated with fishing practices: the generation of discards. However, it is equally important to assess the environmental impacts to ensure the sustainability of such systems, and LCA is an appropriate tool for this purpose. Additionally, safety considerations must be addressed, particularly regarding the potential transfer of pollutants from marine biomass to final products, due to the presence of ENPs in marine environments. Research has shown that ENPs can accumulate in marine organisms and transfer across the trophic chain through diet, which has been recognised as a major contributor to the higher accumulation levels of ENPs compared to the aqueous exposure pathway, raising concerns about their potential impact on human health. Consequently, integrating marine biomass into biorefinery systems may introduce specific challenges, especially since some valorisation processes involve organic-rich phases that could lead to the accumulation of lipophilic ENPs. Therefore, a thorough evaluation and management of these risks are essential for the safe and sustainable implementation of biorefinery systems.

Although LCA is an important tool for assessing environmental impacts and identifying the main hotspots along the value chain of biorefinery systems or other emerging technologies, challenges still remain. One of these challenges relates to the inclusion of ENPs in the system analysis, mainly due to the lack of representative CFs that adequately account for the ecotoxicity and human toxicity of a wide range of ENPs. Moreover, to the best of our knowledge, there is no specific CF for the seawater compartment. Therefore, to fully account for the impact of ENPs and assess their potential risks, further efforts are required to develop effective CFs that enable the inclusion of ENPs in LCA, as well as the proper definition and characterisation of various groups of manufactured nanomaterials, which are currently lacking, despite being released into the environment and potentially posing long-term effects on human health.

Acknowledgements Carla Lopes would like to acknowledge the funding from the European Union's Horizon 2020 research and innovation program under the Marie Skłodowska-Curie grant agreement No. 101032922 (SUSTIMAR). Véronique Adam, Luis Mauricio Ortiz-Galvez, Beatrice Salieri, Blanca Suarez Merino and Cyrille Durand would also like to thank the MACRAMÉ project receives funding from the European Union's Horizon Europe research and innovation programme (Grant agreement No. 101092686) and from the State Secretariat for Education, Research and Innovation (SERI) No. 23.00141.

References

Abdel-latif HMR, Dawood MAO, Mahmoud SF, Shukry M, Noreldin AE, Ghetas HA, Khallaf MA (2021) Copper oxide nanoparticles Alter serum biochemical indices, induce histopathological alterations, and modulate transcription of cytokines, Hsp70, and oxidative stress genes in Oreochromis Niloticus. Animals 11:1–21. https://doi.org/10.3390/ani11030652

Abtahi SMH, Trevisan R, di Giulio R, Murphy CJ, Saleh NB, Vikesland PJ (2019) Implications of aspect ratio on the uptake and Nanotoxicity of Gold nanomaterials. NanoImpact 14:100153. https://doi.org/10.1016/j.impact.2019.100153

Ahlgren S, Björklund A, Ekman A, Karlsson H, Berlin J, Börjesson P, Ekvall T, Finnveden G, Janssen M, Strid I (2015) Review of methodological choices in LCA of biorefinery systems – key issues and recommendations. Biofuels Bioprod Biorefin 9(5):606–619. https://doi.org/10.1002/bbb.1563

Aiken GR, Hsu-Kim H, Ryan JN (2011) Influence of dissolved organic matter on the environmental fate of metals, nanoparticles, and colloids. Environ Sci Technol 45:3196–3201. https://doi.org/10.1021/es103992s

Aman Y, Schmauck-Medina T, Hansen M, Morimoto RI, Simon AK, Bjedov I, Palikaras K et al (2021) Autophagy in healthy aging and diseased. Nature Aging 1:634–650. https://doi.org/10.1038/s43587-021-00098-4

Amponsah L, Chuck C, Parsons S (2024) Life cycle assessment of a marine biorefinery producing protein, bioactives and polymeric packaging material. Int J Life Cycle Assess 29(2):174–191. https://doi.org/10.1007/s11367-023-02239-w

Auguste M, Ciacci C, Balbi T, Brunelli A, Caratto V, Marcomini A, Cuppini R, Canesi L (2018) Effects of Nanosilver on Mytilus Galloprovincialis Hemocytes and early embryo development. Aquat Toxicol 203:107–116. https://doi.org/10.1016/j.aquatox.2018.08.005

Azelee NIW, Noor NM, Rasid ZIA, Suhaimi SH, Salamun N, Jasman SM, Manas NHA, Hasham@ Hisam R (2023) Marine waste for nutraceutical and cosmeceutical production. Valorizat Wastes Sustain Develop Waste Wealth 241–272. https://doi.org/10.1016/B978-0-323-95417-4.00010-X

Baalousha M, Cornelis G, Kuhlbusch T, Lynch I, Nickel C, Peijnenburg W, van den Brink N (2016) Modeling nanomaterial fate and uptake in the environment: current knowledge and future trends. Environ Sci Nano 3:323–345. https://doi.org/10.1039/C5EN00207A

Babaei M, Behzadi M, Seong M, Je I, Ali S (2022) Trophic transfer and toxicity of silver nanoparticles along a phytoplankton-zooplankton-fish food chain. Sci Total Environ 842:156807. https://doi.org/10.1016/j.scitotenv.2022.156807

Bae I, Osatomi K, Yoshida A, Osako K, Yamaguchi A, Hara K (2008) Biochemical properties of acid-soluble collagens extracted from the skins of underutilised fishes. Food Chem 108(1):49–54. https://doi.org/10.1016/j.foodchem.2007.10.039

Barr WJ, Landis AE (2018) Comparative life cycle assessment of a commercial algal multi-product biorefinery and wild caught fishery for small pelagic fish. Int J Life Cycle Assess 23(5):1141–1150. https://doi.org/10.1007/s11367-017-1395-7

Bashiri B, Cropotova J, Kvangarsnes K, Gavrilova O, Vilu R (2024) Environmental and economic life cycle assessment of enzymatic hydrolysis-based fish protein and oil extraction. Resources 13(5):61. https://doi.org/10.3390/resources13050061

Bellingeri A, Casabianca S, Capellacci S, Faleri C, Paccagnini E, Lupetti P, Koelmans AA, Penna A, Corsi I (2020) Impact of polystyrene nanoparticles on marine diatom *Skeletonema Marinoi* chain assemblages and consequences on their ecological role in marine ecosystems. Environ Pollut 262:114268. https://doi.org/10.1016/j.envpol.2020.114268

Beloin-Saint-Pierre D, Turner DA, Salieri B, Haarman A, Hischier R (2018) How suitable is LCA for nanotechnology assessment? Overview of current methodological pitfalls and potential solutions: 65th LCA discussion forum, Swiss Federal Institute of Technology, Zürich, May 24, 2017. Int J Life Cycle Assess 23(1):191–196. https://doi.org/10.1007/s11367-017-1399-3

Bennett DH, Kastenberg WE, Mckone TE (1999) General formulation of characteristic time for persistent chemicals in a multimedia environment. Environ Sci Technol 33(3):503–509. https://doi.org/10.1021/es980556a

Bergami E, Bocci E, Vannuccini ML, Monopoli M, Salvati A, Dawson KA, Corsi I (2016) Nano-sized polystyrene affects feeding, behavior and physiology of brine shrimp *Artemia franciscana* larvae. Ecotoxicol Environ Saf 123:18–25. https://doi.org/10.1016/j.ecoenv.2015.09.021

Bergami E, Pugnalini S, Vannuccini ML, Manfra L, Faleri C, Savorelli F, Dawson KA, Corsi I (2017) Long-term toxicity of surface-charged polystyrene nanoplastics to marine planktonic species *Dunaliella tertiolecta* and *Artemia franciscana*. Aquat Toxicol 189:159–169. https://doi.org/10.1016/j.aquatox.2017.06.008

Bergerson JA, Brandt A, Cresko J, Carbajales-Dale M, MacLean HL, Matthews HS, McCoy S et al (2020) Life cycle assessment of emerging technologies: evaluation techniques at different stages of market and technical maturity. J Ind Ecol 24:11–25. https://doi.org/10.1111/jiec.12954

Berman H, Henrick K, Nakamura H (2003) Announcing the worldwide protein data bank. Nat Struct Mol Biol 10(12):980–980. https://doi.org/10.1038/nsb1203-980

Bian SW, Mudunkotuwa IA, Rupasinghe T, Grassian VH (2011) Aggregation and dissolution of 4 nm ZnO nanoparticles in aqueous environments: influence of pH, ionic strength, size, and adsorption of humic acid. Langmuir 27(10):6059–6068. https://doi.org/10.1021/la200570n

Bobori D, Dimitriadi A, Karasiali S, Tsoumaki-Tsouroufli P, Mastora M, Kastrinaki G, Feidantsis K, Printzi A, Koumoundouros G, Kaloyianni M (2020) Common mechanisms activated in the tissues of aquatic and terrestrial animal models after TiO_2 nanoparticles exposure. Environ Int 138:105611. https://doi.org/10.1016/j.envint.2020.105611

Bouallegui Y, Ben Younes R, Bellamine H, Oueslati R (2017) Histopathology and analyses of inflammation intensity in the gills of mussels exposed to silver nanoparticles: role of nanoparticle size, exposure time, and uptake pathways. Toxicol Mechan Methods 27:582–591. https://doi.org/10.1080/15376516.2017.1337258

Buist HE, Hischier R, Westerhout J, Brouwer DH (2017) Derivation of health effect factors for nanoparticles to be used in LCIA. NanoImpact 7:41–53. https://doi.org/10.1016/j.impact.2017.05.002

Burley SK, Bhikadiya C, Bi C, Bittrich S, Chao H, Chen L, Craig PA et al (2023) RCSB Protein Data Bank (RCSB.org): delivery of experimentally-determined PDB structures alongside one million computed structure models of proteins from artificial intelligence/machine learning. Nucleic Acids Res 51(D1):D488–D508. https://doi.org/10.1093/nar/gkac1077

Butt HJ, Jaschke M, Ducker W (1995) Measuring surface forces in aqueous electrolyte solution with the atomic force microscope. Bioelectrochem Bioenerg 38(1):191–201. https://doi.org/10.1016/0302-4598(95)01800-T

Caldeira C, Vlysidis A, Fiore G, De Laurentiis V, Vignali G, Sala S (2020) Sustainability of food waste biorefinery: A review on valorisation pathways, techno-economic constraints, and environmental assessment. Bioresour Technol 312:123575. https://doi.org/10.1016/J.BIORTECH.2020.123575

Cano AM, Maul JD, Saed M, Irin F, Shah SA, Green MJ, French AD, Klein DM, Crago J, Canas-Carrell JE (2018) Trophic transfer and accumulation of multiwalled carbon nanotubes in the presence of copper ions in *Daphnia magna* and fathead minnow (*Pimephales promelas*). Environ Sci Technol 52:794–800. https://doi.org/10.1021/acs.est.7b03522

Cerbelaud M, Videcoq A, Abelard P, Pagnoux C, Rossignol F, Ferrando R (2008) Heteroaggregation between Al_2O_3 submicrometer particles and SiO_2 nanomaterials: experiment and simulation. Langmuir 24:3001–3008. https://doi.org/10.1021/la702104u

Chalamaiah M, Dinesh Kumar B, Hemalatha R, Jyothirmayi T (2012) Fish protein hydrolysates: proximate composition, amino acid composition, antioxidant activities and applications: a review. Food Chem 135(4):3020–3038. https://doi.org/10.1016/j.foodchem.2012.06.100

Chen KL, Elimelech M (2007) Influence of humic acid on the aggregation kinetics of fullerene (C_{60}) nanomaterials in monovalent and divalent electrolyte solutions. J Colloid Interface Sci 309:126–134. https://doi.org/10.1016/j.jcis.2007.01.074

Chowdhury S, Saikia SK (2020) Oxidative stress in fish: A review. J Sci Res 12:145–160. https://doi.org/10.3329/jsr.v12i1.41716

Christenson HK, Claesson PM (2001) Direct Measurments of the force between hydrophobic surfaces in water. Adv Colloid Interf Sci 91:391–436. https://doi.org/10.1016/S0001-8686(00)00036-1

Clément L, Hurel C, Marmier N (2013) Toxicity of TiO_2 nanoparticles to Cladocerans, algae, rotifers and plants—effects of size and crystalline structure. Chemosphere 90:1083–1090. https://doi.org/10.1016/j.chemosphere.2012.09.013

Climent Barba F, Grasham O, Puri DJ, Blacker AJ (2022) A simple techno-economic assessment for scaling-up the enzymatic hydrolysis of MSW pulp. Front Energy Res 10. https://doi.org/10.3389/fenrg.2022.788534

Connolly M, Martinez-Morcillo S, Kalman J, Navas J-M, Bleeker E, Fernandez-Cruz M-L (2023) Considerations for bioaccumulation studies in fish with nanomaterials. Chemosphere 312:137299. https://doi.org/10.1016/j.chemosphere.2022.137299

Corsi I, Bellingeri A, Eliso MC, Grassi G, Liberatori G, Murano C, Sturba L, Vannuccini ML, Bergami E (2021) Eco-interactions of engineered nanomaterials in the marine environment: towards an eco-design framework. Nano 11:1903. https://doi.org/10.3390/nano11081903

Croteau MN, Misra SK, Luoma SN, Valsami-Jones E (2011) Silver bioaccumulation dynamics in a freshwater invertebrate after aqueous and dietary exposures to Nanosized and ionic ag. Environ Sci Technol 45(15):6600–6607. https://doi.org/10.1021/es200880c

Croteau MN, Misra SK, Luoma SN, Valsami-Jones E (2014) Bioaccumulation and toxicity of CuO nanoparticles by a freshwater invertebrate after waterborne and Dietborne exposures. Environ Sci Technol 48(18):10929–10937. https://doi.org/10.1021/es5018703

de Souza NRD, Matt L, Sedrik R, Vares L, Cherubini F (2023) Integrating ex-ante and prospective life-cycle assessment for advancing the environmental impact analysis of emerging bio-based technologies. Sustain Prod Consum 43:319–332. https://doi.org/10.1016/j.spc.2023.11.002

Deng Y, Li J, Qiu M, Yang F, Zhang J, Yuan C (2017) Deriving characterization factors on freshwater ecotoxicity of graphene oxide nanomaterial for life cycle impact assessment. Int J Life Cycle Assess 22(2):222–236. https://doi.org/10.1007/s11367-016-1151-4

Derjaguin BV, Landau LD (1941) Theory of the stability of strongly charged lyophilic sols and of the adhesion of strongly charged particles in solutions of electrolytes. Prog Surf Sci 43(1–4):30–59. https://doi.org/10.1016/0079-6816(93)90013-L

Diegoli S, Manciulea AL, Begum S, Jones IP, Lead JR, Preece JA (2008) Interaction between manufactured gold nanomaterials and naturally occurring organic macromolecules. Sci Total Environ 402(1):51–61. https://doi.org/10.1016/j.scitotenv.2008.04.023

Dong F, Mohd Zaidi NF, Valsami-Jones E, Kreft JU (2017) Time-resolved toxicity study reveals the dynamic interactions between uncoated silver nanoparticles and bacteria. Nanotoxicology 11(5):637–646. https://doi.org/10.1080/17435390.2017.1342010

Dube E, Okuthe GE (2023) Engineered nanoparticles in aquatic systems: toxicity and mechanism of toxicity in fish. Emerging Contaman 9(2):100212. https://doi.org/10.1016/j.emcon.2023.100212

Ducker WA, Senden TJ, Pashley RM (1991) Direct measurement of colloidal forces using an atomic force microscope. Nature 353:239–241. https://doi.org/10.1038/353239a0

Ducker WA, Senden TJ, Pashley RM (1992) Measurement of forces in liquids using a force microscope. Langmuir 8(7):1831–1836. https://doi.org/10.1021/la00043a024

Eckelman MJ, Mauter MS, Isaac JA, Elimeleck M (2012) New perspectives on nanomaterial aquatic Ecotoxicity: production impacts exceed direct exposure impacts for carbon nanotubes. Environ Sci Technol 46(5):2902–2910. https://doi.org/10.1021/es203409a

Eliso MC, Bergami E, Manfra L, Spagnuolo A, Corsi I (2020) Toxicity of Nanoplastics during the embryogenesis of the ascidian Ciona Robusta (Phylum Chordata). Nanotoxicology 14(10):1415–1431. https://doi.org/10.1080/17435390.2020.1838650

Ettrup K, Kounina A, Hansen SF, Meesters JAJ, Vea EB, Laurent A (2017) Development of comparative toxicity potentials of TiO_2 nanoparticles for use in life cycle assessment. Environ Sci Technol 51(7):4027–4037. https://doi.org/10.1021/acs.est.6b05049

European Commission (2013) Regulation (EU) no 1380/2013, amending council regulations (EC) no 1954/2003 and (EC) no 1224/2009 and repealing council regulations (EC) no 2371/2002 and (EC) no 639/2004 and council decision 2004/585/EC

European Commission (2022). Commission recommendation of 10 June 2022 on the definition of nanomaterial (text with EEA relevance) 2022/C 229/01C/2022/3689

Fadare OO, Wan B, Guo LH, Xin Y, Qin W, Yang Y (2019) Humic acid alleviates the toxicity of polystyrene Nanoplastic particles to *Daphnia magna*. Environ Sci Nano 6:1466–1477. https://doi.org/10.1039/C8EN01457D

Fadare OO, Wan B, Liu K, Yang Y, Zhao L, Guo L (2020) Eco-Corona vs. protein Corona: effects of humic substances on Corona formation and Nanoplastic particle toxicity in *Daphnia magna*. Environ Sci Technol 54(13):8001–8009. https://doi.org/10.1021/acs.est.0c00615

Fantke P, Huijbregts M, Margni M, Hauschild M, Jolliet O, McKone T, Rosenbaum R, van de Meent D (2015) USEtox 2.0 model user manual (version 2). Usetox.org

FAO, Food and Agriculture Organization of the United Nations (2022) The state of world fisheries and aquaculture 2022. https://doi.org/10.4060/cc0461en

Ferraro V, Cruz IB, Jorge RF, Malcata FX, Pintado ME, Castro PML (2010) Valorisation of natural extracts from marine source focused on marine by-products: A review. Food Res Int 43(9):2221–2233. https://doi.org/10.1016/j.foodres.2010.07.034

Fransman W, Buist H, Kuijpers E, Walser T, Meyer D, Zondervan-van den Beuken E, Westerhout J, Klein Entink RH, Brouwer DH (2017) Comparative human health impact assessment of engineered nanomaterials in the framework of life cycle assessment. Risk Anal 37(7):1358–1374. https://doi.org/10.1111/risa.12703

Frazao C, Simes DC, Coelho R, Alves D, Williamson MK, Price PA, Cancela ML, Carrondo MA (2005a) Osteocalcin from fish Argyrosomus regius. https://doi.org/10.2210/pdb1VZM/pdb

Frazao C, Simes DC, Coelho R, Alves D, Williamson MK, Price PA, Cancela ML, Carrondo MA (2005b) Structural evidence of a fourth Gla residue in fish osteocalcin: biological implications. Biochemistry 44(4):1234–1242. https://doi.org/10.1021/bi048336z

Fritz G, Schadler V, Willenbacher N, Wagner NJ (2002) Electrosteric stabilization of colloidal dispersions. Langmuir 18(16):6381–6390. https://doi.org/10.1021/la015734j

Fu PP, Xia Q, Hwang HM, Ray PC, Yu H (2014) Mechanisms of Nanotoxicity: generation of reactive oxygen species. J Food Drug Anal 22(1):64–75. https://doi.org/10.1016/j.jfda.2014.01.005

Gallego-Rios SE, Peñuela GA, Martinez-Lopez E (2021) Updating the use of biochemical biomarkers in fish for the evaluation of alterations produced by pharmaceutical products. Environ Toxicol Pharmacol 88:103756. https://doi.org/10.1016/j.etap.2021.103756

Galloway T, Lewis C, Dolciotti I, Johnston BD, Moger J, Regoli F (2010) Sublethal toxicity of Nano-titanium dioxide and carbon nanotubes in a sediment dwelling marine Polychaete. Environ Pollut 158(5):1748–1755. https://doi.org/10.1016/j.envpol.2009.11.013

Gao R, Yu Q, Shen Y, Chu Q, Chen G, Fen S, Yang M, Yuan L, McClements DJ, Sun Q (2021) Production, bioactive properties, and potential applications of fish protein hydrolysates: developments and challenges. Trends Food Sci Technol 110:687–699. https://doi.org/10.1016/j.tifs.2021.02.031

García-Santiago X, Franco-Uría A, Antelo LT, Vázquez JA, Pérez-Martín R, Moreira MT, Feijoo G (2021) Eco-efficiency of a marine biorefinery for valorization of cartilaginous fish biomass. J Ind Ecol 25(3):789–801. https://doi.org/10.1111/jiec.13066

Garner KL, Suh S, Keller AA (2017) Assessing the risk of engineered nanomaterials in the environment: development and application of the nanoFate model. Environ Sci Technol 51(10):5541–5551. https://doi.org/10.1021/acs.est.6b05279

Gaspar TR, Chi RJ, Parrow MW, Ringwood AH (2018) Cellular bioreactivity of micro- and Nanoplastic particles in oysters. Front Mar Sci 5:345. https://doi.org/10.3389/fmars.2018.00345

Geppert M, Sigg L, Schirmer K (2021) Toxicity and translocation of Ag, CuO, ZnO and TiO_2 nanoparticles upon exposure to fish intestinal epithelial cells. Environ Sci Nano 8:2249–2260. https://doi.org/10.1039/D1EN00050K

Giménez B, Alemán A, Montero P, Gómez-Guillén MC (2009) Antioxidant and functional properties of gelatin hydrolysates obtained from skin of sole and squid. Food Chem 114(3):976–983. https://doi.org/10.1016/J.FOODCHEM.2008.10.050

Giusti A, Atluri R, Tsekovska R, Gajewicz A, Apostolova MD, Battistelli CL, Bleeker EAJ et al (2019) Nanomaterial grouping: existing approaches and future recommendations. NanoImpact 16(100182):100182. https://doi.org/10.1016/j.impact.2019.100182

Golberg A, Polikovsky M, Epstein M, Slegers PM, Drabik D, Kribus A (2021) Hybrid solar-seaweed biorefinery for co-production of biochemicals, biofuels, electricity, and water: thermodynamics, life cycle assessment, and cost-benefit analysis. Energy Convers Manag 246:114679. https://doi.org/10.1016/J.ENCONMAN.2021.114679

Gomes T, Araújo O, Pereira R, Almeida AC, Cravo A, Bebianno MJ (2013) Genotoxicity of copper oxide and silver nanoparticles in the mussel *Mytilus galloprovincialis*. Marine Environ Res 84:51–59. https://doi.org/10.1016/j.marenvres.2012.11.009

Gondikas A, Gallego-Urrea J, Halbach M, Derrien N, Hassellöv M (2020) Nanomaterial fate in seawater: A rapid sink or intermittent stabilization? Front Environ Sci 8. https://doi.org/10.3389/fenvs.2020.00151

Guinée J (2016) Life cycle sustainability assessment: what is it and what are its challenges? In: Clift R, Druckman A (eds) Taking stock of industrial ecology. Springer, Cham. https://doi.org/10.1007/978-3-319-20571-7_3

Halim NRA, Yusof HM, Sarbon NM (2016) Functional and bioactive properties of fish protein hydolysates and peptides: A comprehensive review. Trends Food Sci Technol 51:24–33. https://doi.org/10.1016/j.tifs.2016.02.007

Handy RD, Kammer FVD, Lead JR, Hassellöv M, Owen R, Crane M (2008) The ecotoxicology and chemistry of manufactured nanoparticles. Ecotoxicology 17:287–314. https://doi.org/10.1007/s10646-008-0199-8

Henriques A, Vázquez JA, Valcarcel J, Mendes R, Bandarra NM, Pires C (2021) Characterization of protein hydrolysates from fish discards and by-products from the north-West Spain fishing fleet as potential sources of bioactive peptides. Mar Drugs 19(6). https://doi.org/10.3390/md19060338

Hoek EMV, Agarwal GK (2006) Extended DLVO interactions between spherical particles and rough surfaces. J Colloid Interface Sci 298(1):50–58. https://doi.org/10.1016/j.jcis.2005.12.031

Hosseinzadeh-Bandbafha, H., Tabatabaei, M., Aghbashlo, M., Rehan, M., & Nizami, A.S. (2020). Determining key issues in life-cycle assessment of waste biorefineries. Waste Biorefinery: Integrating Biorefineries for Waste Valorisation, 515–555. https://doi.org/10.1016/B978-0-12-818228-4.00019-8, Elsevier

Hubrecht RC, Carter E (2019) The 3Rs and humane experimental technique: implementing change. Animals (Basel) 9(10). https://doi.org/10.3390/ani9100754

Iribarren D, Moreira MT, Feijoo G (2010) Implementing by-product management into the life cycle assessment of the mussel sector. Resour Conserv Recycl 54(12):1219–1230. https://doi.org/10.1016/J.RESCONREC.2010.03.017

Ishak NH, Sarbon NM (2018) A review of protein hydrolysates and bioactive peptides deriving from wastes generated by fish processing. Food Bioprocess Technol 11(1):2–16. https://doi.org/10.1007/s11947-017-1940-1

ISO, International Organization for Standardization (2006a) ISO 14040:2006—Environmental management—Life cycle assessment—Principles and framework. https://www.iso.org/obp/ui/en/#iso:std:iso:14040:ed-2:v1:en

ISO, International Organization for Standardization (2006b) ISO 14044:2006—Environmental management—Life cycle assessment—Requirements and guidelines. https://www.iso.org/obp/ui/en/#iso:std:iso:14044:ed-1:v1:en

Jeong CB, Won EJ, Kang HM, Lee MC, Hwang DS, Zhou B, Souissi SJ, Lee JS (2016) Microplastic size-dependent toxicity, oxidative stress induction, and p-JNK and p-P38 activation in the

Monogonont rotifer (*Brachionus Koreanus*). Environ Sci Technol 50(16):8849–8857. https://doi.org/10.1021/acs.est.6b01441

Jeong C-B, Kang H-M, Lee M-C, Kim D-H, Han J, Hwang D-S, Souissi S, Lee S-J, Shin K-H, Park HG, Lee J-S (2017) Adverse effects of microplastics and oxidative stress-induced MAPK/Nrf2 pathway-mediated defense mechanisms in the marine copepod *Paracyclopina nana*. Sci Rep 7:1–11. https://doi.org/10.1038/srep41323

Jia J, Wang Z, Yue T, Su G, Teng C, Yan B (2020) Crossing biological barriers by engineered nanoparticles. Chem Res Toxicol 33(5):1055–1060. https://doi.org/10.1021/acs.chemrestox.9b00483

Johnson MS, Ates M, Arslan Z, Farah IO, Bogatu C (2017) Assessment of crystal morphology on uptake, particle dissolution, and toxicity of nanoscale titanium dioxide on Artemia Salina. J Nanotoxicol Nanomed 2:11–27. https://doi.org/10.4018/JNN.2017010102

Johnston BD, Scown TM, Moger J, Cumberland SA, Baalousha M, Linge K, van Aerle R, Jarvis K, Lead JR, Tyler CR (2010) Bioavailability of nanoscale metal oxides TiO_2, CeO_2, and ZnO to fish. Environ Sci Technol 44(3):1144–1151. https://doi.org/10.1021/es901971a

Jolliet O et al (2016) Environmental life cycle assessment. CRC Press, Boca Raton/London/New York. ISBN 978-1-4398-8766-0

Kach DJ, Ward JE (2008) The role of marine aggregates in the ingestion of picoplankton-size particles by suspension-feeding Molluscs. Mar Biol 153:797–805. https://doi.org/10.1007/s00227-007-0852-4

Kaya H, Duysak M, Akbulut M, Yilmaz S, Gürkan M, Arslan Z, Demir V, Ates M (2017) Effects of subchronic exposure to zinc nanoparticles on tissue accumulation, serum biochemistry, and Hispathological changes in tilapia (*Oreochromis Niloticus*). Environ Toxicol 32:1213–1225. https://doi.org/10.1002/tox.22318

Khan FR, Misra SK, Garca-Alonso J, Smith BD, Strekopytov S, Rainbow PS, Luoma SN, Valsami-Jones E (2012) Bioaccumulation dynamics and modeling in an estuarine invertebrate following aqueous exposure to nanosized and dissolved silver. Environ Sci Technol 46(14):7621–7628. https://doi.org/10.1021/es301253s

Khan FR, Paul KB, Dybowska AD, Valsami-Jones E, Lead JR, Stone V, Fernandes TF (2015) Accumulation dynamics and acute toxicity of silver nanoparticles to *Daphnia magna* and *Lumbriculus variegatus*: implications for metal modeling approaches. Environ Sci Technol 49(7):4389–4397. https://doi.org/10.1021/es506124x

Kiehbadroudinezhad M, Hosseinzadeh-Bandbafha H, Varjani S, Wang Y, Peng W, Pan J, Aghbashlo M, Tabatabaei M (2023) Marine shell-based biorefinery: A sustainable solution for aquaculture waste valorization. Renew Energy 206:623–634. https://doi.org/10.1016/j.renene.2023.02.057

Kim SK, Mendis E (2006) Bioactive compounds from marine processing byproducts—a review. Food Res Int 39(4):383–393. https://doi.org/10.1016/J.FOODRES.2005.10.010

Kittler S, Greulich C, Diendorf J, Köller M, Epple M (2010) Toxicity of silver nanoparticles increases during storage because of slow dissolution under release of silver ions. Chem Mater 22(16):4548–4554. https://doi.org/10.1021/cm100023p

Klaine SJ, Alvarez PJJ, Batley GE, Fernandes TF, Handy RD, Lyon DY, Mahendra S, McLaughlin M, Lead JR (2008) Nanomaterials in the environment: behavior, fate, bioavailability, and effects. Environ Toxicol Chem 27(9):1825–1851. https://doi.org/10.1897/08-090.1

Kokossis AC, Tsakalova M, Pyrgakis K (2014) Design of integrated biorefineries. Comput Chem Eng 81:40–56. https://doi.org/10.1016/j.compchemeng.2015.05.021

Kreyling WG, Holzwarth U, Haberl N, Kozempel J, Hirn S, Wenk A, Schleh C, Schäffler M, Lipka J, Semmler-Behnke M, Gibson N (2017) Quantitative biokinetics of titanium dioxide nanoparticles after intravenous injection in rats: part 1. Nanotoxicology 11(4):434–442. https://doi.org/10.1080/17435390.2017.1306892

Kumah EA, Fopa RD, Harati S, Boadu P, Zohoori FV, Pak T (2023) Human and environmental impacts of nanoparticles: a scoping review of the current literature. BMC Public Health 23(1):1059. https://doi.org/10.1186/s12889-023-15958-4

Kurita K (2006) Chitin and chitosan: functional biopolymers from marine crustaceans. Mar Biotechnol 8(3):203–226. https://doi.org/10.1007/s10126-005-0097-5

Lammel T, Thit A, Mouneyrac C, Baun A, Sturve J, Selck H (2019) Trophic transfer of CuO NPs and dissolved cu from sediment to Worms to fish – a proof-of-concept study. Environ Sci Nano 6:1140–1155. https://doi.org/10.1039/c9en00093c

Lead JR, Batley GE, Alvarez PJJ, Croteau M-N, Handy RD, McLaughlin MJ, Judy JD, Schirmer K (2018) Nanomaterials in the environment: behavior, fate, bioavailability, and effects - an updated review. Environ Toxicol Chem 37(87):2029–2063. https://doi.org/10.1002/etc.4147

Lee G, Lee B, Kim KT (2021) Mechanisms and effects of zinc oxide nanoparticle transformations on toxicity to zebrafish embryos. Environ Sci Nano 8(6):1690–1700. https://doi.org/10.1039/D1EN00305D

Li M, Gazang C, Ge H, Li J, Huang T, Gao H, Zhao Y, Mao X, Du J, Wu M, Ma J (2021) The atmospheric travel distance of persistent organic pollutants-revisit and application in climate change impact on long-rang transport potential. Atmos Res 255:105558. https://doi.org/10.1016/j.atmosres.2021.105558

Li M, Zhang Y, Feng S, Zhang X, Xi Y, Xiang X (2022) Bioaccumulation and biomagnification effects of Nano-TiO$_2$ in the aquatic food chain. Ecotoxicology 31:1023–1034. https://doi.org/10.1007/s10646-022-02572-0

Liu HH, Cohen Y (2014) Multimedia environmental distribution of engineered nanomaterials. Environ Sci Technol 48(6):3281–3292. https://doi.org/10.1021/es405132z

Liu Y, Lyu Y, Tian J, Zhao J, Ye N, Zhang Y, Chen L (2021) Review of waste biorefinery development towards a circular economy: from the perspective of a life cycle assessment. Renew Sust Energ Rev 139:110716. https://doi.org/10.1016/J.RSER.2021.110716

Lively JA, & McKenzie J (2023) Discards and bycatch: A review of wasted fishing. Adv Mar Biol 95:1–26. https://doi.org/10.1016/bs.amb.2023.07.001

Lopes C, Antelo LT, Franco-Uría A, Alonso AA, Pérez-Martín R (2015) Valorisation of fish by-products against waste management treatments – comparison of environmental impacts. Waste Manag 46:103–112. https://doi.org/10.1016/j.wasman.2015.08.017

López-Herrada E, Gallardo-Rodríguez JJ, López-Rosales L, Cerón-García MC, Sánchez-Mirón A, García-Camacho F (2023) Life-cycle assessment of a microalgae-based fungicide under a biorefinery approach. Bioresour Technol 383:129244. https://doi.org/10.1016/j.biortech.2023.129244

Lu K, Dong S, Petersen EJ, Niu J, Chang X, Wang P, Lin S, Gao S, Mao L (2017) Biological uptake, distribution, and depuration of radio-labeled graphene in adult zebrafish: effects of graphene size and natural organic matter. ACS Nano 11(3):2872–2885. https://doi.org/10.1021/acsnano.6b07982

Lundqvist M, Stigler J, Elia G, Lynch I, Cedervall T, Dawson KA (2008) Nanoparticle size and surface properties determine the protein Corona with possible implications for biological impacts. Proc Nat Acad Sci USA 105(38):14265–14270. https://doi.org/10.1073/pnas.0805135105

Lynch I, Dawson KA (2008) Protein-nanomaterial Interactions. NanoToday 3(1–2):40–47. https://doi.org/10.1016/S1748-0132(08)70014-8

Manfra L, Rotini A, Bergami E, Grassi G, Faleri C, Corsi I (2017) Comparative Ecotoxicity of polystyrene nanoparticles in natural seawater and reconstituted seawater using the rotifer Brachionus Plicatilis. Ecotoxicol Environ Saf 145:557–563. https://doi.org/10.1016/j.ecoenv.2017.07.068

Mao BH, Tsai JC, Chen CW, Yan SJ, Wang YJ (2016) Mechanisms of silver nanoparticle-induced toxicity and important role of autophagy. Nanotoxicology 10:1021–1040. https://doi.org/10.1080/17435390.2016.1189614

Mawed SA, Marini C, Alagawany M, Farag MR, Reda RM, El-Saadony MT, Elhady WM, Magi GE, Di Cerbo A, El-Nagar WG (2022) Zinc oxide nanoparticles (ZnO-NPs) suppress fertility by activating autophagy, apoptosis, and oxidative stress in the developing oocytes of female zebrafish. Antioxidants 11(8):1567. https://doi.org/10.3390/antiox11081567

Meesters JAJ, Koelmans AA, Quik JTK, Hendriks AJ, Van De Meent D (2014) Multimedia modeling of engineered nanoparticles with simpleBox4nano: model definition and evaluation. Environ Sci Technol 48(10):5726–5736. https://doi.org/10.1021/es500548h

Meesters JAJ, Quik JTK, Koelmans AA, Hendriks AJ, Van De Meent D (2016) Multimedia environmental fate and speciation of engineered nanoparticles: A probabilistic modeling approach. Environ Sci Nano 3(4):715–727. https://doi.org/10.1039/c6en00081a

Mei N, Hedberg J, Ekvall MT, Kelpsiene E, Hansson L, Cedervall T, Blomberg E, Odnevall I (2021) Transfer of cobalt nanoparticles in a simplified food web: from algae to zooplankton to fish. Appl Nano 2(3):184–205. https://doi.org/10.3390/applnano2030014

Miller FJ, Asgharian B, Schroeter JD, Price O (2016) Improvements and additions to the multiple path particle dosimetry model. J Aerosol Sci 99:14–26. https://doi.org/10.1016/j.jaerosci.2016.01.018

Miseljic M, Olsen SI (2014) Life-cycle assessment of engineered nanomaterials: A literature review of assessment status. J Nanopart Res 16(6). https://doi.org/10.1007/s11051-014-2427-x

Mizani M, Aminlari M, Khodabandeh M (2005) An effective method for producing a nutritive protein extract powder from shrimp-head waste. Food Sci Technol Int 11(1):49–54. https://doi.org/10.1177/1082013205051271

Moncada J, Cardona CA, Rincón LE (2015) Design and analysis of a second and third generation biorefinery: the case of castorbean and microalgae. Bioresour Technol 198:836–843. https://doi.org/10.1016/j.biortech.2015.09.077

Moni SM, Mahmud R, High K, Carbajales-Dale M (2019) Life cycle assessment of emerging technologies: a review. J Ind Ecol 24(1):52–63. https://doi.org/10.1111/jiec.12965

Monopoli MP, Pitek AS, Lynch I, & Dawson KA (2013) Formation and characterization of the nanoparticle-protein corona. Methods Mol Biol 1025:137–155. https://doi.org/10.1007/978-1-62703-462-3_11

Monteiro A, Paquincha D, Martins F, Queirós RP, Saraiva JA, Švarc-Gajić J, Nastić N, Delerue-Matos C, Carvalho AP (2018) Liquid by-products from fish canning industry as sustainable sources of ω3 lipids. J Environ Manag 219:9–17. https://doi.org/10.1016/J.JENVMAN.2018.04.102

Mountraki A, Tsakalova M, Panteli A, Papoutsi AI, Kokossis AC (2016) Integrated waste Management in Multiproduct Biorefineries: systems optimization and analysis of a real-life industrial plant. Ind Eng Chem Res 55(12):3478–3492. https://doi.org/10.1021/acs.iecr.5b03431

Nikdehghan N, Kashiri H, Hedayati AA (2018) CuO nanoparticles-induced micronuclei and DNA damage in Cyprinus Carpio. Bioflux 11(3):925–936

Nirmal NP, Santivarangkna C, Rajput MS, Benjakul S, Maqsood S (2022) Valorization of fish byproducts: sources to end-product applications of bioactive protein hydrolysate. Compr Rev Food Sci Food Saf 21(2):1803–1842. https://doi.org/10.1111/1541-4337.12917

Noventa S, Hacker C, Correia A, Drago C, Galloway T (2018) Gold nanoparticles ingested by oyster larvae are internalized by cells through an alimentary endocytic pathway. Nanotoxicology 12:901–913. https://doi.org/10.1080/17435390.2018.1487601

OECD, Organisation for Economic Co-operation and Development (2018) Case study on grouping and read-across for nanomaterials − genotoxicity of nano-TiO2 (Series on testing and assessment no. 292), 1–56. https://www.oecd.org/officialdocuments/publicdisplaydocumentpdf/?cote=ENV/JM/MONO(2018)28&docLanguage=En

Opršal J, Knotek P, Zickler GA, Sigg L, Schirmer K, Pouzar M, Geppert M (2021) Cytotoxicity, accumulation and translocation of silver and silver sulfide nanoparticles in contact with Rainbow trout intestinal cells. Aquat Toxicol 237:105869. https://doi.org/10.1016/j.aquatox.2021.105869

Ortizo RGG, Sharma V, Tsai ML, Wang JX, Sun PP, Nargotra P, Kuo CH, Chen CW, Dong CD (2023) Extraction of novel bioactive peptides from fish protein hydrolysates by enzymatic reactions. Appl Sci 13(9). https://doi.org/10.3390/app13095768

Peijnenburg WJGM, Baalousha M, Chen J, Chaudry Q, von der Kammer F, Kuhlbusch TAJ, Lead JR et al (2015) A review of the properties and processes determining the fate of engineered nanomaterials in the aquatic environment. Crit Rev Environ Sci Technol 45(19):2084–2134. https://doi.org/10.1080/10643389.2015.1010430

Pérez Roda MA, Gilman E, Huntington T, Kennelly SJ, Suuronen P, Chaloupka M, Medley PAH (2019) A third assessment of global marine fisheries discards. Food and Agriculture Organization of the United Nations

Petrova I, Tolstorebrov I, Eikevik TM (2018) Production of fish protein hydrolysates step by step: technological aspects, equipment used, major energy costs and methods of their minimizing. Inter Aquat Res 10(3):223–241. https://doi.org/10.1007/s40071-018-0207-4

Phadke GG, Rathod NB, Ozogul F, Elavarasan K, Karthikeyan M, Shin KH, Kim SK (2021) Exploiting of secondary raw materials from fish processing industry as a source of bioactive peptide-rich protein hydrolysates. Mar Drugs 19(9). https://doi.org/10.3390/md19090480

Phenrat T, Saleh N, Sirk K, Tilton RD, Lowry GV (2007) Aggregation and sedimentation of aqueous nanoscale zerovalent iron dispersions. Environ Sci Technol 41(1):284–290. https://doi.org/10.1021/es061349a

Phenrat T, Saleh N, Sirk K, Kim HJ, Tilton RD, Lowry GV (2008) Stabilization of aqueous nanoscale zerovalent iron dispersions by anionic polyelectrolytes: adsorbed anionic polyelectrolyte layer properties and their effect on aggregation and sedimentation. J Nanopart Res 10:795–814. https://doi.org/10.1007/s11051-007-9315-6

Pini M, Salieri B, Ferrari AM, Nowack B, Hischier R (2016) Human health characterization factors of nano-TiO_2 for indoor and outdoor environments. Int J Life Cycle Assess 21:1452–1462. https://doi.org/10.1007/s11367-016-1115-8

Pu Y, Tang F, Adam P-M, Laratte B, Ionescu RE (2016) Fate and characterization factors of nanoparticles in seventeen subcontinental freshwaters: A case study on copper nanoparticles. Environ Sci Technol 50(17):9370–9379. https://doi.org/10.1021/acs.est.5b06300

Pu Y, Laratte B, Marks RS, Ionescu RE (2017) Impact of copper nanoparticles on porcine neutrophils: ultrasensitive characterization factor combining chemiluminescence information and USEtox assessment model. Mater Today Commun 11:68–75. https://doi.org/10.1016/j.mtcomm.2017.02.008

Reeves JF, Davies SJ, Dodd NJF, Jha AN (2008) Hydroxyl radicals (·OH) are associated with titanium dioxide (TiO_2) nanoparticle-induced cytotoxicity and oxidative DNA damage in fish cells. Mutat Res 640(1–2):113–122. https://doi.org/10.1016/j.mrfmmm.2007.12.010

Righi S (2019) Life cycle assessments of waste-based biorefineries—a critical review. In: Basosi R, Cellura M, Longo S, Parisi M (eds) Life cycle assessment of energy systems and sustainable energy technologies. Green Energy and Technology/Springer, pp 139–154. https://doi.org/10.1007/978-3-319-93740-3_9

Rocco L, Santonastaso M, Mottola F, Costagliola D, Suero T, Pacifico S, Stingo V (2015) Genotoxicity assessment of TiO_2 nanoparticles in the teleost danio Rerio. Ecotoxicol Environ Saf 113:223–230. https://doi.org/10.1016/j.ecoenv.2014.12.012

Rocha TL, Gomes T, Sousa VS, Mestre NC, Bebianno MJ (2015) Ecotoxicological impact of engineered nanomaterials in bivalve Molluscs: an overview. Mar Environ Res 111:74–88. https://doi.org/10.1016/j.marenvres.2015.06.013

Rodriguez-Garcia G, Zimmermann B, Weil M (2014) Nanotoxicity and life cycle assessment: first attempt towards the determination of characterization factors for carbon nanotubes, *IOP Conf. Series*. Mater Sci Eng 64:012029. https://doi.org/10.1088/1757-899X/64/1/012029

Rodriguez-Hernandez AG, Vazquez-Duhalt R, Huerta-Saquero A (2020) Nanoparticle-plasma membrane interactions: thermodynamics, toxicity and cellular response. Curr Med Chem 27(20):3330–3345. https://doi.org/10.2174/0929867325666181112090648

Romeo D, Hischier R, Nowack B, Jolliet O, Fantke P, Wick P (2022a) *In vitro*-based human toxicity effect factors: challenges and opportunities for nanomaterial impact assessment. Environ Sci Nano 9(6):1913–1925. https://doi.org/10.1039/d1en01014j

Romeo D, Hischier R, Nowack B, Wick P (2022b) Approach toward *in vitro*-based human toxicity effect factors for the life cycle impact assessment of inhaled low-solubility particles. Environ Sci Technol 56(12):8552–8560. https://doi.org/10.1021/acs.est.2c01816

Rosenbaum RK, Bachmann TM, Gold LS, Huijbregts MAJ, Jolliet O, Juraske R, Koehler A, Larsen HF, MacLeod M, Margni M, McKone TE, Payet J, Schuhmacher M, Van De Meent D, Hauschild MZ (2008) USEtox – the UNEP-SETAC toxicity model: recommended characterisation factors for human toxicity and freshwater ecotoxicity in life cycle impact assessment. Int J Life Cycle Assess 13(7):532–546. https://doi.org/10.1007/s11367-008-0038-4

Saavedra del Oso M, Mauricio-Iglesias M, Hospido A, Steubing B (2023) Prospective LCA to provide environmental guidance for developing waste-to-PHA biorefineries. J Clean Prod 383:135331. https://doi.org/10.1016/J.JCLEPRO.2022.135331

Sadhukhan J, Gadkari S, Martinez-Hernandez E, Ng KS, Shemfe M, Torres-Garcia E, Lynch J (2019) Novel macroalgae (seaweed) biorefinery systems for integrated chemical, protein, salt, nutrient and mineral extractions and environmental protection by green synthesis and life cycle sustainability assessments. Green Chem 21(10):2635–2655. https://doi.org/10.1039/c9gc00607a

Sahu BB, Paikaray NK, Paikaray A, Agnibesh A, Mohapatra S, Jayasankar P (2016) Fish waste bio-refinery products: its application in organic farming. Inter J Environ Agricult Biotechnol 1(4):837–843. https://doi.org/10.22161/ijeab/1.4.30

Salieri B, Righi S, Pasteris A, Olsen SI (2015) Freshwater ecotoxicity characterisation factor for metal oxide nanoparticles: A case study on titanium dioxide nanoparticle. Sci Total Environ 505:494–502. https://doi.org/10.1016/j.scitotenv.2014.09.107

Salieri B, Turner DA, Nowack B, Hischier R (2018) Life cycle assessment of nanomaterials: where are we? NanoImpact 10:108–120. https://doi.org/10.1016/j.impact.2017.12.003

Salieri B, Hischier R, Quik JTK, & Jolliet O (2019) Fate modelling of nanoparticle releases in LCA: An integrative approach towards "USEtox4Nano." J Clean Prod 206:701–712. https://doi.org/10.1016/J.JCLEPRO.2018.09.187

Salieri B, Kaiser JP, Rösslein M, Nowack B, Hischier R, Wick P (2020) Relative potency factor approach enables the use of in vitro information for estimation of human effect factors for nanoparticle toxicity in life-cycle impact assessment. Nanotoxicology 14(2):275–286. https://doi.org/10.1080/17435390.2019.1710872

Sander S, Mosley LM, Hunter KA (2004) Investigation of Interparticle forces in natural waters: effects of adsorbed humic acids on iron oxide and alumina surface properties. Environ Sci Technol 38(18):4791–4796. https://doi.org/10.1021/es049602z

Selck H, Handy RD, Fernandes TF, Klaine SJ, Petersen EJ (2016) Nanomaterials in the aquatic environment: a European Union – United States perspective on the status of Ecotoxicity testing, research priorities, and challenges ahead. Environ Toxicol Chem 35(5):1055–1067. https://doi.org/10.1002/etc.3385

Sendra M, Saco A, Yeste MP, Romero A, Novoa B, Figueras A (2020) Nanoplastics: from tissue accumulation to cell translocation into *Mytillus galloprovincialis* Hemocytes. Resilience of immune cells exposed to Nanoplastics and Nanoplastics plus *vibrio Splendidus* combination. J Hazard Mater 388:121788. https://doi.org/10.1016/j.jhazmat.2019.121788

Skjolding LM, Winther-Nielsen M, Baun A (2014) Trophic transfer of differently functionalized zinc oxide nanoparticles from crustaceans (daphnia magna) to zebrafish (danio Rerio). Aquat Toxicol 157:101–108. https://doi.org/10.1016/j.aquatox.2014.10.005

Speshock J, Erold N, Sadoski DK, Maurer E, Braydich-Stolle LK, Brady J, Hussain S (2016) Differential organ toxicity in the adult zebra fish following exposure to acute sub-lethal doses of 10 nm silver nanoparticles. Front Nanosci Nanotechnol 2. https://doi.org/10.15761/FNN.1000119

Sung HK, Jo E, Kim E, Yoo SK, Lee JW, Kim PJ, Kim Y, Eom IC (2018) Analysis of Gold and silver nanoparticles internalized by zebrafish (*Danio Rerio*) using single particle-inductively coupled plasma-mass spectrometry. Chemosphere 209:815–822. https://doi.org/10.1016/j.chemosphere.2018.06.149

Tallec K, Huvet A, di Poi C, Gonzalez-Fernandez C, Lambert C, Petton B, le Goïc N, Berchel M, Soudant P, Paul-Pont I (2018) Nanoplastics impaired oyster free living stages, gametes and embryos. Environ Pollut 242(Part B):1226–1235. https://doi.org/10.1016/j.envpol.2018.08.020

Thirukumaran R, Anu Priya VK, Krishnamoorthy S, Ramakrishnan P, Moses JA, Anandharamakrishnan C (2022) Resource recovery from fish waste: prospects and the usage of intensified extraction technologies. Chemosphere 299:134361. https://doi.org/10.1016/j.chemosphere.2022.134361

Thonemann N, Schulte A, Maga D (2020) How to conduct prospective life cycle assessment for emerging technologies? A systematic review and methodological guidance. Sustain For 12(3):1192. https://doi.org/10.3390/su12031192

Toncelli C, Mylona K, Kalantzi I, Tsiola A, Pitta P, Tsapakis M, Pergantis SA (2017) Silver nanoparticles in seawater: A dynamic mass balance at part per trillion silver concentrations. Sci Total Environ 601–602:15–21. https://doi.org/10.1016/j.scitotenv.2017.05.148

Torre CD, Bergami E, Salvati A, Faleri C, Cirino P, Dawson KA, Corsi I (2014) Accumulation and Embryotoxicity of polystyrene nanoparticles at early stage of development of sea urchin embryos *Paracentrotus lividus*. Environ Sci Technol 48(20):12302–12311. https://doi.org/10.1021/es502569w

Torre CD, Balbi T, Grassi G, Frenzilli G, Bernardeschi M, Smerilli A, Guidi P et al (2015) Titanium dioxide nanoparticles modulate the toxicological response to cadmium in the gills of *Mytilus galloprovincialis*. J Hazard Mater 297:92–100. https://doi.org/10.1016/j.jhazmat.2015.04.072

Tsang MP, Li D, Garner KL, Keller AA, Suh S, Sonnemann GW (2017) Modeling human health characterization factors for indoor nanomaterial emissions in life cycle assessment: a case-study of titanium dioxide. Environ Sci Nano 4(8):1705–1721. https://doi.org/10.1039/c7en00251c

Tsoy N, Steubing B, van der Giesen C, Guinée J (2020) Upscaling methods used in ex ante life cycle assessment of emerging technologies: a review. Int J Life Cycle Assess 25:1680–1692. https://doi.org/10.1007/s11367-020-01796-8

Ubando AT, Felix CB, Chen WH (2020) Biorefineries in circular bioeconomy: a comprehensive review. Bioresour Technol 299:122585. https://doi.org/10.1016/J.BIORTECH.2019.122585

van der Giesen C, Cucurachi S, Guinée J, Kramer GJ, Tukker A (2020) A critical view on the current application of LCA for new technologies and recommendations for improved practice. J Clean Prod 259:120904. https://doi.org/10.1016/J.JCLEPRO.2020.120904

Varó I, Perini A, Torreblanca A, Garcia Y, Bergami E, Vannuccini ML, Corsi I (2019) Time-dependent effects of polystyrene nanoparticles in brine shrimp *Artemia franciscana* at physiological, biochemical and molecular levels. Sci Total Environ 675:570–580. https://doi.org/10.1016/j.scitotenv.2019.04.157

Vázquez JA, Rodríguez-Amado I, Montemayor MI, Fraguas J, Del González MP, Murado MA (2013) Chondroitin sulfate, hyaluronic acid and chitin/chitosan production using marine waste sources: characteristics, applications and eco-friendly processes: a review. Mar Drugs 11(3):747–774. https://doi.org/10.3390/md11030747

Vázquez JA, Blanco M, Massa AE, Amado IR, Pérez-Martín RI (2017) Production of fish protein hydrolysates from *scyliorhinus canicula* discards with antihypertensive & antioxidant activities by enzymatic hydrolysis & mathematical optimization using response surface methodology. Mar Drugs 15(10):306. https://doi.org/10.3390/md15100306

Vázquez JA, Sotelo CG, Sanz N, Pérez-Martín RI, Rodríguez-Amado I, Valcarcel J (2019) Valorization of aquaculture by-products of salmonids to produce enzymatic hydrolysates: process optimization, chemical characterization and evaluation of bioactives. Mar Drugs 17(12):676. https://doi.org/10.3390/md17120676

Vázquez JA, Fraguas J, Mirón J, Valcárcel J, Pérez-Martín RI, Antelo LT (2020) Valorisation of fish discards assisted by enzymatic hydrolysis and microbial bioconversion: lab and pilot plant studies and preliminary sustainability evaluation. J Clean Prod 246:119027. https://doi.org/10.1016/j.jclepro.2019.119027

Vázquez JA, Pedreira A, Durán S, Cabanelas D, Souto-Montero P, Martínez P, Mulet M, Pérez-Martín RI, Valcarcel J (2022) Biorefinery for tuna head wastes: production of protein hydrolysates, high-quality oils, minerals and bacterial peptones. J Clean Prod 357:131909. https://doi.org/10.1016/j.jclepro.2022.131909

Venkata Mohan S, Nikhil GN, Chiranjeevi P, Nagendranatha Reddy C, Rohit MV, Kumar AN, Sarkar O (2016) Waste biorefinery models towards sustainable circular bioeconomy: critical review and future perspectives. Bioresour Technol 215:2–12. https://doi.org/10.1016/j.biortech.2016.03.130

Veríssimo NV, Mussagy CU, Oshiro AA, Mendonça CMN, Santos-Ebinuma VDC, Pessoa A, Oliveira RPDS, Pereira JFB (2021) From green to blue economy: marine biorefineries for a sustainable ocean-based economy. Green Chem 23(23):9377–9400. https://doi.org/10.1039/d1gc03191k

Verwey EJW, Overbeek JTG (1948) Theory of the stability of lyophilic colloids. Elsevier, Amsterdam

Vevers W, Jha AN (2008) Genotoxic and cytotoxic potential of titanium dioxide (TiO$_2$) nanoparticles on fish cells *in vitro*. Ecotoxicology 17:410–420. https://doi.org/10.1007/s10646-008-0226-9

Vicari T, Dagostim AC, Klingelfus T, Galvan GL, Monteiro PS, da Silva Pereira L, Silva de Assis HC, Cestari MM (2018) Co-exposure to titanium dioxide nanoparticles (NpTiO$_2$) and Lead at environmentally relevant concentrations in the Neotropical fish species *Hoplias intermedius*. Toxicol Rep 5:1032–1043. https://doi.org/10.1016/j.toxrep.2018.09.001

Vidya PV, Chitra KC (2018) Evaluation of genetic damage in *Oreochromis mossambicus* exposed to selected nanoparticles by using micronucleus and comet bioassays. Croatian J Fish 76(3):115–124. https://doi.org/10.2478/cjf-2018-0015

Villares M, Işildar A, von der Giesen C, Guinée J (2017) Does ex ante application enhance the usefulness of LCA? A case study on an emerging technology for metal recovery from e-waste. Int J Life Cycle Assess 22:1618–1633. https://doi.org/10.1007/s11367-017-1270-6

Vlysidis A, Binns M, Webb C, Theodoropoulos C (2011) A techno-economic analysis of biodiesel biorefineries: assessment of integrated designs for the co-production of fuels and chemicals. Energy 36(8):4671–4683. https://doi.org/10.1016/j.energy.2011.04.046

Ward JE, Kach DJ (2009) Marine aggregates facilitate ingestion of nanoparticles by suspension-feeding bivalves. Mar Environ Res 68(3):137–142. https://doi.org/10.1016/j.marenvres.2009.05.002

Xiang Q-Q, Wang D, Zhang J-L, Ding C-Z, Luo X, Tao J, Ling J, Shea D, Chen L-Q (2020) Effect of silver nanoparticles on gill membranes of common carp: modification of fatty acid Profil, lipid peroxidation and membrane fluidity. Environ Pollut 256:113504. https://doi.org/10.1016/j.envpol.2019.113504

Yang X, Gondikas AP, Marinakos SM, Auffan M, Liu J, Hsu-Kim H, Meyer JN (2012) Mechanism of silver nanoparticle toxicity is dependent on dissolved silver and surface coating in *Caenorhabditis elegans*. Environ Sci Technol 46(2):1119–1127. https://doi.org/10.1021/es202417t

Zaky AS (2021) Introducing a marine biorefinery system for the integrated production of biofuels, high-value-chemicals, and co-products: a path forward to a sustainable future. PRO 9(10). https://doi.org/10.3390/pr9101841

Zande M, van der Jemec Kokalj A, Spurgeon DJ, Loureiro S, Silva PV, Khodaparast Z, Drobne D et al (2020) The gut barrier and the fate of engineered nanomaterials: a view from comparative physiology. Environ Sci Nano 7:1874–1898. https://doi.org/10.1039/D0EN00174K

Zhang Y, Pu S, Lv X, Gao Y, Ge L (2020) Global trends and prospects in microplastic research: A bibliometric analysis. J Hazard Mater 400:123110. https://doi.org/10.1016/j.jhazmat.2020.123110

Zhu X, Wang J, Zhang X, Chang Y, Chen Y (2010) Trophic transfer of TiO$_2$ nanoparticles from daphnia to zebrafish in a simplified freshwater food chain. Chemosphere 79(9):928–933. https://doi.org/10.1016/j.chemosphere.2010.03.022

Open Access This chapter is licensed under the terms of the Creative Commons Attribution-NonCommercial-NoDerivatives 4.0 International License (http://creativecommons.org/licenses/by-nc-nd/4.0/), which permits any noncommercial use, sharing, distribution and reproduction in any medium or format, as long as you give appropriate credit to the original author(s) and the source, provide a link to the Creative Commons license and indicate if you modified the licensed material. You do not have permission under this license to share adapted material derived from this chapter or parts of it.

The images or other third party material in this chapter are included in the chapter's Creative Commons license, unless indicated otherwise in a credit line to the material. If material is not included in the chapter's Creative Commons license and your intended use is not permitted by statutory regulation or exceeds the permitted use, you will need to obtain permission directly from the copyright holder.

Index

A
Accumulations, 29, 63, 70, 71, 121, 176, 202–204, 206, 219, 221, 224–225, 231, 234, 284, 285, 292, 296–298, 318, 320, 353, 367, 369–377, 380, 382, 385, 387, 389, 407, 408, 441, 445, 446, 448, 450, 463, 489, 564, 575, 576, 581
Advanced in vitro models, 2, 7–11, 18, 48, 155
Allergies, 93, 98, 126, 132
Asthma, 93, 97, 115–123, 130–132

B
Bioaccumulation, 2, 14, 202, 259, 261, 264, 287, 300, 372, 373, 380, 389, 405, 407–408, 412, 421, 441, 445–447, 452, 453, 572, 575, 576
Biorefinery, 560–581
Blood-brain barrier (BBB), 219–228, 230–236, 238, 239, 340, 367, 369, 374, 375, 382, 390

C
Caenorhabditis elegans, 11, 275–302, 424
CORAL software, 537, 540, 553
Correlation intensity index (CII), 552, 555

D
Density Functional Theory (DFT), 182, 513, 515, 517–520, 523
DNA damages, 3, 6, 70, 90, 91, 99, 102, 145, 180, 262–264, 280, 282, 283, 291, 294, 296, 298, 320, 342, 348–351, 376, 378, 385, 387, 389, 453, 489, 574
DNA methylation, 88, 90–95, 97, 100, 102, 103
Drosophila, 11, 12, 314, 338–355

E
Eco-corona, 425, 574
Ecotoxicology, 13, 14, 235, 259, 279, 295, 408–411, 446, 460
Engineered nanomaterials, 115–132, 194, 366, 404, 412, 424, 429, 579, 581
Engineered nanoparticles (ENPs), 281, 412, 560–581
Environment, 2, 30, 57, 89, 116, 169, 194, 219, 256, 276, 317, 340, 367, 404, 440, 482, 512, 536, 564
Environmental risk assessment (ERA), 15, 404, 411–413, 429, 430
Epigenetic toxicity, 88, 103
Experimental protocol, 41–47
Exposure, 3, 29, 59, 89, 116, 162, 194, 219, 255, 279, 314, 340, 366, 405, 444, 489, 519, 540, 564

G
Genotoxicity, 2, 3, 6–7, 18, 37, 40, 88, 103, 145, 148–150, 153, 179, 180, 225, 298, 348, 366, 378, 380, 384–385, 389, 408, 430, 489, 491, 494, 574, 576
Glyphosate, 513, 519–523

© The Editor(s) (if applicable) and The Author(s), under exclusive license to Springer Nature Switzerland AG 2025
E. Alfaro-Moreno, F. Murphy (eds.), *Nanosafety*,
https://doi.org/10.1007/978-3-031-93871-9

I

Immune responses, 5, 6, 56–62, 65–67, 69–72, 74, 77, 93, 95, 97–101, 117, 118, 126, 128, 130, 132, 145, 147, 155, 167, 180, 199, 289, 292, 296, 297, 314, 316, 318–321, 324, 325, 327, 339, 373, 377, 405, 410, 451, 575
Immunomodulation, 95–98, 323
Immunotoxicity, 2, 4–6, 18, 72, 74, 76, 88, 100, 103, 172, 297, 319–323, 378–379, 389, 430, 482
Index of ideality of correlation (IIC), 552, 555
Inflammatory responses, 4, 5, 36, 37, 59, 63, 67–69, 71, 73, 89, 90, 92, 96, 100, 103, 117, 146, 147, 150, 165, 172, 204, 224, 323, 324, 340, 342, 373, 377, 380, 450, 493, 574, 576
Intestinal organoids, 9, 202–204
Invertebrate models, 11, 252–266, 314, 367
In vitro digestion, 194, 201
In vitro lung, 150, 155

L

Life cycle analysis, 413
Life cycle assessment (LCA), 560–581
Lung, 8, 30, 59, 93, 115, 219, 323, 344, 371, 445, 577

M

Marine biomass, 560–581
Metal nanooxide
Molecular docking, 181, 182, 482–504, 517
Monte Carlo method, 539

N

Nanomaterials, 2, 29, 57, 88, 132, 162, 194, 219, 260, 281, 317, 338, 366, 404, 440, 482, 512, 532, 571
Nanomaterial safety, 349
Nanoparticles (NPs), 3, 5, 7, 14, 59, 66, 69, 88–103, 116–120, 122–123, 126, 127, 148–150, 152, 162, 181, 183, 194, 197–199, 202, 203, 206, 207, 219, 252–266, 282–284, 286–289, 291, 294–302, 318, 323, 338–354, 366–390, 404–406, 410–414, 420, 422, 425, 426, 428–430, 445, 446, 448, 452, 453, 482, 484–491, 493–498, 501–504, 519, 533, 538, 540–542, 545–551, 572, 577

Nanoplastics, 122, 123, 194, 262, 266, 349, 350, 353, 355, 404, 407, 414, 415, 418, 421, 429, 440–465
Nanosafety, 2–18, 33, 204–208, 219–239, 259, 262, 266, 281–301, 349, 404–430, 482, 512
Nanotechnologies, 2, 10, 11, 13–15, 18, 29, 155, 194, 207, 260, 342, 354, 374, 412, 424–430, 441, 512, 532–537
Nanotoxicology, 12, 17, 18, 29–48, 127, 143, 179, 180, 281–302, 314–331, 341–345, 389, 404, 412, 427, 430, 482–504
Nanotubes, 14, 16, 32, 264
Neurobehavior, 265, 266, 378, 385
Non-coding RNAs (ncRNAs), 92, 100

O

Organs-on-chips (OoCs), 30, 36, 225–226
Oxidative stress, 3, 4, 6, 12, 30, 37, 71, 73, 91, 92, 95, 102, 103, 145, 153, 163–165, 170, 180, 183, 201, 224, 237, 260, 280, 284–293, 295–298, 320, 342–344, 347, 350, 351, 354, 376–389, 405, 408, 410, 411, 449–451, 453, 574, 576, 577

P

Planarians, 11, 13, 252–266
Plastics strategy, 460

Q

Quantitative structure-activity relationship (QSAR), 16, 182, 518, 533

S

Safe-and-sustainable-by-design (SSbD), 425, 426
Safe-by-design (SbD), 426, 431
Safety assessments, 18, 74, 76, 143, 162–183, 194–208, 319, 338, 341–345, 384, 423, 425, 440
Scoring functions, 182, 484–488, 490, 495, 502, 503

T

Test methods, 34, 37–41, 48, 153, 412
3Rs, 343, 577

Toxicity, 2, 29, 60, 88, 116, 163, 194, 224, 254, 279, 314, 338, 366, 405, 441, 483, 533, 563

U
Uptakes, 4, 14, 42, 43, 69, 71, 88, 95–97, 119, 120, 162, 171, 179, 195–199, 202–204, 219, 260, 264, 282–285, 288, 289, 301, 326, 342, 349–351, 354, 368–376, 387, 389, 416, 425, 428, 445, 573

Z
Zebrafish, 14, 256, 366, 453, 575

GPSR Compliance

The European Union's (EU) General Product Safety Regulation (GPSR) is a set of rules that requires consumer products to be safe and our obligations to ensure this.

If you have any concerns about our products, you can contact us on ProductSafety@springernature.com

In case Publisher is established outside the EU, the EU authorized representative is:

Springer Nature Customer Service Center GmbH
Europaplatz 3
69115 Heidelberg, Germany

Batch number: 08794372

Printed by Printforce, the Netherlands